BIG IDEAS
MATH®
COURSE 2

A Bridge to Success

TEACHING EDITION

Ron Larson
Laurie Boswell

BIG IDEAS LEARNING®

Erie, Pennsylvania
BigIdeasLearning.com

Big Ideas Learning, LLC
1762 Norcross Road
Erie, PA 16510-3838
USA

For product information and customer support, contact Big Ideas Learning
at **1-877-552-7766** or visit us at ***BigIdeasLearning.com***.

Cover

Songquan Deng/Shutterstock.com

Printed in the U.S.A.

ISBN 13: 978-1-68033-123-3
ISBN 10: 1-68033-123-X

2 3 4 5 6 7 8 9 10 WEB 17 16 15

AUTHORS

Ron Larson is a professor of mathematics at Penn State Erie, The Behrend College, where he has taught since receiving his Ph.D. in mathematics from the University of Colorado. Dr. Larson is well known as the lead author of a comprehensive program for mathematics that spans middle school, high school, and college courses. His high school and Advanced Placement books are published by Holt McDougal. Ron's numerous professional activities keep him in constant touch with the needs of students, teachers, and supervisors. Ron and Laurie Boswell began writing together in 1992. Since that time, they have authored over two dozen textbooks. In their collaboration, Ron is primarily responsible for the pupil edition and Laurie is primarily responsible for the teaching edition of the text.

Laurie Boswell is the Head of School and a mathematics teacher at the Riverside School in Lyndonville, Vermont. Dr. Boswell received her Ed.D. from the University of Vermont in 2010. She is a recipient of the Presidential Award for Excellence in Mathematics Teaching. Laurie has taught math to students at all levels, elementary through college. In addition, Laurie was a Tandy Technology Scholar, and served on the NCTM Board of Directors from 2002 to 2005. She currently serves on the board of NCSM, and is a popular national speaker. Along with Ron, Laurie has co-authored numerous math programs.

ABOUT THE BOOK

Big Ideas Math: A Bridge to Success was developed using the same research-based strategy of a balanced approach that has become synonymous with the *Big Ideas Math* series. This approach opens doors to abstract thought, reasoning, and inquiry as students persevere to answer the Essential Questions that introduce each section.

From start to finish, this program was designed with the student in mind. Students are subtly introduced to "Habits of Mind" that help them internalize concepts for a greater depth of understanding. These habits serve students well not only in mathematics, but across all curricula throughout their academic careers. *Big Ideas Math: A Bridge to Success* exposes students to highly motivating and relevant problems. Woven throughout the program are the depth and rigor students need to prepare for career-readiness and other college-level courses.

We consider the Big Ideas Math series to be the crowning jewel of 30 years of achievement in writing educational materials.

Ron Larson

Laurie Boswell

TEACHER REVIEWERS

Lisa Amspacher
Milton Hershey School
Hershey, PA

Mary Ballerina
Orange County Public Schools
Orlando, FL

Lisa Bubello
School District of Palm
Beach County
Lake Worth, FL

Sam Coffman
North East School District
North East, PA

Kristen Karbon
Troy School District
Rochester Hills, MI

Laurie Mallis
Westglades Middle School
Coral Springs, FL

Dave Morris
Union City Area
School District
Union City, PA

Bonnie Pendergast
Tolleson Union High
School District
Tolleson, AZ

Valerie Sullivan
Lamoille South
Supervisory Union
Morrisville, VT

Becky Walker
Appleton Area School District
Appleton, WI

Zena Wiltshire
Dade County Public Schools
Miami, FL

STUDENT REVIEWERS

Mike Carter
Matthew Cauley
Amelia Davis
Wisdom Dowds
John Flatley
Nick Ganger

Hannah Iadeluca
Paige Lavine
Emma Louie
David Nichols
Mikala Parnell
Jordan Pashupathi

Stephen Piglowski
Robby Quinn
Michael Rawlings
Garrett Sample
Andrew Samuels
Addie Sedelmyer
Tyler Steffy
Erin Taylor
Reid Wilson

CONSULTANTS

● **Patsy Davis**
Educational Consultant
Knoxville, Tennessee

● **Bob Fulenwider**
Mathematics Consultant
Bakersfield, California

● **Linda Hall**
Mathematics Assessment Consultant
Norman, Oklahoma

● **Ryan Keating**
Special Education Advisor
Gilbert, Arizona

● **Michael McDowell**
Project-Based Instruction Specialist
Fairfax, California

● **Sean McKeighan**
Interdisciplinary Advisor
Norman, Oklahoma

● **Bonnie Spence**
Differentiated Instruction Consultant
Missoula, Montana

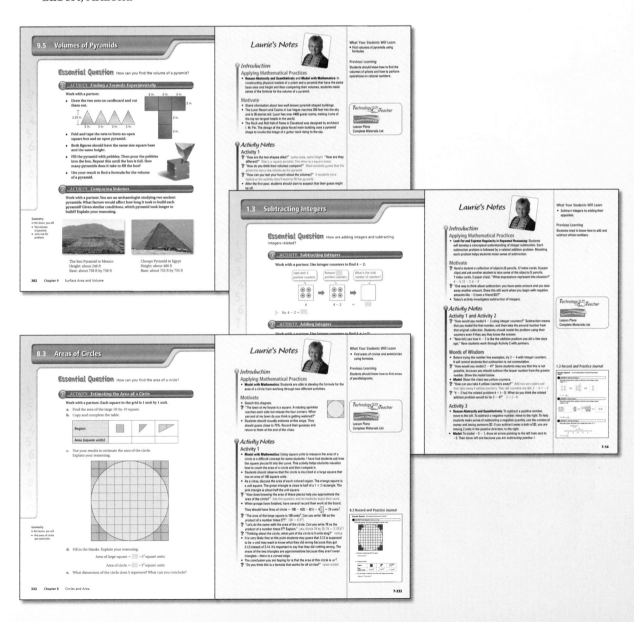

1 Integers

What You Learned Before......................1

Section 1.1 Integers and Absolute Value
Activity...2
Lesson..4

Section 1.2 Adding Integers
Activity...8
Lesson...10

Section 1.3 Subtracting Integers
Activity..14
Lesson...16
Study Help/Graphic Organizer............20
1.1–1.3 Quiz.................................21

Section 1.4 Multiplying Integers
Activity..22
Lesson...24

Section 1.5 Dividing Integers
Activity..28
Lesson...30
1.4–1.5 Quiz.................................34
Chapter Review..............................35
Chapter Test.................................38
Cumulative Assessment....................39

"I like talking about math, and working with a partner allows me to do that."

Rational Numbers

What You Learned Before.................43

Section 2.1 Rational Numbers

Activity...............................44

Lesson................................46

Section 2.2 Adding Rational Numbers

Activity...............................50

Lesson................................52

Study Help/Graphic Organizer...........56

2.1–2.2 Quiz.........................57

Section 2.3 Subtracting Rational Numbers

Activity...............................58

Lesson................................60

Section 2.4 Multiplying and Dividing Rational Numbers

Activity...............................64

Lesson................................66

2.3–2.4 Quiz.........................70

Chapter Review.......................71

Chapter Test.........................74

Cumulative Assessment................75

" With my eBook, I get to decide when I use technology and when I use print. "

3 Expressions and Equations

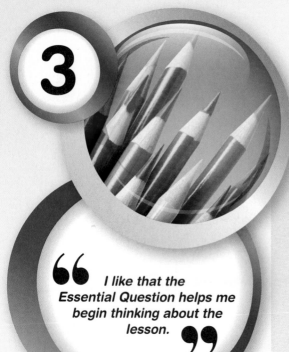

> *I like that the Essential Question helps me begin thinking about the lesson.*

What You Learned Before..................... 79

Section 3.1 Algebraic Expressions
Activity... 80
Lesson... 82

Section 3.2 Adding and Subtracting Linear Expressions
Activity... 86
Lesson... 88
Extension: Factoring Expressions.... 92
Study Help/Graphic Organizer......... 94
3.1–3.2 Quiz.................................... 95

Section 3.3 Solving Equations Using Addition or Subtraction
Activity... 96
Lesson... 98

Section 3.4 Solving Equations Using Multiplication or Division
Activity... 102
Lesson... 104

Section 3.5 Solving Two-Step Equations
Activity... 108
Lesson... 110
3.3–3.5 Quiz.................................. 114
Chapter Review.............................. 115
Chapter Test.................................. 118
Cumulative Assessment................. 119

Inequalities

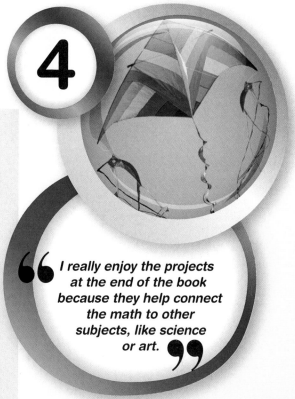

What You Learned Before 123

Section 4.1 **Writing and Graphing Inequalities**

Activity 124

Lesson 126

Section 4.2 **Solving Inequalities Using Addition or Subtraction**

Activity 130

Lesson 132

Study Help/Graphic Organizer 136

4.1–4.2 Quiz 137

Section 4.3 **Solving Inequalities Using Multiplication or Division**

Activity 138

Lesson 140

Section 4.4 **Solving Two-Step Inequalities**

Activity 146

Lesson 148

4.3–4.4 Quiz 152

Chapter Review 153

Chapter Test 156

Cumulative Assessment 157

" I really enjoy the projects at the end of the book because they help connect the math to other subjects, like science or art. "

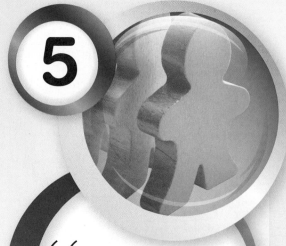

5 Ratios and Proportions

"I like Newton and Descartes! The cartoons are funny and I like that they model the math that we are learning."

What You Learned Before 161

Section 5.1 **Ratios and Rates**
Activity 162
Lesson 164

Section 5.2 **Proportions**
Activity 170
Lesson 172
Extension: Graphing Proportional Relationships 176

Section 5.3 **Writing Proportions**
Activity 178
Lesson 180
Study Help/Graphic Organizer 184
5.1–5.3 Quiz 185

Section 5.4 **Solving Proportions**
Activity 186
Lesson 188

Section 5.5 **Slope**
Activity 192
Lesson 194

Section 5.6 **Direct Variation**
Activity 198
Lesson 200
5.4–5.6 Quiz 204
Chapter Review 205
Chapter Test 208
Cumulative Assessment 209

Percents

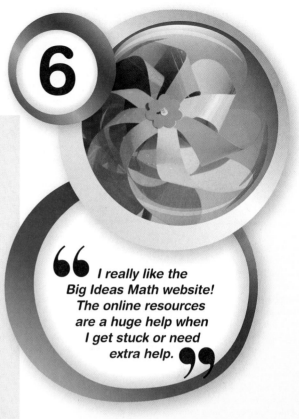

What You Learned Before.................213

Section 6.1 **Percents and Decimals**
Activity..214
Lesson...216

Section 6.2 **Comparing and Ordering Fractions, Decimals, and Percents**
Activity..220
Lesson...222

Section 6.3 **The Percent Proportion**
Activity..226
Lesson...228

Section 6.4 **The Percent Equation**
Activity..232
Lesson...234
Study Help/Graphic Organizer.........238
6.1–6.4 Quiz................................239

Section 6.5 **Percents of Increase and Decrease**
Activity..240
Lesson...242

Section 6.6 **Discounts and Markups**
Activity..246
Lesson...248

Section 6.7 **Simple Interest**
Activity..252
Lesson...254
6.5–6.7 Quiz................................258
Chapter Review............................259
Chapter Test................................264
Cumulative Assessment................265

I really like the Big Ideas Math website! The online resources are a huge help when I get stuck or need extra help.

Constructions and Scale Drawings

What You Learned Before 269

Section 7.1 **Adjacent and Vertical Angles**

Activity 270

Lesson 272

Section 7.2 **Complementary and Supplementary Angles**

Activity 276

Lesson 278

Section 7.3 **Triangles**

Activity 282

Lesson 284

Extension: Angle Measures of Triangles 288

Study Help/Graphic Organizer 290

7.1–7.3 Quiz 291

Section 7.4 **Quadrilaterals**

Activity 292

Lesson 294

Section 7.5 **Scale Drawings**

Activity 298

Lesson 300

7.4–7.5 Quiz 306

Chapter Review 307

Chapter Test 310

Cumulative Assessment 311

I like the real-life application exercises because they show me how I can use the math in my own life.

BIG IDEAS MATH.
COURSE 2
A Bridge to Success

Circles and Area

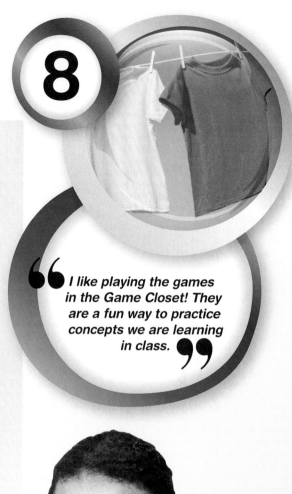

What You Learned Before................... 315

Section 8.1 **Circles and Circumference**

Activity.................................... 316

Lesson...................................... 318

Section 8.2 **Perimeters of Composite Figures**

Activity.................................... 324

Lesson...................................... 326

Study Help/Graphic Organizer........ 330

8.1–8.2 Quiz............................. 331

Section 8.3 **Areas of Circles**

Activity.................................... 332

Lesson...................................... 334

Section 8.4 **Areas of Composite Figures**

Activity.................................... 338

Lesson...................................... 340

8.3–8.4 Quiz............................. 344

Chapter Review........................ 345

Chapter Test............................ 348

Cumulative Assessment............ 349

I like playing the games in the Game Closet! They are a fun way to practice concepts we are learning in class.

9 Surface Area and Volume

> With the BigIdeasMath.com website I don't have to worry if I forget my book or my workbook at school.

What You Learned Before................353

Section 9.1 **Surface Areas of Prisms**

Activity................354

Lesson................356

Section 9.2 **Surface Areas of Pyramids**

Activity................362

Lesson................364

Section 9.3 **Surface Areas of Cylinders**

Activity................368

Lesson................370

Study Help/Graphic Organizer..........374

9.1–9.3 Quiz................375

Section 9.4 **Volumes of Prisms**

Activity................376

Lesson................378

Section 9.5 **Volumes of Pyramids**

Activity................382

Lesson................384

Extension: Cross Sections of Three-Dimensional Figures................388

9.4–9.5 Quiz................390

Chapter Review................391

Chapter Test................394

Cumulative Assessment................395

Probability and Statistics

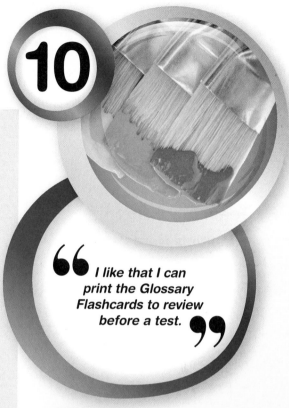

What You Learned Before.................399

Section 10.1 **Outcomes and Events**
Activity..................................400
Lesson...................................402

Section 10.2 **Probability**
Activity..................................406
Lesson...................................408

Section 10.3 **Experimental and Theoretical Probability**
Activity..................................412
Lesson...................................414

Section 10.4 **Compound Events**
Activity..................................420
Lesson...................................422

Section 10.5 **Independent and Dependent Events**
Activity..................................428
Lesson...................................430
Extension: Simulations.........436
Study Help/Graphic Organizer........438
10.1–10.5 Quiz...................439

Section 10.6 **Samples and Populations**
Activity..................................440
Lesson...................................442
Extension: Generating Multiple Samples..............................446

Section 10.7 **Comparing Populations**
Activity..................................448
Lesson...................................450
10.6–10.7 Quiz...................454
Chapter Review..................455
Chapter Test.......................460
Cumulative Assessment...........461

I like that I can print the Glossary Flashcards to review before a test.

Appendix A: My Big Ideas Projects

Section A.1 **Literature Project**...........A2
Section A.2 **History Project**...............A4
Section A.3 **Art Project**......................A6
Section A.4 **Science Project**...............A8

Key Vocabulary Index A11
Student Index A12
Additional Answers A21
Mathematics Reference Sheet B1

PROGRAM OVERVIEW

Print *Also available online*

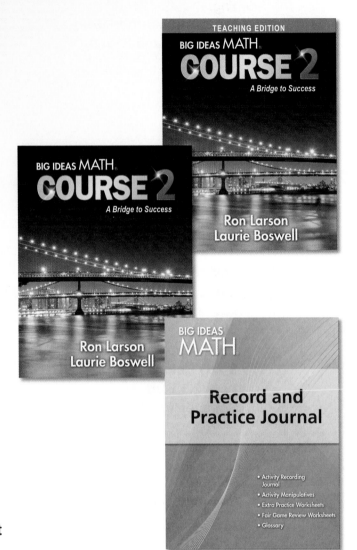

- **Pupil Edition**

- **Teaching Edition**
  - **Laurie's Notes**

- **Record and Practice Journal**
  Available in English and Spanish

- **Assessment Book**
  - **Pre-Course Test**
  - **Quizzes**
  - **Chapter Tests**
  - **Cumulative Assessment**
  - **Alternative Assessment**
  - **End-of-Course Tests**

- **Resources by Chapter**
  - **Family and Community Involvement**
    Available in English and Spanish
  - **Start Thinking! and Warm Up**
  - **Extra Practice**
  - **Enrichment and Extension**
  - **Puzzle Time**
  - **Technology Connection**
  - **Projects with Rubrics**

Technology

Student Edition
With complete English and Spanish Audio

- **Dynamic eBook App**
  - **Lesson Tutorial Videos**
  - **Dynamic Investigations**
- **Online Home Edition**
  - **Multi-Language Glossary**
  - **Basic Skills Handbook**
  - **Skills Review Handbook**
  - **Game Closet**

Dynamic Classroom
- **Interactive Manipulatives**
- **Answer Presentation Tool**
- **Mini-Assessment**
- **Closure Activity**

Dynamic Assessment and Progress Monitoring Tool
- **Homework and Assessment Creation**
- **Progress Monitoring**
- **Direct Ties to Remediation**
- **All-In-One Reporting**
- **Online Chat Tutor**

Dynamic Teaching Tools
- **Interactive Whiteboard Lesson Library**
  - **Includes standard and customizable lessons**
  - **Compatible with SMART®, Promethean®, and Mimeo® technology**
- **Exam**View® **Assessment Suite**
- **Real-Life STEM Videos**
- **Editable Online Resources**
  - **Lesson Plans**
  - **Pacing Guides**
  - **Assessment Book**
  - **Resources by Chapter**

SCOPE AND

Regular Pathway

Grade 6

Ratios and Proportional Relationships	– Understand Ratio Concepts; Use Ratio Reasoning
The Number System	– Perform Fraction and Decimal Operations; Understand Rational Numbers
Expressions and Equations	– Write, Interpret, and Use Expressions, Equations, and Inequalities
Geometry	– Solve Problems Involving Area, Surface Area, and Volume
Statistics and Probability	– Summarize and Describe Distributions; Understand Variability

Grade 7

Ratios and Proportional Relationships	– Analyze Proportional Relationships
The Number System	– Perform Rational Number Operations
Expressions and Equations	– Generate Equivalent Expressions; Solve Problems Using Linear Equations and Inequalities
Geometry	– Understand Geometric Relationships; Solve Problems Involving Angles, Surface Area, and Volume
Statistics and Probability	– Analyze and Compare Populations; Find Probabilities of Events

Grade 8

The Number System	– Approximate Real Numbers; Perform Real Number Operations
Expressions and Equations	– Use Radicals and Integer Exponents; Connect Proportional Relationships and Lines; Solve Systems of Linear Equations
Functions	– Define, Evaluate, and Compare Functions; Model Relationships
Geometry	– Understand Congruence and Similarity; Apply the Pythagorean Theorem; Apply Volume Formulas
Statistics and Probability	– Analyze Bivariate Data

SEQUENCE

Compacted Pathway

 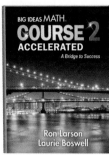 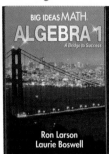

Grade 6

Ratios and Proportional Relationships	– Understand Ratio Concepts; Use Ratio Reasoning
The Number System	– Perform Fraction and Decimal Operations; Understand Rational Numbers
Expressions and Equations	– Write, Interpret, and Use Expressions, Equations, and Inequalities
Geometry	– Solve Problems Involving Area, Surface Area, and Volume
Statistics and Probability	– Summarize and Describe Distributions; Understand Variability

Grade 7 Accelerated

Number and Quantity	– Analyze Proportional Relationships; Perform Real Number Operations; Use Radicals and Integer Exponents
Algebra	– Generate Equivalent Expressions; Connect Proportional Relationships and Lines; Solve Problems Using Linear Equations and Inequalities
Geometry	– Understand Geometric Relationships and Similarity; Solve Problems Involving Angles, Surface Area, and Volume
Statistics and Probability	– Analyze and Compare Populations; Find Probabilities of Events

Grade 8 Algebra 1

Number and Quantity	– Use Rational Exponents; Perform Real Number Operations
Algebra	– Solve Linear and Quadratic Equations; Solve Inequalities and Systems of Equations
Functions	– Define, Evaluate, and Compare Functions; Write Sequences; Model Relationships
Geometry	– Apply the Pythagorean Theorem
Statistics and Probability	– Represent and Interpret Data; Analyze Bivariate Data

LEARNING PROGRESSION

Regular Pathway

Domain	Grade 6	Grade 7	Grade 8
Number and Quantity			
Ratios and Proportional Relationships	Use ratios to solve problems. *Chapter 5*	Use proportional relationships to solve problems. *Chapters 5, 6*	
The Number System	Perform operations with multi-digit numbers and find common factors and multiples. *Chapter 1* Divide fractions by fractions. *Chapter 2* Extend understanding of numbers to the rational number system. *Chapter 6*	Perform operations with rational numbers. *Chapter 2*	Extend understanding of numbers to the real number system. *Chapter 7*
Algebra			
Expressions and Equations	Perform arithmetic with algebraic expressions. *Chapter 3* Solve one-variable equations and inequalities. *Chapter 7* Analyze relationships between dependent and independent variables. *Chapter 7*	Write equivalent expressions. *Chapter 3* Use numerical and algebraic expressions, equations, and inequalities to solve problems. *Chapters 3, 4*	Understand the connections between proportional relationships, lines, and linear equations. *Chapter 4* Solve linear equations and systems of linear equations. *Chapter 5* Work with radicals and integer exponents. *Chapters 7, 10*
Functions			
Functions			Define, evaluate, and compare functions, and use functions to model relationships between quantities. *Chapter 6*

MIDDLE SCHOOL MATH

Domain	Grade 6	Grade 7	Grade 8
Geometry			
Geometry	Solve real-world and mathematical problems involving area, surface area, and volume. *Chapters 4, 8*	Draw, construct, and describe geometrical figures and describe the relationships between them. *Chapters 7–9* Solve problems involving angle measure, area, surface area, and volume. *Chapters 7–9*	Understand congruence and similarity. *Chapters 2, 3* Use the Pythagorean Theorem. *Chapter 7* Solve problems involving volume of cylinders, cones, and spheres. *Chapter 8*
Statistics and Probability			
Statistics and Probability	Develop understanding of statistical variability and summarize and describe distributions. *Chapters 9, 10*	Make inferences about a population, compare two populations, and use probability models. *Chapter 10*	Investigate patterns of association in bivariate data. *Chapter 9*

EMBEDDED MATHEMATICAL PRACTICES THROUGH A BALANCED APPROACH TO INSTRUCTION

STUDENT DIRECTED DISCOVERY

TEACHER DIRECTED INSTRUCTION

The *Big Ideas Math* program balances conceptual understanding with procedural fluency. Embedded Mathematical Practices in grade-level content promote a greater understanding of how mathematical concepts are connected to each other and to real-life, helping turn mathematical learning into an engaging and meaningful way to see and explore the real world. The program provides young people with the skills and knowledge they need to succeed academically in credit-bearing, college entry courses and in workforce training programs.

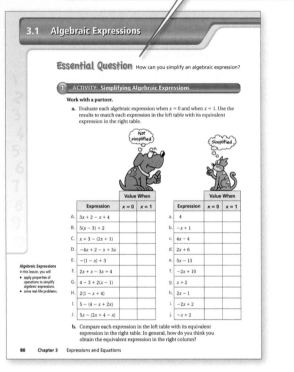

Essential Question How can you simplify an algebraic expression?

3.1 Algebraic Expressions

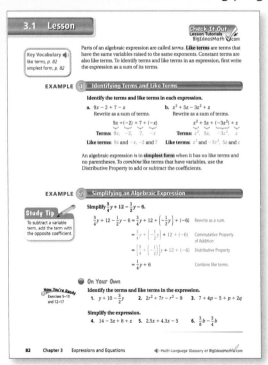

Research shows that students benefit from a program that includes equal exposure to discovery and direct instruction. By beginning each lesson with an inquiry-based activity, *Big Ideas Math* allows students to explore, question, explain, and persevere as they seek to answer Essential Questions that encourage abstract thought.

These rich activities are followed by direct instruction lessons, allowing for procedural fluency, modeling, and the opportunity to use clear, precise mathematical language.

DAILY SUPPORT FOR TEACHERS

Applying Mathematical Practices
- **Look for and Make Use of Structure:** Mathematically proficient students are able to see an algebraic expression being composed of several terms. In this lesson, students will simplify algebraic expressions by combining like terms.

The *Big Ideas Math* Teaching Edition is unique in its organization. It provides teachers with complete support as they teach the program. Using the side-by-side pages, teachers have access to the full student page as they teach and can reference insights from master educator Laurie Boswell.

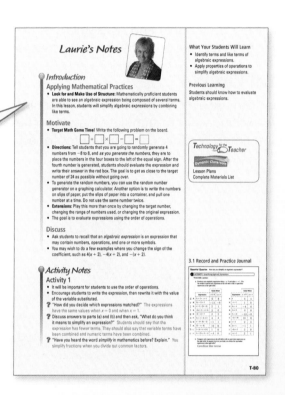

Each student page is accompanied by a support page. Laurie includes connections to previous learning, support for the Mathematical Practices, and closure opportunities.

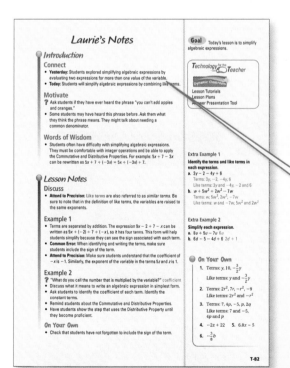

Connect
- **Yesterday:** Students explored simplifying algebraic expressions by evaluating two expressions for more than one value of the variable.
- **Today:** Students will simplify algebraic expressions by combining like terms.

The Teaching Edition also provides Differentiated Instruction, Response to Intervention, and English Language Learner support.

CUSTOMIZED INSTRUCTION

Print Option

The print Teaching Edition provides teachers with help through Laurie's Notes and other features that help manage the classroom. The Chapter Resource Book, Skills Review Handbook, and Assessment Book complete a teaching array that makes it easy for the teacher to differentiate, assess, and teach.

Digital Option

Teachers can use 21st century technology tools found throughout the program to provide exciting ways to stimulate learning. These tools provide innovative electronic activities, timely feedback, and measures for accurate assessment.

Blended Option

Teachers will find that using the blended option provides them a multitude of ways to teach, differentiate, and assess. Teachers and students can customize their teaching and learning by blending the power of creative technology tools with the accessibility of print resources. Rich content and the combination of creative print and online resources allows for an engaging and challenging approach to teaching mathematics.

The Dynamic Classroom

Regardless of the option you choose for your students, *The Dynamic Classroom* will be one of your most valued tools in the *Big Ideas Math* program. This powerful tool can be used with interactive whiteboards and includes the following:

- What You Learned Before
- Start Thinking!
- Warm Ups
- Record and Practice Journal pages
- Virtual Manipulatives
- *On Your Own* Exercises
- Extra Examples
- Mini Assessments
- Closure Activities
- Graphic Organizers
- Mathematical Practices
- Answer Presentation Tool
- Chapter Reviews
- Chapter Tests
- Cumulative Assessments

PERSONALIZED LEARNING

The *Big Ideas Math* program offers teachers and students many ways to personalize and enrich the learning experience of all levels of learners.

Lesson Tutorials Online

Two- to three-minute lesson tutorials provide colorful visuals and audio support for every example in the textbook. The Lesson Tutorials are valuable for students who miss a class, need a second explanation, or just need some help with a homework assignment. Parents can also use the tutorials to stay connected or to provide additional help at home.

The Dynamic Student Edition

This unique tool, available online or as an eBook App, provides students with embedded 21st century learning resources. Students have the opportunity to interact with the underlying mathematics in a number of ways, including engaging tutorials, interactive manipulatives, flashcards, vocabulary support, and games that enhance the learning experience and promote mathematical understanding.

Differentiated Instruction

Through print and digital resources, the *Big Ideas Math* program completely supports the 3-Tier Response to Intervention model. Using research-based instructional strategies, teachers can reach, challenge and motivate each student with germane, high quality instruction targeted to individual needs.

Tier 3:
Customized Learning Intervention
- Intensive Intervention Lessons
- Activities

Tier 2:
Strategic Intervention

- Lesson Tutorials
- Basic Skills Handbook
- Skills Review Handbook
- Differentiated Instruction
- Game Closet

Tier 1:
Daily Intervention

- Record and Practice Journal
- Fair Game Review
- Graphic Organizers
- Vocabulary Support
- Mini Assessments
- Game Closet
- Lesson Tutorials
- On Your Own

CONTINUOUS PREPARATION FOR

In the Textbook

Activities

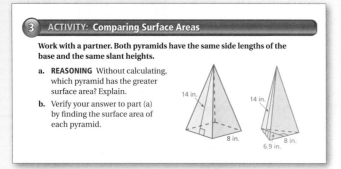
State assessments require a higher level of thinking. In the Activities, students are asked to explain their reasoning.

Exercises

17. **PROBLEM SOLVING** You are making an umbrella that is shaped like a regular octagonal pyramid.

a. Estimate the amount of fabric that you need to make the umbrella.

b. The fabric comes in rolls that are 72 inches wide. You don't want to cut the fabric "on the bias." Find out what this means. Then draw a diagram of how you can cut the fabric most efficiently.

c. How much fabric is wasted?

5 ft

4 ft

State assessments may ask for multiple representations. The *Big Ideas Math* program provides students opportunities to use multiple approaches to solve problems.

Quizzes and Tests

9. **SKYLIGHT** You are making a skylight that has 12 triangular pieces of glass and a slant height of 3 feet. Each triangular piece has a base of 1 foot. *(Section 9.2)*

a. How much glass will you need to make the skylight?

b. Can you cut the 12 glass triangles from a sheet of glass that is 4 feet by 8 feet? If so, draw a diagram showing how this can be done.

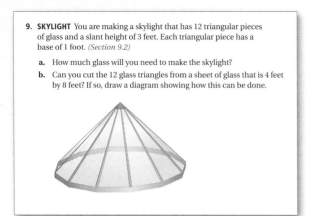

The assessments in the textbook include problems that extend concepts.

Cumulative Assessments

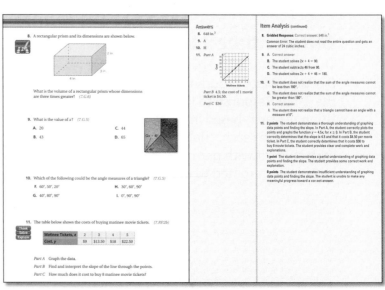

Cumulative Assessments include questions in multiple formats to help prepare students for high-stakes assessments. A detailed Item Analysis for each question in the Pupil Edition is available in the Teaching Edition.

HIGH-STAKES ASSESSMENTS

In the Technology

Online Test Practice

Numeric Response

1. You are adding a three-sided pyramid to your climbing wall. You make the three lateral faces of the pyramid from a 4-foot by 8-foot sheet of plywood. How many square feet of plywood are left over?

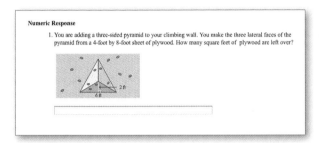

Students will find self-grading quizzes and tests on the student side of the website.

Performance Tasks

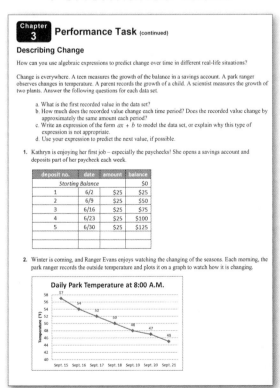

Chapter 3 **Performance Task** (continued)

Describing Change

How can you use algebraic expressions to predict change over time in different real-life situations?

Change is everywhere. A teen measures the growth of the balance in a savings account. A park ranger observes changes in temperature. A parent records the growth of a child. A scientist measures the growth of two plants. Answer the following questions for each data set.

a. What is the first recorded value in the data set?
b. How much does the recorded value change each time period? Does the recorded value change by approximately the same amount each period?
c. Write an expression of the form $ax + b$ to model the data set, or explain why this type of expression is not appropriate.
d. Use your expression to predict the next value, if possible.

Performance Tasks are available at *BigIdeasMath.com*.

Resources

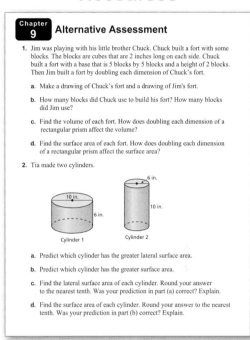

Chapter 9 **Alternative Assessment**

1. Jim was playing with his little brother Chuck. Chuck built a fort with some blocks. The blocks are cubes that are 2 inches long on each side. Chuck built a fort with a base that is 5 blocks by 5 blocks and a height of 2 blocks. Then Jim built a fort by doubling each dimension of Chuck's fort.

a. Make a drawing of Chuck's fort and a drawing of Jim's fort.

b. How many blocks did Chuck use to build his fort? How many blocks did Jim use?

c. Find the volume of each fort. How does doubling each dimension of a rectangular prism affect the volume?

d. Find the surface area of each fort. How does doubling each dimension of a rectangular prism affect the surface area?

2. Tia made two cylinders.

a. Predict which cylinder has the greater lateral surface area.

b. Predict which cylinder has the greater surface area.

c. Find the lateral surface area of each cylinder. Round your answer to the nearest tenth. Was your prediction in part (a) correct? Explain.

d. Find the surface area of each cylinder. Round your answer to the nearest tenth. Was your prediction in part (b) correct? Explain.

Other resources include the Assessment Book, Resources by Chapter, and Additional Support at *BigIdeasMath.com*.

Technology Enhanced Items

The *Big Ideas Math* assessment package includes Technology Enhanced items.

PACING GUIDE

Regular:

Chapters 1–10   154 Days

Accelerated:

Chapters 1–16:   152 Days

	Regular	Accelerated
Scavenger Hunt	**1 Day**	**1 Day**
Chapter 1	**14 Days**	**4 Days**
Chapter Opener	1 Day	1 Day
Section 1.1	2 Days	1 Day
Section 1.2	2 Days	0.5 Day
Section 1.3	2 Days	0.5 Day
Study Help/Quiz	1 Day	0 Days
Section 1.4	2 Days	0.5 Day
Section 1.5	2 Days	0.5 Day
Chapter Review/Chapter Tests	2 Days	0 Days
Chapter 2	**12 Days**	**5 Days**
Chapter Opener	1 Day	0.5 Day
Section 2.1	2 Days	0.5 Day
Section 2.2	2 Days	0.5 Day
Study Help/Quiz	1 Day	0 Days
Section 2.3	2 Days	0.5 Day
Section 2.4	2 Days	1 Day
Chapter Review/Chapter Tests	2 Days	2 Days
Chapter 3	**15 Days**	**12 Days**
Chapter Opener	1 Day	1 Day
Section 3.1	2 Days	0.5 Day
Section 3.2	3 Days	1.5 Days
Study Help/Quiz	1 Day	0 Days
Section 3.3	2 Days	1 Day
Section 3.4	2 Days	1 Day
Section 3.5	2 Days	1 Day
Topic 1		1 Day
Topic 2		2 Days
Topic 3		1 Day
Chapter Review/Chapter Tests	2 Days	2 Days

	Regular	Accelerated
Chapter 4	**12 Days**	**7 Days**
Chapter Opener	**1 Day**	**1 Day**
Section 4.1	2 Days	1 Day
Section 4.2	2 Days	0 Days
Study Help/Quiz	**1 Day**	**1 Day**
Section 4.3	2 Days	1 Day
Section 4.4	2 Days	1 Day
Chapter Review/Chapter Tests	**2 Days**	**2 Days**
Chapter 5	**17 Days**	**11 Days**
Chapter Opener	**1 Day**	**1 Day**
Section 5.1	2 Days	1 Day
Section 5.2	2 Days	1 Day
Section 5.3	3 Days	2 Days
Study Help/Quiz	**1 Day**	**1 Day**
Section 5.4	2 Days	1 Day
Section 5.5	2 Days	1 Day
Section 5.6	2 Days	1 Day
Chapter Review/Chapter Tests	**2 Days**	**2 Days**
Chapter 6	**18 Days**	**11 Days**
Chapter Opener	**1 Day**	**1 Day**
Section 6.1	2 Days	1 Day
Section 6.2	2 Days	1 Day
Section 6.3	2 Days	1 Day
Section 6.4	2 Days	1 Day
Study Help/Quiz	**1 Day**	**1 Day**
Section 6.5	2 Days	1 Day
Section 6.6	2 Days	1 Day
Section 6.7	2 Days	1 Day
Chapter Review/Chapter Tests	**2 Days**	**2 Days**
Chapter 7	**17 Days**	**11 Days**
Chapter Opener	**1 Day**	**1 Day**
Section 7.1	2 Days	1 Day
Section 7.2	2 Days	1 Day
Section 7.3	4 Days	3 Days
Study Help/Quiz	**1 Day**	**1 Day**
Section 7.4	3 Days	1 Day
Section 7.5	2 Days	1 Day
Chapter Review/Chapter Tests	**2 Days**	**2 Days**

PACING GUIDE (CONTINUED)

	Regular	Accelerated
Chapter 8	**12 Days**	**7 Days**
Chapter Opener	1 Day	1 Day
Section 8.1	2 Days	1 Day
Section 8.2	2 Days	1 Day
Study Help/Quiz	1 Day	0 Days
Section 8.3	2 Days	1 Day
Section 8.4	2 Days	1 Day
Chapter Review/Chapter Tests	2 Days	2 Days
Chapter 9	**16 Days**	**10 Days**
Chapter Opener	1 Day	1 Day
Section 9.1	2 Days	1 Day
Section 9.2	3 Days	2 Days
Section 9.3	2 Days	1 Day
Study Help/Quiz	1 Day	0 Day
Section 9.4	2 Days	1 Day
Section 9.5	3 Days	2 Days
Chapter Review/Chapter Tests	2 Days	2 Days
Chapter 10	**20 Days**	**13 Days**
Chapter Opener	1 Day	1 Day
Section 10.1	2 Days	1 Day
Section 10.2	2 Days	2 Days
Section 10.3	2 Days	2 Days
Section 10.4	2 Days	1 Day
Section 10.5	3 Days	1 Day
Study Help/Quiz	1 Day	1 Day
Section 10.6	3 Days	1 Day
Section 10.7	2 Days	1 Day
Chapter Review/Chapter Tests	2 Days	2 Days
Chapter 11	**0 Days**	**10 Days**
Chapter Opener		1 Day
Section 11.1		1 Day
Section 11.2		1 Day
Section 11.3		1 Day
Section 11.4		1 Day
Study Help/Quiz		1 Day
Section 11.5		1 Day
Section 11.6		1 Day
Section 11.7		2 Days
Chapter Review/Chapter Tests		2 Days

	Regular	Accelerated
Chapter 12	**0 Days**	**8 Days**
Chapter Opener		**1 Day**
Section 12.1		1 Day
Section 12.2		1 Day
Section 12.3		2 Days
Section 12.4		1 Day
Chapter Review/Chapter Tests		**2 Days**
Chapter 13	**0 Days**	**12 Days**
Chapter Opener		**1 Day**
Section 13.1		1 Day
Section 13.2		2 Days
Section 13.3		1 Day
Study Help/Quiz		**1 Day**
Section 13.4		1 Day
Section 13.5		1 Day
Section 13.6		1 Day
Section 13.7		1 Day
Chapter Review/Chapter Tests		**2 Days**
Chapter 14	**0 Days**	**9 Days**
Chapter Opener		**1 Day**
Section 14.1		1 Day
Section 14.2		1 Day
Section 14.3		1 Day
Section 14.4		2 Days
Section 14.5		1 Day
Chapter Review/Chapter Tests		**2 Days**
Chapter 15	**0 Days**	**8 Days**
Chapter Opener		**1 Day**
Section 15.1		1 Day
Section 15.2		1 Day
Section 15.3		1 Day
Section 15.4		2 Days
Chapter Review/Chapter Tests		**2 Days**
Chapter 16	**0 Days**	**11 Days**
Chapter Opener		**1 Day**
Section 16.1		1 Day
Section 16.2		1 Day
Section 16.3		1 Day
Section 16.4		1 Day
Study Help/Quiz		**1 Day**
Section 16.5		1 Day
Section 16.6		1 Day
Section 16.7		1 Day
Chapter Review/Chapter Tests		**2 Days**

Mathematical Practices

Make sense of problems and persevere in solving them.
- Multiple representations are presented to help students move from concrete to representative and into abstract thinking
- *Essential Questions* help students focus and analyze
- *In Your Own Words* provide opportunities for students to look for meaning and entry points to a problem

Reason abstractly and quantitatively.
- Visual problem solving models help students create a coherent representation of the problem
- Opportunities for students to decontextualize and contextualize problems are presented in every lesson

Construct viable arguments and critique the reasoning of others.
- *Error Analysis*; *Different Words, Same Question*; and *Which One Doesn't Belong* features provide students the opportunity to construct arguments and critique the reasoning of others
- *Inductive Reasoning* activities help students make conjectures and build a logical progression of statements to explore their conjecture

Model with mathematics.
- Real-life situations are translated into diagrams, tables, equations, and graphs to help students analyze relations and to draw conclusions
- Real-life problems are provided to help students learn to apply the mathematics that they are learning to everyday life

Use appropriate tools strategically.
- *Graphic Organizers* support the thought process of what, when, and how to solve problems
- A variety of tool papers, such as graph paper, number lines, and manipulatives, are available as students consider how to approach a problem
- Opportunities to use the web, graphing calculators, and spreadsheets support student learning

Attend to precision.
- *On Your Own* questions encourage students to formulate consistent and appropriate reasoning
- Cooperative learning opportunities support precise communication

Look for and make use of structure.
- *Inductive Reasoning* activities provide students the opportunity to see patterns and structure in mathematics
- Real-world problems help students use the structure of mathematics to break down and solve more difficult problems

Look for and express regularity in repeated reasoning.
- Opportunities are provided to help students make generalizations
- Students are continually encouraged to check for reasonableness in their solutions

Go to *BigIdeasMath.com* for more information on the Mathematical Practices.

Mathematical Content

Chapter Coverage

Ratios and Proportional Relationships

- Analyze proportional relationships and use them to solve real-world and mathematical problems.

The Number System

- Apply and extend previous understandings of operations with fractions to add, subtract, multiply, and divide rational numbers.

Expressions and Equations

- Use properties of operations to generate equivalent expressions.
- Solve real-life and mathematical problems using numerical and algebraic expressions and equations.

Geometry

- Draw, construct, and describe geometrical figures and describe the relationships between them.
- Solve real-life and mathematical problems involving angle measure, area, surface area, and volume.

Statistics and Probability

- Use random sampling to draw inferences about a population.
- Draw informal comparative inferences about two populations.
- Investigate chance processes and develop, use, and evaluate probability models.

How to Use Your Math Book

- Read the **Essential Question** in the activity.

  Discuss the  question with your partner.

  Work with a partner to decide **What Is Your Answer?**

  Now you are ready to do the Practice problems.

- Find the (Key Vocabulary 🔊) words, **highlighted in yellow**.

  Read their definitions. Study the concepts in each 🔍 **Key Idea**.
  If you forget a definition, you can look it up online in the

  🔊 Multi-Language Glossary at BigIdeasMath✓com.

- After you study each **EXAMPLE**, do the exercises in the ⬤ **On Your Own**.

  Now You're Ready to do the exercises that correspond to the example.

  As you study, look for a **Study Tip** ✏ or a **Common Error** ⚠ .

- The exercises are divided into 3 parts.

   ✓ **Vocabulary and Concept Check**

   **Practice and Problem Solving**

   **Fair Game Review**

  If an exercise has a ① next to it, look back at Example 1 for help with that exercise.

  More help is available at  **Check It Out** Lesson Tutorials BigIdeasMath✓com .

- To help study for your test, use the following.

   **Quiz**   **Study Help**

   **Chapter Review**   **Chapter Test**

SCAVENGER HUNT

Use this *Scavenger Hunt* to find where things are in **Chapter 1**.

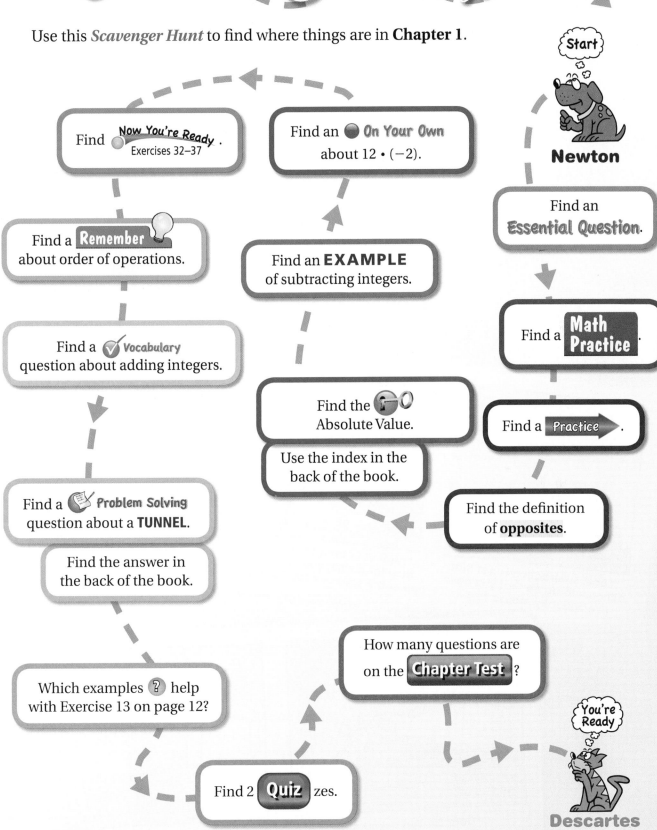

Start

Newton

Find an **Essential Question**.

Find a **Math Practice**.

Find a **Practice**.

Find the definition of **opposites**.

Find an ● **On Your Own** about $12 \cdot (-2)$.

Find an **EXAMPLE** of subtracting integers.

Find the Absolute Value.

Use the index in the back of the book.

Find **Now You're Ready**. Exercises 32–37

Find a **Remember** about order of operations.

Find a ✓ **Vocabulary** question about adding integers.

Find a **Problem Solving** question about a **TUNNEL**.

Find the answer in the back of the book.

Which examples ❓ help with Exercise 13 on page 12?

How many questions are on the **Chapter Test**?

Find 2 **Quiz**zes.

You're Ready

Descartes

1 Integers

1.1   Integers and Absolute Value

1.2   Adding Integers

1.3   Subtracting Integers

1.4   Multiplying Integers

1.5   Dividing Integers

"Look, subtraction is not that difficult. Imagine that you have five squeaky mouse toys."

"After your friend Fluffy comes over for a visit, you notice that one of the squeaky toys is missing."

"Now, you go over to Fluffy's and retrieve the missing squeaky mouse toy. It's easy."

"Dear Sir: You asked me to 'find' the opposite of −1."

"I didn't know it was missing."

What Your Students Have Learned

- Fluently multiply.
- Divide whole numbers, find quotients with and without remainders.
- Create and solve problems with whole number operations.
- Fluently divide.
- Identify and represent integers.
- Order and compare integers.
- Identify and describe absolute values of integers.

What Your Students Will Learn

- Use and justify rules for addition, subtraction, multiplication, and division of integers.
- Find the absolute value of integers.
- Add, subtract, multiply, and divide integers.

Pacing Guide for Chapter 1

Chapter Opener	
Regular	1 Day
Accelerated	1 Day
Section 1	
Regular	2 Days
Accelerated	1 Day
Section 2	
Regular	2 Days
Accelerated	0.5 Day
Section 3	
Regular	2 Days
Accelerated	0.5 Day
Study Help / Quiz	
Regular	1 Day
Accelerated	0 Days
Section 4	
Regular	2 Days
Accelerated	0.5 Day
Section 5	
Regular	2 Days
Accelerated	0.5 Day
Chapter Review/ Chapter Tests	
Regular	2 Days
Accelerated	0 Days*
Total Chapter 1	
Regular	14 Days
Accelerated	4 Days
Year-to-Date	
Regular	15 Days
Accelerated	5 Days

*Assess Chapter 1 with Chapter 2

Technology for the *Teacher*

BigIdeasMath.com
Chapter at a Glance
Complete Materials List
Parent Letters: English and Spanish

What Your Students Have Learned

- Use the Commutative and Associative properties to simplify expressions.
- Use the properties of zero and one to simplify expressions.

Additional Topics for Review

- Compare and Order Integers
- Operations with Whole Numbers
- Distributive Property
- Mental Math Strategies

Try It Yourself

1–6. See Additional Answers.

Record and Practice Journal
Fair Game Review

1. $7 + y$;
   $2 + (5 + y) = (2 + 5) + y$
       Assoc. Prop. of Add.
       $= 7 + y$
       Add 2 and 5.

2. $c + 10$;
   $(c + 1) + 9 = c + (1 + 9)$
       Assoc. Prop. of Add.
       $= c + 10$
       Add 1 and 9.

3. $n + 3.7$;
   $(2.3 + n) + 1.4$
       $= (n + 2.3) + 1.4$
           Comm. Prop. of Add.
       $= n + (2.3 + 1.4)$
           Assoc. Prop. of Add.
       $= n + 3.7$
           Add 2.3 and 1.4.

4. $12 + d$;
   $7 + (d + 5) = 7 + (5 + d)$
           Comm. Prop. of Add.
       $= (7 + 5) + d$
           Assoc. Prop. of Add.
       $= 12 + d$
           Add 7 and 5.

5. $70t$;
   $10(7t) = (10 \cdot 7)t$
           Assoc. Prop. of Mult.
       $= 70t$
           Multiply 10 and 7.

6–12. See Additional Answers.

Math Background Notes

Vocabulary Review

- Commutative Property of Addition
- Commutative Property of Multiplication
- Associative Property of Addition
- Associative Property of Multiplication
- Addition Property of Zero
- Multiplication Property of Zero
- Multiplication Property of One

Commutative and Associative Properties

- Discuss the meaning of the word *commute*. Talk about *commuters* who go to work and back home, switching the direction.
- Discuss the meaning of the word *associate*, putting two things together. You *associate* dogs with barking, Arizona with the Grand Canyon, and sports cars with speed.
- Remind students that it is okay to *just combine numbers* for only the operations of addition and multiplication. If the first problem was $6 - (14 - x)$, combining the numbers would result in a wrong answer.
- Remind students that grouping symbols, such as parentheses, tell you to perform the operation inside them first.
- Ask students to write the Commutative and Associative Properties using variables.

Properties of Zero and One

- Multiplying by zero, in any form, produces a product of 0.
- Multiplying by one, in any form, does not change the value of the quantity.

  Remind students that 1 can be represented in different ways: 1, $\left(\dfrac{1}{2} + \dfrac{1}{2}\right)$, or $\dfrac{3}{3}$.

  This is also called the Multiplicative Identity.

- Ask students to write the Properties of Zero and One using variables.

Reteaching and Enrichment Strategies

If students need help. . .	If students got it. . .
Record and Practice Journal • Fair Game Review Skills Review Handbook Lesson Tutorials	Game Closet at *BigIdeasMath.com* Start the next section

What You Learned Before

Commutative and Associative Properties

Example 1 **a. Simplify the expression 6 + (14 + x).**

$6 + (14 + x) = (6 + 14) + x$	Associative Property of Addition
$= 20 + x$	Add 6 and 14.

b. Simplify the expression (3.1 + x) + 7.4.

$(3.1 + x) + 7.4 = (x + 3.1) + 7.4$	Commutative Property of Addition
$= x + (3.1 + 7.4)$	Associative Property of Addition
$= x + 10.5$	Add 3.1 and 7.4.

c. Simplify the expression 5(12y).

$5(12y) = (5 \cdot 12)y$	Associative Property of Multiplication
$= 60y$	Multiply 5 and 12.

Try It Yourself

Simplify the expression. Explain each step.

1. $3 + (b + 8)$ **2.** $(d + 4) + 6$ **3.** $6(5p)$

Properties of Zero and One

Example 2 **a. Simplify the expression 6 · 0 · q.**

$6 \cdot 0 \cdot q = (6 \cdot 0) \cdot q$	Associative Property of Multiplication
$= 0 \cdot q = 0$	Multiplication Property of Zero

b. Simplify the expression 3.6 · s · 1.

$3.6 \cdot s \cdot 1 = 3.6 \cdot (s \cdot 1)$	Associative Property of Multiplication
$= 3.6 \cdot s$	Multiplication Property of One
$= 3.6s$	

Try It Yourself

Simplify the expression. Explain each step.

4. $13 \cdot m \cdot 0$ **5.** $1 \cdot x \cdot 29$ **6.** $(n + 14) + 0$

Essential Question How can you use integers to represent the velocity and the speed of an object?

On these two pages, you will investigate vertical motion (up or down).

- Speed tells how fast an object is moving, but it does not tell the direction.
- Velocity tells how fast an object is moving, and it also tells the direction.

    When velocity is positive, the object is moving up.

    When velocity is negative, the object is moving down.

1 ACTIVITY: Falling Parachute

Work with a partner. You are gliding to the ground wearing a parachute. The table shows your height above the ground at different times.

Time (seconds)	0	1	2	3
Height (feet)	90	75	60	45

a. Describe the pattern in the table. How many feet do you move each second? After how many seconds will you land on the ground?

b. What integer represents your speed? Give the units.

c. Do you think your velocity should be represented by a positive or negative integer? Explain your reasoning.

d. What integer represents your velocity? Give the units.

2 ACTIVITY: Rising Balloons

Work with a partner. You release a group of balloons. The table shows the height of the balloons above the ground at different times.

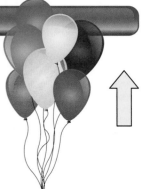

Integers

In this lesson, you will
- define the absolute value of a number.
- find absolute values of numbers.
- solve real-life problems.

Time (seconds)	0	1	2	3
Height (feet)	8	12	16	20

a. Describe the pattern in the table. How many feet do the balloons move each second? After how many seconds will the balloons be at a height of 40 feet?

b. What integer represents the speed of the balloons? Give the units.

c. Do you think the velocity of the balloons should be represented by a positive or negative integer? Explain your reasoning.

d. What integer represents the velocity of the balloons? Give the units.

Laurie's Notes

Introduction

Applying Mathematical Practices

- **Attend to Precision.** In this lesson, students will use integers to describe the velocity of an object. Although integers have not been defined formally, students have heard of negative numbers. In discussing speed and velocity, ask students for examples, being sure that appropriate units are stated.

Motivate

- Ask students if they have ever watched the launch of a NASA shuttle. The velocity of the shuttle describes both its speed and its direction. Today's activity looks at the velocities of various objects.
- **Model:** Use a handkerchief to make a simple parachute by stapling or taping a piece of string or yarn to each corner of the handkerchief. Tie a paper clip to the loose ends of the string or yarn.
- Have two students stand on chairs. One student drops a paper clip without the parachute and the other drops the paper clip with the opened parachute.
- Both paper clips have speed, which can be measured in ft/sec. They also have velocity, moving down, so the velocity is negative and the units would still be ft/sec.
- Write the definitions of speed and velocity on the board and ask for (or share) examples of each.
    - **Examples of speed:** a pitcher throws a 94 mi/h fastball; a manatee swims 4 mi/h; a football travels 15 ft/sec
    - **Examples of velocity:** a NASA shuttle has an orbital velocity of 17,500 mi/h; you walk to school at 8 ft/sec; a feather falls at -2 ft/sec

Activity Notes

Activity 1

- **?** "Do you see any pattern(s) in the table? Describe the pattern(s)." The heights are going down by 15. "Is there a name for this collection of numbers: 90, 75, 60, 45, ...?" multiples of 15
- **?** "When will you land on the ground? Explain how you know." You land after 6 seconds. Students may say that $6 \times 15 = 90$ so after 6 seconds the person is on the ground.
- Discuss the difference between the speed of 15 feet per second and the velocity of -15 feet per second.
- **Attend to Precision:** Be sure to point out that in the table, each of the quantities have units associated with them. The labeling of units is a good habit to develop.

Activity 2

- This activity is similar to Activity 1 except the balloons are rising versus a parachute falling. The speed and velocity will be positive.
- **?** "What was the height of the balloons when they were released? Explain your reasoning." 8 feet; the height when time equals 0 seconds

---

What Your Students Will Learn

- Define the absolute value of a number.
- Find and compare absolute values of numbers.

Previous Learning

Students should know how to compare, order, and graph integers.

Technology for the Teacher

Dynamic Classroom

Lesson Plans
Complete Materials List

1.1 Record and Practice Journal

Essential Question How can you use integers to represent the velocity and the speed of an object?

On these three pages, you will investigate vertical motion (up or down).

- Speed tells how fast an object is moving, but it does not tell the direction.
- Velocity tells how fast an object is moving, and it also tells the direction.

    When velocity is positive, the object is moving up.

    When velocity if negative, the object is moving down.

1 ACTIVITY: Falling Parachute

Work with a partner. You are gliding to the ground wearing a parachute. The table shows your height above the ground at different times.

Time (seconds)	0	1	2	3
Height (feet)	90	75	60	45

a. Describe the pattern in the table. How many feet do you move each second? After how many seconds will you land on the ground?
After each second, your height decreases by 15 feet; 15 ft; 6 sec

b. What integer represents your speed? Give the units.
15 ft/sec

c. Do you think your velocity should be represented by a positive or negative integer? Explain your reasoning.
negative

d. What integer represents your velocity? Give the units.
−15 ft/sec

Vocabulary

On a long strip of paper, mark integers from −10 through 10, with zero in the middle. Hold the strip vertically so that −10 touches the ground. Discuss that when an object is moving from 0 to 10, the *velocity* is positive. Point out the numbers 1 through 10 on the strip are positive integers. Then discuss that an object moving from zero down to the ground has a negative velocity. Again refer to the number strip, pointing out that −1 through −10 are negative integers.

1.1 Record and Practice Journal

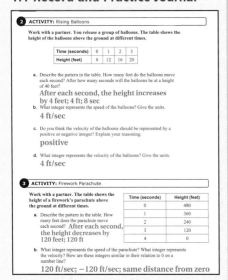

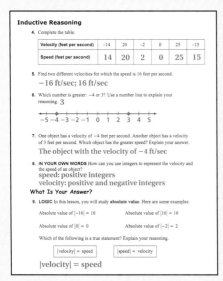

Laurie's Notes

Activity 3

- **Think-Pair-Share:** Students should read each question independently and then work in pairs to answer the questions. When they have answered the questions, the pair should compare their answers with another group and discuss any discrepancies.
- Use curved arrows to show the change in time and the change in height.

	Time (in seconds)	Height (in feet)	
+1	0	480	−120
+1	1	360	−120
+1	2	240	−120
+1	3	120	−120
	4	0	

Inductive Reasoning

- In Question 4, if students have connected the relationship between speed and velocity, they will be able to complete the table. To reinforce labeling of answers, have students write *ft/sec* with each entry in the table.
- Questions 5–7 are developing the notion of absolute value—there are two velocities that have a speed of 16 ft/sec, namely 16 ft/sec and −16 ft/sec.
- In Question 6, it is common for students to say −4 > 3. Remind students that a number farther to the right on a number line is greater, so 3 > −4.
- In Question 7, the sign (negative or positive) is not considered because the question concerns speed and the direction does not matter. An object moving 4 ft/sec has a greater speed than an object moving 3 ft/sec.

What Is Your Answer?

- In Question 9, speed is the **absolute value** of velocity.

Words of Wisdom

- A formal definition of absolute value will be presented in the lesson.
- Do not let students suggest that absolute value simply means to take away the negative sign. This could cause problems in the future when students work with variables or variable expressions within the absolute value symbols. Direct discussions toward the idea that you want to know how far (distance) a number is from zero.

Closure

- **Communication:** Give examples of velocities of two objects (A and B), where Velocity A > Velocity B but Speed A < Speed B.
- Examples are shown below.

Velocity Object A	Velocity Object B	Compare Velocities	Compare Speeds
4 ft/sec	−5 ft/sec	Vel (A) > Vel (B)	Sp (A) < Sp (B)
10 ft/sec	−15 ft/sec	Vel (A) > Vel (B)	Sp (A) < Sp (B)
−5 ft/sec	−15 ft/sec	Vel (A) > Vel (B)	Sp (A) < Sp (B)

ACTIVITY: Firework Parachute

Work with a partner. The table shows the height of a firework's parachute above the ground at different times.

Math Practice

Use Clear Definitions

What information can you use to support your answer?

Time (seconds)	Height (feet)
0	480
1	360
2	240
3	120
4	0

a. Describe the pattern in the table. How many feet does the parachute move each second?

b. What integer represents the speed of the parachute? What integer represents the velocity? How are these integers similar in their relation to 0 on a number line?

Inductive Reasoning

4. Copy and complete the table.

Velocity (feet per second)	-14	20	-2	0	25	-15
Speed (feet per second)						

5. Find two different velocities for which the speed is 16 feet per second.

6. Which number is greater: -4 or 3? Use a number line to explain your reasoning.

7. One object has a velocity of -4 feet per second. Another object has a velocity of 3 feet per second. Which object has the greater speed? Explain your answer.

What Is Your Answer?

8. **IN YOUR OWN WORDS** How can you use integers to represent the velocity and the speed of an object?

9. **LOGIC** In this lesson, you will study *absolute value*. Here are some examples:

$$|-16| = 16 \qquad |16| = 16 \qquad |0| = 0 \qquad |-2| = 2$$

Which of the following is a true statement? Explain your reasoning.

$$|\,\text{velocity}\,| = \text{speed} \qquad\qquad |\,\text{speed}\,| = \text{velocity}$$

Practice

Use what you learned about absolute value to complete Exercises 4–11 on page 6.

The following numbers are **integers:**

$$\ldots, -3, -2, -1, 0, 1, 2, 3, \ldots$$

Key Idea

Absolute Value

Words The **absolute value** of an integer is the distance between the number and 0 on a number line. The absolute value of a number a is written as $|a|$.

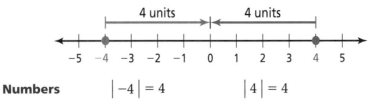

Numbers $|-4| = 4$ $\quad\quad |4| = 4$

EXAMPLE 1 Finding Absolute Value

Find the absolute value of 2.

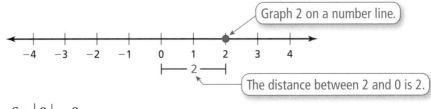

Graph 2 on a number line.

The distance between 2 and 0 is 2.

So, $|2| = 2$.

EXAMPLE 2 Finding Absolute Value

Find the absolute value of -3.

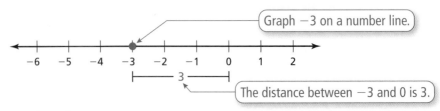

Graph -3 on a number line.

The distance between -3 and 0 is 3.

So, $|-3| = 3$.

On Your Own

Now You're Ready
Exercises 4–19

Find the absolute value.

1. $|7|$　　　　2. $|-1|$　　　　3. $|-5|$　　　　4. $|14|$

　　Multi-Language Glossary at BigIdeasMath.com

Laurie's Notes

Introduction

Connect

- **Yesterday:** Students explored speed and velocity in the activity. They used the relationship between speed and velocity to find speed (the absolute value of velocity) when the velocity was known.
- **Today:** Have students think about the distance a number is from 0. This distance is always positive, just like the speed of an object.

Motivate

- Have two students (A and B) stand at the front of the room with a piece of string between them. You hold a piece of paper with the number 0 written on it. State that the distance between A and B is 10 units.
- **?** Position yourself at the midpoint of A and B.
  - "If A is 5, what number does B represent?" −5
  - "Who is closer to me (meaning 0)?" Neither; both are the same distance.
  - "How far away from me is each person?" 5 units
- **?** Move closer to A so that if A is 3, B would be approximately −7.
  - "If A is 3, what number does B represent?" −7
  - "Who is closer to me (meaning 0)?" A
  - "How far away from me is each person?" A is 3 units. B is 7 units.
- **?** Move closer to B so that if B is −2, A would be approximately 8.
  - "If B is −2, what number does A represent?" 8
  - "Who is closer to me (meaning 0)?" B
  - "How far away from me is each person?" A is 8 units. B is 2 units.
- **?** Without the string, ask students, "What number or numbers are 6 units from 0?" 6 and −6

Lesson Notes

Key Idea

- When students state that "absolute values are always positive," try to clarify their statement. Students may make incorrect assumptions when there are numeric or variable expressions within absolute value symbols.
- At this stage of development, stress the geometric definition of absolute value. **Absolute value** is the distance a number is from zero.

Example 1 and Example 2

- Work through Examples 1 and 2 and then ask the following:
  - **?** "What is the absolute value of 12?" 12
  - **?** "What is the absolute value of −8?" 8
- **Common Error:** When students plot 2 on the number line, they may make a scale on the number line and fail to put a dot on the number 2. Remind students that plotting a point involves actually putting a dot on the number line at the point they are plotting. Remind them as well that the numbers go below the number line, not above it.

Extra Example 1

Find the absolute value of 6. 6

Extra Example 2

Find the absolute value of −11. 11

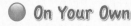 **On Your Own**

1. 7   2. 1

3. 5   4. 14

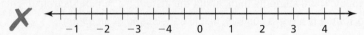
Extra Example 3

Compare $|-9|$ and 7. $|-9| > 7$

 On Your Own

5. $|-2| > -1$

6. $-7 < |6|$

7. $|10| < 11$

8. $9 = |-9|$

Extra Example 4

Seawater freezes at $-2°C$. Is the freezing point of honey (from Example 4) or seawater closer to the freezing point of water, $0°C$? seawater

 On Your Own

9. airplane fuel; Because $|-53| < |55|$, the freezing point of airplane fuel is closer to $0°C$, the freezing point of water.

Differentiated Instruction

Kinesthetic

Stand in front of the classroom with two students, one on each side. Tell the students that you represent zero, your left (the students' right) is the positive direction, and your right (the students' left) is the negative direction. Have the student on your left walk three paces away from you. Say, "This student represents $+3$." Have the student on the right walk three paces from you. Say, "This student represents -3." Ask, "How far is each student from me?" (3 paces away). Say, "Positive 3 and negative 3 are the same distance from zero, so they have the same absolute value, which is 3."

Example 3

- Students may mix up or forget the inequality symbols.
- **Common Error:** Students may label the number line left-to-right for both positive and negative integers as shown below. Explain that negative integers are labeled from 0, such that 1 and -1 are both one unit from 0.

$$\times \quad \xleftarrow{} \overset{-1 \quad -2 \quad -3 \quad -4 \quad 0 \quad 1 \quad 2 \quad 3 \quad 4}{\rule{0pt}{0pt}} \xrightarrow{}$$

- **Attend to Precision:** When students write their answers, check to see that the absolute value symbols are included. It would be wrong to write $1 < -4$. The correct answer is $1 < |-4|$.

On Your Own

- Make sure that students understand that when you write the notation for the absolute value, it means *take the absolute value of the number inside the symbols*.

Example 4

- Have students discuss different liquids that they know freeze, such as water, ice cream, and chocolate. Probe to see if students know that water freezes at $0°C$ and $32°F$.
- **FYI:** Citrus fruit trees can sustain damage from low temperatures. Depending upon the temperature, the leaves, wood, or fruit can be damaged, causing economic problems for both the growers and the consumers.
- In part (a), discuss the substances listed in the table.
  - **?** "Why is the point representing -3 closer to 0 than -10?" -3 is 3 units to the left of 0, and -10 is 10 units to the left of 0. Because 3 is less than 10, -3 is closer to 0 than -10.
  - **?** "Why is it important for airplane fuel to have a very low freezing point?" Planes flying in very cold temperatures need fuel to remain in liquid form.
- In part (b), to answer the question "Which substance has a freezing point closer to the freezing point of water?" be sure students understand that the absolute value of the freezing point is how you measure the distance to 0.

Words of Wisdom

- **Construct Viable Arguments and Critique the Reasoning of Others:** Too often students will say, "Because it is," or "It's obvious." Look for a reference to absolute value when students explain their reasoning.

Closure

- **Exit Ticket:** The freezing point of vinegar is $-2°C$. Is the freezing point of vinegar or honey closer to the freezing point of water? Explain your reasoning. vinegar; Because $|-2| < |-3|$, the freezing point of vinegar is closer to $0°C$, the freezing point of water.

EXAMPLE 3 Comparing Values

Remember

A number line can be used to compare and order integers. Numbers to the left are less than numbers to the right. Numbers to the right are greater than numbers to the left.

Compare 1 and $\left|-4\right|$.

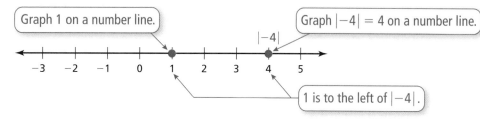

Graph 1 on a number line.

Graph $\left|-4\right| = 4$ on a number line.

$\left|-4\right|$

1 is to the left of $\left|-4\right|$.

∴ So, $1 < \left|-4\right|$.

On Your Own

Now You're Ready
Exercises 20–25

Copy and complete the statement using <, >, or =.

5. $\left|-2\right|$ ▢ -1

6. -7 ▢ $\left|6\right|$

7. $\left|10\right|$ ▢ 11

8. 9 ▢ $\left|-9\right|$

EXAMPLE 4 Real-Life Application

Substance	Freezing Point (°C)
Butter	35
Airplane fuel	−53
Honey	−3
Mercury	−39
Candle wax	55

The *freezing point* is the temperature at which a liquid becomes a solid.

a. Which substance in the table has the lowest freezing point?

b. Is the freezing point of mercury or butter closer to the freezing point of water, 0°C?

a. Graph each freezing point.

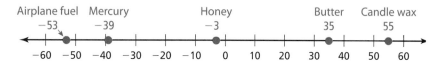

Airplane fuel −53 Mercury −39 Honey −3 Butter 35 Candle wax 55

∴ Airplane fuel has the lowest freezing point, −53°C.

b. The freezing point of water is 0°C, so you can use absolute values.

Mercury: $\left|-39\right| = 39$ **Butter:** $\left|35\right| = 35$

∴ Because 35 is less than 39, the freezing point of butter is closer to the freezing point of water.

On Your Own

9. Is the freezing point of airplane fuel or candle wax closer to the freezing point of water? Explain your reasoning.

 Vocabulary and Concept Check

1. **VOCABULARY** Which of the following numbers are integers?

$$9, 3.2, -1, \frac{1}{2}, -0.25, 15$$

2. **VOCABULARY** What is the absolute value of an integer?

3. **WHICH ONE DOESN'T BELONG?** Which expression does *not* belong with the other three? Explain your reasoning.

| $|6|$ | 6 | -6 | $|-6|$ |

 Practice and Problem Solving

Find the absolute value.

 4. $|9|$ **5.** $|-6|$ **6.** $|-10|$ **7.** $|10|$

8. $|-15|$ **9.** $|13|$ **10.** $|-7|$ **11.** $|-12|$

12. $|5|$ **13.** $|-8|$ **14.** $|0|$ **15.** $|18|$

16. $|-24|$ **17.** $|-45|$ **18.** $|60|$ **19.** $|-125|$

Copy and complete the statement using <, >, or =.

20. $2 \quad\boxed{}\quad |-5|$ **21.** $|-4| \quad\boxed{}\quad 7$ **22.** $-5 \quad\boxed{}\quad |-9|$

23. $|-4| \quad\boxed{}\quad -6$ **24.** $|-1| \quad\boxed{}\quad |-8|$ **25.** $|5| \quad\boxed{}\quad |-5|$

ERROR ANALYSIS Describe and correct the error.

26.

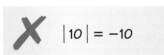

$|10| = -10$

27.

$|-5| < 4$

28. **SAVINGS** You deposit $50 in your savings account. One week later, you withdraw $20. Write each amount as an integer.

29. **ELEVATOR** You go down 8 floors in an elevator. Your friend goes up 5 floors in an elevator. Write each amount as an integer.

Order the values from least to greatest.

30. $8, |3|, -5, |-2|, -2$ **31.** $|-6|, -7, 8, |5|, -6$

32. $-12, |-26|, -15, |-12|, |10|$ **33.** $|-34|, 21, -17, |20|, |-11|$

Simplify the expression.

34. $|-30|$ **35.** $-|4|$ **36.** $-|-15|$

Assignment Guide and Homework Check

Level	Day 1 Activity Assignment	Day 2 Lesson Assignment	Homework Check
Basic	4–11, 46–50	1–3, 15–35 odd	2, 17, 25, 29, 31
Average	4–11, 46–50	1–3, 17–27 odd, 30–36 even, 37–43 odd	2, 17, 25, 30, 41
Advanced	4–11, 46–50	1–3, 20–44 even	24, 30, 38, 40
Accelerated	1–11, 20–44 even, 46–50		24, 30, 38, 40

Common Errors

- **Exercises 4–19** Students may think that the absolute value of a number is its opposite and say $|6| = -6$. Use a number line to show them that the absolute value is a number's distance from 0, so it is always a positive number or zero.
- **Exercises 20–27** When comparing absolute values of negative integers, students may not find the absolute values and instead compare the integers themselves.
- **Exercise 22** A student may find $-5 > |-9|$. The student likely did not find the absolute value of -9 to be 9 and instead compared -5 to -9.
- **Exercise 41** Students may write 14 and 18, rather than -14 and -18, for the diver's positions. These students did not account for the fact that the diver is *below* sea level. Use a vertical scale by turning a number line so that it runs vertically. Point out to students that the exercise defines sea level as 0 on the number line, so the diver's positions are negative.

1.1 Record and Practice Journal

Find the absolute value.

1. $|-1|$
   1

2. $|-14|$
   14

3. $|0|$
   0

4. $|6|$
   6

Complete the statement using <, >, or =.

5. $6 \_\_ |-2|$
   >

6. $-7 \_\_ |-8|$
   <

7. $|-9| \_\_ 5$
   >

8. $|-2| \_\_ 2$
   =

Order the values from least to greatest.

9. $4, |7|, -1, |-3|, -4$
   $-4, -1, |-3|, 4, |7|$

10. $|2|, -3, |-5|, -1, 6$
    $-3, -1, |2|, |-5|, 6$

11. You download 12 new songs to your MP3 player. Then you delete 5 old songs. Write each amount as an integer.
    $12, -5$

Practice and Problem Solving

4. 9
5. 6
6. 10
7. 10
8. 15
9. 13
10. 7
11. 12
12. 5
13. 8
14. 0
15. 18
16. 24
17. 45
18. 60
19. 125
20. $2 < |-5|$
21. $|-4| < 7$
22. $-5 < |-9|$
23. $|-4| > -6$
24. $|-1| < |-8|$
25. $|5| = |-5|$
26. The absolute value of a number cannot be negative. $|10| = 10$
27. Because $|-5| = 5$, the statement is incorrect. $|-5| > 4$
28. $50, -20$
29. $-8, 5$
30. $-5, -2, |-2|, |3|, 8$
31. $-7, -6, |5|, |-6|, 8$
32. $-15, -12, |10|, |-12|, |-26|$
33. $-17, |-11|, |20|, 21, |-34|$
34. 30
35. -4
36. -15

Practice and Problem Solving

37. a. MATE

b. TEAM

38. *Sample answer:* -4

39. $n \geq 0$

40. $n \leq 0$

41. See *Taking Math Deeper.*

42. Loihi

43. a. Player 3

b. Player 2

c. Player 1

44. true; If a number x is negative, then its absolute value is its opposite, $-x$.

45. false; The absolute value of zero is zero, which is neither positive nor negative.

Fair Game Review

46. 51   **47.** 144

48. 398   **49.** 3170

50. A

Mini-Assessment

Find the absolute value of the integer.

1. 6 6   **2.** -13 13

3. -17 17   **4.** 0 0

5. You deposit $125 in your checking account. One month later, you withdraw $65. Write each amount as an integer. 125; -65

Taking Math Deeper

Exercise 41

In this problem, negative numbers are used on a vertical scale to indicate *positions* that are below sea level. The numbers on the number line describe *position* (a negative number), not *depth* below sea level (a positive number).

 a. Draw and label a vertical number line.

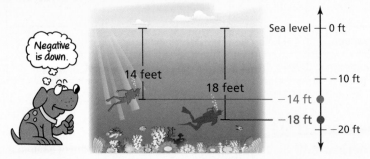

 b. Which integer is greater?

Of the two integers -14 and -18, -14 is greater because it is *higher* on a vertical number line.

c. Which integer has the greater absolute value?

$\left| -18 \right| = 18$ is greater than $\left| -14 \right| = 14$.

Compare this absolute value with the depth of that diver. The numbers are the same. The diver is 18 feet below sea level.

 In mathematics, many concepts require understanding. Some, however, simply require acceptance. No one really knows why negative means "left" or "down" on a number line. We are not sure of the reason for choosing "right" to be positive, but it could have been something as simple as the fact that René Descartes was right-handed.

Project

Draw a picture that illustrates a real-life use of negative integers. Write a paragraph that explains how negative numbers are used in your picture.

Reteaching and Enrichment Strategies

If students need help. . .	If students got it. . .
Resources by Chapter • Practice A and Practice B • Puzzle Time Record and Practice Journal Practice Differentiating the Lesson Lesson Tutorials Skills Review Handbook	Resources by Chapter • Enrichment and Extension • Technology Connection Start the next section

37. **PUZZLE** Use a number line.

    a. Graph and label the following points on a number line: $A = -3$, $E = 2$, $M = -6$, $T = 0$. What word do the letters spell?

    b. Graph and label the absolute value of each point in part (a). What word do the letters spell now?

38. **OPEN-ENDED** Write a negative integer whose absolute value is greater than 3.

REASONING Determine whether $n \geq 0$ or $n \leq 0$.

39. $n + \left| -n \right| = 2n$ 40. $n + \left| -n \right| = 0$

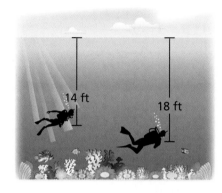

41. **CORAL REEF** The depths of two scuba divers exploring a living coral reef are shown.

    a. Write an integer for the position of each diver relative to sea level.

    b. Which integer in part (a) is greater?

    c. Which integer in part (a) has the greater absolute value? Compare this absolute value with the depth of that diver.

42. **VOLCANOES** The *summit elevation* of a volcano is the elevation of the top of the volcano relative to sea level. The summit elevation of the volcano Kilauea in Hawaii is 1277 meters. The summit elevation of the underwater volcano Loihi in the Pacific Ocean is -969 meters. Which summit is closer to sea level?

43. **MINIATURE GOLF** The table shows golf scores, relative to *par*.

    a. The player with the lowest score wins. Which player wins?

    b. Which player is at par?

    c. Which player is farthest from par?

Player	Score
1	+5
2	0
3	−4
4	−1
5	+2

 Determine whether the statement is *true* or *false*. Explain your reasoning.

44. If $x < 0$, then $\left| x \right| = -x$.

45. The absolute value of every integer is positive.

Fair Game Review *What you learned in previous grades & lessons*

Add. (*Skills Review Handbook*)

46. $19 + 32$ 47. $50 + 94$ 48. $181 + 217$ 49. $1149 + 2021$

50. **MULTIPLE CHOICE** Which value is *not* a whole number? (*Skills Review Handbook*)

    (**A**) -5 (**B**) 0 (**C**) 4 (**D**) 113

1.2 Adding Integers

Essential Question Is the sum of two integers *positive*, *negative*, or *zero*? How can you tell?

1 ACTIVITY: Adding Integers with the Same Sign

Work with a partner. Use integer counters to find −4 + (−3).

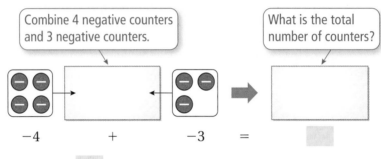

Combine 4 negative counters and 3 negative counters.

What is the total number of counters?

−4   +   −3   =

So, −4 + (−3) = [ ].

2 ACTIVITY: Adding Integers with Different Signs

Work with a partner. Use integer counters to find −3 + 2.

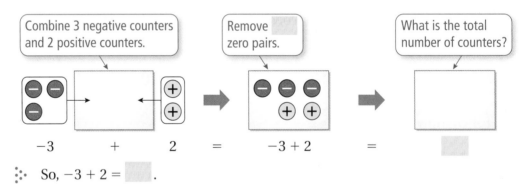

Combine 3 negative counters and 2 positive counters.

Remove [ ] zero pairs.

What is the total number of counters?

−3   +   2   =   −3 + 2   =

So, −3 + 2 = [ ].

Integers

In this lesson, you will

- add integers.
- show that the sum of a number and its opposite is 0.
- solve real-life problems.

3 ACTIVITY: Adding Integers with Different Signs

Work with a partner. Use a number line to find 5 + (−3).

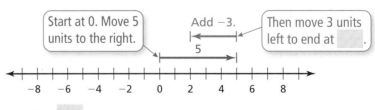

Start at 0. Move 5 units to the right.

Add −3.

Then move 3 units left to end at [ ].

So, 5 + (−3) = [ ].

Laurie's Notes

Introduction

Applying Mathematical Practices
- **Look For and Express Regularity in Repeated Reasoning.** In this lesson, students will use color integer counters to develop a conceptual understanding of integer addition. When the signs are the same, it is a combining process and the sum has the same sign. When the signs are different, it is a combining process *and* you remove some zero pairs.

Motivate
- **?** "What is the net result of an 8-yard loss in football followed by a 10-yard gain?" 2-yard gain
- **?** "What is the net result of scoring 25 points in a video game then losing 40 points?" a loss of 15 points
- Today's activity is about how integers are added.

Demonstrate
- If this is the student's first experience with integer counters, define a *yellow* counter as positive 1 ($+1$) and a *red* counter as negative 1 (-1).
- Counters of opposite color "neutralize" each other, so the net result of such a pair is zero. This is called a *zero pair*.
- **Model:** Show students that there are many ways to model a single integer. For example, the number 2 can be represented by two yellow counters or by three yellow counters with one red counter (2 plus 1 zero pair).

Activity Notes

Activity 1 and Activity 2
- **Management Tip:** Store integer counters in self-locking bags. Put 15–20 counters in each bag.
- Students should use counters even if they say they know the answer.
- **Model:** A student volunteer could model Activity 1 and Activity 2 at the overhead projector saying aloud what he or she is doing with the counters.
- **Common Error:** You may hear students say that -3 is greater than 2. You should respond "Gee, 2 is farther to the right on the number line than -3. Are you sure -3 is greater?" Remind students that the number farther to the right on the number line is greater.
- The use of parentheses around the integer -3 is for clarity. Sometimes people write $-4 + {}^-3$, with the raised negative sign.

Activity 3
- Numbers are being represented by *directed line segments*. Positive numbers point to the right and negative numbers point to the left.
- **Connection:** The amount that the two directed line segments overlap is the same as the number of zero pairs that would result if the same problem were modeled using integer counters.

What Your Students Will Learn
- Add integers with the same sign or with different signs.
- Show that the sum of a number and its opposite is 0 using the Additive Inverse Property.

Previous Learning
Students need to know how to add and subtract whole numbers.

Technology for the *Teacher*

Dynamic Classroom

Lesson Plans
Complete Materials List

1.2 Record and Practice Journal

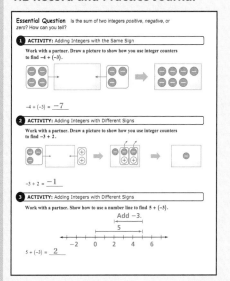

Essential Question Is the sum of two integers *positive*, *negative*, or *zero*? How can you tell?

1 ACTIVITY: Adding Integers with the Same Sign

Work with a partner. Draw a picture to show how you use integer counters to find $-4 + (-3)$.

$-4 + (-3) = \underline{-7}$

2 ACTIVITY: Adding Integers with Different Signs

Work with a partner. Draw a picture to show how you use integer counters to find $-3 + 2$.

$-3 + 2 = \underline{-1}$

3 ACTIVITY: Adding Integers with Different Signs

Work with a partner. Show how to use a number line to find $5 + (-3)$.

$5 + (-3) = \underline{2}$

Differentiated Instruction

Visual

Use the number line to review and reinforce the Commutative Property of Addition. Draw a number line on the board. Model the expressions $-5 + 2$ and $2 + (-5)$. Students should see from the movement on the number line that the sum (result) is the same. The initial direction of movement does not matter.

1.2 Record and Practice Journal

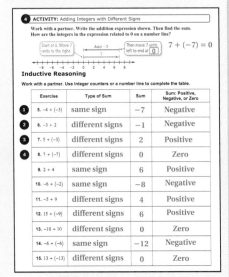

4 ACTIVITY: Adding Integers with Different Signs

Work with a partner. Write the addition expression shown. Then find the sum. How are the integers in the expression related to 0 on a number line?

$7 + (-7) = 0$

Inductive Reasoning

Work with a partner. Use integer counters or a number line to complete the table.

	Exercise	Type of Sum	Sum	Sum: Positive, Negative, or Zero
1	5. $-4 + (-3)$	same sign	-7	Negative
2	6. $-3 + 2$	different signs	-1	Negative
3	7. $5 + (-3)$	different signs	2	Positive
4	8. $7 + (-7)$	different signs	0	Zero
	9. $2 + 4$	same sign	6	Positive
	10. $-6 + (-2)$	same sign	-8	Negative
	11. $-5 + 9$	different signs	4	Positive
	12. $15 + (-9)$	different signs	6	Positive
	13. $-10 + 10$	different signs	0	Zero
	14. $-6 + (-6)$	same sign	-12	Negative
	15. $13 + (-13)$	different signs	0	Zero

What Is Your Answer?

16. **IN YOUR OWN WORDS** Is the sum of two integers *positive, negative,* or *zero*? How can you tell?
Sample answer: The sum will have the same sign as the integer with the greater absolute value.

17. **STRUCTURE** Write a general rule for adding Sample answers given.

a. two integers with the same sign.
To add two positive integers, add normally. To add two negative integers, ignore the signs and add. Then make the answer negative.

b. two integers with different signs.
Subtract the lesser absolute value from the greater absolute value. Use the sign of the integer with the greater absolute value.

c. two integers that vary in sign.
The sum is zero.

T-9

Laurie's Notes

Activity 4

❓ "How would you use the number line to show $-7 + 7$?" Ask a student volunteer to show the solution. Start at zero. Move 7 units to the left to get to -7. Then move 7 units to the right. End at 0.

• Students should recognize that the order in which you add two numbers does not matter (Commutative Property of Addition). You can start at 7 and then move 7 units to the left.

Words of Wisdom

• You can use a standard deck of playing cards to generate random addition problems. Define red = negative, black = positive, Jacks = 11, Queens = 12, Kings = 13, Aces = 1, Jokers = 0.

Inductive Reasoning

• Students should work with partners to complete the table. Note that Questions 5–8 are the problems completed in Activities 1–4.

• The goal of Questions 5–15 is to develop some understanding about the two types of addition problems: integers with the same sign and integers with different signs.

• **Look for and Express Regularity in Repeated Reasoning:** It is important for students to record the *Type of Sum*. Mathematically proficient students will observe that when the signs are the same, you "just add." When the signs are different, the strategy is different, and students may not be articulate in describing the pattern.

• If time permits, you might give a problem such as $47 + (-58)$. Clearly, you do not want to model this with color integer counters. Students should be able to describe the strategy of putting 47 yellow and 58 red together. There would be 47 zero pairs with 11 red remaining representing a sum of -11.

What Is Your Answer?

• In Questions 16 and 17, the sum of two integers can be positive, negative, or zero. If the two integers have the same sign, the sign of the sum is the same as the integers. If the integers have different signs, then the sum is the sign of the integer with the greater absolute value. (Formal rules will be presented in the lesson.)

❓ **Extension:** Use the integer counters or a number line to model the sum of three integers.

• "What is the sum of $3 + (-2) + 5$?" 6

• "What is the sum of $(-4) + 2 + (-5)$?" -7

Closure

• "If the sum of two integers is negative, are both integers negative? How do you know?" Both integers could be negative, but they may not be. The sum of 4 and -5 is negative, but both integers are not negative.

• "If the sum of two integers is positive, are both integers positive? How do you know?" Both integers could be positive, but they may not be. The sum of -4 and 5 is positive, but both integers are not positive.

4 ACTIVITY: Adding Integers with Different Signs

Math Practice

Make Conjectures

How can the relationship between the integers help you write a rule?

Work with a partner. Write the addition expression shown. Then find the sum. How are the integers in the expression related to 0 on a number line?

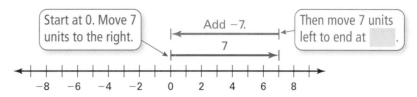

Start at 0. Move 7 units to the right.

Add −7.

7

Then move 7 units left to end at [ ].

Inductive Reasoning

Work with a partner. Use integer counters or a number line to complete the table.

	Exercise	Type of Sum	Sum	Sum: Positive, Negative, or Zero
①	5. −4 + (−3)	Integers with the same sign		
②	6. −3 + 2			
③	7. 5 + (−3)			
④	8. 7 + (−7)			
	9. 2 + 4			
	10. −6 + (−2)			
	11. −5 + 9			
	12. 15 + (−9)			
	13. −10 + 10			
	14. −6 + (−6)			
	15. 13 + (−13)			

What Is Your Answer?

16. **IN YOUR OWN WORDS** Is the sum of two integers *positive*, *negative*, or *zero*? How can you tell?

17. **STRUCTURE** Write general rules for adding (a) two integers with the same sign, (b) two integers with different signs, and (c) two integers that vary only in sign.

Practice

Use what you learned about adding integers to complete Exercises 8–15 on page 12.

Check It Out
Lesson Tutorials
BigIdeasMath.com

Key Vocabulary
opposites, *p. 10*
additive inverse, *p. 10*

Key Idea

Adding Integers with the Same Sign

Words Add the absolute values of the integers. Then use the common sign.

Numbers $2 + 5 = 7$ $-2 + (-5) = -7$

EXAMPLE 1 **Adding Integers with the Same Sign**

Find $-2 + (-4)$. Use a number line to check your answer.

$-2 + (-4) = -6$ Add $|-2|$ and $|-4|$.

Use the common sign.

∴ The sum is -6.

Check

The Meaning of a Word

Opposite

When you walk across a street, you are moving to the **opposite** side of the street.

On Your Own

Add.

1. $7 + 13$ **2.** $-8 + (-5)$ **3.** $-20 + (-15)$

Two numbers that are the same distance from 0, but on opposite sides of 0, are called **opposites.** For example, -3 and 3 are opposites.

Key Ideas

Adding Integers with Different Signs

Words Subtract the lesser absolute value from the greater absolute value. Then use the sign of the integer with the greater absolute value.

Numbers $8 + (-10) = -2$ $-13 + 17 = 4$

Additive Inverse Property

Words The sum of an integer and its **additive inverse,** or opposite, is 0.

Numbers $6 + (-6) = 0$ $-25 + 25 = 0$ **Algebra** $a + (-a) = 0$

Multi-Language Glossary at BigIdeasMath.com

Laurie's Notes

Introduction

Connect

- **Yesterday:** Students used integer counters and a number line to add integers of the same sign and of different signs.
- **Today:** Students will add integers without the use of a visual or concrete model.

Motivate

? "Is the sum of 58 and −72 positive or negative? How do you know?"
Negative. *Sample answers:*

Using Counters: Some students may say that there would be more red counters (72) than yellow counters (58), so the sum is negative.

Using a Number Line: Some students may describe a number line: if you go back (left) 72 units and then forward (right) 58 units, you won't get back to 0, so the sum is negative.

Using Definitions: Some students may remember that the sign of the integer with the greater absolute value (72) is negative, so the sum is negative.

Technology for the **Teacher**

Dynamic Classroom

Lesson Tutorials
Lesson Plans
Answer Presentation Tool

Lesson Notes

Example 1

- **Model with Mathematics:** As you discuss the example, refer to the models from the Activity.
  - When the signs are the *same,* the counters will be the *same color*.
  - When the signs are the *same,* both directed line segments will be going in the *same direction*.

Key Ideas

- Discuss the definition of opposites.
- **?** **Construct Viable Arguments and Critique the Reasoning of Others:** "When you add two integers with different signs, how do you know if the sum is positive or negative?" Students should be using the concept of absolute value even if they don't use the precise language. You want to hear something about the size of the number, meaning its absolute value.
- **?** **Look for and Express Regularity in Repeated Reasoning:** Write these problems on the board: $14 + (-8) = ?$ and $(-14) + 8 = ?$. Ask, "How are the problems alike? How are they different?" *Sample answer:*
  Alike: They each use the numbers 14 and 8. They both consist of two different signs, which are being added together.
  Different: In the first problem, 14 is positive and 8 is negative. In the second problem, 14 is negative and 8 is positive.
- Define how to add integers with different signs.
- Define the Additive Inverse Property. This is a special case of adding integers with different signs.
- **?** "How many zero pairs are there when you add $(-5) + 5$?" 5

Extra Example 1
Find $-3 + (-12)$. -15

On Your Own

1. 20
2. −13
3. −35

Extra Example 2

Add.

a. $-11 + 6$ -5

b. $12 + (-5)$ 7

c. $4 + (-4)$ 0

Extra Example 3

Find the change in the account balance for August.

August Transactions	
Deposit	$35
Deposit	$40
Withdrawal	−$25

increased $50

On Your Own

4. 9	**5.** −1
6. 0	**7.** −$20

Differentiated Instruction

Vocabulary

Write a table of opposites on the board. Encourage students to add to the list.

Word	Opposite
little	big
forward	backward

Ask the English learners to write the words in their native languages in another column and to share them with the class. Explain to the class that in mathematics, every nonzero number has an opposite. Every pair of opposites consists of a positive number and a negative number. Ask students to name some pairs of opposite numbers. Write opposite numbers and the words that represent them in the table.

Example 2

- For each part of this example, have student volunteers explain how it was computed.
- **Part (a):** Because 5 has the lesser absolute value, you subtract it from the absolute value of -10. In general, use the sign of the number with the greater absolute value. In this case, the answer will be negative.

Words of Wisdom

- **Model with Mathematics:** Be sure students understand that the subtraction of the two absolute values is connected to the zero pairs that get removed when using integer counters, or it is the overlapping distance when the number line is used.
- If students are not getting correct answers, use integer counters or the number line to model several additional examples.

Example 3

- Review the properties of addition (associative, commutative, zero) from a previous course. Explain to students that these rules also apply to integers.
- **Financial Literacy:** Provide a brief description of what *deposit* and *withdrawal* mean in banking. A *deposit* is when you add money to an account, and a *withdrawal* is when you take money out of an account. Checkbooks are one context where addition of opposites occurs. Depositing $100 and writing a check for $100 results in a zero change in the balance.
- **Model:** Using *play* money and a student volunteer, act out the following.
  Hand $50 to a banker (*deposit*).
  Ask for $40 back (*withdrawal*).
  Hand the banker $75 (*deposit*).
  Ask for $50 back (*withdrawal*).
- **?** "What is the change in the balance in your account?" $35
- **?** "Would the change in the balance be the same if you added the two deposits first, added the two withdrawals next, and then found the sum of the two answers?" yes; $(50 + 75) + (-40 + (-50)) = 125 + (-90) = 35$
- **FYI:** The Addition Property of Zero states that the sum of any number and zero is that number.

Closure

- What do you know about the sum $A + B$? Explain your reasoning.

- Two integers have different signs. Their sum is -8. What are possible values for the two integers? *Sample answers:* -9 and 1, -10 and 2, 3 and -11, 4 and -12

EXAMPLE 2 · Adding Integers with Different Signs

a. Find 5 + (−10).

$$5 + (-10) = -5 \qquad |-10| > |5|. \text{ So, subtract } |5| \text{ from } |-10|.$$

Use the sign of −10.

::· The sum is −5.

b. Find −3 + 7.

$$-3 + 7 = 4 \qquad |7| > |-3|. \text{ So, subtract } |-3| \text{ from } |7|.$$

Use the sign of 7.

::· The sum is 4.

c. Find −12 + 12.

$$-12 + 12 = 0 \qquad \text{The sum is 0 by the Additive Inverse Property.}$$

−12 and 12 are opposites.

::· The sum is 0.

EXAMPLE 3 · Adding More Than Two Integers

The list shows four bank account transactions in July. Find the change C in the account balance.

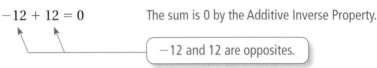

JULY TRANSACTIONS	
Withdrawal	-$40
Deposit	$50
Deposit	$75
Withdrawal	-$50

> **Study Tip**
>
> A deposit of $50 and a withdrawal of $50 represent opposite quantities, +50 and −50, which have a sum of 0.

Find the sum of the four transactions.

$$C = -40 + 50 + 75 + (-50) \qquad \text{Write the sum.}$$
$$= -40 + 75 + 50 + (-50) \qquad \text{Commutative Property of Addition}$$
$$= -40 + 75 + [50 + (-50)] \qquad \text{Associative Property of Addition}$$
$$= -40 + 75 + 0 \qquad \text{Additive Inverse Property}$$
$$= 35 + 0 \qquad \text{Add } -40 \text{ and } 75.$$
$$= 35 \qquad \text{Addition Property of Zero}$$

::· Because $C = 35$, the account balance increased $35 in July.

On Your Own

Now You're Ready
Exercises 8–23
and 28–39

Add.

4. −2 + 11   **5.** 9 + (−10)   **6.** −31 + 31

7. WHAT IF? In Example 3, the deposit amounts are $30 and $40. Find the change C in the account balance.

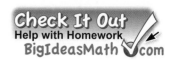

 ## Vocabulary and Concept Check

1. **WRITING** How do you find the additive inverse of an integer?

2. **NUMBER SENSE** Is $3 + (-4)$ the same as $-4 + 3$? Explain.

Tell whether the sum is *positive*, *negative*, or *zero* without adding. Explain your reasoning.

3. $-8 + 20$ 4. $30 + (-30)$ 5. $-10 + (-18)$

Tell whether the statement is *true* or *false*. Explain your reasoning.

6. The sum of two negative integers is always negative.

7. An integer and its absolute value are always opposites.

 ## Practice and Problem Solving

Add.

① ②   8. $6 + 4$ 9. $-4 + (-6)$ 10. $-2 + (-3)$ 11. $-5 + 12$

12. $5 + (-7)$ 13. $8 + (-8)$ 14. $9 + (-11)$ 15. $-3 + 13$

16. $-4 + (-16)$ 17. $-3 + (-1)$ 18. $14 + (-5)$ 19. $0 + (-11)$

20. $-10 + (-15)$ 21. $-13 + 9$ 22. $18 + (-18)$ 23. $-25 + (-9)$

ERROR ANALYSIS Describe and correct the error in finding the sum.

24.

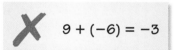

$9 + (-6) = -3$

25.

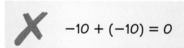

$-10 + (-10) = 0$

26. **TEMPERATURE** The temperature is $-3°F$ at 7:00 A.M. During the next 4 hours, the temperature increases $21°F$. What is the temperature at 11:00 A.M.?

27. **BANKING** Your bank account has a balance of $-\$12$. You deposit $\$60$. What is your new balance?

Tell how the Commutative and Associative Properties of Addition can help you find the sum mentally. Then find the sum.

③ 28. $9 + 6 + (-6)$ 29. $-8 + 13 + (-13)$ 30. $9 + (-17) + (-9)$

31. $7 + (-12) + (-7)$ 32. $-12 + 25 + (-15)$ 33. $6 + (-9) + 14$

Add.

34. $13 + (-21) + 16$ 35. $22 + (-14) + (-35)$ 36. $-13 + 27 + (-18)$

37. $-19 + 26 + 14$ 38. $-32 + (-17) + 42$ 39. $-41 + (-15) + (-29)$

Assignment Guide and Homework Check

Level	Day 1 Activity Assignment	Day 2 Lesson Assignment	Homework Check
Basic	8–15, 50–54	1–7, 17–35 odd	17, 27, 31, 35
Average	8–15, 50–54	1–7, 21, 23–25, 27, 32–46 even	21, 27, 32, 38
Advanced	8–15, 50–54	1–7, 24, 30–48 even, 49	30, 36, 40, 46, 48
Accelerated	1–15, 24–48 even, 49–54		30, 36, 40, 46, 48

Common Errors

- **Exercises 8–23, 28–39** Students may try to ignore the signs and just add the integers. Remind them of the meaning of absolute value. Make sure they understand that they should use the sign of the number that is farther from zero. Also remind them of the Key Ideas, and how the signs of the integers determine if they need to add or subtract the integers.
- **Exercise 48** Students may not realize that each height measurement is given in reference to the previous point. Tell them to determine the measurement in relation to point *A*, which would be zero on a number line.

1.2 Record and Practice Journal

Add.

1. $-9 + 2$ -7
2. $5 + (-5)$ 0
3. $-12 + (-6)$ -18
4. $-10 + 19 + 5$ 14
5. $-11 + (-20) + 9$ -22
6. $-7 + 7 + (-8)$ -8

Use mental math to solve the equation.

7. $x + (-5) = 4$ $x = 9$
8. $y + 6 = -2$ $y = -8$
9. $-10 = -7 + z$ $z = -3$

10. The table shows the change in your hair length over a year.

Month	January	February	August	September	December
Change in hair length (inches)	2	-1	3	-4	3

   a. What is the total change in your hair length at the end of the year? **3 inches**

   b. Is your hair longer in January or December? Explain your reasoning. **December**

   c. When is your hair the longest? Explain your reasoning. **August**

Vocabulary and Concept Check

1. Change the sign of the integer.

2. yes; The sums are the same by the Commutative Property of Addition.

3. positive; 20 has the greater absolute value and is positive.

4. zero; 30 and −30 are additive inverses.

5. negative; The common sign is a negative sign.

6. true; To add integers with the same sign, add the absolute values and use the common sign.

7. false; A positive integer and its absolute value are equal, not opposites.

Practice and Problem Solving

8. 10   9. −10

10. −5   11. 7

12. −2   13. 0

14. −2   15. 10

16. −20   17. −4

18. 9   19. −11

20. −25   21. −4

22. 0   23. −34

24. The wrong sign is used. $9 + (-6) = 3$

25. −10 and −10 are not opposites. $-10 + (-10) = -20$

26. 18°F   27. $48

28. Use the Associative Property to add 6 and −6 first. 9

29. Use the Associative Property to add 13 and −13 first. −8

Practice and Problem Solving

30. *Sample answer:* Use the Commutative Property to switch the last two terms. -17

31. *Sample answer:* Use the Commutative Property to switch the last two terms. -12

32. *Sample answer:* Use the Commutative Property to switch the last two terms. -2

33. *Sample answer:* Use the Commutative Property to switch the last two terms. 11

34. 8   **35.** -27

36. -4 **37.** 21

38. -7 **39.** -85

40. 0

41. *Sample answer:*
$-26 + 1; -12 + (-13)$

42. -1 **43.** -3

44. 9   **45.** $d = -10$

46. $b = 2$ **47.** $m = -7$

48. See Additional Answers.

49. See *Taking Math Deeper*.

Fair Game Review

50. 31   **51.** 8

52. 114   **53.** 183

54. D

Mini-Assessment

Add.

1. $10 + (-12)$ -2

2. $-7 + (-5)$ -12

3. $-17 + 25$ 8

4. $65 + (-99)$ -34

5. The temperature is $-2°$F at 6 A.M. During the next three hours, the temperature increases $15°$F. What is the temperature at 9 A.M.? $13°$F

Taking Math Deeper

Exercise 49

In this puzzle, students get a chance to apply integer addition with *Guess, Check, and Revise.*

 Solve the straightforward part of the puzzle.

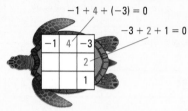

Make a list of the numbers from -4 to 4. Cross off the numbers you have used.

$-4 \ \cancel{-3} \ -2 \ \cancel{-1} \ 0 \ \cancel{1} \ \cancel{2} \ 3 \ \cancel{4}$

Use the strategy *Guess, Check, and Revise* to complete the square.

You can create other magic squares by repeating the same digit in each number in the magic square $(-11, 44, -33,$ etc.$)$.

The emperor Yu-Huang was a legendary figure in China, in the same sense that King Arthur was a legendary figure in Europe. He was called the Jade Emperor, and there are many stories about him. In addition to the story of the magic square and the turtle, the Jade Emperor is credited with creating the Chinese Zodiac in which each sequence of 12 years is given the name of an animal, such as the "Year of the Snake" or the "Year of the Rat."

Project

Create a new magic square using integers. Decide which squares will contain numbers and which squares will be blank. Complete your magic square. Then switch puzzles with a classmate and complete the puzzle you receive. Check your answers.

Reteaching and Enrichment Strategies

If students need help. . .	If students got it. . .
Resources by Chapter • Practice A and Practice B • Puzzle Time Record and Practice Journal Practice Differentiating the Lesson Lesson Tutorials Skills Review Handbook	Resources by Chapter • Enrichment and Extension • Technology Connection Start the next section

40. SCIENCE A lithium atom has positively charged protons and negatively charged electrons. The sum of the charges represents the charge of the lithium atom. Find the charge of the atom.

Lithium Atom

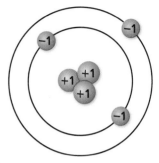

41. OPEN-ENDED Write two integers with different signs that have a sum of -25. Write two integers with the same sign that have a sum of -25.

ALGEBRA Evaluate the expression when $a = 4$, $b = -5$, and $c = -8$.

42. $a + b$ **43.** $-b + c$ **44.** $|a + b + c|$

MENTAL MATH Use mental math to solve the equation.

45. $d + 12 = 2$ **46.** $b + (-2) = 0$ **47.** $-8 + m = -15$

48. PROBLEM SOLVING Starting at point A, the path of a dolphin jumping out of the water is shown.

     a. Is the dolphin deeper at point C or point E? Explain your reasoning.

     b. Is the dolphin higher at point B or point D? Explain your reasoning.

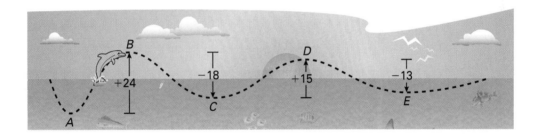

49. **Puzzle** According to a legend, the Chinese Emperor Yu-Huang saw a magic square on the back of a turtle. In a *magic square*, the numbers in each row and in each column have the same sum. This sum is called the *magic sum*.

Copy and complete the magic square so that each row and each column has a magic sum of 0. Use each integer from -4 to 4 exactly once.

Fair Game Review What you learned in previous grades & lessons

Subtract. *(Skills Review Handbook)*

50. $69 - 38$ **51.** $82 - 74$ **52.** $177 - 63$ **53.** $451 - 268$

54. MULTIPLE CHOICE What is the range of the numbers below? *(Skills Review Handbook)*

     12, 8, 17, 12, 15, 18, 30

    (A) 12   **(B)** 15   **(C)** 18   **(D)** 22

Essential Question How are adding integers and subtracting integers related?

1 **ACTIVITY: Subtracting Integers**

Work with a partner. Use integer counters to find 4 − 2.

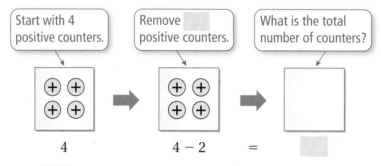

| Start with 4 positive counters. | Remove [ ] positive counters. | What is the total number of counters? |

4   4 − 2   =

So, 4 − 2 = [ ].

2 **ACTIVITY: Adding Integers**

Work with a partner. Use integer counters to find 4 + (−2).

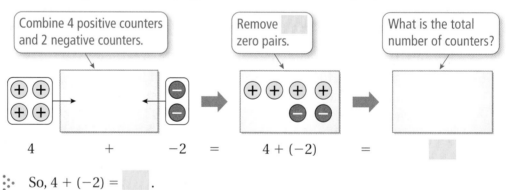

| Combine 4 positive counters and 2 negative counters. | Remove [ ] zero pairs. | What is the total number of counters? |

4   +   −2   =   4 + (−2)   =

So, 4 + (−2) = [ ].

Integers

In this lesson, you will
- subtract integers.
- solve real-life problems.

3 **ACTIVITY: Subtracting Integers**

Work with a partner. Use a number line to find −3 − 1.

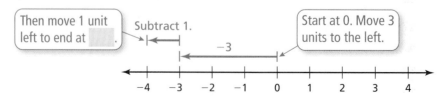

| Then move 1 unit left to end at [ ]. | Subtract 1. | Start at 0. Move 3 units to the left. |

So, −3 − 1 = [ ].

Laurie's Notes

Introduction

Applying Mathematical Practices

- **Look For and Express Regularity in Repeated Reasoning:** Students will develop a conceptual understanding of integer subtraction. Each subtraction problem is followed by a related addition problem. Modeling each problem helps students make sense of subtraction.

Motivate

- ❓ Hand a student a collection of objects (8 pencils, 12 index cards, 9 paper clips) and ask another student to take some of the objects (5 pencils, 7 index cards, 3 paper clips). "What expressions represent this situation?" $8 - 5, 12 - 7, 9 - 3$

- ❓ "One way to think about subtraction: you have some amount and you take away another amount. Does this still work when you begin with negative amounts like -3 (owe a friend \$3)?"

- Today's activity investigates subtraction of integers.

Activity Notes

Activity 1 and Activity 2

- ❓ "How would you model $4 - 2$ using integer counters?" Subtraction means that you model the first number, and then take the second number from that original collection. Students should model the problem using their counters even if they say they know the answer.

- "Now let's see how $4 - 2$ is like the addition problem you did a few days ago." Have students work through Activity 2 with partners.

Words of Wisdom

- Before trying the number line examples, try $2 - 4$ with integer counters. It will remind students that subtraction is *not* commutative.

- ❓ "How would you model $2 - 4$?" Some students may say that this is not possible, because you should subtract the lesser number from the greater number. Show the model below.

- **Model:** Show the class two yellow counters.

- ❓ "How can you take 4 yellow counters away?" Add two zero pairs and then take away 4 yellow counters. Two red counters are left. $2 - 4 = -2$

- ❓ "$4 - 2$ had the related problem $4 + (-2)$. What do you think the related addition problem would be for $2 - 4$?" $2 + (-4)$

Activity 3

- **Reason Abstractly and Quantitatively:** To subtract a positive number, move to the left. To subtract a negative number, move to the right. To help students make sense of subtracting a negative quantity, use the context of money and owing someone \$2. If you subtract away a debt of \$2, you are moving 2 units in the positive direction, to the right.

- **Model:** To model $-3 - 1$, draw an arrow pointing to the left from zero to -3. Then move *left one* because you are *subtracting positive 1*.

What Your Students Will Learn

- Subtract integers by adding their opposites.

Previous Learning

Students need to know how to add and subtract whole numbers.

Technology for the *Teacher*

Dynamic Classroom

Lesson Plans
Complete Materials List

1.3 Record and Practice Journal

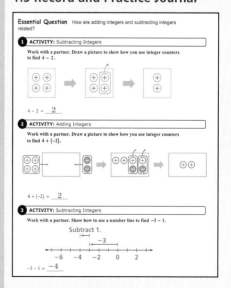

Auditory

When you subtract a number, you add its opposite. So when you subtract a number using a number line, you move in the opposite direction you would move if adding the number. This is why you move left when subtracting a positive number and move right when subtracting a negative number.

Activity 4

- Look at the related addition problem $-3 + (-1)$. Draw an arrow from 0 to -3 to represent -3. Now move to the left one because you are adding 1 in the negative direction (-1). Draw the arrow and write, "Add -1."
- Have students work with partners to write an addition expression.
- You just wrote sum and difference expressions that meant the same thing.

Inductive Reasoning

- Students should work with partners to find the sums. Note that Questions 5–8 are the problems completed in Activities 1–4.
- The goal is to develop some understanding about subtraction and the related addition problem.
- **Look for and Express Regularity in Repeated Reasoning:** It is important for students to record the *Operation: Add or Subtract*. Mathematically proficient students will observe that there is a pattern in the table. There are pairs of problems, one subtraction and one addition.

What Is Your Answer?

- In Questions 17 and 18, subtraction is the same as adding the opposite.
- **Extension:** "Use the integer counters or a number line to model $(8 - 4) - 2$ and $8 - (4 - 2)$. Are the results the same?" No, subtraction is not associative.
- **FYI:** The Associative Property of Addition states that the value of a sum does not depend on how the numbers are grouped. This does not apply for subtraction, as illustrated by the example above.

Closure

- Explain how you would use integer counters to model $4 - 6$.
  Add 6 red counters to the 4 yellow counters and remove the 4 zero pairs. The result is 2 red counters.

1.3 Record and Practice Journal

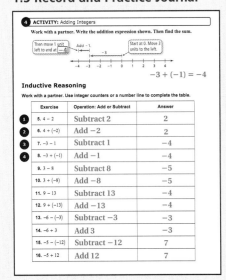

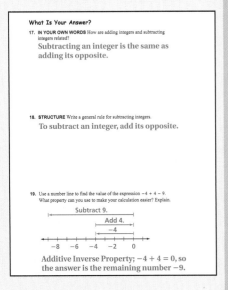

4   ACTIVITY: Adding Integers

Math Practice

Make Sense of Quantities

What integers will you use in your addition expression?

Work with a partner. Write the addition expression shown. Then find the sum.

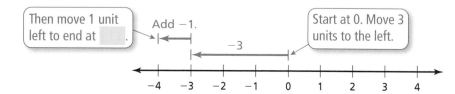

Then move 1 unit left to end at ▢ .

Add −1.

−3

Start at 0. Move 3 units to the left.

Inductive Reasoning

Work with a partner. Use integer counters or a number line to complete the table.

Exercise	Operation: Add or Subtract	Answer
5. $4 - 2$	Subtract 2	
6. $4 + (-2)$		
7. $-3 - 1$		
8. $-3 + (-1)$		
9. $3 - 8$		
10. $3 + (-8)$		
11. $9 - 13$		
12. $9 + (-13)$		
13. $-6 - (-3)$		
14. $-6 + 3$		
15. $-5 - (-12)$		
16. $-5 + 12$		

What Is Your Answer?

17. IN YOUR OWN WORDS How are adding integers and subtracting integers related?

18. STRUCTURE Write a general rule for subtracting integers.

19. Use a number line to find the value of the expression $-4 + 4 - 9$. What property can you use to make your calculation easier? Explain.

Practice

Use what you learned about subtracting integers to complete Exercises 8–15 on page 18.

Check It Out
Lesson Tutorials
BigIdeasMath ✓com

 **Key Idea**

Subtracting Integers

Words To subtract an integer, add its opposite.

Numbers $3 - 4 = 3 + (-4) = -1$

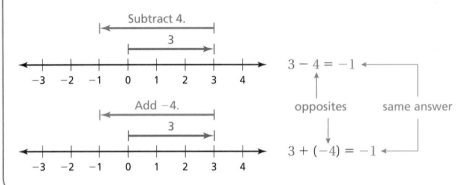

$3 - 4 = -1$

opposites   same answer

$3 + (-4) = -1$

EXAMPLE ① **Subtracting Integers**

a. Find 3 − 12.

$$3 - 12 = 3 + (-12) \qquad \text{Add the opposite of 12.}$$
$$= -9 \qquad \text{Add.}$$

⋮ The difference is -9.

b. Find −8 − (−13).

$$-8 - (-13) = -8 + 13 \qquad \text{Add the opposite of } -13.$$
$$= 5 \qquad \text{Add.}$$

⋮ The difference is 5.

c. Find 5 − (−4).

$$5 - (-4) = 5 + 4 \qquad \text{Add the opposite of } -4.$$
$$= 9 \qquad \text{Add.}$$

⋮ The difference is 9.

● **On Your Own**

Now You're Ready
Exercises 8–23

Subtract.

1. $8 - 3$ **2.** $9 - 17$ **3.** $-3 - 3$

4. $-14 - 9$ **5.** $9 - (-8)$ **6.** $-12 - (-12)$

Laurie's Notes

Introduction

Connect

- **Yesterday:** Students used integer counters and a number line to subtract integers.
- **Today:** Students will use the idea that a subtraction problem can be rewritten as an addition problem.

Motivate

Lesson Tutorials
Lesson Plans
Answer Presentation Tool

- ❓ "Is my age (teacher) minus your age (point to a student) the same as your age minus my age?" Most students will understand that an older person's age minus a younger person's age is a positive number and subtracting in the other order is a negative number.
- Draw the two number lines shown.

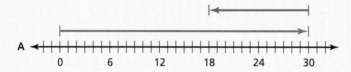

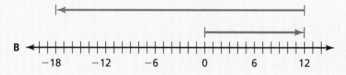

- Assume the teacher is 30 and the student is 12. Take away the context (age of two people) and simply look at the two number lines.
- ❓ "What two addition problems are modeled in each diagram?"
  A: 30 + (−12); B: 12 + (−30)
- Now add the context of the problem and write the following on the board.
  30 − 12 = 30 + (−12) = 18
  12 − 30 = 12 + (−30) = −18
- This introduction states the relationship between a subtraction problem and its related addition problem. It reminds students that subtraction is not commutative and that they need to be careful of order when subtracting.
- **Vocabulary Review:** Ask students what the word *opposite* means.

Lesson Notes

Key Idea

- Each subtraction problem can be rewritten as an addition problem. Remind students that they already know how to add integers.

Example 1

- Work through each part of the example. Pointing to a classroom number line may be helpful.
- Use a different color when rewriting the problem as "add the opposite."
  3 − 12 = 3 + (−12)

Extra Example 1
Subtract.
a. 8 − 10 −2
b. −3 − (−6) 3
c. 2 − (−4) 6

On Your Own

1. 5	**2.** −8
3. −6	**4.** −23
5. 17	**6.** 0

Extra Example 2

Evaluate $-11 - 5 - (-8)$. -8

 On Your Own

7. -33	**8.** -33
9. -4	**10.** -2
11. 15	**12.** -59

Extra Example 3

Which continent has the greater range of elevations? South America

	South America	Europe
Highest Elevation	6960 m	5642 m
Lowest Elevation	-40 m	-92 m

On Your Own

13. 5710 meters

English Language Learners

Class Activity

Reinforce the meaning of the words *difference*, *subtract*, *positive*, and *negative*.

1. Ask students to name two numbers whose difference is 0. Students should realize that the only possibility is a number subtracted from itself.
2. Ask students to name two numbers whose difference is 3. Students should realize that if they start with a number greater than 3, they need to subtract a positive number. If they begin with a number less than 3, they need to subtract a negative number.
3. Ask students to name two numbers whose difference is -3. Starting with a number greater than -3 means you need to subtract a positive number. Starting with a number less than -3 means you need to subtract a negative number.

Laurie's Notes

On Your Own

- **Questions 1–6:**
  - For students who are having difficulty, have them record the problem on the board. They should say aloud, "Add the opposite," and state what that means for the particular problem.
  - **Construct Viable Arguments and Critique the Reasoning of Others:** Ask students if it is possible to determine when the difference of two negative numbers will be positive and when the difference of two negative numbers will be negative.

Example 2

- Caution students to work slowly.
- Subtraction must be performed in order from left to right.

Example 3

- **Vocabulary:** You may need to review the meanings of *elevation* and *range*.
- **Fun Fact:** The highest point in Hawaii is Mauna Kea at 4208 meters above sea level. The lowest points in Hawaii are at sea level, where the coast of Hawaii meets the Pacific Ocean.

Words of Wisdom

- **Model with Mathematics:** Make a colorful number line that stretches the length of your board. Use two 3-inch wide strips of different colored paper. Positive integers are on one color and negative integers are on the other. Label only the integers, but make a hash mark at $\frac{1}{2}$ between each of the integers. This will be useful later in the year.

Closure

- **Writing:** Your friend is home sick today. Imagine you are on the telephone with him or her. How would you explain how to subtract integers? Be sure to use an example.

EXAMPLE ② **Subtracting Integers**

Evaluate $-7 - (-12) - 14$.

$$
\begin{aligned}
-7 - (-12) - 14 &= -7 + 12 - 14 &&\text{Add the opposite of } -12.\\
&= 5 - 14 &&\text{Add } -7 \text{ and } 12.\\
&= 5 + (-14) &&\text{Add the opposite of } 14.\\
&= -9 &&\text{Add.}
\end{aligned}
$$

∴ So, $-7 - (-12) - 14 = -9$.

On Your Own

Now You're Ready
Exercises 27–32

Evaluate the expression.

7. $-9 - 16 - 8$

8. $-4 - 20 - 9$

9. $0 - 9 - (-5)$

10. $-8 - (-6) - 0$

11. $15 - (-20) - 20$

12. $-14 - 9 - 36$

EXAMPLE ③ **Real-Life Application**

Which continent has the greater range of elevations?

	North America	Africa
Highest Elevation	6198 m	5895 m
Lowest Elevation	−86 m	−155 m

To find the range of elevations for each continent, subtract the lowest elevation from the highest elevation.

North America

$$
\begin{aligned}
\text{range} &= 6198 - (-86)\\
&= 6198 + 86\\
&= 6284 \text{ m}
\end{aligned}
$$

Africa

$$
\begin{aligned}
\text{range} &= 5895 - (-155)\\
&= 5895 + 155\\
&= 6050 \text{ m}
\end{aligned}
$$

∴ Because 6284 is greater than 6050, North America has the greater range of elevations.

On Your Own

13. The highest elevation in Mexico is 5700 meters, on Pico de Orizaba. The lowest elevation in Mexico is −10 meters, in Laguna Salada. Find the range of elevations in Mexico.

Vocabulary and Concept Check

1. **WRITING** How do you subtract one integer from another?

2. **OPEN-ENDED** Write two integers that are opposites.

3. **DIFFERENT WORDS, SAME QUESTION** Which is different? Find "both" answers.

Find the difference of 3 and −2.	What is 3 less than −2?
How much less is −2 than 3?	Subtract −2 from 3.

MATCHING Match the subtraction expression with the corresponding addition expression.

4. $9 - (-5)$ 5. $-9 - 5$ 6. $-9 - (-5)$ 7. $9 - 5$

   A. $-9 + 5$ B. $9 + (-5)$ C. $-9 + (-5)$ D. $9 + 5$

Practice and Problem Solving

Subtract.

8. $4 - 7$ 9. $8 - (-5)$ 10. $-6 - (-7)$ 11. $-2 - 3$

12. $5 - 8$ 13. $-4 - 6$ 14. $-8 - (-3)$ 15. $10 - 7$

16. $-8 - 13$ 17. $15 - (-2)$ 18. $-9 - (-13)$ 19. $-7 - (-8)$

20. $-6 - (-6)$ 21. $-10 - 12$ 22. $32 - (-6)$ 23. $0 - 20$

24. **ERROR ANALYSIS** Describe and correct the error in finding the difference $7 - (-12)$.

$$\bcancel{\quad} \quad 7 - (-12) = 7 + (-12) = -5$$

25. **SWIMMING POOL** The floor of the shallow end of a swimming pool is at −3 feet. The floor of the deep end is 9 feet deeper. Which expression can be used to find the depth of the deep end?

$$-3 + 9 \qquad -3 - 9 \qquad 9 - 3$$

26. **SHARKS** A shark is at −80 feet. It swims up and jumps out of the water to a height of 15 feet. Write a subtraction expression for the vertical distance the shark travels.

Evaluate the expression.

27. $-2 - 7 + 15$ 28. $-9 + 6 - (-2)$ 29. $12 - (-5) - 8$

30. $-87 - 5 - 13$ 31. $-6 - (-8) + 6$ 32. $-15 - 7 - (-11)$

Assignment Guide and Homework Check

Level	Day 1 Activity Assignment	Day 2 Lesson Assignment	Homework Check
Basic	8–15, 50–56	1–7, 17–23 odd, 24, 25–29 odd	17, 19, 25, 29
Average	8–15, 50–56	1–7, 20–26 even, 27–31 odd, 39, 40, 42, 44	20, 22, 27, 39, 42
Advanced	8–15, 50–56	1–7, 24, 28–48 even	30, 32, 38, 42, 44
Accelerated	1–15, 24–48 even, 50–56		30, 32, 38, 42, 44

For Your Information

- **Exercise 3** In the *Different Words, Same Question* exercise, three of the four choices pose the same question using different words. The remaining choice poses a different question. So there are two answers.

Common Errors

- **Exercises 8–23** Students may change the sign of the first number or forget to change the problem from subtraction to addition when changing the sign of the second number. Remind them that the first number is a starting point and will never change. Also remind students that the sign of the second number and the operation change.
- **Exercises 27–32** Students may try to do the addition first instead of working left to right. Remind them that the order of operations does not put addition before subtraction, but that addition *and* subtraction are performed from left to right.
- **Exercise 40** Students may try to add $-4 + 11$ instead of subtract $-4 - 11$ because they do not recognize that *change in elevation* means a range (subtraction). Use a number line rotated vertically to help students see the meaning of change in elevation.

1.3 Record and Practice Journal

Subtract.

1. $3 - 8$ 2. $6 - (-7)$ 3. $-10 - 9$ 4. $-5 - (-4)$
   -5 13 -19 -1

Evaluate the expression.

5. $11 - (-2) + 14$ 6. $-16 - (-12) + (-8)$ 7. $6 - 17 - 4$
   27 -12 -15

Use mental math to solve the equation.

8. $6 - x = 10$ 9. $y - (-10) = 2$ 10. $z - 17 = -14$
   $x = -4$ $y = -8$ $z = 3$

11. You begin a hike in Death Valley, California, at an elevation of –86 meters. You hike to a point of elevation at 45 meters. What is your change in elevation?
   131 meters

12. You sell T-shirts for a fundraiser. It costs $112 to have the T-shirts made. You make $98 in sales. What is your profit?
   $-\$14$

Vocabulary and Concept Check

1. You add the integer's opposite.
2. *Sample answer:* 3, −3
3. What is 3 less than −2?; −5; 5
4. D   5. C
6. A   7. B

Practice and Problem Solving

8. −3   9. 13
10. 1   11. −5
12. −3   13. −10
14. −5   15. 3
16. −21   17. 17
18. 4   19. 1
20. 0   21. −22
22. 38   23. −20
24. The *opposite* of −12 should be added.
   $7 - (-12) = 7 + 12 = 19$
25. $-3 - 9$
26. $15 - (-80)$
27. 6   28. −1
29. 9   30. −105
31. 8   32. −11

33. $m = 14$ **34.** $w = 4$

35. $c = 15$ **36.** -5

37. 2 **38.** -17

39. 3 **40.** -15 m

41. *Sample answer:* $x = -2$, $y = -1$; $x = -3$, $y = -2$

42. See *Taking Math Deeper.*

43. sometimes; It's positive only if the first integer is greater.

44. sometimes; It's positive only if the first integer is greater.

45. always; It's always positive because the first integer is always greater.

46. never; It's never positive because the first integer is never greater.

47. all values of a and b

48. when a and b both have the same sign, or $a = 0$, or $b = 0$

49. when a and b have the same sign and $|a| \geq |b|$ or $b = 0$

Fair Game Review

50. -20 **51.** -45

52. 40 **53.** 468

54. 1476 **55.** 2378

56. C

Mini-Assessment

Subtract.

1. $6 - 10$ -4

2. $-14 - 16$ -30

3. $-9 - (-4)$ -5

4. $-26 - (-35)$ 9

5. The top of a flag pole is 15 feet high. The base is at -3 feet. Find the length of the flag pole. 18 feet

Taking Math Deeper

Exercise 42

The exercise reviews the concept of the range of a data set. Students should know that the range of a set is the difference between the greatest number and the least number in the set.

 a. Find the range of temperatures for each month.

	Jan	Feb	Mar	Apr	May	Jun	Jul	Aug	Sep	Oct	Nov	Dec
High (°F)	56	57	56	72	82	92	84	85	73	64	62	53
Low (°F)	-35	-38	-24	-15	1	29	34	31	19	-6	-21	-36
Range (°F)	91	95	80	87	81	63	50	54	54	70	83	89

$56 - (-35) = 91$

 Help me see it.

$$56 - (-35) = 56 + 35$$
$$= 91$$

Add 56 and 35.

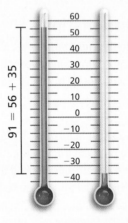

$91 = 56 + 35$

 b. Find the all-time high and the all-time low temperatures.

All-time high: 92°F

All-time low: -38°F

c. Find the range of the all-time high and the all-time low.

$$92 - (-38) = 92 + 38$$
$$= 130°F$$

Project

Create a chart showing the high and low temperatures in your town for each month of the year. Give the range of temperatures for each month.

Reteaching and Enrichment Strategies

If students need help. . .	If students got it. . .
Resources by Chapter • Practice A and Practice B • Puzzle Time Record and Practice Journal Practice Differentiating the Lesson Lesson Tutorials Skills Review Handbook	Resources by Chapter • Enrichment and Extension • Technology Connection Start the next section

MENTAL MATH Use mental math to solve the equation.

33. $m - 5 = 9$
34. $w - (-3) = 7$
35. $6 - c = -9$

ALGEBRA Evaluate the expression when $k = -3$, $m = -6$, and $n = 9$.

36. $4 - n$
37. $m - (-8)$

38. $-5 + k - n$
39. $|m - k|$

40. PLATFORM DIVING The figure shows a diver diving from a platform. The diver reaches a depth of 4 meters. What is the change in elevation of the diver?

41. OPEN-ENDED Write two different pairs of negative integers, x and y, that make the statement $x - y = -1$ true.

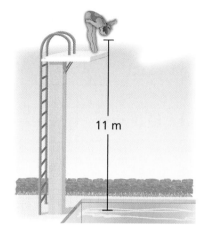

11 m

42. TEMPERATURE The table shows the record monthly high and low temperatures for a city in Alaska.

	Jan	Feb	Mar	Apr	May	Jun	Jul	Aug	Sep	Oct	Nov	Dec
High (°F)	56	57	56	72	82	92	84	85	73	64	62	53
Low (°F)	−35	−38	−24	−15	1	29	34	31	19	−6	−21	−36

    a. Find the range of temperatures for each month.

    b. What are the all-time high and all-time low temperatures?

    c. What is the range of the temperatures in part (b)?

REASONING Tell whether the difference between the two integers is *always*, *sometimes*, or *never* positive. Explain your reasoning.

43. two positive integers
44. two negative integers

45. a positive integer and a negative integer
46. a negative integer and a positive integer

 For what values of a and b is the statement true?

47. $|a - b| = |b - a|$
48. $|a + b| = |a| + |b|$
49. $|a - b| = |a| - |b|$

 Fair Game Review What you learned in previous grades & lessons

Add. *(Section 1.2)*

50. $-5 + (-5) + (-5) + (-5)$
51. $-9 + (-9) + (-9) + (-9) + (-9)$

Multiply. *(Skills Review Handbook)*

52. 8×5
53. 6×78
54. 36×41
55. 82×29

56. MULTIPLE CHOICE Which value of n makes the value of the expression $4n + 3$ a composite number? *(Skills Review Handbook)*

    Ⓐ 1   Ⓑ 2   Ⓒ 3   Ⓓ 4

You can use an **idea and examples chart** to organize information about a concept. Here is an example of an idea and examples chart for absolute value.

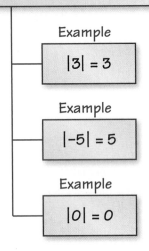

Absolute Value: the distance between a number and 0 on the number line

Example

$|3| = 3$

Example

$|-5| = 5$

Example

$|0| = 0$

On Your Own

Make idea and examples charts to help you study these topics.

1. integers

2. adding integers

   a. with the same sign

   b. with different signs

3. Additive Inverse Property

4. subtracting integers

After you complete this chapter, make idea and examples charts for the following topics.

5. multiplying integers

   a. with the same sign

   b. with different signs

6. dividing integers

   a. with the same sign

   b. with different signs

"I made an idea and examples chart to give my owner ideas for my birthday next week."

Sample Answers

1.

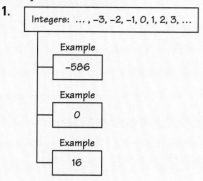

Integers: ... , –3, –2, –1, 0, 1, 2, 3, ...

Example
–586

Example
0

Example
16

2a.

Adding integers with the same sign:
Add the absolute values of the integers.
Then use the common sign.

Example
16 + 17 = 33

Example
–5 + (–4) = –9

Example
–55 + (–45) = –100

2b.

Adding integers with different signs:
Subtract the lesser absolute value from the
greater absolute value. Then use the sign of
the integer with the greater absolute value.

Example
8 + (–2) = 6

Example
–8 + 2 = –6

Example
–97 + 19 = –78

3.

Additive Inverse Property: The sum of an
integer and its *additive inverse*, or opposite, is 0.

Example
5 + (–5) = 0

Example
–100 + 100 = 0

Example
16 + (–16) = 0

4.   Available at *BigIdeasMath.com.*

List of Organizers
Available at *BigIdeasMath.com*

Comparison Chart
Concept Circle
Example and Non-Example Chart
Formula Triangle
Four Square
Idea (Definition) and Examples Chart
Information Frame
Information Wheel
Notetaking Organizer
Process Diagram
Summary Triangle
Word Magnet
Y Chart

About this Organizer

An **Idea and Examples Chart** can be used to organize information about a concept. Students fill in the top rectangle with a term and its definition or description. Students fill in the rectangles that follow with examples to illustrate the term. Each sample answer shows 3 examples, but students can show more or fewer examples. Idea and examples charts are useful for concepts that can be illustrated with more than one type of example.

Technology for the *Teacher*

Editable Graphic Organizer

Answers

1. $\left|-8\right| > 3$

2. $7 = \left|-7\right|$

3. $-6, -4, 3, \left|-4\right|, \left|-5\right|$

4. $-10, -8, \left|-9\right|, 12, \left|-15\right|$

5. -11

6. 12

7. -6

8. 0

9. 1

10. 13

11. a. $-10, -7$

    b. -7

    c. -10

12. yes; They raised $1129.

13. $130°F$

Technology for the *Teacher*

Online Assessment
Assessment Book
ExamView® Assessment Suite

Alternative Quiz Ideas

100% Quiz	Math Log
Error Notebook	Notebook Quiz
Group Quiz	**Partner Quiz**
Homework Quiz	Pass the Paper

Partner Quiz

- Partner quizzes are to be completed by students working in pairs. Student pairs can be selected by the teacher, by students, through a random process, or any way that works for your class.
- Students are permitted to use their notebooks and other appropriate materials.
- Each pair submits a draft of the quiz for teacher feedback. Then they revise their work and turn it in for a grade.
- When the pair is finished they can submit one paper, or each can submit their own.
- Teachers can give feedback in a variety of ways. It is important that the teacher does not reteach or provide the solution. The teacher can tell students which questions they have answered correctly, if they are on the right track, or if they need to rethink a problem.

Reteaching and Enrichment Strategies

If students need help. . .	If students got it. . .
Resources by Chapter • Practice A and Practice B • Puzzle Time Lesson Tutorials *BigIdeasMath.com*	Resources by Chapter • Enrichment and Extension • Technology Connection Game Closet at *BigIdeasMath.com* Start the next section

Copy and complete the statement using <, >, or =.   *(Section 1.1)*

1. $\left| -8 \right|$ ▢ 3

2. 7 ▢ $\left| -7 \right|$

Order the values from least to greatest.   *(Section 1.1)*

3. $-4, \left| -5 \right|, \left| -4 \right|, 3, -6$

4. $12, -8, \left| -15 \right|, -10, \left| -9 \right|$

Evaluate the expression.   *(Section 1.2 and Section 1.3)*

5. $-3 + (-8)$

6. $-4 + 16$

7. $3 - 9$

8. $-5 - (-5)$

Evaluate the expression when $a = -2$, $b = -8$, and $c = 5$.   *(Section 1.2 and Section 1.3)*

9. $4 - a - c$

10. $\left| b - c \right|$

11. EXPLORING Two climbers explore a cave.   *(Section 1.1)*

    a. Write an integer for the position of each climber relative to the surface.

    b. Which integer in part (a) is greater?

    c. Which integer in part (a) has the greater absolute value?

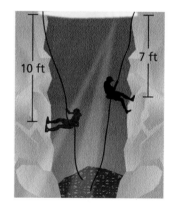

12. SCHOOL CARNIVAL The table shows the income and expenses for a school carnival. The school's goal was to raise $1100. Did the school reach its goal? Explain.   *(Section 1.2)*

Games	Concessions	Donations	Flyers	Decorations
$650	$530	$52	−$28	−$75

13. TEMPERATURE Temperatures in the Gobi Desert reach −40°F in the winter and 90°F in the summer. Find the range of the temperatures.   *(Section 1.3)*

1.4 Multiplying Integers

Essential Question
Is the product of two integers *positive*, *negative*, or *zero*? How can you tell?

1 ACTIVITY: Multiplying Integers with the Same Sign

Work with a partner. Use repeated addition to find 3 · 2.

Recall that multiplication is repeated addition. 3 · 2 means to add 3 groups of 2.

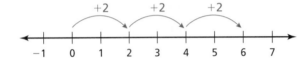

Now you can write

$3 \cdot 2 = \boxed{} + \boxed{} + \boxed{}$

$= \boxed{}.$

So, $3 \cdot 2 = \boxed{}$.

2 ACTIVITY: Multiplying Integers with Different Signs

Work with a partner. Use repeated addition to find 3 · (−2).

3 · (−2) means to add 3 groups of −2.

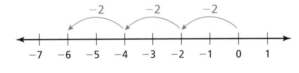

Now you can write

$3 \cdot (-2) = \boxed{} + \boxed{} + \boxed{}$

$= \boxed{}.$

So, $3 \cdot (-2) = \boxed{}$.

3 ACTIVITY: Multiplying Integers with Different Signs

Work with a partner. Use a table to find −3 · 2.

Integers

In this lesson, you will
- multiply integers.
- solve real-life problems.

Describe the pattern of the products in the table. Then complete the table.

2	·	2	=	4
1	·	2	=	2
0	·	2	=	0
−1	·	2	=	
−2	·	2	=	
−3	·	2	=	

So, $-3 \cdot 2 = \boxed{}$.

Laurie's Notes

Introduction

Applying Mathematical Practices

- **Look for and Express Regularity in Repeated Reasoning:** In this lesson, students will use the repeated addition model of multiplication and inductive reasoning to develop a conceptual understanding of integer multiplication.

Motivate

- Play *Guess My Rule*. Write the first 4 terms of a sequence on the board. Ask students to give the next few terms and guess the rule. Here are some possibilities.
  - 2, 4, 8, 16, . . . 32, 64, 128; The rule is to multiply by 2, powers of 2, or doubling (any of those 3 answers is acceptable).
  - 0, 4, 8, 12, . . . 16, 20, 24; The rule is adding 4, multiples of 4, or counting by 4s (any of those 3 answers is acceptable).
  - −6, −3, 0, 3, . . . 6, 9, 12; The rule is adding 3.
- **?** Ask students if any of the patterns seem different than the others. In the sequences above, the third one involves negative numbers.

Discuss

- **?** "Do you remember *skip counting* in elementary school?"
  If no one remembers, explain to students that skip counting is a fast way to count by a number other than 1.
- Skip counting is one way to show multiplication. For example, skip counting by 5s yields 5, 10, 15, 20, 25, ... You can think of the terms of the sequence formed by skip counting by 5s as $5 \times 1, 5 \times 2, 5 \times 3$, etc.

Activity Notes

Activity 1 and Activity 2

- **Model with Mathematics:** This is the *repeated addition model* of multiplication.
- Have students draw a number line to represent 3 groups of 2.
- **?** **Connection:** "I noticed that $3 \times 2 = 2 \times 3$. What property is this?"
  The Commutative Property of Multiplication
- In Activity 2, make sure students understand why the arrows are moving to the left, instead of moving to the right.
- Ask a student to read the last result in Activity 2, namely that $3 \cdot (-2) = -6$. So, a positive number times a negative number is a negative product.

Activity 3

- Make sure that students recognize the pattern—the first factor is decreasing by 1, the second factor is constant, and the product is decreasing by 2.
- Ask a student to read the last result, namely that $-3 \cdot 2 = -6$. So, a negative number times a positive number is a negative product.

What Your Students Will Learn

- Multiply integers with the same sign and with different signs.
- Evaluate expressions with whole number exponents by using repeated multiplication.

Previous Learning

Students need to know how to multiply whole numbers.

Lesson Plans
Complete Materials List

1.4 Record and Practice Journal

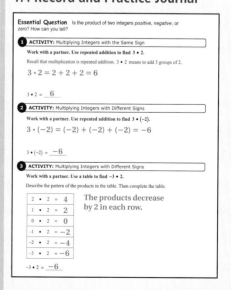

Essential Question Is the product of two integers *positive*, *negative*, or *zero*? How can you tell?

1 **ACTIVITY:** Multiplying Integers with the Same Sign

Work with a partner. Use repeated addition to find $3 \cdot 2$.

Recall that multiplication is repeated addition. $3 \cdot 2$ means to add 3 groups of 2.

$$3 \cdot 2 = 2 + 2 + 2 = 6$$

$3 \cdot 2 = \underline{6}$

2 **ACTIVITY:** Multiplying Integers with Different Signs

Work with a partner. Use repeated addition to find $3 \cdot (-2)$.

$$3 \cdot (-2) = (-2) + (-2) + (-2) = -6$$

$3 \cdot (-2) = \underline{-6}$

3 **ACTIVITY:** Multiplying Integers with Different Signs

Work with a partner. Use a table to find $-3 \cdot 2$.

Describe the pattern of the products in the table. Then complete the table.

$2 \cdot 2 =$	4	The products decrease by 2 in each row.
$1 \cdot 2 =$	2	
$0 \cdot 2 =$	0	
$-1 \cdot 2 =$	-2	
$-2 \cdot 2 =$	-4	
$-3 \cdot 2 =$	-6	

$-3 \cdot 2 = \underline{-6}$

Visual

Use integer counters to demonstrate that the product of two integers with different signs is negative.

$$3(-4) = (-4) + (-4) + (-4)$$
$$= -12$$

1.4 Record and Practice Journal

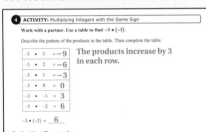

4 ACTIVITY: Multiplying Integers with the Same Sign

Work with a partner. Use a table to find $-3 \cdot (-2)$.

Describe the pattern of the products in the table. Then complete the table.

$-3 \cdot 3 = -9$	The products increase by 3
$-3 \cdot 2 = -6$	in each row.
$-3 \cdot 1 = -3$	
$-3 \cdot 0 = 0$	
$-3 \cdot -1 = 3$	
$-3 \cdot -2 = 6$	

$-3 \cdot (-2) = \underline{6}$

Inductive Reasoning

Work with a partner. Complete the table.

	Exercise	Type of Product	Product	Product: Positive or Negative
1	5. $3 \cdot 2$	same sign	6	Positive
2	6. $3 \cdot (-2)$	different signs	-6	Negative
3	7. $-3 \cdot 2$	different signs	-6	Negative
4	8. $-3 \cdot (-2)$	same sign	6	Positive
	9. $6 \cdot 3$	same sign	18	Positive
	10. $2 \cdot (-5)$	different signs	-10	Negative
	11. $-6 \cdot 5$	different signs	-30	Negative
	12. $-5 \cdot (-3)$	same sign	15	Positive

What Is Your Answer?

13. Write two integers whose product is 0.
 Sample answer: 3 and 0

14. **IN YOUR OWN WORDS** Is the product of two integers *positive, negative, or zero*? How can you tell?
 It can be positive, negative, or zero.

15. **STRUCTURE** Write a general rule for multiplying

 a. two integers with the same sign.
 Multiply the absolute values and make the product positive.

 b. two integers with different signs.
 Multiply the absolute values and make the product negative.

Laurie's Notes

Activity 4

- **Connection:** Activity 1 showed that the product of two positive integers is positive. Activity 2 and Activity 3 showed that the product of a positive and a negative (or a negative and a positive) is negative.
- **?** "Are there any other combinations to consider?" the product of two negatives
- Tell students: "Let's look at the product of two negatives." Students should recognize the patterns: the first factor is constant, the second factor is decreasing by 1, and the product is increasing by 3.
- **Extension:** Use the patterns developed to find the product of three numbers, such as $3(-2)(-4)$. 24

Inductive Reasoning

- Students should work with partners to find the products. The goal is for the students to recognize the bigger pattern. When the factors have the same signs, the product is positive. When the factors have different signs, the product is negative.
- Note that Questions 5–8 are the problems completed in Activities 1–4.
- **Look for and Express Regularity in Repeated Reasoning:** It is important for students to record the *Type of Product*. Mathematically proficient students will observe that there is a pattern in the table. When both factors have the same sign, the product is positive. When the factors have different signs, the product is negative.

Words of Wisdom

- **Common Error:** Students may make mistakes with addition. Review with them that a negative integer added to a negative integer has a negative sum. (Remind students that red counters added to red counters equal red counters.)

What Is Your Answer?

- Students may have a good sense of how to predict the sign of the product; however, they often use language such as: "two positives make a positive and two negatives make a positive." This language should be avoided.

Closure

- "Today we learned that a negative integer multiplied by a negative integer is positive. Be sure to mentally check all of your steps so that you are not confusing anything."

Work with a partner. Use a table to find $-3 \cdot (-2)$.

Describe the pattern of the products in the table. Then complete the table.

-3	$\cdot$	3	$=$	-9
-3	$\cdot$	2	$=$	-6
-3	$\cdot$	1	$=$	-3
-3	$\cdot$	0	$=$	
-3	$\cdot$	-1	$=$	
-3	$\cdot$	-2	$=$	

So, $-3 \cdot (-2) =$ [ ] .

Inductive Reasoning

Work with a partner. Complete the table.

Exercise	Type of Product	Product	Product: Positive or Negative
5. $3 \cdot 2$	Integers with the same sign		
6. $3 \cdot (-2)$			
7. $-3 \cdot 2$			
8. $-3 \cdot (-2)$			
9. $6 \cdot 3$			
10. $2 \cdot (-5)$			
11. $-6 \cdot 5$			
12. $-5 \cdot (-3)$			

What Is Your Answer?

13. Write two integers whose product is 0.

14. **IN YOUR OWN WORDS** Is the product of two integers *positive*, *negative*, or *zero*? How can you tell?

15. **STRUCTURE** Write general rules for multiplying (a) two integers with the same sign and (b) two integers with different signs.

Practice Use what you learned about multiplying integers to complete Exercises 8–15 on page 26.

Check It Out
Lesson Tutorials
BigIdeasMath ✓com

 **Key Ideas**

Multiplying Integers with the Same Sign

Words The product of two integers with the same sign is positive.

Numbers $2 \cdot 3 = 6$ $\qquad$ $-2 \cdot (-3) = 6$

Multiplying Integers with Different Signs

Words The product of two integers with different signs is negative.

Numbers $2 \cdot (-3) = -6$ $\qquad$ $-2 \cdot 3 = -6$

EXAMPLE 1 **Multiplying Integers with the Same Sign**

Find $-5 \cdot (-6)$.

The integers have the same sign.

$$-5 \cdot (-6) = 30$$

The product is positive.

:•: The product is 30.

EXAMPLE 2 **Multiplying Integers with Different Signs**

Multiply.

a. $3(-4)$ $\qquad\qquad\qquad$ **b.** $-7 \cdot 4$

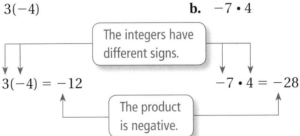

The integers have different signs.

$$3(-4) = -12 \qquad\qquad -7 \cdot 4 = -28$$

The product is negative.

:•: The product is -12. $\qquad$:•: The product is -28.

On Your Own

Now You're Ready
Exercises 8–23

Multiply.

1. $5 \cdot 5$ $\qquad\qquad\qquad\qquad$ **2.** $4(11)$

3. $-1(-9)$ $\qquad\qquad\qquad\quad$ **4.** $-7 \cdot (-8)$

5. $12 \cdot (-2)$ $\qquad\qquad\qquad$ **6.** $4(-6)$

7. $-10(-6)(0)$ $\qquad\qquad\quad$ **8.** $-7 \cdot (-5) \cdot (-4)$

Laurie's Notes

Introduction

Connect
- **Yesterday:** Students used repeated addition on a number line to develop a sense of integer multiplication.
- **Today:** Students will find products of integers with the same sign and products of integers with different signs.

Motivate
- **?** "Will someone summarize what we learned yesterday?"
- Listen for informal language such as "two negatives make a positive." While it is understood that the remark was made about multiplying two negative numbers, some students may incorrectly remember the comment when they are adding two negatives later on.
- **Vocabulary Review:** Ask students to define *factor, product,* and *Commutative Property of Multiplication.*

Lesson Notes

Key Ideas
- **Attend to Precision:** Write the rules for the two cases of multiplying integers. Discuss how multiplication can be represented. The multiplication dot is shown in the book, and parentheses are used to surround a negative integer. The parentheses are used for clarity so that the negative sign is not confused with the operation of subtraction.

Example 1
- Work through each part of the example.
- Say, "You know that 5 times 6 is 30, and because both integers in this example are negative (-5 and -6), the product is 30."
- **Attend to Precision:** Students should use correct language in reading the problems. They should say, "Negative 5 times negative 6 equals 30." If students say "minus 5," remind students that minus is an operation.

Example 2
- **Look for and Make Use of Structure:** Point out to students how multiplication is represented differently in the two problems. Before doing part (a), ask if there is another way the problem could be written.
- The goal is for students to be comfortable with all of the ways in which multiplication is represented.

On Your Own
- Students should work independently and check their work.
- Alternately, you could write the problems on index cards. Ask two students to sort the cards into two piles: integers with the same sign, and integers with different signs. Ask, "What is true about all of the products (or answers) in this pile?" Point to one of the piles and then repeat the same question for the other pile. Ask for volunteers to do the problems aloud.
- When multiplying more than two numbers, remind students that they can rearrange factors using properties of multiplication.

Technology for the **Teacher**

Dynamic Classroom

Lesson Tutorials
Lesson Plans
Answer Presentation Tool

Extra Example 1
Find $-8 \cdot (-12)$. 96

Extra Example 2
Multiply.
a. $9(-7)$ -63
b. $-6 \cdot 6$ -36

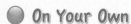

On Your Own

1. 25	**2.** 44
3. 9	**4.** 56
5. -24	**6.** -24
7. 0	**8.** -140

Extra Example 3
Evaluate the expression.
a. $(-8)^2$ 64
b. -9^2 -81
c. -5^3 -125

Example 3

- Students should know the meaning of exponents. Write the expression 5^2 on the board and ask students to tell you what it means.
- **Vocabulary Review:** 5 is the *base* and 2 is the *exponent*. The exponent tells you how many factors of the base (how many times you will see the base number) will be multiplied.
- So, $5 \times 5 = 25$. It is read "5 raised to the second power" or "5 squared."
- **Common Error:** When a negative number is raised to a power, the number must be written within parentheses. In part (b), the example is read "the opposite of 5 squared." If you wanted to raise -5 to the second power, it would be written $(-5)^2$. For the given problem, the order of operations says to square the number and then take its opposite. Part (c) shows how to raise a negative integer to a power.
- **Extension:** "When you raise a negative number to a power, is the answer always positive?" No. If the exponent is odd, the answer is negative.

On Your Own

9. 9 **10.** -8

11. -49 **12.** -216

On Your Own

- Students should work with partners.
- **Construct Viable Arguments and Critique the Reasoning of Others:** Caution students about Questions 11 and 12.
- **Common Error:** Students sometimes multiply the exponent by the base, particularly if the exponent is greater than 2.

Example 4

- There is no scale written on the vertical axis.
- "Is it possible to determine the number of taxis the company began with?" From the graph, you could estimate 300 taxis to start: each horizontal line is 50 taxis, and the first bar is 6 increments tall.
- Read the verbal model: total change = change per year × number of years.
- **Extension:** "At the same rate, how many years before there are no taxis?" 6

Extra Example 4

A football jersey is marked down $10 each week for 3 weeks. Find the total change in the price of the football jersey. $-\$30$

On Your Own

- Note that you do *not* need to know the initial population of manatees in order to answer the question.

On Your Own

13. -45 manatees

Closure

- Write the number -4 on the board. Ask students to write each of the following and then share their responses.
  - "Write a multiplication problem that has -4 as one of the factors, and has a negative product." *Sample answer:* $(-4)(-1)(-1) = -4$
  - "Write a second multiplication problem that has -4 as one of the factors, and has a positive product." *Sample answer:* $(-4)(-1)(1) = 4$

English Language Learners

Vocabulary

Make sure that the students understand the mathematical meanings of the words *positive* and *negative*. In math, *positive* means a number greater than zero and *negative* means a number less than zero. Explain that positive and negative do not mean good and bad.

EXAMPLE 3   Using Exponents

a. Evaluate $(-2)^2$.

$$(-2)^2 = (-2) \cdot (-2)$$ Write $(-2)^2$ as repeated multiplication.

$$= 4$$ Multiply.

b. Evaluate -5^2.

$$-5^2 = -(5 \cdot 5)$$ Write 5^2 as repeated multiplication.

$$= -25$$ Multiply.

c. Evaluate $(-4)^3$.

$$(-4)^3 = (-4) \cdot (-4) \cdot (-4)$$ Write $(-4)^3$ as repeated multiplication.

$$= 16 \cdot (-4)$$ Multiply.

$$= -64$$ Multiply.

Study Tip

Place parentheses around a negative number to raise it to a power.

On Your Own

Now You're Ready
Exercises 32–37

Evaluate the expression.

9. $(-3)^2$ **10.** $(-2)^3$ **11.** -7^2 **12.** -6^3

EXAMPLE 4   Real-Life Application

The bar graph shows the number of taxis a company has in service. The number of taxis decreases by the same amount each year for 4 years. Find the total change in the number of taxis.

The bar graph shows that the number of taxis in service decreases by 50 each year. Use a model to solve the problem.

Taxis in Service

50 fewer taxis

Number of taxis

Year

$$\text{total change} = \text{change per year} \cdot \text{number of years}$$

$$= -50 \cdot 4$$

Use -50 for the change per year because the number *decreases* each year.

$$= -200$$

The total change in the number of taxis is -200.

On Your Own

13. A manatee population decreases by 15 manatees each year for 3 years. Find the total change in the manatee population.

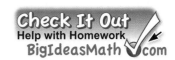
Check It Out
Help with Homework
BigIdeasMath ✓com

 Vocabulary and Concept Check

1. **WRITING** What can you conclude about the signs of two integers whose product is (a) positive and (b) negative?

2. **OPEN-ENDED** Write two integers whose product is negative.

Tell whether the product is *positive* or *negative* without multiplying. Explain your reasoning.

3. $4(-8)$
4. $-5(-7)$
5. $-3 \cdot 12$

Tell whether the statement is *true* or *false*. Explain your reasoning.

6. The product of three positive integers is positive.

7. The product of three negative integers is positive.

 Practice and Problem Solving

Multiply.

① ② 8. $6 \cdot 4$
9. $7(-3)$
10. $-2(8)$
11. $-3(-4)$

12. $-6 \cdot 7$
13. $3 \cdot 9$
14. $8 \cdot (-5)$
15. $-1 \cdot (-12)$

16. $-5(10)$
17. $-13(0)$
18. $-9 \cdot 9$
19. $15(-2)$

20. $-10 \cdot 11$
21. $-6 \cdot (-13)$
22. $7(-14)$
23. $-11 \cdot (-11)$

24. **JOGGING** You burn 10 calories each minute you jog. What integer represents the change in your calories after you jog for 20 minutes?

25. **WETLANDS** About 60,000 acres of wetlands are lost each year in the United States. What integer represents the change in wetlands after 4 years?

Multiply.

26. $3 \cdot (-8) \cdot (-2)$
27. $6(-9)(-1)$
28. $-3(-5)(-4)$

29. $(-5)(-7)(-20)$
30. $-6 \cdot 3 \cdot (-2)$
31. $3 \cdot (-12) \cdot 0$

Evaluate the expression.

③ 32. $(-4)^2$
33. $(-1)^3$
34. -8^2

35. -6^2
36. $-5^2 \cdot 4$
37. $-2 \cdot (-3)^3$

ERROR ANALYSIS Describe and correct the error in evaluating the expression.

38.
✗ $-2(-7) = -14$

39.
✗ $-10^2 = 100$

Assignment Guide and Homework Check

Level	Day 1 Activity Assignment	Day 2 Lesson Assignment	Homework Check
Basic	8–15, 49–53	1–7, 17–39 odd	2, 21, 25, 33
Average	8–15, 49–53	1–7, 20–23, 30–34 even, 39, 40, 45, 46	2, 21, 32, 45
Advanced	8–15, 49–53	1–7, 30–46 even, 47, 48	7, 34, 40, 47
Accelerated	1–15, 30–46 even, 47–53		7, 34, 40, 47

Common Errors

- **Exercises 8–23** Students may not remember that a negative number multiplied by a negative number is positive. Tell them that it is similar to multiplying by -1, which means to take the opposite. For example, $-6(-13) = (-1 \cdot 6)(-13) = -1[6 \cdot (-13)] = -1(-78) = 78$.
- **Exercises 26–31** Students may multiply all the numbers together ignoring the signs and then place the incorrect sign in front. For example, a student might say $-7(-3)(-5) = 105$. Tell them to multiply only two integers at a time, determine the sign, and then multiply by the last number.
- **Exercises 32–37** Students may erroneously interpret -8^2 as $(-8)(-8)$ instead of $-1(8^2)$. Remind them that the negative sign means multiplication by -1 and that exponents are evaluated before multiplication.

1.4 Record and Practice Journal

Multiply.

1. $8 \cdot 9$
   72
2. $7(-7)$
   -49
3. $-10 \cdot 4$
   -40
4. $-5(-6)$
   30
5. $12 \cdot (-1) \cdot (-2)$
   24
6. $-10(-3)(-7)$
   -210
7. $-20 \cdot 0 \cdot (-4)$
   0
8. $-4 \cdot 8 \cdot 3$
   -96

Evaluate the expression.

9. $(-8)^2$
   64
10. -11^2
    -121
11. $9 \cdot (-5)^2$
    225
12. $(-2)^1 \cdot (-6)$
    48

13. You lose 5 points for every wrong answer in a trivia game. What integer represents the change in your points after answering 8 questions wrong?
    -40

Vocabulary and Concept Check

1. **a.** They are the same.
   b. They are different.
2. *Sample answer:* $2, -3$
3. negative; different signs
4. positive; same signs
5. negative; different signs
6. true; The product of the first two positive integers is positive. The product of the result and the third positive integer is positive.
7. false; The product of the first two negative integers is positive. The product of the positive result and the third negative integer is negative.

Practice and Problem Solving

8.	24	9.	-21
10.	-16	11.	12
12.	-42	13.	27
14.	-40	15.	12
16.	-50	17.	0
18.	-81	19.	-30
20.	-110	21.	78
22.	-98	23.	121
24.	-200	25.	$-240,000$
26.	48	27.	54
28.	-60	29.	-700
30.	36	31.	0
32.	16	33.	-1
34.	-64	35.	-36
36.	-100	37.	54

Practice and Problem Solving

38. The product should be positive. $-2(-7) = 14$

39. The answer should be negative. $-10^2 = -(10 \cdot 10) = -100$

40. -6 **41.** 32

42. 38

43. $-7500, 37,500$

44. $1792, -7168$

45. -12

46. See *Taking Math Deeper*.

47. a. $153; 141; 129$

   b. The price drops $12 every month.

   c. no; yes; In August, you have $135 but the cost is $141. In September, you have $153 and the cost is only $129.

48. -25

Fair Game Review

49. 3 **50.** 8

51. 14 **52.** 17

53. D

Mini-Assessment

Multiply.

1. $-4(-5)$ 20

2. $3(-3)$ -9

3. $-1(-12)$ 12

4. $-2(15)$ -30

5. You have $900 in a checking account. You pay a $60 cell phone bill each month using this account. The account balance is given by $900 + (-60t)$, where t is the time in months. What is the balance of the account after 4 months? $660

Taking Math Deeper

Exercise 46

This is a classic type of problem in mathematics. A real-life measurement (such as height) is modeled by an expression. The height h (in feet) depends on the time t (in minutes).

$$h = 22,000 + (-480t) = 22,000 - 480t$$

 First, help students understand the model.

$$h = 22,000 - 480t$$

Starting height is 22,000 feet.   Plane descends 480 feet each minute.

 a. Copy and complete the table.

Time (minutes)	5	10	15	20
Height (feet)	19,600	17,200	14,800	12,400

 b. When does the plane land?

About 46 minutes

$$\frac{22,000}{480} \approx 45.83 \text{ minutes}$$

The **height** of a plane is called its **altitude**.

Descent rates vary greatly. The rate of 480 feet per minute in this problem is low. Descent rates between 500 and 1500 feet per minute are more common. After take-off, an ascent rate of 1000 to 2000 feet per minute is common.

Project

Draw a graph showing the height of the plane in 5-minute intervals from the time it begins the descent at 22,000 feet until it lands.

Reteaching and Enrichment Strategies

If students need help. . .	If students got it. . .
Resources by Chapter • Practice A and Practice B • Puzzle Time Record and Practice Journal Practice Differentiating the Lesson Lesson Tutorials Skills Review Handbook	Resources by Chapter • Enrichment and Extension • Technology Connection Start the next section

ALGEBRA Evaluate the expression when $a = -2$, $b = 3$, and $c = -8$.

40. ab

41. $|a^2c|$

42. $-ab^3 - ac$

NUMBER SENSE Find the next two numbers in the pattern.

43. $-12, 60, -300, 1500, \ldots$

44. $7, -28, 112, -448, \ldots$

45. GYM CLASS You lose four points each time you attend gym class without sneakers. You forget your sneakers three times. What integer represents the change in your points?

46. MODELING The height of an airplane during a landing is given by $22{,}000 + (-480t)$, where t is the time in minutes.

a. Copy and complete the table.

b. Estimate how many minutes it takes the plane to land. Explain your reasoning.

Time (minutes)	5	10	15	20
Height (feet)				

47. INLINE SKATES In June, the price of a pair of inline skates is $165. The price changes each of the next 3 months.

a. Copy and complete the table.

Month	Price of Skates
June	165 $\qquad$ = \$165
July	$165 + \ (-12) = \$\_\_\_$
August	$165 + 2(-12) = \$\_\_\_$
September	$165 + 3(-12) = \$\_\_\_$

b. Describe the change in the price of the inline skates for each month.

c. The table at the right shows the amount of money you save each month to buy the inline skates. Do you have enough money saved to buy the inline skates in August? September? Explain your reasoning.

Amount Saved	
June	$35
July	$55
August	$45
September	$18

48. **Reasoning** Two integers, a and b, have a product of 24. What is the least possible sum of a and b?

Fair Game Review What you learned in previous grades & lessons

Divide. *(Skills Review Handbook)*

49. $27 \div 9$

50. $48 \div 6$

51. $56 \div 4$

52. $153 \div 9$

53. MULTIPLE CHOICE What is the prime factorization of 84?
(Skills Review Handbook)

(A) $2^2 \times 3^2$

(B) $2^3 \times 7$

(C) $3^3 \times 7$

(D) $2^2 \times 3 \times 7$

1.5 Dividing Integers

Essential Question
Is the quotient of two integers *positive*, *negative*, or *zero*? How can you tell?

1 ACTIVITY: Dividing Integers with Different Signs

Work with a partner. Use integer counters to find $-15 \div 3$.

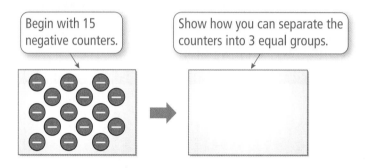

Begin with 15 negative counters.

Show how you can separate the counters into 3 equal groups.

Because there are ▢ negative counters in each group, $-15 \div 3 = $ ▢.

2 ACTIVITY: Rewriting a Product as a Quotient

Work with a partner. Rewrite the product $3 \cdot 4 = 12$ as a quotient in two different ways.

First Way

12 is equal to 3 groups of ▢.

So, $12 \div 3 = $ ▢.

Second Way

12 is equal to 4 groups of ▢.

So, $12 \div 4 = $ ▢.

3 ACTIVITY: Dividing Integers with Different Signs

Integers

In this lesson, you will
- divide integers.
- solve real-life problems.

Work with a partner. Rewrite the product $-3 \cdot (-4) = 12$ as a quotient in two different ways. What can you conclude?

First Way

$12 \div \left(\boxed{} \right) = \boxed{}$

Second Way

$12 \div \left(\boxed{} \right) = \boxed{}$

In each case, when you divide a ▢ integer by a ▢ integer, you get a ▢ integer.

Laurie's Notes

Introduction

Applying Mathematical Practices

- **Look for and Express Regularity in Repeated Reasoning:** In this lesson, students will use the relationship between multiplication and division to develop a conceptual understanding of integer division.

Motivate

- ❓ "What do you know about football?" Guide students to discuss the length of the field. A football field is 100 yards long, plus two 10-yard end zones, for a total length of 120 yards.
- ❓ "If I told you the area of the football field, could you tell me the width of the football field?" The goal is to have students think about the area formula ($A = \ell w$) and realize that if they know the area and one dimension, they can divide to find the other dimension.
- ❓ "The area is 6400 yd² and the length is 120 yd. What is the width?" $53\frac{1}{3}$ yd

Discuss

- ❓ "What are fact families? Give some examples for multiplication and division." Fact families show the inverse relationship between multiplication and division.

  Sample answers: $2 \times 3 = 6$, $3 \times 2 = 6$, $6 \div 2 = 3$, and $6 \div 3 = 2$
  $6 \times 8 = 48$, $8 \times 6 = 48$, $48 \div 6 = 8$, and $48 \div 8 = 6$

Activity Notes

Activity 1

- ❓ Place 15 red integer counters on the overhead, arranged in a 3×5 array. "What integer is being modeled?" -15
- ❓ Use a ruler to separate the counters into 3 groups of 5. "What division problem does this suggest?" $-15 \div 3$
- There are 5 red counters in every group. The quotient is -5. This is the *grouping model* of division.

Activity 2

- **Look for and Make Use of Structure:** Relate this example back to the football problem and fact families, because length $\times$ width = area, area $\div$ length = width, and area $\div$ width = length.

Activity 3

- ❓ "What is the product of two negative integers?" a positive integer
- **Look for and Make Use of Structure:** You can rewrite a multiplication problem as a division problem, but be sure to pay attention to the signs.

What Your Students Will Learn

- Divide integers with the same sign and with different signs.
- Evaluate expressions involving division by using substitution and the order of operations.

Previous Learning

Students need to know how to divide whole numbers.

Lesson Plans
Complete Materials List

1.5 Record and Practice Journal

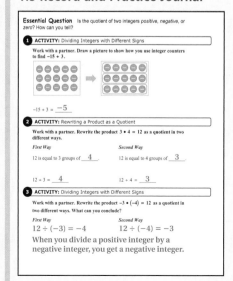

Differentiated Instruction

Visual

Have students find the mean of two negative integers. Then tell the students to graph the two numbers and the mean on a number line. Have students share their results with the class. If any student has a mean that is zero or positive, identify the error.

Activity 4

- **Look for and Make Use of Structure:** Let's look at one last related problem, $3 \cdot (-4) = -12$. In this example, it is the first way (negative ÷ negative) that is new. The second way (dividing integers with different signs) is similar to Activity 3.

Words of Wisdom

? "How are Activity 2 and Activity 3 alike? How are Activity 2 and Activity 4 alike?" In each problem, you rewrote the integer multiplication problem as an integer division problem.

Inductive Reasoning

- Students should work with partners to find the quotients. The goal is for the students to recognize the bigger pattern. When the dividend and divisor have the same signs, the quotient is positive. When the dividend and divisor have different signs, the quotient is negative.
- Note that Questions 5–8 are the problems completed in Activities 1–4.
- **Look for and Express Regularity in Repeated Reasoning:** It is important for students to record the *Type of Quotient*. Mathematically proficient students will observe that there is a pattern in the table. When the dividend and divisor have the same sign, the quotient is positive. When the dividend and divisor have different signs, the quotient is negative.
- **Extension:**
    ? "Is division commutative, meaning do $18 \div 9$ and $9 \div 18$ have the same quotient?" no
    ? "What is the relationship between the two solutions?" They are reciprocals.

What Is Your Answer?

- Students may have a good sense of how to predict the sign of the quotient; however, they often use language (as they do with multiplication) such as: two positives make a positive and two negatives make a positive. This language should be avoided.

1.5 Record and Practice Journal

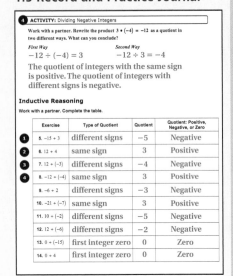

Closure

- "Today you learned that a negative integer divided by a negative integer is positive. Be sure to mentally check all of your steps so that you are not confusing anything."

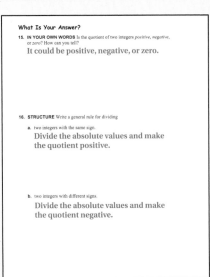

ACTIVITY: Dividing Negative Integers

Work with a partner. Rewrite the product $3 \cdot (-4) = -12$ as a quotient in two different ways. What can you conclude?

First Way

$-12 \div \left(\boxed{} \right) = \boxed{}$

Second Way

$-12 \div \left(\boxed{} \right) = \boxed{}$

⁘ When you divide a ▢ integer by a ▢ integer, you get a ▢ integer. When you divide a ▢ integer by a ▢ integer, you get a ▢ integer.

Inductive Reasoning

Work with a partner. Complete the table.

	Exercise	Type of Quotient	Quotient	Quotient: Positive, Negative, or Zero
①	**5.** $-15 \div 3$	Integers with different signs		
②	**6.** $12 \div 4$			
③	**7.** $12 \div (-3)$			
④	**8.** $-12 \div (-4)$			
	9. $-6 \div 2$			
	10. $-21 \div (-7)$			
	11. $10 \div (-2)$			
	12. $12 \div (-6)$			
	13. $0 \div (-15)$			
	14. $0 \div 4$			

What Is Your Answer?

15. IN YOUR OWN WORDS Is the quotient of two integers *positive*, *negative*, or *zero*? How can you tell?

16. STRUCTURE Write general rules for dividing (a) two integers with the same sign and (b) two integers with different signs.

Practice

Use what you learned about dividing integers to complete Exercises 8–15 on page 32.

1.5 Lesson

Check It Out
Lesson Tutorials
BigIdeasMath ✓com

🔑 Key Ideas

Remember

Division by 0 is undefined.

Dividing Integers with the Same Sign

Words The quotient of two integers with the same sign is positive.

Numbers $8 \div 2 = 4$ $-8 \div (-2) = 4$

Dividing Integers with Different Signs

Words The quotient of two integers with different signs is negative.

Numbers $8 \div (-2) = -4$ $-8 \div 2 = -4$

EXAMPLE 1 **Dividing Integers with the Same Sign**

Find $-18 \div (-6)$.

The integers have the same sign.

$$-18 \div (-6) = 3$$

The quotient is positive.

∴ The quotient is 3.

EXAMPLE 2 **Dividing Integers with Different Signs**

Divide.

a. $75 \div (-25)$ **b.** $\dfrac{-54}{6}$

The integers have different signs.

$$75 \div (-25) = -3 \qquad \dfrac{-54}{6} = -9$$

The quotient is negative.

∴ The quotient is -3.   ∴ The quotient is -9.

On Your Own

Now You're Ready
Exercises 8–23

Divide.

1. $14 \div 2$ **2.** $-32 \div (-4)$ **3.** $-40 \div (-8)$

4. $0 \div (-6)$ **5.** $\dfrac{-49}{7}$ **6.** $\dfrac{21}{-3}$

Laurie's Notes

Introduction

Connect
- **Yesterday:** Students used fact families and what they knew about multiplication of integers to develop a sense of integer division.
- **Today:** Students will use the idea that when the signs of the dividend and divisor are the same, the quotient is positive; when the signs of the dividend and divisor are different, the quotient is negative.

Motivate
- **?** "Will someone summarize what we learned yesterday?"
- **Listen:** Again, watch for informal language such as "two negatives make a positive." While it is understood that the remark was made about dividing two negative numbers, some students may incorrectly remember the comment when they are adding or subtracting two negatives later on.
- **Vocabulary Review:** Ask students to define *dividend, divisor, quotient, Commutative Property,* and *division involving zero.*

Lesson Notes

Key Ideas
- **Attend to Precision:** Students should know that to check a division problem, you multiply the quotient by the divisor, and the answer is the dividend.
- **Summary of division involving zero:** You can divide 0 by a nonzero number and the answer is 0. You cannot divide a number by 0. Later in this lesson, connect this concept to "0 cannot be in the denominator when division is represented in fraction form."

Example 1
- Work through the example.
- Say, "We know that 18 divided by 6 is 3, and because both integers are negative for this example (-18 and -6), the quotient is positive 3."
- **Attend to Precision:** Be sure that students use correct language in reading the problems. When they read the problem they should say, "Negative 18 divided by negative 6 equals 3." If students say "minus 18," remind them that minus is an operation.

Example 2
- **Look for and Make Use of Structure:** Point out to students how division is represented differently in the two problems. Before doing part (a), you may want to ask if the problem could be written another way.
- The goal is for students to be comfortable with all of the ways in which division is represented.

On Your Own
- Students should work independently and check their work.

Goal
Today's lesson is dividing integers.

Technology for the **Teacher**

Dynamic Classroom

Lesson Tutorials
Lesson Plans
Answer Presentation Tool

Extra Example 1
Find $-48 \div (-6)$. 8

Extra Example 2
Divide.

a. $\dfrac{84}{-4}$ -21

b. $-39 \div 3$ -13

On Your Own
1. 7	**2.** 8
3. 5	**4.** 0
5. -7	**6.** -7

Extra Example 3

Evaluate $\dfrac{x-8}{y^2}$ when $x = 4$ and $y = 2$.
-1

On Your Own

7. 3

8. -4

9. 2

Extra Example 4

The morning high tide at a beach is 57 inches. Six hours later, the afternoon low tide is 12 inches. What is the mean hourly change in the height? -7.5 in.

On Your Own

10. -6 ft/h

English Language Learners

Visual

Show students that the rules for multiplication can be used to understand the rules for division. For example, to evaluate $16 \div (-8) = ?$, rewrite it using multiplication, $-8 \times ? = 16$. Students should be able to determine that the answer to the multiplication problem is -2, and the answer to the division problem is -2. So, the quotient of two integers with different signs is negative. Use the same approach to demonstrate the other three cases: $-16 \div 8 = ?$, $16 \div 8 = ?$, and $-16 \div (-8) = ?$.

T-31

Example 3

- Students should know the order of operations and the meaning of exponents. Students will need to use order of operations. Instead of telling students "remember the order of operations," you want to see what the students remember without prompting.

- **?** Write the problem and ask what it means to "evaluate." To evaluate a numerical expression means to perform the operations to find the value of the expression.

- **Common Error:** If students forget the order of operations, they will perform the operations left to right. Solicit responses as to what operations should be done, in the correct order, and why.

On Your Own

- Students should work with partners.

- **?** **Extension:** "Can Question 8 be rewritten as $a + 6/2$?" No. There is an implied order of operations by the division bar and, therefore, it would need to be written as $(a + 6)/2$.

Example 4

- Discuss how the word *mean* is used in this context. Students often only think of computing a mean by adding values and then dividing by the number of values. So, if students think of *mean* as adding values, it could lead to a problem.

- In this problem, the total change in height is found by finding the difference of the final height and the initial height.

On Your Own

- There is a decrease in the water level, so the hourly change in height is negative. If the tide is coming in, the height would increase, and so the hourly change would be positive.

Closure

- How are the rules for multiplication and division of integers related? Why? The rules are the same because the operations are inverses. You can use fact families to rewrite a division problem as a multiplication problem, and vice versa.

EXAMPLE 3 **Evaluating an Expression**

Evaluate $10 - x^2 \div y$ when $x = 8$ and $y = -4$.

$$10 - x^2 \div y = 10 - 8^2 \div (-4)$$ Substitute 8 for x and -4 for y.

$$= 10 - 8 \cdot 8 \div (-4)$$ Write 8^2 as repeated multiplication.

$$= 10 - 64 \div (-4)$$ Multiply 8 and 8.

$$= 10 - (-16)$$ Divide 64 by -4.

$$= 26$$ Subtract.

Remember

Use order of operations when evaluating an expression.

On Your Own

Now You're Ready
Exercises 28–31

Evaluate the expression when $a = -18$ and $b = -6$.

7. $a \div b$ **8.** $\dfrac{a + 6}{3}$ **9.** $\dfrac{b^2}{a} + 4$

EXAMPLE 4 **Real-Life Application**

You measure the height of the tide using the support beams of a pier. Your measurements are shown in the picture. What is the mean hourly change in the height?

59 inches at 2 P.M. →
8 inches at 8 P.M. →

Use a model to solve the problem.

$$\text{mean hourly change} = \frac{\text{final height} - \text{initial height}}{\text{elapsed time}}$$

$$= \frac{8 - 59}{6}$$ Substitute. The elapsed time from 2 P.M. to 8 P.M. is 6 hours.

$$= \frac{-51}{6}$$ Subtract.

$$= -8.5$$ Divide.

∴ The mean change in the height of the tide is -8.5 inches per hour.

On Your Own

10. The height of the tide at the Bay of Fundy in New Brunswick decreases 36 feet in 6 hours. What is the mean hourly change in the height?

 Vocabulary and Concept Check

1. **WRITING** What can you tell about two integers when their quotient is positive? negative? zero?

2. **VOCABULARY** A quotient is undefined. What does this mean?

3. **OPEN-ENDED** Write two integers whose quotient is negative.

4. **WHICH ONE DOESN'T BELONG?** Which expression does *not* belong with the other three? Explain your reasoning.

$$\frac{10}{-5} \qquad \frac{-10}{5} \qquad \frac{-10}{-5} \qquad -\left(\frac{10}{5}\right)$$

Tell whether the quotient is *positive* or *negative* without dividing.

5. $-12 \div 4$

6. $\dfrac{-6}{-2}$

7. $15 \div (-3)$

 Practice and Problem Solving

Divide, if possible.

 8. $4 \div (-2)$

9. $21 \div (-7)$

10. $-20 \div 4$

11. $-18 \div (-3)$

12. $\dfrac{-14}{7}$

13. $\dfrac{0}{6}$

14. $\dfrac{-15}{-5}$

15. $\dfrac{54}{-9}$

16. $-33 \div 11$

17. $-49 \div (-7)$

18. $0 \div (-2)$

19. $60 \div (-6)$

20. $\dfrac{-56}{14}$

21. $\dfrac{18}{0}$

22. $\dfrac{65}{-5}$

23. $\dfrac{-84}{-7}$

ERROR ANALYSIS Describe and correct the error in finding the quotient.

24.

$$\bcancel{\qquad} \quad \frac{-63}{-9} = -7$$

25.

$$\bcancel{\qquad} \quad 0 \div (-5) = -5$$

26. **ALLIGATORS** An alligator population in a nature preserve in the Everglades decreases by 60 alligators over 5 years. What is the mean yearly change in the alligator population?

27. **READING** You read 105 pages of a novel over 7 days. What is the mean number of pages you read each day?

ALGEBRA Evaluate the expression when $x = 10$, $y = -2$, and $z = -5$.

28. $x \div y$

29. $\dfrac{10y^2}{z}$

30. $\left| \dfrac{xz}{-y} \right|$

31. $\dfrac{-x^2 + 6z}{y}$

Assignment Guide and Homework Check

Level	Day 1 Activity Assignment	Day 2 Lesson Assignment	Homework Check
Basic	8–15, 42–45	1–7, 17–31 odd	19, 23, 27, 29
Average	8–15, 42–45	1–7, 20–23, 25, 29–39 odd	22, 29, 37, 39
Advanced	8–15, 42–45	1–7, 24, 28–40 even, 41	28, 30, 36, 38
Accelerated	1–15, 24–40 even, 41–45		28, 30, 36, 38

Common Errors

- **Exercises 8–23** In problems involving zero, students may just say that the quotient is undefined. Remind students that when 0 is the dividend, it means $\square \cdot -2 = 0$, where $\square = 0$. Also, when 0 is the divisor, it means $\square \cdot 0 = 18$, where the answer is undefined.
- **Exercises 28–31 and 34–35** Students may forget to follow the order of operations. Review the order of operations, especially the left-to-right rule in evaluating multiplication/division and addition/subtraction.
- **Exercises 32 and 33** Students may not remember how to find the mean of several numbers. They may get confused by the negative numbers and subtract instead of add. Remind students of the definition of mean.

1. They have the same sign. They have different signs. The dividend is zero.
2. The divisor is zero.
3. *Sample answer:* $-4, 2$
4. $\dfrac{-10}{-5}$, which equals 2. All the others equal -2.
5. negative
6. positive
7. negative

Practice and Problem Solving

8. -2		**9.** -3	
10. -5		**11.** 6	
12. -2		**13.** 0	
14. 3		**15.** -6	
16. -3		**17.** 7	
18. 0		**19.** -10	
20. -4		**21.** undefined	
22. -13		**23.** 12	

24. The quotient should be positive. $\dfrac{-63}{-9} = 7$
25. The quotient should be 0. $0 \div (-5) = 0$
26. -12 alligators
27. 15 pages

28. -5	**29.** -8
30. 25	**31.** 65

1.5 Record and Practice Journal

Divide, if possible.

1. $3 \div (-1)$
   -3
2. $8 \div 2$
   4
3. $-10 \div 5$
   -2
4. $-21 \div (-7)$
   3
5. $\dfrac{48}{-6}$
   -8
6. $\dfrac{-13}{-13}$
   1
7. $\dfrac{0}{3}$
   0
8. $\dfrac{-55}{11}$
   -5

Evaluate the expression.

9. $-63 \div (-7) + 6$
   15
10. $-5 - 12 \div 3$
    -9
11. $-8 \cdot 7 + 33 \div (-11)$
    -59

12. The table shows the number of yards a football player runs in each quarter of a game. Find the mean number of yards the player runs per quarter.

Quarter	1	2	3	4
Yards	-2	14	-18	-6

-3 yards

Practice and Problem Solving

32. 3   **33.** 5

34. −10   **35.** 4

36. −8, 4; Divide the previous number by −2 to obtain the next number.

37. −400 ft/min

38. See *Taking Math Deeper*.

39. 5

40. 20 people

41. *Sample answer:* −20, −15, −10, −5, 0; Start with −10, then pair −15 with −5 and −20 with 0. The sum of the integers must be 5(−10) = −50.

Fair Game Review

42–44. See Additional Answers for number lines.

42. −6, −1, $|2|$, 4, $|−10|$

43. −8, −3, $|0|$, 3, $|−4|$

44. −7, −5, −2, $|−2|$, $|5|$

45. B

Mini-Assessment

Divide.

1. $−16 ÷ (−4)$ 4

2. $−22 ÷ 11$ −2

3. $35 ÷ (−5)$ −7

4. $−36 ÷ (−6)$ 6

5. You play a video game for 15 minutes. You lose 75 points. What integer represents the mean change in points per minute? −5

Taking Math Deeper

Exercise 38

This problem reviews the concept of mean (average) that students learned last year. The difference here is that the data set contains negative numbers. When students studied mean previously, all of the numbers in the data set were positive.

 a. Find the total score.

$$−2 + (−6) + (−7) + (−3) = −18$$

 b. Find the mean score per round.

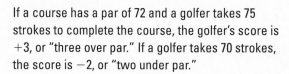

Divide the total score by the number of rounds. $\dfrac{−18}{4} = −4.5$

Scorecard	
Round 1	−2
Round 2	−6
Round 3	−7
Round 4	−3

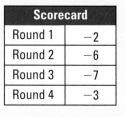

Low is good.

 In golf, par is the number of strokes it should take to complete a hole. An 18-hole course may contain four par 3s, ten par 4s, and four par 5s with a total par score of 72.

If a course has a par of 72 and a golfer takes 75 strokes to complete the course, the golfer's score is +3, or "three over par." If a golfer takes 70 strokes, the score is −2, or "two under par."

Golf tournaments usually have 4 rounds, for a total par of 4(72) = 288. As an extension, ask students to determine the golfer's total score.

$288 + (−18) = 270$

Project

Make a table showing the winning scores for each Masters Tournament since 2000. Graph the winning scores for each year. Describe any patterns you notice.

Reteaching and Enrichment Strategies

If students need help...	If students got it...
Resources by Chapter • Practice A and Practice B • Puzzle Time Record and Practice Journal Practice Differentiating the Lesson Lesson Tutorials Skills Review Handbook	Resources by Chapter • Enrichment and Extension • Technology Connection Start the next section

Find the mean of the integers.

32. 3, −10, −2, 13, 11

33. −26, 39, −10, −16, 12, 31

Evaluate the expression.

34. −8 − 14 ÷ 2 + 5

35. 24 ÷ (−4) + (−2) • (−5)

36. PATTERN Find the next two numbers in the pattern −128, 64, −32, 16, ... Explain your reasoning.

37. SNOWBOARDING A snowboarder descends a 1200-foot hill in 3 minutes. What is the mean change in elevation per minute?

38. GOLF The table shows a golfer's score for each round of a tournament.

   a. What was the golfer's total score?

   b. What was the golfer's mean score per round?

Scorecard	
Round 1	−2
Round 2	−6
Round 3	−7
Round 4	−3

39. TUNNEL The Detroit-Windsor Tunnel is an underwater highway that connects the cities of Detroit, Michigan, and Windsor, Ontario. How many times deeper is the roadway than the bottom of the ship?

40. AMUSEMENT PARK The regular admission price for an amusement park is $72. For a group of 15 or more, the admission price is reduced by $25. How many people need to be in a group to save $500?

41. **Number Sense** Write five different integers that have a mean of −10. Explain how you found your answer.

 Fair Game Review *What you learned in previous grades & lessons*

Graph the values on a number line. Then order the values from least to greatest. *(Section 1.1)*

42. −6, 4, |2|, −1, |−10|

43. 3, |0|, |−4|, −3, −8

44. |5|, −2, −5, |−2|, −7

45. MULTIPLE CHOICE What is the value of 4 • 3 + (12 ÷ 2)^2? *(Skills Review Handbook)*

   (A) 15   **(B)** 48   **(C)** 156   **(D)** 324

Evaluate the expression.   *(Section 1.4 and Section 1.5)*

1. $-7(6)$

2. $-1(-10)$

3. $\dfrac{-72}{-9}$

4. $-24 \div 3$

5. $-3 \cdot 4 \cdot (-6)$

6. $(-3)^3$

Evaluate the expression when $a = 4$, $b = -6$, and $c = -12$.   *(Section 1.4 and Section 1.5)*

7. c^2

8. bc

9. $\dfrac{ab}{c}$

10. $\dfrac{|c - b|}{a}$

11. **SPEECH** In speech class, you lose 3 points for every 30 seconds you go over the time limit. Your speech is 90 seconds over the time limit. What integer represents the change in your points? *(Section 1.4)*

12. **MOUNTAIN CLIMBING** On a mountain, the temperature decreases by 18°F every 5000 feet. What integer represents the change in temperature at 20,000 feet? *(Section 1.4)*

13. **GAMING** You play a video game for 15 minutes. You lose 165 points. What is the mean change in points per minute? *(Section 1.5)*

14. **DIVING** You dive 21 feet from the surface of a lake in 7 seconds.   *(Section 1.4 and Section 1.5)*

    a. What is the mean change in your position in feet per second?

    b. You continue diving. What is your position relative to the surface after 5 more seconds?

15. **HIBERNATION** A female grizzly bear weighs 500 pounds. After hibernating for 6 months, she weighs only 200 pounds. What is the mean change in weight per month? *(Section 1.5)*

Alternative Assessment Options

Math Chat   Student Reflective Focus Question
Structured Interview   Writing Prompt

Math Chat

- Have students work in pairs. One student describes the rule for multiplying two integers with the same sign and the rule for multiplying two integers with different signs. The student should include examples. The other student should probe for more information. Students then switch roles and repeat the process for dividing two integers with the same sign and dividing two integers with different signs.
- The teacher should walk around the classroom listening to the pairs and asking questions to ensure understanding.

Study Help Sample Answers

Remind students to complete Graphic Organizers for the rest of the chapter.

5a.

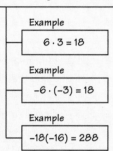

Multiplying integers with the same sign: The product of two integers with the same sign is positive.

Example
$6 \cdot 3 = 18$

Example
$-6 \cdot (-3) = 18$

Example
$-18(-16) = 288$

5b.

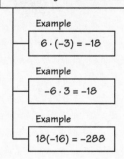

Multiplying integers with different signs: The product of two integers with different signs is negative.

Example
$6 \cdot (-3) = -18$

Example
$-6 \cdot 3 = -18$

Example
$18(-16) = -288$

6. Available at *BigIdeasMath.com*.

Reteaching and Enrichment Strategies

If students need help. . .	If students got it. . .

Resources by Chapter
  • Practice A and Practice B
  • Puzzle Time
Lesson Tutorials
BigIdeasMath.com
	Resources by Chapter
	  • Enrichment and Extension
	  • Technology Connection
	Game Closet at *BigIdeasMath.com*
	Start the Chapter Review

Answers

1. -42
2. 10
3. 8
4. -8
5. 72
6. -27
7. 144
8. 72
9. 2
10. $\dfrac{3}{2}$
11. -9
12. -72
13. -11
14. **a.** -3
    b. -36 ft
15. -50 lb/mo

For the Teacher
Additional Review Options
- *BigIdeasMath.com*
- Online Assessment
- Game Closet at *BigIdeasMath.com*
- Vocabulary Help
- Resources by Chapter

Answers
1. 3
2. 9
3. 17
4. 8
5. Mississippi River in Illinois
6. -27
7. -10
8. 25
9. -34

Review of Common Errors

Exercises 1–5
- Students may think they can find the absolute value of a number by changing its sign and incorrectly find $|6| = -6$.

Exercises 6–9
- Students may ignore the signs and just add the integers.
- Remind students of the Key Ideas and how the signs of the integers determine if the student needs to add or subtract the integers.

Exercises 10–14
- Students may change the sign of the first number or forget to change the problem from subtraction to addition when changing the sign of the second number.

Exercises 15–22
- Students may not remember that a negative number multiplied or divided by a negative number is positive.

Exercises 23–25
- Remind students of the order of operations.

Exercises 26–28
- Remind students of the definition of mean.

Exercise 29
- Students may not recognize that division should be used to find the answer.
- Point out to students that all of the shirts are priced the same and that the total price of the shirts is $30.60.

Review Key Vocabulary

integer, *p. 4*
absolute value, *p. 4*

opposites, *p. 10*
additive inverse, *p. 10*

Review Examples and Exercises

1.1 Integers and Absolute Value *(pp. 2–7)*

Find the absolute value of −2.

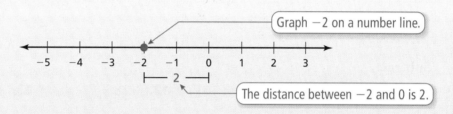

Graph −2 on a number line.

The distance between −2 and 0 is 2.

So, $|{-2}| = 2$.

Exercises

Find the absolute value.

1. $|3|$ **2.** $|{-9}|$ **3.** $|{-17}|$ **4.** $|8|$

5. ELEVATION The elevation of Death Valley, California, is −282 feet. The Mississippi River in Illinois has an elevation of 279 feet. Which is closer to sea level?

1.2 Adding Integers *(pp. 8–13)*

Find 6 + (−14).

$6 + (-14) = -8$ $|{-14}| > |6|$. So, subtract $|6|$ from $|{-14}|$.

Use the sign of −14.

The sum is −8.

Exercises

Add.

6. $-16 + (-11)$ **7.** $-15 + 5$ **8.** $100 + (-75)$ **9.** $-32 + (-2)$

1.3 Subtracting Integers *(pp. 14–19)*

Subtract.

a. $7 - 19 = 7 + (-19)$ Add the opposite of 19.

            $= -12$ Add.

    The difference is -12.

b. $-6 - (-10) = -6 + 10$ Add the opposite of -10.

               $= 4$ Add.

    The difference is 4.

Exercises

Subtract.

10. $8 - 18$ **11.** $-16 - (-5)$ **12.** $-18 - 7$ **13.** $-12 - (-27)$

14. GAME SHOW Your score on a game show is -300. You answer the final question incorrectly, so you lose 400 points. What is your final score?

1.4 Multiplying Integers *(pp. 22–27)*

a. Find $-7 \cdot (-9)$.

The integers have the same sign.

$$-7 \cdot (-9) = 63$$

The product is positive.

    The product is 63.

b. Find $-6(14)$.

The integers have different signs.

$$-6(14) = -84$$

The product is negative.

    The product is -84.

Exercises

Multiply.

15. $-8 \cdot 6$ **16.** $10(-7)$ **17.** $-3 \cdot (-6)$ **18.** $-12(5)$

Review Game

Integer Operations

Materials per Group
- 52 index cards, numbered 1 through 52
- paper for each group member
- pencil for each group member

Directions

This game is like the card game *War*. Divide the class into equally sized groups. One person in each group shuffles and deals the index cards for that group. Eight rounds are played. For each round, each person is dealt two cards. A different operation is used in each round, as follows.

> Round 1—addition
> Round 2—subtraction
> Round 3—multiplication
> Round 4—division
> Round 5—addition
> Round 6—subtraction
> Round 7—multiplication
> Round 8—division

Using the two cards dealt for the round, in Rounds 1 through 4, players evaluate an expression using the number on the first card, the operation for the round, and the number on the second card. In Rounds 5 through 8, players evaluate an expression using the number on the first card, the operation for the round, and *the negative of* the number on the second card.

The person with the greatest result in their group wins the round and takes all the other players' cards. In the event of a tie, the players involved receive one additional card, and the operation for the round is performed using *all three* cards. Players continue in this manner until there is one winner.

If a group runs out of cards before the end of Round 8, each player records how many cards he or she has collected, and the cards are collected from the players. The cards are then reshuffled and reused.

Who Wins?

The player with the most cards after Round 8 wins. In the event of a tie, the rounds are repeated, starting with Round 1, until there is a winner.

For the Student
Additional Practice
- Lesson Tutorials
- Multi-Language Glossary
- Self-Grading Progress Check
- *BigIdeasMath.com*
  Dynamic Student Edition
  Student Resources

Answers

10. -10		**11.** -11	
12. -25		**13.** 15	
14. -700 points			
15. -48		**16.** -70	
17. 18		**18.** -60	
19. -2		**20.** 7	
21. -5		**22.** -12	
23. -2		**24.** 2	
25. 3		**26.** -1	
27. -48		**28.** $-\$48$	
29. 5 shirts			

My Thoughts on the Chapter

What worked. . .

What did not work. . .

What I would do differently. . .

1.5 Dividing Integers (pp. 28–33)

a. Find $30 \div (-10)$.

> The integers have different signs.

$$30 \div (-10) = -3$$

> The quotient is negative.

∴ The quotient is -3.

b. Find $\dfrac{-72}{-9}$.

> The integers have the same sign.

$$\frac{-72}{-9} = 8$$

> The quotient is positive.

∴ The quotient is 8.

Exercises

Divide.

19. $-18 \div 9$ **20.** $\dfrac{-42}{-6}$ **21.** $\dfrac{-30}{6}$ **22.** $84 \div (-7)$

Evaluate the expression when $x = 3$, $y = -4$, and $z = -6$.

23. $z \div x$ **24.** $\dfrac{xy}{z}$ **25.** $\dfrac{z - 2x}{y}$

Find the mean of the integers.

26. $-3, -8, 12, -15, 9$ **27.** $-54, -32, -70, -25, -65, -42$

28. PROFITS The table shows the weekly profits of a fruit vendor. What is the mean profit for these weeks?

Week	1	2	3	4
Profit	$-\$125$	$-\$86$	$\$54$	$-\$35$

29. RETURNS You return several shirts to a store. The receipt shows that the amount placed back on your credit card is $-\$30.60$. Each shirt is $-\$6.12$. How many shirts did you return?

Find the absolute value.

1. $|-9|$

2. $|64|$

3. $|-22|$

Copy and complete the statement using <, >, or =.

4. $4 \;\rule{1cm}{0.4pt}\; |-8|$

5. $|-7| \;\rule{1cm}{0.4pt}\; -12$

6. $-7 \;\rule{1cm}{0.4pt}\; |3|$

Evaluate the expression.

7. $-6 + (-11)$

8. $2 - (-9)$

9. $-9 \cdot 2$

10. $-72 \div (-3)$

Evaluate the expression when $x = 5$, $y = -3$, and $z = -2$.

11. $\dfrac{y + z}{x}$

12. $\dfrac{x - 5z}{y}$

Find the mean of the integers.

13. $11, -7, -14, 10, -5$

14. $-32, -41, -39, -27, -33, -44$

15. **NASCAR** A driver receives -25 points for each rule violation. What integer represents the change in points after 4 rule violations?

16. **GOLF** The table shows your scores, relative to *par*, for nine holes of golf. What is your total score for the nine holes?

Hole	1	2	3	4	5	6	7	8	9	Total
Score	+1	-2	-1	0	-1	+3	-1	-3	+1	?

17. **VISITORS** In a recent 10-year period, the change in the number of visitors to U.S. national parks was about $-11,150,000$ visitors.

a. What was the mean yearly change in the number of visitors?

b. During the seventh year, the change in the number of visitors was about 10,800,000. Explain how the change for the 10-year period can be negative.

Test Item References

Chapter Test Questions	Section to Review
1–6	1.1
7, 16	1.2
8	1.3
9, 15	1.4
10–14, 17	1.5

Test-Taking Strategies

Remind students to quickly look over the entire test before they start so that they can budget their time. They should not spend too much time on any single problem. Urge students to try to work on a part of each problem because partial credit is better than none. Teach students to use the Stop and Think strategy before answering. **Stop** and carefully read the question, and **Think** about what the answer should look like.

Common Errors

- **Exercise 2** Students may think that the absolute value of a number is always its opposite and write $|64| = -64$. Use a number line to show students that absolute value is a number's distance from 0, so it is always positive or zero.
- **Exercises 4 and 5** When comparing absolute values of negative integers, students may not find absolute values first and instead just compare the integers. Remind students to find the absolute values first.
- **Exercises 7–12** Students may ignore the signs of the integers when simplifying. Remind students of the Key Ideas for addition, subtraction, multiplication, and division of integers and that the signs of the integers will affect their answers.
- **Exercises 13 and 14** Students may not remember how to find the mean of several numbers. They may get confused by the negative numbers and subtract instead of add. Remind students of the definition of mean.
- **Exercise 16** Students may not know the meaning of the word *par*, so explain it to them.

Reteaching and Enrichment Strategies

If students need help...	If students got it...
Resources by Chapter • Practice A and Practice B • Puzzle Time Record and Practice Journal Practice Differentiating the Lesson Lesson Tutorials *BigIdeasMath.com* Skills Review Handbook	Resources by Chapter • Enrichment and Extension • Technology Connection Game Closet at *BigIdeasMath.com* Start Cumulative Assessment

Answers

1. 9
2. 64
3. 22
4. $4 < |-8|$
5. $|-7| > -12$
6. $-7 < |3|$
7. -17
8. 11
9. -18
10. 24
11. -1
12. -5
13. -1
14. -36
15. -100
16. -3
17. a. $-1{,}115{,}000$ visitors
    b. During other years, there were more significant changes in visitors in the negative direction.

Technology for the *Teacher*

Online Assessment
Assessment Book
ExamView® Assessment Suite

Test-Taking Strategies

Available at *BigIdeasMath.com*

After Answering Easy Questions, Relax

Answer Easy Questions First

Estimate the Answer

Read All Choices before Answering

Read Question before Answering

Solve Directly or Eliminate Choices

Solve Problem before Looking at Choices

Use Intelligent Guessing

Work Backwards

About this Strategy

When taking a multiple choice test, be sure to read each question carefully and thoroughly. Before answering a question, determine exactly what is being asked, then eliminate the wrong answers and select the best choice.

Answers

1. C
2. H
3. C
4. 25
5. G

Item Analysis

1. **A.** The student treats all numbers as gains and finds their sum.

   B. The student finds the correct difference but thinks it is a gain instead of a loss.

   C. Correct answer

   D. The student treats all numbers as losses and finds their sum.

2. **F.** The student does not perform the operation correctly.

   G. The student does not perform the operation correctly.

   H. Correct answer

   I. The student does not perform the operation correctly.

3. **A.** The student does not evaluate a^2 correctly and does not find the absolute value.

   B. The student does not find the absolute value.

   C. Correct answer

   D. The student does not evaluate a^2 correctly.

4. **Gridded Response:** Correct answer: 25

   Common Error: The student thinks that $17 - (-8)$ is equivalent to $17 - 8$, getting an answer of 9.

5. **F.** The student thinks that $-(-5) = -5$.

   G. Correct answer

   H. The student does not follow the order of operations.

   I. The student does not follow the order of operations.

Technology for the *Teacher*

Performance Tasks

Online Assessment

Assessment Book

ExamView® Assessment Suite

Cumulative Assessment Icons

 Gridded Response

 Short Response (2-point rubric)

 Extended Response (4-point rubric)

1. A football team gains 2 yards on the first play, loses 5 yards on the second play, loses 3 yards on the third play, and gains 4 yards on the fourth play. What is the team's overall gain or loss for all four plays?

   A. a gain of 14 yards   **C.** a loss of 2 yards

   B. a gain of 2 yards   **D.** a loss of 14 yards

2. Which expression is *not* equal to the number 0?

   F. $5 - 5$ **H.** $6 - (-6)$

   G. $-7 + 7$ **I.** $-8 - (-8)$

Test-Taking Strategy

Solve Directly or Eliminate Choices

You ripped out $(-1)^2 + (-2)(-3)$ whiskers. How many did you rip out?

Ⓐ −5   Ⓑ 5   Ⓒ −7   Ⓓ 7

Yeow, why the biggest number?

"You can eliminate **A** and **C.** Then, solve directly to determine that the correct answer is **D.**"

3. What is the value of the expression below when $a = -2$, $b = 3$, and $c = -5$?

   $$\left| a^2 - 2ac + 5b \right|$$

   A. -9 **C.** 1

   B. -1 **D.** 9

4. What is the value of the expression below?

   $$17 - (-8)$$

5. Sam was evaluating an expression in the box below.

   $$(-2)^3 \cdot 3 - (-5) = 8 \cdot 3 - (-5)$$
   $$= 24 + 5$$
   $$= 29$$

   What should Sam do to correct the error that he made?

   F. Subtract 5 from 24 instead of adding.

   G. Rewrite $(-2)^3$ as -8.

   H. Subtract -5 from 3 before multiplying by $(-2)^3$.

   I. Multiply -2 by 3 before raising the quantity to the third power.

6. What is the value of the expression below when $x = 6$, $y = -4$, and $z = -2$?

$$\frac{x - 2y}{-z}$$

A. -7

B. -1

C. 1

D. 7

7. What is the missing number in the sequence below?

$$39, 24, 9, \underline{\quad}, -21$$

8. You are playing a game using the spinner shown. You start with a score of 0 and spin the spinner four times. When you spin blue or green, you add the number to your score. When you spin red or orange, you subtract the number from your score. Which sequence of colors represents the greatest score?

F. red, green, green, red

G. orange, orange, green, blue

H. red, blue, orange, green

I. blue, red, blue, red

9. Which expression represents a negative integer?

A. $5 - (-6)$

B. $(-3)^3$

C. $-12 \div (-6)$

D. $(-2)(-4)$

10. Which expression has the greatest value when $x = -2$ and $y = -3$?

F. $-xy$

G. xy

H. $x - y$

I. $-x - y$

Item Analysis (continued)

6. **A.** The student thinks that $-(-2) = -2$.

   B. The student thinks that $(-2)(-4) = -8$.

   C. The student thinks that $(-2)(-4) = -8$ and $-(-2) = -2$.

   D. Correct answer

7. **Gridded Response:** Correct answer: -6

   Common Error: The student incorrectly finds a number halfway between 9 and -21 by subtracting 9 from -21 to get -30, half of which is -15.

8. **F.** The student adds the red value instead of subtracting.

   G. Correct answer

   H. The student adds the absolute value of each number.

   I. The student chooses the value with the greatest absolute value, not the greatest value.

9. **A.** The student does not perform the operation correctly.

   B. Correct answer

   C. The student does not perform the operation correctly.

   D. The student does not perform the operation correctly.

10. **F.** The student does not perform the operation correctly.

    G. Correct answer

    H. The student does not perform the operation correctly.

    I. The student does not perform the operation correctly.

6. D

7. -6

8. G

9. B

10. G

Answers

11. B

12. G

13. D

14. *Part A* Start at 0. Then move 2 to the left and then 3 more to the left, which results in a position of -5.

   Part B Start at 0. Then move 2 to the right and then 5 to the left, which results in a position of -3.

15. H

Item Analysis (continued)

11. **A.** The student thinks that $-(-3) = -3$.

   B. Correct answer

   C. The student thinks that $-5 \cdot (-4)^2 = 80$ and $-(-3) = -3$.

   D. The student thinks that $-5 \cdot (-4)^2 = 80$.

12. **F.** The student does not have a clear understanding of the properties.

   G. Correct answer

   H. The student does not have a clear understanding of the properties.

   I. The student does not have a clear understanding of the properties.

13. **A.** The student finds the mode instead of the mean.

   B. The student finds the range instead of the mean. The student also makes multiple errors in finding the range, using -8 and 1 instead of -8 and 4, thinking that the range between -8 and 1 is 7, and thinking that the range is negative.

   C. The student finds the median instead of the mean.

   D. Correct answer

14. **2 points** The student demonstrates a thorough understanding of adding and subtracting integers using a number line. The student demonstrates how to add using a number line, writes the expression, and gets the correct answer $-2 + (-3) = -5$. The student demonstrates how to subtract using a number line, writes the expression, and gets the correct answer $2 - 5 = -3$.

   1 point The student demonstrates a partial understanding of adding and subtracting integers using a number line. The student shows some knowledge of adding and subtracting integers, but is not successful in determining the correct answers.

   0 points The student demonstrates insufficient understanding of adding and subtracting integers using a number line.

15. **F.** The student incorrectly evaluates the numerator as $(-3 - 2)^2$ instead of $-3 - (2)^2$.

   G. The student incorrectly evaluates the power as $(-2)^2$ instead of 2^2.

   H. Correct answer

   I. The student incorrectly evaluates the numerator as $(-3 - 2)^2$ instead of $-3 - (2)^2$ and uses the wrong sign in the answer.

11. What is the value of the expression below?

$$-5 \cdot (-4)^2 - (-3)$$

A. -83 **C.** 77

B. -77 **D.** 83

12. Which property does the equation below represent?

$$-80 + 30 + (-30) = -80 + [30 + (-30)]$$

F. Commutative Property of Addition

G. Associative Property of Addition

H. Additive Inverse Property

I. Addition Property of Zero

13. What is the mean of the data set in the box below?

$$-8, -6, -2, 0, -6, -8, 4, -7, -8, 1$$

A. -8 **C.** -6

B. -7 **D.** -4

14. Consider the number line shown below.

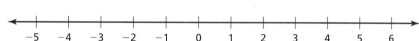

Part A   Use the number line to explain how to add -2 and -3.

Part B   Use the number line to explain how to subtract 5 from 2.

15. What is the value of the expression below?

$$\frac{-3 - 2^2}{-1}$$

F. -25 **H.** 7

G. -1 **I.** 25

2 Rational Numbers

2.1 Rational Numbers

2.2 Adding Rational Numbers

2.3 Subtracting Rational Numbers

2.4 Multiplying and Dividing Rational Numbers

"On the count of 5, I'm going to give you half of my dog biscuits."

"1, 2, 3, 4, $4\frac{1}{2}$, $4\frac{3}{4}$, $4\frac{7}{8}$,..."

"I entered a contest for dog biscuits."

"I was notified that the number of biscuits I won was in the three-digit range."

What Your Students Have Learned

- Fluently multiply and divide whole numbers.
- Add and subtract fractions with unlike denominators.
- Multiply fractions or whole numbers by fractions.
- Divide unit fractions by whole numbers and whole numbers by unit fractions.
- Add, subtract, multiply, and divide decimals up to hundredths.
- Divide fractions.
- Fluently add, subtract, multiply, and divide decimals.
- Describe quantities with positive and negative numbers.

What Your Students Will Learn

- Add, subtract, multiply, and divide rational numbers.
- Apply properties of operations as strategies to perform operations with rational numbers.
- Convert rational numbers to decimals using long division.

Pacing Guide for Chapter 2

Chapter Opener	
Regular	1 Day
Accelerated	0.5 Day
Section 1	
Regular	2 Days
Accelerated	0.5 Day
Section 2	
Regular	2 Days
Accelerated	0.5 Day
Study Help / Quiz	
Regular	1 Day
Accelerated	0 Days
Section 3	
Regular	2 Days
Accelerated	0.5 Day
Section 4	
Regular	2 Days
Accelerated	1 Day
Chapter Review/ Chapter Tests	
Regular	2 Days
Accelerated	2 Days
Total Chapter 2	
Regular	12 Days
Accelerated	5 Days
Year-to-Date	
Regular	27 Days
Accelerated	10 Days

Technology for the *Teacher*

BigIdeasMath.com
Chapter at a Glance
Complete Materials List
Parent Letters: English and Spanish

What Your Students Have Learned

- Write decimals as fractions and fractions as decimals.
- Add and subtract fractions with unlike denominators.
- Multiply and divide fractions with unlike denominators.

Additional Topics for Review

- Place Value
- Writing Mixed Numbers as Improper Fractions
- Dividing Numbers Using the Standard Algorithm
- Simplifying Fractions

Try It Yourself

1. $\dfrac{51}{100}$ 2. $\dfrac{731}{1000}$

3. 0.6 4. 0.875

5. $\dfrac{9}{10}$ 6. $\dfrac{3}{5}$

7. $\dfrac{27}{70}$ 8. $\dfrac{17}{20}$

Record and Practice Journal
Fair Game Review

1. $\dfrac{13}{50}$ 2. $\dfrac{79}{100}$

3. $\dfrac{571}{1000}$ 4. $\dfrac{423}{500}$

5. 0.375 6. 0.4

7. 0.6875 8. 0.85

9. $\dfrac{3}{5}$ 10. $\dfrac{17}{72}$

11–18. See Additional Answers.

Math Background Notes

Vocabulary Review

- Denominator
- Least Common Multiple
- Common Denominator
- Least Common Denominator (LCD)
- Reciprocal
- Divisor

Writing Decimals and Fractions

- Students should know how to convert between decimals and fractions.
- You may need to review place values to the right of the decimal place with students prior to completing Example 1.

Adding and Subtracting Fractions

- Students should know how to add and subtract fractions.
- Remind students that adding and subtracting fractions requires a common denominator.
- You should review the least common multiple with students. This concept will help some students to find a common denominator.
- Using the least common multiple of the denominators will produce the least common denominator. Remind students that there are many common denominators to choose from. Some choices will require students to simplify the fraction at the end.

Multiplying and Dividing Fractions

- Students should know how to multiply and divide fractions.
- Remind students that the rules for multiplying and dividing fractions are different from the rules for adding and subtracting fractions. Multiplying and dividing fractions does not require a common denominator.
- **Teaching Tip:** Most students will remember the process to divide fractions. If your students are comfortable with the process, encourage them to describe it using math vocabulary. Instead of "change the sign and flip the second fraction," encourage "multiply by the reciprocal of the divisor."

Reteaching and Enrichment Strategies

If students need help...	If students got it...
Record and Practice Journal • Fair Game Review Skills Review Handbook Lesson Tutorials	Game Closet at *BigIdeasMath.com* Start the next section

What You Learned Before

"Let's play a game. The goal is to say a positive rational number that is less than the other pet's number... You go first."

This feels like a setup.

Writing Decimals and Fractions

Example 1 Write 0.37 as a fraction.

$$0.37 = \frac{37}{100}$$

Example 2 Write $\frac{2}{5}$ as a decimal.

$$\frac{2}{5} = \frac{2 \cdot 2}{5 \cdot 2} = \frac{4}{10} = 0.4$$

Try It Yourself
Write the decimal as a fraction or the fraction as a decimal.

1. 0.51 **2.** 0.731 **3.** $\frac{3}{5}$ **4.** $\frac{7}{8}$

Adding and Subtracting Fractions

Example 3 Find $\frac{1}{3} + \frac{1}{5}$.

$$\frac{1}{3} + \frac{1}{5} = \frac{1 \cdot 5}{3 \cdot 5} + \frac{1 \cdot 3}{5 \cdot 3}$$

$$= \frac{5}{15} + \frac{3}{15}$$

$$= \frac{8}{15}$$

Example 4 Find $\frac{1}{4} - \frac{2}{9}$.

$$\frac{1}{4} - \frac{2}{9} = \frac{1 \cdot 9}{4 \cdot 9} - \frac{2 \cdot 4}{9 \cdot 4}$$

$$= \frac{9}{36} - \frac{8}{36}$$

$$= \frac{1}{36}$$

Multiplying and Dividing Fractions

Example 5 Find $\frac{5}{6} \cdot \frac{3}{4}$.

$$\frac{5}{6} \cdot \frac{3}{4} = \frac{5 \cdot \overset{1}{\cancel{3}}}{\underset{2}{\cancel{6}} \cdot 4}$$

$$= \frac{5}{8}$$

Example 6 Find $\frac{2}{3} \div \frac{9}{10}$.

$$\frac{2}{3} \div \frac{9}{10} = \frac{2}{3} \cdot \frac{10}{9}$$ ← Multiply by the reciprocal of the divisor.

$$= \frac{2 \cdot 10}{3 \cdot 9}$$

$$= \frac{20}{27}$$

Try It Yourself
Evaluate the expression.

5. $\frac{1}{4} + \frac{13}{20}$ **6.** $\frac{14}{15} - \frac{1}{3}$ **7.** $\frac{3}{7} \cdot \frac{9}{10}$ **8.** $\frac{4}{5} \div \frac{16}{17}$

Essential Question

How can you use a number line to order rational numbers?

The Meaning of a Word ● Rational

The word **rational** comes from the word *ratio*. Recall that you can write a ratio using fraction notation.

If you sleep for 8 hours in a day, then the ratio of your sleeping time to the total hours in a day can be written as $\dfrac{8\text{ h}}{24\text{ h}}$.

A **rational number** is a number that can be written as the ratio of two integers.

$$2 = \frac{2}{1} \qquad -3 = \frac{-3}{1} \qquad -\frac{1}{2} = \frac{-1}{2} \qquad 0.25 = \frac{1}{4}$$

---

1 ACTIVITY: Ordering Rational Numbers

Work in groups of five. Order the numbers from least to greatest.

- Use masking tape and a marker to make a number line on the floor similar to the one shown.

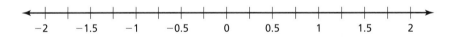

- Write the numbers on pieces of paper. Then each person should choose one.

- Stand on the location of your number on the number line.

- Use your positions to order the numbers from least to greatest.

Rational Numbers

In this lesson, you will
- understand that a rational number is an integer divided by an integer.
- convert rational numbers to decimals.

a. $-0.5,\ 1.25,\ -\dfrac{1}{3},\ 0.5,\ -\dfrac{5}{3}$

b. $-\dfrac{7}{4},\ 1.1,\ \dfrac{1}{2},\ -\dfrac{1}{10},\ -1.3$

c. $-1.4,\ -\dfrac{3}{5},\ \dfrac{9}{2},\ \dfrac{1}{4},\ 0.9$

d. $\dfrac{5}{4},\ 0.75,\ -\dfrac{5}{4},\ -0.8,\ -1.1$

Laurie's Notes

Introduction

Applying Mathematical Practices

- **Make Sense of Problems and Persevere in Solving Them:** In this lesson, students will write fractions as decimals and vice versa. Students should always check the reasonableness of their answers. For instance, $\frac{7}{11}$ is greater than $\frac{1}{2}$. So, when you write $\frac{7}{11}$ as a decimal, the result should be greater than 0.5.

Motivate

- A key skill for both activities today will be the ability to compare fractions and decimals. Try a warm up where students need to fill in the following table.

Fraction	$\frac{1}{2}$		$\frac{3}{5}$		$\frac{3}{4}$	
Decimal		0.1		0.8		1.4

- Check for understanding of the process of converting between these two forms of numbers.
- Students have studied operations with whole numbers, fractions, decimals, and integers.
- ? "Do you think there is a number halfway between -3 and -4? What is that number?"
- Explain that in this chapter, they will perform operations on numbers such as -3.5. Define rational numbers.

Activity Notes

Activity 1

- **Model with Mathematics:** In preparing for this activity, be sure to leave sufficient space between the number line marks so that students are able to stand at their locations comfortably. If there is enough space in the classroom, make multiple number lines on the floor. Consider different orientations so that students may see some as vertical number lines and some as horizontal number lines. If space is limited, pairs of students could do the same problem on the board or on a piece of paper.
- **Construct Viable Arguments:** When the first set of numbers has been located, spend time having students give their reasoning as to why they located the numbers as they did. For instance, how did they know that $-\frac{5}{3}$ was to the left of -0.5 versus to the right of it?
- So that all students have an opportunity to use the number line on the floor, rotate groups at the end of each set of numbers.
- **Extension:** If time permits, ask students to name a decimal between two fractions $\left(-\frac{1}{2} \text{ and } -\frac{3}{4}\right)$ and to name a fraction between two decimals $(-0.6 \text{ and } -0.7)$.

What Your Students Will Learn

- Understand that a rational number is an integer divided by an integer.
- Convert rational numbers to decimals using long division.
- Write decimals as fractions in simplest form using equivalent fractions.
- Order rational numbers by converting them to decimals and then graphing them on a number line.

Previous Learning

Students should know how to convert between common fractions (halves, fourths, fifths, and tenths) and decimals. They should also be able to graph common fractions and decimals on a number line.

Technology for the *Teacher*

Dynamic Classroom

Lesson Plans
Complete Materials List

2.1 Record and Practice Journal

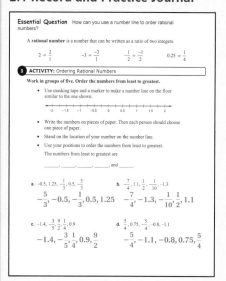

English Language Learners

Vocabulary

Let English language learners know that the pronunciation of rational numbers that end in –th such as fourths and fifths is sometimes difficult for native English speakers.

2.1 Record and Practice Journal

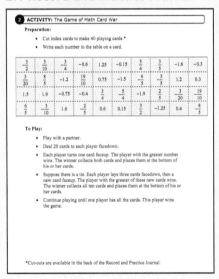

2 ACTIVITY: The Game of Math Card War

Preparation:

- Cut index cards to make 40 playing cards.*
- Write each number in the table on a card.

$-\frac{3}{2}$	$\frac{3}{10}$	$-\frac{3}{4}$	-0.6	1.25	-0.15	$\frac{5}{4}$	$\frac{3}{5}$	-1.6	-0.3
$\frac{3}{20}$	$\frac{8}{10}$	-1.2	$\frac{19}{10}$	0.75	-1.5	$\frac{6}{5}$	$\frac{3}{5}$	1.2	0.3
1.5	1.9	-0.75	-0.4	$\frac{3}{4}$	$-\frac{5}{4}$	-1.9	$\frac{2}{5}$	$-\frac{3}{20}$	$\frac{19}{10}$
$\frac{6}{5}$	$-\frac{3}{10}$	1.6	$-\frac{2}{5}$	0.6	0.15	$\frac{3}{2}$	-1.25	0.4	$-\frac{8}{5}$

To Play:

- Play with a partner.
- Deal 20 cards to each player facedown.
- Each player turns one card faceup. The player with the greater number wins. The winner collects both cards and places them at the bottom of his or her cards.
- Suppose there is a tie. Each player lays three cards facedown, then a new card faceup. The player with the greater of these new cards wins. The winner collects all ten cards and places them at the bottom of his or her cards.
- Continue playing until one player has all the cards. This player wins the game.

*Cut-outs are available in the back of the Record and Practice Journal.

What Is Your Answer?

3. **IN YOUR OWN WORDS** How can you use a number line to order rational numbers? Give an example.

Sample answer: A number line can be used to organize rational numbers from least to greatest based on their order from left to right on the line.

The numbers are in order from least to greatest. Fill in the blank spaces with rational numbers. **Sample answers are given.**

4. $-\frac{1}{2}$, $\boxed{-\frac{1}{4}}$, $\frac{1}{3}$, $\frac{3}{4}$, $\frac{7}{5}$, $\boxed{2}$

5. $-\frac{5}{2}$, $\boxed{-2}$, -1.9, $\boxed{-\frac{3}{2}}$, $\frac{2}{3}$, $\boxed{\frac{1}{2}}$

6. $-\frac{1}{3}$, $\boxed{-\frac{1}{5}}$, -0.1, $\boxed{0.1}$, $\frac{4}{5}$, $\boxed{\frac{5}{4}}$

7. -3.4, $\boxed{}$, -1.5, $\boxed{1.1}$, 2.2, $\boxed{2.5}$
   -2.1

Laurie's Notes

Activity 2

- You may want to make the game cards ahead of time or have students create the game cards.
- **Management Tip:** To preserve cards for multiple uses, make cards on colored cardstock and store individual sets in sealable plastic bags.
- The card game *War* is familiar to many students. The question asked each play is, "Which number is greater?" The player with the greater value collects both cards. If the cards have an equivalent value, there is a tie. As stated in the text, each player lays 3 cards face down and then 1 card face up. The player with the card of greater value collects all of the cards.
- **Comparing Cards:** The key component of this activity is when students actually compare the two rational numbers. Discuss with students how they will compare the numbers. When both numbers are positive or the signs are different, students will have less difficulty. If both numbers are negative, students need to remember that the farther the number is to the right on the number line, the greater its value.

  For example, $-\frac{3}{5} > -0.75$ because $-\frac{3}{5}$ is to the right of -0.75 on a number line.
- To start play, give students the opportunity to preview the cards. Explain the rules and let students begin. If one group finishes early, have them shuffle the cards and play again.
- **Extension:** The cards can also be used to play the game *Memory*. Put the fraction cards in one group and the decimal cards in another group. Place all cards face down in two grids. Students select one card from each group. If the cards match, (meaning they are equivalent), then the student keeps the cards. If they do not match, the cards are put back face down. A deck of 40 cards is too many! Reduce the deck to 24 (12 in each group). Make sure the equivalent decimals and fractions are in each deck.

What Is Your Answer?

- Listen for the big idea, namely that the farther to the right the number is on the number line, the greater the value of that number.
- **Construct Viable Arguments and Critique the Reasoning of Others:** For Questions 4–7, students should work with partners. Have students share their results and their reasoning. Answers will vary, so the explanation is important to hear.

Closure

- Which is greater: *A* or *B*? All have *B* as the greater number.

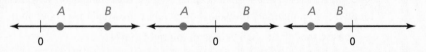

Preparation:

- Cut index cards to make 40 playing cards.
- Write each number in the table on a card.

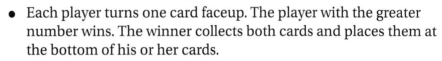

Math Practice

Consider Similar Problems

What are some ways to determine which number is greater?

To Play:

- Play with a partner.
- Deal 20 cards to each player facedown.
- Each player turns one card faceup. The player with the greater number wins. The winner collects both cards and places them at the bottom of his or her cards.
- Suppose there is a tie. Each player lays three cards facedown, then a new card faceup. The player with the greater of these new cards wins. The winner collects all ten cards and places them at the bottom of his or her cards.
- Continue playing until one player has all the cards. This player wins the game.

$-\dfrac{3}{2}$	$\dfrac{3}{10}$	$-\dfrac{3}{4}$	-0.6	1.25	-0.15	$\dfrac{5}{4}$	$\dfrac{3}{5}$	-1.6	-0.3
$\dfrac{3}{20}$	$\dfrac{8}{5}$	-1.2	$\dfrac{19}{10}$	0.75	-1.5	$-\dfrac{6}{5}$	$-\dfrac{3}{5}$	1.2	0.3
1.5	1.9	-0.75	-0.4	$\dfrac{3}{4}$	$-\dfrac{5}{4}$	-1.9	$\dfrac{2}{5}$	$-\dfrac{3}{20}$	$-\dfrac{19}{10}$
$\dfrac{6}{5}$	$-\dfrac{3}{10}$	1.6	$-\dfrac{2}{5}$	0.6	0.15	$\dfrac{3}{2}$	-1.25	0.4	$-\dfrac{8}{5}$

What Is Your Answer?

3. IN YOUR OWN WORDS How can you use a number line to order rational numbers? Give an example.

The numbers are in order from least to greatest. Fill in the blank spaces with rational numbers.

4. $-\dfrac{1}{2},$ ⬜ $,\dfrac{1}{3},$ ⬜ $,\dfrac{7}{5},$ ⬜

5. $-\dfrac{5}{2},$ ⬜ $,-1.9,$ ⬜ $,-\dfrac{2}{3},$ ⬜

6. $-\dfrac{1}{3},$ ⬜ $,-0.1,$ ⬜ $,\dfrac{4}{5},$ ⬜

7. $-3.4,$ ⬜ $,-1.5,$ ⬜ $,2.2,$ ⬜

Practice

Use what you learned about ordering rational numbers to complete Exercises 28–30 on page 48.

Check It Out
Lesson Tutorials
BigIdeasMath com

Key Vocabulary
rational number,
  p. 46
terminating decimal,
  p. 46
repeating decimal,
  p. 46

 Key Idea

Rational Numbers

A **rational number** is a number that can be written as $\dfrac{a}{b}$ where a and b are integers and $b \neq 0$.

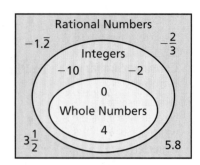

Because you can divide any integer by any nonzero integer, you can use long division to write fractions and mixed numbers as decimals. These decimals are also rational numbers and will either *terminate* or *repeat*.

A **terminating decimal** is a decimal that ends.

$$1.5, \ -0.25, \ 10.625$$

A **repeating decimal** is a decimal that has a pattern that repeats.

$$-1.333\ldots = -1.\overline{3}$$
$$0.151515\ldots = 0.\overline{15}$$

Use *bar notation* to show which of the digits repeat.

EXAMPLE 1 **Writing Rational Numbers as Decimals**

a. Write $-2\dfrac{1}{4}$ as a decimal.

Notice that $-2\dfrac{1}{4} = -\dfrac{9}{4}$.

Divide 9 by 4.

$$
\begin{array}{r}
2.25 \\
4\overline{)9.00} \\
-8 \\
\hline
1\,0 \\
-8 \\
\hline
20 \\
-20 \\
\hline
0
\end{array}
$$

The remainder is 0. So, it is a terminating decimal.

∴ So, $-2\dfrac{1}{4} = -2.25$.

b. Write $\dfrac{5}{11}$ as a decimal.

Divide 5 by 11.

$$
\begin{array}{r}
0.4545 \\
11\overline{)5.0000} \\
-4\,4 \\
\hline
60 \\
-55 \\
\hline
50 \\
-44 \\
\hline
60 \\
-55 \\
\hline
5
\end{array}
$$

The remainder repeats. So, it is a repeating decimal.

∴ So, $\dfrac{5}{11} = 0.\overline{45}$.

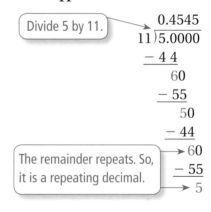

On Your Own

Now You're Ready
Exercises 11–18

Write the rational number as a decimal.

1. $-\dfrac{6}{5}$ **2.** $-7\dfrac{3}{8}$ **3.** $-\dfrac{3}{11}$ **4.** $1\dfrac{5}{27}$

🔊 Multi-Language Glossary at BigIdeasMath✓com

Laurie's Notes

Introduction

Connect

- **Yesterday:** Students ordered rational numbers.
- **Today:** Students will extend this knowledge to include repeating decimals.

Motivate

- Ask students to form a "name fraction," where the numerator is the number of letters in their first name and the denominator is the number of letters in their last name.
- Before class, go through your class roster and select two students whose name fractions are nearly equivalent, but one is a terminating decimal and the other is a repeating decimal. Discuss writing the fractions as decimals.
- When you look at the repeating decimal, share that today's lesson is about writing rational numbers, which may be repeating decimals.

Lesson Notes

Key Idea

- **Discuss:** Students have worked with fractions and decimals before. Explain that when negative fractions and decimals are included, we refer to these numbers as rational numbers. Also point out that the definition includes the word *can*, meaning that the rational numbers do not have to be written in the form $\frac{a}{b}$, but they *can* be.
- Define terminating and repeating decimals. Give examples of each.
- **Common Error:** Some students will write $\frac{1}{3}$ as 0.333 and think that is sufficient. They do not realize what the repeat bar represents.
- **Big Idea:** In this lesson students should gain an understanding that every quotient of integers (with a non-zero divisor) is a rational number.

Example 1

- ❓ **"How do you write a fraction as a decimal?"** Listen for 3 methods: 1) benchmark fractions you know, 2) write the fraction as an equivalent fraction with a denominator as a power of 10 and use the place value, or 3) divide the numerator by the denominator.
- **Make Sense of Problems:** Mathematically proficient students are able to plan a solution. Choosing between methods may help students be more efficient and accurate when writing fractions as decimals.
- Complete part (a) as a class. The first step is to write the mixed number as the equivalent improper fraction. Then divide the numerator by the denominator. Point out that the negative sign is simply placed in the answer after the calculations are complete.
- Complete part (b) as a class. Remind students that you always divide the numerator by the denominator, regardless of the size of the numbers!

On Your Own

- **Neighbor Check:** Have students work independently and then have their neighbors check their work. Have students discuss any discrepancies.

Goal Today's lesson is writing fractions as decimals, including **repeating decimals**, and writing decimals as fractions in simplest form.

Lesson Tutorials
Lesson Plans
Answer Presentation Tool

Extra Example 1

Write the rational number as a decimal.

a. $4\frac{3}{16}$ 4.1875

b. $-3\frac{4}{9}$ $-3.\overline{4}$

On Your Own

1. -1.2 2. -7.375

3. $-0.\overline{27}$ 4. $1.\overline{185}$

Extra Example 2

Write -2.625 as a mixed number in simplest form. $-2\frac{5}{8}$

On Your Own

5. $-\frac{7}{10}$

6. $\frac{1}{8}$

7. $-3\frac{1}{10}$

8. $-10\frac{1}{4}$

Extra Example 3

Order the rational numbers $-\frac{5}{9}$, $-1\frac{3}{4}$, $-\frac{13}{8}$, and -0.6 from least to greatest. $-1\frac{3}{4}$, $-\frac{13}{8}$, -0.6, $-\frac{5}{9}$

On Your Own

9. All of the sea creatures (anglerfish, squid, shark, and whale) are deeper than the dolphin.

Differentiated Instruction

Auditory

Writing terminating decimals as rational numbers is easier if the students read the decimal using place value as opposed to reading the digits.

Terminating decimal:
-0.26

Read as place value:
negative twenty-six hundredths

Read as digits:
negative zero point two six

Example 2

? "How do you write a decimal as a fraction?" Look at the place value of the last digit in the decimal and that will be the denominator.

• Work through Example 2.

? "How was the fraction simplified?" Both the numerator and the denominator were divided by a common factor of 2.

• Be sure to discuss the Study Tip.

• **Extension:** Write -0.026 and -2.6 as fractions. This helps students focus on the importance of place value and where the last digit is located.

On Your Own

• **Neighbor Check:** Have students work independently and then have their neighbors check their work. Have students discuss any discrepancies.

• In Questions 7 and 8, the whole number portion of the decimal can be a problem.

Example 3

• Discuss the unit of measure, kilometers.

• Work through the problem. When doing this problem in class, draw the number line vertically and identify sea level.

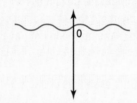

On Your Own

• **Neighbor Check:** Have students work independently and then have their neighbors check their work. Have students discuss any discrepancies.

• **Extension:** If calculators are available to students, explore the repeating patterns for certain sets of fractions (thirds, ninths, elevenths, etc.).

Closure

• **Exit ticket:**

• Write $-\frac{5}{6}$ as a decimal. $-0.8\overline{3}$

• Write -0.56 as a fraction. $-\frac{56}{100} = -\frac{14}{25}$

EXAMPLE (2) **Writing a Decimal as a Fraction**

Write −0.26 as a fraction in simplest form.

Study Tip

If p and q are integers, then $-\dfrac{p}{q} = \dfrac{-p}{q} = \dfrac{p}{-q}$.

$$-0.26 = -\frac{26}{100}$$

> Write the digits after the decimal point in the numerator.

> The last digit is in the hundredths place. So, use 100 in the denominator.

$$= -\frac{13}{50} \qquad \text{Simplify.}$$

On Your Own

Now You're Ready
Exercises 20–27

Write the decimal as a fraction or a mixed number in simplest form.

5. −0.7 **6.** 0.125 **7.** −3.1 **8.** −10.25

EXAMPLE (3) **Ordering Rational Numbers**

Creature	Elevation (kilometers)
Anglerfish	$-\dfrac{13}{10}$
Squid	$-2\dfrac{1}{5}$
Shark	$-\dfrac{2}{11}$
Whale	-0.8

The table shows the elevations of four sea creatures relative to sea level. Which of the sea creatures are deeper than the whale? Explain.

Write each rational number as a decimal.

$$-\frac{13}{10} = -1.3$$

$$-2\frac{1}{5} = -2.2$$

$$-\frac{2}{11} = -0.\overline{18}$$

Then graph each decimal on a number line.

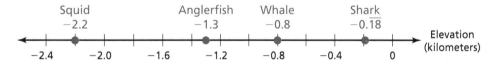

Both −2.2 and −1.3 are less than −0.8. So, the squid and the anglerfish are deeper than the whale.

On Your Own

Now You're Ready
Exercises 28–33

9. WHAT IF? The elevation of a dolphin is $-\dfrac{1}{10}$ kilometer. Which of the sea creatures in Example 3 are deeper than the dolphin? Explain.

✓ Vocabulary and Concept Check

1. **VOCABULARY** Is the quotient of two integers always a rational number? Explain.

2. **WRITING** Are all terminating and repeating decimals rational numbers? Explain.

Tell whether the number belongs to each of the following number sets:
rational numbers, integers, whole numbers.

3. -5
4. $-2.1\overline{6}$
5. 12
6. 0

Tell whether the decimal is *terminating* or *repeating*.

7. $-0.4848\ldots$
8. -0.151
9. 72.72
10. $-5.2\overline{36}$

Practice and Problem Solving

Write the rational number as a decimal.

① 11. $\dfrac{7}{8}$
12. $\dfrac{1}{11}$
13. $-\dfrac{7}{9}$
14. $-\dfrac{17}{40}$

15. $1\dfrac{5}{6}$
16. $-2\dfrac{17}{18}$
17. $-5\dfrac{7}{12}$
18. $8\dfrac{15}{22}$

19. **ERROR ANALYSIS** Describe and correct the error in writing the rational number as a decimal.

$$\times \qquad -\dfrac{7}{11} = -0.6\overline{3}$$

Write the decimal as a fraction or a mixed number in simplest form.

② 20. -0.9
21. 0.45
22. -0.258
23. -0.312

24. -2.32
25. -1.64
26. 6.012
27. -12.405

Order the numbers from least to greatest.

③ 28. $-\dfrac{3}{4}, 0.5, \dfrac{2}{3}, -\dfrac{7}{3}, 1.2$
29. $\dfrac{9}{5}, -2.5, -1.1, -\dfrac{4}{5}, 0.8$
30. $-1.4, -\dfrac{8}{5}, 0.6, -0.9, \dfrac{1}{4}$

31. $2.1, -\dfrac{6}{10}, -\dfrac{9}{4}, -0.75, \dfrac{5}{3}$
32. $-\dfrac{7}{2}, -2.8, -\dfrac{5}{4}, \dfrac{4}{3}, 1.3$
33. $-\dfrac{11}{5}, -2.4, 1.6, \dfrac{15}{10}, -2.25$

34. **COINS** You lose one quarter, two dimes, and two nickels.

   a. Write the amount as a decimal.

   b. Write the amount as a fraction in simplest form.

35. **HIBERNATION** A box turtle hibernates in sand at $-1\dfrac{5}{8}$ feet. A spotted turtle hibernates at $-1\dfrac{16}{25}$ feet. Which turtle is deeper?

Assignment Guide and Homework Check

Level	Day 1 Activity Assignment	Day 2 Lesson Assignment	Homework Check
Basic	28–30, 48–52	1–10, 15–23 odd, 31–35 odd	15, 21, 31, 35
Average	28–30, 48–52	1–10, 15–23 odd, 31–35 odd, 43–45	15, 21, 31, 44
Advanced	28–30, 48–52	1–10, 19, 32–42 even, 45–47	28, 36, 42, 46
Accelerated	1–10, 19, 28–42 even, 45–52		28, 36, 42, 46

Common Errors

- **Exercises 11–18** Students may forget to carry the negative sign through the division operation. Tell them to create a space for the final answer and to write the sign of the number in the space at the beginning.
- **Exercises 20–27** Students may try to put the decimal number over the denominator. Remind them to remove the decimal point before they write it as a fraction. They can also write the whole number in front of the fraction while they are reducing it.
- **Exercises 28–33** Students may just order the fractions or decimals without the negative signs. Remind them that some numbers are negative and will be less than the positive numbers.

2.1 Record and Practice Journal

Write the rational number as a decimal.

1. $-\dfrac{9}{10}$
-0.9

2. $-4\dfrac{2}{3}$
$-4.\overline{6}$

3. $1\dfrac{7}{16}$
1.4375

Write the decimal as a fraction or mixed number in simplest form.

4. -0.84
$-\dfrac{21}{25}$

5. 5.22
$5\dfrac{11}{50}$

6. -1.716
$-1\dfrac{179}{250}$

Order the numbers from least to greatest.

7. $\dfrac{1}{5}, 0.1, -\dfrac{1}{2}, -0.25, 0.3$
$-\dfrac{1}{2}, -0.25, 0.1, \dfrac{1}{5}, 0.3$

8. $-1.6, \dfrac{5}{2}, -\dfrac{7}{8}, 0.9, -\dfrac{6}{5}$
$-1.6, -\dfrac{6}{5}, -\dfrac{7}{8}, 0.9, \dfrac{5}{2}$

9. $-\dfrac{2}{3}, \dfrac{5}{9}, 0.5, -1.3, -\dfrac{10}{3}$
$-\dfrac{10}{3}, -1.3, -\dfrac{2}{3}, 0.5, \dfrac{5}{9}$

10. The table shows the position of each runner relative to when the first place finisher crossed the finish line. Who finished in second place? Who finished in fifth place?

Runner	A	B	C	D	E	F
Meters	−1.264	$-\dfrac{5}{4}$	−1.015	−0.480	$-\dfrac{14}{25}$	$-\dfrac{13}{8}$

Runner D; Runner B

Vocabulary and Concept Check

1. no; The denominator cannot be 0.

2. yes; These decimals can be written as $\dfrac{a}{b}$ where a and b are integers and $b \neq 0$.

3. rational numbers, integers

4. rational numbers

5. rational numbers, integers, whole numbers

6. rational numbers, integers, whole numbers

7. repeating

8. terminating

9. terminating

10. repeating

Practice and Problem Solving

11. 0.875

12. $0.\overline{09}$

13. $-0.\overline{7}$

14. -0.425

15. $1.8\overline{3}$

16. $-2.9\overline{4}$

17. $-5.58\overline{3}$

18. $8.68\overline{1}$

19. The bar should be over both digits to the right of the decimal point.
$-\dfrac{7}{11} = -0.\overline{63}$

20. $-\dfrac{9}{10}$

21. $\dfrac{9}{20}$

22. $-\dfrac{129}{500}$

23. $-\dfrac{39}{125}$

24. $-2\dfrac{8}{25}$

25. $-1\dfrac{16}{25}$

26. $6\dfrac{3}{250}$

27. $-12\dfrac{81}{200}$

28. $-\dfrac{7}{3}, -\dfrac{3}{4}, 0.5, \dfrac{2}{3}, 1.2$

Practice and Problem Solving

29. $-2.5, -1.1, -\frac{4}{5}, 0.8, \frac{9}{5}$

30. $-\frac{8}{5}, -1.4, -0.9, \frac{1}{4}, 0.6$

31. $-\frac{9}{4}, -0.75, -\frac{6}{10}, \frac{5}{3}, 2.1$

32. $-\frac{7}{2}, -2.8, -\frac{5}{4}, 1.3, \frac{4}{3}$

33. $-2.4, -2.25, -\frac{11}{5}, \frac{15}{10}, 1.6$

34. **a.** -0.55

    b. $-\frac{11}{20}$

35. spotted turtle

36. $-2.2 > -2.42$

37. $-1.82 < -1.81$

38. $\frac{15}{8} = 1\frac{7}{8}$ **39.** $-4\frac{6}{10} > -4.65$

40–45. See Additional Answers.

46. See *Taking Math Deeper*.

47. See Additional Answers.

Fair Game Review

48. $\frac{31}{35}$ **49.** $\frac{7}{30}$

50. 4.72 **51.** 21.15

52. D

Mini-Assessment

Write the rational number as a decimal.

1. $\frac{8}{9}$ $0.\overline{8}$ **2.** $-\frac{11}{10}$ -1.1

3. $\frac{4}{125}$ 0.032 **4.** $-\frac{13}{15}$ $-0.8\overline{6}$

5. When your cousin was born, she was $21\frac{4}{5}$ inches long. When your friend was born, he was $21\frac{5}{6}$ inches long. Who was longer at birth?

your friend

Taking Math Deeper

Exercise 46

Students have already learned that it is easier to order numbers in decimal form than in fraction form. This problem gives students practice with this skill using negative numbers. The challenge in this problem is that students need to decide what place values to use for all four decimals.

 Write each number as a decimal.

Week	1	2	3	4
Change (inches)	$-\frac{7}{5}$	$-1\frac{5}{11}$	-1.45	$-1\frac{91}{200}$
Decimal	-1.4000	$-1.45\overline{45}$	-1.4500	-1.4550

 Graph the numbers on a number line.

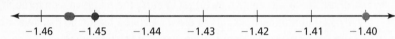

This problem is more difficult than it appears.

 Write the numbers in order from least to greatest.

$$-1\frac{91}{200} \qquad -1\frac{5}{11} \qquad -1.45 \qquad -\frac{7}{5}$$

The U.S. Geological Survey (USGS) records the water levels at various locations in the United States. You can track these measurements by going to www.usgs.org.

Project

Create a chart showing the water levels at various locations in the Great Lakes during the same week. What is the range in water levels? Why do you think the levels vary?

Reteaching and Enrichment Strategies

If students need help. . .	If students got it. . .
Resources by Chapter • Practice A and Practice B • Puzzle Time Record and Practice Journal Practice Differentiating the Lesson Lesson Tutorials Skills Review Handbook	Resources by Chapter • Enrichment and Extension • Technology Connection Start the next section

Copy and complete the statement using <, >, or =.

36. -2.2 ⬚ -2.42

37. -1.82 ⬚ -1.81

38. $\dfrac{15}{8}$ ⬚ $1\dfrac{7}{8}$

39. $-4\dfrac{6}{10}$ ⬚ -4.65

40. $-5\dfrac{3}{11}$ ⬚ $-5.\overline{2}$

41. $-2\dfrac{13}{16}$ ⬚ $-2\dfrac{11}{14}$

42. OPEN-ENDED Find one terminating decimal and one repeating decimal between $-\dfrac{1}{2}$ and $-\dfrac{1}{3}$.

Player	Hits	At Bats
Eva	42	90
Michelle	38	80

43. SOFTBALL In softball, a batting average is the number of hits divided by the number of times at bat. Does Eva or Michelle have the higher batting average?

44. PROBLEM SOLVING You miss 3 out of 10 questions on a science quiz and 4 out of 15 questions on a math quiz. Which quiz has a higher percent of correct answers?

45. SKATING Is the half pipe deeper than the skating pool? Explain.

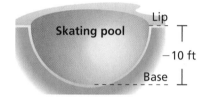
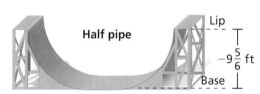

46. ENVIRONMENT The table shows the changes from the average water level of a pond over several weeks. Order the numbers from least to greatest.

Week	1	2	3	4
Change (inches)	$-\dfrac{7}{5}$	$-1\dfrac{5}{11}$	-1.45	$-1\dfrac{91}{200}$

47. *Critical Thinking* Given: a and b are integers.

a. When is $-\dfrac{1}{a}$ positive?

b. When is $\dfrac{1}{ab}$ positive?

 Fair Game Review What you learned in previous grades & lessons

Add or subtract. *(Skills Review Handbook)*

48. $\dfrac{3}{5} + \dfrac{2}{7}$

49. $\dfrac{9}{10} - \dfrac{2}{3}$

50. $8.79 - 4.07$

51. $11.81 + 9.34$

52. MULTIPLE CHOICE In one year, a company has a profit of $-\$2$ million. In the next year, the company has a profit of $\$7$ million. How much more profit did the company make the second year? *(Section 1.3)*

    Ⓐ $2 million   Ⓑ $5 million   Ⓒ $7 million   Ⓓ $9 million

2.2 Adding Rational Numbers

Essential Question How can you use what you know about adding integers to add rational numbers?

1 ACTIVITY: Adding Rational Numbers

Work with a partner. Use a number line to find the sum.

a. $2.7 + (-3.4)$

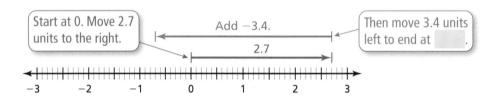

Start at 0. Move 2.7 units to the right.

Add -3.4.

2.7

Then move 3.4 units left to end at [ ].

⋮ So, $2.7 + (-3.4) =$ [ ].

b. $1.3 + (-1.5)$

c. $-2.1 + 0.8$

d. $-1\dfrac{1}{4} + \dfrac{3}{4}$

e. $\dfrac{3}{10} + \left(-\dfrac{3}{10}\right)$

2 ACTIVITY: Adding Rational Numbers

Work with a partner. Use a number line to find the sum.

a. $-1\dfrac{2}{5} + \left(-\dfrac{4}{5}\right)$

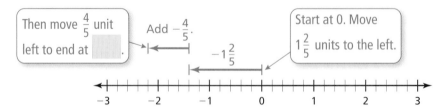

Then move $\dfrac{4}{5}$ unit left to end at [ ].

Add $-\dfrac{4}{5}$.

$-1\dfrac{2}{5}$

Start at 0. Move $1\dfrac{2}{5}$ units to the left.

⋮ So, $-1\dfrac{2}{5} + \left(-\dfrac{4}{5}\right) =$ [ ].

b. $-\dfrac{7}{10} + \left(-1\dfrac{7}{10}\right)$

c. $-1\dfrac{2}{3} + \left(-1\dfrac{1}{3}\right)$

d. $-0.4 + (-1.9)$

e. $-2.3 + (-0.6)$

Rational Numbers

In this lesson, you will
- add rational numbers.
- solve real-life problems.

Laurie's Notes

What Your Students Will Learn
- Add rational numbers with the same sign or with different signs.
- Evaluate expressions involving rational numbers by using substitution.

Previous Learning
Students should be comfortable adding positive fractions, positive decimals, and integers.

Introduction

Applying Mathematical Practices
- **Reason Abstractly and Quantitatively:** Mathematically proficient students are able to use a number line to represent the sum of two rational numbers. Representing a sum on a number line means students must attend to the meaning of addition and to the meaning of a rational number.

Motivate
❓ Pose a series of contextual questions that will help students think about negative rational numbers. These questions should suggest why you need to be able to add rational numbers. Examples:
  - "If finding a dollar and a quarter represented $1.25, how would losing a dollar and a quarter be represented?" $-\$1.25$
  - "If a half mile above sea level is represented as $\frac{1}{2}$, how would a half mile below sea level be represented?" $-\frac{1}{2}$
  - "Represent a loss of $5\frac{1}{2}$ yards on a play in football." $-5\frac{1}{2}$
  - "Represent a drop in temperature of 4.2°." -4.2
- Discuss the use of number lines as a model for addition. Ask students to describe how addition is modeled on a number line.

Activity Notes

Activity 1
- **Reason Abstractly and Quantitatively:** The number line helps students see that the rules for adding rational numbers shouldn't be different from the rules for adding integers.
- Remind students that the first number is represented on the number line by a ray starting at 0. The second number starts at the end of that ray.
- In part (a), you can move 3.4 units to the left in stages. Moving 2 units to the left puts you at 0.7. Moving 1 more unit to the left puts you at -0.3 (not -0.7). Finally, another 0.4 unit left puts you at -0.7.
- **Teaching Tip:** Suggest to students that they use the tick marks on the number line to help them perform the moves in stages.
❓ "When adding rational numbers with different signs, can you predict the sign of the sum? Explain." Yes; the sum has the same sign as the number with the greater absolute value.

Activity 2
- This activity involves adding rational numbers with the same sign. Students should recall that when adding two negative numbers using the number line model, both rays will be drawn in the same direction.
❓ "What is $-1\frac{2}{5}$ as a decimal?" -1.4 "What is $-\frac{4}{5}$ as a decimal?" -0.8
❓ "When adding rational numbers with the same sign, can you predict the sign of the sum? Explain." Yes; the sign is the same.

Technology *for the* Teacher

Dynamic Classroom

Lesson Plans
Complete Materials List

2.2 Record and Practice Journal

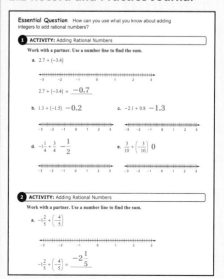

English Language Learners

Vocabulary

Use color-coded examples to help students understand the vocabulary used in this section: denominator, common denominator, least common denominator, improper fraction, and mixed number.

Activity 3

- Have students write the problem modeled on each number line. Ask what clues helped them figure out the problem. State the solution.
- **Common Error:** Students say that $1.5 + (-2.3) = -1.2$, meaning that students will subtract the lesser digit from the greater digit regardless of how the problem is written. (They subtract 1 from 2 and 0.3 from 0.5.) Take time to look at the number line model.

What Is Your Answer?

- For Question 5, students should work with partners. Let students wrestle with the question first, then offer a hint if needed. Five of the six fractions have a common denominator of 24, as does the desired sum of $\frac{3}{4}$. Write each of the fractions as an equivalent fraction with a denominator of 24. The fraction $\left(-\frac{5}{7}\right)$ is not needed to solve the puzzle, so it is not rewritten.

2.2 Record and Practice Journal

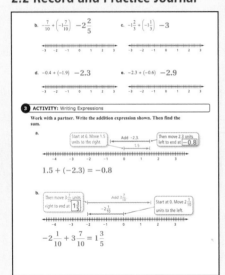

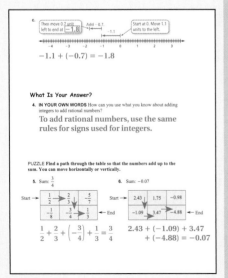

Closure

- Write an addition problem using two rational numbers with different signs whose sum is positive. Find the sum. *Sample answer:* $-3.6 + 7.5 = 3.9$
- Write an addition problem using two rational numbers with different signs whose sum is negative. Find the sum. *Sample answer:* $\frac{1}{4} + \left(-\frac{3}{8}\right) = -\frac{1}{8}$

ACTIVITY: Writing Expressions

Work with a partner. Write the addition expression shown. Then find the sum.

Math Practice

Use Operations

What operation is represented in each number line? How does this help you write an expression?

a.

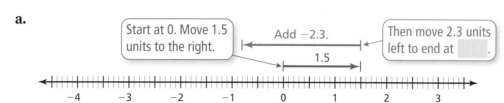

b.

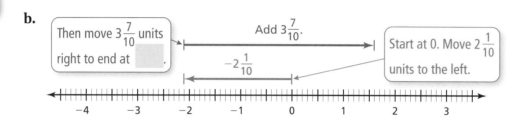

c.

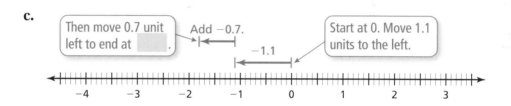

What Is Your Answer?

4. IN YOUR OWN WORDS How can you use what you know about adding integers to add rational numbers?

PUZZLE Find a path through the table so that the numbers add up to the sum. You can move horizontally or vertically.

5. Sum: $\frac{3}{4}$

$\frac{1}{2}$	$\frac{2}{3}$	$-\frac{5}{7}$
$-\frac{1}{8}$	$-\frac{3}{4}$	$\frac{1}{3}$

Start → (top left) ← End (bottom right)

6. Sum: -0.07

2.43	1.75	-0.98
-1.09	3.47	-4.88

Start → (top left) ← End (bottom right)

Practice

Use what you learned about adding rational numbers to complete Exercises 4–6 on page 54.

🔑 Key Idea

Adding Rational Numbers

Words To add rational numbers, use the same rules for signs as you used for integers.

Numbers $-\dfrac{1}{3} + \dfrac{1}{6} = \dfrac{-2}{6} + \dfrac{1}{6} = \dfrac{-2+1}{6} = \dfrac{-1}{6} = -\dfrac{1}{6}$

EXAMPLE 1 Adding Rational Numbers

Study Tip

In Example 1, notice how $-\dfrac{8}{3}$ is written as $-\dfrac{8}{3} = \dfrac{-8}{3} = \dfrac{-16}{6}$.

Find $-\dfrac{8}{3} + \dfrac{5}{6}$.

Estimate $-3 + 1 = -2$

$-\dfrac{8}{3} + \dfrac{5}{6} = \dfrac{-16}{6} + \dfrac{5}{6}$ Rewrite using the LCD (least common denominator).

$= \dfrac{-16 + 5}{6}$ Write the sum of the numerators over the common denominator.

$= \dfrac{-11}{6}$ Add.

$= -1\dfrac{5}{6}$ Write the improper fraction as a mixed number.

∴ The sum is $-1\dfrac{5}{6}$.   **Reasonable?** $-1\dfrac{5}{6} \approx -2$ ✓

EXAMPLE 2 Adding Rational Numbers

Find $-4.05 + 7.62$.

$-4.05 + 7.62 = 3.57$ $|7.62| > |-4.05|$. So, subtract $|-4.05|$ from $|7.62|$.

⬆ Use the sign of 7.62.

∴ The sum is 3.57.

🔵 On Your Own

Now You're Ready
Exercises 4–12

Add.

1. $-\dfrac{7}{8} + \dfrac{1}{4}$

2. $-6\dfrac{1}{3} + \dfrac{20}{3}$

3. $2 + \left(-\dfrac{7}{2}\right)$

4. $-12.5 + 15.3$

5. $-8.15 + (-4.3)$

6. $0.65 + (-2.75)$

Laurie's Notes

Introduction

Connect

- **Yesterday:** Students explored how to add rational numbers.
- **Today:** Students will formalize the process completed yesterday, and they will add rational numbers.

Motivate

? Ask students whether the following questions are *true* or *false*.

$$\frac{\lozenge}{8} + \frac{\triangle}{8} = \frac{\lozenge + \triangle}{16}$$ false, unless the symbols are opposites

$$\frac{\lozenge}{8} + \frac{\triangle}{8} = \frac{\lozenge + \triangle}{8}$$ true

$$0.4 + 0.34 = 0.38$$ false

Lesson Notes

Key Idea

- **Representation:** Take time to talk about how negative fractions are represented, meaning where the negative sign is written. All of the following are equivalent: $-\frac{2}{3} = \frac{-2}{3} = \frac{2}{-3}$.

- **Discuss:** Emphasize the intermediate step: $\frac{-2}{6} + \frac{1}{6} = \frac{-2 + 1}{6}$. This will help the students a great deal.

Example 1

? "What type of fraction is $-\frac{8}{3}$?" improper

? "What would $-\frac{8}{3}$ be as a mixed number?" $-2\frac{2}{3}$

- Note the *Study Tip* when working through the example.
- Be sure to tell students to check for reasonableness of their answers.

Example 2

? Write the problem and ask, "Should the final answer be *positive* or *negative*? Why?"

On Your Own

- Have three pairs of students complete one of the first three fraction problems at the board, while the other students try the problems at their desks. Have students explain their work at the board.
- **?** Ask questions such as, "How do you know your answer is reasonable?"
- Have three different pairs of students complete one of the last three problems at the boards, while the other students try the problems at their desks.

Goal Today's lesson is adding rational numbers.

Technology for the Teacher

Dynamic Classroom

Lesson Tutorials
Lesson Plans
Answer Presentation Tool

Extra Example 1

Find $\frac{4}{5} + \left(-\frac{3}{10}\right)$. $\frac{1}{2}$

Extra Example 2

Find $-3.92 + (-6.89)$. -10.81

On Your Own

1. $-\frac{5}{8}$
2. $\frac{1}{3}$
3. $-1\frac{1}{2}$
4. 2.8
5. -12.45
6. -2.1

Laurie's Notes

Extra Example 3

Evaluate $x + 2y$ when $x = -3.5$ and $y = 1.7$. -0.1

Extra Example 4

The table shows the changes in the value of a stock during one week. Find the stock's gain or loss for the week.

Day	Change in value (dollars)
Monday	3.45
Tuesday	−12.90
Wednesday	−5.02
Thursday	29.31
Friday	−9.44

gain of $5.40

On Your Own

7. $-\dfrac{1}{2}$ 8. 2
9. gain of $770 million

Differentiated Instruction

Auditory

Have students verbally describe the process for adding rational numbers, as well as rewriting and simplifying fractions and mixed numbers. Students should use the word *negative* when referring to the opposite of a positive number. The words *subtract* and *minus* refer to the arithmetic operation. Encourage students to use the words *numerator, denominator, least common denominator*, and *improper fraction*.

Example 3

❓ "What does it mean to *evaluate* an expression?" Substitute the given value for the variable(s) and do the arithmetic.

- **Reason Abstractly and Quantitatively:** When substituting $\dfrac{1}{4}$ for x, make sure students understand that $2\left(\dfrac{1}{4}\right)$ is a multiplication problem. It is not the mixed number $2\dfrac{1}{4}$.

Example 4

- Ask a volunteer to read the problem. Check to see that students are comfortable with the vocabulary.
- Before beginning, tell students that the properties of addition also apply to rational numbers.
- ❓ "What is the relationship between 1.7 and −1.7?" They are opposites.
- Explain how 1.7 and −1.7 represent a gain and a loss of $1.7 billion, which combine to make $0. Discuss other quantities that combine to make 0 for standard 7.NS.1a.
- **Attend to Precision:** When students finish the computation they will have −0.3 billion. Students need to convert −0.3 billion to millions.

On Your Own

- **Neighbor Check:** Have students work independently and then have their neighbors check their work. Have students discuss any discrepancies.

Closure

- Ask students to explain how addition of rational numbers is similar to addition of integers. *Sample answer:* The sign of the sum is the sign of the number with the greater absolute value.

EXAMPLE 3 Evaluating Expressions

Evaluate $2x + y$ when $x = \frac{1}{4}$ and $y = -\frac{3}{2}$.

$$2x + y = 2\left(\frac{1}{4}\right) + \left(-\frac{3}{2}\right)$$ Substitute $\frac{1}{4}$ for x and $-\frac{3}{2}$ for y.

$$= \frac{1}{2} + \left(\frac{-3}{2}\right)$$ Multiply.

$$= \frac{1 + (-3)}{2}$$ Write the sum of the numerators over the common denominator.

$$= -1$$ Simplify.

EXAMPLE 4 Real-Life Application

Year	Profit (billions of dollars)
2008	−1.7
2009	−4.75
2010	1.7
2011	0.85
2012	3.6

The table shows the annual profits (in billions of dollars) of a financial company from 2008 to 2012. Positive numbers represent *gains*, and negative numbers represent *losses*. Which statement describes the profit over the five-year period?

(A) gain of $0.3 billion   **(B)** gain of $30 million

(C) loss of $3 million   **(D)** loss of $300 million

To determine whether there was a gain or a loss, find the sum of the profits.

five-year profit $= -1.7 + (-4.75) + 1.7 + 0.85 + 3.6$ Write the sum.

$$= -1.7 + 1.7 + (-4.75) + 0.85 + 3.6$$ Comm. Prop. of Add.

$$= 0 + (-4.75) + 0.85 + 3.6$$ Additive Inv. Prop.

$$= -4.75 + 0.85 + 3.6$$ Add. Prop. of Zero

$$= -3.9 + 3.6$$ Add −4.75 and 0.85.

$$= -0.3$$ Add −3.9 and 3.6.

The five-year profit is −$0.3 billion. So, the company has a five-year loss of $0.3 billion, or $300 million.

⋮⋅ The correct answer is **(D)**.

On Your Own

Now You're Ready
Exercises 15–17

Evaluate the expression when $a = \frac{1}{2}$ and $b = -\frac{5}{2}$.

7. $b + 4a$

8. $|a + b|$

9. WHAT IF? In Example 4, the 2013 profit is $1.07 billion. State the company's gain or loss over the six-year period in millions of dollars.

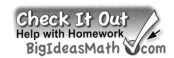
 Vocabulary and Concept Check

1. **WRITING** Explain how to find the sum $-8.46 + 5.31$.

2. **OPEN-ENDED** Write an addition expression using fractions that equals $-\frac{1}{2}$.

3. **DIFFERENT WORDS, SAME QUESTION** Which is different? Find "both" answers.

Add -4.5 and 3.5.	What is the distance between -4.5 and 3.5?
What is -4.5 increased by 3.5?	Find the sum of -4.5 and 3.5.

 Practice and Problem Solving

Add. Write fractions in simplest form.

4. $\frac{11}{12} + \left(-\frac{7}{12}\right)$ **5.** $-1\frac{1}{5} + \left(-\frac{3}{5}\right)$ **6.** $-4.2 + 3.3$

7. $-\frac{9}{14} + \frac{2}{7}$ **8.** $4 + \left(-1\frac{2}{3}\right)$ **9.** $\frac{15}{4} + \left(-4\frac{1}{3}\right)$

10. $-3.1 + (-0.35)$ **11.** $12.48 + (-10.636)$ **12.** $20.25 + (-15.711)$

ERROR ANALYSIS Describe and correct the error in finding the sum.

13.

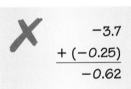

$$\begin{array}{r} -3.7 \\ + (-0.25) \\ \hline -0.62 \end{array}$$

14.

$$-\frac{5}{8} + \frac{1}{8} = \frac{-5 + 1}{8} = \frac{-6}{8} = -\frac{3}{4}$$

Evaluate the expression when $x = \frac{1}{3}$ and $y = -\frac{7}{4}$.

15. $x + y$ **16.** $3x + y$ **17.** $-x + |y|$

18. **BANKING** Your bank account balance is –$20.85. You deposit $15.50. What is your new balance?

19. **HOT DOGS** You eat $\frac{3}{10}$ of a pack of hot dogs.

    Your friend eats $\frac{1}{5}$ of the pack of hot dogs.

    What fraction of the pack of hot dogs do

    you and your friend eat?

Assignment Guide and Homework Check

Level	Day 1 Activity Assignment	Day 2 Lesson Assignment	Homework Check
Basic	4–6, 29–33	1–3, 7–25 odd	7, 11, 15, 17, 23
Average	4–6, 29–33	1–3, 7–16, 17–27 odd	8, 12, 15, 17, 21
Advanced	4–6, 29–33	1–3, 8–22 even, 23–28	8, 12, 16, 20, 24
Accelerated	1–6, 8–22 even, 23–33		8, 12, 16, 20, 24

Common Errors

- **Exercises 7–9** Students may try to identify the sign of the answer before finding a common denominator. Remind them that they need to find the common denominator first.
- **Exercises 10–12** Students may forget to line up the decimal points when they add decimals. Remind them that the decimal points must be lined up before adding. Students may want to use half-inch graph paper to help keep the numbers and decimal points aligned.

Vocabulary and Concept Check

1. Because $|-8.46| > |5.31|$, subtract $|5.31|$ from $|-8.46|$ and the sign is negative.

2. *Sample answer:* $-\frac{1}{4} + \left(-\frac{1}{4}\right)$

3. What is the distance between -4.5 and 3.5?; 8; -1

Practice and Problem Solving

4. $\frac{1}{3}$

5. $-1\frac{4}{5}$

6. -0.9

7. $-\frac{5}{14}$

8. $2\frac{1}{3}$

9. $-\frac{7}{12}$

10. -3.45

11. 1.844

12. 4.539

13. The decimals are not lined up correctly; Line up the decimals; -3.95

14. The sum of the numerators is incorrect.
$$-\frac{5}{8} + \frac{1}{8} = \frac{-5+1}{8} = \frac{-4}{8} = -\frac{1}{2}$$

15. $-1\frac{5}{12}$

16. $-\frac{3}{4}$

17. $1\frac{5}{12}$

18. $-\$5.35$

19. $\frac{1}{2}$

2.2 Record and Practice Journal

Add. Write fractions in simplest form.

1. $-\frac{4}{5} + \frac{3}{20}$
$-\frac{13}{20}$

2. $-8 + \left(-\frac{6}{7}\right)$
$-8\frac{6}{7}$

3. $1\frac{2}{15} + \left(-3\frac{1}{2}\right)$
$-2\frac{11}{30}$

4. $-\frac{1}{6} + \left(-\frac{5}{12}\right)$
$-\frac{7}{12}$

5. $\frac{9}{10} + (-3)$
$-2\frac{1}{10}$

6. $-5\frac{3}{4} + \left(-4\frac{5}{6}\right)$
$-10\frac{7}{12}$

7. $0.46 + (-0.642)$
-0.182

8. $0.13 + (-5.7)$
-5.57

9. $-2.57 + (-3.48)$
-6.05

10. Before a race, you start $4\frac{5}{8}$ feet behind your friend. At the halfway point, you are $3\frac{2}{3}$ feet ahead of your friend. What is the change in distance between you and your friend from the beginning of the race?
$8\frac{7}{24}$ feet

20. $-\dfrac{7}{8}$ **21.** $-9\dfrac{3}{4}$

22. -2.6

23. The sum is an integer when the sum of the fractional parts of the numbers adds up to an integer.

24. See Additional Answers.

25. less than; The water level for the three-month period compared to the normal level is $-1\dfrac{7}{16}$.

26. $450

27. no; This is only true when a and b have the same sign.

28. See *Taking Math Deeper*.

Fair Game Review

29. Commutative Property of Addition; 7

30. Associative Property of Multiplication; 81

31. Associative Property of Addition; $1\dfrac{1}{8}$

32. Commutative Property of Multiplication; $\dfrac{8}{45}$

33. A

Mini-Assessment

Add. Write fractions in simplest form.

1. $2\dfrac{4}{5}+\left(-\dfrac{12}{15}\right)$ 2

2. $-\dfrac{3}{4}+\left(-\dfrac{8}{9}\right)$ $-1\dfrac{23}{36}$

3. $15.48+(-17.23)$ -1.75

4. $-3.89+(-5.34)$ -9.23

5. Your bank account balance is $-\$15.50$. You deposit $75. What is your new balance? $59.50

Taking Math Deeper

Exercise 28

You can evaluate this expression by adding each of the 19 fractions as written. However, this would be tedious. To save time, notice that all the denominators are the same. So, you could find the sum of the numerators and write it over the common denominator, 20.

 One way to quickly find the sum of the numerators is by finding the sums of consecutive terms (1st and 2nd terms, 3rd and 4th terms, and so on.)

$$\underbrace{19+(-18)}_{1}+\underbrace{17+(-16)}_{1}+\underbrace{15+(-14)}_{1}+\underbrace{13+(-12)}_{1}+$$
$$\underbrace{11+(10)}_{1}+\underbrace{9+(-8)}_{1}+\underbrace{7+(-6)}_{1}+\underbrace{5+(-4)}_{1}+\underbrace{3+(-2)}_{1}+\underbrace{1}_{1}$$

There are nine pairs whose sum is 1, and the term 1 is remaining. So, the sum of the numerators is 10.

 Another way to quickly find the sum of the numerators is by pairing terms at the beginning of the expression with terms at the end (1st term and 19th term, 2nd term and 18th term, and so on).

$$\underbrace{19+1+(-18)+(-2)}_{0}+\underbrace{17+3+(-16)+(-4)}_{0}+$$
$$\underbrace{15+5+(-14)+(-6)}_{0}+\underbrace{13+7+(-12)+(-8)}_{0}+\underbrace{11+9+(-10)}_{0}$$

This also shows that the sum of the numerators is 10.

 Answer the question.

So, the value of the expression is $\dfrac{10}{20}$, or $\dfrac{1}{2}$.

Properties of Addition

Project

Write an addition problem with more than 10 terms that can be more easily solved by rearranging terms. Trade problems with a friend and solve.

Reteaching and Enrichment Strategies

If students need help. . .	If students got it. . .

Resources by Chapter
  • Practice A and Practice B
  • Puzzle Time
Record and Practice Journal Practice
Differentiating the Lesson
Lesson Tutorials
Skills Review Handbook
	Resources by Chapter
	  • Enrichment and Extension
	  • Technology Connection
	Start the next section

Add. Write fractions in simplest form.

20. $6 + \left(-4\frac{3}{4}\right) + \left(-2\frac{1}{8}\right)$

21. $-5\frac{2}{3} + 3\frac{1}{4} + \left(-7\frac{1}{3}\right)$

22. $10.9 + (-15.6) + 2.1$

23. NUMBER SENSE When is the sum of two negative mixed numbers an integer?

24. WRITING You are adding two rational numbers with different signs. How can you tell if the sum will be *positive*, *negative*, or *zero*?

25. RESERVOIR The table at the left shows the water level (in inches) of a reservoir for three months compared to the yearly average. Is the water level for the three-month period greater than or less than the yearly average? Explain.

June	July	August
$-2\frac{1}{8}$	$1\frac{1}{4}$	$-\frac{9}{16}$

26. BREAK EVEN The table at the right shows the annual profits (in thousands of dollars) of a county fair from 2008 to 2012. What must the 2012 profit be (in hundreds of dollars) to break even over the five-year period?

Year	Profit (thousands of dollars)
2008	2.5
2009	1.75
2010	−3.3
2011	−1.4
2012	?

27. REASONING Is $|a + b| = |a| + |b|$ for all rational numbers a and b? Explain.

28. **Repeated Reasoning** Evaluate the expression.

$$\frac{19}{20} + \left(\frac{-18}{20}\right) + \frac{17}{20} + \left(\frac{-16}{20}\right) + \cdots + \left(\frac{-4}{20}\right) + \frac{3}{20} + \left(\frac{-2}{20}\right) + \frac{1}{20}$$

 Fair Game Review What you learned in previous grades & lessons

Identify the property. Then simplify. *(Skills Review Handbook)*

29. $8 + (-3) + 2 = 8 + 2 + (-3)$

30. $2 \cdot (4.5 \cdot 9) = (2 \cdot 4.5) \cdot 9$

31. $\frac{1}{4} + \left(\frac{3}{4} + \frac{1}{8}\right) = \left(\frac{1}{4} + \frac{3}{4}\right) + \frac{1}{8}$

32. $\frac{3}{7} \cdot \frac{4}{5} \cdot \frac{14}{27} = \frac{3}{7} \cdot \frac{14}{27} \cdot \frac{4}{5}$

33. MULTIPLE CHOICE The regular price of a photo album is $18. You have a coupon for 15% off. How much is the discount? *(Skills Review Handbook)*

    A $2.70   **B** $3   **C** $15   **D** $15.30

You can use a **process diagram** to show the steps involved in a procedure. Here is an example of a process diagram for adding rational numbers.

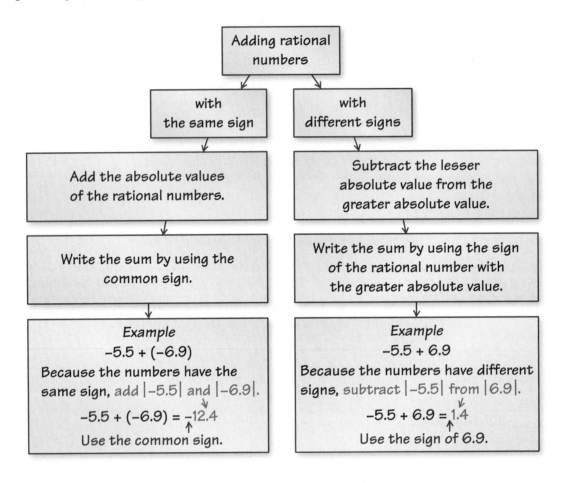

Adding rational numbers

with the same sign → Add the absolute values of the rational numbers. → Write the sum by using the common sign. → **Example** −5.5 + (−6.9) Because the numbers have the same sign, add |−5.5| and |−6.9|. −5.5 + (−6.9) = −12.4 Use the common sign.

with different signs → Subtract the lesser absolute value from the greater absolute value. → Write the sum by using the sign of the rational number with the greater absolute value. → **Example** −5.5 + 6.9 Because the numbers have different signs, subtract |−5.5| from |6.9|. −5.5 + 6.9 = 1.4 Use the sign of 6.9.

On Your Own

Make a process diagram with examples to help you study the topic.

1. writing rational numbers as decimals

After you complete this chapter, make process diagrams with examples for the following topics.

2. subtracting rational numbers

3. multiplying rational numbers

4. dividing rational numbers

"**Does this process diagram accurately show how a cat claws furniture?**"

Sample Answers

1.

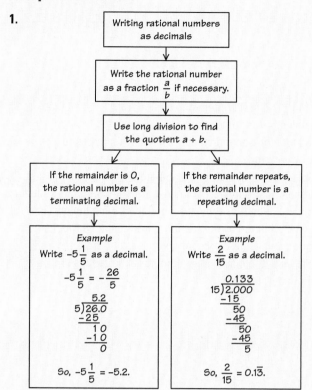

List of Organizers
Available at *BigIdeasMath.com*

Comparison Chart
Concept Circle
Example and Non-Example Chart
Formula Triangle
Four Square
Idea (Definition) and Examples Chart
Information Frame
Information Wheel
Notetaking Organizer
Process Diagram
Summary Triangle
Word Magnet
Y Chart

About this Organizer

A **Process Diagram** can be used to show the steps involved in a procedure. Process diagrams are particularly useful for illustrating procedures with two or more steps, and they can have one or more branches. As shown, students' process diagrams can consist of a single flowchart-type diagram, with example(s) included in the last box to illustrate the steps that precede it. Or, the diagram can have two parallel flowcharts, in which the procedure is stepped out in one chart and an example illustrating each step is shown in the other chart.

Editable Graphic Organizer

Answers

1. -0.15

2. $-1.8\overline{3}$

3. $-\dfrac{13}{40}$

4. $-1\dfrac{7}{25}$

5. $-\dfrac{1}{3}, -0.2, 0.4, 1.3, \dfrac{5}{3}$

6. $-\dfrac{4}{3}, -1.2, -0.8, 0.3, \dfrac{4}{9}$

7. $-1\dfrac{7}{40}$

8. $-1\dfrac{7}{12}$

9. -3.2

10. -6.84

11. $\dfrac{1}{4}$

12. 1

13. $1\dfrac{1}{4}$

14. $1\dfrac{1}{4}$

15. Stock B; Because -3.72 is less than -3.68.

16. $\dfrac{1}{2}$

17. yes; He gained a total of $54\dfrac{3}{4}$ yards, which is greater than 50 yards.

Alternative Quiz Ideas

100% Quiz	Math Log
Error Notebook	Notebook Quiz
Group Quiz	Partner Quiz
Homework Quiz	**Pass the Paper**

Pass the Paper

- Work in groups of four. The first student copies the problem and completes the first step, explaining his or her work.
- The paper is passed and the second student works through the next step, also explaining his or her work.
- This process continues until the problem is completed.
- The second member of the group starts the next problem. Students should be allowed to question and debate as they are working through the quiz.
- Student groups can be selected by the teacher, by students, through a random process, or any way that works for your class.
- The teacher walks around the classroom listening to the groups and asks questions to ensure understanding.

Reteaching and Enrichment Strategies

If students need help. . .	If students got it. . .
Resources by Chapter • Practice A and Practice B • Puzzle Time Lesson Tutorials *BigIdeasMath.com*	Resources by Chapter • Enrichment and Extension • Technology Connection Game Closet at *BigIdeasMath.com* Start the next section

Check It Out
Progress Check
BigIdeasMath ✓com

Write the rational number as a decimal.   *(Section 2.1)*

1. $-\dfrac{3}{20}$

2. $-\dfrac{11}{6}$

Write the decimal as a fraction or a mixed number in simplest form.   *(Section 2.1)*

3. -0.325

4. -1.28

Order the numbers from least to greatest.   *(Section 2.1)*

5. $-\dfrac{1}{3}, -0.2, \dfrac{5}{3}, 0.4, 1.3$

6. $-\dfrac{4}{3}, -1.2, 0.3, \dfrac{4}{9}, -0.8$

Add. Write fractions in simplest form.   *(Section 2.2)*

7. $-\dfrac{4}{5} + \left(-\dfrac{3}{8}\right)$

8. $-\dfrac{13}{6} + \dfrac{7}{12}$

9. $-5.8 + 2.6$

10. $-4.28 + (-2.56)$

Evaluate the expression when $x = \dfrac{3}{4}$ **and** $y = -\dfrac{1}{2}$**.**   *(Section 2.2)*

11. $x + y$

12. $2x + y$

13. $x + |y|$

14. $|-x + y|$

15. STOCK The value of Stock A changes $-\$3.68$, and the value of Stock B changes $-\$3.72$. Which stock has the greater loss? Explain.   *(Section 2.1)*

16. LEMONADE You drink $\dfrac{2}{7}$ of a pitcher of lemonade. Your friend drinks $\dfrac{3}{14}$ of the pitcher. What fraction of the pitcher do you and your friend drink?   *(Section 2.2)*

17. FOOTBALL The table shows the statistics of a running back in a football game. Did he gain more than 50 yards total? Explain.   *(Section 2.2)*

Quarter	1	2	3	4	Total
Yards	$-8\dfrac{1}{2}$	23	$42\dfrac{1}{2}$	$-2\dfrac{1}{4}$	?

2.3 Subtracting Rational Numbers

Essential Question How can you use what you know about subtracting integers to subtract rational numbers?

1 ACTIVITY: Subtracting Rational Numbers

Work with a partner. Use a number line to find the difference.

a. $-1\frac{1}{2} - \frac{1}{2}$

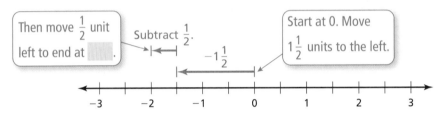

⋮ So, $-1\frac{1}{2} - \frac{1}{2} = $ ▢ .

b. $\frac{6}{10} - 1\frac{3}{10}$

c. $-1\frac{1}{4} - 1\frac{3}{4}$

d. $-1.9 - 0.8$

e. $0.2 - 0.7$

2 ACTIVITY: Finding Distances on a Number Line

Work with a partner.

a. Plot -3 and 2 on the number line. Then find $-3 - 2$ and $2 - (-3)$. What do you notice about your results?

Rational Numbers

In this lesson, you will
- subtract rational numbers.
- solve real-life problems.

b. Plot $\frac{3}{4}$ and 1 on the number line. Then find $\frac{3}{4} - 1$ and $1 - \frac{3}{4}$. What do you notice about your results?

c. Choose any two points a and b on a number line. Find the values of $a - b$ and $b - a$. What do the absolute values of these differences represent? Is this true for any pair of rational numbers? Explain.

Laurie's Notes

Introduction

Applying Mathematical Practices

- **Reason Abstractly and Quantitatively:** Mathematically proficient students are able to use a number line to represent the difference of two rational numbers. Representing a difference on a number line means students must attend to the meaning of subtraction and to the meaning of a rational number.

Motivate

- ❓ "Do you know what an ATM is?" Students should describe features of an automated teller machine.
- Ask students to describe uses and advantages of ATMs. They may have some misconceptions.
- ATMs can be used to deposit money (adding to an account) and to withdraw money (subtracting from an account). Students will investigate transactions in a checkbook.

Activity Notes

Activity 1

- **Reason Abstractly and Quantitatively:** The number line helps students see that the rules for subtracting rational numbers shouldn't be different from the rules for subtracting integers.
- **Teaching Tip:** Suggest to students that they use the tick marks on the number line to help them perform the moves in stages.
- When students have finished, ask volunteers to share their work. Display work at a document camera, if possible.
- Each of these problems involved subtracting a positive number.

Activity 2

- The problems in this activity explore $a - b$ and $b - a$.
- You may need to remind students that when you subtract a positive number n, you move n units to the left. When you subtract a negative number n, you move n units to the right. This is necessary to complete part (a).
- When students have finished the first two parts, ask volunteers to share their work. Display work at a document camera, if possible.
- ❓ "What did you notice about the results in parts (a) and (b)?" Listen for students saying the answers are opposites. They might also say that subtraction is not commutative.
- Students may need guided questioning for part (c). It is common for students to ask if a and b have to be positive, or if they have to be whole numbers. Repeat that both a and b can be any number on the number line.

What Your Students Will Learn

- Subtract rational numbers with the same sign and with different signs.
- Find the distance between two rational numbers using a number line.

Previous Learning

Students should be comfortable subtracting positive fractions, positive decimals, and integers.

Technology for the Teacher

Dynamic Classroom

Lesson Plans
Complete Materials List

2.3 Record and Practice Journal

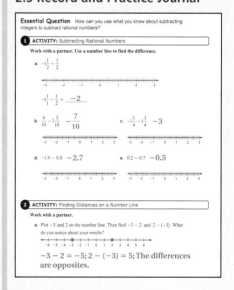

Essential Question How can you use what you know about subtracting integers to subtract rational numbers?

1 ACTIVITY: Subtracting Rational Numbers

Work with a partner. Use a number line to find the difference.

a. $-1\frac{1}{2} - \frac{1}{2}$

$-1\frac{1}{1} - \frac{1}{2} = \underline{-2}$

b. $\frac{6}{10} - 1\frac{3}{10}$ $\underline{-\frac{7}{10}}$

c. $-1\frac{1}{4} - 1\frac{3}{4}$ $\underline{-3}$

d. $-1.9 - 0.8$ $\underline{-2.7}$

e. $0.2 - 0.7$ $\underline{-0.5}$

2 ACTIVITY: Finding Distances on a Number Line

Work with a partner.

a. Plot -3 and 2 on the number line. Then find $-3 - 2$ and $2 - (-3)$. What do you notice about your results?

$-3 - 2 = -5; 2 - (-3) = 5;$ The differences are opposites.

Analyzing Word Problems

Give students a copy of the word problems that are triple-spaced and have wide margins. The additional space gives students plenty of room to add notes. Demonstrate how to underline key words, phrases, and numbers; write down equivalent words in English (or their native language); and draw lines between elements to make the meaning clearer.

2.3 Record and Practice Journal

b. Plot $\frac{3}{4}$ and 1 on the number line. Then find $\frac{3}{4} - 1$ and $1 - \frac{3}{4}$. What do you notice about your results?

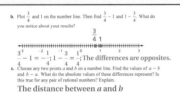

$$\frac{3}{4} - 1 = -\frac{1}{4}; \ 1 - \frac{3}{4} = \frac{1}{4}; \ \text{The differences are opposites.}$$

c. Choose any two points a and b on a number line. Find the values of $a - b$ and $b - a$. What do the absolute values of these differences represent? Is this true for any pair of rational numbers? Explain.

The distance between a and b

3 ACTIVITY: Financial Literacy

Work with a partner. The table shows the balance in a checkbook.

- Deposits and interest are amounts added to the account.
- Amounts shown in parentheses are taken from the account.

Date	Check #	Transaction	Amount	Balance
--	--	Previous Balance	--	100.00
1/02/2013	124	Groceries	(34.57)	65.43
1/07/2013		Check deposit	875.50	940.93
1/11/2013		ATM withdrawal	(40.00)	900.93
1/14/2013	125	Electric company	(78.43)	822.50
1/17/2013		Music store	(10.55)	811.95
1/18/2013	126	Shoes	(47.21)	764.74
1/22/2013		Check deposit	125.00	889.74
1/24/2013		Interest	2.12	891.86
1/25/2013	127	Cell phone	(59.99)	831.87
1/26/2013	128	Clothes	(65.54)	766.33
1/30/2013	129	Cable company	(75.00)	691.33

You can find the balance in the second row two different ways.

$100.00 - 34.57 = 65.43$ Subtract 34.57 from 100.00.

$100.00 + (-34.57) = 65.43$ Add −34.57 to 100.00.

a. Complete the balance column of the table on the previous page.

b. How did you find the balance in the twelfth row?

Sample answer: **Subtract 75.00 from the previous balance of 766.33.**

c. Use a different way to find the balance in part (b).

Sample answer: **Add −75.00 to the previous balance of 766.33.**

What Is Your Answer?

4. **IN YOUR OWN WORDS** How can you use what you know about subtracting integers to subtract rational numbers?

To subtract rational numbers, use the same rules for signs used for integers.

5. Give two real-life examples of subtracting rational numbers that are not integers.

Check students' work.

Laurie's Notes

Activity 3

- **Financial Literacy:** Begin with a discussion on how a checkbook and a debit card are used. In order to know your balance at any time, it is necessary to keep a running balance. Checks written must be subtracted from the balance, and deposits are added to the balance. Interest earned is also added to the balance.
- The activity provides additional practice with decimal addition and subtraction. Be sure that students recall the need to align decimal points. Working with partners, students can check their balances after each transaction.
- Discuss part (c). The check written for $59.99 can be thought of in two ways: balance − $59.99 or balance + (−$59.99).
- Talk about the phrase "in the red," which in accounting means a negative balance. The deposits and interest are "in the black" and are positive. They are being added to the balance and the balance grows (or increases). The checks written are "in the red" and are negative. They are being subtracted from the balance and the balance shrinks (or decreases).

What Is Your Answer?

- Ask students to share their real-life examples for Question 5.

Closure

- **Writing:** Explain how $a - b$ and $a + (-b)$ are equivalent. Create an example to further illustrate what you are explaining.

ACTIVITY: Financial Literacy

Work with a partner. The table shows the balance in a checkbook.

- Black numbers are amounts added to the account.
- Red numbers are amounts taken from the account.

Date	Check #	Transaction	Amount	Balance
--	--	Previous balance	--	100.00
1/02/2013	124	Groceries	34.57	
1/07/2013		Check deposit	875.50	
1/11/2013		ATM withdrawal	40.00	
1/14/2013	125	Electric company	78.43	
1/17/2013		Music store	10.55	
1/18/2013	126	Shoes	47.21	
1/22/2013		Check deposit	125.00	
1/24/2013		Interest	2.12	
1/25/2013	127	Cell phone	59.99	
1/26/2013	128	Clothes	65.54	
1/30/2013	129	Cable company	75.00	

Math Practice

Interpret Results

What does your answer represent? Does your answer make sense?

You can find the balance in the **second row** two different ways.

$$100.00 - 34.57 = 65.43 \qquad \text{Subtract 34.57 from 100.00.}$$
$$100.00 + (-34.57) = 65.43 \qquad \text{Add } -34.57 \text{ to 100.00.}$$

a. Copy the table. Then complete the balance column.

b. How did you find the balance in the **twelfth row**?

c. Use a different way to find the balance in part (b).

What Is Your Answer?

4. **IN YOUR OWN WORDS** How can you use what you know about subtracting integers to subtract rational numbers?

5. Give two real-life examples of subtracting rational numbers that are not integers.

Practice ➤ Use what you learned about subtracting rational numbers to complete Exercises 3–5 on page 62.

 **Key Idea**

Subtracting Rational Numbers

Words To subtract rational numbers, use the same rules for signs as you used for integers.

Numbers $\dfrac{2}{5} - \left(-\dfrac{1}{5}\right) = \dfrac{2}{5} + \dfrac{1}{5} = \dfrac{2+1}{5} = \dfrac{3}{5}$

EXAMPLE 1 **Subtracting Rational Numbers**

Find $-4\dfrac{1}{7} - \left(-\dfrac{6}{7}\right)$.

Estimate $-4 - (-1) = -3$

$$-4\dfrac{1}{7} - \left(-\dfrac{6}{7}\right) = -4\dfrac{1}{7} + \dfrac{6}{7}$$ Add the opposite of $-\dfrac{6}{7}$.

$$= -\dfrac{29}{7} + \dfrac{6}{7}$$ Write the mixed number as an improper fraction.

$$= \dfrac{-29 + 6}{7}$$ Write the sum of the numerators over the common denominator.

$$= \dfrac{-23}{7}$$ Add.

$$= -3\dfrac{2}{7}$$ Write the improper fraction as a mixed number.

The difference is $-3\dfrac{2}{7}$.   **Reasonable?** $-3\dfrac{2}{7} \approx -3$ ✓

EXAMPLE 2 **Subtracting Rational Numbers**

Find $12.8 - 21.6$.

$12.8 - 21.6 = 12.8 + (-21.6)$ Add the opposite of 21.6.

$$= -8.8$$ $|-21.6| > |12.8|$. So, subtract $|12.8|$ from $|-21.6|$.

The difference is -8.8.   Use the sign of -21.6.

On Your Own

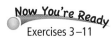 **Now You're Ready**
Exercises 3–11

1. $\dfrac{1}{3} - \left(-\dfrac{1}{3}\right)$ **2.** $-3\dfrac{1}{3} - \dfrac{5}{6}$ **3.** $4\dfrac{1}{2} - 5\dfrac{1}{4}$

4. $-8.4 - 6.7$ **5.** $-20.5 - (-20.5)$ **6.** $0.41 - (-0.07)$

Laurie's Notes

Introduction

Connect

- **Yesterday:** Students explored how to subtract rational numbers.
- **Today:** Students will formalize the process completed yesterday, and they will subtract rational numbers.

Motivate

- Draw two "dots" on the board (locate them horizontally) and ask students about the distance between them. They may say to measure it with a ruler.
- Draw a line through the points, extending the line beyond the points. Put a tick mark to the left of both dots and label it 0.
- **?** "Besides using a ruler, is there another way we could find the distance between the two points?" Listen for scaling the number line and then subtracting the lesser number from the greater number.
- **?** "Would this work if 0 is between the two points? if 0 is to the right of both points?" Students will be less certain that subtraction will work. You may need to try simple integer values where students can count.
- Tell students you will return to distance on a number line in the lesson.

Lesson Notes

Key Idea

- Write the Key Idea. Work through the example.
- **Teaching Tip:** Use a different color when you *add the opposite*.

Example 1

- **?** "How do you subtract integers?" The statement "add the opposite" should be familiar. Once the problem is written as an addition problem, students should recall the rules for integer addition.
- **?** "The problem is now $-4\frac{1}{7} + \frac{6}{7}$. Will the sum be positive or negative?" Negative; the mixed number is negative and has a greater absolute value.
- **FYI:** Some students may choose to solve by first rewriting $-4\frac{1}{7}$ as $-3\frac{8}{7}$.
- **Connection:** The rule for subtracting rational numbers is the same as the rule for subtracting integers. The challenge will be working with fractions!

Example 2

- Write the problem. Students should see that the difference will be negative.
- Rewrite the problem by "adding the opposite."
- Continue to work through the problem as shown.

On Your Own

- **Neighbor Check:** Have students work independently and then have their neighbors check their work. Have students discuss any discrepancies.

Goal Today's lesson is subtracting rational numbers.

Lesson Tutorials
Lesson Plans
Answer Presentation Tool

Extra Example 1
Find $-8\frac{2}{3} - 6\frac{1}{6}$. $-14\frac{5}{6}$

Extra Example 2
Find $-3.75 - (-0.96)$. -2.79

 **On Your Own**

1. $\frac{2}{3}$ 2. $-4\frac{1}{6}$

3. $-\frac{3}{4}$ 4. -15.1

5. 0 6. 0.48

Extra Example 3

Find the distance between the two numbers on the number line.

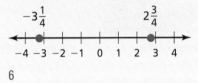

6

Extra Example 4

You have $101.62 in your savings account. You withdraw $45.41 to pay your cell phone bill. Do you have enough money left to buy a video game that costs $49.99?

Yes

 On Your Own

7. 7.8

8. no

Differentiated Instruction

Visual

Some students will incorrectly subtract decimals, especially when the second number has more decimal places than the first. Encourage students to use zeros so that the two numbers have the same number of decimal places. For example, 2.35 − 1.457 should be written as

$$
\begin{array}{r}
2.350 \\
-\ 1.457 \\
\hline
0.893
\end{array}
\quad \text{instead of} \quad
\begin{array}{r}
2.35 \\
-\ 1.457 \\
\hline
0.907 \ \text{✗}
\end{array}
$$

T-61

Laurie's Notes

Example 3

- **Big Idea:** This problem connects subtraction to finding distances on a number line, an idea that was investigated in the Motivate.
- **Model with Mathematics:** The vertical number line provides a visual model that helps students make a reasonable estimate.
- **?** "We want to find the distance between two points. Will the order in which we do the subtraction matter?" Listen for students to state that subtraction is not commutative. However, since we will take the absolute value, you can subtract in either order because the results are opposites.
- Work through the problem, referring to the number line as you work.

Example 4

- To help visualize this problem, fill a glass bowl with water. Float a toy boat in the water so that the distance above the water level is visible.
- **?** "If you know the height of the boat above the water and the depth of the boat below the water, how can you find the total height of the boat?" Students will probably say to add the two together. It is also acceptable to subtract the lowest point relative to sea level (a negative number) from the highest point (a positive number).

On Your Own

- **Neighbor Check:** Have students work independently and then have their neighbors check their work. Have students discuss any discrepancies.

Words of Wisdom

- Students often think that 2.1 feet is equivalent to 2 feet, 1 inch. Have students explore which is greater: 2.1 feet or 2 feet, 1 inch.

$0.1 \text{ ft} \times \dfrac{12 \text{ in.}}{1 \text{ ft}} = 1.2 \text{ in.}$, so 2.1 feet is greater than 2 feet, 1 inch.

Closure

- Ask students to explain how subtraction of rational numbers is similar to subtraction of integers. *Sample answer:* The sign of the difference is the sign of the number with the greater absolute value.

The distance between any two numbers on a number line is the absolute value of the difference of the numbers.

EXAMPLE 3 **Finding Distances Between Numbers on a Number Line**

Find the distance between the two numbers on the number line.

To find the distance between the numbers, first find the difference of the numbers.

$$-2\frac{2}{3} - 2\frac{1}{3} = -2\frac{2}{3} + \left(-2\frac{1}{3}\right)$$ Add the opposite of $2\frac{1}{3}$.

$$= -\frac{8}{3} + \left(-\frac{7}{3}\right)$$ Write the mixed numbers as improper fractions.

$$= \frac{-15}{3}$$ Add.

$$= -5$$ Simplify.

Because $|-5| = 5$, the distance between $-2\frac{2}{3}$ and $2\frac{1}{3}$ is 5.

EXAMPLE 4 **Real-Life Application**

Clearance: 11 ft 8 in.

In the water, the bottom of a boat is 2.1 feet below the surface, and the top of the boat is 8.7 feet above it. Towed on a trailer, the bottom of the boat is 1.3 feet above the ground. Can the boat and trailer pass under the bridge?

Step 1: Find the height h of the boat.

$$h = 8.7 - (-2.1)$$ Subtract the lowest point from the highest point.

$$= 8.7 + 2.1$$ Add the opposite of -2.1.

$$= 10.8$$ Add.

Step 2: Find the height t of the boat and trailer.

$$t = 10.8 + 1.3$$ Add the trailer height to the boat height.

$$= 12.1$$ Add.

Because 12.1 feet is greater than 11 feet 8 inches, the boat and trailer cannot pass under the bridge.

On Your Own

Now You're Ready
Exercises 13–15

7. Find the distance between -7.5 and -15.3 on a number line.

8. **WHAT IF?** In Example 4, the clearance is 12 feet 1 inch. Can the boat and trailer pass under the bridge?

2.3 Exercises

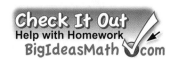

✓ Vocabulary and Concept Check

1. **WRITING** Explain how to find the difference $-\dfrac{4}{5} - \dfrac{3}{5}$.

2. **WHICH ONE DOESN'T BELONG?** Which expression does *not* belong with the other three? Explain your reasoning.

$$-\dfrac{5}{8} - \dfrac{3}{4} \qquad -\dfrac{3}{4} + \dfrac{5}{8} \qquad -\dfrac{5}{8} + \left(-\dfrac{3}{4}\right) \qquad -\dfrac{3}{4} - \dfrac{5}{8}$$

Practice and Problem Solving

Subtract. Write fractions in simplest form.

 3. $\dfrac{5}{8} - \left(-\dfrac{7}{8}\right)$

4. $-1\dfrac{1}{3} - 1\dfrac{2}{3}$

5. $-1 - 2.5$

6. $-5 - \dfrac{5}{3}$

7. $-8\dfrac{3}{8} - 10\dfrac{1}{6}$

8. $-\dfrac{1}{2} - \left(-\dfrac{5}{9}\right)$

9. $5.5 - 8.1$

10. $-7.34 - (-5.51)$

11. $6.673 - (-8.29)$

12. **ERROR ANALYSIS** Describe and correct the error in finding the difference.

$$\times \quad \dfrac{3}{4} - \dfrac{9}{2} = \dfrac{3 - 9}{4 - 2} = \dfrac{-6}{2} = -3$$

Find the distance between the two numbers on a number line.

13. $-2\dfrac{1}{2}, -5\dfrac{3}{4}$

14. $-2.2, 8.4$

15. $-7, -3\dfrac{2}{3}$

16. **SPORTS DRINK** Your sports drink bottle is $\dfrac{5}{6}$ full. After practice, the bottle is $\dfrac{3}{8}$ full. Write the difference of the amounts after practice and before practice.

17. **SUBMARINE** The figure shows the depths of a submarine.

   a. Find the vertical distance traveled by the submarine.

   b. Find the mean hourly vertical distance traveled by the submarine.

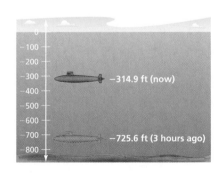

Evaluate.

18. $2\dfrac{1}{6} - \left(-\dfrac{8}{3}\right) + \left(-4\dfrac{7}{9}\right)$

19. $6.59 + (-7.8) - (-2.41)$

20. $-\dfrac{12}{5} + \left|-\dfrac{13}{6}\right| + \left(-3\dfrac{2}{3}\right)$

Assignment Guide and Homework Check

Level	Day 1 Activity Assignment	Day 2 Lesson Assignment	Homework Check
Basic	3–5, 31–35	1, 2, 7–21 odd, 12	7, 11, 13, 17
Average	3–5, 31–35	1, 2, 9–12, 15–19 odd, 22–26 even	10, 17, 19, 22
Advanced	3–5, 31–35	1, 2, 8–30 even	8, 10, 18, 22, 24
Accelerated	1–5, 8–30 even, 31–35		8, 10, 18, 22, 24

Common Errors

- **Exercises 6–8** Students may try to identify the sign of the answer before finding a common denominator. Remind them that they need to find the common denominator first.
- **Exercises 5, 9–11** Students may forget to line up the decimal points when they subtract decimals. Remind them that the decimal points must be lined up before subtracting. Students may want to use half-inch graph paper to help keep the numbers and decimal points aligned.
- **Exercise 8** Students may not know where to put the negative sign in the fraction. Remind them that the negative can go in the numerator or the denominator (although the numerator is usually best when doing calculations), but not both.

2.3 Record and Practice Journal

Subtract. Write fractions in simplest form.

1. $\frac{4}{9} - \left(-\frac{2}{9}\right)$
$\frac{2}{3}$

2. $-2\frac{3}{7} - 1\frac{2}{3}$
$-4\frac{2}{21}$

3. $-2.35 - (-1.27)$
-1.08

Find the distance between the two numbers on a number line.

4. $-3\frac{1}{4}, -6\frac{1}{2}$
$3\frac{1}{4}$

5. $-1.5, 2.8$
4.3

6. $-4, -7\frac{1}{3}$
$3\frac{1}{3}$

Evaluate.

7. $2\frac{1}{2} + \left(-\frac{7}{6}\right) - 1\frac{3}{4}$
$-\frac{5}{12}$

8. $2.37 - (-1.55) - 2.48$
1.44

9. Your friend drinks $\frac{2}{3}$ of a bottle of water. You drink $\frac{5}{7}$ of a bottle of water. Find the difference of the amounts of water left in each bottle.
$\frac{1}{21}$

Practice and Problem Solving

21. The difference is an integer when (1) the decimals have the same sign and the digits to the right of the decimal point are the same, or (2) the decimals have different signs and the sum of the decimal parts of the numbers add up to 1.

22. No, the cook needs $\frac{1}{12}$ cup more.

23. $-1\frac{7}{8}$ miles

24–26. See *Taking Math Deeper*.

27. *Sample answer:* $x = -1.8$ and $y = -2.4$; $x = -5.5$ and $y = -6.1$

28. sometimes; It is positive only if the first fraction is greater.

29. always; It is always positive because the first decimal is always greater.

30. $5.24 - (8.85) = -3.61$

Fair Game Review

31. 35.88   32. 3

33. $8\frac{2}{3}$ 34. $2\frac{4}{5}$

35. C

Mini-Assessment

Subtract. Write fractions in simplest form.

1. $\frac{1}{2} - \frac{3}{4}$ $-\frac{1}{4}$

2. $2\frac{2}{5} - \left(-\frac{6}{5}\right)$ $3\frac{3}{5}$

3. $-8.4 - 0.9$ -9.3

4. $-12.55 - (-23.08)$ 10.53

5. The temperature in a town is $-4.7°C$. The temperature decreases $5.4°C$. What is the new temperature?
   $-10.1°C$

Taking Math Deeper

Exercises 24–26

This problem gives students a chance to find the sum of a long list of signed numbers. To do this efficiently, students can use the Commutative and Associative Properties of Addition.

 Read and interpret the bar graph.

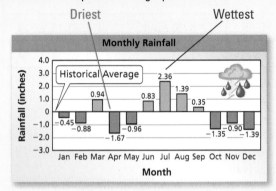

24. Difference $= 2.36 - (-1.67)$
    $= 2.36 + 1.67$
    $= 4.03$ in.

2 Find the sum of the differences.

	-0.45	
	-0.88	
	-1.67	0.94
Sum of	-0.96	0.83
negative	-1.35	2.36
numbers	-0.90	1.39
	-1.39	0.35
	-7.60	5.87

Sum of positive numbers

25. Total sum: $-7.60 + 5.87 = -1.73$ in.

3 Interpret.

26. The total rainfall for the year was 1.73 inches *less* than the historical average.

Reteaching and Enrichment Strategies

If students need help. . .	If students got it. . .
Resources by Chapter • Practice A and Practice B • Puzzle Time Record and Practice Journal Practice Differentiating the Lesson Lesson Tutorials Skills Review Handbook	Resources by Chapter • Enrichment and Extension • Technology Connection Start the next section

21. **REASONING** When is the difference of two decimals an integer? Explain.

22. **RECIPE** A cook has $2\frac{2}{3}$ cups of flour. A recipe calls for $2\frac{3}{4}$ cups of flour. Does the cook have enough flour? If not, how much more flour is needed?

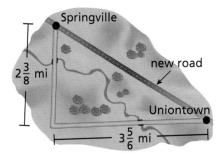

Springville

$2\frac{3}{8}$ mi

new road

Uniontown

$3\frac{5}{6}$ mi

23. **ROADWAY** A new road that connects Uniontown to Springville is $4\frac{1}{3}$ miles long. What is the change in distance when using the new road instead of the dirt roads?

RAINFALL In Exercises 24–26, the bar graph shows the differences in a city's rainfall from the historical average.

24. What is the difference in rainfall between the wettest and the driest months?

25. Find the sum of the differences for the year.

26. What does the sum in Exercise 25 tell you about the rainfall for the year?

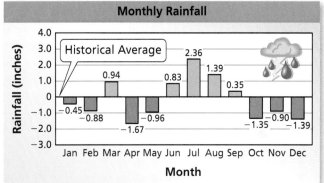

27. **OPEN-ENDED** Write two different pairs of negative decimals, x and y, that make the statement $x - y = 0.6$ true.

REASONING Tell whether the difference between the two numbers is *always*, *sometimes*, or *never* positive. Explain your reasoning.

28. two negative fractions

29. a positive decimal and a negative decimal

30. **Structure** Fill in the blanks to make the solution correct.

$$5.\,\boxed{}4 - \left(\boxed{}.8\,\boxed{}\right) = -3.61$$

 Fair Game Review What you learned in previous grades & lessons

Evaluate. *(Skills Review Handbook)*

31. 5.2×6.9

32. $7.2 \div 2.4$

33. $2\frac{2}{3} \times 3\frac{1}{4}$

34. $9\frac{4}{5} \div 3\frac{1}{2}$

35. **MULTIPLE CHOICE** A sports store has 116 soccer balls. Over 6 months, it sells 8 soccer balls per month. How many soccer balls are in inventory at the end of the 6 months? *(Section 1.3 and Section 1.4)*

Ⓐ −48　　　Ⓑ 48　　　Ⓒ 68　　　Ⓓ 108

Essential Question Why is the product of two negative rational numbers positive?

In Section 1.4, you used a table to see that the product of two negative integers is a positive integer. In this activity, you will find that same result another way.

1   ACTIVITY: Showing $(-1)(-1) = 1$

Work with a partner. How can you show that $(-1)(-1) = 1$?

From the Additive Inverse Property, you know that $1 + (-1) = 0$. If you can show that $(-1)(-1) + (-1) = 0$ is true, then you have shown that $(-1)(-1) = 1$.

Justify each step.

$$(-1)(-1) + (-1) = (-1)(-1) + 1(-1)$$

$$= (-1)[(-1) + 1]$$

$$= (-1)0$$

$$= 0$$

So, $(-1)(-1) = 1$.

2   ACTIVITY: Multiplying by -1

Work with a partner.

a. Graph each number below on three different number lines. Then multiply each number by -1 and graph the product on the appropriate number line.

| | 2 | 8 | -1 |

Rational Numbers

In this lesson, you will
- multiply and divide rational numbers.
- solve real-life problems.

b. How does multiplying by -1 change the location of the points in part (a)? What is the relationship between the number and the product?

c. Graph each number below on three different number lines. Where do you think the points will be after multiplying by -1? Plot the points. Explain your reasoning.

| | $\dfrac{1}{2}$ | 2.5 | $-\dfrac{5}{2}$ |

d. What is the relationship between a rational number $-a$ and the product $-1(a)$? Explain your reasoning.

Laurie's Notes

Introduction

Applying Mathematical Practices

- **Reason Abstractly and Quantitatively:** Mathematically proficient students are able to follow sequential statements about equivalent expressions. They are able to recognize and state the algebraic properties that support each statement.

Motivate

- Display the table used in Section 1.4 to show $-3(-2) = 6$.
- Discuss this approach, an additive interpretation of multiplication.
- **?** "Does this approach make sense for a problem involving fractions, such as $-3\frac{1}{2}\left(-2\frac{3}{4}\right)$?" Students may not be sure how to create a table involving fractions that would show a pattern.
- Explain that they will not use the table approach today. The first three activities use an analytic approach to show that the product of two negatives is positive. This approach requires students to read carefully and recognize the application of the Distributive Property.

Activity Notes

Activity 1

- Activity 1 is an alternate way to show that $(-1)(-1) = 1$.
- **Teaching Tip:** You may want to do a related example before students begin this activity to get them thinking about properties.

$$4 + [2 + (-2)] = 4 + 0 \qquad \text{Additive Inverse Property}$$
$$= 4 \qquad \text{Addition Property of Zero}$$

- **Reason Abstractly and Quantitatively:** Students should read the introduction carefully as it describes the strategy that will be used to show $(-1)(-1) = 1$. When finished, discuss how to show $(-2)(-2) = 4$.

Activity 2

- This activity explores the result of multiplying numbers by -1.
- Part (b) asks two questions. In the first question, students recognize that the point they plot is a reflection (flipped) over 0. The second question is recognizing the relationship between the number and the product (they are opposites).
- **Big Idea:** Multiplying by -1 is the same as taking the opposite of a number. This idea is used in Activity 3.

What Your Students Will Learn

- Multiply and divide rational numbers with the same sign and with different signs.

Previous Learning

Students should be comfortable multiplying and dividing positive fractions, positive decimals, and integers.

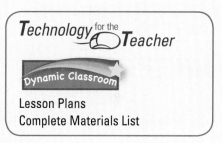

Lesson Plans
Complete Materials List

2.4 Record and Practice Journal

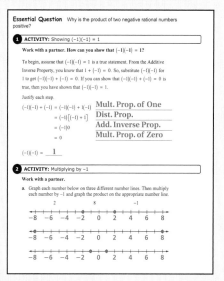

English Language Learners

Simplified Language
Writing stories poses a challenge for English learners. You may want to allow students who struggle with language to outline a story or to create a story using pictures.

2.4 Record and Practice Journal

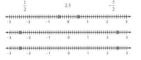

b. How does multiplying by −1 change the location of the points in part (a)? What is the relationship between the number and the product?
reflects point on the other side of 0; opposites

c. Graph each number below on three different number lines. Where do you think the points will be after multiplying by −1? Plot the points. Explain your reasoning.
$\frac{1}{2}$ 2.5 $-\frac{5}{2}$

d. What is the relationship between a rational number −a and the product −1(a)? Explain your reasoning. **They are the same.**

3 ACTIVITY: Understanding the product of Rational Numbers

Work with a partner. Let a and b be positive rational numbers.
a. Because a and b are positive, what do you know about −a and −b?
They represent the opposites of a and b and are negative.
b. Justify each step.
$(-a)(-b) = (-1)(a)(-1)(b)$ — Mult. by −1 is the same as taking the opposite.
$= (-1)(-1)(a)(b)$ — Comm. Prop. of Mult.
$= (1)(a)(b)$ — Result of Activity 1
$= ab$ — Mult. Prop. of One
c. Because a and b are positive, what do you know about the product ab?
It is positive.

d. What does this tell you about products of rational numbers? Explain.
The products of rational numbers follow the same rules as the products of integers.

4 ACTIVITY: Writing a Story

Work with a partner. Write a story that uses addition, subtraction, multiplication, or division of rational numbers. **Check students' work.**
- At least one of the numbers in the story has to be negative and not an integer.
- Draw pictures to help illustrate what is happening in the story.
- Include the solution of the problem in the story.

If you are having trouble thinking of a story, here are some common uses of negative numbers:
- A profit of −$15 is a loss of $15.
- An elevation of −100 feet is a depth of 100 feet below sea level.
- A gain of −5 yards in football is a loss of 5 yards.
- A score of −4 in golf is 4 strokes under par.

What Is Your Answer?

5. IN YOUR OWN WORDS Why is the product of two negative rational numbers positive?
Check students' work.

6. PRECISION Show that $(-2)(-3) = 6$. $(-2)(-3) = (-1)(2)(-1)(3)$
$= (-1)(-1)(2)(3)$
$= 1(2)(3)$
$= 6$

7. How can you show that the product of a negative rational number and a positive rational number is negative?
$(-a)(b) = (-1)(a)(b)$
$= -ab$

T-65

Laurie's Notes

Activity 3
- This activity is similar in style to Activity 1.
- **Construct Viable Arguments:** Give students time to work through the steps in this activity. Resist jumping in too soon to give answers. Remind students to read carefully and observe what changes have taken place in each step.
- Discuss the results. If a and b are positive numbers, then $(-a)(-b)$ is the product of two negative numbers. The conclusion is that this product equals ab, which is positive because it is the product of two positive numbers. This won't be surprising to students because of their work with integers. This extends that knowledge to all rational numbers.
- Reasoning abstractly is challenging for many students and this activity may be less convincing to students than observing a pattern with integers. It is important for students to have the opportunity to make sense of an abstract approach. Students will be required to construct these types of proofs on their own in future courses.

Activity 4
- Read through the directions together as a class.
- Have students work in pairs so that brainstorming can occur. Both students should be actively engaged, with one doing the writing while the other draws a diagram to illustrate the problem.
- The four examples showing where negative numbers are commonly used should help students get started.
- Provide at least 20–25 minutes for the brainstorming and writing process. Students' stories should include computations and a final solution.
- **Discuss:** As time allows, have pairs of students share their stories. To help students see when each operation is used, make a table on the board with four columns, one for each operation. Record the context used for each operation.
- **Interdisciplinary:** Some of your language arts colleagues may want to review the students' stories. Speak with them about different possibilities.

What Is Your Answer?
- **Attend to Precision:** Question 6 is connected to Activity 3. Students should recognize and use a similar strategy to the one used in Activity 3.

Closure
- **Writing:** Have students write brief scenarios for all four operations (addition, subtraction, multiplication, and division). Note: The scenarios *do not* need to be connected to one another. Instead of creating a whole story, students just need to write four sentences.

Work with a partner. Let *a* and *b* be positive rational numbers.

a. Because *a* and *b* are positive, what do you know about $-a$ and $-b$?

b. Justify each step.

$$(-a)(-b) = (-1)(a)(-1)(b)$$

$$= (-1)(-1)(a)(b)$$

$$= (1)(a)(b)$$

$$= ab$$

c. Because *a* and *b* are positive, what do you know about the product *ab*?

d. What does this tell you about products of rational numbers? Explain.

4   **ACTIVITY: Writing a Story**

Work with a partner. Write a story that uses addition, subtraction, multiplication, or division of rational numbers.

- At least one of the numbers in the story has to be negative and *not* an integer.
- Draw pictures to help illustrate what is happening in the story.
- Include the solution of the problem in the story.

Math Practice

Specify Units

What units are in your story?

If you are having trouble thinking of a story, here are some common uses of negative numbers:

- A profit of $-\$15$ is a loss of \$15.
- An elevation of -100 feet is a depth of 100 feet below sea level.
- A gain of -5 yards in football is a loss of 5 yards.
- A score of -4 in golf is 4 strokes under par.

What Is Your Answer?

5. IN YOUR OWN WORDS Why is the product of two negative rational numbers positive?

6. PRECISION Show that $(-2)(-3) = 6$.

7. How can you show that the product of a negative rational number and a positive rational number is negative?

Practice

Use what you learned about multiplying rational numbers to complete Exercises 7–9 on page 68.

Key Idea

Remember

The *reciprocal* of $\dfrac{a}{b}$ is $\dfrac{b}{a}$.

Multiplying and Dividing Rational Numbers

Words To multiply or divide rational numbers, use the same rules for signs as you used for integers.

Numbers

$$-\frac{2}{7} \cdot \frac{1}{3} = \frac{-2 \cdot 1}{7 \cdot 3} = \frac{-2}{21} = -\frac{2}{21}$$

$$-\frac{1}{2} \div \frac{4}{9} = \frac{-1}{2} \cdot \frac{9}{4} = \frac{-1 \cdot 9}{2 \cdot 4} = \frac{-9}{8} = -\frac{9}{8}$$

EXAMPLE 1 **Dividing Rational Numbers**

Find $-5\dfrac{1}{5} \div 2\dfrac{1}{3}$.   **Estimate** $-5 \div 2 = -2\dfrac{1}{2}$

$$-5\frac{1}{5} \div 2\frac{1}{3} = -\frac{26}{5} \div \frac{7}{3}$$ Write mixed numbers as improper fractions.

$$= \frac{-26}{5} \cdot \frac{3}{7}$$ Multiply by the reciprocal of $\dfrac{7}{3}$.

$$= \frac{-26 \cdot 3}{5 \cdot 7}$$ Multiply the numerators and the denominators.

$$= \frac{-78}{35}, \text{ or } -2\frac{8}{35}$$ Simplify.

∴ The quotient is $-2\dfrac{8}{35}$.   **Reasonable?** $-2\dfrac{8}{35} \approx -2\dfrac{1}{2}$ ✓

EXAMPLE 2 **Multiplying Rational Numbers**

Find $-2.5 \cdot 3.6$.

$$
\begin{array}{r}
-2.5 \\
\times\ 3.6 \\
\hline
1\,5\,0 \\
7\,5\,0 \\
\hline
-9.0\,0
\end{array}
$$

The decimals have different signs.

The product is negative.

∴ The product is -9.

Laurie's Notes

Introduction

Connect

- **Yesterday:** Students used an analytic approach to show that the product of two negative numbers is positive.
- **Today:** Students will learn the rules for multiplying and dividing rational numbers.

Discuss

- Before beginning the formal lesson, it would be helpful to review rules for multiplying and dividing integers.
  - same signs ⟶ product/quotient is positive
  - different signs ⟶ product/quotient is negative

Lesson Notes

Key Idea

- Write the definition for multiplication and division of rational numbers. Note that the sign of the fraction is written with the numerator when the computation is performed.

Example 1

- **Discuss:** Before starting the first example, take time to discuss estimating products and quotients. This will help students check their answers.
- Work through the problem. Do not skip the initial estimate.
- Remind students that when multiplying or dividing fractions, mixed numbers must be written as improper fractions.
- **Discuss:** There are several important skills involved in this example. Identify each skill with students so that vocabulary is reviewed and each process is made clear.

Example 2

- This example involves multiplying decimals *and* signed numbers.
- Write the example and ask how they might estimate an answer.
- **Extension:** If time permits, repeat this example by converting the decimals to fractions:

$$-2\frac{5}{10} \times 3\frac{6}{10} = -2\frac{1}{2} \times 3\frac{3}{5}$$

$$= \frac{-5}{2} \times \frac{18}{5}$$

$$= \frac{-90}{10}$$

$$= -9$$

Lesson Tutorials
Lesson Plans
Answer Presentation Tool

Extra Example 1

Find $3\frac{1}{4} \div \left(-1\frac{1}{8}\right)$. $-2\frac{8}{9}$

Extra Example 2

Find $-4.8(-5.2)$. 24.96

Laurie's Notes

Extra Example 3

Find $\frac{10}{3} \cdot \left(-3\frac{3}{5}\right) \cdot (-3)$. 36

 On Your Own

1. $2\frac{2}{5}$ **2.** $-\frac{1}{8}$

3. -9.18 **4.** 3.78

5. $-7\frac{7}{8}$ **6.** 72

Extra Example 4

Find the mean of -25.63, 37.15, 18.92, and -44.28. -3.46

On Your Own

7. $58.65

Differentiated Instruction

Inclusion

Remind students that one difference between multiplying decimals and dividing decimals is the placement of the decimal point. In multiplication, the decimal point is placed after the decimals are multiplied. In division, the placement of the decimal point is determined before dividing.

Example 3

- **Look for and Make Use of Structure:** Write the problem. Discuss possible strategies for performing the computation. Students should recognize that algebraic properties can be used to perform the operations in a more efficient manner than how the problem is presented.
- This problem provides a good review of several algebraic properties. The more frequently we refer to these properties by name, the more fluent students become in using them.

On Your Own

- Have three pairs of students choose one question from 1–3 to complete at the board. Have the other students try these problems at their desks. Have the pairs of students explain their work at the board.
- **?** "How can you check that your answer is reasonable?" Use estimation.
- Have three different pairs of students choose one question from 4–6 to complete at the board. Have the other students try these problems at their desks. Have the pairs of students explain their work at the board. Students may need to be reminded that multiplication can be represented using parentheses around one or both of the factors.

Example 4

- **Financial Literacy:** This example uses stock prices to review decimal addition, subtraction, and division. Remind students that the word *mean* is the same as the arithmetic average.
- Explain the stock context and what each column of the table means.

On Your Own

- Predict whether the mean change will be *positive* or *negative*. Explain your reasoning. positive; Students should recognize that the mean of the four stocks is the sum of the change in the first three stocks ($-$333.63$) and the change in Stock D (568.23), divided by four. This will be positive because $-$333.63 + 568.23 is positive.

Closure

- **Exit Ticket:**

  $-2\frac{1}{3} \times 3\frac{2}{3}$ $-8\frac{5}{9}$ $(-0.5)(-4.2) \div 0.03$ 70

EXAMPLE ③ **Multiplying More Than Two Rational Numbers**

Find $-\dfrac{1}{7} \cdot \left[\dfrac{4}{5} \cdot (-7) \right]$.

You can use properties of multiplication to make the product easier to find.

$-\dfrac{1}{7} \cdot \left[\dfrac{4}{5} \cdot (-7) \right] = -\dfrac{1}{7} \cdot \left(-7 \cdot \dfrac{4}{5} \right)$ Commutative Property of Multiplication

$= -\dfrac{1}{7} \cdot (-7) \cdot \dfrac{4}{5}$ Associative Property of Multiplication

$= 1 \cdot \dfrac{4}{5}$ Multiplicative Inverse Property

$= \dfrac{4}{5}$ Multiplication Property of One

∴ The product is $\dfrac{4}{5}$.

On Your Own

Now You're Ready
Exercises 10–30

Multiply or divide. Write fractions in simplest form.

1. $-\dfrac{6}{5} \div \left(-\dfrac{1}{2} \right)$

2. $\dfrac{1}{3} \div \left(-2\dfrac{2}{3} \right)$

3. $1.8(-5.1)$

4. $-6.3(-0.6)$

5. $-\dfrac{2}{3} \cdot 7\dfrac{7}{8} \cdot \dfrac{3}{2}$

6. $-7.2 \cdot 0.1 \cdot (-100)$

EXAMPLE ④ **Real-Life Application**

Account Positions	⟳		
Stock	**Original Value**	**Current Value**	**Change**
A	600.54	420.15	−180.39
B	391.10	518.38	127.28
C	380.22	99.70	−280.52

An investor owns Stocks A, B, and C. What is the mean change in the value of the stocks?

$$\text{mean} = \frac{-180.39 + 127.28 + (-280.52)}{3} = \frac{-333.63}{3} = -111.21$$

∴ The mean change in the value of the stocks is −$111.21.

On Your Own

7. **WHAT IF?** The change in the value of Stock D is $568.23. What is the mean change in the value of the four stocks?

 Vocabulary and Concept Check

1. **WRITING** How is multiplying and dividing rational numbers similar to multiplying and dividing integers?

2. **NUMBER SENSE** Find the reciprocal of $-\frac{2}{5}$.

Tell whether the expression is *positive* or *negative* without evaluating.

3. $-\frac{3}{10} \times \left(-\frac{8}{15}\right)$ 4. $1\frac{1}{2} \div \left(-\frac{1}{4}\right)$ 5. -6.2×8.18 6. $\frac{-8.16}{-2.72}$

 Practice and Problem Solving

Multiply.

7. $-1\left(\frac{4}{5}\right)$ 8. $-1\left(-3\frac{1}{2}\right)$ 9. $-0.25(-1)$

Divide. Write fractions in simplest form.

① 10. $-\frac{7}{10} \div \frac{2}{5}$ 11. $\frac{1}{4} \div \left(-\frac{3}{8}\right)$ 12. $-\frac{8}{9} \div \left(-\frac{8}{9}\right)$ 13. $-\frac{1}{5} \div 20$

14. $-2\frac{4}{5} \div (-7)$ 15. $-10\frac{2}{7} \div \left(-4\frac{4}{11}\right)$ 16. $-9 \div 7.2$ 17. $8 \div 2.2$

18. $-3.45 \div (-15)$ 19. $-0.18 \div 0.03$ 20. $8.722 \div (-3.56)$ 21. $12.42 \div (-4.8)$

Multiply. Write fractions in simplest form.

② ③ 22. $-\frac{1}{4} \times \left(-\frac{4}{3}\right)$ 23. $\frac{5}{6}\left(-\frac{8}{15}\right)$ 24. $-2\left(-1\frac{1}{4}\right)$

25. $-3\frac{1}{3} \cdot \left(-2\frac{7}{10}\right)$ 26. $0.4 \times (-0.03)$ 27. $-0.05 \times (-0.5)$

28. $-8(0.09)(-0.5)$ 29. $\frac{5}{6} \cdot \left(-4\frac{1}{2}\right) \cdot \left(-2\frac{1}{5}\right)$ 30. $\left(-1\frac{2}{3}\right)^3$

ERROR ANALYSIS Describe and correct the error.

31.
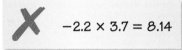
$-2.2 \times 3.7 = 8.14$

32.
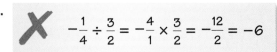
$-\frac{1}{4} \div \frac{3}{2} = -\frac{4}{1} \times \frac{3}{2} = -\frac{12}{2} = -6$

33. **HOUR HAND** The hour hand of a clock moves $-30°$ every hour. How many degrees does it move in $2\frac{1}{5}$ hours?

34. **SUNFLOWER SEEDS** How many 0.75-pound packages can you make with 6 pounds of sunflower seeds?

Assignment Guide and Homework Check

Level	Day 1 Activity Assignment	Day 2 Lesson Assignment	Homework Check
Basic	7–9, 47–51	1–6, 11–33 odd	15, 19, 25, 27, 33
Average	7–9, 47–51	1–6, 14–34 even, 42–44	14, 20, 26, 30
Advanced	7–9, 47–51	1–6, 18–24 even, 32–46 even	18, 20, 24, 42, 44
Accelerated	1–9, 18–46 even, 47–51		18, 20, 24, 42, 44

Common Errors

- **Exercises 10–15** Students may use the reciprocal of the first fraction instead of the second, or they might forget to write a mixed number as an improper fraction before finding the reciprocal. Review multiplying and dividing fractions and the definition of reciprocal.
- **Exercises 16–21** Students may mix up the dividend and divisor. Remind them that the first number is the dividend and the second is the divisor.
- **Exercises 16–21** Students may forget to shift the decimal point when dividing or they might move the decimal point the wrong number of places. Remind students to use estimation to check their answer and the placement of the decimal.
- **Exercises 35–40** Students may forget to follow the order of operations. Tell them to write parentheses around the multiplication or division parts so that they remember to evaluate them first.

2.4 Record and Practice Journal

Multiply or divide. Write fractions in simplest form.

1. $-\frac{8}{9}\left(-\frac{18}{25}\right)$ $\frac{16}{25}$

2. $-4\left(\frac{9}{16}\right)$ $-2\frac{1}{4}$

3. $-3\frac{3}{7} \times 2\frac{1}{2}$ $-8\frac{4}{7}$

4. $-\frac{2}{3} \div \frac{5}{9}$ $-1\frac{1}{5}$

5. $\frac{7}{13} \div (-2)$ $-\frac{7}{26}$

6. $-5\frac{5}{8} \div \left(-4\frac{7}{12}\right)$ $1\frac{5}{22}$

7. $-1.39 \times (-6.8)$ 9.452

8. $-10 \div 0.22$ $-45.\overline{45}$

9. $-12.166 \div (-1.54)$ 7.9

10. In a game of tug of war, your team changes $-1\frac{3}{10}$ feet in position every 10 seconds. What is your change in position after 30 seconds?
$-3\frac{9}{10}$ ft

Vocabulary and Concept Check

1. The same rules for signs of integers are applied to rational numbers.

2. $-\frac{5}{2}$ 3. positive

4. negative 5. negative

6. positive

Practice and Problem Solving

7. $-\frac{4}{5}$ 8. $3\frac{1}{2}$

9. 0.25 10. $-1\frac{3}{4}$

11. $-\frac{2}{3}$ 12. 1

13. $-\frac{1}{100}$ 14. $\frac{2}{5}$

15. $2\frac{5}{14}$ 16. -1.25

17. $3.\overline{63}$ 18. 0.23

19. -6 20. -2.45

21. -2.5875 22. $\frac{1}{3}$

23. $-\frac{4}{9}$ 24. $2\frac{1}{2}$

25. 9 26. -0.012

27. 0.025 28. 0.36

29. $8\frac{1}{4}$ 30. $-4\frac{17}{27}$

31. The answer should be negative. $-2.2 \times 3.7 = -8.14$

32. The wrong fraction was inverted.
$$-\frac{1}{4} \div \frac{3}{2} = -\frac{1}{4} \times \frac{2}{3}$$
$$= -\frac{2}{12}$$
$$= -\frac{1}{6}$$

33. $-66°$

34. 8 packages

35. −19.59 **36.** 1.3

37. −22.667 **38.** $-4\frac{14}{15}$

39. $-5\frac{11}{24}$ **40.** $-1\frac{11}{36}$

41. *Sample answer:* $-\frac{9}{10}, \frac{2}{3}$

42. $191\frac{11}{12}$ yd

43. $3\frac{5}{8}$ gal

44. See *Taking Math Deeper.*

45. −1.28 sec

46. a. −0.02 in.

　　 b. See Additional Answers.

Fair Game Review

47. −1.5 **48.** −5.4

49. $4\frac{1}{2}$ **50.** $-8\frac{5}{18}$

51. D

Taking Math Deeper

Exercise 44

Problems like this one beg for a diagram. It would be easy to misinterpret what "width" is referring to without drawing a diagram and labeling it.

 Draw a diagram. Label the known and unknown lengths.

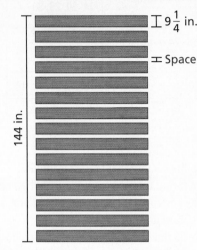

 Find the total width of the boards by multiplying by 15.

$$15\left(9\frac{1}{4}\right) = 15\left(\frac{37}{4}\right) = \frac{555}{4} = 138\frac{3}{4} \text{ in.}$$

3 Subtract the width of the boards from 144 inches.

$$144 - 138\frac{3}{4} = 5\frac{1}{4} \text{ in.}$$

There are 14 spaces, so divide $5\frac{1}{4}$ by 14.

$$5\frac{1}{4} \div 14 = \frac{21}{4} \div 14$$
$$= \frac{21}{4} \cdot \frac{1}{14}$$
$$= \frac{3}{8} \text{ in.} \quad \text{Space}$$

Mini-Assessment

Multiply or divide. Write fractions in simplest form.

1. $-\frac{6}{7}\left(-\frac{5}{2}\right)$ $2\frac{1}{7}$

2. $6\frac{1}{2} \div \left(-2\frac{3}{4}\right)$ $-2\frac{4}{11}$

3. $3.5(-7.65)$ -26.775

4. $-0.25 \div (-0.05)$ 5

5. The cell phone company will add −$2.74 to your next bill for each of the 4 months you were overcharged. How much will be added to your next bill? −$10.96

Reteaching and Enrichment Strategies

If students need help...	If students got it...
Resources by Chapter 　• Practice A and Practice B 　• Puzzle Time Record and Practice Journal Practice Differentiating the Lesson Lesson Tutorials Skills Review Handbook	Resources by Chapter 　• Enrichment and Extension 　• Technology Connection Start the next section

Evaluate.

35. $-4.2 + 8.1 \times (-1.9)$

36. $2.85 - 6.2 \div 2^2$

37. $-3.64 \cdot |-5.3| - 1.5^3$

38. $1\frac{5}{9} \div \left(-\frac{2}{3}\right) + \left(-2\frac{3}{5}\right)$

39. $-3\frac{3}{4} \times \frac{5}{6} - 2\frac{1}{3}$

40. $\left(-\frac{2}{3}\right)^2 - \frac{3}{4}\left(2\frac{1}{3}\right)$

41. OPEN-ENDED Write two fractions whose product is $-\frac{3}{5}$.

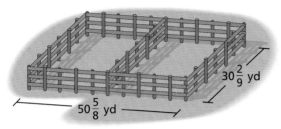

$30\frac{2}{9}$ yd

$50\frac{5}{8}$ yd

42. FENCING A farmer needs to enclose two adjacent rectangular pastures. How much fencing does the farmer need?

43. GASOLINE A 14.5-gallon gasoline tank is $\frac{3}{4}$ full. How many gallons will it take to fill the tank?

44. PRECISION A section of a boardwalk is made using 15 boards. Each board is $9\frac{1}{4}$ inches wide. The total width of the section is 144 inches. The spacing between each board is equal. What is the width of the spacing between each board?

45. RUNNING The table shows the changes in the times (in seconds) of four teammates. What is the mean change?

Teammate	Change
1	-2.43
2	-1.85
3	0.61
4	-1.45

46. *Critical Thinking* The daily changes in the barometric pressure for four days are -0.05, 0.09, -0.04, and -0.08 inches.

　a. What is the mean change?

　b. The mean change after five days is -0.01 inch. What is the change on the fifth day? Explain.

 Fair Game Review What you learned in previous grades & lessons

Add or subtract. *(Section 2.2 and Section 2.3)*

47. $-6.2 + 4.7$

48. $-8.1 - (-2.7)$

49. $\frac{9}{5} - \left(-2\frac{7}{10}\right)$

50. $-4\frac{5}{6} + \left(-3\frac{4}{9}\right)$

51. MULTIPLE CHOICE What are the coordinates of the point in Quadrant IV? *(Skills Review Handbook)*

　Ⓐ $(-4, 1)$　　Ⓑ $(-3, -3)$

　Ⓒ $(0, -2)$　　Ⓓ $(3, -3)$

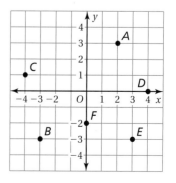

Subtract. Write fractions in simplest form.   *(Section 2.3)*

1. $\dfrac{2}{7} - \left(\dfrac{6}{7}\right)$

2. $\dfrac{12}{7} - \left(-\dfrac{2}{9}\right)$

3. $9.1 - 12.9$

4. $5.647 - (-9.24)$

Find the distance between the two numbers on the number line.   *(Section 2.3)*

5.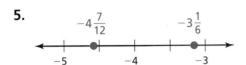

6.

Divide. Write fractions in simplest form.   *(Section 2.4)*

7. $\dfrac{2}{3} \div \left(-\dfrac{5}{6}\right)$

8. $-8\dfrac{5}{9} \div \left(-1\dfrac{4}{7}\right)$

9. $-8.4 \div 2.1$

10. $32.436 \div (-4.24)$

Multiply. Write fractions in simplest form.   *(Section 2.4)*

11. $\dfrac{5}{8} \times \left(-\dfrac{4}{15}\right)$

12. $-2\dfrac{3}{8} \times \dfrac{8}{5}$

13. $-9.4 \times (-4.7)$

14. $-100(-0.6)(0.01)$

15. PARASAILING A parasail is at 200.6 feet above the water. After 5 minutes, the parasail is at 120.8 feet above the water. What is the change in height of the parasail?   *(Section 2.3)*

16. TEMPERATURE Use the thermometer shown. How much did the temperature drop from 5:00 P.M. to 10:00 P.M.?   *(Section 2.3)*

17. LATE FEES You were overcharged $4.52 on your cell phone bill 3 months in a row. The cell phone company says that it will add −$4.52 to your next bill for each month you were overcharged. On the next bill, you see an adjustment of −13.28. Is this amount correct? Explain.   *(Section 2.4)*

18. CASHEWS How many $1\dfrac{1}{4}$-pound packages can you make with $7\dfrac{1}{2}$ pounds of cashews?   *(Section 2.4)*

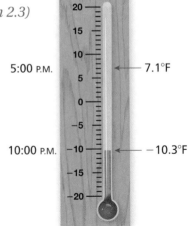

Alternative Assessment Options

Math Chat   **Student Reflective Focus Question**

Structured Interview   Writing Prompt

Student Reflective Focus Question

Ask students to summarize the rules for adding, subtracting, multiplying, and dividing rational numbers. Be sure that they include examples. Select students at random to present their summaries to the class.

Study Help Sample Answers

Remind students to complete Graphic Organizers for the rest of the chapter.

2.

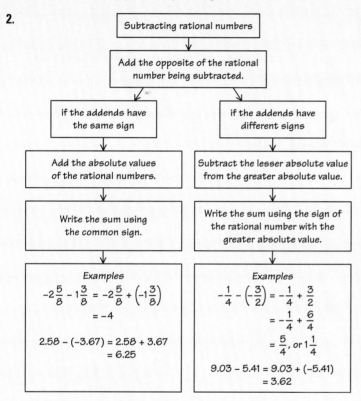

3–4. Available at *BigIdeasMath.com.*

Answers

1. $-\dfrac{4}{7}$ 2. $1\dfrac{59}{63}$

3. -3.8 4. 14.887

5. $1\dfrac{5}{12}$ 6. 6.2

7. $-\dfrac{4}{5}$ 8. $5\dfrac{4}{9}$

9. -4 10. -7.65

11. $-\dfrac{1}{6}$ 12. $-3\dfrac{4}{5}$

13. 44.18 14. 0.6

15. -79.8 ft   16. $17.4°F$

17. No, because $3 \times (-4.52) = -13.56$.

18. 6 packages

Reteaching and Enrichment Strategies

If students need help. . .	If students got it. . .
Resources by Chapter • Practice A and Practice B • Puzzle Time Lesson Tutorials *BigIdeasMath.com*	Resources by Chapter • Enrichment and Extension • Technology Connection Game Closet at *BigIdeasMath.com* Start the Chapter Review

Technology for the *Teacher*

Online Assessment
Assessment Book
ExamView® Assessment Suite

For the Teacher
Additional Review Options

- *BigIdeasMath.com*
- Online Assessment
- Game Closet at *BigIdeasMath.com*
- Vocabulary Help
- Resources by Chapter

Answers

1. $-0.5\overline{3}$ 2. 0.625

3. $-2.1\overline{6}$ 4. 1.4375

5. $-\dfrac{3}{5}$ 6. $-\dfrac{7}{20}$

7. $-5\dfrac{4}{5}$ 8. $24\dfrac{23}{100}$

Review of Common Errors

Exercises 1–4
- Students may forget to carry the negative sign through the division operation.

Exercises 5–8
- Students may use the wrong numerator.

Exercises 9–14
- When adding and subtracting decimals, students may forget to line up the decimal points.

Exercises 9–15
- Students may forget to find a common denominator. Remind students that adding and subtracting fractions always requires a common denominator.

Exercise 16–25
- Students may place the decimal point incorrectly in their answers. Remind students of the rules for multiplying and dividing decimals. Also, remind students to use estimation to check their answers.

Exercises 16–26
- When dividing fractions, students may use the reciprocal of the first fraction instead of the reciprocal of the second fraction.

Check It Out
Vocabulary Help
BigIdeasMath ✓com

Review Key Vocabulary

rational number, *p. 46*
terminating decimal, *p. 46*

repeating decimal, *p. 46*

Review Examples and Exercises

2.1 **Rational Numbers** *(pp. 44–49)*

a. **Write** $4\frac{3}{5}$ **as a decimal.**

Notice that $4\frac{3}{5} = \frac{23}{5}$.

> Divide 23 by 5.

$$
\begin{array}{r}
4.6 \\
5\overline{)23.0} \\
-20 \\
\hline
3\ 0 \\
-3\ 0 \\
\hline
0
\end{array}
$$

> The remainder is 0. So, it is a terminating decimal.

∴ So, $4\frac{3}{5} = 4.6$.

b. **Write** -0.14 **as a fraction in simplest form.**

$-0.14 = -\dfrac{14}{100}$

> Write the digits after the decimal point in the numerator.

> The last digit is in the hundredths place. So, use 100 in the denominator.

$= -\dfrac{7}{50}$ Simplify.

Exercises

Write the rational number as a decimal.

1. $-\dfrac{8}{15}$ **2.** $\dfrac{5}{8}$ **3.** $-\dfrac{13}{6}$ **4.** $1\dfrac{7}{16}$

Write the decimal as a fraction or a mixed number in simplest form.

5. -0.6 **6.** -0.35 **7.** -5.8 **8.** 24.23

2.2 Adding Rational Numbers (pp. 50–55)

Find $-\dfrac{7}{2} + \dfrac{5}{4}$.

$$-\dfrac{7}{2} + \dfrac{5}{4} = \dfrac{-14}{4} + \dfrac{5}{4}$$ Rewrite using the LCD (least common denominator).

$$= \dfrac{-14 + 5}{4}$$ Write the sum of the numerators over the common denominator.

$$= \dfrac{-9}{4}$$ Add.

$$= -2\dfrac{1}{4}$$ Write the improper fraction as a mixed number.

The sum is $-2\dfrac{1}{4}$.

Exercises

Add. Write fractions in simplest form.

9. $\dfrac{9}{10} + \left(-\dfrac{4}{5}\right)$

10. $-4\dfrac{5}{9} + \dfrac{8}{9}$

11. $-1.6 + (-2.4)$

2.3 Subtracting Rational Numbers (pp. 58–63)

Find $-4\dfrac{2}{5} - \left(-\dfrac{3}{5}\right)$.

$$-4\dfrac{2}{5} - \left(-\dfrac{3}{5}\right) = -4\dfrac{2}{5} + \dfrac{3}{5}$$ Add the opposite of $-\dfrac{3}{5}$.

$$= -\dfrac{22}{5} + \dfrac{3}{5}$$ Write the mixed number as an improper fraction.

$$= \dfrac{-22 + 3}{5}$$ Write the sum of the numerators over the common denominator.

$$= \dfrac{-19}{5}, \text{ or } -3\dfrac{4}{5}$$ Simplify.

The difference is $-3\dfrac{4}{5}$.

Exercises

Subtract. Write fractions in simplest form.

12. $-\dfrac{5}{12} - \dfrac{3}{10}$

13. $3\dfrac{3}{4} - \dfrac{7}{8}$

14. $3.8 - (-7.45)$

15. TURTLE A turtle is $20\dfrac{5}{6}$ inches below the surface of a pond. It dives to a depth of $32\dfrac{1}{4}$ inches. What is the change in the turtle's position?

Review Game

Rational Numbers

Materials

- questions from the chapter's homework, quizzes, examples, or tests
- 5 index cards, each with a letter in the word HORSE written on it, for each group

Directions

Divide the class into groups. Ask a group one of the questions. If the group is correct, the game continues to the next group. If they answer incorrectly, they receive an H. Each wrong answer in a group will result in that group receiving the next letter in the word HORSE. When a group has all 5 letters, they are out of the game. Choose questions with a wide range of difficulty to control how long the game takes.

Who Wins?

The last group with 4 or fewer letters wins.

For the Student
Additional Practice

- Lesson Tutorials
- Multi-Language Glossary
- Self-Grading Progress Check
- *BigIdeasMath.com*
  Dynamic Student Edition
  Student Resources

Answers

9. $\dfrac{1}{10}$ 10. $-3\dfrac{2}{3}$

11. -4 12. $-\dfrac{43}{60}$

13. $2\dfrac{7}{8}$ 14. 11.25

15. $-11\dfrac{5}{12}$ inches

16. $-\dfrac{3}{4}$ 17. $-1\dfrac{3}{11}$

18. -2 19. 6.16

20. $\dfrac{28}{81}$ 21. $-\dfrac{16}{45}$

22. 57.23 23. -23.67

24. 5 25. 16

26. -75 ft

My Thoughts on the Chapter

What worked. . .

Teacher Tip

Not allowed to write in your teaching edition? Use sticky notes to record your thoughts.

What did not work. . .

What I would do differently. . .

Multiplying and Dividing Rational Numbers *(pp. 64–69)*

a. Find $-4\dfrac{1}{6} \div 1\dfrac{1}{3}$.

$$-4\frac{1}{6} \div 1\frac{1}{3} = -\frac{25}{6} \div \frac{4}{3}$$ Write mixed numbers as improper fractions.

$$= \frac{-25}{6} \cdot \frac{3}{4}$$ Multiply by the reciprocal of $\dfrac{4}{3}$.

$$= \frac{-25 \cdot 3}{6 \cdot 4}$$ Multiply the numerators and the denominators.

$$= \frac{-25}{8}, \text{ or } -3\frac{1}{8}$$ Simplify.

∴ The quotient is $-3\dfrac{1}{8}$.

b. Find $-1.6 \cdot 2.4$.

$$\begin{array}{r} -1.6 \\ \times\ 2.4 \\ \hline 64 \\ 320 \\ \hline -3.84 \end{array}$$

The decimals have different signs.

The product is negative.

∴ The product is -3.84.

Exercises

Divide. Write fractions in simplest form.

16. $\dfrac{9}{10} \div \left(-\dfrac{6}{5}\right)$ **17.** $-\dfrac{4}{11} \div \dfrac{2}{7}$ **18.** $6.4 \div (-3.2)$ **19.** $-15.4 \div (-2.5)$

Multiply. Write fractions in simplest form.

20. $-\dfrac{4}{9}\left(-\dfrac{7}{9}\right)$ **21.** $\dfrac{8}{15}\left(-\dfrac{2}{3}\right)$ **22.** $-5.9(-9.7)$

23. $4.5(-5.26)$ **24.** $-\dfrac{2}{3} \cdot \left(2\dfrac{1}{2}\right) \cdot (-3)$ **25.** $-1.6 \cdot (0.5) \cdot (-20)$

26. SUNKEN SHIP The elevation of a sunken ship is -120 feet. Your elevation is $\dfrac{5}{8}$ of the ship's elevation. What is your elevation?

Check It Out
Test Practice
BigIdeasMath ✓com

Write the rational number as a decimal.

1. $\dfrac{7}{40}$

2. $-\dfrac{1}{9}$

3. $-\dfrac{21}{16}$

4. $\dfrac{36}{5}$

Write the decimal as a fraction or a mixed number in simplest form.

5. -0.122

6. 0.33

7. -4.45

8. -7.09

Add or subtract. Write fractions in simplest form.

9. $-\dfrac{4}{9} + \left(-\dfrac{23}{18}\right)$

10. $\dfrac{17}{12} - \left(-\dfrac{1}{8}\right)$

11. $9.2 + (-2.8)$

12. $2.86 - 12.1$

Multiply or divide. Write fractions in simplest form.

13. $3\dfrac{9}{10} \times \left(-\dfrac{8}{3}\right)$

14. $-1\dfrac{5}{6} \div 4\dfrac{1}{6}$

15. $-4.4 \times (-6.02)$

16. $-5 \div 1.5$

17. $-\dfrac{3}{5} \cdot \left(2\dfrac{2}{7}\right) \cdot \left(-3\dfrac{3}{4}\right)$

18. $-6 \cdot (-0.05) \cdot (-0.4)$

19. ALMONDS How many 2.25-pound containers can you make with 24.75 pounds of almonds?

20. FISH The elevation of a fish is -27 feet.

    a. The fish decreases its elevation by 32 feet, and then increases its elevation by 14 feet. What is its new elevation?

    b. Your elevation is $\dfrac{2}{5}$ of the fish's new elevation. What is your elevation?

21. RAINFALL The table shows the rainfall (in inches) for three months compared to the yearly average. Is the total rainfall for the three-month period greater than or less than the yearly average? Explain.

November	December	January
-0.86	2.56	-1.24

22. BANK ACCOUNTS Bank Account A has $750.92, and Bank Account B has $675.44. Account A changes by $-\$216.38$, and Account B changes by $-\$168.49$. Which account has the greater balance? Explain.

Test Item References

Chapter Test Questions	Section to Review
1–8	2.1
9, 11, 21, 22	2.2
10, 12, 20(a)	2.3
13–19, 20(b)	2.4

Test-Taking Strategies

Remind students to quickly look over the entire test before they start so that they can budget their time. On tests, it is really important for students to **Stop** and **Think**. When students hurry on a test dealing with signed numbers, they often make "sign" errors. Sometimes it helps to represent each problem with a number line to ensure that they are thinking through the process.

Common Errors

- **Exercises 1–4** Students may forget to carry the negative sign through the division operation. Tell them to create a space for the final answer and to write the sign of the number in the space at the beginning.
- **Exercises 9 and 10** Students may forget to find a common denominator. Remind students that adding and subtracting fractions always requires a common denominator.
- **Exercise 14** Students may use the reciprocal of the first fraction instead of the second, or they might forget to write a mixed number as an improper fraction before finding the reciprocal. Review multiplying and dividing fractions and the definition of reciprocal.
- **Exercises 15 and 16** Students may place the decimal point incorrectly in their answers. Remind students of the rules for multiplying and dividing decimals. Also, remind students to use estimation to check their answers.

Reteaching and Enrichment Strategies

If students need help. . .	If students got it. . .
Resources by Chapter • Practice A and Practice B • Puzzle Time Record and Practice Journal Practice Differentiating the Lesson Lesson Tutorials *BigIdeasMath.com* Skills Review Handbook	Resources by Chapter • Enrichment and Extension • Technology Connection Game Closet at *BigIdeasMath.com* Start Cumulative Assessment

Technology for the *Teacher*

Online Assessment
Assessment Book
ExamView® Assessment Suite

After Answering Easy Questions, Relax

Answer Easy Questions First

Estimate the Answer

Read All Choices before Answering

Read Question before Answering

Solve Directly or Eliminate Choices

Solve Problem before Looking at
    Choices

Use Intelligent Guessing

Work Backwards

About this Strategy

When taking a multiple choice test, be sure to read each question carefully and thoroughly. After reading the question, estimate the answer before trying to solve.

Answers

1. A
2. H
3. −18
4. C
5. I

Item Analysis

1. **A.** Correct answer

   B. The student correctly finds José's height at 5 years old, which was 41 inches, but then reverses the relationship between José and Sean.

   C. When multiplying the rate of growth by the number of elapsed years, the student multiplies only the whole number parts to get $16\frac{3}{4}$.

   D. When multiplying the rate of growth by the number of elapsed years, the student multiplies only the whole number parts to get $16\frac{3}{4}$. After using this to find José's height at 5 years old, the student also reverses the relationship between José and Sean.

2. **F.** The student does not perform the operation correctly.

   G. The student does not perform the operation correctly.

   H. Correct answer

   I. The student does not perform the operation correctly.

3. **Gridded Response:** Correct answer: −18

   Common Error: The student thinks that each number is 2 times the previous number, rather than −2 times the previous number, and gets an answer of 18.

4. **A.** The student thinks that the absolute value of each individual number is negative and finds the sum of −2 and −2.5.

   B. The student correctly simplifies the expression inside the absolute value bars, but then thinks that the absolute value means to take the opposite.

   C. Correct answer

   D. The student takes the absolute value of each individual number and finds the sum of 2 and 2.5.

5. **F.** The student correctly subtracts $\frac{3}{8}$ from $-\frac{7}{4}$, but forgets to find the absolute value.

   G. The student incorrectly finds the sum of the two numbers.

   H. The student incorrectly finds the sum of the two numbers and then finds the absolute value.

   I. Correct answer

Technology for the **Teacher**

Performance Tasks
Online Assessment
Assessment Book
ExamView® Assessment Suite

1. When José and Sean were each 5 years old, José was $1\frac{1}{2}$ inches taller than Sean. José grew at an average rate of $2\frac{3}{4}$ inches per year from the time that he was 5 years old until the time he was 13 years old. José was 63 inches tall when he was 13 years old. How tall was Sean when he was 5 years old?

   A. $39\frac{1}{2}$ in.

   B. $42\frac{1}{2}$ in.

   C. $44\frac{3}{4}$ in.

   D. $47\frac{3}{4}$ in.

2. Which expression represents a positive integer?

   F. -6^2

   G. $(-3)^3$

   H. $(-5)^2$

   I. -2^3

3. What is the missing number in the sequence below?

   $$\frac{9}{16}, \ -\frac{9}{8}, \ \frac{9}{4}, \ -\frac{9}{2}, \ 9, \ \underline{\hspace{1cm}}$$

4. What is the value of the expression below?

   $$\left| -2 - (-2.5) \right|$$

   A. -4.5

   B. -0.5

   C. 0.5

   D. 4.5

5. What is the distance between the two numbers on the number line?

   $-\frac{7}{4}$ ... $\frac{3}{8}$

   | | | | | | | | | |
   -2   -1   0   1   2

   F. $-2\frac{1}{8}$

   G. $-1\frac{3}{8}$

   H. $1\frac{3}{8}$

   I. $2\frac{1}{8}$

6. Sandra was evaluating an expression in the box below.

$$-4\frac{3}{4} \div 2\frac{1}{5} = -\frac{19}{4} \div \frac{11}{5}$$

$$= \frac{-4}{19} \cdot \frac{5}{11}$$

$$= \frac{-4 \cdot 5}{19 \cdot 11}$$

$$= \frac{-20}{209}$$

What should Sandra do to correct the error that she made?

A. Rewrite $-\frac{19}{4}$ as $-\frac{4}{19}$ and multiply by $\frac{11}{5}$.

B. Rewrite $\frac{11}{5}$ as $\frac{5}{11}$ and multiply by $-\frac{19}{4}$.

C. Rewrite $\frac{11}{5}$ as $-\frac{5}{11}$ and multiply by $-\frac{19}{4}$.

D. Rewrite $-4\frac{3}{4}$ as $-\frac{13}{4}$ and multiply by $\frac{5}{11}$.

7. What is the value of the expression below when $q = -2$, $r = -12$, and $s = 8$?

$$\frac{-q^2 - r}{s}$$

F. -2 **H.** 1

G. -1 **I.** 2

8. You are stacking wooden blocks with the dimensions shown below. How many blocks do you need to stack to build a block tower that is $7\frac{1}{2}$ inches tall?

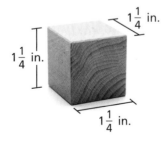

$1\frac{1}{4}$ in.

$1\frac{1}{4}$ in.

$1\frac{1}{4}$ in.

Item Analysis (continued)

6. **A.** The student thinks that you divide two fractions by multiplying the reciprocal of the dividend by the divisor.

   B. Correct answer

   C. The student thinks that you divide two fractions by multiplying the dividend by the opposite of the reciprocal of the divisor.

   D. The student incorrectly rewrites the negative mixed number.

7. **F.** The student subtracts 12 in the numerator.

   G. The student subtracts 12 from 4 in the numerator.

   H. Correct answer

   I. The student thinks that the opposite of the square of -2 is 4.

8. **Gridded Response:** Correct answer: 6

   Common Error: The student divides the integer parts and the fractional parts of the mixed numbers and gets an answer of $7 + 2 = 9$.

Answers

6. B

7. H

8. 6 blocks

9. B

10. G

11. *Part A* 1.7
 Part B 4
 Part C $0.9\overline{6}$
 Part D 8.7

12. C

Item Analysis (continued)

9. A. The student finds $\frac{1}{2}$ of the sum of the base and the height.

 B. Correct answer

 C. The student finds the sum of the base and the height.

 D. The student finds the product of the base and the height.

10. F. The student thinks that the cube of -2 is 8.

 G. Correct answer

 H. The student finds the square of -4 and then thinks that the cube of -2 is 8.

 I. The student finds the square of -4.

11. 4 points The student demonstrates a thorough understanding of interpreting rational numbers on a number line and a thorough conceptual understanding of the four operations using rational numbers. In Part A, the student correctly recognizes that the two greatest values, T and U, have the greatest sum, which is approximately 1.7. In Part B, the student correctly recognizes that the two values that are the farthest apart, U and R, have the greatest difference, which is approximately 4. In Part C, the student correctly recognizes that the two values that have the same sign and also the greatest magnitude, R and S, have the greatest product, which is approximately $0.9\overline{6}$. In Part D, the student correctly recognizes that the two values that have the same sign and also the greatest ratio, R and S, have the greatest quotient, which is approximately 8.7. The student provides clear and complete explanations of the reasoning used.

3 points The student demonstrates an understanding of interpreting rational numbers on a number line and a good conceptual understanding of the four operations using rational numbers, but the student's work and explanations demonstrate an essential but less than thorough understanding.

2 points The student demonstrates an understanding of interpreting rational numbers on a number line and a partial conceptual understanding of the four operations using rational numbers. The student's work and explanations demonstrate a lack of essential understanding.

1 point The student demonstrates a partial understanding of interpreting rational numbers on a number line and a limited conceptual understanding of the four operations using rational numbers. The student's response is incomplete and exhibits many flaws.

0 points The student provided no response, a completely incorrect or incomprehensible response, or a response that demonstrates insufficient understanding of interpreting rational numbers on a number line and an insufficient conceptual understanding of the four operations using rational numbers.

12. A. The student thinks that $\frac{-0.4}{-1} = -0.4$ and that $-0.4 + 0.8 = -1.2$.

 B. The student thinks that $\frac{-0.4}{-0.2} = -2$.

 C. Correct answer

 D. The student thinks that $-0.4 + 0.8 = -1.2$.

9. What is the area of a triangle with a base length of $2\frac{1}{2}$ inches and a height of 2 inches?

   A. $2\frac{1}{4}$ in.^{2} C. $4\frac{1}{2}$ in.^{2}

   B. $2\frac{1}{2}$ in.^{2} D. 5 in.^{2}

10. What is the value of the expression below?
$$\frac{-4^2 - (-2)^3}{4}$$

   F. -6 H. 2

   G. -2 I. 6

11. Four points are graphed on the number line below.

   Part A Choose the two points whose values have the greatest sum. Approximate this sum. Explain your reasoning.

   Part B Choose the two points whose values have the greatest difference. Approximate this difference. Explain your reasoning.

   Part C Choose the two points whose values have the greatest product. Approximate this product. Explain your reasoning.

   Part D Choose the two points whose values have the greatest quotient. Approximate this quotient. Explain your reasoning.

12. What number belongs in the box to make the equation true?
$$\frac{-0.4}{\square} + 0.8 = -1.2$$

   A. -1 C. 0.2

   B. -0.2 D. 1

3 Expressions and Equations

3.1 Algebraic Expressions

3.2 Adding and Subtracting Linear Expressions

3.3 Solving Equations Using Addition or Subtraction

3.4 Solving Equations Using Multiplication or Division

3.5 Solving Two-Step Equations

"I can't find my algebra tiles, so I am painting some of my dog biscuits."

"Now I will be able to solve the equation $2x + (-2) = 2$."

"I hope the paint is edible."

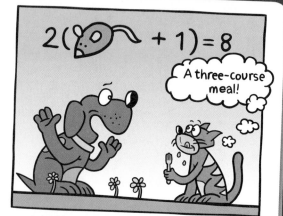

"Descartes, if you solve for 🐭 in the equation, what do you get?"

"A three-course meal!"

What Your Students Have Learned

- Use parentheses, brackets, or braces in numerical expressions, and evaluate expressions with these symbols.
- Write and interpret numerical expressions without evaluating them.
- Use and interpret simple equations.
- Write and evaluate numerical expressions involving whole-number exponents.
- Read, write, and evaluate algebraic expressions.
- Apply the properties of operations to generate equivalent expressions.
- Factor out the greatest common factor (GCF) in algebraic and numerical expressions.
- Identify equivalent expressions.
- Determine if a value is a solution of an equation.
- Solve one-step equations.

What Your Students Will Learn

- Add, subtract, factor, and expand linear expressions with rational coefficients.
- Understand that rewriting expressions in different forms can show how the quantities are related.
- Write, graph, and solve one-step equations (including negative numbers).
- Solve two-step equations.
- Compare algebraic solutions to arithmetic solutions.

Pacing Guide for Chapter 3

Chapter Opener	
Regular	1 Day
Accelerated	1 Day
Section 1	
Regular	2 Days
Accelerated	0.5 Day
Section 2	
Regular	3 Days
Accelerated	1.5 Days
Study Help / Quiz	
Regular	1 Day
Accelerated	0 Days
Section 3	
Regular	2 Days
Accelerated	1 Day
Section 4	
Regular	2 Days
Accelerated	1 Day
Section 5	
Regular	2 Days
Accelerated	1 Day
Topic 1	
Regular	0 Days
Accelerated	1 Day
Topic 2	
Regular	0 Days
Accelerated	2 Days
Topic 3	
Regular	0 Days
Accelerated	1 Day
Chapter Review/ Chapter Tests	
Regular	2 Days
Accelerated	2 Days
Total Chapter 3	
Regular	15 Days
Accelerated	12 Days
Year-to-Date	
Regular	42 Days
Accelerated	22 Days

Technology for the *Teacher*

BigIdeasMath.com
Chapter at a Glance
Complete Materials List
Parent Letters: English and Spanish

What Your Students Have Learned

- Evaluate algebraic expressions by using substitution and the order of operations.
- Write algebraic expressions using key terms and phrases.

Additional Topics for Review

- Powers and Exponents
- Order of Operations
- Greatest Common Factor (GCF)
- The Distributive Property

Try It Yourself

1. $-1\frac{1}{2}$ 2. -12

3. 2 4. -14

5. $3q + 5$ 6. $n - 9$

7. $6p$ 8. $8 \div h$

9. $3t + 4$ 10. $7c - 2$

Record and Practice Journal
Fair Game Review

1. 6 2. 12

3. 20 4. $\frac{19}{4}$ or $4\frac{3}{4}$

5. 12 6. -1

7. 24 ft^2 8. $8 + y$

9. $p - 6$ 10. $7m$

11. $11c - 8$

12. $r - \dfrac{r}{2}$

13. $9(z + 4)$

Math Background Notes

Vocabulary Review

- Evaluate
- Expression
- Substitute
- Order of Operations

Evaluating Expressions

- Students should know how to evaluate expressions. Because this skill is fairly new to them, you may want to allot slightly more time for practice.
- Remind students that evaluate means to find a value.
- To evaluate an expression, students will need to substitute values into the expression.
- **Teaching Tip:** English Language Learners may find the vocabulary and writing involved in these opening exercises challenging. To ease the transition, help them make analogies. In the same way a sugar substitute replaces sugar, substituting numbers into an expression will replace the variables until you are able to evaluate.
- After they substitute, remind students that they must follow the correct order of operations as they simplify. You may wish to review order of operations prior to completing the examples.

Writing Algebraic Expressions

- Students should know how to write expressions.
- Review key words with students. For example, words such as sum and product alert students to use addition and multiplication.
- **Common Error:** Many students will read Example 3, part (b) and incorrectly write the expression $8 - 3x$. Remind students that they cannot subtract 8 from a number unless they have a number to start with. Always write the number first and then add or subtract as indicated.

Reteaching and Enrichment Strategies

If students need help. . .	If students got it. . .
Record and Practice Journal • Fair Game Review Skills Review Handbook Lesson Tutorials	Game Closet at *BigIdeasMath.com* Start the next section

What You Learned Before

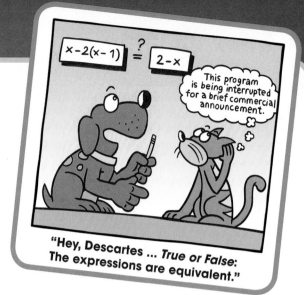

"Hey, Descartes ... True or False: The expressions are equivalent."

Evaluating Expressions

Example 1 Evaluate $6x + 2y$ when $x = -3$ and $y = 5$.

$6x + 2y = 6(-3) + 2(5)$ Substitute -3 for x and 5 for y.

$\quad\quad\quad = -18 + 10$ Using order of operations, multiply 6 and -3, and 2 and 5.

$\quad\quad\quad = -8$ Add -18 and 10.

Example 2 Evaluate $6x^2 - 3(y + 2) + 8$ when $x = -2$ and $y = 4$.

$6x^2 - 3(y + 2) + 8 = 6(-2)^2 - 3(4 + 2) + 8$ Substitute -2 for x and 4 for y.

$\quad\quad\quad\quad\quad\quad\quad = 6(-2)^2 - 3(6) + 8$ Using order of operations, evaluate within the parentheses.

$\quad\quad\quad\quad\quad\quad\quad = 6(4) - 3(6) + 8$ Using order of operations, evaluate the exponent.

$\quad\quad\quad\quad\quad\quad\quad = 24 - 18 + 8$ Using order of operations, multiply 6 and 4, and 3 and 6.

$\quad\quad\quad\quad\quad\quad\quad = 14$ Subtract 18 from 24. Add the result to 8.

Try It Yourself

Evaluate the expression when $x = -\dfrac{1}{4}$ and $y = 3$.

1. $2xy$ **2.** $12x - 3y$ **3.** $-4x - y + 4$ **4.** $8x - y^2 - 3$

Writing Algebraic Expressions

Example 3 Write the phrase as an algebraic expression.

  a. the sum of twice a number m and four   **b.** eight less than three times a number x

                $2m + 4$ $3x - 8$

Try It Yourself

Write the phrase as an algebraic expression.

  5. five more than three times a number q **6.** nine less than a number n

  7. the product of a number p and six   **8.** the quotient of eight and a number h

  9. four more than three times a number t **10.** two less than seven times a number c

Essential Question How can you simplify an algebraic expression?

---

1 ACTIVITY: Simplifying Algebraic Expressions

Work with a partner.

a. Evaluate each algebraic expression when $x = 0$ and when $x = 1$. Use the results to match each expression in the left table with its equivalent expression in the right table.

	Expression	Value When $x = 0$	Value When $x = 1$
A.	$3x + 2 - x + 4$		
B.	$5(x - 3) + 2$		
C.	$x + 3 - (2x + 1)$		
D.	$-4x + 2 - x + 3x$		
E.	$-(1 - x) + 3$		
F.	$2x + x - 3x + 4$		
G.	$4 - 3 + 2(x - 1)$		
H.	$2(1 - x + 4)$		
I.	$5 - (4 - x + 2x)$		
J.	$5x - (2x + 4 - x)$		

	Expression	Value When $x = 0$	Value When $x = 1$
a.	4		
b.	$-x + 1$		
c.	$4x - 4$		
d.	$2x + 6$		
e.	$5x - 13$		
f.	$-2x + 10$		
g.	$x + 2$		
h.	$2x - 1$		
i.	$-2x + 2$		
j.	$-x + 2$		

Algebraic Expressions

In this lesson, you will

- apply properties of operations to simplify algebraic expressions.
- solve real-life problems.

b. Compare each expression in the left table with its equivalent expression in the right table. In general, how do you think you obtain the equivalent expression in the right column?

Laurie's Notes

What Your Students Will Learn
- Identify terms and like terms of algebraic expressions.
- Apply properties of operations to simplify algebraic expressions.

Introduction
Applying Mathematical Practices
- **Look for and Make Use of Structure:** Mathematically proficient students are able to see an algebraic expression being composed of several terms. In this lesson, students will simplify algebraic expressions by combining like terms.

Motivate
- **Target Math Game Time!** Write the following problem on the board.

$$\boxed{} + \boxed{} \times \boxed{} - \boxed{} = \boxed{}$$

- **Directions:** Tell students that you are going to randomly generate 4 numbers from -8 to 8, and *as you generate the numbers,* they are to place the numbers in the four boxes to the left of the equal sign. After the fourth number is generated, students should evaluate the expression and write their answer in the red box. The goal is to get as close to the target number of 24 as possible without going over.
- To generate the random numbers, you can use the random number generator on a graphing calculator. Another option is to write the numbers on slips of paper, put the slips of paper into a container, and pull one number at a time. Do not use the same number twice.
- **Extensions:** Play this more than once by changing the target number, changing the range of numbers used, or changing the original expression.
- The goal is to evaluate expressions using the order of operations.

Discuss
- Ask students to recall that an *algebraic expression* is an expression that may contain numbers, operations, and one or more symbols.
- You may wish to do a few examples where you change the sign of the coefficient, such as $4(x + 2)$, $-4(x + 2)$, and $-(x + 2)$.

Activity Notes
Activity 1
- It will be important for students to use the order of operations.
- Encourage students to write the expression, then rewrite it with the value of the variable substituted.
- ❓ "How did you decide which expressions matched?" The expressions have the same values when $x = 0$ and when $x = 1$.
- ❓ Discuss answers to parts (a) and (b) and then ask, "What do you think it means to simplify an expression?" Students should say that the expression has fewer terms. They should also say that variable terms have been combined and numeric terms have been combined.
- ❓ "Have you heard the word *simplify* in mathematics before? Explain." You simplify fractions when you divide out common factors.

Previous Learning
Students should know how to evaluate algebraic expressions.

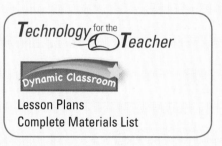

Lesson Plans
Complete Materials List

3.1 Record and Practice Journal

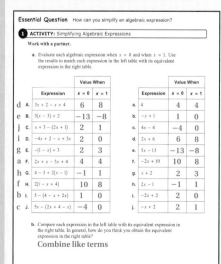

Essential Question How can you simplify an algebraic expression?

1 ACTIVITY: Simplifying Algebraic Expressions

Work with a partner.

a. Evaluate each algebraic expression when $x = 0$ and when $x = 1$. Use the results to match each expression in the left table with its equivalent expression in the right table.

	Expression	$x=0$	$x=1$		Expression	$x=0$	$x=1$
d A.	$3x + 2 - x + 4$	6	8	a.	4	4	4
e B.	$5(x-3) + 2$	-13	-8	b.	$-x + 1$	1	0
j C.	$x + 3 - (2x + 1)$	2	1	c.	$4x - 4$	-4	0
i D.	$-4x + 2 - x + 3x$	2	0	d.	$2x + 6$	6	8
g E.	$-(1-x) + 3$	2	3	e.	$5x - 13$	-13	-8
a F.	$2x + x - 3x + 4$	4	4	f.	$-2x + 10$	10	8
h G.	$4 - 3 + 2(x-1)$	-1	1	g.	$x + 2$	2	3
f H.	$2(1 - x + 4)$	10	8	h.	$2x - 1$	-1	1
b I.	$5 - (4 - x + 2x)$	1	0	i.	$-2x + 2$	2	0
c J.	$5x - (2x + 4 - x)$	-4	0	j.	$-x + 2$	2	1

b. Compare each expression in the left table with its equivalent expression in the right table. In general, how do you think you obtain the equivalent expression in the right table?

Combine like terms

Differentiated Instruction

Inclusion

When simplifying expressions with parentheses such as $3(x + 2) - 5$, students should use the Distributive Property first. This follows the order of operations.

Activity 2

- This activity will give you insight into how students think about simplifying algebraic expressions. The previous activity contained only like variable terms, so students will likely not consider expressions with unlike variable terms such as $3x + 2y - 4x + 5$.
- **Construct Viable Arguments and Critique the Reasoning of Others:** When partners have finished, have them exchange their lessons with another pair of students. Students should read the lessons they receive and write critiques to the authors of the lessons. They should provide feedback on what is clear in the lesson and what can be made more clear.
- Have students complete the exercises of the lesson they received.

What Is Your Answer?

- Students should be able to answer Question 3 independently.
- Discuss student answers for Question 4.

Closure

- **Exit Ticket:** Which of the following expressions would simplify to $3x + 4$? If the expression does not simplify to $3x + 4$, what does it simplify to?

  A. $-6 + 3x + 2$ $3x - 4$

  B. $3(x + 4)$ $3x + 12$

  C. $2x + 8 + x - 2$ $3x + 6$

  D. $-5x + 4 + 8x$ $3x + 4$

3.1 Record and Practice Journal

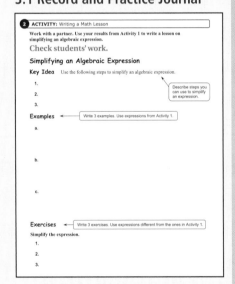

What Is Your Answer?

3. **IN YOUR OWN WORDS** How can you simplify an algebraic expression? Give an example that demonstrates your procedure.

   Sample answer: You can simplify an algebraic expression by: (1) using the Distributive Property to eliminate parentheses, (2) use the Commutative and Associative Properties of Addition to get the variable terms together and the terms without variables together, (3) combine the variable terms and combine the terms without variables.

4. **REASONING** Why would you want to simplify an algebraic expression? Discuss several reasons.

   Check students' work.

2 ACTIVITY: Writing a Math Lesson

Work with a partner. Use your results from Activity 1 to write a lesson on simplifying an algebraic expression.

Math Practice

Communicate Precisely

What can you do to make sure that you are communicating exactly what is needed in the Key Idea?

Describe steps you can use to simplify an expression.

Write 3 examples. Use expressions from Activity 1.

Write 3 exercises. Use expressions different from the ones in Activity 1.

Simplifying an Algebraic Expression

Key Idea Use the following steps to simplify an algebraic expression.
1.
2.
3.

Examples

a.

b.

c.

Exercises

Simplify the expression.
1.
2.
3.

What Is Your Answer?

3. **IN YOUR OWN WORDS** How can you simplify an algebraic expression? Give an example that demonstrates your procedure.

4. **REASONING** Why would you want to simplify an algebraic expression? Discuss several reasons.

Practice Use what you learned about simplifying algebraic expressions to complete Exercises 12–14 on page 84.

Key Vocabulary 🔊
like terms, *p. 82*
simplest form, *p. 82*

Parts of an algebraic expression are called *terms*. **Like terms** are terms that have the same variables raised to the same exponents. Constant terms are also like terms. To identify terms and like terms in an expression, first write the expression as a sum of its terms.

EXAMPLE 1 Identifying Terms and Like Terms

Identify the terms and like terms in each expression.

a. $9x - 2 + 7 - x$
Rewrite as a sum of terms.

$$9x + (-2) + 7 + (-x)$$

Terms: $9x, \quad -2, \quad 7, \quad -x$

Like terms: $9x$ and $-x$, -2 and 7

b. $z^2 + 5z - 3z^2 + z$
Rewrite as a sum of terms.

$$z^2 + 5z + (-3z^2) + z$$

Terms: $z^2, \quad 5z, \quad -3z^2, \quad z$

Like terms: z^2 and $-3z^2$, $5z$ and z

An algebraic expression is in **simplest form** when it has no like terms and no parentheses. To *combine* like terms that have variables, use the Distributive Property to add or subtract the coefficients.

EXAMPLE 2 Simplifying an Algebraic Expression

Study Tip ✏️
To subtract a variable term, add the term with the opposite coefficient.

Simplify $\frac{3}{4}y + 12 - \frac{1}{2}y - 6$.

$$\frac{3}{4}y + 12 - \frac{1}{2}y - 6 = \frac{3}{4}y + 12 + \left(-\frac{1}{2}y\right) + (-6) \qquad \text{Rewrite as a sum.}$$

$$= \frac{3}{4}y + \left(-\frac{1}{2}y\right) + 12 + (-6) \qquad \text{Commutative Property of Addition}$$

$$= \left[\frac{3}{4} + \left(-\frac{1}{2}\right)\right]y + 12 + (-6) \qquad \text{Distributive Property}$$

$$= \frac{1}{4}y + 6 \qquad \text{Combine like terms.}$$

● On Your Own

Now You're Ready
Exercises 5–10
and 12–17

Identify the terms and like terms in the expression.

1. $y + 10 - \frac{3}{2}y$ **2.** $2r^2 + 7r - r^2 - 9$ **3.** $7 + 4p - 5 + p + 2q$

Simplify the expression.

4. $14 - 3z + 8 + z$ **5.** $2.5x + 4.3x - 5$ **6.** $\frac{3}{8}b - \frac{3}{4}b$

Laurie's Notes

Introduction

Connect
- **Yesterday:** Students explored simplifying algebraic expressions by evaluating two expressions for more than one value of the variable.
- **Today:** Students will simplify algebraic expressions by combining like terms.

Motivate
- **?** Ask students if they have ever heard the phrase "you can't add apples and oranges."
- Some students may have heard this phrase before. Ask them what they think the phrase means. They might talk about needing a common denominator.

Words of Wisdom
- Students often have difficulty with simplifying algebraic expressions. They must be comfortable with integer operations and be able to apply the Commutative and Distributive Properties. For example: $5x + 7 - 3x$ can be rewritten as $5x + 7 + (-3x) = 5x + (-3x) + 7$.

Lesson Notes

Discuss
- **Attend to Precision: Like terms** are also referred to as *similar terms*. Be sure to note that in the definition of like terms, the variables are raised to the same exponents.

Example 1
- Terms are separated by addition. The expression $9x - 2 + 7 - x$ can be written as $9x + (-2) + 7 + (-x)$, so it has four terms. This form will help students simplify because they can see the sign associated with each term.
- **Common Error:** When identifying and writing the terms, make sure students include the sign of the term.
- **Attend to Precision:** Make sure students understand that the coefficient of $-x$ is -1. Similarly, the exponent of the variable in the terms $5z$ and z is 1.

Example 2
- **?** "What do you call the number that is multiplied by the variable?" coefficient
- Discuss what it means to write an algebraic expression in simplest form.
- Ask students to identify the coefficient of each term. Identify the constant terms.
- Remind students about the Commutative and Distributive Properties.
- Have students show the step that uses the Distributive Property until they become proficient.

On Your Own
- Check that students have not forgotten to include the sign of the term.

Goal Today's lesson is to simplify algebraic expressions.

Technology for the Teacher

Dynamic Classroom

Lesson Tutorials
Lesson Plans
Answer Presentation Tool

Extra Example 1
Identify the terms and like terms in each expression.
- **a.** $3y - 2 - 4y + 6$
  Terms: $3y$, -2, $-4y$, 6
  Like terms: $3y$ and $-4y$, -2 and 6
- **b.** $w + 5w^2 + 2w^2 - 7w$
  Terms: w, $5w^2$, $2w^2$, $-7w$
  Like terms: w and $-7w$, $5w^2$ and $2w^2$

Extra Example 2
Simplify each expression.
- **a.** $8u + 5u - 7u$ $6u$
- **b.** $6d - 5 - 4d + 6$ $2d + 1$

On Your Own

1. Terms: y, 10, $-\frac{3}{2}y$
   Like terms: y and $-\frac{3}{2}y$

2. Terms: $2r^2$, $7r$, $-r^2$, -9
   Like terms: $2r^2$ and $-r^2$

3. Terms: 7, $4p$, -5, p, $2q$
   Like terms: 7 and -5, $4p$ and p

4. $-2z + 22$ 5. $6.8x - 5$

6. $-\frac{3}{8}b$

English learners will benefit from understanding that a term is a number, a variable, or the product of a number and a variable. Like terms are terms that have identical variable parts.

3 and 16 are like terms because they contain no variable.

$4x$ and $7x$ are like terms because they have the same variable x.

$5a$ and $5b$ are not like terms because they have different variables.

Extra Example 3

Simplify $12g + 4 - 5g$. $7g + 4$

 On Your Own

7. $3q - 1$

8. $5g - 8$

9. $-3x + 8$

Extra Example 4

Each person in Example 4 buys a ticket, a small drink, and a small popcorn. Write an expression in simplest form that represents the amount of money the group spends at the movies. $12.25x$

 On Your Own

10. $14x$;

$7.50x + 3.50x + 3x$

$= (7.50 + 3.50 + 3)x$

$= 14x$

Laurie's Notes

Example 3

- Students have not had a lot of practice with a fractional factor in the Distributive Property.

- **Teaching Tip:** Use arrows to show the $-\frac{1}{2}$ being distributed over the $6n$ and the 4.

$$-\frac{1}{2}(6n + 4)$$

On Your Own

- Students should write the original expression, followed by each step in the simplifying process.
- **Common Error:** In Questions 7, 8, and 9, students often forget to distribute the constant over *both* of the terms inside the parentheses.
- **Neighbor Check:** Have students work independently and then have their neighbors check their work. Have students discuss any discrepancies.

Example 4

- Discuss the information provided in the side column. You might ask how these prices compare to the prices at a local movie theater.
- Work through the example with students. Students are writing an algebraic expression, so it is important to identify what the variable represents in the problem. Do not skip this step!
- When you finish the problem, ask students what the total cost would be for a group of 4 people. $\$14.25 \times 4 = \$57$

On Your Own

- Ask a student to share his or her answer.

Closure

- Simplify the following algebraic expressions.
  a. $5x - 8 + 2x^2 + 7x$ $2x^2 + 12x - 8$
  b. $4n + 6(n - 4)$ $10n - 24$

EXAMPLE ③ **Simplifying an Algebraic Expression**

Simplify $-\dfrac{1}{2}(6n + 4) + 2n.$

$$-\frac{1}{2}(6n + 4) + 2n = -\frac{1}{2}(6n) + \left(-\frac{1}{2}\right)(4) + 2n \qquad \text{Distributive Property}$$

$$= -3n + (-2) + 2n \qquad \text{Multiply.}$$

$$= -3n + 2n + (-2) \qquad \begin{array}{l}\text{Commutative Property} \\ \text{of Addition}\end{array}$$

$$= (-3 + 2)n + (-2) \qquad \text{Distributive Property}$$

$$= -n - 2 \qquad \text{Simplify.}$$

On Your Own

Now You're Ready
Exercises 18–20

Simplify the expression.

7. $3(q + 1) - 4$ **8.** $-2(g + 4) + 7g$ **9.** $7 - 4\left(\dfrac{3}{4}x - \dfrac{1}{4}\right)$

EXAMPLE ④ **Real-Life Application**

Each person in a group buys a ticket, a medium drink, and a large popcorn. Write an expression in simplest form that represents the amount of money the group spends at the movies. Interpret the expression.

Evening Tickets $7.50

REFRESHMENTS

Drinks
Small $1.75
Medium $2.75
Large $3.50

Popcorn
Small $3.00
Large $4.00

Words   Each is   each medium is and each large is $4.
ticket $7.50,   drink $2.75,   popcorn

Variable   The same number of each item is purchased. So, x can represent the number of tickets, the number of medium drinks, and the number of large popcorns.

Expression $7.50\,x$ $+$ $2.75\,x$ $+$ $4\,x$

$$7.50x + 2.75x + 4x = (7.50 + 2.75 + 4)x \qquad \text{Distributive Property}$$

$$= 14.25x \qquad \text{Add coefficients.}$$

∴ The expression $14.25x$ indicates that the total cost per person is $14.25.

On Your Own

10. **WHAT IF?** Each person buys a ticket, a large drink, and a small popcorn. How does the expression change? Explain.

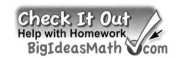

Vocabulary and Concept Check

1. **WRITING** Explain how to identify the terms of $3y - 4 - 5y$.

2. **WRITING** Describe how to combine like terms in the expression $3n + 4n - 2$.

3. **VOCABULARY** Is the expression $3x + 2x - 4$ in simplest form? Explain.

4. **REASONING** Which algebraic expression is in simplest form? Explain.

$5x - 4 + 6y$	$4x + 8 - x$
$3(7 + y)$	$12n - n$

Practice and Problem Solving

Identify the terms and like terms in the expression.

① 5. $t + 8 + 3t$ 6. $3z + 4 + 2 + 4z$ 7. $2n - n - 4 + 7n$

8. $-x - 9x^2 + 12x^2 + 7$ 9. $1.4y + 5 - 4.2 - 5y^2 + z$ 10. $\frac{1}{2}s - 4 + \frac{3}{4}s + \frac{1}{8} - s^3$

11. **ERROR ANALYSIS** Describe and correct the error in identifying the like terms in the expression.

$$3x - 5 + 2x^2 + 9x = 3x + 2x^2 + 9x - 5$$

Like Terms: $3x$, $2x^2$, and $9x$

Simplify the expression.

② 12. $12g + 9g$ 13. $11x + 9 - 7$ 14. $8s - 11s + 6$

15. $4.2v - 5 - 6.5v$ 16. $8 + 4a + 6.2 - 9a$ 17. $\frac{2}{5}y - 4 + 7 - \frac{9}{10}y$

③ 18. $4(b - 6) + 19$ 19. $4p - 5(p + 6)$ 20. $-\frac{2}{3}(12c - 9) + 14c$

21. **HIKING** On a hike, each hiker carries the items shown. Write an expression in simplest form that represents the weight carried by x hikers. Interpret the expression.

4.6 lb

3.4 lb   2.2 lb

Assignment Guide and Homework Check

Level	Day 1 Activity Assignment	Day 2 Lesson Assignment	Homework Check
Basic	12–14, 30–32	1–4, 5–11 odd, 15–23 odd	7, 17, 19, 21
Average	12–14, 30–32	1–4, 7–9, 11, 15, 16, 18, 19–27 odd, 28	9, 19, 23, 27
Advanced	12–14, 30–32	1–4, 6–10 even, 11, 16–28 even	8, 10, 18, 20, 26
Accelerated	1–4, 6–10 even, 11–14, 16–28 even, 30–32		8, 10, 18, 20, 26

Common Errors

- **Exercises 5–10** When identifying and writing terms, make sure students include the sign of the term. Students may find it helpful to write the original problem using addition. For example, $3n - 10 + 2n = 3n + (-10) + 2n$.
- **Exercises 5–10** Students may confuse like variables with like terms. Remind them that the same variables must be raised to the same exponents for terms to be like terms. The terms $3x$ and $7x^2$ are not like terms because one has an exponent of 1 and the other has an exponent of 2.
- **Exercises 13–20** The subtraction operation can confuse students. It is not obvious to them why it is okay to rewrite $6t - 24 + 3t$ as $6t + 3t - 24$. Tell students to write the original problem using addition, and then use the Commutative Property.
$$6t - 24 + 3t = 6t + (-24) + 3t = 6t + 3t + (-24)$$
$$= 6t + 3t - 24 = 9t - 24$$
- **Exercises 18–20** Students often forget to distribute the constant over *both* of the terms inside the parentheses. Remind them of the Distributive Property, $a(b + c) = ab + ac$.

3.1 Record and Practice Journal

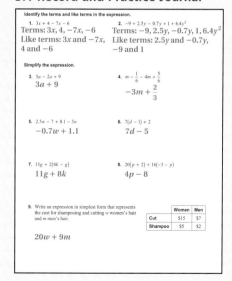

Vocabulary and Concept Check

1. Terms of an expression are separated by addition. Rewrite the expression as $3y + (-4) + (-5y)$. The terms in the expression are $3y$, -4, and $-5y$.

2. Use the Distributive Property to add the coefficients of the like terms $3n$ and $4n$.
$$3n + 4n - 2 = (3 + 4)n - 2$$
$$= 7n - 2$$

3. no; The like terms $3x$ and $2x$ should be combined.
$$3x + 2x - 4 = (3 + 2)x - 4$$
$$= 5x - 4$$

4. $5x - 4 + 6y$; There are no like terms or parentheses in the expression.

Practice and Problem Solving

5. Terms: t, 8, $3t$; Like terms: t and $3t$

6. Terms: $3z$, 4, 2, $4z$; Like terms: $3z$ and $4z$, 4 and 2

7. Terms: $2n$, $-n$, -4, $7n$; Like terms: $2n$, $-n$, and $7n$

8. Terms: $-x$, $-9x^2$, $12x^2$, 7; Like terms: $-9x^2$ and $12x^2$

9. Terms: $1.4y$, 5, -4.2, $-5y^2$, z; Like terms: 5 and -4.2

10. Terms: $\frac{1}{2}s$, -4, $\frac{3}{4}s$, $\frac{1}{8}$, $-s^3$; Like terms: $\frac{1}{2}s$ and $\frac{3}{4}s$, -4 and $\frac{1}{8}$

11. $2x^2$ is not a like term, because x is squared. The like terms are $3x$ and $9x$.

12. $21g$

13. $11x + 2$

14. $-3s + 6$

15. $-2.3v - 5$

16. $14.2 - 5a$

17. $3 - \frac{1}{2}y$

Practice and Problem Solving

18. $4b - 5$ **19.** $-p - 30$

20. $6c + 6$

21. $10.2x$; The weight carried by each hiker is 10.2 pounds.

22. -9; -9; *Sample answer:* Simplifying the expression first is easier because you only have to substitute once instead of substituting three times.

23. yes; Both expressions simplify to $11x^2 + 3y$.

24–28. See Additional Answers.

29. See *Taking Math Deeper.*

Fair Game Review

30. 14.5 in., 14.8 in., 15 in., 15.3 in., 15.8 in.

31. 0.52 m, 0.545 m, 0.55 m, 0.6 m, 0.65 m

32. C

Mini-Assessment

Identify the terms and like terms in the expression.

1. $4r + 2 - 6 + 3r$

Terms: $4r$, 2, -6, $3r$
Like terms: $4r$ and $3r$, 2 and -6

2. $5h^2 - 3h^2 - 4h + 3h + 7$

Terms: $5h^2$, $-3h^2$, $-4h$, $3h$, 7
Like terms: $5h^2$ and $-3h^2$, $-4h$ and $3h$

Simplify the expression.

3. $6m + 7 - 3m - 1$ $3m + 6$

4. $3(5b + 2) - 4$ $15b + 2$

5. Write an expression in simplest form that represents the perimeter of the polygon. $(3x + 9)$ m

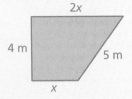

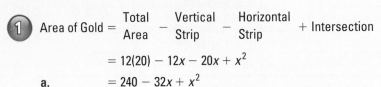

Taking Math Deeper

Exercise 29

In this problem, students have to realize that when you subtract the two red strips, you have subtracted their intersection twice.

① Area of Gold = $\dfrac{\text{Total}}{\text{Area}} - \dfrac{\text{Vertical}}{\text{Strip}} - \dfrac{\text{Horizontal}}{\text{Strip}} + \text{Intersection}$

$= 12(20) - 12x - 20x + x^2$

a. $= 240 - 32x + x^2$

Notice that the intersection of the two red strips is subtracted twice, so it must be added back into the expression once.

② When $x = 3$, the area of the gold foil is

$$\text{Area} = 240 - 32x + x^2$$
$$= 240 - 32(3) + 3^2$$
$$= 240 - 96 + 9$$
b. $\qquad = 153 \text{ in.}^2$

England is only part of the UK

③ **c.** This pattern is used as the flag of England.

| England | Historical Greece | Georgia |

Note: The flag for the United Kingdom (England, Scotland, Wales, and Northern Ireland) is different and has a criss-cross pattern.

Reteaching and Enrichment Strategies

If students need help...	If students got it...
Resources by Chapter • Practice A and Practice B • Puzzle Time Record and Practice Journal Practice Differentiating the Lesson Lesson Tutorials Skills Review Handbook	Resources by Chapter • Enrichment and Extension • Technology Connection Start the next section

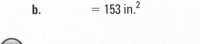

22. **STRUCTURE** Evaluate the expression $-8x + 5 - 2x - 4 + 5x$ when $x = 2$ before and after simplifying. Which method do you prefer? Explain.

23. **REASONING** Are the expressions $8x^2 + 3(x^2 + y)$ and $7x^2 + 7y + 4x^2 - 4y$ equivalent? Explain your reasoning.

24. **CRITICAL THINKING** Which solution shows a correct way of simplifying $6 - 4(2 - 5x)$? Explain the errors made in the other solutions.

(A) $6 - 4(2 - 5x) = 6 - 4(-3x) = 6 + 12x$

(B) $6 - 4(2 - 5x) = 6 - 8 + 20x = -2 + 20x$

(C) $6 - 4(2 - 5x) = 2(2 - 5x) = 4 - 10x$

(D) $6 - 4(2 - 5x) = 6 - 8 - 20x = -2 - 20x$

25. **BANNER** Write an expression in simplest form that represents the area of the banner.

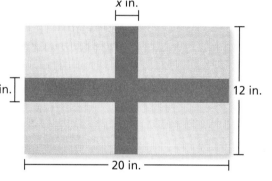

3 ft

(3 + x) ft

	Car	Truck
Wash	$8	$10
Wax	$12	$15

26. **CAR WASH** Write an expression in simplest form that represents the earnings for washing and waxing x cars and y trucks.

MODELING Draw a diagram that shows how the expression can represent the area of a figure. Then simplify the expression.

27. $5(2 + x + 3)$

28. $(4 + 1)(x + 2x)$

29. You apply gold foil to a piece of red poster board to make the design shown.

   a. Write an expression in simplest form that represents the area of the gold foil.

   b. Find the area of the gold foil when $x = 3$.

   c. The pattern at the right is called "St. George's Cross." Find a country that uses this pattern as its flag.

   x in.

   x in.

   12 in.

   20 in.

Fair Game Review What you learned in previous grades & lessons

Order the lengths from least to greatest. *(Skills Review Handbook)*

30. 15 in., 14.8 in., 15.8 in., 14.5 in., 15.3 in.

31. 0.65 m, 0.6 m, 0.52 m, 0.55 m, 0.545 m

32. **MULTIPLE CHOICE** A bird's nest is 12 feet above the ground. A mole's den is 12 inches below the ground. What is the difference in height of these two positions? *(Section 1.3)*

(A) 24 in.   (B) 11 ft   (C) 13 ft   (D) 24 ft

Essential Question How can you use algebra tiles to add or subtract algebraic expressions?

Key: = variable   = −variable   = zero pair

= 1   = −1   = zero pair

1 ACTIVITY: Writing Algebraic Expressions

Work with a partner. Write an algebraic expression shown by the algebra tiles.

a.

b.

c.

d.

2 ACTIVITY: Adding Algebraic Expressions

Work with a partner. Write the sum of two algebraic expressions modeled by the algebra tiles. Then use the algebra tiles to simplify the expression.

a.

b.

Linear Expressions

In this lesson, you will

- apply properties of operations to add and subtract linear expressions.
- solve real-life problems.

c.

d.

Laurie's Notes

Introduction

Applying Mathematical Practices

- **Reason Abstractly and Quantitatively:** Algebra tiles help students make sense of algebraic expressions by modeling them and finding sums and differences. Algebra tiles are a concrete representation, deepening student understanding of the meaning of each expression.

Discuss

- **FYI:** Show students a collection of yellow integer-tiles and one green variable-tile. Define the yellow integer-tile as having dimensions 1 by 1 with an area of 1 square unit and the variable-tile as having dimensions 1 by x with an area of x square units. The Record and Practice Journal has algebra tiles and they are also available commercially. Be sure to point out to students that the variable-tile is NOT an integral length, meaning you should not be able to *measure* the length of the variable-tile by lining up yellow integer-tiles. The length of the tile is a variable—x!
- Display a collection of tiles, say 1 variable-tile, 3 yellow integer-tiles ($+3$) and 2 red integer-tiles (-2). Say, "These algebra tiles represent an algebraic expression and just as you simplify algebraic expressions, you are going to simplify expressions modeled by the algebra tiles."

Activity Notes

Activity 1

- **Management Tip:** Distribute a set of algebra tiles to each pair of students. Presort them in baggies for easy distribution and collection.
- Even though the collection of tiles is shown, encourage students to make the collection shown with their own algebra tiles.
- Remind students that any letter can be used to represent a variable.
- Some students may write expressions that represent each algebra tile such as "$x + x - 1 - 1$" for part (b). Ask them "Is your expression in simplest form? If not, how can you write it in simplest form?"
- Ask for volunteers to share their results.
- Students may write $3 + x$ for part (a). Explain that it is more common to state the x-term first, as $x + 3$. The Commutative Property of Addition assures that $3 + x$ and $x + 3$ are equivalent.

Activity 2

- Have each partner represent one of the expressions using their tiles. To add, have them combine their tiles together in the common work space. Then simplify by removing any zero pairs.
- Note that all expressions in Activity 2 have positive coefficients.
- Ask for volunteers to explain how they used the tiles to simplify.
- **Reason Abstractly and Quantitatively:** Handling the tiles helps students understand that $x + x = 2x$ and not x^2. Students who have worked with algebra tiles should not make that mistake.

What Your Students Will Learn

- Use a vertical or horizontal method and apply properties of operations to add or subtract linear expressions.

Previous Learning

Students should know how to simplify algebraic expressions.

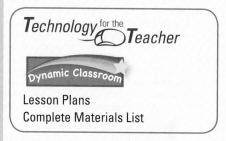

Lesson Plans
Complete Materials List

3.2 Record and Practice Journal

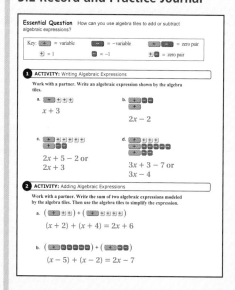

Have students verbally describe the difference of two algebraic expressions, such as

$$2x - (x + 1).$$

This expression is read as "two *x* minus the quantity *x* plus one." Remind students that "the quantity *x* plus one" means that *x* and 1 are grouped together.

3.2 Record and Practice Journal

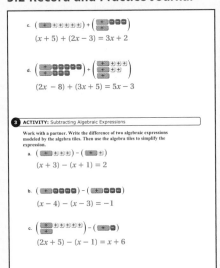

c. (...) + (...)
$(x + 5) + (2x - 3) = 3x + 2$

d. (...) + (...)
$(2x - 8) + (3x + 5) = 5x - 3$

3 ACTIVITY: Subtracting Algebraic Expressions

Work with a partner. Write the difference of two algebraic expressions modeled by the algebra tiles. Then use the algebra tiles to simplify the expression.

a. (...) – (...)
$(x + 3) - (x + 1) = 2$

b. (...) – (...)
$(x - 4) - (x - 3) = -1$

c. (...) – (...)
$(2x + 5) - (x - 1) = x + 6$

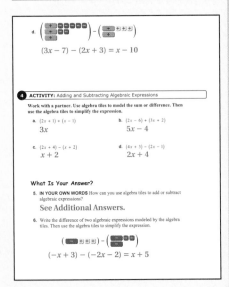

d. (...) – (...)
$(3x - 7) - (2x + 3) = x - 10$

4 ACTIVITY: Adding and Subtracting Algebraic Expressions

Work with a partner. Use algebra tiles to model the sum or difference. Then use the algebra tiles to simplify the expression.

a. $(2x + 1) + (x - 1)$
$3x$

b. $(2x - 6) + (3x + 2)$
$5x - 4$

c. $(2x + 4) - (x + 2)$
$x + 2$

d. $(4x + 3) - (2x - 1)$
$2x + 4$

What Is Your Answer?

5. **IN YOUR OWN WORDS** How can you use algebra tiles to add or subtract algebraic expressions?
   See Additional Answers.

6. Write the difference of two algebraic expressions modeled by the algebra tiles. Then use the algebra tiles to simplify the expression.
   (...) – (...)
   $(-x + 3) - (-2x - 2) = x + 5$

Laurie's Notes

Activity 3

- Review the meaning of subtraction and how it is performed using integer tiles. Begin by representing the first expression on the work space. Students find the difference by removing the algebra tiles in the second expression from the algebra tiles in the first expression.
- For parts (c) and (d), students will need to add zero pairs in order to subtract the second expression.
- Ask for volunteers to share their results. This should include modeling at least one of the problems at the document camera or overhead projector so that the language of the subtraction is heard.
- **Reason Abstractly and Quantitatively:** Using the algebra tiles, it should be clear that $3x - 2x = 1x$. Without the algebra tiles, students may incorrectly reason that $3x - 2x = 1$. They can lose track of what the expressions $3x$ and $2x$ represent and see them only as symbols, not quantities.

Activity 4

- Now students start with the expressions and create a model using algebra tiles. To model $x - 1$ in part (a), students can think of the equivalent expression $x + (-1)$.
- If time permits, have students model these problems so that classmates hear the language associated with performing these operations.
- Students may need help with part (d). Model $4x + 3$. When students go to remove $2x - 1$, remind them that $2x - 1$ can be written as $2x + (-1)$.

What Is Your Answer?

- In Question 6, students need to add zero pairs to the first expression in order to "take away" the algebra tiles in the second expression. This is also the first problem that uses the negative variable-tile.

Closure

- Use algebra tiles to create the following model and simplify it.

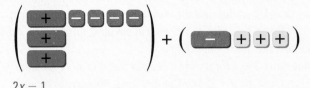

$$2x - 1$$

3 ACTIVITY: Subtracting Algebraic Expressions

Math Practice

Use Expressions

What do the tiles represent? How does this help you write an expression?

Work with a partner. Write the difference of two algebraic expressions modeled by the algebra tiles. Then use the algebra tiles to simplify the expression.

a. $\left(\boxed{+} \; \boxed{+} \boxed{+} \boxed{+} \right) - \left(\boxed{+} \; \boxed{+} \right)$

b. $\left(\boxed{+} \; \boxed{-} \boxed{-} \boxed{-} \boxed{-} \right) - \left(\boxed{+} \; \boxed{-} \boxed{-} \boxed{-} \right)$

c. $\left(\begin{array}{l} \boxed{+} \\ \boxed{+} \end{array} \; \boxed{+} \boxed{+} \boxed{+} \boxed{+} \boxed{+} \right) - \left(\boxed{+} \; \boxed{-} \right)$

d. $\left(\begin{array}{l} \boxed{+} \\ \boxed{+} \\ \boxed{+} \end{array} \; \begin{array}{l} \boxed{-} \boxed{-} \boxed{-} \boxed{-} \boxed{-} \\ \boxed{-} \boxed{-} \end{array} \right) - \left(\begin{array}{l} \boxed{+} \\ \boxed{+} \end{array} \; \boxed{+} \boxed{+} \boxed{+} \right)$

4 ACTIVITY: Adding and Subtracting Algebraic Expressions

Work with a partner. Use algebra tiles to model the sum or difference. Then use the algebra tiles to simplify the expression.

a. $(2x + 1) + (x - 1)$

b. $(2x - 6) + (3x + 2)$

c. $(2x + 4) - (x + 2)$

d. $(4x + 3) - (2x - 1)$

What Is Your Answer?

5. **IN YOUR OWN WORDS** How can you use algebra tiles to add or subtract algebraic expressions?

6. Write the difference of two algebraic expressions modeled by the algebra tiles. Then use the algebra tiles to simplify the expression.

$\left(\boxed{-} \; \boxed{+} \boxed{+} \boxed{+} \right) - \left(\begin{array}{l} \boxed{-} \\ \boxed{-} \end{array} \; \boxed{-} \boxed{-} \boxed{-} \right)$

Practice ➤ Use what you learned about adding and subtracting algebraic expressions to complete Exercises 6 and 7 on page 90.

Check It Out
Lesson Tutorials
BigIdeasMath ✓com

Key Vocabulary ◀))
linear expression,
  p. 88

A **linear expression** is an algebraic expression in which the exponent of the variable is 1.

Linear Expressions	$-4x$	$3x + 5$	$5 - \dfrac{1}{6}x$
Nonlinear Expressions	x^2	$-7x^3 + x$	$x^5 + 1$

You can use a vertical or a horizontal method to add linear expressions.

EXAMPLE ① **Adding Linear Expressions**

Find each sum.

a. $(x - 2) + (3x + 8)$

Vertical method: Align
like terms vertically and add.

$$\begin{array}{r} x - 2 \\ + 3x + 8 \\ \hline 4x + 6 \end{array}$$

b. $(-4y + 3) + (11y - 5)$

Horizontal method: Use properties of operations to group
like terms and simplify.

$$(-4y + 3) + (11y - 5) = -4y + 3 + 11y - 5 \qquad \text{Rewrite the sum.}$$

$$= -4y + 11y + 3 - 5 \qquad \text{Commutative Property of Addition}$$

$$= (-4y + 11y) + (3 - 5) \qquad \text{Group like terms.}$$

$$= 7y - 2 \qquad \text{Combine like terms.}$$

EXAMPLE ② **Adding Linear Expressions**

Find $2(-7.5z + 3) + (5z - 2)$.

$$2(-7.5z + 3) + (5z - 2) = -15z + 6 + 5z - 2 \qquad \text{Distributive Property}$$

$$= -15z + 5z + 6 - 2 \qquad \text{Commutative Property of Addition}$$

$$= -10z + 4 \qquad \text{Combine like terms.}$$

 On Your Own

Now You're Ready
Exercises 8–16

Find the sum.

1. $(x + 3) + (2x - 1)$

2. $(-8z + 4) + (8z - 7)$

3. $(4 - n) + 2(-5n + 3)$

4. $\dfrac{1}{2}(w - 6) + \dfrac{1}{4}(w + 12)$

◀)) Multi-Language Glossary at BigIdeasMath ✓com

Laurie's Notes

Introduction

Connect

- **Yesterday:** Students used algebra tiles to develop an understanding of how to add and subtract algebraic expressions.
- **Today:** Students will use a horizontal or vertical format to add and subtract linear expressions.

Motivate

- Draw a vertical line on the middle of your board. On the left write "These are" and on the right write "These are not." On each side write examples of expressions that are linear (left side) and are not linear (right side).
- Explain that you are not giving names to either side yet. You are just trying to have them be good detectives in thinking about what characteristics they see.
- **?** "Can you give other examples of what you think would be on the left or right?"
- **?** "What feature(s) distinguish the expressions that **are**, from the expressions that **are not**?" Listen for a reference to exponents (right side) and the lack of exponents (left side).

Discuss

- Tell students that the expressions on the left are examples of **linear expressions**, which are algebraic expressions in which the exponent of the variable is 1. This is not a precise definition because $\frac{1}{x}$ has a variable with an exponent of 1 and it is not a linear expression. However, this description is appropriate for this grade level.

Lesson Notes

Example 1

- **Connection:** When you add (or subtract) whole numbers, you use the place values of the numbers. The same is true when you add (or subtract) decimals—lining up the decimal points assures that this happens. Lining up place values is similar to lining up like terms. Make this connection for students as you begin to work these problems.
- Using the vertical method, students should see the connection to adding two whole numbers.
- **Teaching Tip:** Before adding, rewrite $x - 2$ as $x + (-2)$.
- The Commutative Property of Addition is used to change the order of the terms so that like terms are adjacent to one another.

Example 2

- **Attend to Precision:** Ask for a volunteer to read the problem. Listen for, "Two times the quantity negative seven point five z plus 3, plus the quantity five z minus two." Students should be able to read this.
- **?** "What is the first step?" Simplify $2(-7.5z + 3)$.

Lesson Tutorials
Lesson Plans
Answer Presentation Tool

Extra Example 1

Find each sum.

a. $(-2x + 2) + (4x - 7)$ $\quad 2x - 5$

b. $(7y - 5) + (3y + 8)$ $\quad 10y + 3$

Extra Example 2

Find $(7w - 6) + 5(-2.4w + 1)$

$-5w - 1$

On Your Own

1. $3x + 2$ **2.** -3

3. $-11n + 10$

4. $\frac{3}{4}w$

Extra Example 3

Find each difference.

a. $(-3x + 7) - (4x - 8)$ $-7x + 15$

b. $-3(2y - 9) - (5y + 4)$ $-11y + 23$

Extra Example 4

The original price of a coffee table is *d* dollars. You use a coupon and buy the table for $(d - 4)$ dollars. You paint the table and sell it for $(3d + 1)$ dollars. Write an expression that represents your earnings from buying and selling the coffee table. Interpret the expression.

$(3d + 1) - (d - 4)$; You earn $(2d + 5)$ dollars.

On Your Own

5. $2m - 15$

6. $-8c - 25$

7. $4

Laurie's Notes

Example 3

? "How do you think you subtract linear expressions?" Subtract like terms.

? Write part (a) and ask, "Can you subtract the quantity $(-x + 6)$ by removing the parentheses? No, you must subtract each term in the linear expression. So, you add the opposite.

- My experience is that students make more errors when subtracting linear expressions using the vertical method unless they take the time to rewrite the problem as shown where *adding the opposite* is obvious. As stated in the Study Tip, to find the opposite of a linear expression you can multiply the expression by -1.

- **Reason Abstractly and Quantitatively:** It may be helpful to rewrite $(5x + 6) - (-x + 6)$ as $(5x + 6) + [-(-x + 6)]$ and then $(5x + 6) + (-1)(-x + 6)$. This is the Multiplication Property of -1.

- Write part (b). This may be easier for students to understand than part (a) because the constant 2 is written in the problem, whereas the constant 1 in part (a) is not written. When students rewrite the problem as *add the opposite*, they can see that -2 needs to be distributed.

- **Look for and Make Use of Structure:** Using the Commutative Property to rewrite $7y + 5 - 8y + 6$ as $7y - 8y + 5 + 6$ is not obvious to all students. Take time to probe for understanding. Subtracting $8y$ is the same as adding the opposite of $8y$. You may need to work through these extra steps so students make sense of how the order of the terms can be changed.

? "Do you prefer the vertical or horizontal method? Why?" Answers will vary.

Example 4

- Have a quick discussion about how to calculate the earnings when buying something and reselling it.

- Ask for a volunteer to read the problem. Remind students that the variable *d* is unknown, and you are not writing an expression for the selling price.

? "What is the value of the coupon? Explain." $2, because you purchase the hat for $(d - 2)$ dollars.

- Write the verbal model and substitute the linear expressions.

? "This is a subtraction problem. What is our next step?" Add the opposite.

? "If you pay $(d - 2)$ dollars for an item and earn $(d - 2)$ dollars back, what does this mean?" If students are having difficulty interpreting this problem, substitute a value for *d*, such as $20, then explain.

- Students can verify that the selling price of $(2d - 4)$ dollars is twice that of the purchase price $(d - 2)$ by multiplying by 2. You could decide to have a quick review of factoring by factoring 2 out of the selling price, or wait and do this as an introduction to 3.2 Extension.

Closure

- **Exit Ticket:** Find the sum or difference.

  $2(3x - 4) + (2x - 5)$ $8x - 13$ $2(3x - 4) - (2x - 5)$ $4x - 3$

To subtract one linear expression from another, add the opposite of each term in the expression. You can use a vertical or a horizontal method.

EXAMPLE ③ **Subtracting Linear Expressions**

Study Tip

To find the opposite of a linear expression, you can multiply the expression by −1.

Find each difference.

a. $(5x + 6) - (-x + 6)$ **b.** $(7y + 5) - 2(4y - 3)$

a. Vertical method: Align like terms vertically and subtract.

$$
\begin{array}{r}
(5x + 6) \\
- (-x + 6)
\end{array}
\quad \boxed{\text{Add the opposite.}} \quad
\begin{array}{r}
5x + 6 \\
+\ \ x - 6 \\
\hline
6x
\end{array}
$$

b. Horizontal method: Use properties of operations to group like terms and simplify.

$$
\begin{aligned}
(7y + 5) - 2(4y - 3) &= 7y + 5 - 8y + 6 & &\text{Distributive Property} \\
&= 7y - 8y + 5 + 6 & &\text{Commutative Property of Addition} \\
&= (7y - 8y) + (5 + 6) & &\text{Group like terms.} \\
&= -y + 11 & &\text{Combine like terms.}
\end{aligned}
$$

EXAMPLE ④ **Real-Life Application**

The original price of a cowboy hat is d dollars. You use a coupon and buy the hat for $(d - 2)$ dollars. You decorate the hat and sell it for $(2d - 4)$ dollars. Write an expression that represents your earnings from buying and selling the hat. Interpret the expression.

$$
\begin{aligned}
\text{earnings} &= \text{selling price} - \text{purchase price} & &\text{Use a model.} \\
&= (2d - 4) - (d - 2) & &\text{Write the difference.} \\
&= (2d - 4) + (-d + 2) & &\text{Add the opposite.} \\
&= 2d - d - 4 + 2 & &\text{Group like terms.} \\
&= d - 2 & &\text{Combine like terms.}
\end{aligned}
$$

⋮ You earn $(d - 2)$ dollars. You also paid $(d - 2)$ dollars, so you doubled your money by selling the hat for twice as much as you paid for it.

On Your Own

Exercises 19–24

Find the difference.

5. $(m - 3) - (-m + 12)$ **6.** $-2(c + 2.5) - 5(1.2c + 4)$

7. WHAT IF? In Example 4, you sell the hat for $(d + 2)$ dollars. How much do you earn from buying and selling the hat?

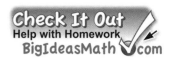

 Vocabulary and Concept Check

VOCABULARY Determine whether the algebraic expression is a linear expression. Explain.

1. $x^2 + x + 1$

2. $-2x - 8$

3. $x - x^4$

4. WRITING Describe two methods for adding or subtracting linear expressions.

5. DIFFERENT WORDS, SAME QUESTION Which is different? Find "both" answers.

Subtract x from $3x - 1$.

Find $3x - 1$ decreased by x.

What is x more than $3x - 1$?

What is the difference of $3x - 1$ and x?

 Practice and Problem Solving

Write the sum or difference of two algebraic expressions modeled by the algebra tiles. Then use the algebra tiles to simplify the expression.

6.

7.

Find the sum.

8. $(n + 8) + (n - 12)$

9. $(7 - b) + (3b + 2)$

10. $(2w - 9) + (-4w - 5)$

11. $(2x - 6) + 4(x - 3)$

12. $5(-3.4k - 7) + (3k + 21)$

13. $(1 - 5q) + 2(2.5q + 8)$

14. $3(2 - 0.9h) + (-1.3h - 4)$

15. $\frac{1}{3}(9 - 6m) + \frac{1}{4}(12m - 8)$

16. $-\frac{1}{2}(7z + 4) + \frac{1}{5}(5z - 15)$

17. BANKING You start a new job. After w weeks, you have $(10w + 120)$ dollars in your savings account and $(45w + 25)$ dollars in your checking account. Write an expression that represents the total in both accounts.

18. FIREFLIES While catching fireflies, you and a friend decide to have a competition. After m minutes, you have $(3m + 13)$ fireflies and your friend has $(4m + 6)$ fireflies.

  a. Write an expression that represents the number of fireflies you and your friend caught together.

  b. The competition ends after 5 minutes. Who has more fireflies?

Assignment Guide and Homework Check

Level	Day 1 Activity Assignment	Day 2 Lesson Assignment	Homework Check
Basic	6 and 7, 32–35	1–5, 9–29 odd	9, 13, 17, 21
Average	6 and 7, 32–35	1–5, 9–25 odd, 26–30 even	9, 13, 17, 23, 26
Advanced	6 and 7, 32–35	1–5, 10–24 even, 25, 26–30 even	10, 16, 18, 24, 28
Accelerated	1–7, 10–24 even, 25, 26–30 even, 32–35		10, 16, 18, 24, 28

Common Errors

- **Exercise 30** Students may count the corner tiles twice. Remind them that the corner tiles are the end of one length and the beginning of another, and should not be counted twice.
- **Exercise 31** Students may try to make distance negative. Remind them that distance is always positive.

3.2 Record and Practice Journal

Find the sum or difference.

1. $(x - 2) + (x + 6)$

   $2x + 4$

2. $(2n - 4) - (4n - 3)$

   $-2n - 1$

3. $2(-3y - 1) + (2y + 7)$

   $-4y + 5$

4. $(1 - 3k) - 4(2 + 2.5k)$

   $-13k - 7$

5. $(6g - 9) + \frac{1}{3}(15 - 9g)$

   $3g - 4$

6. $\frac{1}{2}(2r + 4) - \frac{1}{4}(16 - 8r)$

   $3r - 2$

7. You earn $(4x + 12)$ points after completing x levels of a video game and then lose $(2x - 5)$ points. Write an expression that represents the total number of points you have now.

   $2x + 17$

Vocabulary and Concept Check

1. not linear; An exponent of a variable is not equal to 1.

2. linear; The exponent of the variable is equal to 1.

3. not linear; An exponent of a variable is not equal to 1.

4. Vertical method: Align like terms vertically and add or subtract the opposite. Horizontal method: Group like terms using properties of operations and simplify.

5. What is x more than $3x - 1$?; $4x - 1$; $2x - 1$

Practice and Problem Solving

6. *Sample answer:*
   $(2x - 6) + (x + 5) = 3x - 1$

7. *Sample answer:*
   $(2x + 7) - (2x - 4) = 11$

8. $2n - 4$

9. $2b + 9$

10. $-2w - 14$

11. $6x - 18$

12. $-14k - 14$

13. 17

14. $-4h + 2$

15. $m + 1$

16. $-2\frac{1}{2}z - 5$

17. $55w + 145$

18. **a.** $7m + 19$

   b. you

19. $-3g - 4$

20. $9d + 3$

21. $-12y + 20$

22. $14n - 29$

23. $-2c$

24. $x + 10\frac{1}{2}$

25. See Additional Answers.

26. **a.** 7 fireflies per minute

 b. 19 fireflies

27. no; If the variable terms are opposites, the sum is a numerical expression.

28. $8n$

29. $0.25x + 0.15$

30. See *Taking Math Deeper.*

31. $|x - 3|$, or equivalently $|-x + 3|$; 0; 6

Fair Game Review

32. $-\dfrac{7}{15}$ 33. $\dfrac{2}{5}$

34. $2\dfrac{2}{15}$ 35. D

Mini-Assessment

Find the sum or difference.

1. $(5m + 3) + (-8m + 8)$ $-3m + 11$

2. $(4 - x) + (2x + 5)$ $x + 9$

3. $(8x - 3) - (2x + 6)$ $6x - 9$

4. $(2 - 7y) - 3(y - 9)$ $-10y + 29$

5. A rectangle has side lengths $(x + 5)$ meters and $(2x - 1)$ meters. Write an expression in simplest form that represents the perimeter of the rectangle.

 $6x + 8$ meters

Taking Math Deeper

Exercise 30

It is easy to count tiles more than once in this problem and then write an incorrect expression. You can avoid this pitfall by drawing a diagram.

 Let w represent the width of the room in feet. The expression $(w + 10)$ represents the length of the room. Draw the room using 1-foot-by-1-foot tiles.

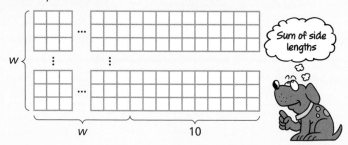

 Using the diagram, you can see that if you find the sum $w + w + (w + 10) + (w + 10)$, then you will count each corner tile twice. So, you must subtract 4 from this sum.

$$w + w + (w + 10) + (w + 10) - 4$$
$$= w + w + w + w + 10 + 10 - 4$$
$$= 4w + 16$$

 Another way to find the sum is to keep track when you are counting each corner tile. Starting at the bottom and adding side lengths counterclockwise, you can write

$$w + 10 + (w - 1) + (w + 9) + (w - 2)$$
$$= w + w + w + w + 10 - 1 + 9 - 2$$
$$= 4w + 16.$$

So, an expression for the number of tiles along the outside of the room is $4w + 16$.

Project

Research the costs of at least 3 different types of floor tiles. Choose a reasonable value for the width and find how much more it would cost to tile the room with the most expensive tile than with the least expensive tile.

Reteaching and Enrichment Strategies

If students need help. . .	If students got it. . .
Resources by Chapter • Practice A and Practice B • Puzzle Time Record and Practice Journal Practice Differentiating the Lesson Lesson Tutorials Skills Review Handbook	Resources by Chapter • Enrichment and Extension • Technology Connection Start the next section

Find the difference.

19. $(-2g + 7) - (g + 11)$

20. $(6d + 5) - (2 - 3d)$

21. $(4 - 5y) - 2(3.5y - 8)$

22. $(2n - 9) - 5(-2.4n + 4)$

23. $\frac{1}{8}(-8c + 16) - \frac{1}{3}(6 + 3c)$

24. $\frac{3}{4}(3x + 6) - \frac{1}{4}(5x - 24)$

25. ERROR ANALYSIS Describe and correct the error in finding the difference.

$$\begin{aligned}(4m + 9) - 3(2m - 5) &= 4m + 9 - 6m - 15 \\ &= 4m - 6m + 9 - 15 \\ &= -2m - 6\end{aligned}$$

26. STRUCTURE Refer to the expressions in Exercise 18.

    a. How many fireflies are caught each minute during the competition?

    b. How many fireflies are caught before the competition starts?

27. LOGIC Your friend says the sum of two linear expressions is always a linear expression. Is your friend correct? Explain.

28. GEOMETRY The expression $17n + 11$ represents the perimeter (in feet) of the triangle. Write an expression that represents the measure of the third side.

29. TAXI Taxi Express charges $2.60 plus $3.65 per mile, and Cab Cruiser charges $2.75 plus $3.90 per mile. Write an expression that represents how much more Cab Cruiser charges than Taxi Express.

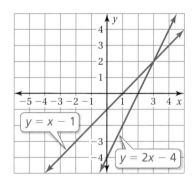

30. MODELING A rectangular room is 10 feet longer than it is wide. One-foot-by-one-foot tiles cover the entire floor. Write an expression that represents the number of tiles along the outside of the room.

31. Reasoning Write an expression in simplest form that represents the vertical distance between the two lines shown. What is the distance when $x = 3$? when $x = -3$?

Fair Game Review What you learned in previous grades & lessons

Evaluate the expression when $x = -\frac{4}{5}$ and $y = \frac{1}{3}$. *(Section 2.2)*

32. $x + y$

33. $2x + 6y$

34. $-x + 4y$

35. MULTIPLE CHOICE What is the surface area of a cube that has a side length of 5 feet? *(Skills Review Handbook)*

    (A) 25 ft^2
    (B) 75 ft^2
    (C) 125 ft^2
    (D) 150 ft^2

Check It Out
Lesson Tutorials
BigIdeasMathcom

Key Vocabulary
factoring an expression, *p. 92*

When **factoring an expression**, you write the expression as a product of factors. You can use the Distributive Property to factor expressions.

EXAMPLE 1 Factoring Out the GCF

Factor $24x - 18$ using the GCF.

Find the GCF of $24x$ and 18 by writing their prime factorizations.

$$24x = 2 \cdot 2 \cdot 2 \cdot 3 \cdot x$$
$$18 = 2 \cdot 3 \cdot 3$$

Circle the common prime factors.

So, the GCF of $24x$ and 18 is $2 \cdot 3 = 6$. Use the GCF to factor the expression.

$$24x - 18 = 6(4x) - 6(3)$$ Rewrite using GCF.
$$= 6(4x - 3)$$ Distributive Property

⁞· So, $24x - 18 = 6(4x - 3)$.

You can also use the Distributive Property to factor out any rational number from an expression.

EXAMPLE 2 Factoring Out a Fraction

Factor $\frac{1}{2}$ out of $\frac{1}{2}x + \frac{3}{2}$.

Write each term as a product of $\frac{1}{2}$ and another factor.

$$\frac{1}{2}x = \frac{1}{2} \cdot x$$ Think: $\frac{1}{2}x$ is $\frac{1}{2}$ times what?

$$\frac{3}{2} = \frac{1}{2} \cdot 3$$ Think: $\frac{3}{2}$ is $\frac{1}{2}$ times what?

Linear Expressions
In this extension, you will
• factor linear expressions.

Use the Distributive Property to factor out $\frac{1}{2}$.

$$\frac{1}{2}x + \frac{3}{2} = \frac{1}{2} \cdot x + \frac{1}{2} \cdot 3$$ Rewrite the expression.

$$= \frac{1}{2}(x + 3)$$ Distributive Property

⁞· So, $\frac{1}{2}x + \frac{3}{2} = \frac{1}{2}(x + 3)$.

Laurie's Notes

Introduction

Connect
- **Yesterday:** Students found the sums and differences of linear expressions.
- **Today:** Students will factor linear expressions.

Discuss
- In today's lesson students will be factoring expressions. The connection you want them to make is that the Distributive Property states there is an equality between two expressions: $a(b + c) = ab + ac$. Students should be comfortable *distributing the factor of a*. Here they will *factor out the a*.
- Students often view the equal sign in the Distributive Property as an arrow pointing to the right. Remind them that the arrow also points to the left!

Lesson Notes

Example 1
- **?** "Do $24x$ and 18 have any common factors?" 2, 3, and 6
- Now students may find it odd that you did not just ask about 24 and 18. Instead you included the variable factor. Explain that an algebraic term has factors, just like numbers.
- Write the prime factorizations of $24x$ and 18. Students will be comfortable seeing 6 as the GCF of $24x$ and 18. The variable x has not changed that.
- Say, "You want to rewrite $24x - 18$ as a product." Get students started by writing $24x - 18 = \underline{\quad}(\underline{\quad} - \underline{\quad})$
- **?** "How can you use the Distributive Property?" Listen for students to mention that the GCF is the factor you want to remove.
- **Look for and Make Use of Structure:** When you finish, be sure that students recognize that this is the Distributive Property, with the arrow pointing left!

$$6(4x - 3) = 24x - 18 \qquad \text{Distributive Property} \longrightarrow \text{expanding}$$
$$24x - 18 = 6(4x - 3) \qquad \text{Distributive Property} \longleftarrow \text{factoring}$$

Example 2
- **Look for and Make Use of Structure:** Writing each term as a product helps students see the common factor. This is particularly true with fractions.
- **?** "How can you write $\frac{1}{2}x$ as a product?" *Sample answer:* $\frac{1}{2} \cdot x$
- **?** "How can you write $\frac{3}{2}$ as a product?" *Sample answer:* $\frac{1}{2} \cdot 3$
- Because there is a common factor of $\frac{1}{2}$ in each term, you can factor it out.

$$\frac{1}{2}(x + 3) = \frac{1}{2}x + \frac{3}{2} \qquad \text{Distributive Property} \longrightarrow \text{expanding}$$
$$\frac{1}{2}x + \frac{3}{2} = \frac{1}{2}(x + 3) \qquad \text{Distributive Property} \longleftarrow \text{factoring}$$

What Your Students Will Learn
- Factor linear expressions by factoring out the Greatest Common Factor (GCF) or the coefficient of the variable.

Goal Today's lesson is factoring linear expressions.

Technology for the Teacher

Dynamic Classroom

Lesson Tutorials
Lesson Plans
Answer Presentation Tool

Extra Example 1
Factor $6x - 27$ using the GCF.
$3(2x - 9)$

Extra Example 2
Factor $\frac{1}{5}$ out of $\frac{1}{5}x - \frac{4}{5}$.
$\frac{1}{5}(x - 4)$

Record and Practice Journal
Extension 3.2 Practice

1. $7(1 + 4)$ 2. $25(1 + 2)$
3. $7(b - 1)$ 4. $8(a - 2)$
5. $4(2x + 3)$ 6. $12(y + 2t)$
7. $10(w + 5z)$ 8. $2(5v + 6u)$
9. $3(3a + 5b)$
10. $\frac{1}{2}(a - 1)$ 11. $\frac{1}{4}(d - 3)$
12. $\frac{5}{6}\left(s + \frac{4}{5}\right)$ 13. $\frac{3}{10}\left(y - \frac{4}{3}\right)$
14. $1.1(x + 9)$ 15. $3.4(c + 3)$
16. $-2(3x - 5)$
17. $-\frac{1}{3}\left(-y + \frac{9}{2}\right)$
18. $(2x + 3)$ ft

Extra Example 3

Factor -4 out of $-12r - 20$

$-4(3r + 5)$

Practice

1. $3(3 + 7)$

2. $16(2 - 3)$

3. $2(4x + 1)$

4. $3(y - 8)$

5. $4(5z - 2)$

6. $5(3w + 13)$

7. $4(9a + 4b)$

8. $7(3m - 7n)$

9. $\frac{1}{3}(b - 1)$

10. $\frac{3}{8}(d + 2)$

11. $2.2(x + 2)$

12. $4\left(h - \frac{3}{4}\right)$

13. $-\frac{1}{2}(x - 12)$

14. $-\frac{1}{4}(2x + 5y)$

15. $(3x - 8)$ ft

16. See Additional Answers.

17. *Sample answer:* $2x - 1$ and x, $2x$ and $x - 1$

Example 3

- This example looks at factoring out a negative number. If students are comfortable with the two directions of the Distributive Property identified in the last two examples, then they should be able to discuss the role of the negative factor in this example.
- As a prompt, write $-4p + 10 = -2(\underline{\hspace{1cm}} + \underline{\hspace{1cm}})$
- ❓ "-2 times what is $-4p$?" $\ 2p$
- ❓ "-2 times what is 10?" $\ -5$
- Fill in the blanks, $-4p + 10 = -2(2p + (-5))$, which can be written as $-2(2p - 5)$.

Closure

- Match the algebraic expression on the left with its factored form on the right.

1. $12x + 6$ C
2. $12x - 6$ B
3. $-12x - 6$ D
4. $-12x + 6$ A

   A. $-6(2x - 1)$
   B. $6(2x - 1)$
   C. $6(2x + 1)$
   D. $-6(2x + 1)$

Mini-Assessment

1. Factor $21w + 28$ using the GCF.

   $7(3w + 4)$

2. Factor $\frac{1}{3}$ out of $\frac{1}{3}p - \frac{2}{3}$.

   $\frac{1}{3}(p - 2)$

3. Factor -6 out of $-12y - 42$.

   $-6(2y + 7)$

EXAMPLE 3   Factoring Out a Negative Number

Factor -2 out of $-4p + 10$.

Math Practice

View as Components

How does rewriting each term as a product help you see the common factor?

Write each term as a product of -2 and another factor.

$$-4p = -2 \cdot 2p \qquad \text{Think: } -4p \text{ is } -2 \text{ times what?}$$

$$10 = -2 \cdot (-5) \qquad \text{Think: } 10 \text{ is } -2 \text{ times what?}$$

Use the Distributive Property to factor out -2.

$$-4p + 10 = -2 \cdot 2p + (-2) \cdot (-5) \qquad \text{Rewrite the expression.}$$

$$= -2[2p + (-5)] \qquad \text{Distributive Property}$$

$$= -2(2p - 5) \qquad \text{Simplify.}$$

⠶ So, $-4p + 10 = -2(2p - 5)$.

● Practice

Factor the expression using the GCF.

1. $9 + 21$ **2.** $32 - 48$ **3.** $8x + 2$ **4.** $3y - 24$

5. $20z - 8$ **6.** $15w + 65$ **7.** $36a + 16b$ **8.** $21m - 49n$

Factor out the coefficient of the variable.

9. $\frac{1}{3}b - \frac{1}{3}$ **10.** $\frac{3}{8}d + \frac{3}{4}$ **11.** $2.2x + 4.4$ **12.** $4h - 3$

13. Factor $-\frac{1}{2}$ out of $-\frac{1}{2}x + 6$.   **14.** Factor $-\frac{1}{4}$ out of $-\frac{1}{2}x - \frac{5}{4}y$.

15. WRESTLING A square wrestling mat has a perimeter of $(12x - 32)$ feet. Write an expression that represents the side length of the mat (in feet).

16. MAKING A DIAGRAM A table is 6 feet long and 3 feet wide. You extend the table by inserting two identical table *leaves*. The longest side length of each rectangular leaf is 3 feet. The extended table is rectangular with an area of $(18 + 6x)$ square feet.

   a. Make a diagram of the table and leaves.

   b. Write an expression that represents the length of the extended table. What does x represent?

17. STRUCTURE The area of the trapezoid is $\left(\frac{3}{4}x - \frac{1}{4}\right)$ square centimeters. Write two different pairs of expressions that represent possible lengths of the bases.

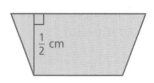

You can use a **four square** to organize information about a topic. Each of the four squares can be a category, such as *definition, vocabulary, example, non-example, words, algebra, table, numbers, visual, graph,* or *equation.* Here is an example of a four square for like terms.

Definition	Examples
Terms that have the same variables raised to the same exponents	2 and −3, 3x and −7x, x^2 and $6x^2$

Like Terms

Words	Non-Examples
To *combine* like terms that have variables, use the Distributive Property to add or subtract the coefficients.	y and 4, 3x and −4y, $6x^2$ and 2x

On Your Own

Make four squares to help you study these topics.

1. simplest form

2. linear expression

3. factoring expressions

After you complete this chapter, make four squares for the following topics.

4. equivalent equations

5. solving equations using addition or subtraction

6. solving equations using multiplication or division

7. solving two-step equations

"My four square shows that my new red skateboard is faster than my old blue skateboard."

Sample Answers

1.

Definition	Words

An algebraic expression is in *simplest form* when it has:
  1. no like terms and
  2. no parentheses.
	To write an algebraic expression in simplest form:
	Step 1: Rewrite as a sum.
	Step 2: Use the Distributive Property on parentheses, if necessary.
	Step 3: Rearrange terms.
	Step 4: Combine like terms.

$$\text{Simplest form}$$

Example	Example
$5x^2 + 6x - 3x^2 + 8 - x$ $= 5x^2 + 6x + (-3x^2) + 8 + (-1x)$ $= 5x^2 + (-3x^2) + 6x + (-1x) + 8$ $= [5 + (-3)]x^2 + [6 + (-1)]x + 8$ $= 2x^2 + 5x + 8$	$9 - 3\left(\frac{2}{3}m - \frac{1}{3}\right) + 3m$ $= 9 + (-3)\left(\frac{2}{3}m\right) + \left(-\frac{1}{3}\right) + 3m$ $= 9 + (-3)\left(\frac{2}{3}m\right) + (-3)\left(-\frac{1}{3}\right) + 3m$ $= 9 + (-2m) + 1 + 3m$ $= (-2m) + 3m + 9 + 1$ $= (-2 + 3)m + (9 + 1)$ $= m + 10$

2.

Definition	Examples
An algebraic expression in which the exponent of the variable is 1	$-7x,\ 2x + 3,\ 8 - \frac{1}{4}x$ Non-examples: $x^3,\ -5x^2 + x,\ x^7 - 9$

$$\text{Linear expression}$$

Example	Example
Adding linear expressions: $(7 - w) + 3(-2w + 4)$ $= 7 + (-1w) + 3(-2w) + 3(4)$ $= 7 + (-1w) + (-6w) + 12$ $= (-1w) + (-6w) + 7 + 12$ $= -7w + 19$	Subtracting linear expressions: $(4y + 7) - (y - 8)$ $\begin{array}{r} 4y + 7 \\ -(1y - 8) \end{array} \Rightarrow \begin{array}{r} 4y + 7 \\ + (-1y) + 8 \\ \hline 3y + 15 \end{array}$

3.

Words	Example
Write the expression as a product of factors. You can use the Distributive Property.	Factor $12a - 30$ using the GCF. $12a = ②\cdot 2 \cdot ③\cdot a$ $30 = ②\cdot ③\cdot 5$ $GCF = 2 \cdot 3 = 6$ $12a - 30 = 6(2a) - 6(5)$ $\qquad\qquad = 6(2a - 5)$

$$\text{Factoring expressions}$$

Example	Example
Factor $\frac{1}{4}$ out of $\frac{1}{4}r + \frac{3}{4}$. $\frac{1}{4}r = \frac{1}{4}\cdot r$ $\frac{3}{4} = \frac{1}{4}\cdot 3$ $\frac{1}{4}r + \frac{3}{4} = \frac{1}{4}\cdot r + \frac{1}{4}\cdot 3$ $\qquad\qquad = \frac{1}{4}(r + 3)$	Factor -7 out of $-21p + 28$. $-21p = -7 \cdot 3p$ $28 = -7 \cdot (-4)$ $-21p + 28$ $\qquad = -7(3p) + (-7)(-4)$ $\qquad = -7(3p - 4)$

List of Organizers
Available at *BigIdeasMath.com*

Comparison Chart
Concept Circle
Example and Non-Example Chart
Formula Triangle
Four Square
Idea (Definition) and Examples Chart
Information Frame
Information Wheel
Notetaking Organizer
Process Diagram
Summary Triangle
Word Magnet
Y Chart

About this Organizer

A **Four Square** can be used to organize information about a topic. Students write the topic in the "bubble" in the middle of the four square. Then students write concepts related to the topic in the four squares surrounding the bubble. Any concept related to the topic can be used. Encourage students to include concepts that will help them learn the topic. Students can place their four squares on note cards to use as a quick study reference.

Technology for the **Teacher**
Editable Graphic Organizer

Answers

1. Terms: $11x$, $2x$;
   Like terms: $11x$ and $2x$

2. Terms: $9x$, $-5x$
   Like terms: $9x$ and $-5x$

3. Terms: $21x$, 6, $-x$, -5;
   Like terms: $21x$ and $-x$;
   6 and -5

4. Terms: $8x$, 14, $-3x$, 1;
   Like terms: $8x$ and $-3x$;
   14 and 1

5. $8x$

6. $-7 + 7x$

7. $2x + 6$

8. $5x + 4$

9. $4s + 4$

10. $12t - 1$

11. $-13k + 8$

12. $\frac{7}{12}q$

13. $3n - 10$

14. $9h + 2$

15. $5(c - 3)$

16. $\frac{2}{9}(j + 3)$

17. $2.4(n + 4)$

18. $-6(z - 2)$

19. $32.67x$

20. $3n + 1$; The total number of apples you and your friend picked is one more than 3 full baskets.

21. $8w$

Alternative Quiz Ideas

100% Quiz	Math Log
Error Notebook	Notebook Quiz
Group Quiz	Partner Quiz
Homework Quiz	Pass the Paper

Group Quiz

Students work in groups. Give each group a large index card. Each group writes five questions that they feel evaluate the material they have been studying. On a separate piece of paper, students solve the problems. When they are finished, they exchange cards with another group. The new groups work through the questions on the card.

Reteaching and Enrichment Strategies

If students need help. . .	If students got it. . .
Resources by Chapter	Resources by Chapter
• Practice A and Practice B	• Enrichment and Extension
• Puzzle Time	• Technology Connection
Lesson Tutorials	Game Closet at *BigIdeasMath.com*
BigIdeasMath.com	Start the next section

3.1–3.2　Quiz

Check It Out
Progress Check
BigIdeasMath✓com

Identify the terms and like terms in the expression. *(Section 3.1)*

1. $11x + 2x$

2. $9x - 5x$

3. $21x + 6 - x - 5$

4. $8x + 14 - 3x + 1$

Simplify the expression. *(Section 3.1)*

5. $2(3x + x)$

6. $-7 + 3x + 4x$

7. $2x + 4 - 3x + 2 + 3x$

8. $7x + 6 + 3x - 2 - 5x$

Find the sum or difference. *(Section 3.2)*

9. $(s + 12) + (3s - 8)$

10. $(9t + 5) + (3t - 6)$

11. $(2 - k) + 3(-4k + 2)$

12. $\frac{1}{4}(q - 12) + \frac{1}{3}(q + 9)$

13. $(n - 8) - (-2n + 2)$

14. $-3(h - 4) - 2(-6h + 5)$

Factor out the coefficient of the variable. *(Section 3.2)*

15. $5c - 15$

16. $\frac{2}{9}j + \frac{2}{3}$

17. $2.4n + 9.6$

18. $-6z + 12$

Paint
$21.79

Interior
Latex Paint
Orange
1 gallon

Brush
$3.99

Paint roller
$6.89

19. PAINTING You buy the same number of brushes, rollers, and paint cans. Write an expression in simplest form that represents the total amount of money you spend for painting supplies. *(Section 3.1)*

20. APPLES A basket holds n apples. You pick $2n - 3$ apples, and your friend picks $n + 4$ apples. Write an expression that represents the number of apples you and your friend picked. Interpret the expression. *(Section 3.2)*

21. EXERCISE Write an expression in simplest form for the perimeter of the exercise mat. *(Section 3.1)*

w

$3w$

3.3 Solving Equations Using Addition or Subtraction

Essential Question How can you use algebra tiles to solve addition or subtraction equations?

1 ACTIVITY: Solving Equations

Work with a partner. Use algebra tiles to model and solve the equation.

a. $x - 3 = -4$

Model the equation $x - 3 = -4$.

To get the variable tile by itself, remove the [ ] tiles on the left side by adding [ ] [ ] tiles to each side.

How many *zero pairs* can you remove from each side? [ ]
Circle them.

The remaining tile shows the value of x.

So, $x =$ [ ].

b. $z - 6 = 2$ **c.** $p - 7 = -3$ **d.** $-15 = t - 5$

2 ACTIVITY: Solving Equations

Work with a partner. Use algebra tiles to model and solve the equation.

a. $-5 = n + 2$

Model the equation $-5 = n + 2$.

Remove the [ ] tiles on the right side by adding [ ] [ ] tiles to each side.

How many *zero pairs* can you remove from the right side? [ ]
Circle them.

The remaining tiles show the value of n.

So, $n =$ [ ].

b. $y + 10 = -5$ **c.** $7 + b = -1$ **d.** $8 = 12 + z$

Solving Equations

In this lesson, you will
- write simple equations.
- solve equations using addition or subtraction.
- solve real-life problems.

Laurie's Notes

Introduction

Applying Mathematical Practices

- **Model with Mathematics:** Algebra tiles can help students make sense of equations. Algebra tiles are a concrete representation, deepening student understanding of what it means to solve an equation.

Motivate

- Show students a collection of algebra tiles and ask them what the collection represents.
- **?** "Can the collection be simplified? (Can you remove zero pairs?)"
- **?** "What is the expression represented by the collection?"
- **Model:** As a class, model the equations $x + 3 = 7$ and $x + 2 = 5$ using algebra tiles. These do not require a zero pair to solve and will help remind students how to solve equations using algebra tiles.

Activity Notes

Activity 1

- **?** There are two points to make at the beginning. First, ask students what it means to solve an equation. Second, mention that students need to think of $x - 3$ as $x + (-3)$ when using algebra tiles. (You can only *add* a positive tile or a negative tile.) to find the value of the variable that makes the equation true
- **?** "To get the variable tile by itself, what do you have to do to both sides of the equation?" Students may mention removing 3 red tiles from each side or adding 3 yellow tiles to each side.
- **Reason Abstractly and Quantitatively:** Subtracting -3 is equivalent to adding 3. In the activity, the approach is to add 3 to each side. Removing the red tiles is intuitive to students when the symbolic representation is introduced, but it is adding 3 (inverse operations) that will make sense. Mathematically proficient students recognize the equivalence.
- After adding 3 to each side, the green variable-tile is equal to -1.
- Students may say they can use mental math to solve. It is the tactile experience of adding and removing tiles that you want them to experience.

Activity 2

- **Representation:** While the equations $-5 = n + 2$ and $n + 2 = -5$ are the same to mathematics teachers, students may see these as very different equations. Students even see $2 + n = -5$ as a different equation. Take time to discuss the equivalence of all three equations.
- Ask students why $x = 4$ is equivalent to $4 = x$. Student conceptions and misconceptions show up when the original equation is modified.
- **?** When students finish part (a) ask, "What did you do to both sides of the equation in order to solve?" Add two red tiles.
- **?** **Reason Abstractly and Quantitatively:** "Adding -2 is equivalent to what?" Subtracting 2

What Your Students Will Learn

- Write simple equations using key terms and phrases.
- Solve equations using the Addition Property of Equality or the Subtraction Property of Equality.

Previous Learning

Students have solved equations with whole numbers.

Technology for the *Teacher*

Dynamic Classroom

Lesson Plans
Complete Materials List

3.3 Record and Practice Journal

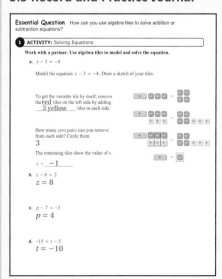

Essential Question How can you use algebra tiles to solve addition or subtraction equations?

1 ACTIVITY: Solving Equations

Work with a partner. Use algebra tiles to model and solve the equation.

a. $x - 3 = -4$

Model the equation $x - 3 = -4$. Draw a sketch of your tiles.

To get the variable tile by itself, remove the red tiles on the left side by adding __3 yellow__ tiles to each side.

How many zero pairs can you remove from each side? Circle them.
3

The remaining tiles show the value of x.

$x = \underline{-1}$

b. $z - 6 = 2$
$z = 8$

c. $p - 7 = -3$
$p = 4$

d. $-15 = t - 5$
$t = -10$

Differentiated Instruction

Kinesthetic

When working out solutions, ask two students to assist you at the board or overhead. Assign one student to the left side of the equation and the other student to the right side. Each student is responsible for performing the operations on his/her side of the equation. Emphasize that in order for both sides of the equation to remain equal, both students must perform the same operation at the same time to solve the equation.

3.3 Record and Practice Journal

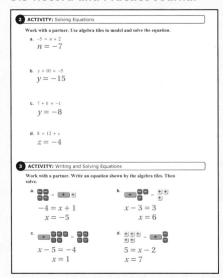

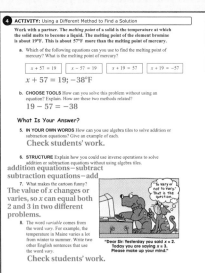

Laurie's Notes

Activity 3

- **Model with Mathematics:** Students are asked to write "an" equation, not "the" equation. This is because there are many correct answers. Students can use any variable and different forms of the equation. For example, part (b) could be written as $n - 3 = 3$ or $n + (-3) = 3$.
- ? Ask students to share the different equations they wrote.
- **Extension:** Share common tasks that *undo* one another. Examples: tying and untying your shoes, filling and emptying a glass, and opening and closing a door

Activity 4

- At this point, students should be able to figure out how to solve an equation like this. If they have difficulty, they can sketch algebra tiles on a piece of paper and find the solution, despite the large numbers.
- It is important to compare algebraic and arithmetic solutions. Explain to students that setting up equations is important since the equations will get more and more complex. Solving arithmetically may seem easy now to students.

What Is Your Answer?

- **Think-Pair-Share:** Students should read each question independently and then work in pairs to answer the questions. When they have answered the questions, the pair should compare their answers with another group and discuss any discrepancies.

Closure

- Translate the following model into symbols and explain in words how it could be solved.

Sample answer: $x - 3 = 1$; Add three yellow tiles to each side. The result will be $x = 4$.

Math Practice

Interpret Results

How can you add tiles to make zero pairs? Explain how this helps you solve the equation.

③ ACTIVITY: Writing and Solving Equations

Work with a partner. Write an equation shown by the algebra tiles. Then solve.

a. [−][−] = [+] [+]
    [−][−]

b. [+] [−][−] = [+][+]
        [−]   [+]

c. [+] [−][−][−] = [−][−]
        [−][−]   [−][−]

d. [+][+][+] = [+] [−]
    [+][+]   [−]

④ ACTIVITY: Using a Different Method to Find a Solution

Work with a partner. The *melting point* of a solid is the temperature at which the solid melts to become a liquid. The melting point of the element bromine is about 19°F. This is about 57°F more than the melting point of mercury.

a. Which of the following equations can you use to find the melting point of mercury? What is the melting point of mercury?

$$x + 57 = 19 \qquad x - 57 = 19 \qquad x + 19 = 57 \qquad x + 19 = -57$$

b. **CHOOSE TOOLS** How can you solve this problem without using an equation? Explain. How are these two methods related?

What Is Your Answer?

5. **IN YOUR OWN WORDS** How can you use algebra tiles to solve addition or subtraction equations? Give an example of each.

6. **STRUCTURE** Explain how you could use inverse operations to solve addition or subtraction equations without using algebra tiles.

7. What makes the cartoon funny?

8. The word *variable* comes from the word *vary*. For example, the temperature in Maine varies a lot from winter to summer.

"To vary or not to vary." That is the question.

"Dear Sir: Yesterday you said x = 2. Today you are saying x = 3. Please make up your mind."

Write two other English sentences that use the word *vary*.

Practice ➤ Use what you learned about solving addition or subtraction equations to complete Exercises 5–8 on page 100.

Check It Out
Lesson Tutorials
BigIdeasMath ✓com

Key Vocabulary 🔊
equivalent equations, p. 98

Two equations are **equivalent equations** if they have the same solutions. The Addition and Subtraction Properties of Equality can be used to write equivalent equations.

 **Key Ideas**

Addition Property of Equality

Words Adding the same number to each side of an equation produces an equivalent equation.

Algebra If $a = b$, then $a + c = b + c$.

Subtraction Property of Equality

Words Subtracting the same number from each side of an equation produces an equivalent equation.

Algebra If $a = b$, then $a - c = b - c$.

Remember

Addition and subtraction are inverse operations.

EXAMPLE 1 **Solving Equations**

a. Solve $x - 5 = -1$.

$$x - 5 = -1$$ Write the equation.

Undo the subtraction. → $\underline{+5 \quad +5}$ Addition Property of Equality

$$x = 4$$ Simplify.

⋮ The solution is $x = 4$.

Check

$x - 5 = -1$

$4 - 5 \overset{?}{=} -1$

$-1 = -1$ ✓

b. Solve $z + \dfrac{3}{2} = \dfrac{1}{2}$.

$$z + \frac{3}{2} = \frac{1}{2}$$ Write the equation.

Undo the addition. → $-\dfrac{3}{2} \quad -\dfrac{3}{2}$ Subtraction Property of Equality

$$z = -1$$ Simplify.

⋮ The solution is $z = -1$.

Check

$z + \dfrac{3}{2} = \dfrac{1}{2}$

$-1 + \dfrac{3}{2} \overset{?}{=} \dfrac{1}{2}$

$\dfrac{1}{2} = \dfrac{1}{2}$ ✓

● **On Your Own**

Now You're Ready
Exercises 5–20

Solve the equation. Check your solution.

1. $p - 5 = -2$ **2.** $w + 13.2 = 10.4$ **3.** $x - \dfrac{5}{6} = -\dfrac{1}{6}$

Laurie's Notes

Introduction

Connect
- **Yesterday:** Students used algebra tiles to model solving equations.
- **Today:** Students will formalize the process using the Addition and Subtraction Properties of Equality.

Motivate
- Have two students stand at the front of the room and write an "=" on the board between them. Hand each the same number of items (i.e., pencils, paper clips, etc.). The students should verify that they have the same number of items. Then give each two more of the same item. Verify that the number of items they have is equal. Finally, take four of the items from each student. Verify that they have the same amount.
- **Discuss:** This is the essence of the two properties used today—as long as each side of the equation has the same amount added to it or subtracted from it, the two sides of the equation are still equal.

Lesson Notes

Key Ideas
- Discuss how the activity in the introduction modeled the two properties.
- **?** "What are inverse operations?" Inverse operations undo one another.
- Ask students to give examples of inverse operations. They may say addition and subtraction or multiplication and division. They may even offer actions such as opening and closing a door.

Example 1
- Work through each part as a class. Notice that a vertical format is used. Use color to show the quantity being added to or subtracted from each side.
- **?** **Discuss:** The equations in parts (a) and (b) have the variable on the left.
  - "Would part (a) have the same solution if it was written as $-1 = x - 5$?" yes
  - "Would part (b) have the same solution if it was written as $\frac{1}{2} = z + \frac{3}{2}$?" yes

On Your Own
- After students have completed Example 1, they should be able to do these questions independently.

Words of Wisdom
- **Struggling Students:** If students have difficulty with the *On Your Own* questions, assess whether it is algebraic (how to solve equations) or computational (how to add or subtract rational numbers). Use this information to guide your instruction. Provide colored pencils so students can record the quantity being added to or subtracted from each side.
- Encourage students to be neat and to keep their equal signs lined up.

Technology for the Teacher

Dynamic Classroom

Lesson Tutorials
Lesson Plans
Answer Presentation Tool

Extra Example 1
a. Solve $t + 6 = -5$. -11
b. Solve $y - \frac{4}{5} = -\frac{2}{5}$. $\frac{2}{5}$

On Your Own
1. $p = 3$
2. $w = -2.8$
3. $x = \frac{2}{3}$

Laurie's Notes

Extra Example 2

You spent $7.25 this week. This is $3.65 less than you spent last week. Write and solve an equation to find the amount s you spent last week.
$s - 3.65 = 7.25$, $10.90

 On Your Own

4. $P - 145.25 = 120.50$

Extra Example 3

You have -1 point after Level 2 of a video game. Your score is 24 points less than your friend's score. Write and solve an equation to find your friend's score after Level 2. $-1 = f - 24$, 23 points

 On Your Own

5. 15 points

English Language Learners

Vocabulary

In this section, students learn to use *inverse* (or *opposite*) operations to solve equations. Students use addition to solve a subtraction equation and use subtraction to solve an addition equation. Review these pairs of words that are essential to understanding mathematics. Give students one word of a pair and ask them to provide the opposite.

odd, even	positive, negative
add, subtract	sum, difference
multiply, divide	product, quotient
plus, minus	

Example 2

- **Financial Literacy:** Discuss the word *profit* and how it is computed: income − expenses = profit.
- The second sentence contains key information. When translated into symbols, students can tell that "this profit" refers to "the profit this week."
- The color-coding in this text is very helpful in assisting students as they translate from words to symbols. Students may not recognize that "is" translates to "equals," so give a quick example. (Evan is $5\frac{1}{2}$ feet tall means the same as $E = 5.5$.)

On Your Own

- **Neighbor Check:** Have students work independently and then have their neighbors check their work. Have students discuss any discrepancies.

Example 3

- This example includes a line graph as a way to present information about the problem. Take time to have students *read and interpret* the information in the line graph.
- **?** Here are some questions to ask about the graph.
  - "What information is displayed on each axis of the line graph?" The horizontal axis shows the level of a video game, and the vertical axis shows the number of points scored.
  - "Were the scores ever tied?" Yes, at the very start and at some point in Level 3.
  - "Who was ahead after Level 2?" your friend
  - "What does '33 points' on the line graph mean?" It is the difference of your score and your friend's score after Level 4.
  - "Describe each player's performance from start to finish." *Sample answer:* Your friend did better than you at the beginning, but after Level 2 your score increased and your friend's score decreased. You ended up with 33 more points than your friend.
- **Model with Mathematics:** Take time to discuss the verbal model and how it translates information from the line graph. Mathematically proficient students are able to identify important quantities in a graph and make use of them to solve problems.

On Your Own

- Encourage students to write the key words and phrases using colored pencils and then translate the words to symbols.

Closure

- **Exit Ticket:**
  $p - 3.5 = -1.3$ 2.2 $-4.2 + m = 8.6$ 12.8

EXAMPLE **2** **Writing an Equation**

A company has a profit of $750 this week. This profit is $900 more than the profit P last week. Which equation can be used to find P?

(A) $750 = 900 - P$ **(B)** $750 = P + 900$

(C) $900 = P - 750$ **(D)** $900 = P + 750$

Words   The profit this week is $900 more than the profit last week.

Equation   750   = P +   900

:·:   The equation is $750 = P + 900$. The correct answer is **(B)**.

On Your Own

Now You're Ready
Exercises 22–25

4. A company has a profit of $120.50 today. This profit is $145.25 less than the profit P yesterday. Write an equation that can be used to find P.

EXAMPLE **3** **Real-Life Application**

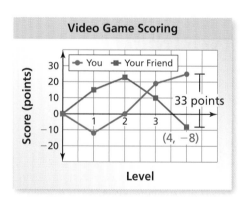

Video Game Scoring

The line graph shows the scoring while you and your friend played a video game. Write and solve an equation to find your score after Level 4.

You can determine the following from the graph.

Words   Your friend's score is 33 points less than your score.

Variable   Let s be your score after Level 4.

Equation -8 = s –   33

$-8 = s - 33$ Write equation.

$\underline{+\ 33} \quad \underline{+\ 33}$ Addition Property of Equality

$25 = s$ Simplify.

:·:   Your score after Level 4 is 25 points.

Reasonable? From the graph, your score after Level 4 is between 20 points and 30 points. So, 25 points is a reasonable answer.

On Your Own

5. **WHAT IF?** You have -12 points after Level 1. Your score is 27 points less than your friend's score. What is your friend's score?

 Vocabulary and Concept Check

1. **VOCABULARY** What property would you use to solve $m + 6 = -4$?

2. **VOCABULARY** Name two inverse operations.

3. **WRITING** Are the equations $m + 3 = -5$ and $m = -2$ equivalent? Explain.

4. **WHICH ONE DOESN'T BELONG?** Which equation does *not* belong with the other three? Explain your reasoning.

$$x + 3 = -1 \qquad x + 1 = -5 \qquad x - 2 = -6 \qquad x - 9 = -13$$

 Practice and Problem Solving

Solve the equation. Check your solution.

① 5. $a - 6 = 13$ **6.** $-3 = z - 8$ **7.** $-14 = k + 6$ **8.** $x + 4 = -14$

9. $c - 7.6 = -4$ **10.** $-10.1 = w + 5.3$ **11.** $\dfrac{1}{2} = q + \dfrac{2}{3}$ **12.** $p - 3\dfrac{1}{6} = -2\dfrac{1}{2}$

13. $g - 9 = -19$ **14.** $-9.3 = d - 3.4$ **15.** $4.58 + y = 2.5$ **16.** $x - 5.2 = -18.73$

17. $q + \dfrac{5}{9} = \dfrac{1}{6}$ **18.** $-2\dfrac{1}{4} = r - \dfrac{4}{5}$ **19.** $w + 3\dfrac{3}{8} = 1\dfrac{5}{6}$ **20.** $4\dfrac{2}{5} + k = -3\dfrac{2}{11}$

21. **ERROR ANALYSIS** Describe and correct the error in finding the solution.

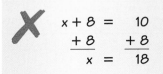

Write the word sentence as an equation. Then solve.

② 22. 4 less than a number n is -15.   **23.** 10 more than a number c is 3.

24. The sum of a number y and -3 is -8.

25. The difference between a number p and 6 is -14.

In Exercises 26–28, write an equation. Then solve.

26. **DRY ICE** The temperature of dry ice is $-109.3°F$. This is $184.9°F$ less than the outside temperature. What is the outside temperature?

27. **PROFIT** A company makes a profit of $1.38 million. This is $2.54 million more than last year. What was the profit last year?

28. **HELICOPTER** The difference in elevation of a helicopter and a submarine is $18\dfrac{1}{2}$ meters. The elevation of the submarine is $-7\dfrac{3}{4}$ meters. What is the elevation of the helicopter?

Assignment Guide and Homework Check

Level	Day 1 Activity Assignment	Day 2 Lesson Assignment	Homework Check
Basic	5–8, 41–45	1–4, 9–29 odd	9, 17, 23, 29
Average	5–8, 41–45	1–4, 13–21 odd, 22–28 even, 31, 33, 34	13, 17, 24, 28
Advanced	5–8, 41–45	1–4, 14–20 even, 21, 22–40 even	18, 30, 34, 36
Accelerated	1–8, 14–20 even, 21, 22–40 even, 41–45		18, 30, 34, 36

Common Errors

- **Exercises 5–20** Students may use the same operation in solving for x instead of the inverse operation. Demonstrate that this will not work to simplify the equation. Students most likely ignored the side with the variable when they made this mistake. Remind them to check their answers in the original equation.
- **Exercises 5–20** Students may add or subtract the number on the side of the equation without the variable. For example, they might write $-14 + 14 = k + 6 + 14$ instead of $-14 - 6 = x + 6 - 6$. Remind students that they are trying to get the variable by itself, so they have to start with the side that the variable is on and use the inverse of that operation.
- **Exercises 29–31** Students may try to use inverse operations to combine like terms. Remind them that inverse operations are used on both sides of the equation.

3.3 Record and Practice Journal

Solve the equation. Check your solution.

1. $y + 12 = -26$

-38

2. $15 + c = -12$

-27

3. $-16 = d + 21$

-37

4. $n + 12.8 = -0.3$

-13.1

5. $1\frac{1}{8} = g - 4\frac{2}{5}$

$5\frac{21}{40}$

6. $-5.47 + k = -14.19$

-8.72

Write the word sentence as an equation. Then solve.

7. 42 less than x is -50.

$x - 42 = -50;\ -8$

8. 32 is the sum of a number z and 9.

$32 = z + 9;\ 23$

9. A clothing company makes a profit of $2.3 million. This is $4.1 million more than last year. What was the profit last year?

$-\$1.8$ million

10. A drop on a wooden roller coaster is $-98\frac{1}{2}$ feet. A drop on a steel roller coaster is $100\frac{1}{4}$ feet lower than the drop on the wooden roller coaster. What is the drop on the steel roller coaster?

$-198\frac{3}{4}$ ft

Vocabulary and Concept Check

1. Subtraction Property of Equality

2. *Sample answer:* addition and subtraction

3. No, $m = -8$ not -2 in the first equation.

4. The equation $x + 1 = -5$ does not belong because its solution is $x = -6$ and the solution of the other equations is $x = -4$.

Practice and Problem Solving

5. $a = 19$

6. $z = 5$

7. $k = -20$

8. $x = -18$

9. $c = 3.6$

10. $w = -15.4$

11. $q = -\dfrac{1}{6}$

12. $p = \dfrac{2}{3}$

13. $g = -10$

14. $d = -5.9$

15. $y = -2.08$

16. $x = -13.53$

17. $q = -\dfrac{7}{18}$

18. $r = -1\dfrac{9}{20}$

19. $w = -1\dfrac{13}{24}$

20. $k = -7\dfrac{32}{55}$

21. The 8 should have been subtracted rather than added.

$$\begin{aligned} x + 8 &= 10 \\ -8 &\ -8 \\ \hline x &= 2 \end{aligned}$$

22. $n - 4 = -15;\ n = -11$

23. $c + 10 = 3;\ c = -7$

24. $y + (-3) = -8;\ y = -5$

25. $p - 6 = -14;\ p = -8$

26. $t - 184.9 = -109.3;\ 75.6°F$

27. $p + 2.54 = 1.38;$ $-\$1.16$ million

28. $h - \left(-7\dfrac{3}{4}\right) = 18\dfrac{1}{2};\ 10\dfrac{3}{4}$ m

29. $x + 8 = 12$; 4 cm

30. $x + 20.4 = 24.2$; 3.8 in.

31. $x + 22.7 = 34.6$; 11.9 ft

32. $305 = h + 153$; 152 ft

33. See *Taking Math Deeper*.

34. $d + 24\frac{1}{3} = 65\frac{3}{5}$; $41\frac{4}{15}$ km

35. $m + 30.3 + 40.8 = 180$; 108.9°

36. $p + 63.43 + 87.15 + 81.96 = 311.62$; more than 79.08

37. -9

38. $2, -2$

39. $6, -6$

40. $13, -13$

 Fair Game Review

41. -56 42. -72

43. -9 44. -6.5

45. B

Mini-Assessment

Solve the equation.

1. $x + 3.6 = -4.75$ $x = -8.35$

2. $-15.8 = y - 24.3$ $y = 8.5$

3. $t - 2\frac{2}{3} = -\frac{5}{2}$ $t = \frac{1}{6}$

4. $-\frac{5}{6} = z + \frac{1}{8}$ $z = -\frac{23}{24}$

5. You withdrew $47.25 from your checking account. Now your balance is $-\$23.75$. Write and solve an equation to find the amount of money in your account before you withdrew the money. $x - 47.25 = -23.75$; $23.50

Taking Math Deeper

Exercise 33

It's surprising how difficult this problem can be for students. There are two reasons for this. One is that you are not given the location of 0 on the vertical number line. The second is that the information is not given in the order it is used.

 Draw a vertical number line. Locate the jumping platform at 0.

 Draw the first jump. Draw the second jump so that the first jump is higher.

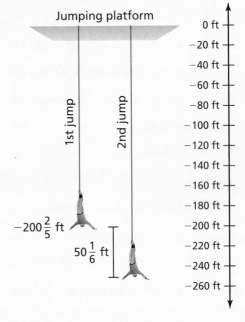

 Subtract to find the height of the second jump.

$$-200\frac{2}{5} - 50\frac{1}{6} = -250 - \frac{2}{5} - \frac{1}{6}$$
$$= -250 - \frac{12}{30} - \frac{5}{30}$$
$$= -250\frac{17}{30} \text{ ft}$$

Project

Research bungee jumping. What safety requirements are necessary for a bungee jumping business?

Reteaching and Enrichment Strategies

If students need help...	If students got it...
Resources by Chapter • Practice A and Practice B • Puzzle Time Record and Practice Journal Practice Differentiating the Lesson Lesson Tutorials Skills Review Handbook	Resources by Chapter • Enrichment and Extension • Technology Connection Start the next section

GEOMETRY Write and solve an equation to find the unknown side length.

29. Perimeter = 12 cm

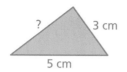

?   3 cm
5 cm

30. Perimeter = 24.2 in.

8.3 in.
?   3.8 in.
8.3 in.

31. Perimeter = 34.6 ft

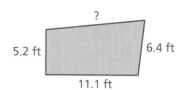

?
5.2 ft   6.4 ft
11.1 ft

In Exercises 32–36, write an equation. Then solve.

305 ft

32. STATUE OF LIBERTY The total height of the Statue of Liberty and its pedestal is 153 feet more than the height of the statue. What is the height of the statue?

33. BUNGEE JUMPING Your first jump is $50\frac{1}{6}$ feet higher than your second jump. Your first jump reaches $-200\frac{2}{5}$ feet. What is the height of your second jump?

34. TRAVEL Boatesville is $65\frac{3}{5}$ kilometers from Stanton. A bus traveling from Stanton is $24\frac{1}{3}$ kilometers from Boatesville. How far has the bus traveled?

35. GEOMETRY The sum of the measures of the angles of a triangle equals 180°. What is the measure of the missing angle?

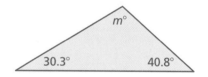

$m°$
30.3°   40.8°

36. SKATEBOARDING The table shows your scores in a skateboarding competition. The leader has 311.62 points. What score do you need in the fourth round to win?

Round	1	2	3	4
Points	63.43	87.15	81.96	?

37. CRITICAL THINKING Find the value of $2x - 1$ when $x + 6 = 2$.

 Find the values of x.

38. $|x| = 2$

39. $|x| - 2 = 4$

40. $|x| + 5 = 18$

 Fair Game Review What you learned in previous grades & lessons

Multiply or divide. *(Section 1.4 and Section 1.5)*

41. -7×8

42. $6 \times (-12)$

43. $18 \div (-2)$

44. $-26 \div 4$

45. MULTIPLE CHOICE A class of 144 students voted for a class president. Three-fourths of the students voted for you. Of the students who voted for you, $\frac{5}{9}$ are female. How many female students voted for you? *(Section 2.4)*

  (A) 50   **(B)** 60   **(C)** 80   **(D)** 108

3.4 Solving Equations Using Multiplication or Division

Essential Question How can you use multiplication or division to solve equations?

---

1 ACTIVITY: Using Division to Solve Equations

Work with a partner. Use algebra tiles to model and solve the equation.

a. $3x = -12$

Model the equation $3x = -12$.

Your goal is to get one variable tile by itself. Because there are ▢ variable tiles, divide the ▢ tiles into ▢ equal groups. Circle the groups.

Keep one of the groups. This shows the value of x.

So, $x = $ ▢.

b. $2k = -8$

c. $-15 = 3t$

d. $-20 = 5m$

e. $4h = -16$

---

2 ACTIVITY: Writing and Solving Equations

Work with a partner. Write an equation shown by the algebra tiles. Then solve.

a.

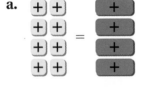

b.

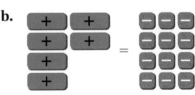

Solving Equations

In this lesson, you will

- solve equations using multiplication or division.
- solve real-life problems.

c.

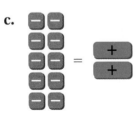

d.

Laurie's Notes

Introduction

Applying Mathematical Practices

- **Model with Mathematics:** Algebra tiles can help students make sense of equations. Algebra tiles are a concrete representation, deepening student understanding of what it means to solve an equation.

Motivate

- **Model:** Display two green variable-tiles and four yellow integer-tiles to the class.
- ❓ "If two green tiles equal four yellow tiles, what does one green tile equal?" two yellow tiles
- ❓ "How did you decide that one green tile equals two yellow tiles?" Divide each side into groups. The number of groups is the number of variable-tiles.

Activity Notes

Activity 1

- Model the first equation as students model the equation at their desks. Write the corresponding algebraic equation represented by the tiles with the first and last step. Encourage students to do the same.
- ❓ **Discuss:** Remind students that the goal is to find the value of just one green variable-tile. "If three green tiles equal 12 red tiles, what is the value of each green tile? How did you find your answer?" 4; To get one green tile, you need three groups. So, divide the 12 red tiles into three equal groups.
- Remind students that variables can be on either side of the equation. If students are more comfortable with variables on the left, they can write part (c) as $3t = -15$.

Activity 2

- **Think-Pair-Share:** After students work on the problems in pairs, ask for volunteers to work the problems for the class. Listen for how students describe the solutions.
- ❓ "Why is it difficult to model the equation $\frac{1}{3}x = 6$ with algebra tiles?"

  You can't show $\frac{1}{3}$ of a green variable-tile, but you can talk about the meaning. If $\frac{1}{3}$ of a green variable-tile is 6, then $\frac{2}{3}$ would be 12, and $\frac{3}{3}$ (or a whole green variable-tile) would be 18.

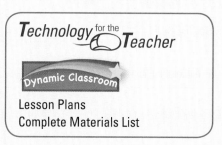

Lesson Plans
Complete Materials List

3.4 Record and Practice Journal

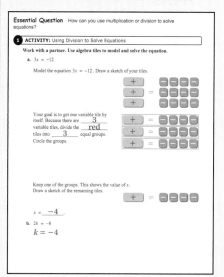

Differentiated Instruction

Visual

To model a division equation, such as $\frac{d}{4} = -3$, use the variable tile to represent the fractional part of a variable.

Then, model the solution.

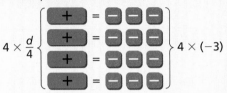

So, $d = -12$.

3.4 Record and Practice Journal

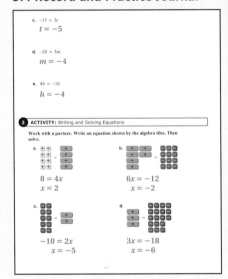

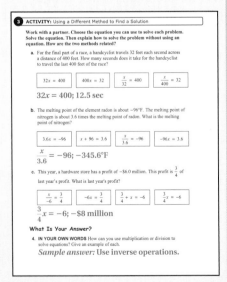

Activity 3

- Students often comment that they solve the problem by just doing a computation. Students might tell you, "I just know what to do" and say having to write and solve an equation makes it harder. You want students to practice the process of identifying what the unknown is, how to represent the unknown with a variable, and then use inverse operations to solve.

- **Make Sense of Problems and Persevere in Solving Them:** Having four possible equations requires students to make sense of each equation and to determine which equation represents the problem. Students need to understand the arithmetic solution and look at the various steps needed to solve algebraically in order to see the relationship.

- It is important to compare algebraic and arithmetic solutions. Explain to students that setting up equations is important since the equations will get more and more complex. Solving arithmetically may seem easy now to students.

What Is Your Answer?

- **Neighbor Check:** Have students work independently and then have their neighbors check their work. Have students discuss any discrepancies.

Closure

- **Exit Ticket:** Solve $\frac{x}{2} = -14$ and $2x = -14$. $-28; -7$

Work with a partner. Choose the equation you can use to solve each problem. Solve the equation. Then explain how to solve the problem without using an equation. How are the two methods related?

a. For the final part of a race, a handcyclist travels 32 feet each second across a distance of 400 feet. How many seconds does it take for the handcyclist to travel the last 400 feet of the race?

$$32x = 400 \qquad 400x = 32$$

$$\frac{x}{32} = 400 \qquad \frac{x}{400} = 32$$

b. The melting point of the element radon is about −96°F. The melting point of nitrogen is about 3.6 times the melting point of radon. What is the melting point of nitrogen?

$$3.6x = -96 \qquad x + 96 = 3.6$$

$$\frac{x}{3.6} = -96 \qquad -96x = 3.6$$

c. This year, a hardware store has a profit of −$6.0 million. This profit is $\frac{3}{4}$ of last year's profit. What is last year's profit?

$$\frac{x}{-6} = \frac{3}{4} \qquad -6x = \frac{3}{4}$$

$$\frac{3}{4} + x = -6 \qquad \frac{3}{4}x = -6$$

What Is Your Answer?

4. **IN YOUR OWN WORDS** How can you use multiplication or division to solve equations? Give an example of each.

Use what you learned about solving equations to complete Exercises 7–10 on page 106.

Check It Out
Lesson Tutorials
BigIdeasMath ✓com

 **Key Ideas**

Multiplication Property of Equality

Words Multiplying each side of an equation by the same number produces an equivalent equation.

Algebra If $a = b$, then $a \cdot c = b \cdot c$.

 **Remember**

Multiplication and division are inverse operations.

Division Property of Equality

Words Dividing each side of an equation by the same number produces an equivalent equation.

Algebra If $a = b$, then $a \div c = b \div c$, $c \neq 0$.

EXAMPLE 1 **Solving Equations**

a. **Solve** $\dfrac{x}{3} = -6$.

$$\dfrac{x}{3} = -6 \qquad \text{Write the equation.}$$

Undo the division. → $3 \cdot \dfrac{x}{3} = 3 \cdot (-6)$ Multiplication Property of Equality

$$x = -18 \qquad \text{Simplify.}$$

∴ The solution is $x = -18$.

Check
$$\dfrac{x}{3} = -6$$
$$\dfrac{-18}{3} \overset{?}{=} -6$$
$$-6 = -6 \checkmark$$

b. **Solve** $18 = -4y$.

$$18 = -4y \qquad \text{Write the equation.}$$

Undo the multiplication. → $\dfrac{18}{-4} = \dfrac{-4y}{-4}$ Division Property of Equality

$$-4.5 = y \qquad \text{Simplify.}$$

∴ The solution is $y = -4.5$.

Check
$$18 = -4y$$
$$18 \overset{?}{=} -4(-4.5)$$
$$18 = 18 \checkmark$$

● **On Your Own**

Solve the equation. Check your solution.

Now You're Ready
Exercises 7–18

1. $\dfrac{x}{5} = -2$ 2. $-a = -24$ 3. $3 = -1.5n$

Laurie's Notes

Introduction

Connect

- **Yesterday:** Students used algebra tiles to model solving equations.
- **Today:** Students will formalize the process of solving equations using the Multiplication and Division Properties of Equality.

Motivate

- Have two students stand at the front of the room. Hand a third student an odd number of index cards without telling the student how many cards he or she has been given.
- Ask the student with the index cards to share them equally between the two students. The student may pause when he or she realizes that there is an odd number of cards. Give the student time to realize that the remaining card needs to be divided into two pieces and that each student will receive one-half of a card.
- Ask the students holding the index cards to verify that they have the same number of cards.

Lesson Notes

Key Ideas

- Write the Properties of Equality on the board.
- ? If you started the class with the index card activity, then ask students which property was modeled in the opening activity. Division Property of Equality
- Remind students of how multiplication and division are represented with a variable, such as $4x$ and $\frac{x}{4}$.

Example 1

- Work through each problem. If possible, use colors to show the multiplication or division on each side of the equation.
- **FYI:** Note that the -6 is written in parentheses in the solution of part (a). When you do this step in class, you may want to write both numbers in parentheses $(3)(-6)$ to avoid students thinking that the multiplication dot is a decimal point.
- ? "Could the problem be represented as $(-6) \cdot 3$ instead of $3 \cdot (-6)$? Why or why not?" yes; This is an example of the Commutative Property of Multiplication.

On Your Own

- If students have difficulty as they work these problems, assess whether it is algebraic (how to solve equations) or computational (how to multiply or divide rational numbers). Use this information to guide your instruction.
- You may want to provide colored pencils to students so that they can highlight the quantity being multiplied or divided on each side.
- Encourage students to be neat and to keep the equal signs lined up.

Goal Today's lesson is solving equations using multiplication or division.

Technology for the Teacher

Dynamic Classroom

Lesson Tutorials
Lesson Plans
Answer Presentation Tool

Extra Example 1

a. Solve $\frac{c}{8} = -7$. -56

b. Solve $-5p = -32$. 6.4

On Your Own

1. $x = -10$
2. $a = 24$
3. $n = -2$

Extra Example 2

Solve $-\dfrac{5}{9}m = 25$. $\quad -45$

 On Your Own

4. $x = -21$

5. $b = -3\dfrac{1}{8}$

6. $h = -24$

Extra Example 3

The record low temperature in Nevada is $-50°F$. The record low temperature in Montana is 1.4 times the record low temperature in Nevada. What is the record low temperature in Montana? $\quad -70°F$

On Your Own

7. $-80°F$

English Language Learners

Graphic Organizer

When solving a one-step equation, students must remember to isolate the variable. Encourage students to make a table in their notebooks that will help them remember which operation to use to solve a one-step equation.

Operation on Variable	Operation to Solve	Example
Addition	Subtraction	$a + 3 = -5$
Subtraction	Addition	$b - 4 = 2$
Multiplication	Division	$c \cdot (-2) = 7$
Division	Multiplication	$\dfrac{d}{-4} = -8$

Laurie's Notes

Example 2

- Explain that $\dfrac{x}{3}$ and $\dfrac{1}{3}x$ are equivalent. Discuss how to multiply a fraction and a whole number: $\dfrac{1}{3}x = \dfrac{1}{3} \cdot \dfrac{x}{1} = \dfrac{x}{3}$. Repeat to show that $\dfrac{4}{5}x = \dfrac{4x}{5}$.

? "What is x being multiplied by?" $\quad -\dfrac{4}{5}$

? "Can you divide both sides by $-\dfrac{4}{5}$?" yes

- **Look for and Make Use of Structure:** Dividing by a fraction is equivalent to multiplying by its reciprocal.
- Students may need a quick review of multiplying fractions.
- **FYI:** You may want to emphasize that you are dividing each side by $-\dfrac{4}{5}$.

   This will emphasize the connection to multiplying by the reciprocal $-\dfrac{5}{4}$, and that both of these processes are equivalent.

Words of Wisdom

- When checking a solution, read it out loud. It is helpful for students to hear (as well as to see) what they are reading.

On Your Own

- **Think-Pair-Share:** Students should read each question independently and then work in pairs to answer the questions. When they have answered the questions, the pair should compare their answers with another group and discuss any discrepancies.

Example 3

- Encourage students to look at the artwork next to the problem. The first sentence contains key information that is translated into the equation.
- **Model with Mathematics:** The color-coding in the text is very helpful in assisting students as they translate from words to symbols. You may want to use color-coding when you do other examples.

On Your Own

- **Think-Pair-Share:** Students should read each question independently and then work in pairs to answer the questions. When they have answered the questions, the pair should compare their answers with another group and discuss any discrepancies.

Closure

- **Writing:** The variable in a one-step equation is being multiplied by $-\dfrac{3}{4}$. Describe how to solve the equation for x. You divide both sides of the equation by $-\dfrac{3}{4}$, which is the same as multiplying by the reciprocal. So, you multiply both sides of the equation by $-\dfrac{4}{3}$ and simplify.

EXAMPLE ② **Solving an Equation Using a Reciprocal**

Solve $-\dfrac{4}{5}x = -8$.

$$-\frac{4}{5}x = -8 \qquad \text{Write the equation.}$$

Multiply each side by $-\dfrac{5}{4}$, the reciprocal of $-\dfrac{4}{5}$. → $-\dfrac{5}{4} \cdot \left(-\dfrac{4}{5}x\right) = -\dfrac{5}{4} \cdot (-8)$ — Multiplicative Inverse Property

$$x = 10 \qquad \text{Simplify.}$$

∴ The solution is $x = 10$.

On Your Own

Now You're Ready
Exercises 19–22

Solve the equation. Check your solution.

4. $-14 = \dfrac{2}{3}x$ 5. $-\dfrac{8}{5}b = 5$ 6. $\dfrac{3}{8}h = -9$

EXAMPLE ③ **Real-Life Application**

Record low temperature in Arizona

The record low temperature in Arizona is 1.6 times the record low temperature in Rhode Island. What is the record low temperature in Rhode Island?

Words   The record low in Arizona is 1.6 times the record low in Rhode Island.

Variable   Let t be the record low in Rhode Island.

Equation -40 $=$ 1.6 $\times$ t

$$-40 = 1.6t \qquad \text{Write equation.}$$

$$-\frac{40}{1.6} = \frac{1.6t}{1.6} \qquad \text{Division Property of Equality}$$

$$-25 = t \qquad \text{Simplify.}$$

∴ The record low temperature in Rhode Island is $-25°$F.

On Your Own

Now You're Ready
Exercises 24–27

7. The record low temperature in Hawaii is −0.15 times the record low temperature in Alaska. The record low temperature in Hawaii is 12°F. What is the record low temperature in Alaska?

 Vocabulary and Concept Check

1. **WRITING** Explain why you can use multiplication to solve equations involving division.

2. **OPEN-ENDED** Turning a light on and then turning the light off are considered to be inverse operations. Describe two other real-life situations that can be thought of as inverse operations.

Describe the inverse operation that will undo the given operation.

3. multiplying by 5   4. subtracting 12   5. dividing by -8 6. adding -6

 Practice and Problem Solving

Solve the equation. Check your solution.

① 7. $3h = 15$

8. $-5t = -45$

9. $\dfrac{n}{2} = -7$

10. $\dfrac{k}{-3} = 9$

11. $5m = -10$

12. $8t = -32$

13. $-0.2x = 1.6$

14. $-10 = -\dfrac{b}{4}$

15. $-6p = 48$

16. $-72 = 8d$

17. $\dfrac{n}{1.6} = 5$

18. $-14.4 = -0.6p$

② 19. $\dfrac{3}{4}g = -12$

20. $8 = -\dfrac{2}{5}c$

21. $-\dfrac{4}{9}f = -3$

22. $26 = -\dfrac{8}{5}y$

23. **ERROR ANALYSIS** Describe and correct the error in finding the solution.

$$-4.2x = 21$$
$$\dfrac{-4.2x}{4.2} = \dfrac{21}{4.2}$$
$$x = 5$$

Write the word sentence as an equation. Then solve.

③ 24. A number divided by -9 is -16.

25. A number multiplied by $\dfrac{2}{5}$ is $\dfrac{3}{20}$.

26. The product of 15 and a number is -75.

27. The quotient of a number and -1.5 is 21.

In Exercises 28 and 29, write an equation. Then solve.

28. **NEWSPAPERS** You make $0.75 for every newspaper you sell. How many newspapers do you have to sell to buy the soccer cleats?

29. **ROCK CLIMBING** A rock climber averages $12\dfrac{3}{5}$ feet per minute.

    How many feet does the rock climber climb in 30 minutes?

Soccer Cleats $36⁰⁰

Assignment Guide and Homework Check

Level	Day 1 Activity Assignment	Day 2 Lesson Assignment	Homework Check
Basic	7–10, 41–45	1–6, 11–29 odd	13, 17, 19, 25, 29
Average	7–10, 41–45	1–6, 11–23 odd, 28, 34–37	13, 17, 19, 28
Advanced	7–10, 41–45	1–6, 16–22 even, 23, 30–40 even, 39	20, 30, 36, 39
Accelerated	1–10, 16–22 even, 23, 30–40 even, 39, 41–45		20, 30, 36, 39

Common Errors

- **Exercises 7–18** When the variable is multiplied by a negative number, students may not remember to keep the negative with the number and will really solve for $-x$ instead of x. Do an example of one of these problems on the board. Solve for $-x$ and ask students if x is by itself. If they do not realize it, remind them that there is a -1 in front of the variable and that they must divide by -1 to "get the variable by itself."
- **Exercises 19–22** Students may not understand why they should multiply by the reciprocal and may try to divide by the reciprocal. Ask students how they would solve the problem without using the reciprocal (divide by the fractional coefficient). Then ask how to divide a number by a fraction (multiply by the reciprocal). It is a short cut to multiply by the reciprocal from the beginning.
- **Exercises 19–22** If students have a difficult time grasping the idea of multiplying by the reciprocal, have them write out each step instead of just multiplying by the reciprocal. Remind students to check their answers.

3.4 Record and Practice Journal

Solve the equation. Check your solution.

1. $\dfrac{d}{5} = -6$

-30

2. $8x = -6$

$-\dfrac{3}{4}$

3. $-15 = \dfrac{z}{-2}$

30

4. $3.2n = -0.8$

-0.25

5. $-\dfrac{3}{10}h = 15$

-50

6. $-1.1k = -1.21$

1.1

Write the word sentence as an equation. Then solve.

7. A number divided by -8 is 7.

$\dfrac{x}{-8} = 7;\ -56$

8. The product of -12 and a number is 60.

$-12x = 60;\ -5$

9. You earn $0.85 for every cup of hot chocolate you sell. How many cups do you need to sell to earn $55.25?

65 cups

 Practice and Problem Solving

7. $h = 5$
8. $t = 9$

9. $n = -14$
10. $k = -27$

11. $m = -2$
12. $t = -4$

13. $x = -8$
14. $b = 40$

15. $p = -8$
16. $d = -9$

17. $n = 8$
18. $p = 24$

19. $g = -16$
20. $c = -20$

21. $f = 6\dfrac{3}{4}$
22. $y = -16\dfrac{1}{4}$

23. They should divide by -4.2.

$$-4.2x = 21$$
$$\dfrac{-4.2x}{-4.2} = \dfrac{21}{-4.2}$$
$$x = -5$$

24. $\dfrac{x}{-9} = -16;\ x = 144$

25. $\dfrac{2}{5}x = \dfrac{3}{20};\ x = \dfrac{3}{8}$

26. $15x = -75;\ x = -5$

27. $\dfrac{x}{-1.5} = 21;\ x = -31.5$

28. $0.75n = 36;\ 48$ newspapers

29. $\dfrac{x}{30} = 12\dfrac{3}{5};\ 378$ ft

30–33. Sample answers are given.

30. **a.** $3x = -9$ **b.** $\frac{x}{2} = -1.5$

31. **a.** $-2x = 4.4$ **b.** $\frac{x}{1.1} = -2$

32. **a.** $5x = -\frac{5}{2}$ **b.** $\frac{x}{2} = -\frac{1}{4}$

33. **a.** $4x = -5$ **b.** $\frac{x}{5} = -\frac{1}{4}$

34. All of them except "multiply each side by $-\frac{2}{3}$."

35. $-1.26n = -10.08$; 8 days

36. $\frac{3}{4}s = 1464$; 1952 students

37. -50 ft

38. See *Taking Math Deeper*.

39. $-5, 5$

40. $1\frac{3}{5}$ days

Fair Game Review

41. -7 42. -9

43. 12 44. -9

45. B

Mini-Assessment
Solve the equation.

1. $7x = -84$ $x = -12$

2. $-0.3y = 2.4$ $y = -8$

3. $\frac{1}{2}m = -\frac{6}{7}$ $m = -1\frac{5}{7}$

4. $4\frac{1}{2} = -\frac{8}{9}k$ $k = -5\frac{1}{16}$

5. A stock has a return of $-\$1.40$ per day. Write and solve an equation to find the number of days until the total return is $-\$12.60$.
   $-1.4x = -12.6$; 9 days

Taking Math Deeper

Exercise 38

This problem is a good example of the type of question that can occur on a standardized test. For this problem, remind students to *read the question carefully.*

 Organize the given information.

	Store A	Store B
Price	$150.60	x

$150.60 is $\frac{5}{6}$ x.

 Write and solve an equation for x.

$150.60 = \frac{5}{6}x$	Write the equation.
$6(150.60) = 5x$	Multiply each side by 6.
$\frac{6(150.60)}{5} = x$	Divide each side by 5.
$\$180.72 = x$	Simplify.

The bike costs $180.72 at Store B.

Not the answer

3 How much do you save?

$180.72 - 150.60 = \$30.12$

You save $30.12 at Store A.

Project

Find the prices of bikes in at least 2 different stores in your area. How much can you save by making your purchase at the store with the lowest price? Do you think it pays to comparison shop? Why or why not?

Reteaching and Enrichment Strategies

If students need help...	If students got it...
Resources by Chapter • Practice A and Practice B • Puzzle Time Record and Practice Journal Practice Differentiating the Lesson Lesson Tutorials Skills Review Handbook	Resources by Chapter • Enrichment and Extension • Technology Connection Start the next section

OPEN-ENDED (a) Write a multiplication equation that has the given solution.
(b) Write a division equation that has the same solution.

30. -3 **31.** -2.2 **32.** $-\dfrac{1}{2}$ **33.** $-1\dfrac{1}{4}$

34. REASONING Which of the methods can you use to solve $-\dfrac{2}{3}c = 16$?

> Multiply each side by $-\dfrac{2}{3}$.   Multiply each side by $-\dfrac{3}{2}$.
>
> Divide each side by $-\dfrac{2}{3}$.   Multiply each side by 3, then divide each side by -2.

35. STOCK A stock has a return of $-\$1.26$ per day. Write and solve an equation to find the number of days until the total return is $-\$10.08$.

36. ELECTION In a school election, $\dfrac{3}{4}$ of the students vote. There are 1464 ballots. Write and solve an equation to find the number of students.

37. OCEANOGRAPHY Aquarius is an underwater ocean laboratory located in the Florida Keys National Marine Sanctuary. Solve the equation $\dfrac{31}{25}x = -62$ to find the value of x.

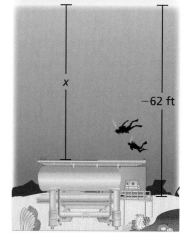

x

-62 ft

38. SHOPPING The price of a bike at Store A is $\dfrac{5}{6}$ the price at Store B. The price at Store A is $\$150.60$. Write and solve an equation to find how much you save by buying the bike at Store A.

39. CRITICAL THINKING Solve $-2|m| = -10$.

40. In four days, your family drives $\dfrac{5}{7}$ of a trip. Your rate of travel is the same throughout the trip. The total trip is 1250 miles. In how many more days will you reach your destination?

Fair Game Review What you learned in previous grades & lessons

Subtract. (Section 1.3)

41. $5 - 12$ **42.** $-7 - 2$ **43.** $4 - (-8)$ **44.** $-14 - (-5)$

45. MULTIPLE CHOICE Of the 120 apartments in a building, 75 have been scheduled to receive new carpet. What fraction of the apartments have not been scheduled to receive new carpet? (Skills Review Handbook)

    Ⓐ $\dfrac{1}{4}$ Ⓑ $\dfrac{3}{8}$ Ⓒ $\dfrac{5}{8}$ Ⓓ $\dfrac{3}{4}$

3.5 Solving Two-Step Equations

Essential Question How can you use algebra tiles to solve a two-step equation?

1 ACTIVITY: Solving a Two-Step Equation

Work with a partner. Use algebra tiles to model and solve $2x - 3 = -5$.

Model the equation $2x - 3 = -5$.

Remove the ⬚ red tiles on the left side by adding ⬚ yellow tiles to each side.

How many *zero pairs* can you remove from each side? ⬚
Circle them.

Because there are ⬚ green tiles, divide the red tiles into ⬚ equal groups. Circle the groups.

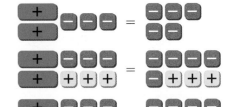

Keep one of the groups. This shows the value of x.

∴ So, $x = $ ⬚ .

2 ACTIVITY: The Math behind the Tiles

Work with a partner. Solve $2x - 3 = -5$ without using algebra tiles. Complete each step. Then answer the questions.

Use the steps in Activity 1 as a guide.

$2x - 3 = -5$	Write the equation.
$2x - 3 + \boxed{} = -5 + \boxed{}$	Add ⬚ to each side.
$2x = \boxed{}$	Simplify.
$\dfrac{2x}{\boxed{}} = \dfrac{\boxed{}}{\boxed{}}$	Divide each side by ⬚ .
$x = \boxed{}$	Simplify.

Solving Equations
In this lesson, you will
- solve two-step equations.
- solve real-life problems.

∴ So, $x = $ ⬚ .

a. Which step is first, adding 3 to each side or dividing each side by 2?

b. How are the above steps related to the steps in Activity 1?

Laurie's Notes

Introduction

Applying Mathematical Practices

- **Model with Mathematics:** Algebra tiles can help students make sense of equations. Algebra tiles are a concrete representation, deepening student understanding of what it means to solve an equation.

Motivate

- Write the number four on a slip of paper and put it in an envelope. Seal the envelope. Write the number 15 on the outside of the envelope.
- Hold the envelope up to your forehead and say, "I'm thinking of a number. When I double the number and add 7, I get an answer of 15." Show that the number 15 is written on the outside of the envelope.
- ❓ "What number did I start with?" 4
- ❓ "Can anyone explain how they know what number I was thinking of?" Listen for students to "undo" your process by working backwards: subtract 7 from 15 (to get 8) and then divide by 2 (to get 4).
- Open the envelope and reveal the four on your slip of paper.
- ❓ "Why didn't you divide by 2 first and then subtract 7?" You need to do the steps in the reverse order to undo the calculations.
- Explain that today you will investigate how to solve equations with two operations, like the number puzzle. These are called two-step equations.

Activity Notes

Activity 1

- Have students write the corresponding algebraic equations that result with each step to connect the model to the algebraic representation.
- **Discuss:** When students have finished, summarize by saying that the goal is to find the value of just one green variable-tile, so it should seem reasonable to "get rid of" the red integer-tiles on the left-hand side. This is referred to as *isolating the variable.*
- ❓ "How did you get the green variable-tiles by themselves?" Add three yellow integer-tiles to each side or remove three red integer-tiles from each side.
- Take time to look at each method discussed in the question above. Adding three yellow tiles is represented by $2x - 3 + 3 = -5 + 3$. Taking three red tiles away from each side results in $2x = -2$.
- **Reason Abstractly and Quantitatively:** Mathematically proficient students recognize the equivalence of adding 3 and subtracting -3.

Activity 2

- ❓ After you complete Activity 2, ask, "What happens if you divide each side by 2 first? Will you get the same answer?"
  $\dfrac{2x - 3}{2} = -\dfrac{5}{2}$ simplifies to $x - \dfrac{3}{2} = -\dfrac{5}{2}$. This introduces fractions into the problem, but you will still get the same answer.
- ❓ "Which method do you prefer?" Most students will want to avoid fractions.

What Your Students Will Learn

- Solve two-step equations by applying properties of equality.
- Combine like terms before solving algebraic equations.

Previous Learning

Students solved one-step equations.

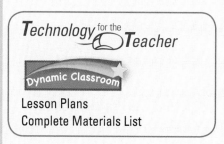

Technology for the Teacher

Dynamic Classroom

Lesson Plans
Complete Materials List

3.5 Record and Practice Journal

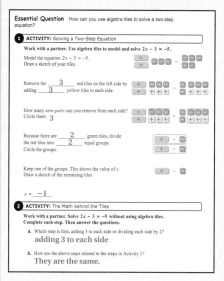

Differentiated Instruction

Auditory

Write the equation $5h + 6 = 1$ on the board or overhead. Ask students to tell you the steps needed to solve the equation. Repeat the steps out loud as you solve the problem. Ask students for another way to solve the problem. Which way is more efficient? Why?

3.5 Record and Practice Journal

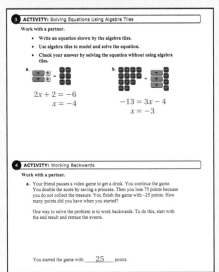

Activity 3

- Ask for volunteers to share their work with the class.
- **Attend to Precision:** Listen to the language that students use when they explain their solutions. If they say, "I'll put two red tiles on each side," ask them to express their steps mathematically. They *should* say, "Add 2 red tiles to each side."
- **Reason Abstractly and Quantitatively:** You want students to be able to connect their manipulation of the tiles with operations they'll record symbolically.
- Students may use different methods to solve part (b). Some students may add four yellow tiles to each side (add the opposite), and others may remove four red tiles from each side (subtract −4).
- After one method of solving an equation is described, be sure to ask if anyone approached the problem in another way so students don't think their method is wrong.

Activity 4

- If you did not do the activity in the introduction, you may want to before this activity. Both use the strategy *Working Backwards*.
- Have students make up their own number puzzles and have their partners work backwards to guess their number.

What Is Your Answer?

- Some students may not have the appropriate language to describe the process yet. Have them focus on creating examples.
- For Question 7, listen for the idea that it is the last operation performed that is "undone" first in the solving process.

Closure

- Match the equation with the first step used in solving the equation.

Equation	First Step in Solving
1) $4x - 3 = -7$	A) Divide by −3.
2) $5 = -2x + 4$	B) Subtract 2.
3) $-3x + 2 = 6$	C) Multiply by $\frac{1}{3}$.
4) $-4 = 3x - 2$	D) Add 3.
	E) Add 2.
	F) Subtract 4.

Answers: 1D, 2F, 3B, 4E

3

ACTIVITY: Solving Equations Using Algebra Tiles

Work with a partner.

- Write an equation shown by the algebra tiles.
- Use algebra tiles to model and solve the equation.
- Check your answer by solving the equation without using algebra tiles.

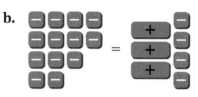

a.   b.

ACTIVITY: Working Backwards

4

Math Practice

Maintain Oversight

How does working backwards help you decide which operation to do first? Explain.

Work with a partner.

a. **Sample:** Your friend pauses a video game to get a drink. You continue the game. You double the score by saving a princess. Then you lose 75 points because you do not collect the treasure. You finish the game with −25 points. How many points did you have when you started?

One way to solve the problem is to work backwards. To do this, start with the end result and retrace the events.

You have −25 points at the end of the game.	**−25**
You lost 75 points for not collecting the treasure, so add 75 to −25.	**−25 + 75 = 50**
You doubled your score for saving the princess, so find half of 50.	**50 ÷ 2 = 25**

So, you started the game with 25 points.

b. You triple your account balance by making a deposit. Then you withdraw $127.32 to buy groceries. Your account is now overdrawn by $10.56. By working backwards, find your account balance before you made the deposit.

What Is Your Answer?

5. **IN YOUR OWN WORDS** How can you use algebra tiles to solve a two-step equation?

6. When solving the equation $4x + 1 = -11$, what is the first step?

7. **REPEATED REASONING** Solve the equation $2x - 75 = -25$. How do your steps compare with the strategy of working backwards in Activity 4?

Practice

Use what you learned about solving two-step equations to complete Exercises 6–11 on page 112.

3.5 Lesson

Check It Out
Lesson Tutorials
BigIdeasMath ✓ com

EXAMPLE 1 Solving a Two-Step Equation

Solve $-3x + 5 = 2$. Check your solution.

$$-3x + 5 = 2 \qquad \text{Write the equation.}$$

Undo the addition. $\longrightarrow \quad \underline{-5 \quad -5} \qquad$ Subtraction Property of Equality

$$-3x = -3 \qquad \text{Simplify.}$$

Undo the multiplication. $\longrightarrow \quad \dfrac{-3x}{-3} = \dfrac{-3}{-3} \qquad$ Division Property of Equality

$$x = 1 \qquad \text{Simplify.}$$

Check

$$-3x + 5 = 2$$
$$-3(1) + 5 \stackrel{?}{=} 2$$
$$-3 + 5 \stackrel{?}{=} 2$$
$$2 = 2 \checkmark$$

∴ The solution is $x = 1$.

On Your Own

Now You're Ready
Exercises 6–17

Solve the equation. Check your solution.

1. $2x + 12 = 4$

2. $-5c + 9 = -16$

3. $3(x - 4) = 9$

EXAMPLE 2 Solving a Two-Step Equation

Solve $\dfrac{x}{8} - \dfrac{1}{2} = -\dfrac{7}{2}$. Check your solution.

Study Tip

You can simplify the equation in Example 2 before solving. Multiply each side by the LCD of the fractions, 8.

$$\dfrac{x}{8} - \dfrac{1}{2} = -\dfrac{7}{2}$$
$$x - 4 = -28$$
$$x = -24$$

$$\dfrac{x}{8} - \dfrac{1}{2} = -\dfrac{7}{2} \qquad \text{Write the equation.}$$

$$\underline{+\dfrac{1}{2} \qquad +\dfrac{1}{2}} \qquad \text{Addition Property of Equality}$$

$$\dfrac{x}{8} = -3 \qquad \text{Simplify.}$$

$$8 \cdot \dfrac{x}{8} = 8 \cdot (-3) \qquad \text{Multiplication Property of Equality}$$

$$x = -24 \qquad \text{Simplify.}$$

Check

$$\dfrac{x}{8} - \dfrac{1}{2} = -\dfrac{7}{2}$$
$$\dfrac{-24}{8} - \dfrac{1}{2} \stackrel{?}{=} -\dfrac{7}{2}$$
$$-3 - \dfrac{1}{2} \stackrel{?}{=} -\dfrac{7}{2}$$
$$-\dfrac{7}{2} = -\dfrac{7}{2} \checkmark$$

∴ The solution is $x = -24$.

On Your Own

Now You're Ready
Exercises 20–25

Solve the equation. Check your solution.

4. $\dfrac{m}{2} + 6 = 10$

5. $-\dfrac{z}{3} + 5 = 9$

6. $\dfrac{2}{5} + 4a = -\dfrac{6}{5}$

Laurie's Notes

Introduction

Connect
- **Yesterday:** Students used algebra tiles to model solving two-step equations.
- **Today:** Students will solve equations by undoing the operations in the reverse order of how the expression would have been evaluated.

Motivate
- ❓ "Four friends each purchase a large beverage and share a $9 pizza. The total bill before tax is $16. What is the cost of each beverage?" $1.75
- Ask students to explain how they solved this problem. Listen for students to mention subtracting the cost of the pizza from the total before dividing by four.

Lesson Notes

Example 1
- ❓ **Vocabulary Review:** "In the expression $-3x + 5$, what is -3 called?" the coefficient
- Work through the example. Before doing each step, ask students what the next step should be.
- Take the time to check the solution so that students see this as important.

On Your Own
- Students may be uncertain of how to solve Question 3 because of the parentheses. Remind students about the Distributive Property.
- Review the methods students use to solve each problem. In Question 3, for example, some students may distribute the three and some may realize they can divide both sides of the equation by three.
- **Challenge:** Ask students to describe two methods for solving Question 1. Some may notice that each number in the equation can be divided by 2.

Example 2
- ❓ "If you knew the value of x, how would you evaluate the expression $\frac{x}{8} - \frac{1}{2}$?" *Sample answer:* Divide the number by 8 and subtract $\frac{1}{2}$; Some students might say that you need to find a common denominator and then subtract the fractions.
- ❓ "What is the first step to solve this equation?" Add $\frac{1}{2}$ to each side.
- ❓ "What is the second step to solve this equation?" Multiply each side by 8.

On Your Own
- **Think-Pair-Share:** Students should read each question independently and then work in pairs to answer the questions. When they have answered the questions, the pair should compare their answers with another group and discuss any discrepancies.

Goal Today's lesson is solving two-step equations.

Technology for the **Teacher**

Dynamic Classroom

Lesson Tutorials
Lesson Plans
Answer Presentation Tool

Extra Example 1
Solve $4t - 7 = -15$. Check your solution. -2

On Your Own
1. $x = -4$
2. $c = 5$
3. $x = 7$

Extra Example 2
Solve $\frac{n}{9} + \frac{2}{3} = -\frac{2}{3}$. Check your solution. -12

On Your Own
4. $m = 8$
5. $z = -12$
6. $a = -\frac{2}{5}$

Laurie's Notes

Extra Example 3

Solve $12.5 = 0.3m - 2.8m$. -5

Extra Example 4

A taxi charges $2.50 plus $2 for every mile traveled. Find the number of miles traveled for a fare of $10.50. 4 miles

 On Your Own

7. $y = 8$

8. $x = -5$

9. $m = 10$

10. 9.5 ft

English Language Learners

Verbal Clues

English learners should become familiar with words and phrases that give clues to the types of operations required. Word problems calling for two-step equations almost always contain words such as *per*, *each*, and *every*. These are clues to quantities that will appear in the equation, usually as the coefficient of the variable. Point out these words in the exercises and identify the terms associated with them in the equations used to solve the problems.

Example 3

- This problem requires students to *combine like terms* as the first step.
- **?** "What do we call $3y$ and $-8y$?" They are like terms.

Example 4

- Note that the unknown value in this problem is the starting height. Roller coasters do not begin on the ground!
- **Reason Abstractly and Quantitatively:** The table helps students develop the ability to translate from words to symbols.

On Your Own

- **Think-Pair-Share**: Students should read each question independently and then work in pairs to answer the questions. When they have answered the questions, the pair should compare their answers with another group and discuss any discrepancies.

Closure

- What does it mean to *isolate the variable term*? Use inverse operations to get the variable by itself.
- What are like or similar terms? Give examples. Terms that can be combined are like terms. Some examples are $3x$ and $-2x$, $5a$ and $-a$, and 5 and -8.
- Explain how the solutions of the two equations are similar.

$$4x - 5 = 7 \qquad \frac{4}{3}x - \frac{5}{3} = \frac{7}{3}$$

You can multiply each term of the second equation by 3 and then the two equations will be the same. The solution of the two equations is the same.

EXAMPLE **3** Combining Like Terms Before Solving

Solve $3y - 8y = 25$.

$3y - 8y = 25$	Write the equation.
$-5y = 25$	Combine like terms.
$y = -5$	Divide each side by -5.

⋮ The solution is $y = -5$.

EXAMPLE **4** Real-Life Application

The height at the top of a roller coaster hill is 10 times the height h of the starting point. The height decreases 100 feet from the top to the bottom of the hill. The height at the bottom of the hill is -10 feet. Find h.

Location	Verbal Description	Expression
Start	The height at the start is h.	h
Top of hill	The height at the top of the hill is 10 times the starting height h.	$10h$
Bottom of hill	The height decreases by 100 feet. So, subtract 100.	$10h - 100$

The height at the bottom of the hill is -10 feet. Solve $10h - 100 = -10$ to find h.

$10h - 100 = -10$	Write equation.
$10h = 90$	Add 100 to each side.
$h = 9$	Divide each side by 10.

⋮ So, the height at the start is 9 feet.

On Your Own

Now You're Ready
Exercises 29–34

Solve the equation. Check your solution.

7. $4 - 2y + 3 = -9$ **8.** $7x - 10x = 15$ **9.** $-8 = 1.3m - 2.1m$

10. **WHAT IF?** In Example 4, the height at the bottom of the hill is -5 feet. Find the height h.

 ## Vocabulary and Concept Check

1. **WRITING** How do you solve two-step equations?

Match the equation with the first step to solve it.

2. $4 + 4n = -12$ 3. $4n = -12$ 4. $\dfrac{n}{4} = -12$ 5. $\dfrac{n}{4} - 4 = -12$

A. Add 4.   B. Subtract 4.   C. Multiply by 4.   D. Divide by 4.

 ## Practice and Problem Solving

Solve the equation. Check your solution.

① 6. $2v + 7 = 3$ 7. $4b + 3 = -9$ 8. $17 = 5k - 2$

9. $-6t - 7 = 17$ 10. $8n + 16.2 = 1.6$ 11. $-5g + 2.3 = -18.8$

12. $2t - 5 = -10$ 13. $-4p + 9 = -5$ 14. $11 = -5x - 2$

15. $4 + 2.2h = -3.7$ 16. $-4.8f + 6.4 = -8.48$ 17. $7.3y - 5.18 = -51.9$

ERROR ANALYSIS Describe and correct the error in finding the solution.

18.
$$-6 + 2x = -10$$
$$-6 + \frac{2x}{2} = -\frac{10}{2}$$
$$-6 + x = -5$$
$$x = 1$$

19.
$$-3x + 2 = -7$$
$$-3x = -9$$
$$-\frac{3x}{3} = \frac{-9}{3}$$
$$x = -3$$

Solve the equation. Check your solution.

② 20. $\dfrac{3}{5}g - \dfrac{1}{3} = -\dfrac{10}{3}$ 21. $\dfrac{a}{4} - \dfrac{5}{6} = -\dfrac{1}{2}$ 22. $-\dfrac{1}{3} + 2z = -\dfrac{5}{6}$

23. $2 - \dfrac{b}{3} = -\dfrac{5}{2}$ 24. $-\dfrac{2}{3}x + \dfrac{3}{7} = \dfrac{1}{2}$ 25. $-\dfrac{9}{4}v + \dfrac{4}{5} = \dfrac{7}{8}$

In Exercises 26–28, write an equation. Then solve.

26. **WEATHER** Starting at 1:00 P.M., the temperature changes −4 degrees per hour. How long will it take to reach −1°?

27. **BOWLING** It costs $2.50 to rent bowling shoes. Each game costs $2.25. You have $9.25. How many games can you bowl?

28. **CELL PHONES** A cell phone company charges a monthly fee plus $0.25 for each text message. The monthly fee is $30.00 and you owe $59.50. How many text messages did you have?

Temperature at 1:00 P.M.

35°F

Assignment Guide and Homework Check

Level	Day 1 Activity Assignment	Day 2 Lesson Assignment	Homework Check
Basic	6–11, 42–46	1–5, 13–31 odd	13, 21, 27, 29
Average	6–11, 42–46	1–5, 13–25 odd, 28–36 even	13, 21, 28, 30
Advanced	6–11, 42–46	1–5, 18–40 even, 41	24, 28, 30, 36, 40
Accelerated	1–11, 18–40 even, 41–46		24, 28, 30, 36, 40

Common Errors

- **Exercises 6–17** Students may divide the coefficient first instead of adding or subtracting first. Tell them that while this is a valid method, they must remember to divide each part of the equation by the coefficient.
- **Exercises 20–25** Students may immediately multiply each term by one of the denominators without thinking if it will help them solve for the variable. Ask them to check if all the denominators would be eliminated.
- **Exercises 32–34** Students may try to add or subtract without distributing. Remind them that when parentheses are present, they either need to use the Distributive Property or they need to undo the multiplication first. All of the exercises can be solved using either method.

3.5 Record and Practice Journal

Solve the equation. Check your solution.

1. $3a - 5 = -14$
-3

2. $10 = -2c + 22$
6

3. $18 = -5b - 17$
-7

4. $-12 = -8z + 12$
3

5. $1.3n - 0.03 = -9$
-6.9

6. $-\frac{5}{11}h + \frac{7}{9} = \frac{2}{9}$
$1\frac{2}{9}$

7. The length of a rectangle is 3 meters less than twice its width.

 a. Write an equation to find the length of the rectangle.
 $\ell = 2w - 3$

 b. The length of the rectangle is 11 meters. What is the width of the rectangle?
 7 meters

Vocabulary and Concept Check

1. Eliminate the constants on the side with the variable. Then solve for the variable using either division or multiplication.

2. B   3. D

4. C   5. A

Practice and Problem Solving

6. $v = -2$ 7. $b = -3$

8. $k = 3\frac{4}{5}$ 9. $t = -4$

10. $n = -1.825$

11. $g = 4.22$

12. $t = -2\frac{1}{2}$ 13. $p = 3\frac{1}{2}$

14. $x = -2\frac{3}{5}$ 15. $h = -3.5$

16. $f = 3.1$ 17. $y = -6.4$

18. The steps are out of order.
$$-6 + 2x = -10$$
$$2x = -4$$
$$\frac{2x}{2} = \frac{-4}{2}$$
$$x = -2$$

19. Each side should be divided by -3, not 3.
$$-3x + 2 = -7$$
$$-3x = -9$$
$$\frac{-3x}{-3} = \frac{-9}{-3}$$
$$x = 3$$

20. $g = -5$ 21. $a = 1\frac{1}{3}$

22. $z = -\frac{1}{4}$ 23. $b = 13\frac{1}{2}$

24. $x = -\frac{3}{28}$ 25. $v = -\frac{1}{30}$

26. $-4x + 35 = -1$;
9 hours (10:00 P.M.)

27. $2.5 + 2.25x = 9.25$; 3 games

28. $30 + 0.25x = 59.5$;
118 text messages

29. $v = -5$ **30.** $t = -13$

31. $d = -12$ **32.** $x = -1$

33. $m = -9$ **34.** $y = -4$

35. *Sample answer:* You travel halfway up a ladder. Then you climb down two feet and are 8 feet above the ground. How long is the ladder? $x = 20$

36. $12\frac{3}{4}$ ft   **37.** the initial fee

38. the coldest surface temperature on the moon

39. See *Taking Math Deeper.*

40. -21 ft

41. decrease the length by 10 cm; $2(25 + x) + 2(12) = 54$

Fair Game Review

42. -34.72 **43.** $-6\frac{2}{3}$

44. $-3\frac{1}{8}$ **45.** 6.2

46. C

Taking Math Deeper

Exercise 39

This problem asks students to answer a question *with* and *without* using algebra.

 Summarize the given information.

1. You caught *x* insects on Saturday.

2. 5 of the insects escaped.

3. The remaining insects form 3 groups of 9 each.

 a. Work backwards.

3. There are $3(9) = 27$ insects remaining.

2. Add the 5 that escaped.

1. You caught $27 + 5$, or 32, insects on Saturday.

 b. Write and solve an equation.

1. *x* insects on Saturday

2. $(x - 5)$ are remaining.

3. $\dfrac{(x - 5)}{3} = 9$

$\dfrac{x - 5}{3} = 9$ Write the equation.

$x - 5 = 27$ Multiply each side by 3.

$x = 32$ Add 5 to each side.

You caught 32 insects on Saturday.

Reteaching and Enrichment Strategies

If students need help. . .	If students got it. . .
Resources by Chapter • Practice A and Practice B • Puzzle Time Record and Practice Journal Practice Differentiating the Lesson Lesson Tutorials Skills Review Handbook	Resources by Chapter • Enrichment and Extension • Technology Connection Start the next section

Solve the equation. Check your solution.

③ 29. $3v - 9v = 30$
30. $12t - 8t = -52$
31. $-8d - 5d + 7d = 72$

32. $6(x - 2) = -18$
33. $-4(m + 3) = 24$
34. $-8(y + 9) = -40$

35. **WRITING** Write a real-world problem that can be modeled by $\frac{1}{2}x - 2 = 8$. Then solve the equation.

36. **GEOMETRY** The perimeter of the parallelogram is 102 feet. Find m.

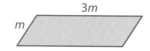

REASONING Exercises 37 and 38 are missing information. Tell what information you need to solve the problem.

37. **TAXI** A taxi service charges an initial fee plus $1.80 per mile. How far can you travel for $12?

38. **EARTH** The coldest surface temperature on the Moon is 57 degrees colder than twice the coldest surface temperature on Earth. What is the coldest surface temperature on Earth?

39. **PROBLEM SOLVING** On Saturday, you catch insects for your science class. Five of the insects escape. The remaining insects are divided into three groups to share in class. Each group has nine insects. How many insects did you catch on Saturday?

   a. Solve the problem by working backwards.

   b. Solve the equation $\frac{x - 5}{3} = 9$. How does the answer compare with the answer to part (a)?

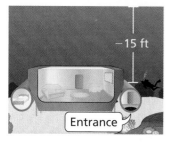

40. **UNDERWATER HOTEL** You must scuba dive to the entrance of your room at Jules' Undersea Lodge in Key Largo, Florida. The diver is 1 foot deeper than $\frac{2}{3}$ of the elevation of the entrance. What is the elevation of the entrance?

41. ⟨**Geometry**⟩ How much should you change the length of the rectangle so that the perimeter is 54 centimeters? Write an equation that shows how you found your answer.

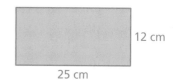

12 cm

25 cm

Fair Game Review *What you learned in previous grades & lessons*

Multiply or divide. *(Section 2.4)*

42. -6.2×5.6
43. $\frac{8}{3} \times \left(-2\frac{1}{2}\right)$
44. $\frac{5}{2} \div \left(-\frac{4}{5}\right)$
45. $-18.6 \div (-3)$

46. **MULTIPLE CHOICE** Which fraction is *not* equivalent to 0.75? *(Skills Review Handbook)*

   Ⓐ $\frac{15}{20}$
   Ⓑ $\frac{9}{12}$
   Ⓒ $\frac{6}{9}$
   Ⓓ $\frac{3}{4}$

Solve the equation. Check your solution. *(Section 3.3, Section 3.4, and Section 3.5)*

1. $-6.5 + x = -4.12$

2. $4\frac{1}{2} + p = -5\frac{3}{4}$

3. $-\frac{b}{7} = 4$

4. $-2w + 3.7 = -0.5$

Write the word sentence as an equation. Then solve. *(Section 3.3 and Section 3.4)*

5. The difference between a number b and 7.4 is -6.8.

6. $5\frac{2}{5}$ more than a number a is $7\frac{1}{2}$.

7. A number x multiplied by $\frac{3}{8}$ is $-\frac{15}{32}$.

8. The quotient of two times a number k and -2.6 is 12.

Write and solve an equation to find the value of x. *(Section 3.3 and Section 3.5)*

9. Perimeter = 26

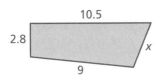

10. Perimeter = 23.59

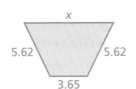

11. Perimeter = 33

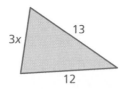

12. BANKING You withdraw $29.79 from your bank account. Now your balance is $-\$20.51$. Write and solve an equation to find the amount of money in your bank account before you withdrew the money. *(Section 3.3)*

13. WATER LEVEL During a drought, the water level of a lake changes $-3\frac{1}{5}$ feet per day. Write and solve an equation to find how long it takes for the water level to change -16 feet. *(Section 3.4)*

14. BASKETBALL A basketball game has four quarters. The length of a game is 32 minutes. You play the entire game except for $4\frac{1}{2}$ minutes. Write and solve an equation to find the mean time you play per quarter. *(Section 3.5)*

15. SCRAPBOOKING The mat needs to be cut to have a 0.5-inch border on all four sides. *(Section 3.5)*

a. How much should you cut from the left and right sides?

b. How much should you cut from the top and bottom?

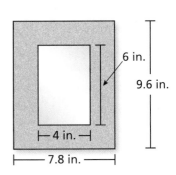

Alternative Assessment Options

Math Chat   Student Reflective Focus Question

Structured Interview   Writing Prompt

Structured Interview

Interviews can occur formally or informally. Ask a student to perform a task and explain it, describing his or her thought process throughout the task. Probe the student for more information. Do not ask leading questions. Keep a rubric or notes.

Teacher Prompts	Student Answers	Teacher Notes
Tell me a story about buying flowers. Include Cost of a vase: $8.50 Cost per rose: $2.25 Total amount spent: $26.50 Number of roses bought: ?	I bought flowers for my aunt for her birthday. I bought a vase that costs $8.50 and roses that cost $2.25 each. I spent a total of $26.50. So, I bought 8 roses.	Student can write and solve a two-step equation.
Add to your story using this phrase. $50 bill and receive ? back	I gave the florist a $50 bill and received $23.50 back in change.	Student can write and solve an equation involving addition or subtraction.

Study Help Sample Answers

Remind students to complete Graphic Organizers for the rest of the chapter.

4.

Words	Algebra
Two equations are equivalent equations if they have the same solutions. You can use the Addition, Subtraction, Multiplication, and Division Properties of Equality to write equivalent equations.	$a = b$ and $a + c = b + c$ $a = b$ and $a - c = b - c$ $a = b$ and $a \cdot c = b \cdot c$ $a = b$ and $\frac{a}{c} = \frac{b}{c}, c \neq 0$

$$\text{Equivalent equations}$$

Examples	Non-Examples
$x - 7 = 2$ and $x - 7 + 7 = 2 + 7$ $2d + 5 = -7$ and $2d + 5 - 5 = -7 - 5$ $24 = \frac{y}{-4}$ and $-4 \cdot 24 = -4 \cdot \frac{y}{-4}$ $3c = -12$ and $\frac{3c}{3} = \frac{-12}{3}$	$x + 7 = 2$ and $x + 7 - 7 = 2 + 7$ $3c = -4$ and $\frac{3c}{3} = 3 \cdot (-4)$ $7 = m + 3$ and $7 - 7 = m + 3 - 3$ $3x + 7 = 3$ and $3x = 7 + 3$

5–7. Available at *BigIdeasMath.com*.

Reteaching and Enrichment Strategies

If students need help...	If students got it...
Resources by Chapter • Practice A and Practice B • Puzzle Time Lesson Tutorials *BigIdeasMath.com*	Resources by Chapter • Enrichment and Extension • Technology Connection Game Closet at *BigIdeasMath.com* Start the Chapter Review

Technology for the *Teacher*

Online Assessment
Assessment Book
ExamView® Assessment Suite

Answers

1. Terms: z, 8, $-4z$;
   Like terms: z and $-4z$

2. Terms: $3n$, 7, $-n$, -3;
   Like terms: $3n$ and $-n$,
   7 and -3

3. Terms: $10x^2$, $-y$, 12, $-3x^2$;
   Like terms: $10x^2$ and $-3x^2$

4. $-4h$

5. $3.5r - 7$

6. $\dfrac{9}{20}x + 12$

7. $3q + 21$

8. $2m - 18$

9. $1.5n - 3.2$

Review of Common Errors

Exercises 1–3

- When identifying and writing terms, make sure students include the sign of the term. They may find it helpful to write the original problem using addition.
- Students may confuse like variables with like terms. Remind them that the same variables must be raised to the same exponents for terms to be like terms.

Exercises 4–9

- The subtraction operation can confuse students. Tell students to write the original problem using addition and then to use the Commutative Property.

Exercises 7–9

- Students often forget to distribute the constant over *both* of the terms inside the parentheses. Remind them of the Distributive Property, $a(b + c) = ab + ac$.

Exercises 16–23

- Students may use the same operation in solving for x instead of the inverse operation.

Exercises 24–32

- When the variable is multiplied by a negative number, students forget to keep the negative with the number and solve for $-x$ instead of x.

Exercises 33–37

- Students might divide the coefficient first instead of adding or subtracting.

3 Chapter Review

Review Key Vocabulary

like terms, *p. 82*
simplest form, *p. 82*
linear expression, *p. 88*

factoring an expression, *p. 92*
equivalent equations, *p. 98*

Review Examples and Exercises

3.1 Algebraic Expressions *(pp. 80–85)*

a. **Identify the terms and like terms in the expression** $6y + 9 + 3y - 7$.

Rewrite as a sum of terms.

$$6y + 9 + 3y + (-7)$$

Terms: $6y$, $\ 9$, $\ 3y$, $\ -7$

Like terms: $6y$ and $3y$, 9 and -7

b. **Simplify** $\dfrac{2}{3}y + 14 - \dfrac{1}{6}y - 8$.

$$\frac{2}{3}y + 14 - \frac{1}{6}y - 8 = \frac{2}{3}y + 14 + \left(-\frac{1}{6}y\right) + (-8) \qquad \text{Rewrite as a sum.}$$

$$= \frac{2}{3}y + \left(-\frac{1}{6}y\right) + 14 + (-8) \qquad \text{Commutative Property of Addition}$$

$$= \left[\frac{2}{3} + \left(-\frac{1}{6}\right)\right]y + 14 + (-8) \qquad \text{Distributive Property}$$

$$= \frac{1}{2}y + 6 \qquad \text{Combine like terms.}$$

Exercises

Identify the terms and like terms in the expression.

1. $z + 8 - 4z$

2. $3n + 7 - n - 3$

3. $10x^2 - y + 12 - 3x^2$

Simplify the expression.

4. $4h - 8h$

5. $6.4r - 7 - 2.9r$

6. $\dfrac{3}{5}x + 19 - \dfrac{3}{20}x - 7$

7. $3(2 + q) + 15$

8. $\dfrac{1}{8}(16m - 8) - 17$

9. $-1.5(4 - n) + 2.8$

Adding and Subtracting Linear Expressions   *(pp. 86–93)*

a. Find $(5z + 4) + (3z - 6)$.

$$
\begin{array}{r}
5z + 4 \\
+\ 3z - 6 \\
\hline
8z - 2
\end{array}
$$ Align like terms vertically and add.

b. Factor $\dfrac{1}{4}$ out of $\dfrac{1}{4}x - \dfrac{3}{4}$.

Write each term as a product of $\dfrac{1}{4}$ and another factor.

$$\frac{1}{4}x = \frac{1}{4} \cdot x \qquad\qquad -\frac{3}{4} = \frac{1}{4} \cdot (-3)$$

Use the Distributive Property to factor out $\dfrac{1}{4}$.

$$\frac{1}{4}x - \frac{3}{4} = \frac{1}{4} \cdot x + \frac{1}{4} \cdot (-3) = \frac{1}{4}(x - 3)$$

So, $\dfrac{1}{4}x - \dfrac{3}{4} = \dfrac{1}{4}(x - 3)$.

Exercises

Find the sum or difference.

10. $(c - 4) + (3c + 9)$

11. $\dfrac{2}{5}(d - 10) - \dfrac{2}{3}(d + 6)$

Factor out the coefficient of the variable.

12. $2b + 8$ **13.** $\dfrac{1}{4}y + \dfrac{3}{8}$ **14.** $1.7j - 3.4$ **15.** $-5p + 20$

Solving Equations Using Addition or Subtraction   *(pp. 96–101)*

Solve $x - 9 = -6$.

Undo the subtraction. →

$$
\begin{array}{rl}
x - 9 = -6 & \text{Write the equation.} \\
\underline{+\ 9\quad +\ 9} & \text{Addition Property of Equality} \\
x = 3 & \text{Simplify.}
\end{array}
$$

Check

$$
\begin{array}{l}
x - 9 = -6 \\
3 - 9 \overset{?}{=} -6 \\
-6 = -6 \ \checkmark
\end{array}
$$

Exercises

Solve the equation. Check your solution.

16. $p - 3 = -4$ **17.** $6 + q = 1$ **18.** $-2 + j = -22$ **19.** $b - 19 = -11$

20. $n + \dfrac{3}{4} = \dfrac{1}{4}$ **21.** $v - \dfrac{5}{6} = -\dfrac{7}{8}$ **22.** $t - 3.7 = 1.2$ **23.** $\ell + 15.2 = -4.5$

Review Game

Expressions and Equations

Materials
- color-coded spinner with operation symbols
- 20 cards numbered −10 through 10, excluding 0
- 2 pencils
- paper

Players: 2

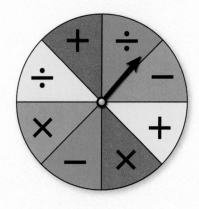

Directions
One player draws three cards and writes each number in one of the shaded boxes in the equation below.

The player then spins the spinner. The spinner is used to determine the operation to be used. The player writes the operation in the unshaded box above. For example, if a player draws −10, −3, and 2, and spins *red* +, the equation may be $-3x + 2 = -10$.

Each player solves the equation, exchanges his or her work with the other player, and checks the other player's work. Each player receives 1 point for the round for a correct solution. Each round can result in a total of 0, 1, or 2 points received between the two players. Players take turns drawing/spinning and writing the equations. Students can play several rounds or for a predetermined amount of time.

Who Wins?
The player with the most points wins.

For the Student
Additional Practice
- Lesson Tutorials
- Multi-Language Glossary
- Self-Grading Progress Check
- *BigIdeasMath.com*
  Dynamic Student Edition
  Student Resources

Answers

10. $4c + 5$ **11.** $-\dfrac{4}{15}d - 8$

12. $2(b + 4)$ **13.** $\dfrac{1}{4}\left(y + \dfrac{3}{2}\right)$

14. $1.7(j - 2)$ **15.** $-5(p - 4)$

16. $p = -1$ **17.** $q = -5$

18. $j = -20$ **19.** $b = 8$

20. $n = -\dfrac{1}{2}$ **21.** $v = -\dfrac{1}{24}$

22. $t = 4.9$ **23.** $\ell = -19.7$

24. $x = -24$ **25.** $y = -49$

26. $z = 3$ **27.** $w = 50$

28. $x = -2$ **29.** $y = -5$

30. $z = 6$ **31.** $w = -0.5$

32. $-16°F$ **33.** $c = 7$

34. $w = -\dfrac{8}{9}$ **35.** $w = -12$

36. $x = -3.5$ **37.** 11 years

My Thoughts on the Chapter

What worked. . .

Teacher Tip

Not allowed to write in your teaching edition? Use sticky notes to record your thoughts.

What did not work. . .

What I would do differently. . .

3.4 Solving Equations Using Multiplication or Division (pp. 102–107)

Solve $\dfrac{x}{5} = -7$.

$\dfrac{x}{5} = -7$	Write the equation.	

Undo the division. ⟶ $5 \cdot \dfrac{x}{5} = 5 \cdot (-7)$ Multiplication Property of Equality

$x = -35$ Simplify.

Check

$\dfrac{x}{5} = -7$

$\dfrac{-35}{5} \overset{?}{=} -7$

$-7 = -7$ ✓

Exercises

Solve the equation. Check your solution.

24. $\dfrac{x}{3} = -8$ **25.** $-7 = \dfrac{y}{7}$ **26.** $-\dfrac{z}{4} = -\dfrac{3}{4}$ **27.** $-\dfrac{w}{20} = -2.5$

28. $4x = -8$ **29.** $-10 = 2y$ **30.** $-5.4z = -32.4$ **31.** $-6.8w = 3.4$

32. TEMPERATURE The mean temperature change is $-3.2°F$ per day for 5 days. What is the total change over the 5-day period?

3.5 Solving Two-Step Equations (pp. 108–113)

Solve $-6y + 7 = -5$. Check your solution.

$-6y + 7 = -5$	Write the equation.
$\underline{-7 \quad -7}$	Subtraction Property of Equality
$-6y = -12$	Simplify.
$\dfrac{-6y}{-6} = \dfrac{-12}{-6}$	Division Property of Equality
$y = 2$	Simplify.

Check

$-6y + 7 = -5$

$-6(2) + 7 \overset{?}{=} -5$

$-12 + 7 \overset{?}{=} -5$

$-5 = -5$ ✓

∴ The solution is $y = 2$.

Exercises

Solve the equation. Check your solution.

33. $-2c + 6 = -8$ **34.** $3(3w - 4) = -20$

35. $\dfrac{w}{6} + \dfrac{5}{8} = -1\dfrac{3}{8}$ **36.** $-3x - 4.6 = 5.9$

37. EROSION The floor of a canyon has an elevation of -14.5 feet. Erosion causes the elevation to change by -1.5 feet per year. How many years will it take for the canyon floor to have an elevation of -31 feet?

Check It Out
Test Practice
BigIdeasMath ✓com

Simplify the expression.

1. $8x - 5 + 2x$

2. $2.5w - 3y + 4w$

3. $3(5 - 2n) + 9n$

4. $\dfrac{5}{7}x + 15 - \dfrac{9}{14}x - 9$

Find the sum or difference.

5. $(3j + 11) + (8j - 7)$

6. $\dfrac{3}{4}(8p + 12) + \dfrac{3}{8}(16p - 8)$

7. $(2r - 13) - (-6r + 4)$

8. $-2.5(2s - 5) - 3(4.5s - 5.2)$

Factor out the coefficient of the variable.

9. $3n - 24$

10. $\dfrac{1}{2}q + \dfrac{5}{2}$

Solve the equation. Check your solution.

11. $7x = -3$

12. $2(x + 1) = -2$

13. $\dfrac{2}{9}g = -8$

14. $z + 14.5 = 5.4$

15. $-14 = 6c$

16. $\dfrac{2}{7}k - \dfrac{3}{8} = -\dfrac{19}{8}$

17. HAIR SALON Write an expression in simplest form that represents the income from w women and m men getting a haircut and a shampoo.

	Women	Men
Haircut	$45	$15
Shampoo	$12	$7

18. RECORD A runner is compared with the world record holder during a race. A negative number means the runner is ahead of the time of the world record holder. A positive number means that the runner is behind the time of the world record holder. The table shows the time difference between the runner and the world record holder for each lap. What time difference does the runner need for the fourth lap to match the world record?

Lap	Time Difference
1	−1.23
2	0.45
3	0.18
4	?

19. GYMNASTICS You lose 0.3 point for stepping out of bounds during a floor routine. Your final score is 9.124. Write and solve an equation to find your score before the penalty.

20. PERIMETER The perimeter of the triangle is 45. Find the value of x.

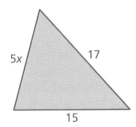

$5x$ 17

15

Test Item References

Chapter Test Questions	Section to Review
1–4, 17	3.1
5–10	3.2
11–16, 19	3.3
11–16	3.4
11–16, 18, 20	3.5

Test-Taking Strategies

Remind students to quickly look over the entire test before they start so that they can budget their time. On tests, it is really important for students to **Stop** and **Think.** When students hurry on a test dealing with signed numbers, they often make "sign" errors. There are equations on the test, so remind students to always check their solutions.

Common Errors

- **Exercises 1–8** The subtraction operation can confuse students. Tell students to write the original problem using addition and then to use the Commutative Property.

- **Exercises 11–16** Students may use the same operations, instead of inverse operations, when solving for the variable. Demonstrate to students that this will not give the correct solution. Also, students may use the properties of equality improperly by adding, subtracting, multiplying, or dividing on one side of the equation only, or by using inverse operations on opposite sides of the equation. Remind students of the properties and to check their answers in the original equation.

Reteaching and Enrichment Strategies

If students need help. . .	If students got it. . .
Resources by Chapter • Practice A and Practice B • Puzzle Time Record and Practice Journal Practice Differentiating the Lesson Lesson Tutorials *BigIdeasMath.com* Skills Review Handbook	Resources by Chapter • Enrichment and Extension • Technology Connection Game Closet at *BigIdeasMath.com* Start Cumulative Assessment

Answers

1. $10x - 5$
2. $6.5w - 3y$
3. $15 + 3n$
4. $\dfrac{1}{14}x + 6$
5. $11j + 4$
6. $12p + 6$
7. $8r - 17$
8. $-18.5s + 28.1$
9. $3(n - 8)$
10. $\dfrac{1}{2}(q + 5)$
11. $x = -\dfrac{3}{7}$
12. $x = -2$
13. $g = -36$
14. $z = -9.1$
15. $c = -2\dfrac{1}{3}$
16. $k = -7$
17. $57w + 22m$
18. 0.6
19. $x - 0.3 = 9.124;\ 9.424$
20. $2\dfrac{3}{5}$

Technology for the *Teacher*

Online Assessment
Assessment Book
ExamView® Assessment Suite

Test-Taking Strategies
Available at *BigIdeasMath.com*

After Answering Easy Questions, Relax
Answer Easy Questions First
Estimate the Answer
Read All Choices before Answering
Read Question before Answering
Solve Directly or Eliminate Choices
Solve Problem before Looking at Choices
Use Intelligent Guessing
Work Backwards

About this Strategy

When taking a multiple choice test, be sure to read each question carefully and thoroughly. After skimming the test and answering the easy questions, stop for a few seconds, take a deep breath, and relax. Work through the remaining questions carefully, using your knowledge and test-taking strategies. Remember, you already completed many of the questions on the test!

Answers

1. B
2. −9
3. G
4. B

Technology for the *Teacher*

Performance Tasks
Online Assessment
Assessment Book
ExamView® Assessment Suite

Item Analysis

1. **A.** A student multiplies instead of divides.
   B. Correct answer
   C. The student exchanges the dividend and divisor.
   D. The student sets the equation equal to the wrong integer.

2. **Gridded Response:** Correct answer: −9

   Common Error: The student incorrectly substitutes and evaluates $\dfrac{(-6)(0) - 0^2}{4}$ to get 0 as an answer.

3. **F.** The student adds −38 and −14.
   G. Correct answer
   H. The student adds 38 and −14.
   I. The student adds 38 and 14.

4. **A.** The student finds the temperature of the first thermometer.
   B. Correct answer
   C. The student finds the median of the temperatures.
   D. The student finds the mean of 8°, 10°, and 12°.

1. Which equation represents the word sentence shown below?

   > The quotient of a number b and 0.3 equals negative 10.

   A. $0.3b = 10$

   C. $\dfrac{0.3}{b} = -10$

   B. $\dfrac{b}{0.3} = -10$

   D. $\dfrac{b}{0.3} = 10$

2. What is the value of the expression below when $c = 0$ and $d = -6$?

   $$\dfrac{cd - d^2}{4}$$

3. What is the value of the expression below?

   $$-38 - (-14)$$

   F. -52

   H. 24

   G. -24

   I. 52

4. The daily low temperatures last week are shown below.

   What is the mean low temperature of last week?

   A. $-2°F$

   C. $8°F$

   B. $6°F$

   D. $10°F$

5. Which equation is equivalent to the equation shown below?

$$-\frac{3}{4}x + \frac{1}{8} = -\frac{3}{8}$$

F. $-\frac{3}{4}x = -\frac{3}{8} - \frac{1}{8}$

G. $-\frac{3}{4}x = -\frac{3}{8} + \frac{1}{8}$

H. $x + \frac{1}{8} = -\frac{3}{8} \cdot \left(-\frac{4}{3}\right)$

I. $x + \frac{1}{8} = -\frac{3}{8} \cdot \left(-\frac{3}{4}\right)$

6. What is the value of the expression below?

$$-0.28 \div (-0.07)$$

7. Karina was solving the equation in the box below.

> $-96 = -6(x - 15)$
>
> $-96 = -6x - 90$
>
> $-96 + 90 = -6x - 90 + 90$
>
> $-6 = -6x$
>
> $\dfrac{-6}{-6} = \dfrac{-6x}{-6}$
>
> $1 = x$

What should Karina do to correct the error that she made?

A. First add 6 to both sides of the equation.

B. First subtract x from both sides of the equation.

C. Distribute the -6 to get $6x - 90$.

D. Distribute the -6 to get $-6x + 90$.

8. The perimeter of the rectangle is 400 inches. What is the value of j? (All measurements are in inches.)

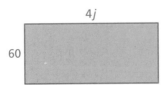

F. 35

G. 85

H. 140

I. 200

Item Analysis (continued)

5. **F.** Correct answer

   G. The student adds $\frac{1}{8}$ to both sides instead of subtracting.

   H. The student creates an equation that is not equivalent to the given equation, because if the student multiplies both sides of the equation by $-\frac{4}{3}$, this would require distributing the multiplication on the left side of the equation, thereby also multiplying $\frac{1}{8}$ by $-\frac{4}{3}$.

   I. The student creates an equation that is not equivalent to the given equation, because if the student multiplies both sides of the equation by $-\frac{3}{4}$, this would require distributing the multiplication on the left side of the equation, thereby also multiplying $\frac{1}{8}$ by $-\frac{3}{4}$. Furthermore, multiplying $-\frac{3}{4}x$ by $-\frac{3}{4}$ does not result in a product of x.

6. **Gridded Response:** Correct answer: 4

   Common Error: The student thinks that the quotient of two negative numbers is a negative number, or the student misplaces the decimal point in the answer.

7. **A.** The student thinks that the inverse operation of multiplying by -6 is adding 6.

   B. The student disregards that the -6 first must be distributed and violates the order of operations.

   C. The student does not distribute the negative sign to the two terms inside the parentheses.

   D. Correct answer

8. **F.** Correct answer

   G. The student solves $60 + 4j = 400$.

   H. The student finds the length, $4j = 140$ inches.

   I. The student adds the length and the width.

Answers

9. B

10. H

11. *Part A*

$116 = 43.50 + 7.25w$

$w = 10$

10 weeks

Part B

$116 - 20 = 43.50 + 8.75w$

$w = 6$

4 weeks sooner because in Part A it takes 10 weeks and in Part B it takes 6 weeks. The difference is $10 - 6 = 4$ weeks.

Item Analysis (continued)

9. **A.** The student incorrectly performs the order of operations.

B. Correct answer

C. The student incorrectly performs the order of operations.

D. The student incorrectly performs the order of operations.

10. **F.** $-2\frac{1}{4} - \left(8\frac{3}{8}\right)$

G. $-(2 + 8)\frac{1 + 3}{4 + 8}$

H. Correct answer

I. $(8 - 2)\frac{3 - 1}{8 - 4}$

11. **4 points** The student demonstrates a thorough understanding of writing and solving two-step equations and of interpreting the results. In Part A, the student correctly writes an equation such as $116 = 43.50 + 7.25w$ and solves to get $w = 10$. The student shows appropriate work and states that it would take 10 weeks to save the money before you can purchase the bicycle. In Part B, the student correctly adjusts the equation written in Part A to get an equation such as $116 - 20 = 43.50 + 8.75w$ and solves to get $w = 6$. The student shows appropriate work and states you could have purchased the bicycle 4 weeks sooner because $10 - 6 = 4$.

3 points The student demonstrates a good understanding of writing and solving two-step equations, and interpreting the results, but the student's work and explanations demonstrate an essential but less than thorough understanding.

2 points The student demonstrates a partial understanding of writing and solving two-step equations and of interpreting the results. The student's work and explanations demonstrate a lack of essential understanding.

1 point The student demonstrates a limited understanding of writing and solving two-step equations and of interpreting the results. The student's response is incomplete and exhibits many flaws.

0 points The student provided no response, a completely incorrect or incomprehensible response, or a response that demonstrates insufficient understanding of writing and solving two-step equations, and interpreting the results.

9. Jacob was evaluating the expression below when $x = -2$ and $y = 4$.

$$3 + x^2 \div y$$

His work is in the box below.

$$3 + x^2 \div y = 3 + (-2^2) \div 4$$
$$= 3 - 4 \div 4$$
$$= 3 - 1$$
$$= 3$$

What should Jacob do to correct the error that he made?

A. Divide 3 by 4 before subtracting.

B. Square -2, then divide.

C. Square then divide.

D. Subtract 4 from 3 before dividing.

10. Which number is equivalent to the expression shown below?

$$-2\frac{1}{4} - \left(-8\frac{3}{8}\right)$$

F. $-10\frac{5}{8}$ **H.** $6\frac{1}{8}$

G. $-10\frac{1}{3}$ **I.** $6\frac{1}{2}$

11. You want to buy the bicycle. You already have $43.50 saved and plan to save an additional $7.25 every week.

Think
Solve
Explain

Part A   Write and solve an equation to find the number of weeks you need to save before you can purchase the bicycle.

Part B   How much sooner could you purchase the bicycle if you had a coupon for $20 off and saved $8.75 every week? Explain your reasoning.

HOT PRICE!!
$116

4 Inequalities

4.1   Writing and Graphing Inequalities

4.2   Solving Inequalities Using Addition or Subtraction

4.3   Solving Inequalities Using Multiplication or Division

4.4   Solving Two-Step Inequalities

"If you reached into your water bowl and found more than $20..."

"And then reached into your cat food bowl and found more than $40..."

Someone else's bowls!

"What would you have?"

"Dear Precious Pet World: Your ad says 'Up to 75% off on selected items.'"

Hey, it didn't say who's doing the selecting.

"I select Yummy Tummy Bacon-Flavored Dog Biscuits."

What Your Students Have Learned

- Use and interpret simple equations.
- Determine whether a value is a solution of an inequality.
- Represent constraints with inequalities and recognize that they can have infinitely many solutions.
- Solve one-step equations and inequalities.

What Your Students Will Learn

- Solve one-step inequalities involving integers and rational numbers.
- Solve two-step inequalities.

Pacing Guide for Chapter 4

Chapter Opener	
Regular	1 Day
Accelerated	1 Day
Section 1	
Regular	2 Days
Accelerated	1 Day
Section 2	
Regular	2 Days
Accelerated	1 Day
Study Help / Quiz	
Regular	1 Day
Accelerated	0 Days
Section 3	
Regular	2 Days
Accelerated	1 Day
Section 4	
Regular	2 Days
Accelerated	1 Day
Chapter Review/ Chapter Tests	
Regular	2 Days
Accelerated	2 Days
Total Chapter 4	
Regular	12 Days
Accelerated	7 Days
Year-to-Date	
Regular	54 Days
Accelerated	29 Days

Technology for the *Teacher*

BigIdeasMath.com
Chapter at a Glance
Complete Materials List
Parent Letters: English and Spanish

What Your Students Have Learned

- Graph inequalities on a number line.
- Compare rational numbers and decimals.

Additional Topics for Review

- Writing Expressions
- Evaluating Expressions
- Solving Two-Step Equations
- Converting Between Fractions and Decimals

Try It Yourself

1.
2.
3.
4.
5. <   6. >
7. <   8. <
9. >   10. <

Record and Practice Journal
Fair Game Review

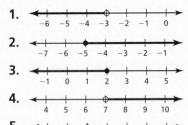

1.
2.
3.
4.
5.
6.
7.

8. >   9. >
10. >   11. <
12. <   13. <

14. your friend; 5.6 ft is about 5 ft and 7 in.

Math Background Notes

Vocabulary Review

- Inequality
- Number Line
- Integers
- Rational Numbers

Graphing Inequalities

- Students should be able to graph inequalities on a number line.
- Remind students that an equation produces a finite number of solutions, but an inequality produces an entire set of solutions. That is why an inequality requires you to shade the number line to describe the solutions.
- Remind students that inequalities containing ≤ or ≥ will require a closed circle. Inequalities containing < or > will require an open circle.
- **Teaching Tip:** Some students have difficulty deciding which side of the number line to shade. Encourage students to pick a test value on each side of the circle. Substitute each test value for *x*. Only one of the resulting inequalities will be true. Shade the number line on the side of the circle from which the valid test value was selected.

Comparing Numbers

- Students should know how to order integers and rational numbers and work with the number line.
- You may want to discuss the scaling of the number line with students. Each tick mark should count the same amount and be equally spaced on the number line.
- **Common Error:** Students may not use a number line and end up with incorrect comparisons. Encourage students to graph the numbers on a number line, as shown in Example 2, to help them correctly compare the numbers.

Reteaching and Enrichment Strategies

If students need help. . .	If students got it. . .
Record and Practice Journal • Fair Game Review Skills Review Handbook Lesson Tutorials	Game Closet at *BigIdeasMath.com* Start the next section

What You Learned Before

Why do I always have to be farther from the fulcrum?

"Move farther back. We still have an inequality."

Graphing Inequalities

Example 1 Graph $x \geq 2$.

Use a closed circle because 2 is a solution.

Shade the number line on the side where you found the solution.

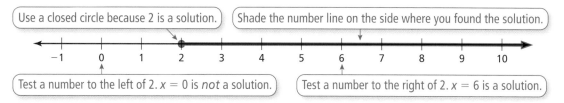

Test a number to the left of 2. $x = 0$ is *not* a solution.

Test a number to the right of 2. $x = 6$ is a solution.

Try It Yourself
Graph the inequality.

1. $x \geq 1$ **2.** $x < 5$ **3.** $x \leq 20$ **4.** $x > 13$

Comparing Numbers

Example 2 Compare $-\dfrac{1}{3}$ and $-\dfrac{5}{6}$.

Graph $-\dfrac{5}{6}$.

Graph $-\dfrac{1}{3}$.

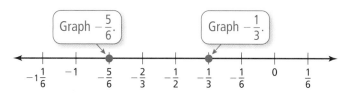

$-\dfrac{5}{6}$ is to the left of $-\dfrac{1}{3}$.

So, $-\dfrac{5}{6} < -\dfrac{1}{3}$.

Try It Yourself
Copy and complete the statement using < or >.

5. $-\dfrac{2}{3} \quad\rule{1cm}{0.4pt}\quad \dfrac{3}{8}$ **6.** $-\dfrac{1}{2} \quad\rule{1cm}{0.4pt}\quad -\dfrac{7}{8}$

7. $-\dfrac{1}{5} \quad\rule{1cm}{0.4pt}\quad \dfrac{1}{10}$ **8.** $-1.4 \quad\rule{1cm}{0.4pt}\quad 1.2$

9. $-2.2 \quad\rule{1cm}{0.4pt}\quad -4.6$ **10.** $-1.9 \quad\rule{1cm}{0.4pt}\quad -1.1$

4.1 Writing and Graphing Inequalities

Essential Question How can you use a number line to represent solutions of an inequality?

1 ACTIVITY: Understanding Inequality Statements

Work with a partner. Read the statement. Circle each number that makes the statement true, and then answer the questions.

a. "You are in at least 5 of the photos."

$$-3 \quad -2 \quad -1 \quad 0 \quad 1 \quad 2 \quad 3 \quad 4 \quad 5 \quad 6$$

- What do you notice about the numbers that you circled?

- Is the number 5 included? Why or why not?

- Write four other numbers that make the statement true.

b. "The temperature is less than −4 degrees Fahrenheit."

$$-7 \quad -6 \quad -5 \quad -4 \quad -3 \quad -2 \quad -1 \quad 0 \quad 1 \quad 2$$

- What do you notice about the numbers that you circled?

- Can the temperature be exactly −4 degrees Fahrenheit? Explain.

- Write four other numbers that make the statement true.

c. "More than 3 students from our school are in the chess tournament."

$$-3 \quad -2 \quad -1 \quad 0 \quad 1 \quad 2 \quad 3 \quad 4 \quad 5 \quad 6$$

- What do you notice about the numbers that you circled?

- Is the number 3 included? Why or why not?

- Write four other numbers that make the statement true.

Inequalities

In this lesson, you will

- write and graph inequalities.
- use substitution to check whether a number is a solution of an inequality.

d. "The balance in a yearbook fund is no more than −$5."

$$-7 \quad -6 \quad -5 \quad -4 \quad -3 \quad -2 \quad -1 \quad 0 \quad 1 \quad 2$$

- What do you notice about the numbers that you circled?

- Is the number −5 included? Why or why not?

- Write four other numbers that make the statement true.

Laurie's Notes

Introduction

Applying Mathematical Practices

- **Attend to Precision:** Mathematically proficient students communicate precisely to others. This is done orally, in writing, and in the graphs they construct.

Motivate

- Ask students to write down their heights in two ways.
  - in feet and inches
  - in inches

  For instance, my height is 5 feet 7 inches or 67 inches. Tell students you are going to ask questions about their height written both ways.
- Read each statement below. Have students stand up (or raise their hands) when the statement is true for their height. Discuss each statement before going on to the next one. For instance, for the first statement ask, "Should someone who is 72 inches tall stand?" yes "Should someone who is 64 inches tall stand?" no
  - Your height is greater than 64 inches ($h > 64''$).
  - Your height is at most 5 feet 2 inches ($h \leq 5'2''$).
  - Your height is at least 63 inches ($h \geq 63''$).

Activity Notes

Activity 1

- Have students work through the four parts with their partners.
- If you discussed each statement in the Motivate activity, then students should have no difficulty answering the questions posed.
- When students finish, have volunteers share their results.
- **Big Idea:** You want students to notice that there is an infinite number of solutions to each statement (though some solutions may not make sense in the context of the problem).
- Another idea to focus on is that there is a boundary point for the set of solutions and you need to pay attention to the problem wording to understand the boundary point.
- **Attend to Precision:** Although only integers are listed as possible solutions, students should recognize that the solutions in parts (b) and (d) also include non-integer values. For example, a solution in part (b) is $-8.3°F$ and a possible balance in part (d) is $-\$7.50$.

What Your Students Will Learn

- Write and graph inequalities.
- Use substitution to check whether numbers are solutions of inequalities.

Previous Learning

Students should know how to graph numbers on a number line, solve single variable equations, and solve single variable inequalities using whole numbers.

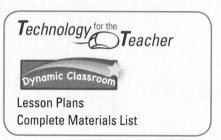

Lesson Plans
Complete Materials List

4.1 Record and Practice Journal

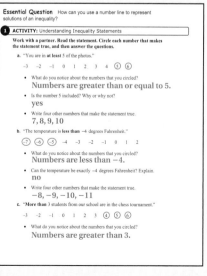

English Language Learners

Vocabulary and Symbols

Students should review the vocabulary and symbols for inequalities. Have students add to their notebooks a table of symbols and what the symbols mean. Students should add to the table as new phrases are used in the chapter.

Symbol	Phrase
$=$	is equal to
$\neq$	is not equal to
$<$	is less than
$\leq$	is less than or equal to
$>$	is greater than
$\geq$	is greater than or equal to

4.1 Record and Practice Journal

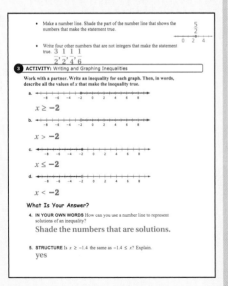

Laurie's Notes

Activity 2

- The questions in this activity have no context.
- Students will consider the boundary point and which direction to look (greater than or less than) for additional solutions. The solutions are not limited to integer values.
- **Attend to Precision:** Students must pay attention to the inequality symbol. In part (a) they need to consider whether the boundary point (-1.5) is a solution.
- When students finish, have volunteers share their results. Pay attention to how students locate the boundary point and how they deal with the strict and non-strict inequality symbols.
- Students may or may not use circles. They may just shade the number line to the left or right of a number or shade through a number.

Activity 3

- The boundary point in each problem is -2. The inequality symbol is what changes for each graph.
- Without giving students any definitions, let them think about the difference between the closed circle and open circle.
- Discuss students' answers for parts (a) and (b), comparing the open and closed circle. Do the same for parts (c) and (d).

What Is Your Answer?

- Some students may not know the language needed to describe the process. They might focus on creating clear examples.
- In Question 5, students may be confused because of the rational number. You could modify the question and ask, is $x \geq 4$ the same as $4 \leq x$?

Closure

- **Writing:** Write an inequality for each graph. Describe all the values of x that make the inequality true.

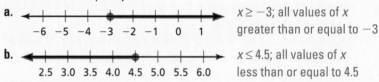

a. $x \geq -3$; all values of x greater than or equal to -3

b. $x \leq 4.5$; all values of x less than or equal to 4.5

2 ACTIVITY: Understanding Inequality Symbols

Work with a partner.

a. Consider the statement "*x* is a number such that $x > -1.5$."

- Can the number be exactly -1.5? Explain.

- Make a number line. Shade the part of the number line that shows the numbers that make the statement true.

- Write four other numbers that are not integers that make the statement true.

b. Consider the statement "*x* is a number such that $x \leq \dfrac{5}{2}$."

- Can the number be exactly $\dfrac{5}{2}$? Explain.

- Make a number line. Shade the part of the number line that shows the numbers that make the statement true.

- Write four other numbers that are not integers that make the statement true.

3 ACTIVITY: Writing and Graphing Inequalities

Math Practice

Check Progress

All the graphs are similar. So, what can you do to make sure that you have correctly written each inequality?

Work with a partner. Write an inequality for each graph. Then, in words, describe all the values of *x* that make the inequality true.

a.

b.

c.

d.

What Is Your Answer?

4. IN YOUR OWN WORDS How can you use a number line to represent solutions of an inequality?

5. STRUCTURE Is $x \geq -1.4$ the same as $-1.4 \leq x$? Explain.

Practice

Use what you learned about writing and graphing inequalities to complete Exercises 4 and 5 on page 128.

Key Vocabulary 🔊
inequality, *p. 126*
solution of an
    inequality, *p. 126*
solution set, *p. 126*
graph of an
    inequality, *p. 127*

An **inequality** is a mathematical sentence that compares expressions. It contains the symbols $<$, $>$, $\leq$, or $\geq$. To write an inequality, look for the following phrases to determine where to place the inequality symbol.

Inequality Symbols				
Symbol	$<$	$>$	$\leq$	$\geq$
Key Phrases	• is less than • is fewer than	• is greater than • is more than	• is less than or equal to • is at most • is no more than	• is greater than or equal to • is at least • is no less than

EXAMPLE ① **Writing an Inequality**

A number q plus 5 is greater than or equal to -7.9. Write this word sentence as an inequality.

A $\underbrace{\text{number } q \text{ plus } 5}_{q+5}$ $\underbrace{\text{is greater than or equal to}}_{\geq}$ -7.9.

An inequality is $q + 5 \geq -7.9$.

On Your Own

Now You're Ready
Exercises 6−9

Write the word sentence as an inequality.

1. A number x is at most -10.
2. Twice a number y is more than $-\dfrac{5}{2}$.

A **solution of an inequality** is a value that makes the inequality true. An inequality can have more than one solution. The set of all solutions of an inequality is called the **solution set**.

Reading

The symbol $\nleq$ means *is not less than or equal to*.

Value of x	$x + 2 \leq -1$	Is the inequality true?
-2	$-2 + 2 \overset{?}{\leq} -1$ $0 \nleq -1$ ✗	no
-3	$-3 + 2 \overset{?}{\leq} -1$ $-1 \leq -1$ ✓	yes
-4	$-4 + 2 \overset{?}{\leq} -1$ $-2 \leq -1$ ✓	yes

🔊 Multi-Language Glossary at BigIdeasMath⎷com

Laurie's Notes

Introduction

Connect

- **Yesterday:** Yesterday students investigated writing and graphing inequalities.
- **Today:** Students will translate inequalities from words to symbols and check to see whether a value is a solution of the inequality.

Motivate

- **Story Time:** You are planning to visit several theme parks and notice in doing your research that some of the rides have height restrictions.

Attraction	Restriction	Inequality
Dinosaur	Minimum is now 40 inches	$h \geq 40$
Primeval Whirl	Must be at least 48 inches	$h \geq 48$
Bay Slide	Must be under 60 inches	$h < 60$

- Ask students to write each as an inequality, where h is the rider's height.
- In today's lesson, they will be translating words to symbols.

Lesson Notes

Discuss

- Write the definition of an inequality.
- Review the four inequality symbols and key phrases or words that suggest each inequality.

Example 1

- ❓ Read the problem and ask, "How do you write a number q plus 5 in symbols?" $q + 5$
- For clarity, notice the use of color to help students translate each portion of the inequality.

On Your Own

- **Think-Pair-Share:** Students should read each question independently and then work in pairs to answer the questions. When they have answered the questions, the pair should compare their answers with another group and discuss any discrepancies.

Discuss

- Discuss what is meant by a solution of an inequality. Inequalities can, and generally do, have more than one solution. All of the solutions are collectively referred to as the **solution set**.
- It is helpful to write the inequality and substitute the value you are checking, as shown in the table.
- **Common Error:** Students will often make the mistake of thinking $-2 \geq -1$, forgetting that relationships are reversed on the negative side of 0; $-2 \leq -1$.

Lesson Tutorials
Lesson Plans
Answer Presentation Tool

Extra Example 1

A number y minus 3 is less than -15.3. Write this word sentence as an inequality.

$y - 3 < -15.3$

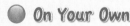 **On Your Own**

1. $x \leq -10$
2. $2y > -\dfrac{5}{2}$

Differentiated Instruction

Auditory

Stress to students the importance of reading a statement and translating it into an expression, equation, or inequality. The word "is" plays an important role in the meaning of the statement. For instance, *six less than a number* translates to $x - 6$, while *six is less than a number* translates to $6 < x$.

Extra Example 2

Tell whether −3 is a solution of each inequality.

a. $y + 6 < 4$ solution

b. $\dfrac{y}{-4} > 4$ not a solution

On Your Own

3. not a solution

4. not a solution

5. solution

Extra Example 3

Graph $p < -3$.

 ## On Your Own

6.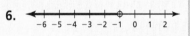

7. ![number line with closed circle at 4, ranging 0 to 8]

8. ![number line with 1.4 labeled, closed circle between 1 and 2, ranging -3 to 3]

9. ![number line with -1/2 labeled, open circle, ranging -2 to 4]

Example 2

? "How do you determine whether −2 is a solution of an inequality?" Substitute −2 for the variable, simplify, and decide whether the resulting inequality is true.

- Work through each example as shown. In part (b), students must recall that the product of two negatives is a positive.

On Your Own

- **Common Error:** In Question 3, when students substitute −5 for x, they may incorrectly see the result $7 > 7$ as a true inequality. Remind students to pay close attention to the inequality symbol.
- Ask volunteers to share their work at the board.

Discuss

- Discuss what is meant by the graph of an inequality. Remind students of the difference between the open and closed circles.

Example 3

- Decide which side of the boundary point to shade by testing one number on each side of the boundary point. It is called a boundary point because all of the values on one side of this value satisfy the inequality while all of the values on the other side of this value do *not* satisfy the inequality.
- Because −8 is not a solution of the inequality, use an open circle for the graph. Use a closed circle only when the boundary point is a solution of the graph (the inequality involves ≤ or ≥).
- **Use Appropriate Tools Strategically:** The graph of an inequality is a helpful visual tool that allows the solution to be seen. It also shows what values are *not* solutions.

On Your Own

- In Question 9, check to see that students locate $-\dfrac{1}{2}$ correctly.
- Ask students to share their graphs at the board.

Closure

- **Writing Prompt:** To decide whether a number is a solution of the inequality, you . . .

EXAMPLE ❷ **Checking Solutions**

Tell whether −2 is a solution of each inequality.

a. $y - 5 \geq -6$

$y - 5 \geq -6$	Write the inequality.
$-2 - 5 \overset{?}{\geq} -6$	Substitute −2 for y.
$-7 \not\geq -6$ ✗	Simplify.

−7 is *not* greater than or equal to −6.

∴ So, −2 is *not* a solution of the inequality.

b. $-5.5y < 14$

$-5.5y < 14$
$-5.5(-2) \overset{?}{<} 14$
$11 < 14$ ✓

11 is less than 14.

∴ So, −2 is a solution of the inequality.

⬤ **On Your Own**

Now You're Ready
Exercises 11–16

Tell whether −5 is a solution of the inequality.

3. $x + 12 > 7$ **4.** $1 - 2p \leq -9$ **5.** $n \div 2.5 \geq -3$

The **graph of an inequality** shows all the solutions of the inequality on a number line. An open circle ○ is used when a number is *not* a solution. A closed circle ● is used when a number is a solution. An arrow to the left or right shows that the graph continues in that direction.

EXAMPLE ❸ **Graphing an Inequality**

Graph $y > -8.$

Use an open circle because −8 is *not* a solution.

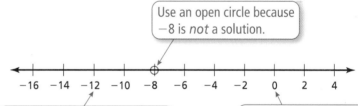

Test a number to the left of −8. $y = -12$ is *not* a solution.

Test a number to the right of −8. $y = 0$ is a solution.

Study Tip

The graph in Example 3 shows that the inequality has *infinitely many* solutions.

Shade the number line on the side where you found the solution.

⬤ **On Your Own**

Now You're Ready
Exercises 17–20

Graph the inequality on a number line.

6. $x < -1$ **7.** $z \geq 4$ **8.** $s \leq 1.4$ **9.** $-\frac{1}{2} < t$

✓ Vocabulary and Concept Check

1. **PRECISION** Should you use an open circle or a closed circle in the graph of the inequality $b \geq -42$? Explain.

2. **DIFFERENT WORDS, SAME QUESTION** Which is different? Write "both" inequalities.

k is less than or equal to -3.	k is no more than -3.
k is at most -3.	k is at least -3.

3. **REASONING** Do $x < 5$ and $5 < x$ represent the same inequality? Explain.

✎ Practice and Problem Solving

Write an inequality for the graph. Then, in words, describe all the values of x that make the inequality true.

4.

5.

Write the word sentence as an inequality.

① 6. A number y is no more than -8.

7. A number w added to 2.3 is more than 18.

8. A number t multiplied by -4 is at least $-\dfrac{2}{5}$.

9. A number b minus 4.2 is less than -7.5.

10. **ERROR ANALYSIS** Describe and correct the error in writing the word sentence as an inequality.

✗ Twice a number x is at most –24.

$$2x \geq -24$$

Tell whether the given value is a solution of the inequality.

② 11. $n + 8 \leq 13; n = 4$ 12. $5h > -15; h = -5$ 13. $p + 1.4 \leq 0.5; p = 0.1$

14. $\dfrac{a}{6} > -4; a = -18$ 15. $-\dfrac{2}{3}s \geq 6; s = -9$ 16. $\dfrac{7}{8} - 3k < -\dfrac{1}{2}; k = \dfrac{1}{4}$

Graph the inequality on a number line.

③ 17. $r \leq -9$ 18. $g > 2.75$ 19. $x \geq -3\dfrac{1}{2}$ 20. $z < 1\dfrac{1}{4}$

21. **FOOD TRUCK** Each day at lunchtime, at least 53 people buy food from a food truck. Write an inequality that represents this situation.

Assignment Guide and Homework Check

Level	Day 1 Activity Assignment	Day 2 Lesson Assignment	Homework Check
Basic	4, 5, 29–32	1–3, 7, 9, 10, 11–27 odd	7, 11, 13, 21, 23
Average	4, 5, 29–32	1–3, 6–20 even, 21– 27 odd	8, 14, 21, 23, 27
Advanced	4, 5, 29–32	1–3, 6–28 even	8, 14, 22, 24, 26
Accelerated	1–5, 6–28 even, 29–32		8, 14, 22, 24, 26

Common Errors

- **Exercises 11–16** Students may not understand when the boundary point is a solution of the inequality. Remind them that the inequality is true for the boundary point only when the inequality symbol is ≤ or ≥.
- **Exercises 17–20** Students may use the wrong type of circle at the boundary point. Remind them to use a closed circle when the boundary point is a solution and an open circle when the boundary point is not a solution.
- **Exercises 17–20** Students may shade on the wrong side of the boundary point. Remind them to test one point on each side of the boundary point.

4.1 Record and Practice Journal

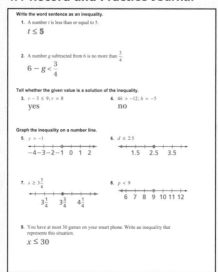

Vocabulary and Concept Check

1. A closed circle would be used because -42 is a solution.

2. k is at least -3; $k \geq -3$; $k \leq -3$

3. no; $x < 5$ is all values of x less than 5. $5 < x$ is all values of x greater than 5.

Practice and Problem Solving

4. $x > 12$; all values of x greater than 12

5. $x \leq -4$; all values of x less than or equal to -4

6. $y \leq -8$

7. $w + 2.3 > 18$

8. $-4t \geq -\dfrac{2}{5}$

9. $b - 4.2 < -7.5$

10. The inequality symbol is reversed; $2x \leq -24$

11. yes

12. no

13. no

14. yes

15. yes

16. no

17. ![number line with closed circle at -9]
$-12\ -11\ -10\ -9\ -8\ -7\ -6$

18. ![number line with open circle at 2.75]
2.75
$-1\ \ 0\ \ 1\ \ 2\ \ 3\ \ 4\ \ 5$

19. ![number line with closed circle at -3½]
$-3\tfrac{1}{2}$
$-6\ -5\ -4\ -3\ -2\ -1\ \ 0$

20. ![number line with open circle at 1¼]
$1\tfrac{1}{4}$
$-2\ -1\ \ 0\ \ 1\ \ 2\ \ 3\ \ 4$

21. $p \geq 53$

22. no   **23.** yes

24. no   **25.** yes

26. a. $1.25x > 35$

   b. yes; It costs $56.25 for 45 trips, which is more than the $35 monthly pass.

27. a. any value that is greater than -2

   b. any value that is less than or equal to -2; $b \leq -2$

   c. They represent the entire set of real numbers; yes

28. See *Taking Math Deeper.*

Fair Game Review

29. $p = 11$ **30.** $w = -3.6$

31. $x = -7$ **32.** C

Mini-Assessment

1. A number a is at least 5. Write this word sentence as an inequality.
$a \geq 5$

2. Four times a number b is at most -4.73. Write this word sentence as an inequality. $4b \leq -4.73$

3. Tell whether -2 is a solution of the inequality $6g - 14 > -21$. not a solution

4. Graph $p \leq -1.7$ on a number line.

-1.7
$-5\ -4\ -3\ -2\ -1\ \ 0\ \ 1\ \ 2$

Taking Math Deeper

Exercise 28

This problem gives students an opportunity to use an inequality to describe a real-life situation.

 Write an expression that represents the girth of the rectangular package.

$$\text{girth} = w + h + w + h$$
$$= 2w + 2h$$

 The combined length ℓ and girth of the package is represented by the expression $\ell + 2w + 2h$. The combined length and girth can be no more than 108 inches.

a. So, an inequality that represents the allowable dimensions is
$$\ell + 2w + 2h \leq 108.$$

 b. You can use the strategy *Guess, Check, and Revise* to find three different sets of allowable dimensions. Organize your results in a table. Below are some examples.

Dimensions (in.)	$\ell + 2w + 2h \overset{?}{\leq} 108$	Volume (in.^{3})
$\ell = 12, w = 5, h = 5$	$32 \leq 108$ ✓	300
$\ell = 20, w = 8, h = 12$	$60 \leq 108$ ✓	1920
$\ell = 30, w = 20, h = 19$	$108 \leq 108$ ✓	11,400

Project

Research the pricing offered by the postal service and delivery companies in your area. What are some of their requirements for shipping packages? What is their pricing based on?

Reteaching and Enrichment Strategies

If students need help...	If students got it...
Resources by Chapter	Resources by Chapter
• Practice A and Practice B	• Enrichment and Extension
• Puzzle Time	• Technology Connection
Record and Practice Journal Practice	Start the next section
Differentiating the Lesson	
Lesson Tutorials	
Skills Review Handbook	

Tell whether the given value is a solution of the inequality.

22. $4k < k + 8$; $k = 3$

23. $\dfrac{w}{3} \geq w - 12$; $w = 15$

24. $7 - 2y > 3y + 13$; $y = -1$

25. $\dfrac{3}{4}b - 2 \leq 2b + 8$; $b = -4$

26. MODELING A subway ride for a student costs $1.25. A monthly pass costs $35.

    a. Write an inequality that represents the number of times you must ride the subway for the monthly pass to be a better deal.

    b. You ride the subway about 45 times per month. Should you buy the monthly pass? Explain.

27. LOGIC Consider the inequality $b > -2$.

    a. Describe the values of b that are solutions of the inequality.

    b. Describe the values of b that are *not* solutions of the inequality. Write an inequality for these values.

    c. What do all the values in parts (a) and (b) represent? Is this true for any inequality?

28. **Critical Thinking** A postal service says that a rectangular package can have a maximum combined length and *girth* of 108 inches. The girth of a package is the distance around the perimeter of a face that does not include the length.

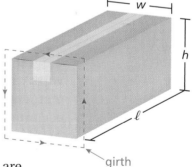

    a. Write an inequality that represents the allowable dimensions for the package.

    b. Find three different sets of allowable dimensions that are reasonable for the package. Find the volume of each package.

 Fair Game Review What you learned in previous grades & lessons

Solve the equation. Check your solution. *(Section 3.3)*

29. $p - 8 = 3$

30. $8.7 + w = 5.1$

31. $x - 2 = -9$

32. MULTIPLE CHOICE Which expression has a value less than -5? *(Section 1.2)*

    A $5 + 8$ **B** $-9 + 5$ **C** $1 + (-8)$ **D** $7 + (-2)$

Essential Question How can you use addition or subtraction to solve an inequality?

1 ACTIVITY: Writing an Inequality

Work with a partner. Members of the Boy Scouts must be less than 18 years old. In 4 years, your friend will still be eligible to be a scout.

a. Which of the following represents your friend's situation? What does x represent? Explain your reasoning.

$x + 4 > 18$ $x + 4 < 18$

$x + 4 \geq 18$ $x + 4 \leq 18$

b. Graph the possible ages of your friend on a number line. Explain how you decided what to graph.

2 ACTIVITY: Writing an Inequality

Work with a partner. Supercooling is the process of lowering the temperature of a liquid or a gas below its freezing point without it becoming a solid. Water can be supercooled to 86°F below its normal freezing point (32°F) and still not freeze.

Inequalities

In this lesson, you will
- solve inequalities using addition or subtraction.
- solve real-life problems.

a. Let x represent the temperature of water. Which inequality represents the temperature at which water can be a liquid or a gas? Explain your reasoning.

$x - 32 > -86$ $x - 32 < -86$

$x - 32 \geq -86$ $x - 32 \leq -86$

b. On a number line, graph the possible temperatures at which water can be a liquid or a gas. Explain how you decided what to graph.

Laurie's Notes

Introduction

Applying Mathematical Practices

- **Model with Mathematics:** Mathematically proficient students are able to make connections between the context of a problem and the graph of the solution. In checking solutions, students verify that solutions make sense (are reasonable) in the context of the problem.

Motivate

- Ask students to name the seven continents. Then have students work with partners to guess a ranked order for the continents based on the coldest temperature recorded on each continent.
- When students finish ranking the continents, have them guess the actual coldest temperature for each continent.
- Tell students that you will return to this data later in the class.

Activity Notes

Activity 1

- Ask students whether any have belonged to organizations such as scouts, 4-H, etc. Students may be aware of age restrictions.
- Partners work together to answer each part. In graphing possible ages, students may intuitively solve the inequality as they would an equation. This process is used in Activity 3.
- **?** "What does x represent?" your friend's age
- **Attend to Precision:** Be sure that students accurately define the variables. The variable x represents your friend's *current* age.
- **?** "What does $x + 4$ represent?" your friend's age in 4 years
- **Model with Mathematics:** In discussing their graphs, students should be able to explain how they chose the domain relative to the limitations stated in the problem and age restrictions in scouting. (The minimum age for a scout is 10.)

Activity 2

- To help students comprehend the problem, discuss the background information for this activity. University of Utah chemists have shown that water does not necessarily have to freeze until its temperature drops to $-55°F$. This is 87 degrees Fahrenheit colder than the well-known standard of $32°F$.
- When students finish, have volunteers share their results to each part.
- **?** "What does the expression $x - 32$ represent?" the number of degrees Fahrenheit below the standard freezing point of 32°F that the water temperature can reach without freezing
- Interpret the graph. It represents the temperatures at which water *can* be a liquid or gas. All numbers not included in the graph are Fahrenheit temperatures at which water must be ice.

What Your Students Will Learn

- Solve inequalities using the Addition Property of Inequality or the Subtraction Property of Inequality.

Previous Learning

Students should know how to graph inequalities, solve single variable equations, and solve single variable inequalities using whole numbers.

Technology for the Teacher

Dynamic Classroom

Lesson Plans
Complete Materials List

4.2 Record and Practice Journal

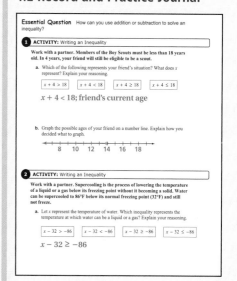

Essential Question How can you use addition or subtraction to solve an inequality?

1 ACTIVITY: Writing an Inequality

Work with a partner. Members of the Boy Scouts must be less than 18 years old. In 4 years, your friend will still be eligible to be a scout.

a. Which of the following represents your friend's situation? What does x represent? Explain your reasoning.

| $x + 4 > 18$ | $x + 4 < 18$ | $x + 4 \geq 18$ | $x + 4 \leq 18$ |

$x + 4 < 18$; friend's current age

b. Graph the possible ages of your friend on a number line. Explain how you decided what to graph.

$$8 \quad 10 \quad 12 \quad 14 \quad 16 \quad 18$$

2 ACTIVITY: Writing an Inequality

Work with a partner. Supercooling is the process of lowering the temperature of a liquid or a gas below its freezing point without it becoming a solid. Water can be supercooled to 86°F below its normal freezing point (32°F) and still not freeze.

a. Let x represent the temperature of water. Which inequality represents the temperature at which water can be a liquid or a gas? Explain your reasoning.

| $x - 32 > -86$ | $x - 32 < -86$ | $x - 32 \geq -86$ | $x - 32 \leq -86$ |

$x - 32 \geq -86$

4.2 Record and Practice Journal

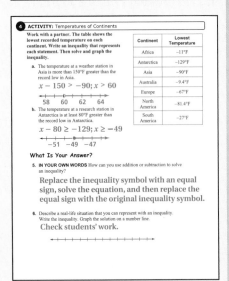

Laurie's Notes

Activity 3

- Students have already graphed the solution, and they have prior experience in solving equations. This activity connects these two skills.
- **?** "How do you think addition and subtraction inequalities are solved?" Listen for the same techniques used in solving equations.
- Check that students perform the steps correctly.

Activity 4

- Students can now check their rankings in the Motivate activity and how well they guessed the coldest recorded temperature for each continent.
- Students will use the data in the table to write the two inequalities in this activity.
- If students have trouble writing the inequalities, encourage them to reread the words and circle key words that are helpful in translating the words to symbols.
- When students finish, ask volunteers to share and discuss their work for each part.
- **?** "Does the graph help you to consider the reasonableness of your answer?" Answers will vary.

What Is Your Answer?

- In Question 5, some students may say to replace the inequality symbol with an equal sign, solve the equation, and then replace the equation with the original inequality symbol.
- **?** "Is it necessary to replace the inequality symbol with an equal sign to solve an inequality?" Answers will vary. This is an opportunity to discuss the Addition and Subtraction Properties of Inequality.

Closure

- Write, solve, and graph an inequality to describe your sister's height now. A ride at the amusement park has a maximum height limit of 48 inches. When your sister grows 3 inches, she will still be able to go on the ride.
  $x + 3 \le 48$; $x \le 45$;

3 ACTIVITY: Solving Inequalities

Math Practice

Interpret Results

What does the solution of the inequality represent?

Work with a partner. Complete the following steps for Activity 1. Then repeat the steps for Activity 2.

- Use your inequality from part (a). Replace the inequality symbol with an equal sign.
- Solve the equation.
- Replace the equal sign with the original inequality symbol.
- Graph this new inequality.
- Compare the graph with your graph in part (b). What do you notice?

4 ACTIVITY: Temperatures of Continents

Work with a partner. The table shows the lowest recorded temperature on each continent. Write an inequality that represents each statement. Then solve and graph the inequality.

a. The temperature at a weather station in Asia is more than 150°F greater than the record low in Asia.

b. The temperature at a research station in Antarctica is at least 80°F greater than the record low in Antarctica.

Continent	Lowest Temperature
Africa	−11°F
Antarctica	−129°F
Asia	−90°F
Australia	−9.4°F
Europe	−67°F
North America	−81.4°F
South America	−27°F

What Is Your Answer?

5. **IN YOUR OWN WORDS** How can you use addition or subtraction to solve an inequality?

6. Describe a real-life situation that you can represent with an inequality. Write the inequality. Graph the solution on a number line.

Practice Use what you learned about solving inequalities to complete Exercises 3–5 on page 134.

Key Ideas

Study Tip

You can solve inequalities in the same way you solve equations. Use inverse operations to get the variable by itself.

Addition Property of Inequality

Words When you add the same number to each side of an inequality, the inequality remains true.

Numbers		**Algebra**	If $a < b$, then $a + c < b + c$.
-4	$<$	3	
$+2$		$+2$	If $a > b$, then $a + c > b + c$.
-2	$<$	5	

Subtraction Property of Inequality

Words When you subtract the same number from each side of an inequality, the inequality remains true.

Numbers		**Algebra**	If $a < b$, then $a - c < b - c$.
-2	$<$	2	
-3		-3	If $a > b$, then $a - c > b - c$.
-5	$<$	-1	

These properties are also true for $\leq$ and $\geq$.

EXAMPLE 1 **Solving an Inequality Using Addition**

Solve $x - 5 < -3$. Graph the solution.

$$x - 5 < -3 \qquad \text{Write the inequality.}$$

Undo the subtraction. $\longrightarrow$ $\underline{+5 \quad +5}$ $\qquad$ Addition Property of Inequality

$$x < 2 \qquad \text{Simplify.}$$

The solution is $x < 2$.

Check:

$x = 0$: $\quad 0 - 5 \overset{?}{<} -3$

$\qquad -5 < -3$ ✓

$x = 5$: $\quad 5 - 5 \overset{?}{<} -3$

$\qquad 0 \not< -3$ ✗

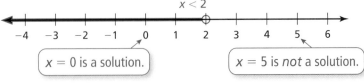

$x = 0$ is a solution.

$x = 5$ is *not* a solution.

On Your Own

Solve the inequality. Graph the solution.

1. $y - 6 > -7$

2. $b - 3.8 \leq 1.7$

3. $-\dfrac{1}{2} > z - \dfrac{1}{4}$

Laurie's Notes

Introduction

Connect

- **Yesterday:** Yesterday students investigated solving addition and subtraction inequalities.
- **Today:** Students will use Properties of Inequality to solve addition and subtraction inequalities.

Motivate

- It's "Did you know?" time for the students. The application problem involves NASA astronauts. Share some astronaut food facts with the students.
- To reduce launch weight during space shuttle flights, water was removed from certain foods and beverages. To eat these items, astronauts rehydrated them with water produced by the shuttle fuel cells. Rehydratable foods include breakfast cereals with milk, scrambled eggs, casseroles, and appetizers such as shrimp cocktail.

Lesson Notes

Key Ideas

- Write the Key Ideas. These properties should look familiar, as they are similar to the Addition and Subtraction Properties of Equality that students have used in solving equations.
- **Teaching Tip:** Summarize these two properties in the following way: George is older than Martha. In two years, George will still be older than Martha.

George's age > Martha's age	If $a > b$,
George's age + 2 > Martha's age + 2	then $a + c > b + c$.

  Two years ago, George was older than Martha.

George's age > Martha's age	If $a > b$,
George's age − 2 > Martha's age − 2	then $a - c > b - c$.

Example 1

- **?** Write the problem. "How do you isolate the variable, meaning get x by itself? Add 5 to each side of the inequality.
- Adding 5 is the inverse operation of subtracting 5.
- Solve, graph, and check.
- Take time to check a solution point. 0 is generally an easy value to work with.
- Also check a point that is not a solution on the graph, and verify that it does not satisfy the inequality.

On Your Own

- These problems integrate review of fraction and decimal operations.

Goal  Today's lesson is solving inequalities using addition or subtraction.

Technology for the **Teacher**

Dynamic Classroom

Lesson Tutorials
Lesson Plans
Answer Presentation Tool

Extra Example 1

Solve $x - 2 \geq -4$. Graph the solution. $x \geq -2$;

On Your Own

1. $y > -1$;

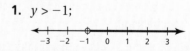

2. $b \leq 5.5$;

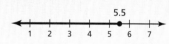

3. $-\dfrac{1}{4} > z$, or $z < -\dfrac{1}{4}$;

Extra Example 2

Solve $14 > x + 8$. Graph the solution.

$x < 6$;

On Your Own

4. $w \le -3$;

5. $2.5 \ge d$, or $d \le 2.5$;

6. $x > \dfrac{3}{4}$;

Differentiated Instruction

Auditory

To help auditory learners comprehend the verbal model in Example 3, point at each part of the verbal model as you explain or ask what it represents in the context of the problem.

Extra Example 3

A person can be no taller than $6\dfrac{5}{12}$ feet to become a fighter pilot for the United States Air Force. Your brother is 6 feet 3 inches tall. Write and solve an inequality that represents how much your brother can grow and still meet the requirement.

$6\dfrac{1}{4} + h \le 6\dfrac{5}{12}$; $h \le \dfrac{1}{6}$; Your brother can grow at most $\dfrac{1}{6}$ foot, or 2 inches.

On Your Own

7. $5.25 + h \le 6.25$; $h \le 1$; Your cousin can grow no more than 1 foot, or 12 inches.

Laurie's Notes

Example 2

❓ Write the inequality and ask, "How do you isolate the variable x?" Subtract 14 from each side of the inequality.
- **Common Error:** When students graph the solution they may look at the inequality symbol $\le$ and shade to the left of -1. It helps to rewrite the solution $-1 \le x$ as $x \ge -1$.

On Your Own

- **Think-Pair-Share:** Students should read each question independently and then work in pairs to answer the questions. When they have answered the questions, the pair should compare their answers with another group and discuss any discrepancies.
- Note that the variable is on the right side of the inequality in Question 5.

Example 3

- Ask a volunteer to read the problem.
- ❓ "The height limit is written as a decimal number of feet. Your friend's height is written in feet and inches. Can you work with the measurements as they are?" no; One of the forms must be changed to the other form.
- ❓ "A height of 6.25 feet is how many inches? Explain."

  75 inches; $6.25 \text{ ft} = 6\dfrac{1}{4} \text{ ft} = 6 \text{ ft} + \dfrac{1}{4} \text{ ft} = 72 \text{ in.} + 3 \text{ in.} = 75 \text{ in.}$
- ❓ "A height of 5 feet 9 inches is how many feet in decimal form? Explain."

  5.75 feet; Because $9 \text{ in.} = \dfrac{3}{4} \text{ ft}$, $5 \text{ ft } 9 \text{ in.} = 5\dfrac{3}{4} \text{ ft} = 5.75 \text{ ft}$.
- **Model with Mathematics:** Writing verbal models is an important step in helping students gain confidence in translating and setting up equations and inequalities. Color code the verbal model, if possible.
- **FYI:** The problem is solved using decimal heights, but it could be solved using inches as well.
- Continue to solve the problem as shown. Interpret the answer, $h \le 0.5$, in terms of a decimal number of feet and in terms of inches.

On Your Own

- Remind students to write a verbal model and then define variables before writing the inequality.

Closure

- **Exit ticket:** Solve and graph the solution for $-5.2 + x > 14.8$

  $x > 20$;

EXAMPLE 2 **Solving an Inequality Using Subtraction**

Solve $13 \leq x + 14$. Graph the solution.

$13 \leq x + 14$		Write the inequality.
$\underline{-\ 14 \qquad -\ 14}$		Subtraction Property of Inequality
$-1 \leq x$		Simplify.

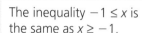 Undo the addition.

∴ The solution is $x \geq -1$.

 **Reading**

The inequality $-1 \leq x$ is the same as $x \geq -1$.

$x \geq -1$

$\longleftarrow\ \overset{}{+}\ \underset{-4}{+}\ \underset{-3}{+}\ \underset{-2}{+}\ \underset{-1}{\bullet}\ \underset{0}{+}\ \underset{1}{+}\ \underset{2}{+}\ \underset{3}{+}\ \underset{4}{+}\ \underset{5}{+}\ \underset{6}{+}\ \longrightarrow$

On Your Own

Now You're Ready
Exercises 3–17

Solve the inequality. Graph the solution.

4. $w + 7 \leq 4$ **5.** $12.5 \geq d + 10$ **6.** $x + \dfrac{3}{4} > 1\dfrac{1}{2}$

EXAMPLE 3 **Real-Life Application**

A person can be no taller than 6.25 feet to become an astronaut pilot for NASA. Your friend is 5 feet 9 inches tall. Write and solve an inequality that represents how much your friend can grow and still meet the requirement.

Words	Current height	plus	amount your friend can grow	is no more than	the height limit.
Variable	Let h be the possible amounts your friend can grow.				
Inequality	5.75	+	h	$\leq$	6.25

5 ft 9 in. = 60 + 9 = 69 in.

$69 \text{ in.} \times \dfrac{1 \text{ ft}}{12 \text{ in.}} = 5.75 \text{ ft}$

$5.75 + h \leq 6.25$		Write the inequality.
$\underline{-\ 5.75 \qquad\quad -\ 5.75}$		Subtraction Property of Inequality
$h \leq 0.5$		Simplify.

∴ So, your friend can grow no more than 0.5 foot, or 6 inches.

On Your Own

7. Your cousin is 5 feet 3 inches tall. Write and solve an inequality that represents how much your cousin can grow and still meet the requirement.

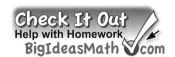
 Vocabulary and Concept Check

1. **REASONING** Is the inequality $c + 3 > 5$ the same as $c > 5 - 3$? Explain.

2. **WHICH ONE DOESN'T BELONG?** Which inequality does *not* belong with the other three? Explain your reasoning.

$$w + \frac{7}{4} > \frac{3}{4}$$

$$w - \frac{3}{4} > -\frac{7}{4}$$

$$w + \frac{7}{4} < \frac{3}{4}$$

$$\frac{3}{4} < w + \frac{7}{4}$$

 Practice and Problem Solving

Solve the inequality. Graph the solution.

① ② **3.** $x + 7 \geq 18$ **4.** $a - 2 > 4$ **5.** $3 \leq 7 + g$

6. $8 + k \leq -3$ **7.** $-12 < y - 6$ **8.** $n - 4 < 5$

9. $t - 5 \leq -7$ **10.** $p + \frac{1}{4} \geq 2$ **11.** $\frac{2}{7} > b + \frac{5}{7}$

12. $z - 4.7 \geq -1.6$ **13.** $-9.1 < d - 6.3$ **14.** $\frac{8}{5} > s + \frac{12}{5}$

15. $-\frac{7}{8} \geq m - \frac{13}{8}$ **16.** $r + 0.2 < -0.7$ **17.** $h - 6 \leq -8.4$

ERROR ANALYSIS Describe and correct the error in solving the inequality or graphing the solution of the inequality.

18.
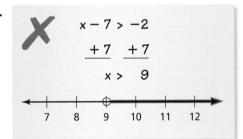

19.
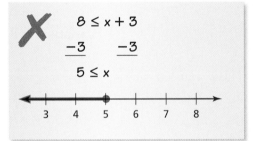

20. AIRPLANE A small airplane can hold 44 passengers. Fifteen passengers board the plane.

   a. Write and solve an inequality that represents the additional number of passengers that can board the plane.

   b. Can 30 more passengers board the plane? Explain.

Assignment Guide and Homework Check

Level	Day 1 Activity Assignment	Day 2 Lesson Assignment	Homework Check
Basic	3–5, 29–33	1, 2, 7–27 odd	7, 11, 21, 25, 27
Average	3–5, 29–33	1, 2, 7–19 odd, 20–28 even	7, 13, 20, 22, 26
Advanced	3–5, 29–33	1, 2, 6–28 even	12, 14, 20, 22, 26
Accelerated	1–5, 6–28 even, 29–33		12, 14, 20, 22, 26

Common Errors

- **Exercises 3–17** Students may use the same operation rather than the inverse operation to isolate the variable. Remind students that when a number is added to the variable, subtract that number from each side. When a number is subtracted from the variable, add that number to each side.
- **Exercises 3–17** Students may shade the number line in the wrong direction when the variable is on the right side of the inequality symbol. Remind them to rewrite the inequality with the variable at the left or to check a value on each side of the boundary point.

4.2 Record and Practice Journal

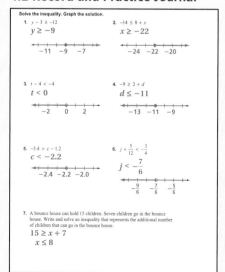

Vocabulary and Concept Check

1. Yes, because of the Subtraction Property of Inequality.

2. $w - \frac{3}{4} > -\frac{7}{4}$ does not belong with the other three because to solve this inequality, you use the Addition Property of Inequality. To solve the other three, you use the Subtraction Property of Inequality.

Practice and Problem Solving

3. $x \geq 11$;

4. $a > 6$;

5. $-4 \leq g$;

6. $k \leq -11$;

7. $-6 < y$;

8. $n < 9$;

9. $t \leq -2$;

10. $p \geq 1\frac{3}{4}$;

11. $-\frac{3}{7} > b$;

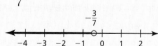

12–20. See Additional Answers.

21. $7 + 7 + x < 28$; $x < 14$ ft

Practice and Problem Solving

22. $x + 3 > 8$; $x > 5$ in.

23. $8 + 8 + 10 + 10 + x \le 51$; $x \le 15$ m

24. 4

25. $x - 3 \ge 5$; $x \ge 8$ ft

26. a. $x + 14 \le 35$; $x \le \$21$

   b. It changes the number added to x. So, the inequality becomes $x + 9.8 \le 35$ and you have more money left.

   c. yes; The cost of the shirt and the pants is $\$32.80$, which is less than $\$35$.

27. See *Taking Math Deeper.*

28. -14

 ## Fair Game Review

29. $x = 9$ **30.** $w = -27$

31. $b = -22$ **32.** $h = 80$

33. A

Mini-Assessment

1. Solve $x - 4 > 11$. Graph the solution.

$x > 15$;

2. Solve $11 \le w + 3.4$. Graph the solution. $w \ge 7.6$;

3. Solve $k - \dfrac{2}{5} < -\dfrac{4}{5}$. Graph the solution. $k < -\dfrac{2}{5}$;

4. The school cafeteria seats 250 students. 203 students are already seated. Write and solve an inequality that represents the additional number of students that can be seated. $203 + s \le 250$; $s \le 47$; At most 47 more students can be seated.

Taking Math Deeper

Exercise 27

Some students may not have had experiences with an electrical circuit that overloads and triggers the circuit breaker. This problem is a nice opportunity to familiarize students with the fact that different appliances use different amounts of electricity.

 Write an inequality.

   Let x = amount (in watts) of additional electricity used.

   1050 = amount (in watts) of electricity used by the portable heater.

   2400 = amount (in watts) of electricity that overloads the circuit.

   $x + 1050 < 2400$

 Solve the inequality.

   a. $x + 1050 < 2400$

   $x < 1350$ watts

 Answer the question.

   b. Yes, there is more than one possibility.

   You can also plug in the following pairs of items without overloading the circuit.

Item	Watts
Aquarium	200
Hair dryer	1200
Television	150
Vacuum cleaner	1100

   aquarium and television: $200 + 150 = 350$ watts

   aquarium and vacuum cleaner: $200 + 1100 = 1300$ watts

   television and vacuum cleaner: $150 + 1100 = 1250$ watts

Project

It might be interesting for students to research other types of items and how much electricity they use. In this example, you can see that the item that generates heat uses a lot of electricity.

Reteaching and Enrichment Strategies

If students need help...	If students got it...
Resources by Chapter • Practice A and Practice B • Puzzle Time Record and Practice Journal Practice Differentiating the Lesson Lesson Tutorials Skills Review Handbook	Resources by Chapter • Enrichment and Extension • Technology Connection Start the next section

Write and solve an inequality that represents x.

21. The perimeter is less than 28 feet.

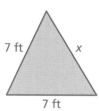

7 ft   x

7 ft

22. The base is greater than the height.

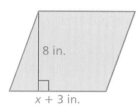

8 in.

x + 3 in.

23. The perimeter is less than or equal to 51 meters.

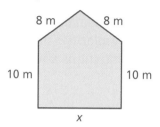

8 m   8 m

10 m   10 m

x

24. REASONING The solution of $d + s > -3$ is $d > -7$. What is the value of s?

25. BIRDFEEDER The hole for a birdfeeder post is 3 feet deep. The top of the post needs to be at least 5 feet above the ground. Write and solve an inequality that represents the required length of the post.

26. SHOPPING You can spend up to $35 on a shopping trip.

   a. You want to buy a shirt that costs $14. Write and solve an inequality that represents the amount of money you will have left if you buy the shirt.

   b. You notice that the shirt is on sale for 30% off. How does this change the inequality?

   c. Do you have enough money to buy the shirt that is on sale and a pair of pants that costs $23? Explain.

27. POWER A circuit overloads at 2400 watts of electricity. A portable heater that uses 1050 watts of electricity is plugged into the circuit.

   a. Write and solve an inequality that represents the additional number of watts you can plug in without overloading the circuit.

   b. In addition to the portable heater, what two other items in the table can you plug in at the same time without overloading the circuit? Is there more than one possibility? Explain.

Item	Watts
Aquarium	200
Hair dryer	1200
Television	150
Vacuum cleaner	1100

28. **Number Sense** The possible values of x are given by $x + 8 \leq 6$. What is the greatest possible value of $7x$?

Fair Game Review What you learned in previous grades & lessons

Solve the equation. Check your solution. *(Section 3.4)*

29. $4x = 36$

30. $\dfrac{w}{3} = -9$

31. $-2b = 44$

32. $60 = \dfrac{3}{4}h$

33. MULTIPLE CHOICE Which fraction is equivalent to –2.4? *(Section 2.1)*

  Ⓐ $-\dfrac{12}{5}$ Ⓑ $-\dfrac{51}{25}$ Ⓒ $-\dfrac{8}{5}$ Ⓓ $-\dfrac{6}{25}$

You can use a **Y chart** to compare two topics. List differences in the branches and similarities in the base of the Y. Here is an example of a Y chart that compares solving equations and solving inequalities.

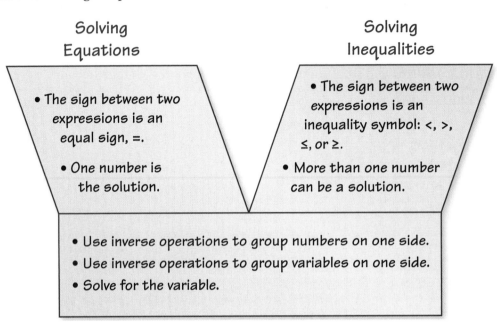

On Your Own

Make Y charts to help you study and compare these topics.

1. writing equations and writing inequalities

2. graphing the solution of an equation and graphing the solution of an inequality

3. graphing inequalities that use > and graphing inequalities that use <

4. graphing inequalities that use > or < and graphing inequalities that use ≥ or ≤

5. solving inequalities using addition and solving inequalities using subtraction

"Hey Descartes, do you have any suggestions for the Y chart I am making?"

After you complete this chapter, make Y charts for the following topics.

6. solving inequalities using multiplication and solving inequalities using division

7. solving two-step equations and solving two-step inequalities

8. Pick two other topics that you studied earlier in this course and make a Y chart to compare them.

Sample Answers

1.

Writing equations
- The sign between two expressions is an equal sign, =.
- One number is the solution.

Writing inequalities
- The sign between two expressions is an inequality symbol: $<$, $>$, $\leq$ or $\geq$.
- More than one number can be a solution.

- Write one expression on the left and one expression on the right.
- Look for key phrases to determine which operation(s) to use: $+$, $-$, $\times$, or $\div$.
- Look for key phrases to determine where to place the equal or inequality sign.

2.

Graphing the solution of an equation
- A solution is represented by a closed circle, ●.
- One number is the solution.

Graphing the solution of an inequality
- The endpoint of the graph can be an open circle, ○, or a closed circle, ●.
- An arrow pointing to the left or the right shows that the graph continues in that direction.
- More than one number can be a solution.

- Solve for the variable.
- Graph the solution on a number line.

3.

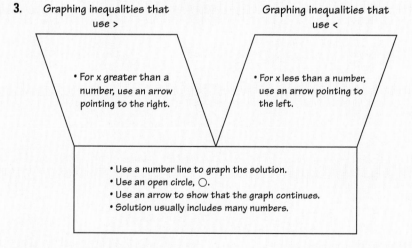

Graphing inequalities that use $>$
- For x greater than a number, use an arrow pointing to the right.

Graphing inequalities that use $<$
- For x less than a number, use an arrow pointing to the left.

- Use a number line to graph the solution.
- Use an open circle, ○.
- Use an arrow to show that the graph continues.
- Solution usually includes many numbers.

4–5. Available at *BigIdeasMath.com*.

List of Organizers

Available at *BigIdeasMath.com*

Comparison Chart
Concept Circle
Definition (Idea) and Example Chart
Example and Non-Example Chart
Formula Triangle
Four Square
Information Frame
Information Wheel
Notetaking Organizer
Process Diagram
Summary Triangle
Word Magnet
Y Chart

About this Organizer

A **Y Chart** can be used to compare two topics. Students list differences between the two topics in the branches of the Y and similarities in the base of the Y. As with an example and non-example chart, a Y chart serves as a good tool for assessing students' knowledge of a pair of topics that have subtle but important differences. You can include blank Y charts on tests or quizzes for this purpose.

Technology for the *Teacher*

Editable Graphic Organizer

Answers

1. $y + 2 > -5$

2. $s - 2.4 \geq 8$

3. yes   **4.** no

5.

6.

7.

$-\frac{1}{3}$

8.

4.2

9. $n \leq -8$;

10. $t > \frac{9}{7}$;

$\frac{9}{7}$

11. $-\frac{7}{4} \geq w$;

$-\frac{7}{4}$

12. $y < -0.8$;

-0.8

13. $s \geq 1.5$

14. a. $j \geq 2$;

$p \geq 25$;

$n \geq 10$;

b. yes; $2.5 \geq 2$; $30 \geq 25$; $20 \geq 10$

15. $-\frac{7}{2} + x > -\frac{3}{2}$; $x > 2$

You must move more than 2 feet in the positive direction.

Alternative Quiz Ideas

100% Quiz	Math Log
Error Notebook	Notebook Quiz
Group Quiz	Partner Quiz
Homework Quiz	Pass the Paper

Homework Quiz

A homework notebook provides an opportunity for teachers to check that students are doing their homework regularly. Students keep homework in their notebooks. They should be told to record the page number, problem number, and copy the problem exactly in their homework notebooks. Each day the teacher walks around and visually checks that homework is completed. Periodically, without advance notice, the teacher tells the students to put everything away except their homework notebooks.

Questions are from students' homework.

1. What are the answers to Exercises 1, 5, 10, and 21 on page 128?
2. What are the answers to Exercises 23 and 27 on page 129?
3. What are the answers to Exercises 3–5 on page 134?
4. What is the answer to Exercise 26 on page 135?

Reteaching and Enrichment Strategies

If students need help. . .	If students got it. . .
Resources by Chapter • Practice A and Practice B • Puzzle Time Lesson Tutorials *BigIdeasMath.com*	Resources by Chapter • Enrichment and Extension • Technology Connection Game Closet at *BigIdeasMath.com* Start the next section

Check It Out
Progress Check
BigIdeasMath ✓com

Write the word sentence as an inequality.   *(Section 4.1)*

1. A number y plus 2 is greater than -5.

2. A number s minus 2.4 is at least 8.

Tell whether the given value is a solution of the inequality.   *(Section 4.1)*

3. $8p < -3; p = -2$

4. $z + 2 > -4; z = -8$

Graph the inequality on a number line.   *(Section 4.1)*

5. $x < -12$

6. $v > \dfrac{5}{4}$

7. $b \geq -\dfrac{1}{3}$

8. $q \leq 4.2$

Solve the inequality. Graph the solution.   *(Section 4.2)*

9. $n + 2 \leq -6$

10. $t - \dfrac{3}{7} > \dfrac{6}{7}$

11. $-\dfrac{3}{4} \geq w + 1$

12. $y - 2.6 < -3.4$

13. STUDYING You plan to study at least 1.5 hours for a geography test. Write an inequality that represents this situation.   *(Section 4.1)*

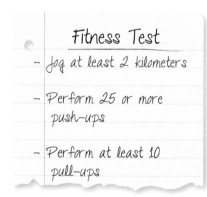

Fitness Test
– Jog at least 2 kilometers
– Perform 25 or more push-ups
– Perform at least 10 pull-ups

14. FITNESS TEST The three requirements to pass a fitness test are shown.   *(Section 4.1)*

    a. Write and graph three inequalities that represent the requirements.

    b. You can jog 2500 meters, perform 30 push-ups, and perform 20 pull-ups. Do you satisfy the requirements of the test? Explain.

15. NUMBER LINE Use tape on the floor to make the number line shown. All units are in feet. You are standing at $-\dfrac{7}{2}$. You want to move to a number greater than $-\dfrac{3}{2}$. Write and solve an inequality that represents the distance you must move.   *(Section 4.2)*

Essential Question How can you use multiplication or division to solve an inequality?

1 ACTIVITY: Using a Table to Solve an Inequality

Work with a partner.

- Copy and complete the table.
- Decide which graph represents the solution of the inequality.
- Write the solution of the inequality.

a. $4x > 12$

x	−1	0	1	2	3	4	5
4x							
4x $\overset{?}{>}$ 12							

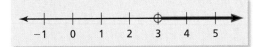 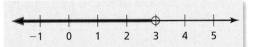

b. $-3x \leq 9$

x	−5	−4	−3	−2	−1	0	1
−3x							
−3x $\overset{?}{\leq}$ 9							

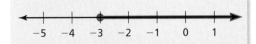 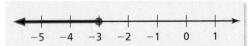

Inequalities

In this lesson, you will
- solve inequalities using multiplication or division.
- solve real-life problems.

2 ACTIVITY: Solving an Inequality

Work with a partner.

a. Solve $-3x \leq 9$ by adding $3x$ to each side of the inequality first. Then solve the resulting inequality.

b. Compare the solution in part (a) with the solution in Activity 1(b).

Laurie's Notes

Introduction

Applying Mathematical Practices

- **Construct Viable Arguments and Critique the Reasoning of Others:** Mathematically proficient students use the results of investigations to observe patterns and make conjectures. Although student conjectures are not always correct, encourage them to try. Prior to this activity, students tend to think that inequalities involving multiplication and division will be solved in the same fashion as the related equations. After completing the four activities, students may revise their thoughts.

Motivate

- Ask a series of questions and record the students' solutions.
  - ❓ "What integers are solutions of $x > 4$?" 5, 6, 7, ...
  - ❓ "What integers are solutions of $-x > 4$, meaning what numbers have an opposite that is greater than 4?" $-5, -6, -7, ...$
  - ❓ "What integers are solutions of $x < -4$?" $-5, -6, -7, ...$
- Leave these 3 problems on the board and refer to them at the end of class.

Activity Notes

Activity 1

- Explain that students are to evaluate one side of the inequality for each value of x and then decide whether the inequality is satisfied. This means, *is the value of x a solution of the inequality?* Students will write *yes* or *no* in the third row of the table to indicate whether the value is a solution or not.
- Using the information in the table, students decide which graph represents the solution. Finally, they write the solution.
- ❓ Discuss the results with your students.
  - "What did you find as the solution for part (a)?" $x > 3$
  - "Is this what you would have expected?" Likely, they will say yes.
  - "What did you find as the solution for part (b)?" $x \geq -3$
  - "Is this what you would have expected?" Likely, they will say no.
- Do not tell students a rule at this point. Simply say that perhaps they need to try a few more problems to help figure out what is going on.

Activity 2

- The directions may seem odd to students. It is not obvious to all students that adding $3x$ to each side of the inequality makes the left side 0.
- **Make Sense of Problems and Persevere in Solving Them:** Resist the urge to jump in too quickly to help students. Tell them to reread the directions carefully. Students should be able to combine what they know about solving multi-step equations with what they learned in Section 4.2 to solve the multi-step inequality.
- Ask a volunteer to share his or her work. Discuss how the process and solution compare to the process and solution in Activity 1 part (b).

What Your Students Will Learn

- Solve inequalities using Case 1 or Case 2 of the Multiplication and Division Properties of Inequality.

Previous Learning

Students should know how to solve equations using multiplication and division. Students should be able to evaluate expressions and decide whether a number is a solution of an inequality.

Lesson Plans
Complete Materials List

4.3 Record and Practice Journal

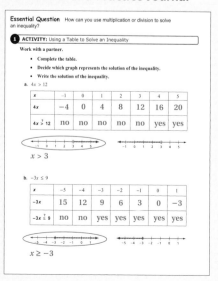

T-138

English Language Learners

Pair Activity

Create index cards with problems similar to those in Activities 2 and 4. Pair English learners with English speakers and give 5 cards to each pair. Have students work together to solve the inequalities. When students have completed the problems, check their work and give them another set of cards.

Laurie's Notes

Activity 3

- Explain that this activity is similar to Activity 1 except the variables are involved in division instead of multiplication.
- Give time for students to work through the two problems.
- ❓ Discuss the results with your students.
  - "What did you find as the solution for part (a)?" $x < 3$
  - "Is this what you would have expected?" Likely, they will say yes.
  - "What did you find as the solution for part (b)?" $x \le -3$
  - "Is this what you would have expected?" Likely, they will say no.
- Again, resist the temptation to simply tell students a rule. Suggest that trying additional problems might help them.

Activity 4

- Give time for students to work through the four problems with their partners.
- ❓ "Did any of the inequalities have solutions that you expected?" Again, listen for the same comments from students as before. The solution of each problem involves the expected number, but the direction of the inequality symbol is reversed.
- After working through all four problems, students should have a sense that solving these inequalities is the same as solving equations except when the coefficient is negative.

What Is Your Answer?

- **Neighbor Check:** Have students work independently and then have their neighbors check their work. Have students discuss any discrepancies.

Closure

- Refer to the three inequalities written at the beginning of class.

$$x > 4 \qquad -x > 4 \qquad x < -4$$

- ❓ "Which inequalities have the same solution?" $-x > 4$ and $x < -4$
- ❓ "Is this consistent with what you discovered in the activities? Explain." yes; Listen for comments about the negative coefficient of x and the switching of the inequality symbol.

4.3 Record and Practice Journal

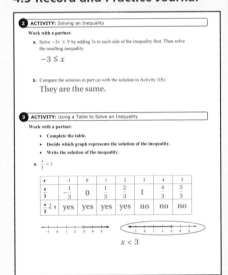

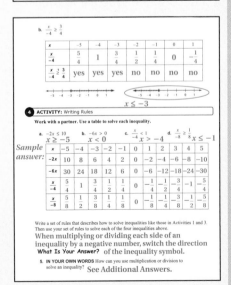

ACTIVITY: Using a Table to Solve an Inequality

Work with a partner.

- Copy and complete the table.

- Decide which graph represents the solution of the inequality.

- Write the solution of the inequality.

a. $\dfrac{x}{3} < 1$

x	−1	0	1	2	3	4	5
$\dfrac{x}{3}$							
$\dfrac{x}{3} \overset{?}{<} 1$							

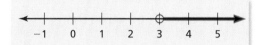   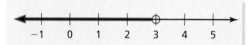

b. $\dfrac{x}{-4} \geq \dfrac{3}{4}$

x	−5	−4	−3	−2	−1	0	1
$\dfrac{x}{-4}$							
$\dfrac{x}{-4} \overset{?}{\geq} \dfrac{3}{4}$							

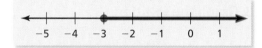   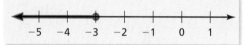

4 **ACTIVITY: Writing Rules**

Work with a partner. Use a table to solve each inequality.

a. $-2x \leq 10$ **b.** $-6x > 0$ **c.** $\dfrac{x}{-4} < 1$ **d.** $\dfrac{x}{-8} \geq \dfrac{1}{8}$

Write a set of rules that describes how to solve inequalities like those in Activities 1 and 3. Then use your rules to solve each of the four inequalities above.

What Is Your Answer?

5. **IN YOUR OWN WORDS** How can you use multiplication or division to solve an inequality?

Practice

Use what you learned about solving inequalities using multiplication or division to complete Exercises 4–9 on page 143.

🔑 Key Idea

Remember

Multiplication and division are inverse operations.

Multiplication and Division Properties of Inequality (Case 1)

Words   When you multiply or divide each side of an inequality by the same *positive* number, the inequality remains true.

Numbers $-4 < 6$ $4 > -6$

$2 \cdot (-4) < 2 \cdot 6$ $\dfrac{4}{2} > \dfrac{-6}{2}$

$-8 < 12$ $2 > -3$

Algebra   If $a < b$ and c is positive, then

$$a \cdot c < b \cdot c \qquad \text{and} \qquad \frac{a}{c} < \frac{b}{c}.$$

If $a > b$ and c is positive, then

$$a \cdot c > b \cdot c \qquad \text{and} \qquad \frac{a}{c} > \frac{b}{c}.$$

These properties are also true for $\leq$ and $\geq$.

EXAMPLE ① **Solving an Inequality Using Multiplication**

Solve $\dfrac{x}{5} \leq -3$. **Graph the solution.**

$$\frac{x}{5} \leq -3 \qquad \text{Write the inequality.}$$

Undo the division. ⟶ $5 \cdot \dfrac{x}{5} \leq 5 \cdot (-3)$ Multiplication Property of Inequality

$$x \leq -15 \qquad \text{Simplify.}$$

∴ The solution is $x \leq -15$.

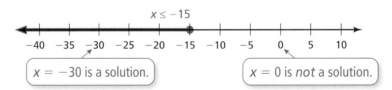

$x \leq -15$

$x = -30$ is a solution. $x = 0$ is *not* a solution.

⚫ On Your Own

Solve the inequality. Graph the solution.

1. $n \div 3 < 1$ **2.** $-0.5 \leq \dfrac{m}{10}$ **3.** $-3 > \dfrac{2}{3}p$

Laurie's Notes

Introduction

- **Yesterday:** Students gained an intuitive understanding of solving inequalities involving multiplication and division.
- **Today:** Students will use the Multiplication and Division Properties of Inequality to solve inequalities.

Motivate

? "Have you heard of Ultimate?" Answers will vary.

- It is a sport played with a flying disc at colleges, high schools, and some middle schools. There are 10 simple rules, one of which is that there aren't any officials! Pretty cool.
- The popularity of the sport has skyrocketed, but there is *at most* one-fifth the numbers of students playing Ultimate as there are playing lacrosse. If there are 26 students playing Ultimate, what is the minimum number playing lacrosse? $\frac{1}{5}x \geq 26$; $x \geq 130$ players

Lesson Notes

Key Idea

- These properties should look familiar, as they are similar to the Multiplication and Division Properties of Equality used in solving equations.
- Note that the properties are restricted to multiplying and dividing by a *positive* number. This is very important.

Example 1

? "How do you isolate the variable, meaning get *x* by itself?" Multiply by 5 on each side of the inequality.

- Multiplying by 5 is the inverse operation of dividing by 5.

On Your Own

- **Think-Pair-Share:** Students should read each question independently and then work in pairs to answer the questions. When they have answered the questions, the pair should compare their answers with another group and discuss any discrepancies.
- Division is represented in different ways in Questions 1 and 2. The second representation is more common in algebra (higher mathematics).
- After solving the inequality in Question 3, the result will be $-\frac{9}{2} > p$. Students can also rewrite this as $p < -\frac{9}{2}$. The direction of the inequality symbol is reversed *only* because the solution is being rewritten with the variable on the left side of the inequality statement.

Lesson Tutorials
Lesson Plans
Answer Presentation Tool

Extra Example 1

Solve $\frac{m}{4} > -4$. Graph the solution.

$m > -16$;

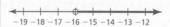

On Your Own

1. $n < 3$;

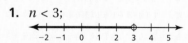

2. $m \geq -5$;

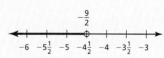

3. $p < -\frac{9}{2}$;

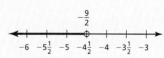

Extra Example 2

Solve $7y \leq -21$. Graph the solution.
$y \leq -3$;

On Your Own

4. $b \geq \frac{1}{2}$;

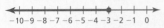

5. $k \leq -2$;

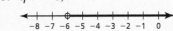

6. $q > -6$;

Differentiated Instruction

Visual

Use the inequality $6 < 9$ to show students why it is necessary to reverse the inequality symbol when multiplying or dividing by a negative number.

Add -3 to each side. The result is $3 < 6$, a true statement.

Subtract -3 from each side. The result is $9 < 12$, a true statement.

Multiply each side by -3. If the inequality is *not* reversed, then the statement $-18 < -27$ is false. By reversing the inequality, the statement $-18 > -27$ is true.

Divide each side by -3. If the inequality is *not* reversed, then the statement $-2 < -3$ is false. By reversing the inequality, the statement $-2 > -3$ is true.

Example 2

? "What operation is being performed on x?" multiplication

? "How do you undo a multiplication problem?" divide

- Solve, graph, and check.

On Your Own

- **Think-Pair-Share:** Students should read each question independently and then work in pairs to answer the questions. When they have answered the questions, the pair should compare their answers with another group and discuss any discrepancies.

- **Construct Viable Arguments and Critique the Reasoning of Others:** Notice that although all of the coefficients are positive, sometimes the constant is negative. This is important in helping students understand when the direction of the inequality symbol is going to be reversed. The focus is on the sign of the coefficient, not the sign of the constant.

- For Question 6, remind students that after solving this inequality, the result will be $-6 < q$. Students can rewrite this as $q > -6$. The direction of the inequality symbol is reversed *only* because the two sides of the inequality are being reversed. Reversing the sign has nothing to do with the negative constant (-6).

- These problems integrate review of decimal operations.

Key Idea

- These properties look identical to what they have been using in the lesson, *except* now the direction of the inequality symbol must be reversed for the inequality to remain true because they are multiplying or dividing by a *negative* quantity!

- The short version of the property: When you multiply or divide by a negative quantity, reverse the direction of the inequality symbol.

- **Common Error:** When students solve $2x < -4$, they sometimes reverse the inequality symbol because there's a negative number in the problem. Reverse the inequality symbol when you multiply or divide each side by a negative number to eliminate a negative coefficient. Do not reverse the inequality symbol just because there is a negative constant.

EXAMPLE ② **Solving an Inequality Using Division**

Solve $6x > -18$. Graph the solution.

$$6x > -18 \qquad \text{Write the inequality.}$$

Undo the multiplication. ⟶ $$\dfrac{6x}{6} > \dfrac{-18}{6} \qquad \text{Division Property of Inequality}$$

$$x > -3 \qquad \text{Simplify.}$$

⋰ The solution is $x > -3$.

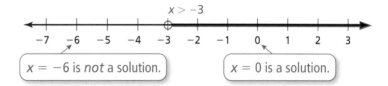

$x = -6$ is *not* a solution. $x = 0$ is a solution.

On Your Own

Now You're Ready
Exercises 10–18

Solve the inequality. Graph the solution.

4. $4b \geq 2$ 5. $12k \leq -24$ 6. $-15 < 2.5q$

🔑 Key Idea

Multiplication and Division Properties of Inequality (Case 2)

Words When you multiply or divide each side of an inequality by the same *negative* number, the direction of the inequality symbol must be reversed for the inequality to remain true.

Common Error ⚠️

A negative sign in an inequality does not necessarily mean you must reverse the inequality symbol.

Only reverse the inequality symbol when you multiply or divide both sides by a negative number.

Numbers $\qquad -4 < 6 \qquad\qquad\qquad 4 > -6$

$$-2 \cdot (-4) > -2 \cdot 6 \qquad \dfrac{4}{-2} < \dfrac{-6}{-2}$$

$$8 > -12 \qquad\qquad -2 < 3$$

Algebra If $a < b$ and c is negative, then

$$a \cdot c > b \cdot c \qquad \text{and} \qquad \dfrac{a}{c} > \dfrac{b}{c}.$$

If $a > b$ and c is negative, then

$$a \cdot c < b \cdot c \qquad \text{and} \qquad \dfrac{a}{c} < \dfrac{b}{c}.$$

These properties are also true for $\leq$ and $\geq$.

EXAMPLE 3 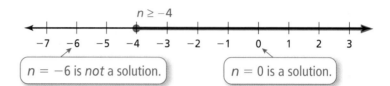 **Solving an Inequality Using Multiplication**

Solve $-\dfrac{3}{2}n \le 6$. Graph the solution.

$$-\dfrac{3}{2}n \le 6 \qquad\qquad \text{Write the inequality.}$$

$$-\dfrac{2}{3} \cdot \left(-\dfrac{3}{2}n\right) \ge -\dfrac{2}{3} \cdot 6 \qquad \begin{array}{l}\text{Use the Multiplication Property of Inequality.}\\ \text{Reverse the inequality symbol.}\end{array}$$

$$n \ge -4 \qquad\qquad \text{Simplify.}$$

⋮ The solution is $n \ge -4$.

$n \ge -4$

$n = -6$ is *not* a solution. $n = 0$ is a solution.

On Your Own

Solve the inequality. Graph the solution.

7. $\dfrac{x}{-3} > -4$

8. $0.5 \le -\dfrac{y}{2}$

9. $-12 \ge \dfrac{6}{5}m$

10. $-\dfrac{2}{5}h \le -8$

EXAMPLE 4 **Solving an Inequality Using Division**

Solve $-3z > -4.5$. Graph the solution.

$$-3z > -4.5 \qquad\qquad \text{Write the inequality.}$$

Undo the multiplication. ⟶ $$\dfrac{-3z}{-3} < \dfrac{-4.5}{-3} \qquad \begin{array}{l}\text{Use the Division Property of Inequality.}\\ \text{Reverse the inequality symbol.}\end{array}$$

$$z < 1.5 \qquad\qquad \text{Simplify.}$$

⋮ The solution is $z < 1.5$.

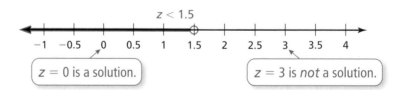

$z < 1.5$

$z = 0$ is a solution. $z = 3$ is *not* a solution.

On Your Own

Now You're Ready
Exercises 27–35

Solve the inequality. Graph the solution.

11. $-5z < 35$

12. $-2a > -9$

13. $-1.5 < 3n$

14. $-4.2 \ge -0.7w$

Laurie's Notes

Example 3

- Write the inequality.
- ❓ "What operation is performed on n?" n is multiplied by $-\frac{3}{2}$.
- ❓ "How do you undo multiplying by $-\frac{3}{2}$?" Divide by $-\frac{3}{2}$.
- ❓ "What is equivalent to dividing by $-\frac{3}{2}$?" Multiplying by $-\frac{2}{3}$.
- Solve as usual, but remember to reverse the direction of the inequality symbol when multiplying each side by $-\frac{2}{3}$.
- When graphing, remember to use a closed circle because the inequality is greater than or equal to.

On Your Own

- **Think-Pair-Share:** Students should read each question independently and then work in pairs to answer the questions. When they have answered the questions, the pair should compare their answers with another group and discuss any discrepancies.

Example 4

- Write the inequality.
- ❓ "What operation is performed on z?" z is multiplied by -3.
- ❓ "How do you undo multiplying by -3?" Divide by -3.
- Solve as usual, but remember to reverse the direction of the inequality symbol when dividing each side by -3. Note that the quotient of two negatives is positive.
- ❓ "Should you use an open or closed circle?" Use an open circle because the inequality is strictly less than.

On Your Own

- **Neighbor Check:** Have students work independently and then have their neighbors check their work. Have students discuss any discrepancies.
- Have students share their work at the board.

Closure

- **Exit Ticket:** Solve and graph.

$\dfrac{x}{-3} \le -9$ $x \ge 27$;

$-8 > 4x$ $x < -2$;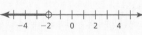

Extra Example 3

Solve $-\frac{4}{3}b > 8$. Graph the solution.

$b < -6$;

On Your Own

7. $x < 12$;

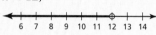

8. $y \le -1$;

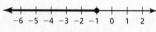

9. $m \le -10$;

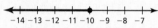

10. $h \ge 20$;

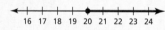

Extra Example 4

Solve $-2w \le -5.2$. Graph the solution.

$w \ge 2.6$;

On Your Own

11. $z > -7$;

12. $a < \frac{9}{2}$;

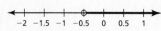

13. $n > -0.5$;

14. $w \ge 6$;

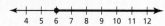

Vocabulary and Concept Check

1. Multiply each side by 3.

2. The first inequality will be divided by a positive number. The second inequality will be divided by a negative number. Because this inequality is divided by a negative number, the direction of the inequality symbol will be reversed.

3. *Sample answer:* $-4x < 16$

Practice and Problem Solving

4. $x < 1$

5. $x \geq -1$

6. $x < -3$

7. $x \leq -35$

8. $x < -\dfrac{2}{5}$

9. $x \leq \dfrac{3}{2}$

10. $n > 10$;

11. $c \leq -36$;
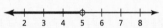

12. $m < 5$;

13. $x < -32$;

14. $w \geq 15$;

15. $k > 2$;
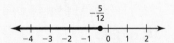

16. $x \leq -\dfrac{5}{12}$;
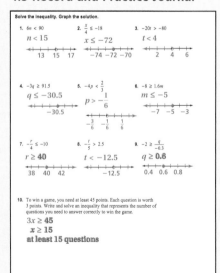

17–23. See Additional Answers.

24. $9.2x \geq 299$; $x \geq 32.5$ h

Assignment Guide and Homework Check

Level	Day 1 Activity Assignment	Day 2 Lesson Assignment	Homework Check
Basic	4–9, 48–52	1–3, 11–33 odd	13, 17, 25, 29, 31
Average	4–9, 48–52	1–3, 11–23 odd, 24–38 even	13, 17, 26, 32, 34
Advanced	4–9, 48–52	1–3, 14–46 even	14, 18, 26, 30, 34
Accelerated	1–9, 14–18 even, 19, 20–46 even, 48–52		14, 18, 26, 30, 34

Common Errors

- **Exercises 10–18** Students may perform the same operation on both sides instead of the opposite operation when solving the inequality. Remind them that solving inequalities is similar to solving equations.
- **Exercises 10–18** When there is a negative in the inequality, students may reverse the direction of the inequality symbol. Remind them that they only reverse the direction when they are multiplying or dividing by a negative number. All of these exercises keep the same inequality symbol.
- **Exercises 10–18** Students may shade the wrong direction when the variable is on the right side of the inequality instead of the left. Remind them to rewrite the inequality by reversing the right and left sides and reversing the inequality symbol.

4.3 Record and Practice Journal

Solve the inequality. Graph the solution.

1. $6n < 90$
$n < 15$

2. $\dfrac{x}{4} \leq -18$
$x \leq -72$

3. $-20r > -80$
$t < 4$

4. $-3q \geq 91.5$
$q \leq -30.5$

5. $-4p < \dfrac{2}{3}$
$p > -\dfrac{1}{6}$

6. $-8 \geq 1.6m$
$m \leq -5$

7. $-\dfrac{r}{4} \leq -10$
$r \geq 40$

8. $-\dfrac{t}{5} > 2.5$
$t < -12.5$

9. $-2 \geq \dfrac{q}{-0.3}$
$q \geq 0.6$

10. To win a game, you need at least 45 points. Each question is worth 3 points. Write and solve an inequality that represents the number of questions you need to answer correctly to win the game.
$3x \geq 45$
$x \geq 15$
at least 15 questions

 Vocabulary and Concept Check

1. **WRITING** Explain how to solve $\frac{x}{3} < -2$.

2. **PRECISION** Explain how solving $4x < -16$ is different from solving $-4x < 16$.

3. **OPEN-ENDED** Write an inequality that you can solve using the Division Property of Inequality where the direction of the inequality symbol must be reversed.

 Practice and Problem Solving

Use a table to solve the inequality.

4. $2x < 2$

5. $-3x \le 3$

6. $-6x > 18$

7. $\frac{x}{-5} \ge 7$

8. $\frac{x}{-1} > \frac{2}{5}$

9. $\frac{x}{3} \le \frac{1}{2}$

Solve the inequality. Graph the solution.

① ② 10. $2n > 20$

11. $\frac{c}{9} \le -4$

12. $2.2m < 11$

13. $-16 > x \div 2$

14. $\frac{1}{6}w \ge 2.5$

15. $7 < 3.5k$

16. $3x \le -\frac{5}{4}$

17. $4.2y \le -12.6$

18. $11.3 > \frac{b}{4.3}$

19. **ERROR ANALYSIS** Describe and correct the error in solving the inequality.

$$\frac{x}{3} < -9$$

$$3 \cdot \frac{x}{3} > 3 \cdot (-9)$$

$$x > -27$$

Write the word sentence as an inequality. Then solve the inequality.

20. The quotient of a number and 4 is at most 5.

21. A number divided by 7 is less than -3.

22. Six times a number is at least -24.

23. The product of -2 and a number is greater than 30.

24. **SMART PHONE** You earn $9.20 per hour at your summer job. Write and solve an inequality that represents the number of hours you need to work in order to buy a smart phone that costs $299.

25. **AVOCADOS** You have $9.60 to buy avocados for a guacamole recipe. Avocados cost $2.40 each.

   a. Write and solve an inequality that represents the number of avocados you can buy.

   b. Are there infinitely many solutions in this context? Explain.

26. **SCIENCE PROJECT** Students in a science class are divided into 6 equal groups with at least 4 students in each group for a project. Write and solve an inequality that represents the number of students in the class.

Solve the inequality. Graph the solution.

③ ④ 27. $-5n \leq 15$

28. $-7w > 49$

29. $-\frac{1}{3}h \geq 8$

30. $-9 < -\frac{1}{5}x$

31. $-3y < -14$

32. $-2d \geq 26$

33. $4.5 > -\frac{m}{6}$

34. $\frac{k}{-0.25} \leq 36$

35. $-2.4 > \frac{b}{-2.5}$

36. **ERROR ANALYSIS** Describe and correct the error in solving the inequality.

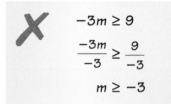

37. **TEMPERATURE** It is currently 0°C outside. The temperature is dropping 2.5°C every hour. Write and solve an inequality that represents the number of hours that must pass for the temperature to drop below −20°C.

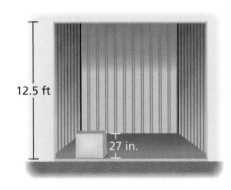

12.5 ft

27 in.

38. **STORAGE** You are moving some of your belongings into a storage facility.

   a. Write and solve an inequality that represents the number of boxes that you can stack vertically in the storage unit.

   b. Can you stack 6 boxes vertically in the storage unit? Explain.

Write and solve an inequality that represents x.

39. Area ≥ 120 cm²

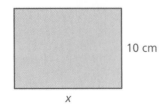

10 cm

x

40. Area < 20 ft²

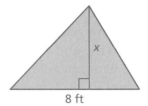

x

8 ft

Common Errors

- **Exercises 27–35** Students may forget to reverse the inequality symbol when multiplying or dividing by a negative number. Remind them of this rule. Encourage students to substitute values into the original inequality to check that the solution is correct.

- **Exercise 41** Students may write an incorrect inequality before solving. They may write $\frac{x}{4} < 100$ because there are four friends. However, the student is included in the trip as well, so there are 5 people going on the trip. The inequality should be $\frac{x}{5} < 100$.

Differentiated Instruction

Visual

Students may question why they are asked to graph the solution of an inequality, but not asked to graph the solution of an equation. The graph of an equation is just a point on the number line. The graph of an inequality provides more information, because of the infinite number of possible solutions.

Practice and Problem Solving

25. **a.** $2.40x \le 9.60$;
   $x \le 4$ avocados

   b. no; You must buy a whole number of avocados.

26. $\frac{x}{6} \ge 4$; $x \ge 24$; There are at least 24 students.

27. $n \ge -3$;

28. $w < -7$;

29. $h \le -24$;

30. $x < 45$;

31. $y > \frac{14}{3}$;

32. $d \le -13$;

33. $m > -27$;

34. $k \ge -9$;

35–39. See Additional Answers.

40. $4x < 20$; $x < 5$ ft

41. $\frac{x}{5} < 100$; $x < \$500$

42. *Sample answer:* Consider the inequality $5 > 3$. If you multiply or divide each side by -1 without reversing the direction of the inequality symbol, you obtain $-5 > -3$, which is not true. So, whenever you multiply or divide an inequality by a negative number, you must reverse the direction of the inequality symbol to obtain a true statement.

43. *Answer should include, but is not limited to:* Use the correct number of months that the novel has been out.

44. See *Taking Math Deeper*.

45. $n \geq -12$ and $n \leq -5$;

46–47. See Additional Answers.

Fair Game Review

48. $w = 8$ **49.** $v = 45$

50. $x = 7$ **51.** $m = 4$

52. B

Mini-Assessment

Solve and graph the inequality.

1. $\frac{b}{6} > -11$ $b > -66$;

2. $4c \leq -28$ $c \leq -7$;

3. $-\frac{6}{7}k > -12$ $k < 14$;

4. $-4b \leq 9.6$ $b \geq -2.4$;

Taking Math Deeper

Exercise 44

Double (or compound) inequalities, like those in Exercises 44–47, can often be written using a single inequality statement.

1 Begin by graphing each inequality.

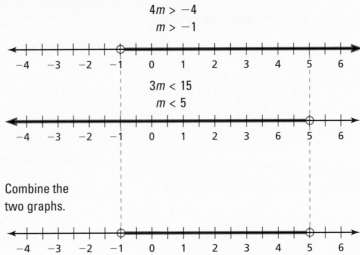

$$4m > -4$$
$$m > -1$$

$$3m < 15$$
$$m < 5$$

2 Combine the two graphs.

3 The numbers that satisfy both inequalities are all numbers greater than -1 and less than 5. If you rewrite $m > -1$ as $-1 < m$, then you can write the statement as a single inequality, $-1 < m < 5$.

Graphs overlap.

Project

Use the newspaper, Internet, TV, radio, or any other source to record and graph at least 10 different uses of inequalities.

Reteaching and Enrichment Strategies

If students need help. . .	If students got it. . .
Resources by Chapter • Practice A and Practice B • Puzzle Time Record and Practice Journal Practice Differentiating the Lesson Lesson Tutorials Skills Review Handbook	Resources by Chapter • Enrichment and Extension • Technology Connection Start the next section

41. AMUSEMENT PARK You and four friends are planning a visit to an amusement park. You want to keep the cost below $100 per person. Write and solve an inequality that represents the total cost of visiting the amusement park.

42. LOGIC When you multiply or divide each side of an inequality by the same negative number, you must reverse the direction of the inequality symbol. Explain why.

43. PROJECT Choose two novels to research.

a. Use the Internet or a magazine to complete the table.

b. Use the table to find and compare the average number of copies sold per month for each novel. Which novel do you consider to be the most successful? Explain.

c. Assume each novel continues to sell at the average rate. Write and solve an inequality that represents the number of months it will take for the total number of copies sold to exceed twice the current number sold.

Author	Name of Novel	Release Date	Current Number of Copies Sold
1.			
2.			

 Describe all numbers that satisfy *both* inequalities. Include a graph with your description.

44. $4m > -4$ and $3m < 15$

45. $\dfrac{n}{3} \geq -4$ and $\dfrac{n}{-5} \geq 1$

46. $2x \geq -6$ and $2x \geq 6$

47. $-\dfrac{1}{2}s > -7$ and $\dfrac{1}{3}s < 12$

Fair Game Review What you learned in previous grades & lessons

Solve the equation. Check your solution. *(Section 3.5)*

48. $-2w + 4 = -12$

49. $\dfrac{v}{5} - 6 = 3$

50. $3(x - 1) = 18$

51. $\dfrac{m + 200}{4} = 51$

52. MULTIPLE CHOICE What is the value of $\dfrac{2}{3} + \left(-\dfrac{5}{7}\right)$? *(Section 2.2)*

    Ⓐ $-\dfrac{3}{4}$ Ⓑ $-\dfrac{1}{21}$ Ⓒ $\dfrac{7}{10}$ Ⓓ $1\dfrac{8}{21}$

4.4 Solving Two-Step Inequalities

Essential Question How can you use an inequality to describe the dimensions of a figure?

1 ACTIVITY: Areas and Perimeters of Figures

Work with a partner.

- Use the given condition to choose the inequality that you can use to find the possible values of the variable. Justify your answer.

- Write four values of the variable that satisfy the inequality you chose.

a. You want to find the values of x so that the area of the rectangle is more than 22 square units.

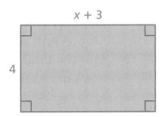

$4x + 12 > 22$ $4x + 3 > 22$

$4x + 12 \geq 22$ $2x + 14 > 22$

b. You want to find the values of x so that the perimeter of the rectangle is greater than or equal to 28 units.

$x + 7 \geq 28$ $4x + 12 \geq 28$ $2x + 14 \geq 28$ $2x + 14 \leq 28$

c. You want to find the values of y so that the area of the parallelogram is fewer than 41 square units.

$5y + 7 < 41$ $5y + 35 < 41$

$5y + 7 \leq 41$ $5y + 35 \leq 41$

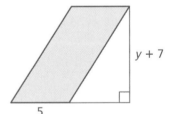

Inequalities

In this lesson, you will

- solve multi-step inequalities.
- solve real-life problems.

d. You want to find the values of z so that the area of the trapezoid is at most 100 square units.

$5z + 30 \leq 100$ $10z + 30 \leq 100$

$5z + 30 < 100$ $10z + 30 < 100$

Laurie's Notes

Introduction

Applying Mathematical Practices

- **Construct Viable Arguments and Critique the Reasoning of Others:** In these activities, students must choose the inequality that solves the stated problem and then justify that choice.

Motivate

- Display a set of tangram pieces arranged as shown. Tell students that the area of the entire square formed by the tangrams is 16 square units.
- Ask students to find the area of each tangram piece.
  gold triangle = 4, green triangle = 2,
  red triangle = 1, turquoise square = 2,
  purple parallelogram = 2

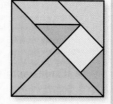

- **?** "Did you use any area formulas?" no; Students likely used fractional relationships.
- Explain that this activity involves area formulas. Review formulas if necessary.

Activity Notes

Activity 1

- This activity reviews perimeters of rectangles and areas of parallelograms and triangles.
- Students know how to use the Distributive Property and combine like terms. Encourage them to use these skills to first write algebraic expressions for the perimeter or area.
- Students are not asked to solve the inequalities, though some may.
- When students finish, ask volunteers to discuss parts (a) and (b). In part (a), some students may have initially written $4 + (x + 3) + 4 + (x + 3)$ while others may have written $2(4) + 2(x + 3)$.
- **Construct Viable Arguments and Critique the Reasoning of Others:** In each part, listen carefully to the student's justification for choosing the inequality. There should be a clear reason for the choice.
- **Extension:** Ask what happens to the shape of the rectangle as x varies.
- In part (c), students need to interpret the figure correctly. The parallelogram has a base of 5 units and a height of $y + 7$ units.
- **Common Error:** Students may forget to use the Distributive Property when they find the expression for area and incorrectly write $5(y + 7) = 5y + 7$.
- **Extension:** Ask what happens to the shape of the parallelogram as y varies.
- In part (d), students could find the sum of the areas of the rectangle and right triangle.
- **Extension:** Ask what happens to the shape of the trapezoid as z varies.

What Your Students Will Learn

- Use the properties of inequalities to solve and graph solutions of multi-step inequalities.

Previous Learning

Students should know how to solve two-step equations.

4.4 Record and Practice Journal

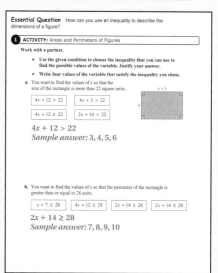

Essential Question How can you use an inequality to describe the dimensions of a figure?

1 ACTIVITY: Areas and Perimeters of Figures

Work with a partner.

- Use the given condition to choose the inequality that you can use to find the possible values of the variable. Justify your answer.
- Write four values of the variable that satisfy the inequality you chose.

a. You want to find the values of x so that the area of the rectangle is more than 22 square units.

| $4x + 12 > 22$ | $4x + 3 > 22$ |
| $4x + 12 \geq 22$ | $2x + 14 > 22$ |

$4x + 12 > 22$
Sample answer: 3, 4, 5, 6

b. You want to find the values of x so that the perimeter of the rectangle is greater than or equal to 28 units.

| $x + 7 \geq 28$ | $4x + 12 \geq 28$ | $2x + 14 \geq 28$ | $2x + 14 \leq 28$ |

$2x + 14 \geq 28$
Sample answer: 7, 8, 9, 10

In each part of the Activities, ask a volunteer to say out loud the key phrase ("is fewer than," "is no more than," etc.) of the given condition that determines the correct inequality symbol to use. Discuss which inequality symbol represents the phrase, and why.

4.4 Record and Practice Journal

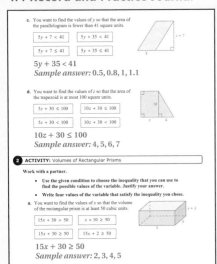

c. You want to find the values of y so that the area of the parallelogram is fewer than 41 square units.

$5y + 7 < 41$	$5y + 35 < 41$
$5y + 7 \le 41$	$5y + 35 \le 41$

$5y + 35 < 41$
Sample answer: 0.5, 0.8, 1, 1.1

d. You want to find the values of z so that the area of the trapezoid is at most 100 square units.

$5z + 30 \le 100$	$10z + 30 \le 100$
$5z + 30 < 100$	$10z + 30 < 100$

$10z + 30 \le 100$
Sample answer: 4, 5, 6, 7

2 ACTIVITY: Volumes of Rectangular Prisms

Work with a partner.

- Use the given condition to choose the inequality that you can use to find the possible values of the variable. Justify your answer.
- Write four values of the variable that satisfy the inequality you chose.

a. You want to find the values of x so that the volume of the rectangular prism is at least 50 cubic units.

$15x + 30 > 50$	$x + 10 \ge 50$
$15x + 30 \ge 50$	$15x + 2 \ge 50$

$15x + 30 \ge 50$
Sample answer: 2, 3, 4, 5

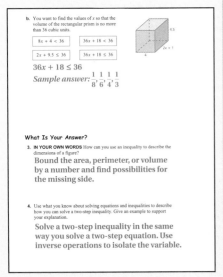

b. You want to find the values of x so that the volume of the rectangular prism is no more than 36 cubic units.

$8x + 4 < 36$	$36x + 18 < 36$
$2x + 9.5 \le 36$	$36x + 18 \le 36$

$36x + 18 \le 36$
Sample answer: $\dfrac{1}{8}, \dfrac{1}{6}, \dfrac{1}{4}, \dfrac{1}{3}$

What Is Your Answer?

3. IN YOUR OWN WORDS How can you use an inequality to describe the dimensions of a figure?

Bound the area, perimeter, or volume by a number and find possibilities for the missing side.

4. Use what you know about solving equations and inequalities to describe how you can solve a two-step inequality. Give an example to support your explanation.

Solve a two-step inequality in the same way you solve a two-step equation. Use inverse operations to isolate the variable.

Laurie's Notes

Activity 2

- This activity reviews volumes of rectangular prisms.
- Students will need to use the Distributive Property to find the volume of each prism.
- **Construct Viable Arguments and Critique the Reasoning of Others:** In each part, listen carefully to the student's justification for the inequality he or she chose. Make sure the student mentions using the Distributive Property.
- Check that students translate "at least" and "no more than" correctly.
- Have students share and discuss values of the variable that satisfy each inequality. This will remind students that each problem has infinitely many solutions.
- **?** "Does x need to be a whole number in each problem?" no
- **Extension:** Ask what happens to the shape of the rectangular prism as x varies in each problem.

What Is Your Answer?

- In Question 4, students are likely to say that solving two-step inequalities is just like solving two-step equations. Make sure they mention how they are different: When the coefficient is negative, you need to multiply or divide each side by a negative quantity and reverse the direction of the inequality symbol.

Closure

- **Exit Ticket:** Solve each inequality.

  $4x + 5 \ge 21$ $x \ge 4$ and $\dfrac{x}{4} - 3 \ge 13$ $x \ge 64$

Work with a partner.

- Use the given condition to choose the inequality that you can use to find the possible values of the variable. Justify your answer.

- Write four values of the variable that satisfy the inequality you chose.

Math Practice

State the Meaning of Symbols

What inequality symbols do the phrases *at least* and *no more than* represent? Explain.

a. You want to find the values of x so that the volume of the rectangular prism is at least 50 cubic units.

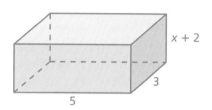

| $15x + 30 > 50$ | $x + 10 \geq 50$ | $15x + 30 \geq 50$ | $15x + 2 \geq 50$ |

b. You want to find the values of x so that the volume of the rectangular prism is no more than 36 cubic units.

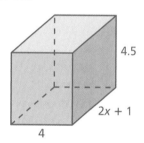

| $8x + 4 < 36$ | $36x + 18 < 36$ | $2x + 9.5 \leq 36$ | $36x + 18 \leq 36$ |

What Is Your Answer?

3. **IN YOUR OWN WORDS** How can you use an inequality to describe the dimensions of a figure?

4. Use what you know about solving equations and inequalities to describe how you can solve a two-step inequality. Give an example to support your explanation.

Practice

Use what you learned about solving two-step inequalities to complete Exercises 3 and 4 on page 150.

You can solve two-step inequalities in the same way you solve two-step equations.

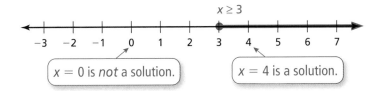

EXAMPLE **1** Solving Two-Step Inequalities

a. Solve $5x - 4 \geq 11$. **Graph the solution.**

$$5x - 4 \geq 11 \qquad \text{Write the inequality.}$$

Step 1: Undo the subtraction. → $\underline{+\,4 \qquad +\,4}$ Addition Property of Inequality

$$5x \geq 15 \qquad \text{Simplify.}$$

Step 2: Undo the multiplication. → $\dfrac{5x}{5} \geq \dfrac{15}{5}$ Division Property of Inequality

$$x \geq 3 \qquad \text{Simplify.}$$

⫶· The solution is $x \geq 3$.

$x \geq 3$

$x = 0$ is *not* a solution.

$x = 4$ is a solution.

b. Solve $\dfrac{b}{-3} + 4 < 13$. **Graph the solution.**

$$\frac{b}{-3} + 4 < 13 \qquad \text{Write the inequality.}$$

Step 1: Undo the addition. → $\underline{-\,4 \qquad -\,4}$ Subtraction Property of Inequality

$$\frac{b}{-3} < 9 \qquad \text{Simplify.}$$

Step 2: Undo the division. → $-3 \cdot \dfrac{b}{-3} > -3 \cdot 9$ Use the Multiplication Property of Inequality. Reverse the inequality symbol.

$$b > -27 \qquad \text{Simplify.}$$

⫶· The solution is $b > -27$.

$b > -27$

On Your Own

Now You're Ready
Exercises 5–10

Solve the inequality. Graph the solution.

1. $6y - 7 > 5$ **2.** $4 - 3d \geq 19$ **3.** $\dfrac{w}{-4} + 8 > 9$

Laurie's Notes

Introduction

Connect

- **Yesterday:** Students developed an intuitive understanding of solving two-step inequalities.
- **Today:** Students will solve and graph two-step inequalities.

Motivate

- Share the following scenario with students.
    - Student A has test scores of 82, 94, 86, and 81.
    - Student B has test scores of 92, 98, 88 and 94.
- Student A wants to achieve an average of at least 85, and Student B wants to achieve an average of at least 90. What must each student score on the 5th and final test to meet their goals?
- **?** "Have any of you ever wondered what score you needed on a particular test to have a certain average?" Answers will vary.
- **?** "Do you think it is mathematically possible for Student A and Student B to achieve their goals?" Answers will vary.
- Tell students you will return to this problem at the end of class.

Lesson Notes

Discuss

- You solve two-step inequalities in much the same way you solve two-step equations. You only need to remember to change the direction of the inequality symbol if you multiply or divide by a negative quantity.
- Recall that solving an equation undoes the evaluating in reverse order. The goal is to isolate the variable.

Example 1

- **?** "What operations are being performed on the left side of the inequality? multiplication and subtraction
- **?** "What is the first step in isolating the variable, meaning getting the x-term by itself?" Add 4 to each side of the inequality.
- Notice that subtracting 4 would have been the last step if evaluating the left side, so its inverse operation is the first step in solving the inequality.
- **?** "To solve for x, what is the last step?" Divide both sides by 5.
- Because you are dividing by a positive quantity, the inequality symbol does not change. The solution is $x \geq 3$. Graph and check.
- Part (b) is solved in a similar fashion.
- **?** "To solve $\dfrac{b}{-3} < 9$, what do you need to do?" Multiply both sides by -3 and change the direction of the inequality symbol.
- Graph and check. Remember to use an open circle because the variable cannot equal -27.
- Point out to students that the number line has increments of 3 units.

On Your Own

- These are all straightforward problems. Students should not be confused by them.

Goal Today's lesson is solving two-step inequalities.

Lesson Tutorials
Lesson Plans
Answer Presentation Tool

Extra Example 1

a. Solve $17 \leq 3y - 4$. Graph the solution. $y \geq 7$;

b. Solve $\dfrac{x}{-2} - 8 \geq -6$. Graph the solution. $x \leq -4$;

On Your Own

1. $y > 2$;

2. $d \leq -5$;

3. $w < -4$;

English Language Learners

Vocabulary

Give English learners the opportunity to use precise language to solve an inequality. Write the inequality $-2x + 4 > 8$ on the board. Have one student come to the board. For each step of the solution, call on another student to give the instruction for solving. The instructions should be given in complete sentences. The instructions for the inequality are:

(1) Subtract 4 from each side.

(2) Simplify.

(3) Divide each side by -2.
    Reverse the inequality symbol.

(4) Simplify.

Extra Example 2

Solve $12 > -2(y - 4)$. Graph the solution. $y > -2$;

Extra Example 3

In Example 3, the contestant wants to lose an average of at least 7 pounds per month. How many pounds must the contestant lose in the fifth month to meet the goal? at least 1 pound

On Your Own

4. $k < 8$;

   number line with open circle at 8, shaded left; marks at 4, 5, 6, 7, 8, 9, 10

5. $n > 2$;

   number line with open circle at 2, shaded right; marks at $-2, -1, 0, 1, 2, 3, 4$

6. $y \geq -14$;

   number line with closed circle at -14, shaded right; marks at $-16, -15, -14, -13, -12, -11, -10$

7. at least 11 pounds

Example 2

- **Look for and Make Use of Structure:** There is another way to solve Example 2. The inequality has two factors on the left side: -7 and $(x + 3)$. Instead of distributing, divide both sides by -7. Dividing by a negative number changes the direction of the inequality symbol.

$-7(x + 3) \leq 28$	Write the inequality.
$\dfrac{-7(x + 3)}{-7} \geq \dfrac{28}{-7}$	Divide each side by -7. Reverse the inequality symbol.
$x + 3 \geq -4$	Simplify.
$\underline{ -3 \quad -3}$	Subtract 3 from each side.
$x \geq -7$	Simplify.

- Discuss each method with students.

Example 3

- Before beginning to solve the example, talk about different total numbers of pounds that average at least 8 pounds per month for 5 months.
- **?** "Would a total of 35 pounds be enough?" no
- **?** "Would a total of 45 pounds be enough?" yes
- Ask students to interpret the information in the Progress Report. They should recognize that 34 pounds were lost in the first four months.
- **?** "What are you trying to solve for in this problem?" the number of pounds the contestant needs to lose in the 5th month to meet the goal
- **?** "What inequality symbol is needed if the goal is "at least" 8 pounds per month?" $\geq$
- You may want to show students a verbal model for the inequality.
- **Common Error:** Students may want to subtract 34 from each side before dealing with the division by 5. The division must be undone first.
- **Connection:** This problem has a different context but is of the same type as the question in the Motivate activity.

On Your Own

- **Common Error:** If students solve Question 5 by distributing the -4, it is very possible they will write $-4n - 40$ instead of $-4n + 40$. For the factor $n - 10$, they need to remember to *add the opposite* so that the initial equation could be written as $-4[n + (-10)] < 32$. Then distribute the -4.
- Students may need guidance on Question 6. Distributing 0.5 results in $-3 \leq 4 + 0.5y$.

Closure

- Have students write and solve the inequalities for each of the students in the Motivate activity.

  Student A: $(82 + 94 + 86 + 81 + a) \div 5 \geq 85$; $a \geq 82$

  Student B: $(92 + 98 + 88 + 94 + b) \div 5 \geq 90$; $b \geq 78$

  It is mathematically possible for both students to reach their goals.

EXAMPLE **2** **Graphing an Inequality**

Which graph represents the solution of $-7(x + 3) \le 28$?

Ⓐ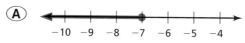

Ⓑ

Ⓒ

Ⓓ

$-7(x + 3) \le 28$	Write the inequality.
$-7x - 21 \le 28$	Distributive Property

Step 1: Undo the subtraction. → $\underline{+\,21 \quad +\,21}$ Addition Property of Inequality

$-7x \le 49$ Simplify.

Step 2: Undo the multiplication. → $\dfrac{-7x}{-7} \ge \dfrac{49}{-7}$ Use the Division Property of Inequality. Reverse the inequality symbol.

$x \ge -7$ Simplify.

∴ The correct answer is Ⓑ.

EXAMPLE **3** **Real-Life Application**

Progress Report	
Month	**Pounds Lost**
1	12
2	9
3	5
4	8

A contestant in a weight-loss competition wants to lose an average of at least 8 pounds per month during a 5-month period. How many pounds must the contestant lose in the fifth month to meet the goal?

Write and solve an inequality. Let x be the number of pounds lost in the fifth month.

$\dfrac{12 + 9 + 5 + 8 + x}{5} \ge 8$ *The phrase at least means greater than or equal to.*

$\dfrac{34 + x}{5} \ge 8$ Simplify.

$5 \cdot \dfrac{34 + x}{5} \ge 5 \cdot 8$ Multiplication Property of Inequality

$34 + x \ge 40$ Simplify.

$x \ge 6$ Subtract 34 from each side.

∴ So, the contestant must lose at least 6 pounds to meet the goal.

Remember

In Example 3, the average is equal to the sum of the pounds lost divided by the number of months.

On Your Own

Now You're Ready
Exercises 12–17

Solve the inequality. Graph the solution.

4. $2(k - 5) < 6$ **5.** $-4(n - 10) < 32$ **6.** $-3 \le 0.5(8 + y)$

7. **WHAT IF?** In Example 3, the contestant wants to lose an average of at least 9 pounds per month. How many pounds must the contestant lose in the fifth month to meet the goal?

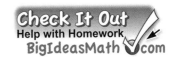

✓ Vocabulary and Concept Check

1. **WRITING** Compare and contrast solving two-step inequalities and solving two-step equations.

2. **OPEN-ENDED** Describe how to solve the inequality $3(a + 5) < 9$.

Practice and Problem Solving

Match the inequality with its graph.

3. $\frac{t}{3} - 1 \geq -3$

A.

B.

C.

4. $5x + 7 \leq 32$

A.

B.

C.

Solve the inequality. Graph the solution.

5. $8y - 5 < 3$

6. $3p + 2 \geq -10$

7. $2 > 8 - \frac{4}{3}h$

8. $-2 > \frac{m}{6} - 7$

9. $-1.2b - 5.3 \geq 1.9$

10. $-1.3 \geq 2.9 - 0.6r$

11. **ERROR ANALYSIS** Describe and correct the error in solving the inequality.

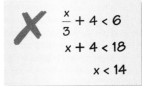

$$\frac{x}{3} + 4 < 6$$
$$x + 4 < 18$$
$$x < 14$$

Solve the inequality. Graph the solution.

12. $5(g + 4) > 15$

13. $4(w - 6) \leq -12$

14. $-8 \leq \frac{2}{5}(k - 2)$

15. $-\frac{1}{4}(d + 1) < 2$

16. $7.2 > 0.9(n + 8.6)$

17. $20 \geq -3.2(c - 4.3)$

10 cm

18. **UNICYCLE** The first jump in a unicycle high-jump contest is shown. The bar is raised 2 centimeters after each jump. Solve the inequality $2n + 10 \geq 26$ to find the number of additional jumps needed to meet or exceed the goal of clearing a height of 26 centimeters.

Assignment Guide and Homework Check

Level	Day 1 Activity Assignment	Day 2 Lesson Assignment	Homework Check
Basic	3, 4, 25–27	1, 2, 5–17 odd, 18, 19, 21	7, 15, 18, 19
Average	3, 4, 25–27	1, 2, 7–17 odd, 18–22 even	7, 15, 18, 20
Advanced	3, 4, 25–27	1, 2, 6–10 even, 11, 12–24 even	10, 14, 20, 22
Accelerated	1–4, 6–10 even, 11, 12–24 even, 25–27		10, 14, 20, 22

Common Errors

- **Exercises 5–10** Students may incorrectly multiply or divide before adding to or subtracting from both sides. Remind them that they should work backward through the order of operations, or that they should start away from the variable and move toward it.
- **Exercises 5–10, 12–17** Students may forget to reverse the inequality symbol when multiplying or dividing by a negative number. Encourage them to write the inequality symbol that they should have in the solution before solving.
- **Exercises 12–17** If students distribute before solving, they may forget to distribute the number to the second term. Remind them that they need to distribute to everything within the parentheses. Encourage students to draw arrows to represent the multiplication.

4.4 Record and Practice Journal

Solve the inequality. Graph the solution.

1. $5 - 3x > 8$
   $x < -1$

2. $-4x - 7 \le 9$
   $x \ge -4$

3. $3 + 4.5x \ge 21$
   $x \ge 4$

4. $-2y - 5 > \dfrac{5}{2}$ $y < -\dfrac{15}{4}$

5. $2(y - 4) < -18$
   $y < -5$

6. $-6 \ge -6(y - 3)$
   $y \ge 4$

7. You borrow $200 from a friend to help pay for a new laptop computer. You pay your friend back $12 per week. Write and solve an inequality to find when you will owe your friend less than $60.
   $200 - 12x < 60$
   $x > 11.6$

   12 weeks

Vocabulary and Concept Check

1. *Sample answer:* They use the same techniques, but when solving an inequality, you must be careful to reverse the inequality symbol when you multiply or divide by a negative number.

2. *Sample answer:* Divide both sides by 3 and then subtract 5 from both sides.

Practice and Problem Solving

3. C   4. A

5. $y < 1$;
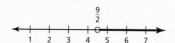

6. $p \ge -4$;

7. $h > \dfrac{9}{2}$;

8. $m < 30$;

9. $b \le -6$;

10. $r \ge 7$;

11. They did not perform the operations in the proper order.

$$\dfrac{x}{3} + 4 < 6$$

$$\dfrac{x}{3} < 2$$

$$x < 6$$

12–17. See Additional Answers.

18. $n \ge 8$ additional jumps

Practice and Problem Solving

19. $x \geq 4$;

20. $d > 6$;

21. $-12x - 38 < -200$;
$x > 13.5$ min

22. See *Taking Math Deeper*.

23. a. $9.5(70 + x) \geq 1000$;

$x \geq 35\frac{5}{19}$, which means
that at least 36 more tickets
must be sold.

b. Because each ticket costs
$1 more, fewer tickets will
be needed for the theater
to earn $1000.

24. $r \geq 8$ units

Fair Game Review

25.

Flutes	7	21	28
Clarinets	4	12	16

$7 : 4$, $21 : 12$, and $28 : 16$

26.

Boys	6	3	30
Girls	10	5	50

$6 : 10$, $3 : 5$, and $30 : 50$

27. A

Mini-Assessment

Solve the inequality. Graph the solution.

1. $2x + 4 < 10$ $x < 3$;

2. $3 \leq \dfrac{y}{-5} + 7$ $y \leq 20$;

3. $-4.2 - 1.1b \leq 2.4$ $b \geq -6$;

4. $\dfrac{2}{3}m + \dfrac{2}{3} \geq -\dfrac{1}{3}$ $m \geq -\dfrac{3}{2}$;

Taking Math Deeper

Exercise 22

Inequality problems can throw students off, simply because of the "inequality." A good way to approach the problem is to imagine that it is an "equality" problem. After the "equation" is written, decide which way the inequality symbol should point.

① Write an equation.

$$75 + 15x = 140$$

| Eaten today | Pounds per bucket | Number of buckets | Total for a day |

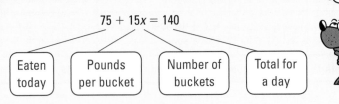

② **a.** Write an inequality and solve it. Recall the whale needs to eat at least 140 pounds of fish each day.

$75 + 15x \geq 140$ Write inequality.

$15x > 65$ Subtract 75 from each side.

$x \geq 4\frac{1}{3}$ Divide each side by 15.

The whale needs to eat at least $4\frac{1}{3}$ more buckets of fish.

③ **b.** If you only have a choice of whole buckets, then the whale should be given 5 more buckets of food. With 4 buckets, the whale would only get $75 + 60 = 135$ pounds of fish.

Project

Select a marine park. Research the amount of fish needed to feed all of the animals at the park. Determine the cost of the fish, not including storage and personnel needed to feed the animals. Using the entrance fee to the park, how many visitors are needed every day to meet the cost? What are some ways that park officials might help offset the daily costs?

Reteaching and Enrichment Strategies

If students need help. . .	If students got it. . .
Resources by Chapter • Practice A and Practice B • Puzzle Time Record and Practice Journal Practice Differentiating the Lesson Lesson Tutorials Skills Review Handbook	Resources by Chapter • Enrichment and Extension • Technology Connection Start the next section

Solve the inequality. Graph the solution.

19. $9x - 4x + 4 \geq 36 - 12$

20. $3d - 7d + 2.8 < 5.8 - 27$

21. **SCUBA DIVER** A scuba diver is at an elevation of -38 feet. The diver starts moving at a rate of -12 feet per minute. Write and solve an inequality that represents how long it will take the diver to reach an elevation deeper than -200 feet.

22. **KILLER WHALES** A killer whale has eaten 75 pounds of fish today. It needs to eat at least 140 pounds of fish each day.

   a. A bucket holds 15 pounds of fish. Write and solve an inequality that represents how many more buckets of fish the whale needs to eat.

   b. Should the whale eat *four* or *five* more buckets of fish? Explain.

23. **REASONING** A student theater charges $9.50 per ticket.

   a. The theater has already sold 70 tickets. Write and solve an inequality that represents how many more tickets the theater needs to sell to earn at least $1000.

   b. The theater increases the ticket price by $1. Without solving an inequality, describe how this affects the total number of tickets needed to earn at least $1000.

24. For what values of r will the area of the shaded region be greater than or equal to 12 square units?

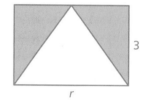

 Fair Game Review What you learned in previous grades & lessons

Find the missing values in the ratio table. Then write the equivalent ratios.
(Skills Review Handbook)

25.

Flutes	7		28
Clarinets	4	12	

26.

Boys	6	3	
Girls	10		50

27. **MULTIPLE CHOICE** What is the volume of the cube?
(Skills Review Handbook)

   A 8 ft³   **B** 16 ft³

   C 24 ft³   **D** 32 ft³

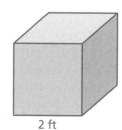

2 ft

Solve the inequality. Graph the solution. *(Section 4.3 and Section 4.4)*

1. $3p \le 18$

2. $2x > -\dfrac{3}{5}$

3. $\dfrac{r}{3} \ge -5$

4. $-\dfrac{z}{8} < 1.5$

5. $3n + 2 \le 11$

6. $-2 < 5 - \dfrac{k}{2}$

7. $1.3m - 3.8 < -1.2$

8. $4.8 \ge 0.3(12 - y)$

Write the word sentence as an inequality. Then solve the inequality. *(Section 4.3)*

9. The quotient of a number and 5 is less than 4.

10. Six times a number is at least -14.

11. PEPPERS You have $18 to buy peppers. Peppers cost $1.50 each. Write and solve an inequality that represents the number of peppers you can buy. *(Section 4.3)*

12. MOVIES You have a gift card worth $90. You want to buy several movies that cost $12 each. Write and solve an inequality that represents the number of movies you can buy and still have at least $30 on the gift card. *(Section 4.4)*

13. CHOCOLATES Your class sells boxes of chocolates to raise $500 for a field trip. You earn $6.25 for each box of chocolates sold. Write and solve an inequality that represents the number of boxes your class must sell to meet or exceed the fundraising goal. *(Section 4.3)*

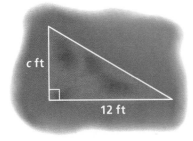

c ft

12 ft

14. FENCE You want to put up a fence that encloses a triangular region with an area greater than or equal to 60 square feet. What is the least possible value of *c*? Explain. *(Section 4.3)*

Alternative Assessment Options

Math Chat **Student Reflective Focus Question**
Structured Interview **Writing Prompt**

Math Chat

- Have students work in pairs. One student describes how to write and graph inequalities, giving examples. The other student probes for more information. Students then switch roles and repeat the process for how to solve one- and two-step inequalities.
- The teacher should walk around the classroom listening to the pairs and asking questions to ensure understanding.

Study Help Sample Answers

Remind students to complete Graphic Organizers for the rest of the chapter.

6.

Solving inequalities using multiplication

- Use the Multiplication Property of Inequality (Case 1): When you multiply each side of an inequality by the same *positive* number, the inequality remains true.

  Example: $\frac{x}{5} > 10$
  $\frac{x}{5} \cdot 5 > 10 \cdot 5$
  $x > 50$

- Use the Multiplication Property of Inequality (Case 2): When you multiply each side of an inequality by the same *negative* number, the direction of the inequality symbol must be reversed for the inequality to remain true.

  Example: $-\frac{1}{2}x \le 10$
  $-\frac{2}{1} \cdot \left(-\frac{1}{2}x\right) \ge -\frac{2}{1} \cdot 10$
  $x \ge -20$

Solving inequalities using division

- Use the Division Property of Inequality (Case 1): When you divide each side of an inequality by the same *positive* number, the inequality remains true.

  Example: $5x > 10$
  $\frac{5x}{5} > \frac{10}{5}$
  $x > 2$

- Use the Division Property of Inequality (Case 2): When you divide each side of an inequality by the same *negative* number, the direction of the inequality symbol must be reversed for the inequality to remain true.

  Example: $-3x \ge 9$
  $\frac{-3x}{-3} \le \frac{9}{-3}$
  $x \le -3$

- Use inverse operations to group numbers on one side.
- Use inverse operations to group variables on one side.
- Solve for the variable.

7–8. Available at *BigIdeasMath.com*.

Reteaching and Enrichment Strategies

If students need help. . .	If students got it. . .
Resources by Chapter • Practice A and Practice B • Puzzle Time Lesson Tutorials *BigIdeasMath.com*	Resources by Chapter • Enrichment and Extension • Technology Connection Game Closet at *BigIdeasMath.com* Start the Chapter Review

Answers

1. $p \le 6$;
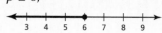

2. $x > -\frac{3}{10}$;
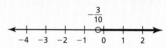

3. $r \ge -15$;

4. $z > -12$;

5. $n \le 3$;
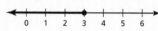

6. $k < 14$;
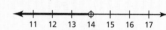

7. $m < 2$;
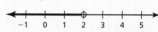

8. $-4 \le y$;

9. $q \div 5 < 4; q < 20$

10. $6t \ge -14; t \ge -\frac{7}{3}$

11. $1.5p \le 18; p \le 12$ peppers

12. $90 - 12c \ge 30; c \le 5$ movies

13. $6.25b \ge 500; b \ge 80$ boxes

14. 10 feet; $\frac{1}{2}(12)c \ge 60 \longrightarrow c \ge 10$; The least value for which $c \ge 10$ is 10.

Technology for the *Teacher*

Online Assessment
Assessment Book
ExamView® Assessment Suite

Answers

1. $w > -3$

2. $y - \dfrac{1}{2} \leq -\dfrac{3}{2}$

3. yes   4. no

5.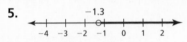

6.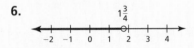

7. $h \geq 42$

Review of Common Errors

Exercises 5 and 6
- Students may use the wrong type of circle at the boundary point. Remind them to use a closed circle when the boundary point is a solution and an open circle when the boundary point is not a solution.

Exercises 5 and 6
- Students may shade on the wrong side of the boundary point. Remind them to test one point on each side of the boundary point.

Exercises 8–10
- Students may use the same operation rather than the inverse operation to isolate the variable. Remind students that when a number is added to the variable, subtract that number from each side. When a number is subtracted from the variable, add that number to each side.

Exercises 10 and 13
- Students may shade the number line in the wrong direction when the variable is on the right side of the inequality symbol. Remind them to rewrite the inequality with the variable at the left or to check a value on each side of the boundary point.

Exercises 11–13
- Students may perform the same operation on both sides instead of the opposite operation when solving the inequality. Remind them that solving inequalities is similar to solving equations.

Exercises 11–13
- When there is a negative in the inequality, students may reverse the direction of the inequality symbol. Remind them that they only reverse the direction when they are multiplying or dividing by a negative number.

Exercises 14–16
- Students may incorrectly multiply or divide before adding to or subtracting from both sides. Remind them that they should work backward through the order of operations, or that they should start away from the variable and move toward it.

Exercises 14–19
- Students may forget to reverse the inequality symbol when multiplying or dividing by a negative number. Encourage them to write the inequality symbol that they should have in the solution before solving.

Exercises 17–19
- If students distribute before solving, they may forget to distribute the number to the second term. Remind them that they need to distribute to everything within the parentheses. Encourage students to draw arrows to represent the multiplication.

4 Chapter Review

Check It Out
Vocabulary Help
BigIdeasMath ✓com

Review Key Vocabulary

inequality, *p. 126*
solution of an inequality, *p. 126*

solution set, *p. 126*
graph of an inequality, *p. 127*

Review Examples and Exercises

4.1 Writing and Graphing Inequalities *(pp. 124–129)*

a. Six plus a number x is at most $-\dfrac{1}{4}$. Write this word sentence as an inequality.

Six plus a number x	is at most	$-\dfrac{1}{4}$.
$6 + x$	$\leq$	$-\dfrac{1}{4}$

∴ An inequality is $6 + x \leq -\dfrac{1}{4}$.

b. Graph $m > 3$.

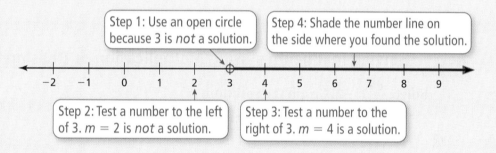

Step 1: Use an open circle because 3 is *not* a solution.

Step 4: Shade the number line on the side where you found the solution.

Step 2: Test a number to the left of 3. $m = 2$ is *not* a solution.

Step 3: Test a number to the right of 3. $m = 4$ is a solution.

Exercises

Write the word sentence as an inequality.

1. A number w is greater than -3.

2. A number y minus $\dfrac{1}{2}$ is no more than $-\dfrac{3}{2}$.

Tell whether the given value is a solution of the inequality.

3. $5 + j > 8; j = 7$

4. $6 \div n \leq -5; n = -3$

Graph the inequality on a number line.

5. $q > -1.3$

6. $s < 1\dfrac{3}{4}$

7. BUMPER CARS You must be at least 42 inches tall to ride the bumper cars at an amusement park. Write an inequality that represents this situation.

4.2 Solving Inequalities Using Addition or Subtraction (pp. 130–135)

Solve −5 < m − 3. Graph the solution.

$$-5 < m - 3 \qquad \text{Write the inequality.}$$

Undo the subtraction. →	$+\ 3 \qquad\ +\ 3$	Addition Property of Inequality

$$-2 < m \qquad \text{Simplify.}$$

The solution is $m > -2$.

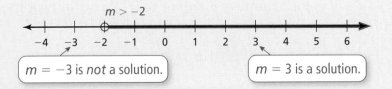

$m > -2$

m = −3 is *not* a solution.   m = 3 is a solution.

Exercises

Solve the inequality. Graph the solution.

8. $d + 12 < 19$ **9.** $t - 4 \leq -14$ **10.** $-8 \leq z + 6.4$

4.3 Solving Inequalities Using Multiplication or Division (pp. 138–145)

Solve $\dfrac{c}{-3} \geq -2$. Graph the solution.

$$\frac{c}{-3} \geq -2 \qquad \text{Write the inequality.}$$

Undo the division. →	$-3 \cdot \dfrac{c}{-3} \ \leq\ -3 \cdot (-2)$	Use the Multiplication Property of Inequality. Reverse the inequality symbol.

$$c \leq 6 \qquad \text{Simplify.}$$

The solution is $c \leq 6$.

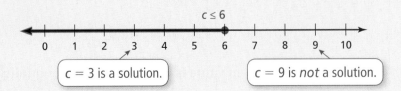

$c \leq 6$

c = 3 is a solution.   c = 9 is *not* a solution.

Exercises

Solve the inequality. Graph the solution.

11. $6q < -18$ **12.** $-\dfrac{r}{3} \leq 6$ **13.** $-4 > -\dfrac{4}{3}s$

Review Game

Inequalities

Materials

- Questions (solving inequalities) written on index cards (one question per card) from the chapter homework, quizzes, examples, or tests—at least as many cards as students
- pencil
- paper
- 10 pre-made number lines per pair of students

Directions

Play in pairs. Each pair is a team. Each player is given a question card and solves the inequality on the card. Both players graph their solutions on the same number line. If the shadings on the graph intersect, then the team earns 1 point for the round. If not, then the team does not earn a point.

After each round, each student on the team passes their card in opposite directions to another team along a set rotation to ensure unique pairs of cards. This continues for ten rounds.

Who wins?

The team with the most points at the end of ten rounds wins.

Alternate Point Scoring: In each round, after all the teams complete their graphs, a coin toss can be used to randomly determine whether a point is earned for intersecting graphs or for non-intersecting graphs.

Answers

8. $d < 7$;

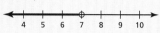

9. $t \le -10$;

10. $z \ge -14.4$;

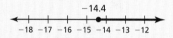

11. $q < -3$;

12. $r \ge -18$;

13. $3 < s$;

14. $x > 4$;

15. $z \ge -8$;

16. $t > -7$;

17. $q < -13$;

18. $p \ge -21$;

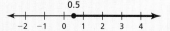

19. $j \ge 0.5$;

My Thoughts on the Chapter

Teacher Tip

Not allowed to write in your teaching edition? Use sticky notes to record your thoughts.

What worked. . .

What did not work. . .

What I would do differently. . .

4.4 **Solving Two-Step Inequalities** *(pp. 146–151)*

a. **Solve** $6x - 8 \leq 10$. **Graph the solution.**

$$6x - 8 \leq 10$$ Write the inequality.

> Step 1: Undo the subtraction.

$$\underline{+8} \quad \underline{+8}$$ Addition Property of Inequality

$$6x \leq 18$$ Simplify.

> Step 2: Undo the multiplication.

$$\frac{6x}{6} \leq \frac{18}{6}$$ Division Property of Inequality

$$x \leq 3$$ Simplify.

∴ The solution is $x \leq 3$.

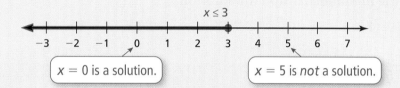

$x \leq 3$

$x = 0$ is a solution. $x = 5$ is *not* a solution.

b. **Solve** $\dfrac{q}{-4} + 7 < 11$. **Graph the solution.**

$$\frac{q}{-4} + 7 < 11$$ Write the inequality.

> Step 1: Undo the addition.

$$\underline{-7} \quad \underline{-7}$$ Subtraction Property of Inequality

$$\frac{q}{-4} < 4$$ Simplify.

> Step 2: Undo the division.

$$-4 \cdot \frac{q}{-4} > -4 \cdot 4$$ Use the Multiplication Property of Inequality. Reverse the inequality symbol.

$$q > -16$$ Simplify.

∴ The solution is $q > -16$.

$q > -16$

$q = -20$ is *not* a solution. $q = -12$ is a solution.

Exercises

Solve the inequality. Graph the solution.

14. $3x + 4 > 16$ **15.** $\dfrac{z}{-2} - 6 \leq -2$ **16.** $-2t - 5 < 9$

17. $7(q + 2) < -77$ **18.** $-\dfrac{1}{3}(p + 9) \leq 4$ **19.** $1.2(j + 3.5) \geq 4.8$

Write the word sentence as an inequality.

1. A number k plus 19.5 is less than or equal to 40.

2. A number q multiplied by $\frac{1}{4}$ is greater than -16.

Tell whether the given value is a solution of the inequality.

3. $n - 3 \leq 4;\ n = 7$

4. $-\frac{3}{7}m < 1;\ m = -7$

5. $-4c \geq 7;\ c = -2$

6. $-2.4m > -6.8;\ m = -3$

Solve the inequality. Graph the solution.

7. $w + 4 \leq 3$

8. $x - 4 > -6$

9. $-\frac{2}{9} + y \leq \frac{5}{9}$

10. $-6z \geq 36$

11. $-5.2 \geq \frac{p}{4}$

12. $4k - 8 \geq 20$

13. $\frac{4}{7} - b \geq -\frac{1}{7}$

14. $-0.6 > -0.3(d + 6)$

15. **GUMBALLS** You have $2.50. Each gumball in a gumball machine costs $0.25. Write and solve an inequality that represents the number of gumballs you can buy.

16. **PARTY** You can spend no more than $100 on a party you are hosting. The cost per guest is $8.

   a. Write and solve an inequality that represents the number of guests you can invite to the party.

   b. What is the greatest number of guests that you can invite to the party? Explain your reasoning.

17. **BASEBALL CARDS** You have $30 to buy baseball cards. Each pack of cards costs $5. Write and solve an inequality that represents the number of packs of baseball cards you can buy and still have at least $10 left.

Test Item References

Chapter Test Questions	Section to Review
1–6	4.1
7–9	4.2
10, 11, 15, 16	4.3
12–14, 17	4.4

Test-Taking Strategies

Remind students to quickly look over the entire test before they start so that they can budget their time. When writing word phrases as inequalities, students can get confused by the subtle differences in wording. Encourage students to think carefully about which inequality symbol is implied by the wording. Have students use the **Stop** and **Think** strategy.

Common Errors

- **Exercises 3–6** Students may not understand when the boundary point is a solution of the inequality. Remind them that the inequality is true for the boundary point only when the inequality symbol is ≤ or ≥.
- **Exercises 7–14** Students may perform the same operation on both sides instead of the opposite operation when solving the inequality. Remind them that solving inequalities is similar to solving equations.
- **Exercises 7–14** Students may reverse the direction of the inequality symbol just because there is a negative number in the inequality. Remind them that you only reverse the direction of the inequality symbol when you multiply each side by a negative number, or when you reverse the right and left side of the inequality.
- **Exercises 7–14** Students may shade in the wrong direction when the variable is on the right side of the inequality instead of the left. Remind them to rewrite the inequality by reversing the right and left sides and reversing the inequality symbol.
- **Exercise 14** If students distribute before solving, they may forget to distribute the number to the second term. Remind them that they need to distribute to everything within the parentheses. Encourage students to draw arrows to represent the multiplication.

Reteaching and Enrichment Strategies

If students need help...	If students got it...
Resources by Chapter • Practice A and Practice B • Puzzle Time Record and Practice Journal Practice Differentiating the Lesson Lesson Tutorials *BigIdeasMath.com* Skills Review Handbook	Resources by Chapter • Enrichment and Extension • Technology Connection Game Closet at *BigIdeasMath.com* Start Cumulative Assessment

Answers

1. $k + 19.5 \le 40$

2. $\frac{1}{4}q > -16$

3. yes   4. no

5. yes   6. yes

7. $w \le -1$;

8. $x > -2$;
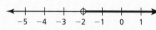

9. $y \le \frac{7}{9}$;
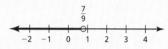

10. $z \le -6$;

11. $p \le -20.8$;

12. $k \ge 7$;

13. $b \le \frac{5}{7}$;

14. $d > -4$;

15. $0.25g \le 2.50$; $g \le 10$ gumballs

16. **a.** $8g \le 100$; $g \le 12.5$

    b. Twelve guests because 12 is the largest whole number that satisfies the inequality.

17. $30 - 5c \ge 10$; $c \le 4$ packs of cards

After Answering Easy Questions, Relax

Answer Easy Questions First

Estimate the Answer

Read All Choices before Answering

Read Question before Answering

Solve Directly or Eliminate Choices

Solve Problem before Looking at Choices

Use Intelligent Guessing

Work Backwards

About this Strategy

When taking a multiple choice test, be sure to read each question carefully and thoroughly. When taking a timed test, it is often best to skim the test and answer the easy questions first. Be careful that you record your answer in the correct position on the answer sheet.

Answers

1. B
2. I
3. A
4. F

Technology for the **Teacher**

Performance Tasks

Online Assessment

Assessment Book

ExamView® Assessment Suite

Item Analysis

1. **A.** The student makes at least two sign errors while performing the operations.

   B. Correct answer

   C. The student makes a sign error while performing the operations.

   D. The student makes one or more sign errors while performing the operations.

2. **F.** The student inverts the product.

   G. The student inverts the first factor before multiplying.

   H. The student inverts the second factor before multiplying.

   I. Correct answer

3. **A.** Correct answer

   B. The student makes sign errors and fails to reverse the inequality symbol while solving the inequality.

   C. The student makes a sign error while solving the inequality.

   D. The student fails to reverse the inequality symbol while solving the inequality.

4. **F.** Correct answer

   G. The student subtracts 6 from each side, and then divides each side by 5.

   H. The student divides each side by 5, and then adds 6 to each side instead of subtracting.

   I. The student subtracts 5 from each side, and then subtracts 6 from each side.

1. What is the value of the expression below when $x = -5$, $y = 3$, and $z = -1$?

$$\frac{x^2 - 3y}{z}$$

A. -34

B. -16

C. 16

D. 34

2. What is the value of the expression below?

$$-\frac{3}{8} \cdot \frac{2}{5}$$

F. $-\dfrac{20}{3}$

G. $-\dfrac{16}{15}$

H. $-\dfrac{15}{16}$

I. $-\dfrac{3}{20}$

3. Which graph represents the inequality below?

$$\frac{x}{-4} - 8 \geq -9$$

A.

B.

C.

D.

4. Which value of p makes the equation below true?

$$5(p + 6) = 25$$

F. -1

G. $3\dfrac{4}{5}$

H. 11

I. 14

5. You set up the lemonade stand. Your profit is equal to your revenue from lemonade sales minus your cost to operate the stand. Your cost is $8. How many cups of lemonade must you sell to earn a profit of $30?

LEMONADE

Lemonade
50¢
per cup

A. 4

C. 60

B. 44

D. 76

6. Which value is a solution of the inequality below?

$$3 - 2y < 7$$

F. -6

H. -2

G. -3

I. -1

7. What value of y makes the equation below true?

$$12 - 3y = -6$$

8. What is the mean distance of the four points from -3?

A. $-\dfrac{1}{2}$

C. 3

B. $2\dfrac{1}{2}$

D. $7\dfrac{1}{8}$

Item Analysis (continued)

Answers

5. D

6. I

7. 6

8. C

5. A. The student writes and solves the equation $(8 - 0.50)x = 30$.

   B. The student writes and solves the equation $0.50x = 30 - 8$.

   C. The student finds the number of cups to earn a revenue of $30.

   D. Correct answer

6. F. Because the inequality symbol is <, the student chooses the least possible answer.

   G. The student fails to reverse the inequality symbol and gets $y < -2$.

   H. The student solves the problem as an equation instead of as an inequality.

   I. Correct answer

7. Gridded response: Correct answer: 6

Common error: The student makes a sign error.

8. A. The student finds the mean of the four points.

   B. The student finds the difference of each number and -3 and then finds the mean of the results.

   C. Correct answer

   D. The student finds the distance (the range) between the greatest and least points.

Answers

9. F

10. $-\dfrac{11}{24}$

11. *Part A* at least 48 more T-shirts; The inequality is $20 + 10t \geq 500$.

   Part B at least 63 T-shirts; The inequality is $8t \geq 500$.

   Part C Your friend; at least 13 more T-shirts; You must sell at least 50 total T-shirts and your friend must sell at least 63 total T-shirts.

12. A

Item Analysis (continued)

9. **F.** Correct answer

   G. The student fails to reverse the inequality symbol when dividing each side by a negative number.

   H. The student fails to reverse the inequality symbol when dividing each side by a negative number and misinterprets the non-strict inequality symbol.

   I. The student makes a sign error in solving the inequality.

10. **Gridded response:** Correct answer: $-\dfrac{11}{24}$

   Common error: The student makes an arithmetic error in finding a common denominator.

11. **4 points** The student demonstrates a thorough understanding of writing and solving inequalities. The student identifies the correct quantities, operations, and inequality symbols, and the solutions are clear, neat, and correct. In Part A, you must sell at least 48 more T-shirts. In Part B, your friend must sell at least 63 T-shirts. In Part C, your friend must sell 13 more T-shirts; He must sell 63, you must sell a total of 50.

   3 points The student demonstrates an essential but less than thorough understanding of writing and solving inequalities. There may be one error made, but subsequent work is consistent with the error.

   2 points The student demonstrates a partial understanding of writing and solving inequalities. The student's work and explanations demonstrate a lack of essential understanding. The student sets up one or more of the three problem situations incorrectly.

   1 point The student demonstrates limited understanding. The student's response is incomplete and exhibits many errors.

   0 points The student provides no response, or a response that is completely incorrect, incomprehensible, or fails to demonstrate sufficient understanding of writing and solving inequalities.

12. **A.** Correct answer

   B. The student makes a sign error in multiplying.

   C. The student fails to handle the signs correctly when subtracting the negative fraction in the expression in the problem statement.

   D. The student inverts the dividend instead of the divisor before multiplying.

9. Martin graphed the solution of the inequality $-4x + 18 > 6$ in the box below.

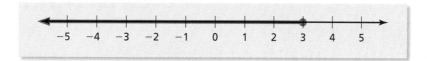

What should Martin do to correct the error that he made?

   F. Use an open circle at 3 and shade to the left of 3.

   G. Use an open circle at 3 and shade to the right of 3.

   H. Use a closed circle and shade to the right of 3.

   I. Use an open circle and shade to the left of -3.

10. What is the value of the expression below?

$$\frac{5}{12} - \frac{7}{8}$$

11. You are selling T-shirts to raise money for a charity. You sell the T-shirts for $10 each.

Part A   You have already sold 2 T-shirts. How many more T-shirts must you sell to raise at least $500? Explain.

Part B   Your friend is raising money for the same charity and has not sold any T-shirts previously. He sells the T-shirts for $8 each. What is the total number of T-shirts he must sell to raise at least $500? Explain.

Part C   Who has to sell more T-shirts in total? How many more? Explain.

12. Which expression is equivalent to the expression below?

$$-\frac{2}{3} - \left(-\frac{4}{9}\right)$$

   A. $-\dfrac{1}{3} + \dfrac{1}{9}$

   C. $-\dfrac{1}{3} - \dfrac{7}{9}$

   B. $-\dfrac{2}{3} \times \left(-\dfrac{1}{3}\right)$

   D. $\dfrac{3}{2} \div \left(-\dfrac{1}{3}\right)$

5 Ratios and Proportions

5.1 Ratios and Rates

5.2 Proportions

5.3 Writing Proportions

5.4 Solving Proportions

5.5 Slope

5.6 Direct Variation

"I am doing an experiment with slope. I want you to run up and down the board 10 times."

"Now with 2 more dog biscuits, do it again and we'll compare your rates."

"Dear Sir: I counted the number of bacon, cheese, and chicken dog biscuits in the box I bought."

"There were 16 bacon, 12 cheese, and only 8 chicken. That's a ratio of 4:3:2. Please go back to the original ratio of 1:1:1."

What Your Students Have Learned

- Generate numerical patterns given rules, identify the relationship, and form ordered pairs.
- Plot points in the first quadrant of the coordinate plane.
- Convert standard measurement units within a measurement system.
- Graph ordered pairs in all four quadrants of the coordinate plane.
- Understand ratios and describe ratio relationships.
- Compare ratios using tables.
- Use ratio reasoning to convert measurement units.
- Understand rates and unit rates.

What Your Students Will Learn

- Find unit rates associated with ratios of fractions, areas, and other quantities in like or different units.
- Decide whether two quantities are proportional using ratio tables and graphs.
- Use graphs to determine whether two ratios form a proportion.
- Identify the constant of proportionality (unit rate) in tables, graphs, equations, diagrams, and verbal descriptions.
- Represent proportional relationships with equations.
- Explain what a point (x, y) means on a proportional graph in context, particularly $(0, 0)$ and $(1, r)$, where r is the unit rate.
- Use proportionality to solve problems.

Pacing Guide for Chapter 5

Chapter Opener	
Regular	1 Day
Accelerated	1 Day
Section 1	
Regular	2 Days
Accelerated	1 Day
Section 2	
Regular	3 Days
Accelerated	2 Days
Section 3	
Regular	2 Days
Accelerated	1 Day
Study Help / Quiz	
Regular	1 Day
Accelerated	1 Day
Section 4	
Regular	2 Days
Accelerated	1 Day
Section 5	
Regular	2 Days
Accelerated	1 Day
Section 6	
Regular	2 Days
Accelerated	1 Day
Chapter Review/ Chapter Tests	
Regular	2 Days
Accelerated	2 Days
Total Chapter 5	
Regular	17 Days
Accelerated	11 Days
Year-to-Date	
Regular	71 Days
Accelerated	40 Days

Technology for the *Teacher*

BigIdeasMath.com
Chapter at a Glance
Complete Materials List
Parent Letters: English and Spanish

What Your Students Have Learned
- Simplify fractions by using factors.
- Identify equivalent fractions.
- Solve one-step equations.

Additional Topics for Review
- Identifying Patterns in Tables
- Operations on Fractions and Decimals
- Coordinate Plane
- Graphing Ordered Pairs
- Equations in Two Variables
- Ratios
- Ratio Tables
- Rates
- Unit Rates
- Converting Measures

Try It Yourself

1. $\frac{1}{12}$ 2. $\frac{1}{3}$

3. $\frac{3}{4}$ 4. $\frac{2}{3}$

5. no   6. no

7. no   8. yes

9. -15 10. 3

11. $\frac{9}{2}$ 12. -28

Record and Practice Journal
Fair Game Review

1. $\frac{1}{6}$ 2. $\frac{2}{3}$

3. $\frac{1}{5}$ 4. $\frac{1}{2}$

5. $\frac{4}{9}$ 6. $\frac{4}{5}$

7–20. See Additional Answers.

Math Background Notes

Vocabulary Review
- Greatest Common Factor
- Equivalent Fractions
- Equation
- Inverse Operations
- Properties of Equality

Simplifying Fractions
- Students should know how to simplify fractions.
- Some students may have learned simplifying fractions as reducing fractions.
- Remind students that you must divide the numerator and the denominator by the same factor. This is equivalent to dividing by one which does not change the value of the fraction but does change the form.

Identifying Equivalent Fractions
- Students should know how to identify equivalent fractions.
- Encourage students to simplify the fraction with the greater numbers. If the fraction simplifies to the second fraction, students can conclude the fractions are equivalent.
- **Teaching Tip:** Some students may have difficulty simplifying fractions. This makes the search for equivalent fractions difficult. Encourage these students to start with the fraction with lesser numbers and multiply the numerator and denominator by the same factor to see if they can produce the second fraction.

Solving Equations
- Students should know how to solve equations.
- Remind students that whatever is performed on one side of the equation must be performed on the other side.
- Remind students that variables can appear on either side of the equal sign.
- **Common Error:** Students may do the same operation on both sides instead of the inverse operation. Remind them that to get the variable alone, they need to use the inverse operation.

Reteaching and Enrichment Strategies

If students need help...	If students got it...
Record and Practice Journal • Fair Game Review Skills Review Handbook Lesson Tutorials	Game Closet at *BigIdeasMath.com* Start the next section

What You Learned Before

"I wonder if our rate is proportional to the slope of the hill."

...or possibly proportional to our stupidity!

Simplifying Fractions

Example 1 Simplify $\dfrac{4}{8}$.

$$\dfrac{4 \div 4}{8 \div 4} = \dfrac{1}{2}$$

Example 2 Simplify $\dfrac{10}{15}$.

$$\dfrac{10 \div 5}{15 \div 5} = \dfrac{2}{3}$$

Identifying Equivalent Fractions

Example 3 Is $\dfrac{1}{4}$ equivalent to $\dfrac{13}{52}$?

$$\dfrac{13 \div 13}{52 \div 13} = \dfrac{1}{4}$$

∴ $\dfrac{1}{4}$ is equivalent to $\dfrac{13}{52}$.

Example 4 Is $\dfrac{30}{54}$ equivalent to $\dfrac{5}{8}$?

$$\dfrac{30 \div 6}{54 \div 6} = \dfrac{5}{9}$$

∴ $\dfrac{30}{54}$ is *not* equivalent to $\dfrac{5}{8}$.

Solving Equations

Example 5 Solve $12x = 168$.

$12x = 168$ Write the equation.

$\dfrac{12x}{12} = \dfrac{168}{12}$ Division Property of Equality

$x = 14$ Simplify.

Check

$12x = 168$

$12(14) \stackrel{?}{=} 168$

$168 = 168$ ✓

Try It Yourself

Simplify.

1. $\dfrac{12}{144}$

2. $\dfrac{15}{45}$

3. $\dfrac{75}{100}$

4. $\dfrac{16}{24}$

Are the fractions equivalent? Explain.

5. $\dfrac{15}{60} \stackrel{?}{=} \dfrac{3}{4}$

6. $\dfrac{2}{5} \stackrel{?}{=} \dfrac{24}{144}$

7. $\dfrac{15}{20} \stackrel{?}{=} \dfrac{3}{5}$

8. $\dfrac{2}{8} \stackrel{?}{=} \dfrac{16}{64}$

Solve the equation. Check your solution.

9. $\dfrac{y}{-5} = 3$

10. $0.6 = 0.2a$

11. $-2w = -9$

12. $\dfrac{1}{7}n = -4$

Essential Question How do rates help you describe real-life problems?

The Meaning of a Word Rate

When you rent snorkel gear at the beach, you should pay attention to the rental **rate**. The rental rate is in dollars per hour.

Snorkel Rentals
$8.75 per hour

Snorkel Rentals
$7.25 per hour

1 ACTIVITY: Finding Reasonable Rates

Work with a partner.

a. Match each description with a verbal rate.

b. Match each verbal rate with a numerical rate.

c. Give a reasonable numerical rate for each description. Then give an unreasonable rate.

Description	*Verbal Rate*	*Numerical Rate*
Your running rate in a 100-meter dash	Dollars per year	$\dfrac{\quad\text{in.}}{\text{yr}}$
The fertilization rate for an apple orchard	Inches per year	$\dfrac{\quad\text{lb}}{\text{acre}}$
The average pay rate for a professional athlete	Meters per second	$\dfrac{\$\quad}{\text{yr}}$
The average rainfall rate in a rain forest	Pounds per acre	$\dfrac{\quad\text{m}}{\text{sec}}$

Ratios and Rates

In this lesson, you will
- find ratios, rates, and unit rates.
- find ratios and rates involving ratios of fractions.

2 ACTIVITY: Simplifying Expressions That Contain Fractions

Work with a partner. Describe a situation where the given expression may apply. Show how you can rewrite each expression as a division problem. Then simplify and interpret your result.

a. $\dfrac{\frac{1}{2}\,\text{c}}{4\text{ fl oz}}$

b. $\dfrac{2\text{ in.}}{\frac{3}{4}\,\text{sec}}$

c. $\dfrac{\frac{3}{8}\,\text{c sugar}}{\frac{3}{5}\,\text{c flour}}$

d. $\dfrac{\frac{5}{6}\,\text{gal}}{\frac{2}{3}\,\text{sec}}$

Laurie's Notes

Introduction

Applying Mathematical Practices

- **Look for and Make Use of Structure:** In working with ratios and rates, students will make connections to their work with fractions.

Motivate

- **Model:** In an area visible to students, set a wind-up toy in motion. If a toy is not available, (quietly) ask a student to walk across the room at a constant speed.
- ❓ "How fast is the toy or student moving?" There will be no exact answer.
- ❓ "How do you measure the *rate* that the toy or student is moving? Which two pieces of information do you need to find the *rate*?"
- Provide measuring tape and a stopwatch. Ask volunteers to compute the rate.
- Discuss why you might use a convenient unit of time (i.e., 5 seconds) versus trying to use a convenient unit of distance (i.e., 10 feet).
- Write the information measured on the board in words, and also as a numerical rate.

Activity Notes

Meaning of the Word

- Begin with a general discussion of rates that students should understand and ask for reasonable values for each: speed limit (65 miles per hour), heart rate (70 beats per minute), and gas mileage (35 miles per gallon).

Activity 1

- Share and discuss sample answers when students have finished.
- ❓ "How would you describe what a rate is to someone who doesn't know?" Listen for a comparison of two quantities where the units are different. The word "per" is used when comparing the two quantities.
- *Note:* All of the rates in this activity are **unit rates**. The denominator references a single unit (i.e., inches per one month, miles per one hour).
- Share with students that rates do not have to be unit rates. Example: $1.89 per 12 ounces and 100 shared minutes per 4 people.

Activity 2

- Before students begin Activity 2, explain that sometimes a rate can have fractions in the numerator, denominator, or both. The point of this activity is to simplify complex fractions. Students should not feel overwhelmed by the sight of a complex fraction.
- **Look for and Make Use of Structure:** The units may prevent some students from *seeing* the division. If students are stuck, then ask how the problem would look without units.
- Point out how the expression in part (c) is different from the others because the units (cups) are the same in the numerator and the denominator.
- After they have finished, have some students share their answers.

What Your Students Will Learn

- Find ratios, rates, and unit rates using division.
- Find ratios and rates involving ratios of fractions using graphs and ratio tables.

Previous Learning

Students have used reasoning about multiplication and division to solve ratio and rate problems. Knowing measurement abbreviations is helpful.

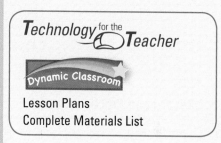

Lesson Plans
Complete Materials List

5.1 Record and Practice Journal

Essential Question How do rates help you describe real-life problems?

① ACTIVITY: Finding Reasonable Rates

Work with a partner.

a. Match each description with a verbal rate.   See Additional Answers.

b. Match each verbal rate with a numerical rate.

c. Give a reasonable numerical rate for each description. Then give an unreasonable rate.

Description	Verbal Rate	Numerical Rate
Your running rate in a 100-meter dash	Dollars per year	$= \boxed{} \frac{\text{in.}}{\text{yr}}$
The fertilization rate for an apple orchard	Inches per year	$= \boxed{} \frac{\text{lb}}{\text{acre}}$
The average pay rate for a professional athlete	Meters per second	$= \boxed{} \frac{\$}{\text{yr}}$
The average rainfall rate in a rainforest	Pounds per acre	$= \boxed{} \frac{\text{m}}{\text{sec}}$

② ACTIVITY: Simplifying Expressions That Contain Fractions

Work with a partner. Describe a situation where the given expression may apply. Show how you can rewrite each expression as a division problem. Then simplify and interpret your result.

a. $\dfrac{\frac{1}{2}\text{ c}}{4\text{ fl oz}}$ $\dfrac{1}{8}$ c/fl oz

b. $\dfrac{2\text{ in.}}{\frac{3}{4}\text{ sec}}$ $\dfrac{8}{3}$, or $2\frac{2}{3}$ in./sec

T-162

Activity 3

- Students use a ratio table in part (b). They should have studied ratio tables in a previous course. If students are not familiar with ratio tables, then they may need extra help with this activity.
- In my experience, students are more successful correctly answering the types of questions in this activity when they organize their work in a ratio table *and* when they label the units. The ratio table is an organizing structure and the units guide the computations.
- Take time to discuss each part of the activity.

Activity 4

- **Look for and Make Use of Structure:** Explain that some real-life problems involve the product of an amount and a rate. The structure you want students to observe is the product of a whole number and a fraction. There is a numerical component and a units component.
- Unit analysis, which again students should have learned in a previous course, is also known as *dimensional analysis*.
- Students may have difficulty thinking of a context for part (c).

What Is Your Answer?

- The estimation rule in Question 6 is very interesting.
- Students will need to try a few examples to see why it works. One example to try: the 40-hour work week.

Closure

- Describe three common rates and give a numerical example of each. Listen to be sure that students are comparing two units using the word "per."

5.1 Record and Practice Journal

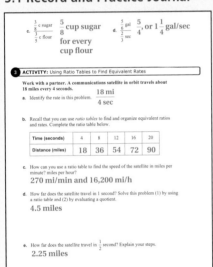

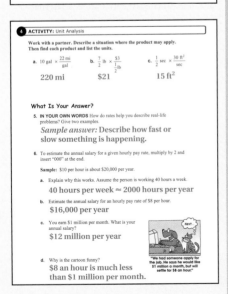

3 ACTIVITY: Using Ratio Tables to Find Equivalent Rates

Work with a partner. A communications satellite in orbit travels about 18 miles every 4 seconds.

a. Identify the rate in this problem.

b. Recall that you can use *ratio tables* to find and organize equivalent ratios and rates. Complete the ratio table below.

Time (seconds)	4	8	12	16	20
Distance (miles)					

c. How can you use a ratio table to find the speed of the satellite in miles per minute? miles per hour?

d. How far does the satellite travel in 1 second? Solve this problem (1) by using a ratio table and (2) by evaluating a quotient.

e. How far does the satellite travel in $\frac{1}{2}$ second? Explain your steps.

4 ACTIVITY: Unit Analysis

Math Practice

View as Components

What is the product of the numbers? What is the product of the units? Explain.

Work with a partner. Describe a situation where the product may apply. Then find each product and list the units.

a. $10 \text{ gal} \times \dfrac{22 \text{ mi}}{\text{gal}}$

b. $\dfrac{7}{2} \text{ lb} \times \dfrac{\$3}{\frac{1}{2} \text{ lb}}$

c. $\dfrac{1}{2} \text{ sec} \times \dfrac{30 \text{ ft}^2}{\text{sec}}$

What Is Your Answer?

5. **IN YOUR OWN WORDS** How do rates help you describe real-life problems? Give two examples.

6. To estimate the annual salary for a given hourly pay rate, multiply by 2 and insert "000" at the end.

   Sample: $10 per hour is about $20,000 per year.

   a. Explain why this works. Assume the person is working 40 hours a week.

   b. Estimate the annual salary for an hourly pay rate of $8 per hour.

   c. You earn $1 million per month. What is your annual salary?

   d. Why is the cartoon funny?

"We had someone apply for the job. He says he would like $1 million a month, but will settle for $8 an hour."

Practice Use what you discovered about ratios and rates to complete Exercises 7–10 on page 167.

Key Vocabulary 🔊
ratio, *p. 164*
rate, *p. 164*
unit rate, *p. 164*
complex fraction,
  p. 165

A **ratio** is a comparison of two quantities using division.

$$\frac{3}{4}, \text{ 3 to 4, 3:4}$$

A **rate** is a ratio of two quantities with different units.

$$\frac{60 \text{ miles}}{2 \text{ hours}}$$

A rate with a denominator of 1 is called a **unit rate**.

$$\frac{30 \text{ miles}}{1 \text{ hour}}$$

EXAMPLE 1 Finding Ratios and Rates

There are 45 males and 60 females in a subway car. The subway car travels 2.5 miles in 5 minutes.

a. Find the ratio of males to females.

$$\frac{\text{males}}{\text{females}} = \frac{45}{60} = \frac{3}{4}$$

⋮⋅ The ratio of males to females is $\frac{3}{4}$.

b. Find the speed of the subway car.

$$2.5 \text{ miles in 5 minutes} = \frac{2.5 \text{ mi}}{5 \text{ min}} = \frac{2.5 \text{ mi} \div 5}{5 \text{ min} \div 5} = \frac{0.5 \text{ mi}}{1 \text{ min}}$$

⋮⋅ The speed is 0.5 mile per minute.

EXAMPLE 2 Finding a Rate from a Ratio Table

The ratio table shows the costs for different amounts of artificial turf. Find the unit rate in dollars per square foot.

	×4	×4	×4	
Amount (square feet)	25	100	400	1600
Cost (dollars)	100	400	1600	6400
	×4	×4	×4	

Use a ratio from the table to find the unit rate.

$$\frac{\text{cost}}{\text{amount}} = \frac{\$100}{25 \text{ ft}^2} \qquad \text{Use the first ratio in the table.}$$

$$= \frac{\$4}{1 \text{ ft}^2} \qquad \text{Simplify.}$$

Remember

The abbreviation ft² means *square feet*.

⋮⋅ So, the unit rate is \$4 per square foot.

Laurie's Notes

Introduction

Connect
- **Yesterday:** Students explored rates.
- **Today:** Students determine rates from words, tables, and graphs.

Motivate
- ❓ "A pitcher for a baseball team is able to throw a fastball approximately 132 feet in 1 second. How fast would this be in miles per hour?" 90 mi/h
- Explain that in this lesson they will be working with equivalent rates written in different units.

Lesson Notes

Discuss
- Remind students of the definitions for *ratio, rate,* and *unit rate.*
- ❓ Ask students to give their own examples of each.

Example 1
- **Common Error:** Be sure students read the question carefully. Order matters when writing a ratio. Writing $\frac{60}{45}$ would be incorrect.
- In part (b), 2.5 miles per 5 minutes is a rate. 0.5 mile per 1 minute is a unit rate.
- ❓ **Extension:** "How would you change 2.5 miles per 5 minutes into a unit rate in miles per hour?" To answer this question, use dimensional analysis introduced in the investigation.

$$\frac{2.5 \text{ miles}}{5 \text{ minutes}} \times \frac{60 \text{ minutes}}{1 \text{ hour}} = 30 \text{ miles per hour}$$

- Besides dividing out a common unit of minutes, divide out a common factor of 5.

Example 2
- Read the problem and write the ratio table. Ask students for assistance in filling in the table.
- ❓ Ask questions to review ratio tables. "Instead of multiplying each column by 4, could you multiply each column by 3?" yes "Could you add the first two columns and say that 125 ft² costs $500?" yes
- ❓ "What is the goal when finding a unit rate?" finding a rate with a denominator of 1
- **Look for and Make Use of Structure:** Explain that any equivalent ratio can be used to find the **unit rate**. Each simplifies to 4 : 1.
- **Common Error:** Students may try to find a unit rate using $\frac{25}{100}$ instead of $\frac{100}{25}$ because of the way the ratio table is set up. The problem asks for the unit rate in *dollars per square foot.* Ask students, "What is more useful to know—how many square feet of sod you can purchase for $1 or the cost of 1 square foot of sod?"

Goal Today's lesson is determining **ratios** and **rates**.

Technology for the **Teacher**

Dynamic Classroom

Lesson Tutorials
Lesson Plans
Answer Presentation Tool

Extra Example 1
a. There are 12 dogs and 15 cats at the pet store. Find the ratio of cats to dogs. $\frac{5}{4}$

b. You bicycle 30 blocks in 20 minutes. Find your speed. 1.5 blocks per minute

Extra Example 2
The table shows the amount of money you can raise by dancing for a charity. Find your unit rate in dollars per hour.

Time (hours)	6	12	18	24
Money (dollars)	$90	$180	$270	$360

$15 per hour

On Your Own

1. $\frac{4}{3}$ 2. $\frac{4}{7}$

3. 4.8 mi per sec

Extra Example 3

The graph shows the distance that you walk. Find your rate in feet per second.

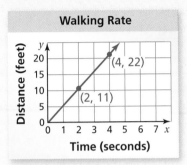

Walking Rate

5.5 feet per second

On Your Own

4. No; the unit rate is still $\frac{1}{2}$ mile per minute because the rates represented by points on the graph are equivalent.

Differentiated Instruction

Visual

Students may see little difference between fractions and ratios. Fractions are one type of ratio, part-to-whole. Ratios include part-to-part and whole-to-whole.

Circle 1   Circle 2

Part-to-whole of Circle 1:

$$\frac{\text{unshaded parts}}{\text{whole number of parts}} = \frac{2}{4}$$

Part-to-part of Circle 2:

$$\frac{\text{shaded parts}}{\text{unshaded parts}} = \frac{5}{3}$$

Whole-to-whole:

$$\frac{\text{parts of Circle 1}}{\text{parts of Circle 2}} = \frac{4}{8}$$

On Your Own

? "How do you calculate a speed?" distance ÷ time

? "Speed is a rate. Why?" You are comparing two quantities with different units.

• Students should work with partners on these three problems.

• **Extension:** Have students estimate the speed in miles per hour.

$$\frac{14.4 \text{ miles}}{3 \text{ seconds}} \times \frac{3600 \text{ seconds}}{1 \text{ hour}} = \frac{17{,}280 \text{ miles}}{\text{hour}}$$

Write

• Define a *complex fraction* and give examples. Including an example with units would be helpful, such as $\dfrac{30 \text{ meters}}{\frac{1}{2} \text{ minute}}$.

Example 3

? Ask general questions about subways—have you ridden on one, how fast do you think they move, how much does it cost to ride on one, and so on. Some students may have no context for these questions.

• Refer to the graph and ask students to identify the quantities on each axis.

? "What information do you need to find the speed of the subway car?" how much time it takes to travel a given distance

• Choose and interpret a point on the line. For instance, the point $\left(\frac{1}{2}, \frac{1}{4}\right)$ indicates that the subway car travels $\frac{1}{4}$ mile in $\frac{1}{2}$ minute.

• **Reason Abstractly and Quantitatively** and **Look for and Make Use of Structure:** At this point, students may very quickly observe that this is equivalent to $\frac{1}{2}$ mile in 1 minute. It is important for students to see the rate written as a complex fraction. The process shown would be the same for a subway car that travels, for instance, $\frac{4}{7}$ mile in $\frac{5}{6}$ minutes.

• Finish working the problem as shown.

On Your Own

• **Construct Viable Arguments and Critique the Reasoning of Others:** There are several ways in which students may explain their reasoning. Take time to hear a variety of approaches.

On Your Own

1. In Example 1, find the ratio of females to males.

2. In Example 1, find the ratio of females to total passengers.

3. The ratio table shows the distance that the *International Space Station* travels while orbiting Earth. Find the speed in miles per second.

Time (seconds)	3	6	9	12
Distance (miles)	14.4	28.8	43.2	57.6

A **complex fraction** has at least one fraction in the numerator, denominator, or both. You may need to simplify complex fractions when finding ratios and rates.

EXAMPLE 3 **Finding a Rate from a Graph**

The graph shows the speed of a subway car. Find the speed in miles per minute. Compare the speed to the speed of the subway car in Example 1.

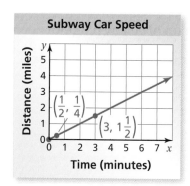

Subway Car Speed

Step 1: Choose and interpret a point on the line.

The point $\left(\dfrac{1}{2}, \dfrac{1}{4}\right)$ indicates that the subway car travels $\dfrac{1}{4}$ mile in $\dfrac{1}{2}$ minute.

Step 2: Find the speed.

$$\dfrac{\text{distance traveled}}{\text{elapsed time}} = \dfrac{\dfrac{1}{4} \;\leftarrow\; \text{miles}}{\dfrac{1}{2} \;\leftarrow\; \text{minutes}}$$

$$= \dfrac{1}{4} \div \dfrac{1}{2} \qquad \text{Rewrite the quotient.}$$

$$= \dfrac{1}{4} \cdot 2 = \dfrac{1}{2} \qquad \text{Simplify.}$$

∴ The speed of the subway car is $\dfrac{1}{2}$ mile per minute.

Because $\dfrac{1}{2}$ mile per minute = 0.5 mile per minute, the speeds of the two subway cars are the same.

On Your Own

4. You use the point $\left(3, 1\dfrac{1}{2}\right)$ to find the speed of the subway car. Does your answer change? Explain your reasoning.

EXAMPLE 4 Solving a Ratio Problem

You mix $\frac{1}{2}$ cup of yellow paint for every $\frac{3}{4}$ cup of blue paint to make 15 cups of green paint. How much yellow paint and blue paint do you use?

Math Practice

Analyze Givens
What information is given in the problem? How does this help you know that the ratio table needs a "total" column? Explain.

Method 1: The ratio of yellow paint to blue paint is $\frac{1}{2}$ to $\frac{3}{4}$. Use a ratio table to find an equivalent ratio in which the total amount of yellow paint and blue paint is 15 cups.

Yellow (cups)	Blue (cups)	Total (cups)
$\frac{1}{2}$	$\frac{3}{4}$	$\frac{1}{2} + \frac{3}{4} = \frac{5}{4}$
2	3	5
6	9	15

$\times 4$ $\times 3$ $\times 4$ $\times 3$

∴ So, you use 6 cups of yellow paint and 9 cups of blue paint.

Method 2: Use the fraction of the green paint that is made from yellow paint and the fraction of the green paint that is made from blue paint. You use $\frac{1}{2}$ cup of yellow paint for every $\frac{3}{4}$ cup of blue paint, so the fraction of the green paint that is made from yellow paint is

yellow → green → $\dfrac{\frac{1}{2}}{\frac{1}{2} + \frac{3}{4}} = \dfrac{\frac{1}{2}}{\frac{5}{4}} = \frac{1}{2} \cdot \frac{4}{5} = \frac{2}{5}.$

Similarly, the fraction of the green paint that is made from blue paint is

blue → green → $\dfrac{\frac{3}{4}}{\frac{1}{2} + \frac{3}{4}} = \dfrac{\frac{3}{4}}{\frac{5}{4}} = \frac{3}{4} \cdot \frac{4}{5} = \frac{3}{5}.$

∴ So, you use $\frac{2}{5} \cdot 15 = 6$ cups of yellow paint and $\frac{3}{5} \cdot 15 = 9$ cups of blue paint.

● **On Your Own**

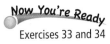
Now You're Ready
Exercises 33 and 34

5. How much yellow paint and blue paint do you use to make 20 cups of green paint?

Laurie's Notes

Example 4

? "What primary colors do you mix to get green?" yellow and blue

? "When you mix $\frac{1}{2}$ cup of yellow paint with $\frac{1}{2}$ cup of blue paint, what do you get?" 1 cup of green paint

- Ask a volunteer to read the problem.

? "How much green paint do you want to make?" 15 cups

- Point out that the ratio table shows the total number of cups of green paint for each combination of yellow paint and blue paint listed.

- Supply the thinking behind the multiplications used in the ratio table: multiply by 4 first so that there is a whole number of cups of green paint, then multiply by 3 so that there are 15 cups of green paint as specified in the problem.

- Encourage students to think about different ways of finding the solution using a ratio table. Have them try something to see whether the results give insight to the next step. For instance, if students double the original amounts of yellow and blue paint, then there would be 1 cup of yellow, $1\frac{1}{2}$ cups of blue, and $2\frac{1}{2}$ cups total. These amounts could be doubled and then tripled, giving the results shown in the table.

- **Model with Mathematics:** Method 2 shows a different approach. You could model it with a tape diagram as follows.

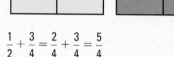

$$\frac{1}{2} + \frac{3}{4} = \frac{2}{4} + \frac{3}{4} = \frac{5}{4}$$

- Every $\frac{5}{4}$ cups of green paint are made up of $\frac{2}{4}$ cup of yellow paint and $\frac{3}{4}$ cup of blue paint. In other words, every 5 parts of green paint are made up of 2 parts yellow and 3 parts blue. So, $\frac{2}{5}$ of the green paint is made up of yellow paint and $\frac{3}{5}$ is made up of blue paint. Point out that these fractions have a sum of 1, as should be expected.

- Find $\frac{2}{5}$ of 15 and $\frac{3}{5}$ of 15 to answer the question.

- **Connection:** After you finish going over Method 2, go back to the ratio table used in Method 1 and show that $\frac{2}{5}$ of each total listed is made up of yellow paint and $\frac{3}{5}$ of each total listed is made up of blue paint.

On Your Own

- If time permits, then ask students to solve the problem using both methods shown in Example 4.

Closure

- **Exit Ticket:** Write 4.8 meters per 3 seconds as a unit rate.
  1.6 meters per second

1.6 meters per second is in answer position

Extra Example 4

You mix $\frac{1}{2}$ cup of red paint for every $\frac{1}{4}$ cup of blue paint to make 12 cups of purple paint. How much red paint and blue paint do you use? 8 cups of red paint and 4 cups of blue paint

On Your Own

5. 8 cups of yellow paint and 12 cups of blue paint

Vocabulary and Concept Check

1. It has a denominator of 1.

2. Unit rates are easier to compare.

3. *Sample answer:* A basketball player runs 10 feet down the court in 2 seconds.

4. $15 per gal

5. $0.10 per fl oz

6. $0.20 per egg

Practice and Problem Solving

7. $72

8. $28

9. 870 MB

10. 57 mi

11. $\dfrac{5}{9}$ 12. $\dfrac{9}{4}$

13. $\dfrac{7}{3}$ 14. $\dfrac{17}{3}$

15. $\dfrac{4}{3}$ 16. $\dfrac{14}{27}$

17. 60 mi/h

18. 32 mi per gal

19. $2.40 per lb

20. $0.80 per can

21. 54 words per min

22. 8.7 m per h

23. 4.5 servings per package

24. 3.6 ft per yr

25. 4.8 MB per min

Assignment Guide and Homework Check

Level	Day 1 Activity Assignment	Day 2 Lesson Assignment	Homework Check
Basic	7–10, 40–43	1–6, 11–27 odd	2, 13, 17, 23, 25
Average	7–10, 40–43	1–6, 11–21 odd, 24–28 even, 29, 31	2, 13, 17, 24, 28
Advanced	7–10, 40–43	1–6, 18–38 even	18, 22, 30, 32, 38
Accelerated	1–10, 18–38 even, 40–43		18, 22, 30, 32, 38

Common Errors

- **Exercises 11–16** Students may put the wrong number in the numerator. Remind them that the first number or object is the numerator and the second is the denominator.
- **Exercises 17–22** Students may find the unit rate but forget to include the units. Remind them that the units are necessary for understanding a unit rate, or any rate.
- **Exercises 23 and 24** When finding the rate from the table, students may put the wrong unit in the numerator. Tell them that the unit in the second row of the table will be the unit in the numerator of the rate.

5.1 Record and Practice Journal

Write the ratio as a fraction in simplest form.

1. 8 to 14
   $$\dfrac{4}{7}$$

2. 36 even : 12 odd
   $$\dfrac{3}{1}$$

3. 42 vanilla to 48 chocolate
   $$\dfrac{7}{8}$$

Find the unit rate.

4. $2.50 for 5 ounces
   $0.50 per ounce

5. 15 degrees in 2 hours
   7.5 degrees per hour

6. 183 miles in 3 hours
   61 miles per hour

Use the ratio table to find the unit rate with the specified units.

7. pounds per box

Boxes	0	1	2	3
Pounds	0	30	60	90

30 pounds per box

8. cost per notebook

Notebooks	0	5	10	15
Cost (dollars)	0	9.45	18.90	28.35

$1.89 per notebook

9. You create 15 centerpieces for a party in 5 hours.

   a. What is the unit rate?
      3 centerpieces per hour

   b. How long will it take you to make 42 centerpieces?
      14 hours

Vocabulary and Concept Check

1. **VOCABULARY** How can you tell when a rate is a unit rate?

2. **WRITING** Why do you think rates are usually written as unit rates?

3. **OPEN-ENDED** Write a real-life rate that applies to you.

Estimate the unit rate.

4. $74.75

Gloss White
PAINT
5 gal

5. $1.19

GRAPE JUICE
12 fl oz

6. $2.35

12 Grade AA Eggs

Practice and Problem Solving

Find the product. List the units.

7. $8 \text{ h} \times \dfrac{\$9}{\text{h}}$

8. $8 \text{ lb} \times \dfrac{\$3.50}{\text{lb}}$

9. $\dfrac{29}{2} \text{ sec} \times \dfrac{60 \text{ MB}}{\text{sec}}$

10. $\dfrac{3}{4} \text{ h} \times \dfrac{19 \text{ mi}}{\frac{1}{4}\text{ h}}$

Write the ratio as a fraction in simplest form.

① 11. 25 to 45

12. 63 : 28

13. 35 girls : 15 boys

14. 51 correct : 9 incorrect

15. 16 dogs to 12 cats

16. $2\dfrac{1}{3}$ feet : $4\dfrac{1}{2}$ feet

Find the unit rate.

17. 180 miles in 3 hours

18. 256 miles per 8 gallons

19. $9.60 for 4 pounds

20. $4.80 for 6 cans

21. 297 words in 5.5 minutes

22. $21\dfrac{3}{4}$ meters in $2\dfrac{1}{2}$ hours

Use the ratio table to find the unit rate with the specified units.

② 23. servings per package

Packages	3	6	9	12
Servings	13.5	27	40.5	54

24. feet per year

Years	2	6	10	14
Feet	7.2	21.6	36	50.4

25. **DOWNLOAD** At 1:00 P.M., you have 24 megabytes of a movie. At 1:15 P.M., you have 96 megabytes. What is the download rate in megabytes per minute?

26. **POPULATION** In 2007, the U.S. population was 302 million people. In 2012, it was 314 million. What was the rate of population change per year?

27. **PAINTING** A painter can paint 350 square feet in 1.25 hours. What is the painting rate in square feet per hour?

③ 28. **TICKETS** The graph shows the cost of buying tickets to a concert.

   a. What does the point (4, 122) represent?

   b. What is the unit rate?

   c. What is the cost of buying 10 tickets?

29. **CRITICAL THINKING** Are the two statements equivalent? Explain your reasoning.

   • The ratio of boys to girls is 2 to 3.

   • The ratio of girls to boys is 3 to 2.

30. **TENNIS** A sports store sells three different packs of tennis balls. Which pack is the best buy? Explain.

31. **FLOORING** It costs $68 for 16 square feet of flooring. How much does it cost for 12 square feet of flooring?

32. **OIL SPILL** An oil spill spreads 25 square meters every $\frac{1}{6}$ hour. How much area does the oil spill cover after 2 hours?

④ 33. **JUICE** You mix $\frac{1}{4}$ cup of juice concentrate for every 2 cups of water to make 18 cups of juice. How much juice concentrate and water do you use?

34. **LANDSCAPING** A supplier sells $2\frac{1}{4}$ pounds of mulch for every $1\frac{1}{3}$ pounds of gravel. The supplier sells 172 pounds of mulch and gravel combined. How many pounds of each item does the supplier sell?

35. **HEART RATE** Your friend's heart beats 18 times in 15 seconds when at rest. While running, your friend's heart beats 25 times in 10 seconds.

   a. Find the heart rate in beats per minute at rest and while running.

   b. How many more times does your friend's heart beat in 3 minutes while running than while at rest?

Common Errors

- **Exercise 30** Students may find the reciprocal of the unit rate and come up with the incorrect conclusion. Remind them that the unit rate represents the rate for one unit, so they should divide by the number of units.
- **Exercises 32–34** Students may have difficulty simplifying the complex fractions that result In the solutions of these exercises. A quick review of complex fractions may be helpful. Be sure to point out that a fraction bar means division, so they can rewrite the complex fraction as *numerator ÷ denominator*.

Practice and Problem Solving

26. 2.4 million people per year

27. 280 square feet per hour

28. **a.** It costs $122 for 4 tickets.

    b. $30.50 per ticket

    c. $305

29. no; Although the relative number of boys and girls are the same, the two ratios are inverses.

30. The 9-pack is the best buy at $2.55 per container.

31. $51

32. 300 square meters

33. 2 cups of juice concentrate, 16 cups of water

34. 108 pounds of mulch, 64 pounds of gravel

35. **a.** rest: 72 beats per minute running: 150 beats per minute

    b. 234 beats

Differentiated Instruction

Auditory

Ask students to recall when they have heard or used expressions with the word *per*, such as dollars per gallon. Make a list of the expressions. Explain that these are all rates and have a denominator of 1. They are called *unit rates*. Ask students how they would find the unit rate from a rate such as $42.50 for 10 gallons of gasoline. Conversely, ask how they would use a unit rate of 25 miles per gallon to determine the number of miles traveled using 10 gallons. Motivate the students by posting the list and adding to it as you work through the chapter.

36. **a.** whole milk

    b. orange juice

37. See *Taking Math Deeper*.

38. **a.** 16 cups of red paint, 10 cups of blue paint

    b. $3\frac{1}{5}$ cups of red paint, 2 cups of blue paint, $\frac{4}{5}$ cup of white paint

39. **a.** you; $\frac{1}{3}$ mile per hour faster

    b. $3\frac{1}{2}$ hours

    c. you; $1\frac{1}{6}$ miles

Fair Game Review

40. >   41. <

42. =   43. B

Mini-Assessment

Write the ratio as a fraction in simplest form.

1. 30 to 50 $\frac{3}{5}$ 2. 3 : 12 $\frac{1}{4}$

Find the unit rate.

3. 165 miles in 3 hours 55 mi/h

4. $9.60 for 8 cans $1.20 per can

5. The graph shows the cost of buying movie tickets.

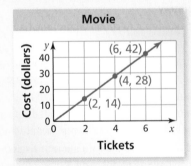

Movie

a. What does the point (4, 28) represent? $28 for 4 tickets

b. What is the unit rate? $7 per ticket

c. What is the cost of buying 9 tickets? $63

Taking Math Deeper

Exercise 37

This is a nice real-life problem that deals with rates. If students search the Internet for "fire hydrant colors" they can get the information about the rates in gallons per minute (GPM).

 a. Perform an Internet search and make a table from the results.

Blue	1500 + GPM	Very good
Green	1000–1499 GPM	Good
Yellow	500–999 GPM	Adequate
Red	Below 500 GPM	Inadequate

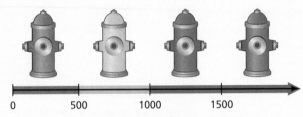 A number line helps display the information.

Rates in gallons per minute (GPM)

 b. Knowing the rate at which the water comes out of the fire hydrant is critical. If a firefighter pumps water out at too high a rate, the system of water pipes in the ground could be stressed and burst.

Reteaching and Enrichment Strategies

If students need help. . .	If students got it. . .
Resources by Chapter • Practice A and Practice B • Puzzle Time Record and Practice Journal Practice Differentiating the Lesson Lesson Tutorials Skills Review Handbook	Resources by Chapter • Enrichment and Extension • Technology Connection Start the next section

36. PRECISION The table shows nutritional information for three beverages.

Beverage	Serving Size	Calories	Sodium
Whole milk	1 c	146	98 mg
Orange juice	1 pt	210	10 mg
Apple juice	24 fl oz	351	21 mg

  a. Which has the most calories per fluid ounce?

  b. Which has the least sodium per fluid ounce?

37. RESEARCH Fire hydrants are painted one of four different colors to indicate the rate at which water comes from the hydrant.

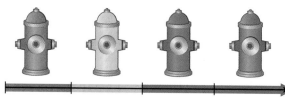

  a. Use the Internet to find the ranges of the rates for each color.

  b. Research why a firefighter needs to know the rate at which water comes out of a hydrant.

38. PAINT You mix $\frac{2}{5}$ cup of red paint for every $\frac{1}{4}$ cup of blue paint to make $1\frac{5}{8}$ gallons of purple paint.

  a. How much red paint and blue paint do you use?

  b. You decide that you want to make a lighter purple paint. You make the new mixture by adding $\frac{1}{10}$ cup of white paint for every $\frac{2}{5}$ cup of red paint and $\frac{1}{4}$ cup of blue paint. How much red paint, blue paint, and white paint do you use to make $\frac{3}{8}$ gallon of lighter purple paint?

39. *Critical Thinking* You and a friend start hiking toward each other from opposite ends of a 17.5-mile hiking trail. You hike $\frac{2}{3}$ mile every $\frac{1}{4}$ hour. Your friend hikes $2\frac{1}{3}$ miles per hour.

Big South Fork Trail
17.5 mi

  a. Who hikes faster? How much faster?

  b. After how many hours do you meet?

  c. When you meet, who hiked farther? How much farther?

Fair Game Review What you learned in previous grades & lessons

Copy and complete the statement using <, >, or =. *(Section 2.1)*

40. $\dfrac{9}{2}$ ⬚ $\dfrac{8}{3}$

41. $-\dfrac{8}{15}$ ⬚ $\dfrac{10}{18}$

42. $\dfrac{-6}{24}$ ⬚ $\dfrac{-2}{8}$

43. MULTIPLE CHOICE Which fraction is greater than $-\frac{2}{3}$ and less than $-\frac{1}{2}$? *(Section 2.1)*

  Ⓐ $-\dfrac{3}{4}$ **Ⓑ** $-\dfrac{7}{12}$ **Ⓒ** $-\dfrac{5}{12}$ **Ⓓ** $-\dfrac{3}{8}$

Essential Question
How can proportions help you decide when things are "fair"?

The Meaning of a Word ● Proportional

When you work toward a goal, your success is usually **proportional** to the amount of work you put in.

An equation stating that two ratios are equal is a **proportion**.

---

1 ACTIVITY: Determining Proportions

Work with a partner. Tell whether the two ratios are equivalent. If they are not equivalent, change the next day to make the ratios equivalent. Explain your reasoning.

a. On the first day, you pay $5 for 2 boxes of popcorn. The next day, you pay $7.50 for 3 boxes.

First Day   Next Day

$$\frac{\$5.00}{2 \text{ boxes}} \overset{?}{=} \frac{\$7.50}{3 \text{ boxes}}$$

b. On the first day, it takes you $3\frac{1}{2}$ hours to drive 175 miles. The next day, it takes you 5 hours to drive 200 miles.

First Day   Next Day

$$\frac{3\frac{1}{2}\text{ h}}{175 \text{ mi}} \overset{?}{=} \frac{5 \text{ h}}{200 \text{ mi}}$$

Proportions

In this lesson, you will
- use equivalent ratios to determine whether two ratios form a proportion.
- use the Cross Products Property to determine whether two ratios form a proportion.

c. On the first day, you walk 4 miles and burn 300 calories. The next day, you walk $3\frac{1}{3}$ miles and burn 250 calories.

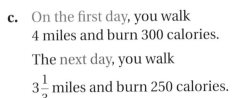

First Day   Next Day

$$\frac{4 \text{ mi}}{300 \text{ cal}} \overset{?}{=} \frac{3\frac{1}{3}\text{ mi}}{250 \text{ cal}}$$

d. On the first day, you paint 150 square feet in $2\frac{1}{2}$ hours. The next day, you paint 200 square feet in 4 hours.

First Day   Next Day

$$\frac{150 \text{ ft}^2}{2\frac{1}{2}\text{ h}} \overset{?}{=} \frac{200 \text{ ft}^2}{4 \text{ h}}$$

Laurie's Notes

Introduction

Applying Mathematical Practices

- **Construct Viable Arguments and Critique the Reasoning of Others:**
  To develop an understanding of proportions, ask students to explain their reasoning. A proportion is an equation stating that two ratios are equivalent. Explanations offered by students need to be connected to this definition.

Motivate

- Ask for two volunteers. Hand student A 8 square tiles and student B 4 square tiles.
- Make up a story as to why student A starts off with more than student B.
- **?** "What is the ratio of student A's tiles to student B's tiles?" 2 : 1
- **?** "What is the ratio of student B's tiles to student A's tiles?" 1 : 2
- **?** "If I give student A 2 more tiles, how many should I give student B so that they still have the same ratio? Explain your reasoning." 1; Student A has twice as many as student B, so you need to give him/her twice as many tiles each time.
- Hand each student 2 more tiles.
- **?** Is the ratio of student A's tiles to student B's tiles still 2 : 1? Explain." No, the ratio is 10 : 6 = 5 : 3 ≠ 2 : 1.
- Ask additional questions if time permits.

Activity Notes

Meaning of the Word

- Explain the meaning of the word proportional. Reference the activity used to motivate today's lesson for examples.

Activity 1

- Note the use of color to distinguish the ratios and help students focus on the writing of the proportions. The ratios on the left contain information from the first day, and the ratios on the right contain information from the second day.
- **Management Tip:** For this activity, students will work in pairs. To allow the activity to run smoothly and to save time in class, plan a partner for each student before class.
- Discuss student explanations.
- **Construct Viable Arguments and Critique the Reasoning of Others:**
  Encourage students to share their strategies. How did they decide whether the ratios were equal or not? There will be different strategies, and it is important to hear a variety.

What Your Students Will Learn

- Use equivalent ratios to determine whether two ratios form a proportion or whether two quantities are proportional.
- Use the Cross Products Property to identify proportional relationships.

Previous Learning

Students have written and simplified ratios.

5.2 Record and Practice Journal

Essential Question How can proportions help you decide when things are "fair"?

1 ACTIVITY: Determining Proportions

Work with a partner. Tell whether the two ratios are equivalent. If they are not equivalent, change the next day to make the ratios equivalent. Explain your reasoning.

a. On the first day, you pay $5 for 2 boxes of popcorn. The next day, you pay $7.50 for 3 boxes.

equivalent

First Day   Next Day

$$\frac{\$5.00}{2 \text{ boxes}} \stackrel{?}{=} \frac{\$7.50}{3 \text{ boxes}}$$

b. On the first day, it takes you $3\frac{1}{2}$ hours to drive 175 miles. The next day, it takes you 5 hours to drive 200 miles.

not equivalent; 250 miles in 5 hours

First Day   Next Day

$$\frac{3\frac{1}{2} \text{ h}}{175 \text{ mi}} \stackrel{?}{=} \frac{5 \text{ h}}{200 \text{ mi}}$$

c. On the first day, you walk 4 miles and burn 300 calories. The next day, you walk $3\frac{1}{3}$ miles and burn 250 calories.

equivalent

First Day   Next Day

$$\frac{4 \text{ mi}}{300 \text{ cal}} \stackrel{?}{=} \frac{3\frac{1}{3} \text{ mi}}{250 \text{ cal}}$$

d. On the first day, you paint 150 square feet in $2\frac{1}{2}$ hours. The next day, you paint 200 square feet in 4 hours.

not equivalent; 240 square feet in 4 hours

First Day   Next Day

$$\frac{150 \text{ ft}^2}{2\frac{1}{2} \text{ h}} \stackrel{?}{=} \frac{200 \text{ ft}^2}{4 \text{ h}}$$

English Language Learners

Word Problems

Most word problems follow a standard format that allows English learners to recognize key words that are integral to writing a mathematical statement of the problem. Most numbers given in a word problem are used. Analyzing the units in the mathematical statement and determining the units of the solution give students confidence that they are on the right path for solving the problem.

5.2 Record and Practice Journal

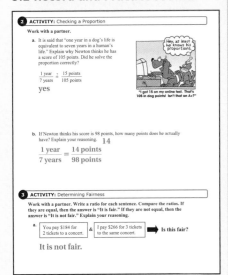

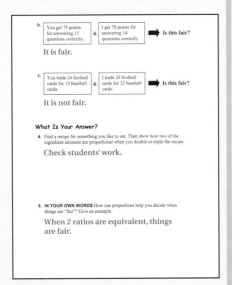

Laurie's Notes

Activity 2

- Help students think through this problem.
- The ratio often heard is 1 dog year : 7 human years. Years may be in the ratio of 1 : 7, but there is no reason why test scores need to be.
- The proportion that is being used is $\dfrac{1 \text{ dog year}}{7 \text{ human years}} = \dfrac{15 \text{ test points}}{105 \text{ dog points}}$.
- While the numeric proportion is a true proportion, the units of $\dfrac{1}{7} = \dfrac{15}{105}$ make no sense.

Activity 3

- Students like to think about fairness when deciding if two ratios are equal. When they finish, have students share their reasoning.
- One common strategy is to compare unit rates.

Words of Wisdom

- All of the work today is to build an understanding of what it means for two ratios to be equal and to develop different strategies for deciding if the ratios are equal.
- **Construct Viable Arguments and Critique the Reasoning of Others:** Student discussion can often be quite interesting as they explain their reasoning. It is important to focus on both the computations and the information that students use to explain their reasoning.

What Is Your Answer?

- Question 4 can be assigned for homework.

Closure

- **Writing Prompt:** One way to decide if 6 hours for $4.80 is the same rate as 10 hours for $8 is . . . *Sample answer:* to write each comparison as a ratio and then find the unit rates.

  $\dfrac{\$4.80}{6 \text{ hours}} = \$0.80 \text{ per hour}$

  $\dfrac{\$8.00}{10 \text{ hours}} = \$0.80 \text{ per hour}$

2 ACTIVITY: Checking a Proportion

Work with a partner.

a. It is said that "one year in a dog's life is equivalent to seven years in a human's life." Explain why Newton thinks he has a score of 105 points. Did he solve the proportion correctly?

$$\frac{1 \text{ year}}{7 \text{ years}} \stackrel{?}{=} \frac{15 \text{ points}}{105 \text{ points}}$$

b. If Newton thinks his score is 98 points, how many points does he actually have? Explain your reasoning.

"I got 15 on my online test. That's 105 in dog points! Isn't that an A+?"

3 ACTIVITY: Determining Fairness

> **Math Practice**
>
> **Justify Conclusions**
> What information can you use to justify your conclusion?

Work with a partner. Write a ratio for each sentence. Compare the ratios. If they are equal, then the answer is "It is fair." If they are not equal, then the answer is "It is not fair." Explain your reasoning.

a.

| You pay $184 for 2 tickets to a concert. | & | I pay $266 for 3 tickets to the same concert. | → Is this fair? |

b.

| You get 75 points for answering 15 questions correctly. | & | I get 70 points for answering 14 questions correctly. | → Is this fair? |

c.

| You trade 24 football cards for 15 baseball cards. | & | I trade 20 football cards for 32 baseball cards. | → Is this fair? |

What Is Your Answer?

4. Find a recipe for something you like to eat. Then show how two of the ingredient amounts are proportional when you double or triple the recipe.

5. IN YOUR OWN WORDS How can proportions help you decide when things are "fair"? Give an example.

> **Practice**
>
> Use what you discovered about proportions to complete Exercises 15–20 on page 174.

Key Vocabulary 🔊
proportion, *p. 172*
proportional, *p. 172*
cross products, *p. 173*

Key Idea

Proportions

Words A **proportion** is an equation stating that two ratios are equivalent. Two quantities that form a proportion are **proportional**.

Numbers $\dfrac{2}{3} = \dfrac{4}{6}$ The proportion is read "2 is to 3 as 4 is to 6."

EXAMPLE 1 **Determining Whether Ratios Form a Proportion**

Tell whether $\dfrac{6}{4}$ and $\dfrac{8}{12}$ form a proportion.

Compare the ratios in simplest form.

$$\dfrac{6}{4} = \dfrac{6 \div 2}{4 \div 2} = \dfrac{3}{2}$$ ← The ratios are *not* equivalent.

$$\dfrac{8}{12} = \dfrac{8 \div 4}{12 \div 4} = \dfrac{2}{3}$$

So, $\dfrac{6}{4}$ and $\dfrac{8}{12}$ do *not* form a proportion.

EXAMPLE 2 **Determining Whether Two Quantities Are Proportional**

Tell whether *x* and *y* are proportional.

Compare each ratio *x* to *y* in simplest form.

$$\dfrac{\frac{1}{2}}{3} = \dfrac{1}{6} \qquad \dfrac{1}{6} \qquad \dfrac{\frac{3}{2}}{9} = \dfrac{1}{6} \qquad \dfrac{2}{12} = \dfrac{1}{6}$$

The ratios are equivalent.

x	y
$\frac{1}{2}$	3
1	6
$\frac{3}{2}$	9
2	12

Reading

Two quantities that are proportional are in a *proportional relationship*.

So, *x* and *y* are proportional.

On Your Own

Now You're Ready
Exercises 5–14

Tell whether the ratios form a proportion.

1. $\dfrac{1}{2}, \dfrac{5}{10}$ **2.** $\dfrac{4}{6}, \dfrac{18}{24}$ **3.** $\dfrac{10}{3}, \dfrac{5}{6}$ **4.** $\dfrac{25}{20}, \dfrac{15}{12}$

5. Tell whether *x* and *y* are proportional.

Birdhouses Built, *x*	1	2	4	6
Nails Used, *y*	12	24	48	72

🔊 Multi-Language Glossary at BigIdeasMath✓com

Laurie's Notes

Introduction

Connect

- **Yesterday:** Students explored pairs of rates and decided if they were equivalent, or fair.
- **Today:** Students will use multiplication and division, and the Cross Products Property to decide if two ratios are equal.

Motivate

- Draw the following on the board: $\dfrac{\square}{\square} = \dfrac{\square}{\square}$

- Ask students to use the numbers 2, 3, 4, and 6 placing one number in each square to make two ratios. They should list all combinations that are different $\left(\text{i.e., } \dfrac{2}{4} = \dfrac{3}{6} \text{ is not different from } \dfrac{3}{6} = \dfrac{2}{4}\right)$.
- ❓ "How did you decide where to place the numbers?" Listen for ideas related to fractions (reducing, using the Cross Products Property, etc.).
- Record student solutions to reference later.

Lesson Notes

Key Idea

- Write the definition of proportion on the board.
- **FYI:** Without units associated with the numeric values, students think of proportions as fractions.
- If students are comfortable with writing equivalent fractions and simplifying fractions, they will generally have a good sense about working with proportions.

Example 1

- The strategy is to write the ratios in simplest form.
- ❓ "What is the relationship between $\dfrac{2}{3}$ and $\dfrac{3}{2}$?" They are reciprocals.

Example 2

- Copy the table of values. Students may see x and y as ordered pairs or see the table as a vertical ratio table.
- ❓ "How can you decide whether x and y are proportional?" Determine whether each ratio $x : y$ is equivalent

On Your Own

- **Construct Viable Arguments and Critique the Reasoning of Others:** Students may come up with different strategies for answering Question 5. Ask them to explain their reasoning.

Goal Today's lesson is comparing ratios using **proportions** and the **Cross Products** Property.

Technology for the Teacher

Dynamic Classroom

Lesson Tutorials
Lesson Plans
Answer Presentation Tool

Extra Example 1

Tell whether $\dfrac{10}{18}$ and $\dfrac{45}{81}$ form a proportion.

yes

Extra Example 2

Tell whether x and y are proportional.

x	y
$\dfrac{1}{2}$	4
1	8
$\dfrac{3}{2}$	12
2	16

yes

On Your Own

1. yes
2. no
3. no
4. yes
5. yes

Laurie's Notes

Key Ideas

- Write the Key Ideas on the board.
- ❓ "Why can't *b* or *d* equal zero?" or "Which number should *b* or *d* not be equal to? Why not?" Zero, because it would lead to division by zero.
- Use the Cross Products Property to verify that each of the solutions written at the beginning of class is a proportion.
- Discuss the Study Tip. It shows why the Cross Products Property is true.

Example 3

- Ask the students to read the problem.
- Work through each method of the solution.
- ❓ **Connection:** When you have finished each method, tie this lesson to equivalent ratios by asking, "If you are swimming at a constant rate and you swam 4 laps in 2.4 minutes, how long should it take you to complete 16 laps?" 4 times as long, or 9.6 min
- ❓ "What does this mean in the context of the problem?" You slowed down after 4 laps.

Closure

- Write an example of two ratios that are equal. Explain how you know they are equal.
- Write an example of two ratios that are not equal. Explain how you know they are not equal.

Extra Example 3

You run the first 3 laps around the gym in 1.5 minutes. You complete 24 laps in 12 minutes. Is the number of laps proportional to your time? The number of laps is proportional to the time.

 On Your Own

6. yes

Differentiated Instruction

Visual

As suggested in the Study Tip, the Cross Products Property is an application of the Multiplication Property of Equality. For those students having trouble following the steps shown in the Study Tip, show the following more detailed steps on the board or overhead.

$\dfrac{a}{b} = \dfrac{c}{d}$	Proportion
$b \cdot d \cdot \dfrac{a}{b} = b \cdot d \cdot \dfrac{c}{d}$	Multiply each side by $b \cdot d$.
$\cancel{b}d \cdot \dfrac{a}{\cancel{b}} = b\cancel{d} \cdot \dfrac{c}{\cancel{d}}$	Divide out common factors.
$d \cdot a = b \cdot c$	Simplify.
$ad = bc$	Commutative Property of Multiplication

 **Key Ideas**

Cross Products

In the proportion $\dfrac{a}{b} = \dfrac{c}{d}$, the products $a \cdot d$ and $b \cdot c$ are called **cross products**.

Cross Products Property

Words The cross products of a proportion are equal.

Numbers	**Algebra**
$\dfrac{2}{3} \!\!\!\bowtie\!\!\! \dfrac{4}{6}$	$\dfrac{a}{b} \!\!\!\bowtie\!\!\! \dfrac{c}{d}$
$2 \cdot 6 = 3 \cdot 4$	$ad = bc$, where $b \neq 0$ and $d \neq 0$

Study Tip

You can use the Multiplication Property of Equality to show that the cross products are equal.

$$\frac{a}{b} = \frac{c}{d}$$

$$\cancel{bd} \cdot \frac{a}{\cancel{b}} = b\cancel{d} \cdot \frac{c}{\cancel{d}}$$

$$ad = bc$$

EXAMPLE ③ **Identifying Proportional Relationships**

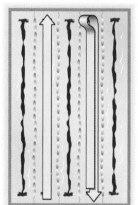

1 length   1 lap

You swim your first 4 laps in 2.4 minutes. You complete 16 laps in 12 minutes. Is the number of laps proportional to your time?

Method 1: Compare unit rates.

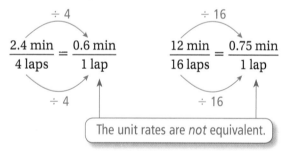

$$\frac{2.4 \text{ min}}{4 \text{ laps}} = \frac{0.6 \text{ min}}{1 \text{ lap}} \qquad \frac{12 \text{ min}}{16 \text{ laps}} = \frac{0.75 \text{ min}}{1 \text{ lap}}$$

The unit rates are *not* equivalent.

∴ So, the number of laps is *not* proportional to the time.

Method 2: Use the Cross Products Property.

$$\frac{2.4 \text{ min}}{4 \text{ laps}} \stackrel{?}{=} \frac{12 \text{ min}}{16 \text{ laps}}$$ Test to see if the rates are equivalent.

$$2.4 \cdot 16 \stackrel{?}{=} 4 \cdot 12$$ Find the cross products.

$$38.4 \neq 48$$ The cross products are *not* equal.

∴ So, the number of laps is *not* proportional to the time.

● **On Your Own**

Now You're Ready
Exercises 15–20

6. You read the first 20 pages of a book in 25 minutes. You read 36 pages in 45 minutes. Is the number of pages read proportional to your time?

 Vocabulary and Concept Check

1. **VOCABULARY** What does it mean for two ratios to form a proportion?

2. **VOCABULARY** What are two ways you can tell that two ratios form a proportion?

3. **OPEN-ENDED** Write two ratios that are equivalent to $\frac{3}{5}$.

4. **WHICH ONE DOESN'T BELONG?** Which ratio does *not* belong with the other three? Explain your reasoning.

$$\frac{4}{10} \qquad \frac{2}{5} \qquad \frac{3}{5} \qquad \frac{6}{15}$$

 Practice and Problem Solving

Tell whether the ratios form a proportion.

① 5. $\frac{1}{3}, \frac{7}{21}$ 6. $\frac{1}{5}, \frac{6}{30}$ 7. $\frac{3}{4}, \frac{24}{18}$ 8. $\frac{2}{5}, \frac{40}{16}$

9. $\frac{48}{9}, \frac{16}{3}$ 10. $\frac{18}{27}, \frac{33}{44}$ 11. $\frac{7}{2}, \frac{16}{6}$ 12. $\frac{12}{10}, \frac{14}{12}$

Tell whether x and y are proportional.

② 13.

x	1	2	3	4
y	7	8	9	10

14.

x	2	4	6	8
y	5	10	15	20

Tell whether the two rates form a proportion.

③ 15. 7 inches in 9 hours; 42 inches in 54 hours

16. 12 players from 21 teams; 15 players from 24 teams

17. 440 calories in 4 servings; 300 calories in 3 servings

18. 120 units made in 5 days; 88 units made in 4 days

19. 66 wins in 82 games; 99 wins in 123 games

20. 68 hits in 172 at bats; 43 hits in 123 at bats

21. **FITNESS** You can do 90 sit-ups in 2 minutes. Your friend can do 135 sit-ups in 3 minutes. Do these rates form a proportion? Explain.

22. **HEART RATES** Find the heart rates of you and your friend. Do these rates form a proportion? Explain.

	Heartbeats	Seconds
You	22	20
Friend	18	15

Assignment Guide and Homework Check

Level	Day 1 Activity Assignment	Day 2 Lesson Assignment	Homework Check
Basic	15–20, 33–37	1–4, 5–13 odd, 21–27 odd	9, 13, 21, 27
Average	15–20, 33–37	1–4, 5–13 odd, 22–30 even	9, 13, 26, 30
Advanced	15–20, 33–37	1–4, 6–14 even, 22–32 even	14, 22, 26, 28, 30
Accelerated	1–4, 6–14 even, 15–20, 22–32 even, 33–37		14, 22, 26, 28, 30

Common Errors

- **Exercises 5–12** Students may have difficulty understanding why you can write the ratio in simplest form. Tell students to compare the ratio to a fraction. Simplifying ratios is the same as writing equivalent fractions.

- **Exercises 15–20** Students may mix up the rates and incorrectly find that they are not proportional. For example, they might write $\dfrac{7 \text{ inches}}{9 \text{ hours}} \overset{?}{=} \dfrac{54 \text{ hours}}{42 \text{ inches}}$. Remind students about writing a rate and help them to identify which unit goes in the numerator.

Vocabulary and Concept Check

1. Both ratios are equal.

2. Compare the ratios in simplest form and compare the cross products.

3. *Sample answer:* $\dfrac{6}{10}, \dfrac{12}{20}$

4. $\dfrac{3}{5}$; The others are equal to $\dfrac{2}{5}$.

Practice and Problem Solving

5. yes		6. yes	
7. no		8. no	
9. yes		10. no	
11. no		12. no	
13. no		14. yes	
15. yes		16. no	
17. no		18. no	
19. yes		20. no	

21. yes; Both can do 45 sit-ups per minute.

22. you: 1.1 beats per second friend: 1.2 beats per second No, the rates are not equivalent.

5.2 Record and Practice Journal

Tell whether the ratios form a proportion.

1. $\dfrac{1}{5}, \dfrac{5}{15}$ **no**

2. $\dfrac{2}{3}, \dfrac{12}{18}$ **yes**

3. $\dfrac{15}{2}, \dfrac{4}{30}$ **no**

4. $\dfrac{56}{21}, \dfrac{8}{3}$ **yes**

5. $\dfrac{5}{8}, \dfrac{62.5}{100}$ **yes**

6. $\dfrac{17}{20}, \dfrac{90.1}{106}$ **yes**

7. $\dfrac{3.2}{4}, \dfrac{16}{24}$ **no**

8. $\dfrac{34}{50}, \dfrac{6.8}{10}$ **yes**

Tell whether the two rates form a proportion.

9. 28 points in 3 games; 112 points in 12 games

yes

10. 32 notes in 4 measures; 12 notes in 2 measures

no

11. You can type 105 words in two minutes. Your friend can type 210 words in four minutes. Are these rates proportional? Explain.

yes

23. yes   24. no

25. yes

26. a. $7 per hour

    b. $9 per hour

    c. no; Your friend earns more money per hour.

27. yes; The ratio of height to base for both triangles is $\frac{4}{5}$.

28. a. x and y, x and z, y and z

    b. 30

29. See *Taking Math Deeper*.

30. a. no

    b. *Sample answer:* If the collection has 50 quarters and 30 dimes, when 10 of each coin are added, the new ratio of quarters to dimes is 3 : 2.

31. no; The ratios are not equivalent;
$\frac{13}{19} \neq \frac{14}{20} \neq \frac{15}{21}$ etc.

32. See Additional Answers.

Fair Game Review

33. -13 34. -17

35. -18 36. -3

37. D

Mini-Assessment

Tell whether the ratios form a proportion.

1. $\frac{4}{12}$, $\frac{5}{15}$ yes   2. $\frac{8}{4}$, $\frac{12}{8}$ no

3. $\frac{14}{17}$, $\frac{42}{51}$ yes   4. $\frac{16}{12}$, $\frac{26}{22}$ no

5. You can do 40 push-ups in 2 minutes. Your friend can do 57 push-ups in 3 minutes. Are these rates proportional?   no

Taking Math Deeper

Exercise 29

In this problem, students are given a mixture of red and yellow pigment and are asked to decide whether they should add more red or more yellow to get a desired shade.

Add red or yellow.

 Summarize the given information. Find each unit rate.

Ratio for desired shade: $\dfrac{7 \text{ parts red}}{2 \text{ parts yellow}} = \dfrac{3\frac{1}{2} \text{ parts red}}{1 \text{ part yellow}}$

Given mixture: $\dfrac{35 \text{ parts red}}{8 \text{ parts yellow}} = \dfrac{4\frac{3}{8} \text{ parts red}}{1 \text{ part yellow}}$

 Interpret the given information.

The ratio for the given mixture needs more yellow to take it down to the desired ratio of $3\frac{1}{2} : 1$ (red to yellow).

 Use a table with *Guess, Check, and Revise* to find how much yellow to add to the given mixture.

Red	Yellow	Ratio
35 qt	8 qt	$\dfrac{35}{8} = \dfrac{4\frac{3}{8} \text{ parts red}}{1 \text{ part yellow}}$
35 qt	9 qt	$\dfrac{35}{9} = \dfrac{3\frac{8}{9} \text{ parts red}}{1 \text{ part yellow}}$
35 qt	10 qt	$\dfrac{35}{10} = \dfrac{3\frac{1}{2} \text{ parts red}}{1 \text{ part yellow}}$

Add 2 quarts of yellow.

0% magenta	50% magenta	100% magenta
100% yellow	50% yellow	0% yellow

Here are some mixtures of yellow and magenta (red).

Reteaching and Enrichment Strategies

If students need help. . .	If students got it. . .

Resources by Chapter
  • Practice A and Practice B
  • Puzzle Time
Record and Practice Journal Practice
Differentiating the Lesson
Lesson Tutorials
Skills Review Handbook
	Resources by Chapter
	  • Enrichment and Extension
	  • Technology Connection
	Start the next section

Tell whether the ratios form a proportion.

23. $\dfrac{2.5}{4}, \dfrac{7}{11.2}$

24. 2 to 4, 11 to $\dfrac{11}{2}$

25. $2 : \dfrac{4}{5}, \dfrac{3}{4} : \dfrac{3}{10}$

26. PAY RATE You earn $56 walking your neighbor's dog for 8 hours. Your friend earns $36 painting your neighbor's fence for 4 hours.

    a. What is your pay rate?

    b. What is your friend's pay rate?

    c. Are the pay rates equivalent? Explain.

27. GEOMETRY Are the heights and bases of the two triangles proportional? Explain.

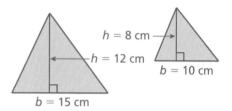

28. BASEBALL A pitcher coming back from an injury limits the number of pitches thrown in bull pen sessions as shown.

    a. Which quantities are proportional?

    b. How many pitches that are not curveballs do you think the pitcher will throw in Session 5?

Session Number, x	Pitches, y	Curveballs, z
1	10	4
2	20	8
3	30	12
4	40	16

29. NAIL POLISH A specific shade of red nail polish requires 7 parts red to 2 parts yellow. A mixture contains 35 quarts of red and 8 quarts of yellow. How can you fix the mixture to make the correct shade of red?

30. COIN COLLECTION The ratio of quarters to dimes in a coin collection is 5 : 3. You add the same number of new quarters as dimes to the collection.

    a. Is the ratio of quarters to dimes still 5 : 3?

    b. If so, illustrate your answer with an example. If not, show why with a "counterexample."

31. AGE You are 13 years old, and your cousin is 19 years old. As you grow older, is your age proportional to your cousin's age? Explain your reasoning.

32. Ratio A is equivalent to Ratio B. Ratio B is equivalent to Ratio C. Is Ratio A equivalent to Ratio C? Explain.

Fair Game Review What you learned in previous grades & lessons

Add or subtract. *(Section 1.2 and Section 1.3)*

33. $-28 + 15$

34. $-6 + (-11)$

35. $-10 - 8$

36. $-17 - (-14)$

37. MULTIPLE CHOICE Which fraction is not equivalent to $\dfrac{2}{6}$? *(Skills Review Handbook)*

    (**A**) $\dfrac{1}{3}$ (**B**) $\dfrac{12}{36}$ (**C**) $\dfrac{4}{12}$ (**D**) $\dfrac{6}{9}$

Recall that you can graph the values from a ratio table.

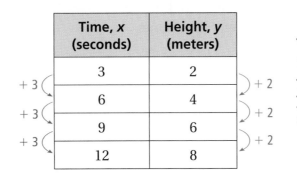

Time, x (seconds)	Height, y (meters)
3	2
6	4
9	6
12	8

$+3$ ... $+2$

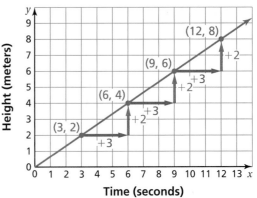

The structure in the ratio table shows why the graph has a constant *rate of change.* You can use the constant rate of change to show that the graph passes through the origin. The graph of every proportional relationship is a line through the origin.

EXAMPLE 1 Determining Whether Two Quantities Are Proportional

Use a graph to tell whether x and y are in a proportional relationship.

a.

x	2	4	6
y	6	8	10

Plot (2, 6), (4, 8), and (6, 10). Draw a line through the points.

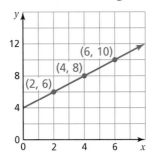

The graph is a line that does not pass through the origin.

⋮ So, x and y are not in a proportional relationship.

b.

x	1	2	3
y	2	4	6

Plot (1, 2), (2, 4), and (3, 6). Draw a line through the points.

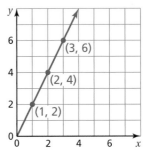

The graph is a line that passes through the origin.

⋮ So, x and y are in a proportional relationship.

Proportions
In this extension, you will
• use graphs to determine whether two ratios form a proportion.
• interpret graphs of proportional relationships.

Practice

Use a graph to tell whether x and y are in a proportional relationship.

1.

x	1	2	3	4
y	3	4	5	6

2.

x	1	3	5	7
y	0.5	1.5	2.5	3.5

Laurie's Notes

Introduction

Connect
- **Yesterday:** Students determined whether two ratios are proportional.
- **Today:** Students will graph proportional relationships.

Motivate
- Share the following information with students. The Mars rover *Curiosity*, a large mobile laboratory, launched from Cape Canaveral on November 26, 2011, and landed successfully on Mars on August 6, 2012. For three minutes right before touchdown, the spacecraft used a parachute and retrorockets to slow its descent. In the final seconds before touchdown, the upper stage acted as a crane, lowering the rover to the surface on a tether.
- Explain that one of the examples in today's lesson will explore the distance *Curiosity* travels over time.

Lesson Notes

Discuss
- Introduce the lesson with the ratio table and graph shown. This is an example in which a constant amount is added to the *x*-values and a different constant amount is added to the *y*-values. The ratios are equivalent: $\frac{3}{2} = \frac{6}{4} = \frac{9}{6} = \frac{12}{8}$.
- State that the graph of every proportional relationship is a line through the origin.
- You could justify that the line goes through the origin using the arrows shown in the graph. You could also work backwards in the table to show that (0, 0) would be in the table. Students will learn more about rate of change and slope in Sections 5.5 and 5.6.

Example 1
- Write the table of values and plot the corresponding ordered pairs for each part of the example.
- **Attend to Precision:** Have students use grid paper instead of making freehand sketches.
- **Connection:** Check the tables for equivalent ratios to justify the answers. In part (a), $\frac{2}{6} \neq \frac{4}{8}$ and $\frac{4}{8} \neq \frac{6}{10}$. In part (b), $\frac{1}{2} = \frac{2}{4} = \frac{3}{6}$.

Extra Example 1

Use a graph to tell whether *x* and *y* are in a proportional relationship.

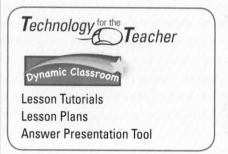

a.

x	2	4	6
y	4	6	8

no

b.

x	3	6	9
y	1	2	3

yes

Record and Practice Journal Extension 5.2 Practice
1–8. See Additional Answers.

Extra Example 2

The graph shows that the distance traveled by a kayak at top speed is proportional to the time traveled. Interpret each plotted point in the graph.

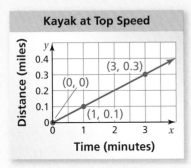

Kayak at Top Speed

Sample Answer: The kayak travels 0 miles in 0 minutes, 0.1 mile in 1 minute, and 0.3 mile in 3 minutes. The unit rate is 0.1 mile per minute.

Practice

1.

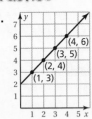

no

2.

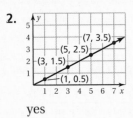

yes

3–8. See Additional Answers.

Mini-Assessment

Use a graph to tell whether *x* and *y* are in a proportional relationship.

1.

x	2	4	6
y	2	3	4

no

2.

x	2	4	6
y	3	6	9

yes

Laurie's Notes

Example 2

- Explain that this example involves the top speed of the Mars rover *Curiosity* discussed earlier.
- ❓ "How do you know that the graph shows a proportional relationship between distance traveled and time traveled?" The line passes through the origin.
- Ask students to read and interpret each plotted point in the graph. Note that in addition to interpreting the points, the unit rate is also found. It is a constant value of 1.5 inches per second.
- Make sure that students understand the Study Tip. If you know (1, *y*), then you know the unit rate *y*.

Practice

- In Exercises 3 and 4, students should be thinking about the labels on the axes as they interpret the plotted points.
- In Exercise 6, *x* and *y* are not in a proportional relationship, but you could still ask students whether there is a constant rate of change, and if so, what it is. This gives a preview of Section 5.5.
- **Connection:** In Exercise 7, students can draw a line through (12, 16) and (0, 0), and then think about where the ordered pair (1, *y*) would be on the line.

Closure

- ❓ Is it possible to have a constant rate of change and not have a proportional relationship? Explain.

  yes; *Sample answer:* Example 1(a) has a constant rate of change, because the points lie on a line. The relationship is not proportional, because the line does not pass through the origin.

EXAMPLE 2 Interpreting the Graph of a Proportional Relationship

The graph shows that the distance traveled by the Mars rover *Curiosity* is proportional to the time traveled. Interpret each plotted point in the graph.

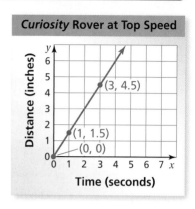

Curiosity Rover at Top Speed

(0, 0): The rover travels 0 inches in 0 seconds.

(1, 1.5): The rover travels 1.5 inches in 1 second. So, the unit rate is 1.5 inches per second.

Study Tip

In the graph of a proportional relationship, you can find the unit rate from the point (1, y).

(3, 4.5): The rover travels 4.5 inches in 3 seconds. Because the relationship is proportional, you can also use this point to find the unit rate.

$$\frac{4.5 \text{ in.}}{3 \text{ sec}} = \frac{1.5 \text{ in.}}{1 \text{ sec}}, \text{ or } 1.5 \text{ inches per second}$$

Practice

Interpret each plotted point in the graph of the proportional relationship.

3.
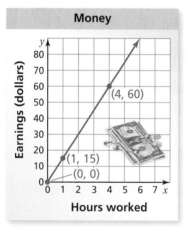
Money

4.
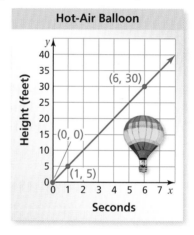
Hot-Air Balloon

Tell whether *x* and *y* are in a proportional relationship. If so, find the unit rate.

5.
x (hours)	1	4	7	10
y (feet)	5	20	35	50

6. Let *y* be the temperature *x* hours after midnight. The temperature is 60°F at midnight and decreases 2°F every $\frac{1}{2}$ hour.

7. **REASONING** The graph of a proportional relationship passes through (12, 16) and (1, y). Find *y*.

8. **MOVIE RENTAL** You pay $1 to rent a movie plus an additional $0.50 per day until you return the movie. Your friend pays $1.25 per day to rent a movie.

   a. Make tables showing the costs to rent a movie up to 5 days.

   b. Which person pays an amount proportional to the number of days rented?

5.3 Writing Proportions

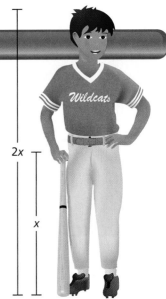

Essential Question How can you write a proportion that solves a problem in real life?

1 ACTIVITY: Writing Proportions

Work with a partner. A rough rule for finding the correct bat length is "the bat length should be half of the batter's height." So, a 62-inch-tall batter uses a bat that is 31 inches long. Write a proportion to find the bat length for each given batter height.

a. 58 inches

b. 60 inches

c. 64 inches

2 ACTIVITY: Bat Lengths

Work with a partner. Here is a more accurate table for determining the bat length for a batter. Find all the batter heights and corresponding weights for which the rough rule in Activity 1 is exact.

Weight of Batter (pounds)	Height of Batter (inches)							
	45–48	49–52	53–56	57–60	61–64	65–68	69–72	Over 72
Under 61	28	29	29					
61–70	28	29	30	30				
71–80	28	29	30	30	31			
81–90	29	29	30	30	31	32		
91–100	29	30	30	31	31	32		
101–110	29	30	30	31	31	32		
111–120	29	30	30	31	31	32		
121–130	29	30	30	31	32	33	33	
131–140	30	30	31	31	32	33	33	
141–150	30	30	31	31	32	33	33	
151–160	30	31	31	32	32	33	33	33
161–170		31	31	32	32	33	33	34
171–180				32	33	33	34	34
Over 180					33	33	34	34

Proportions

In this lesson, you will
- write proportions.
- solve proportions using mental math.

Laurie's Notes

What Your Students Will Learn
- Write proportions using tables.
- Solve proportions using mental math.

Previous Learning
Students have written and simplified ratios and determined if two ratios are equal.

Introduction

Applying Mathematical Practices

- **Reason Abstractly and Quantitatively:** In the first two activities, students work with three quantities: a batter's height and weight and the length of the batter's bat. Activity 1 states a rough rule for determining bat length based on the batter's height. Mathematically proficient students are able to apply the rule and also recognize that other variables are involved in bat selection.

Motivate

- **Management Tip:** You may want to pre-cut several lengths of string prior to this activity so students can join in.
- Ask for a volunteer. Say, "I can estimate the distance around your neck without actually measuring your neck!"
- Use the string to measure the distance around the student's wrist.
- Double this length, and it will be approximately the distance around his/her neck. Have the student verify this length by measuring the distance around his/her own neck.
- **Write:** $\dfrac{\text{distance around wrist}}{\text{distance around neck}} = \dfrac{1}{2}$
- Write a new proportion substituting the length around the wrist:

  $\dfrac{8.5 \text{ inches}}{x \text{ inches}} = \dfrac{1}{2}$

- **Solve:** The distance around the neck is two times 8.5 inches, or 17 inches.

Technology *for the* Teacher

Dynamic Classroom

Lesson Plans
Complete Materials List

Activity Notes

Activity 1

- Borrow a baseball bat for this activity to use as a prop.
- Help students translate the words in the activity into a proportion: $\dfrac{\text{length of bat}}{\text{height of batter}} = \dfrac{1}{2}$. Say, "The ratio of the length of the bat to the height of the batter is 1 : 2. A proportion to determine the bat length for a player 58 inches tall is $\dfrac{1}{2} = \dfrac{b}{58}$."
- You want students to understand that two measures are being compared, and the order in which you write (and say) them does matter.

Activity 2

- This activity promotes reading information from a table.
- The batter's height *and* weight factor into selecting the correct bat length.
- Practice reading information from the table.
- **?** "A 50-inch tall batter weighing 95 pounds should use what length bat?" 30"
- **?** "A 5-foot, 5-inch tall batter weighing 95 pounds should use what length bat?" 32"

5.3 Record and Practice Journal

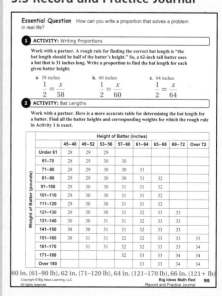

Essential Question How can you write a proportion that solves a problem in real life?

1 ACTIVITY: Writing Proportions

Work with a partner. A rough rule for finding the correct bat length is "the bat length should be half of the batter's height." So, a 62-inch tall batter uses a bat that is 31 inches long. Write a proportion to find the bat length for each given batter height.

a. 58 inches
$\dfrac{1}{2} = \dfrac{x}{58}$

b. 60 inches
$\dfrac{1}{2} = \dfrac{x}{60}$

c. 64 inches
$\dfrac{1}{2} = \dfrac{x}{64}$

2 ACTIVITY: Bat Lengths

Work with a partner. Here is a more accurate table for determining the bat length for a batter. Find all the batter heights and corresponding weights for which the rough rule in Activity 1 is exact.

	Height of Batter (inches)							
Weight of Batter (pounds)	45–48	49–52	53–56	57–60	61–64	65–68	69–72	Over 72
Under 61	28	29	29					
61–70	28	29	30	30				
71–80	28	29	30	30	31			
81–90	29	29	30	30	31	32		
91–100	29	30	30	31	31	32		
101–110	29	30	30	31	31	32		
111–120	29	30	30	31	31	32		
121–130	29	30	30	31	32	33	33	
131–140	30	30	31	31	32	33	33	
141–150	30	30	31	31	32	33	33	
151–160	30	31	31	32	32	33	33	33
161–170		31	31	32	32	33	33	34
171–180			32	32	33	33	34	34
Over 180				33	33	34	34	

60 in. (61–90 lb), 62 in. (71–120 lb), 64 in. (121–170 lb), 66 in. (121+ lb)

Big Ideas Math Red
Record and Practice Journal

95

English Language Learners

Visual Aids

English learners might find it useful to use a general template when writing a proportion problem.

$$\frac{part}{whole} = \frac{part}{whole}$$

5.3 Record and Practice Journal

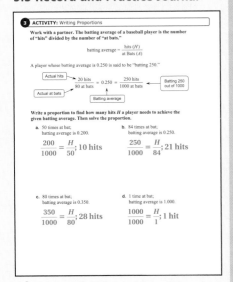

3 ACTIVITY: Writing Proportions

Work with a partner. The batting average of a baseball player is the number of "hits" divided by the number of "at bats."

$$\text{batting average} = \frac{\text{hits } (H)}{\text{at Bats } (A)}$$

A player whose batting average is 0.250 is said to be "batting 250."

Actual hits — 20 hits / 80 at bats = 0.250 = 250 hits / 1000 at bats — Batting 250 out of 1000

Actual at bats / Batting average

Write a proportion to find how many hits H a player needs to achieve the given batting average. Then solve the proportion.

a. 50 times at bat; batting average is 0.200.
$$\frac{200}{1000} = \frac{H}{50}; \ 10 \text{ hits}$$

b. 84 times at bat; batting average is 0.250.
$$\frac{250}{1000} = \frac{H}{84}; \ 21 \text{ hits}$$

c. 80 times at bat; batting average is 0.350.
$$\frac{350}{1000} = \frac{H}{80}; \ 28 \text{ hits}$$

d. 1 time at bat; batting average is 1.000.
$$\frac{1000}{1000} = \frac{H}{1}; \ 1 \text{ hit}$$

What Is Your Answer?

4. IN YOUR OWN WORDS How can you write a proportion that solves a problem in real life?

Check students' work.

5. Two players have the same batting average.

	At Bats	Hits	Batting Average
Player 1	132	45	
Player 2	132	45	

Player 1 gets four hits in the next five at bats. Player 2 gets three hits in the next three at bats.

a. Who has the higher batting average?

Player 1

b. Does this seem fair? Explain your reasoning.

no; Check students' work.

Laurie's Notes

Activity 3

? "Does anyone know what the term *batting average* means and how it is computed?" It is actually more involved than explained in the text. For instance, if a batter walks, has a sacrifice fly, or is hit by a pitch, it is not considered an "at bat."

- After the discussion of batting average, write the formula followed by the example. To have a proportion, the decimal form of the fraction is written.
- Remind students of how ratios are simplified (by dividing out common factors).
- Example: "Determine how many hits a batter has if he has 100 at bats and his batting average is 0.300.

$$\frac{H}{100} = \frac{300}{1000} \rightarrow \text{dividing out a common factor of 10} \rightarrow \frac{H}{100} = \frac{30}{100}$$

- **Common Error:** Students have divided out common factors when simplifying a simple fraction or when multiplying two fractions. A proportion is neither of these! A common error at this point is for students to divide out a factor from the numerator on one side of the equal sign with a factor in the denominator on the other side of the equal sign. For instance, in this example a student would incorrectly write $\frac{H}{1} = \frac{3}{1000}$ and get an answer of 0.003 hit.

- **Look for and Express Regularity in Repeated Reasoning** and **Reason Abstractly and Quantitatively :** Students should be able to use the information given to understand what *batting average* means. Students should also be able to check and compare their answers to each part of the activity for reasonableness. For instance, the batter in part (c) has a better average than, and has been at bat fewer times than, the batter in part (b). So, it should make sense that the batter in part (c) has had more hits than the batter in part (b).

- Have pairs of students show their work at the board.

What Is Your Answer?

- To answer Question 5, students may need to be reminded of how to write a fraction as a decimal. A calculator will be helpful.

Closure

- Write and solve a proportion: A stadium holds approximately 45,000 people during a baseball game. If the ratio of season ticket holders to all tickets is 1 : 3, approximately how many season ticket holders are there? 15,000 season ticket holders

ACTIVITY: Writing Proportions

Work with a partner. The batting average of a baseball player is the number of "hits" divided by the number of "at bats."

$$\text{batting average} = \frac{\text{hits } (H)}{\text{at bats } (A)}$$

A player whose batting average is 0.250 is said to be "batting 250."

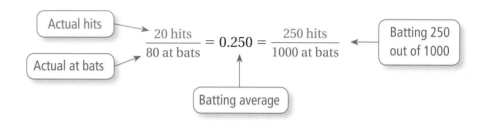

Write a proportion to find how many hits *H* a player needs to achieve the given batting average. Then solve the proportion.

a. 50 times at bat; batting average is 0.200.

b. 84 times at bat; batting average is 0.250.

c. 80 times at bat; batting average is 0.350.

d. 1 time at bat; batting average is 1.000.

What Is Your Answer?

4. IN YOUR OWN WORDS How can you write a proportion that solves a problem in real life?

5. Two players have the same batting average.

	At Bats	Hits	Batting Average
Player 1	132	45	
Player 2	132	45	

Player 1 gets four hits in the next five at bats. Player 2 gets three hits in the next three at bats.

a. Who has the higher batting average?

b. Does this seem fair? Explain your reasoning.

Practice Use what you discovered about proportions to complete Exercises 4–7 on page 182.

One way to write a proportion is to use a table.

	Last Month	This Month
Purchase	2 ringtones	3 ringtones
Total Cost	6 dollars	x dollars

Use the columns or the rows to write a proportion.

Use columns:

$$\frac{2 \text{ ringtones}}{6 \text{ dollars}} = \frac{3 \text{ ringtones}}{x \text{ dollars}}$$

Numerators have the same units.

Denominators have the same units.

Use rows:

$$\frac{2 \text{ ringtones}}{3 \text{ ringtones}} = \frac{6 \text{ dollars}}{x \text{ dollars}}$$

The units are the same on each side of the proportion.

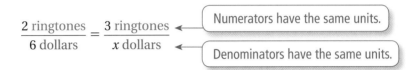

EXAMPLE 1 Writing a Proportion

Black Bean Soup

1.5 cups black beans
0.5 cup salsa
2 cups water
1 tomato
2 teaspoons seasoning

A chef increases the amounts of ingredients in a recipe to make a proportional recipe. The new recipe has 6 cups of black beans. Write a proportion that gives the number x of tomatoes in the new recipe.

Organize the information in a table.

	Original Recipe	New Recipe
Black Beans	1.5 cups	6 cups
Tomatoes	1 tomato	x tomatoes

∴ One proportion is $\dfrac{1.5 \text{ cups beans}}{1 \text{ tomato}} = \dfrac{6 \text{ cups beans}}{x \text{ tomatoes}}$.

On Your Own

Now You're Ready
Exercises 8–11

1. Write a different proportion that gives the number x of tomatoes in the new recipe.

2. Write a proportion that gives the amount y of water in the new recipe.

Laurie's Notes

Introduction

Connect

- **Yesterday:** Students wrote and solved proportions related to baseball that used simple mental math.
- **Today:** Students will write and solve a proportion using mental math.

Motivate

- ❓ "A student is reading a 280-page book. On average he/she reads 8 pages in 3 minutes. How long will it take to read the book?"
- If students can offer strategies, work on this problem now. If not, wait until later in the class.
- There are two equivalent proportions that can be set up. Be sure to use labels.

$$\frac{8 \text{ pages}}{3 \text{ minutes}} = \frac{280 \text{ pages}}{x \text{ minutes}} \quad \text{or} \quad \frac{8 \text{ pages}}{280 \text{ pages}} = \frac{3 \text{ minutes}}{x \text{ minutes}}$$

- The Cross Products Property is not needed. Because $280 \div 8 = 35$, mental math strategies are sufficient.

Lesson Notes

Discuss

- ❓ "What is the information in the table saying?" You purchased 2 ringtones last month for $6. This month, you purchase 3 ringtones, and the cost is unknown.
- Work through the example showing how the labels (ringtones and dollars) are used to identify the numbers.
- **Big Idea:** As identified in the text, when you use the rows or columns from the table, the proportion will be set up correctly. If the table had *not* been provided, it is possible for students to incorrectly set up the proportion:

$$\frac{2 \text{ ringtones}}{x \text{ dollars}} = \frac{3 \text{ ringtones}}{6 \text{ dollars}} \quad \text{or} \quad \frac{2 \text{ ringtones}}{3 \text{ ringtones}} = \frac{x \text{ dollars}}{6 \text{ dollars}}$$

- What students must also see besides the labels is that the information from one month (i.e., 2 ringtones, $6) must be in one ratio *or* in the numerators, and the information from the second month (i.e., 3 ringtones, x) must be in a second ratio *or* in the denominators.
- **Model with Mathematics:** The column and the rows from a correct table ensure this happening.

Example 1

- Organizing the information in a table helps to write the proportion correctly.
- ❓ "Can the rows and columns be interchanged?" yes

On Your Own

- **Think-Pair-Share:** Students should read each question independently and then work in pairs to answer the questions. When they have answered the questions, the pair should compare their answers with another group and discuss any discrepancies.

Goal Today's lesson is writing and solving a proportion using mental math.

Lesson Tutorials
Lesson Plans
Answer Presentation Tool

Extra Example 1

The chef increases the amounts of ingredients in the recipe in Example 1 to make a proportional recipe. The new recipe has 3 cups of salsa. Write a proportion that gives the amount w of water in the new recipe.

Sample answer:

$$\frac{0.5 \text{ cup salsa}}{3 \text{ cups salsa}} = \frac{2 \text{ cups water}}{w \text{ cups water}}$$

On Your Own

1. $\dfrac{1.5 \text{ cups beans}}{6 \text{ cups beans}} = \dfrac{1 \text{ tomato}}{x \text{ tomatoes}}$

2. *Sample answer:*

$$\frac{1.5 \text{ cups beans}}{2 \text{ cups water}} = \frac{6 \text{ cups beans}}{y \text{ cups water}}$$

Extra Example 2

Solve $\dfrac{8}{5} = \dfrac{n}{15}$. 24

Extra Example 3

In Extra Example 1, how much water is in the new recipe? 12 cups water

● On Your Own

3. $d = 32$

4. $z = 5$

5. $x = 7$

6. $\dfrac{48}{95} = \dfrac{f}{950}$; 480 female students

Differentiated Instruction

Kinesthetic

Provide students with counters and two pieces of blank paper. Draw fraction bar lines on each of the papers. Place counters on the papers to represent the two ratios. For each ratio, rearrange the counters in the numerator and denominator in stacks of equal number. If the stacks in each individual ratio are the same size and there are the same number of stacks in each ratio, then the ratios are equal and form a proportion. For instance, in Example 2, the first ratio has 3 stacks of 1 in the numerator and 2 stacks of 1 in the denominator. The second ratio has 3 stacks of 4 in the numerator and 2 stacks of 4 in the denominator. Because there are 3 stacks over 2 stacks in each of the ratios, the ratios form a proportion.

Example 2

- **Reason Abstractly and Quantitatively:** This example has no context. The focus is on the process and how mental math is used in solving the proportion.
- When finished, present the following problems to assess if students can distinguish when mental math is a reasonable approach.
- Tell whether you can easily use mental math to solve these problems:

  a. $\dfrac{3}{7} = \dfrac{x}{27}$ b. $\dfrac{3}{7} = \dfrac{27}{x}$

  Answers:

  a. 7 is not a factor of 27, so mental math is not an easy approach.

  b. 3 is a factor of 27, so mental math can be used.

Example 3

- Work through this example.
- Not all students will know that $4 \times 1.5 = 6$. If students are still struggling with multiplying decimals, you may want to review these rules prior to Example 3.
- Two ideas to help develop fluency:
  $4 \times 15 = 60$, so $4 \times 1.5 = 6$.
  $1 \times 4 = 4$, and half (0.5) of 4 is 2, so add $2 + 4$ to get $4 \times 1.5 = 6$.

On Your Own

- **Neighbor Check:** Have students work independently and then have their neighbors check their work. Have students discuss any discrepancies.

Closure

- **Exit Ticket:** The ratio of quarts to gallons is 4 : 1. If a recipe calls for 14 quarts, how many gallons would be needed? 3.5

EXAMPLE **2** **Solving Proportions Using Mental Math**

Solve $\dfrac{3}{2} = \dfrac{x}{8}$.

Step 1: Think: The product of 2 and what number is 8?

$$\dfrac{3}{2} = \dfrac{x}{8}$$

$$2 \times ? = 8$$

Step 2: Because the product of 2 and 4 is 8, multiply the numerator by 4 to find x.

$$3 \times 4 = 12$$

$$\dfrac{3}{2} = \dfrac{x}{8}$$

$$2 \times 4 = 8$$

∴ The solution is $x = 12$.

EXAMPLE **3** **Solving Proportions Using Mental Math**

In Example 1, how many tomatoes are in the new recipe?

Solve the proportion $\dfrac{1.5}{1} = \dfrac{6}{x}$. ← cups black beans
← tomatoes

Step 1: Think: The product of 1.5 and what number is 6?

$$1.5 \times ? = 6$$

$$\dfrac{1.5}{1} = \dfrac{6}{x}$$

Step 2: Because the product of 1.5 and 4 is 6, multiply the denominator by 4 to find x.

$$1.5 \times 4 = 6$$

$$\dfrac{1.5}{1} = \dfrac{6}{x}$$

$$1 \times 4 = 4$$

∴ So, there are 4 tomatoes in the new recipe.

On Your Own

Now You're Ready
Exercises 16–21

Solve the proportion.

3. $\dfrac{5}{8} = \dfrac{20}{d}$

4. $\dfrac{7}{z} = \dfrac{14}{10}$

5. $\dfrac{21}{24} = \dfrac{x}{8}$

6. A school has 950 students. The ratio of female students to all students is $\dfrac{48}{95}$. Write and solve a proportion to find the number f of students who are female.

 Vocabulary and Concept Check

1. **WRITING** Describe two ways you can use a table to write a proportion.

2. **WRITING** What is your first step when solving $\dfrac{x}{15} = \dfrac{3}{5}$? Explain.

3. **OPEN-ENDED** Write a proportion using an unknown value x and the ratio $5:6$. Then solve it.

 Practice and Problem Solving

Write a proportion to find how many points a student needs to score on the test to get the given score.

4. test worth 50 points; test score of 40%

5. test worth 50 points; test score of 78%

6. test worth 80 points; test score of 80%

7. test worth 150 points; test score of 96%

 Use the table to write a proportion.

① 8.

	Game 1	Game 2
Points	12	18
Shots	14	w

9.

	May	June
Winners	n	34
Entries	85	170

10.

	Today	Yesterday
Miles	15	m
Hours	2.5	4

11.

	Race 1	Race 2
Meters	100	200
Seconds	x	22.4

12. **ERROR ANALYSIS** Describe and correct the error in writing the proportion.

✗

	Monday	Tuesday
Dollars	2.08	d
Ounces	8	16

$$\dfrac{2.08}{16} = \dfrac{d}{8}$$

13. **T-SHIRTS** You can buy 3 T-shirts for $24. Write a proportion that gives the cost c of buying 7 T-shirts.

14. **COMPUTERS** A school requires 2 computers for every 5 students. Write a proportion that gives the number c of computers needed for 145 students.

15. **SWIM TEAM** The school team has 80 swimmers. The ratio of seventh-grade swimmers to all swimmers is $5:16$. Write a proportion that gives the number s of seventh-grade swimmers.

Assignment Guide and Homework Check

Level	Day 1 Activity Assignment	Day 2 Lesson Assignment	Homework Check
Basic	4–7, 26–30	1–3, 9, 11, 12, 13–23 odd	9, 13, 15, 17
Average	4–7, 26–30	1–3, 8–14 even, 19–23	10, 14, 19, 23
Advanced	4–7, 26–30	1–3, 8–24 even, 25	8, 14, 18, 24
Accelerated	1–7, 8–24 even, 25–30		8, 14, 18, 24

Common Errors

- **Exercises 8–11** Students may write half of the proportion using rows and the other half using columns. They will have forgotten to include one of the values. Remind students that they need to pick a method for writing proportions with tables and be consistent throughout the problem.
- **Exercises 16–21** Students may get confused with using mental math to find the value of the variable and try to multiply the numerator and denominator by different numbers. Tell students that they are finding an equivalent ratio, or an equivalent fraction. They are multiplying the original fraction by 1 $\left(\text{or } \dfrac{4}{4}, \text{ for example}\right)$ to find the equivalent fraction, so they must multiply the numerator and denominator by the same number.

5.3 Record and Practice Journal

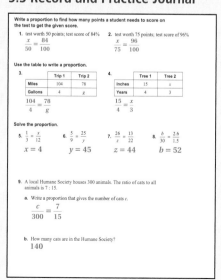

Vocabulary and Concept Check

1. You can use the columns or the rows of the table to write a proportion.

2. Find the number that when multiplied by 5 is 15.

3. *Sample answer:* $\dfrac{x}{12} = \dfrac{5}{6}$; $x = 10$

Practice and Problem Solving

4. $\dfrac{x}{50} = \dfrac{40}{100}$

5. $\dfrac{x}{50} = \dfrac{78}{100}$

6. $\dfrac{x}{80} = \dfrac{80}{100}$

7. $\dfrac{x}{150} = \dfrac{96}{100}$

8. $\dfrac{12 \text{ points}}{14 \text{ shots}} = \dfrac{18 \text{ points}}{w \text{ shots}}$

9. $\dfrac{n \text{ winners}}{85 \text{ entries}} = \dfrac{34 \text{ winners}}{170 \text{ entries}}$

10. $\dfrac{15 \text{ miles}}{2.5 \text{ hours}} = \dfrac{m \text{ miles}}{4 \text{ hours}}$

11. $\dfrac{100 \text{ meters}}{x \text{ seconds}} = \dfrac{200 \text{ meters}}{22.4 \text{ seconds}}$

12. The proportion cannot be written using diagonals of the table. $\dfrac{2.08}{8} = \dfrac{d}{16}$

13. $\dfrac{\$24}{3 \text{ shirts}} = \dfrac{c}{7 \text{ shirts}}$

14. $\dfrac{2 \text{ computers}}{5 \text{ students}} = \dfrac{c \text{ computers}}{145 \text{ students}}$

15. $\dfrac{5 \text{ 7th grade swimmers}}{16 \text{ swimmers}} = \dfrac{s \text{ 7th grade swimmers}}{80 \text{ swimmers}}$

Practice and Problem Solving

16. $z = 5$ **17.** $y = 16$

18. $k = 15$ **19.** $c = 24$

20. $b = 20$ **21.** $g = 14$

22. a. $\dfrac{1 \text{ trombone}}{3 \text{ violas}} = \dfrac{t \text{ trombones}}{9 \text{ violas}}$

b. 3 trombones

23. $\dfrac{1}{200} = \dfrac{19.5}{x}$; Dimensions for the model are in the numerators and the corresponding dimensions for the actual space shuttle are in the denominators.

24. no; The solution of that equation is $x = 1.5$, but using mental math, you can see that the solution of the proportion is $x = 24$.

25. See *Taking Math Deeper*.

Fair Game Review

26. $x = 150$ **27.** $x = 9$

28. $x = 75$ **29.** $x = 140$

30. C

Mini-Assessment

Write a proportion to find how many points a student needs to score on the test to get the given score.

1. Test worth 60 points; test score of 60% $\dfrac{x}{60} = \dfrac{60}{100}$

2. Test worth 50 points; test score of 70% $\dfrac{x}{50} = \dfrac{70}{100}$

3. Test worth 100 points; test score of 85% $\dfrac{x}{100} = \dfrac{85}{100}$

4. Test worth 120 points; test score of 88% $\dfrac{x}{120} = \dfrac{88}{100}$

5. You can buy four DVDs for $48. Write a proportion that gives the cost c of buying six DVDs. $\dfrac{4}{48} = \dfrac{6}{c}$

Taking Math Deeper

Exercise 25

Although this problem does not have difficult or messy mathematics, it is still difficult for many students to know how to start. Emphasize that it is good to just "write things down" and organize the given facts. In this problem, it might help the visual learner to sketch 3 white lockers for every 5 blue lockers.

① Draw a diagram that shows the given values and the unknown values.

3 white lockers for every 5 blue lockers

180 white lockers

② Write and solve a proportion.

$\dfrac{x \text{ blue lockers}}{180 \text{ white lockers}} = \dfrac{5}{3}$ Write a proportion.

$x = 300$ Use mental math.

③ Answer the question.

There are 180 white lockers and 300 blue lockers. So, there are a total of $180 + 300 = 480$ lockers in the school.

Reteaching and Enrichment Strategies

If students need help. . .	If students got it. . .
Resources by Chapter • Practice A and Practice B • Puzzle Time Record and Practice Journal Practice Differentiating the Lesson Lesson Tutorials Skills Review Handbook	Resources by Chapter • Enrichment and Extension • Technology Connection Start the next section

Solve the proportion.

②③ 16. $\dfrac{1}{4} = \dfrac{z}{20}$

17. $\dfrac{3}{4} = \dfrac{12}{y}$

18. $\dfrac{35}{k} = \dfrac{7}{3}$

19. $\dfrac{15}{8} = \dfrac{45}{c}$

20. $\dfrac{b}{36} = \dfrac{5}{9}$

21. $\dfrac{1.4}{2.5} = \dfrac{g}{25}$

22. ORCHESTRA In an orchestra, the ratio of trombones to violas is 1 to 3.

  a. There are 9 violas. Write a proportion that gives the number t of trombones in the orchestra.

  b. How many trombones are in the orchestra?

23. ATLANTIS Your science teacher has a 1 : 200 scale model of the space shuttle *Atlantis*. Which of the proportions can you use to find the actual length x of *Atlantis*? Explain.

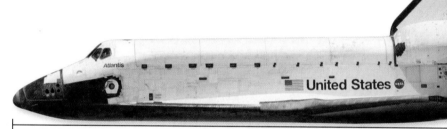

$$\dfrac{1}{200} = \dfrac{19.5}{x} \qquad \dfrac{1}{200} = \dfrac{x}{19.5} \qquad \dfrac{200}{19.5} = \dfrac{x}{1} \qquad \dfrac{x}{200} = \dfrac{1}{19.5}$$

19.5 cm

24. YOU BE THE TEACHER Your friend says "$48x = 6 \cdot 12$." Is your friend right? Explain.

Solve $\dfrac{6}{x} = \dfrac{12}{48}$.

25. Reasoning There are 180 white lockers in the school. There are 3 white lockers for every 5 blue lockers. How many lockers are in the school?

Fair Game Review What you learned in previous grades & lessons

Solve the equation. *(Section 3.4)*

26. $\dfrac{x}{6} = 25$

27. $8x = 72$

28. $150 = 2x$

29. $35 = \dfrac{x}{4}$

30. MULTIPLE CHOICE What is the value of $-\dfrac{9}{4} + \left| -\dfrac{8}{5} \right| - 2\dfrac{1}{2}$? *(Section 2.3)*

  Ⓐ $-6\dfrac{7}{20}$

  Ⓑ $-5\dfrac{7}{20}$

  Ⓒ $-3\dfrac{3}{20}$

  Ⓓ $-2\dfrac{3}{20}$

You can use an **information wheel** to organize information about a concept. Here is an example of an information wheel for ratio.

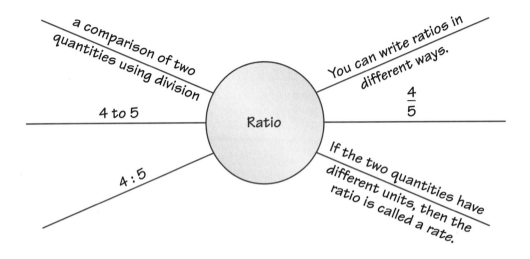

a comparison of two
quantities using division

You can write ratios in
different ways.

4 to 5

Ratio

$\dfrac{4}{5}$

4 : 5

If the two quantities have
different units, then the
ratio is called a rate.

On Your Own

**Make information wheels to help you
study these topics.**

1. rate

2. unit rate

3. proportion

4. cross products

5. graphing proportional relationships

**After you complete this chapter, make
information wheels for the following topics.**

6. solving proportions

7. slope

8. direct variation

"My information wheel summarizes how
cats act when they get baths."

Sample Answers

1.

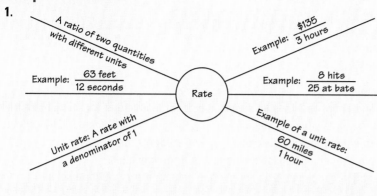

A ratio of two quantities with different units

Example: $\dfrac{63\ feet}{12\ seconds}$

Rate

Example: $\dfrac{\$135}{3\ hours}$

Example: $\dfrac{8\ hits}{25\ at\ bats}$

Unit rate: A rate with a denominator of 1

Example of a unit rate: $\dfrac{60\ miles}{1\ hour}$

2.

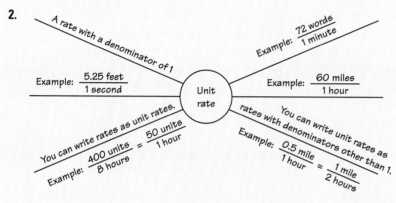

A rate with a denominator of 1

Example: $\dfrac{5.25\ feet}{1\ second}$

Unit rate

Example: $\dfrac{72\ words}{1\ minute}$

Example: $\dfrac{60\ miles}{1\ hour}$

You can write rates as unit rates. Example: $\dfrac{400\ units}{8\ hours} = \dfrac{50\ units}{1\ hour}$

You can write unit rates as rates with denominators other than 1. Example: $\dfrac{0.5\ mile}{1\ hour} = \dfrac{1\ mile}{2\ hours}$

3.

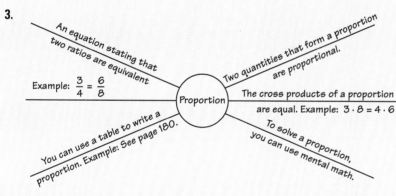

An equation stating that two ratios are equivalent

Example: $\dfrac{3}{4} = \dfrac{6}{8}$

Proportion

Two quantities that form a proportion are proportional.

The cross products of a proportion are equal. Example: $3 \cdot 8 = 4 \cdot 6$

You can use a table to write a proportion. Example: See page 180.

To solve a proportion, you can use mental math.

4.

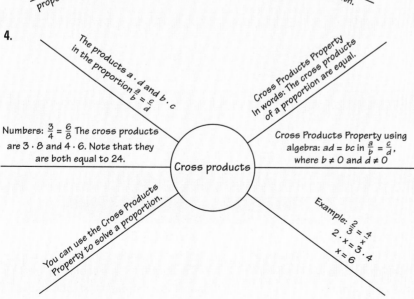

The products $a \cdot d$ and $b \cdot c$ in the proportion $\dfrac{a}{b} = \dfrac{c}{d}$

Numbers: $\dfrac{3}{4} = \dfrac{6}{8}$ The cross products are $3 \cdot 8$ and $4 \cdot 6$. Note that they are both equal to 24.

Cross products

Cross Products Property In words: The cross products of a proportion are equal.

Cross Products Property using algebra: $ad = bc$ in $\dfrac{a}{b} = \dfrac{c}{d}$, where $b \neq 0$ and $d \neq 0$

You can use the Cross Products Property to solve a proportion.

Example: $\dfrac{2}{3} = \dfrac{4}{x}$
$2 \cdot x = 3 \cdot 4$
$x = 6$

5. Available at *BigIdeasMath.com*.

List of Organizers
Available at *BigIdeasMath.com*

Comparison Chart
Concept Circle
Example and Non-Example Chart
Formula Triangle
Four Square
Idea (Definition) and Examples Chart
Information Frame
Information Wheel
Notetaking Organizer
Process Diagram
Summary Triangle
Word Magnet
Y Chart

About this Organizer

An **Information Wheel** can be used to organize information about a concept. Students write the concept in the middle of the "wheel." Then students write information related to the concept on the "spokes" of the wheel. Related information can include, but is not limited to: vocabulary words or terms, definitions, formulas, procedures, examples, and visuals. This type of organizer serves as a good summary tool because any information related to a concept can be included.

Technology for the *Teacher*

Editable Graphic Organizer

Answers

1. $\dfrac{3}{2}$ 2. $\dfrac{15}{8}$

3. $0.99 per song

4. 3.5 gallons per hour

5. yes   6. no

7. yes   8. no

9. yes   10. yes

11. no

12. *Sample answer:* $\dfrac{\$56}{\$42} = \dfrac{h \text{ hours}}{6 \text{ hours}}$

13. *Sample answer:*
    $\dfrac{g \text{ games}}{4 \text{ wins}} = \dfrac{6 \text{ games}}{3 \text{ wins}}$

14. $\dfrac{1}{3}$ MB per second

15. 6 km per second

16. no; Your rate is 5 minutes per level and your friend's rate is 4 minutes per level.

17. $\dfrac{150 \text{ minutes}}{3 \text{ classes}} = \dfrac{x \text{ minutes}}{5 \text{ classes}}$;

    250 minutes

Alternative Quiz Ideas

100% Quiz	Math Log
Error Notebook	Notebook Quiz
Group Quiz	Partner Quiz
Homework Quiz	Pass the Paper

Error Notebook

An error notebook provides an opportunity for students to analyze and learn from their errors. Have students make an error notebook for this chapter. They should work in their notebook a little each day. Give students the following directions.

* Use a notebook and divide the page into three columns.
* Label the first column *problem*, second column *error*, and third column *correction*.
* In the first column, write the exercise in which the errors were made. Record the source of the exercise (homework, quiz, in-class assignment).
* The second column should show the exact error that was made. Include a statement of why you think the error was made. This is where the learning takes place, so it is helpful to use a different color ink for the work in this column.
* The last column contains the corrected problems and comments that will help with future work.
* Separate each problem with horizontal lines.

Reteaching and Enrichment Strategies

If students need help. . .	If students got it. . .
Resources by Chapter • Practice A and Practice B • Puzzle Time Lesson Tutorials *BigIdeasMath.com*	Resources by Chapter • Enrichment and Extension • Technology Connection Game Closet at *BigIdeasMath.com* Start the next section

Write the ratio as a fraction in simplest form.   *(Section 5.1)*

1. 18 red buttons : 12 blue buttons

2. $\frac{5}{4}$ inches to $\frac{2}{3}$ inch

Use the ratio table to find the unit rate with the specified units.   *(Section 5.1)*

3. cost per song

Songs	0	2	4	6
Cost	$0	$1.98	$3.96	$5.94

4. gallons per hour

Hours	3	6	9	12
Gallons	10.5	21	31.5	42

Tell whether the ratios form a proportion.   *(Section 5.2)*

5. $\frac{1}{8}, \frac{4}{32}$

6. $\frac{2}{3}, \frac{10}{30}$

7. $\frac{7}{4}, \frac{28}{16}$

Tell whether the two rates form a proportion.   *(Section 5.2)*

8. 75 miles in 3 hours; 140 miles in 4 hours

9. 12 gallons in 4 minutes; 21 gallons in 7 minutes

10. 150 steps in 50 feet; 72 steps in 24 feet

11. 3 rotations in 675 days; 2 rotations in 730 days

Use the table to write a proportion.   *(Section 5.3)*

12.

	Monday	Tuesday
Dollars	42	56
Hours	6	h

13.

	Series 1	Series 2
Games	g	6
Wins	4	3

14. **MUSIC DOWNLOAD** The amount of time needed to download music is shown in the table. Find the unit rate in megabytes per second.   *(Section 5.1)*

Seconds	6	12	18	24
Megabytes	2	4	6	8

15. **SOUND** The graph shows the distance that sound travels through steel. Interpret each plotted point in the graph of the proportional relationship.   *(Section 5.2)*

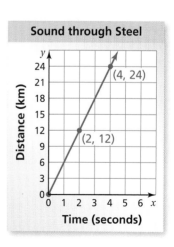

Sound through Steel

16. **GAMING** You advance 3 levels in 15 minutes. Your friend advances 5 levels in 20 minutes. Do these rates form a proportion? Explain.   *(Section 5.2)*

17. **CLASS TIME** You spend 150 minutes in 3 classes. Write and solve a proportion to find how many minutes you spend in 5 classes.   *(Section 5.3)*

Essential Question

How can you use ratio tables and cross products to solve proportions?

1 ACTIVITY: Solving a Proportion in Science

Work with a partner. You can use ratio tables to determine the amount of a compound (like salt) that is dissolved in a solution. Determine the unknown quantity. Explain your procedure.

a. **Salt Water**

Salt Water	1 L	3 L
Salt	250 g	x g

1 liter   3 liters

$$\frac{1\text{ L}}{\boxed{}} = \frac{\boxed{}}{\boxed{}}$$ Write proportion.

$$1 \cdot \boxed{} = \boxed{} \cdot \boxed{}$$ Set cross products equal.

$$\boxed{} = \boxed{}$$ Simplify.

⋮ There are ⬚ grams of salt in the 3-liter solution.

b. **White Glue Solution**

Water	½ cup	1 cup
White Glue	½ cup	x cups

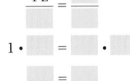

c. **Borax Solution**

Borax	1 tsp	2 tsp
Water	1 cup	x cups

d. **Slime (See recipe.)**

Borax Solution	½ cup	1 cup
White Glue Solution	y cups	x cups

Proportions

In this lesson, you will

- solve proportions using multiplication or the Cross Products Property.
- use a point on a graph to write and solve proportions.

Recipe for SLIME

1. Add ½ cup of water and ½ cup white glue. Mix thoroughly. This is your white glue solution.

2. Add a couple drops of food coloring to the white glue solution. Mix thoroughly.

3. Add 1 teaspoon of borax to 1 cup of water. Mix thoroughly. This is your borax solution (about 1 cup).

4. Pour the borax solution and the glue solution into a separate bowl.

5. Place the slime that forms into a plastic bag. Squeeze the mixture repeatedly to mix it up.

Laurie's Notes

Introduction

Applying Mathematical Practices

- **Model with Mathematics:** Mathematically proficient students use a ratio table as a tool to show relationships between different quantities.

Motivate

- **Model:** Display two containers, each filled with water. The ratio of their volumes should be 2 : 1. Let students watch you put 4 drops of food coloring in the smaller one. Stir the water.

- **?** "How many drops do I need to put in the larger container so that the water is the same darkness as the smaller vessel?" Students should note the ratio of the volumes of the two vessels. It may be helpful to label the volume of each container. Students should have no difficulty in understanding that 8 drops are needed.

- **?** "Suppose that the two volumes are not in the ratio 1 : 2. The smaller container has a volume of 400 milliliters and the larger vessel has a volume of 600 milliliters. If I add 4 drops of food coloring to the smaller vessel, how much should I add to the larger vessel?" 6 drops

- This should get students thinking about a proportion.

- **Discuss:** Ask for volunteers to share their thinking.

Activity Notes

Activity 1

- Write the ratio table.

- **?** "If there are 250 grams of salt in 1 liter of salt water, how many grams of salt are in 3 liters?" Students should quickly answer 750 grams, not just 750. The units are extremely important!

- Set up the proportion from the ratio table and discuss how common units divide out when comparing liters to liters and grams to grams.

- Then use the Cross Products Property to verify the answer.

- **?** "Could the proportion $\dfrac{1\,L}{250\,g} = \dfrac{3\,L}{x\,g}$ be used to solve for x?" yes; Listen for an explanation that 1 liter has 250 grams of salt and you're solving for the number of grams of salt in 3 liters.

- **?** "Is there another way you might solve this proportion without using the Cross Products Property?" mental math: $1 \times 250 = 250$, so $3 \times 250 = 750$

- **Model with Mathematics:** Ask students to solve for x in each of the three parts—white glue solution, borax solution, and Slime. In each part of the activity, a ratio table helps to show relationships between quantities.

- **?** **Extension:** Ask the following:

  - "If you only had $\dfrac{1}{4}$ cup of white glue, how much water should you mix with it to make the white glue solution?" $\dfrac{1}{4}$ cup

  - "If you have 1 tablespoon of borax, how much water do you need to make the borax solution?" 3 cups

What Your Students Will Learn

- Solve proportions using the Multiplication Property of Equality or the Cross Products Property.
- Use a point on a graph to write and solve proportions.

Previous Learning

Students should know how to write and solve proportions using mental math and how to convert measures.

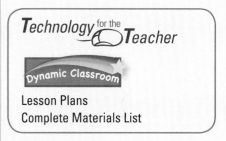

Technology for the **Teacher**

Dynamic Classroom

Lesson Plans
Complete Materials List

5.4 Record and Practice Journal

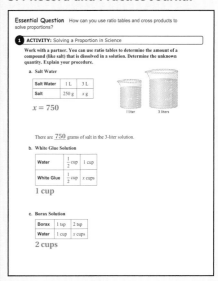

English Language Learners

Visual

When solving a proportion using the Cross Products Property, use a visual X through the equal sign.

$$\frac{4}{3} \diagdown \frac{f}{27}$$ Write the proportion.

$4 \cdot 27 = 3 \cdot f$ Cross Products Property

$108 = 3f$ Multiply.

$36 = f$ Divide.

5.4 Record and Practice Journal

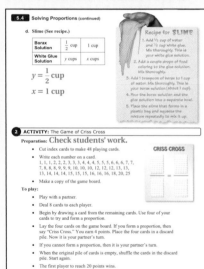

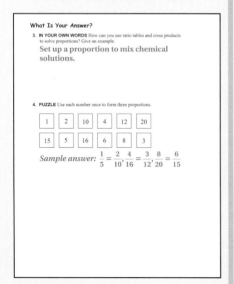

Activity 2

- **Management Tip:** The cards can be photocopied onto heavier weight paper, laminated, cut apart, and stored in locking plastic bags for easy distribution *and* for use next year.
- **Game Notes:**
  - The number of cards in your hand will vary depending upon whether you have been able to make a proportion.
  - The game moves along fairly quickly.
  - If a pair of students finishes early, they can play again.
- After students have played for a period of time, ask what strategies they used to decide if they have a proportion or not. Make the decision on whether or not they had a proportion.
- **Construct Viable Arguments and Critique the Reasoning of Others:** Mathematically proficient students should be expected to explain their reasoning. It is also important for them to hear other students explain their reasoning.
- Some students may only think about equivalent fractions. Others may think about the Cross Products Property.

What Is Your Answer?

Question 4 is a nice recap of the lesson. When students have finished, ask: "Do you think everyone has the same 3 proportions?" Students may likely say yes, forgetting that there are different arrangements of the 4 numbers in a proportion that will form another proportion.

Closure

- Use the information in the table to solve for x.

# of bracelets	3	x
Yellow twine	48 in.	80 in.

$$\frac{3}{48} = \frac{x}{80}$$

$x = 5$

five bracelets

CRISS CROSS

Preparation:

- Cut index cards to make 48 playing cards.
- Write each number on a card.

  1, 1, 1, 2, 2, 2, 3, 3, 3, 4, 4, 4, 5, 5, 5, 6, 6, 6, 7, 7,

  7, 8, 8, 8, 9, 9, 9, 10, 10, 10, 12, 12, 12, 13, 13,

  13, 14, 14, 14, 15, 15, 15, 16, 16, 16, 18, 20, 25

- Make a copy of the game board.

To Play:

- Play with a partner.
- Deal eight cards to each player.
- Begin by drawing a card from the remaining cards. Use four of your cards to try to form a proportion.
- Lay the four cards on the game board. If you form a proportion, then say "Criss Cross." You earn 4 points. Place the four cards in a discard pile. Now it is your partner's turn.
- If you cannot form a proportion, then it is your partner's turn.
- When the original pile of cards is empty, shuffle the cards in the discard pile. Start again.
- The first player to reach 20 points wins.

What Is Your Answer?

3. **IN YOUR OWN WORDS** How can you use ratio tables and cross products to solve proportions? Give an example.

4. **PUZZLE** Use each number once to form three proportions.

1	2	10	4	12	20

15	5	16	6	8	3

Practice ➤ Use what you discovered about solving proportions to complete Exercises 10–13 on page 190.

 **Key Idea**

Solving Proportions

Method 1 Use mental math. *(Section 5.3)*

Method 2 Use the Multiplication Property of Equality. *(Section 5.4)*

Method 3 Use the Cross Products Property. *(Section 5.4)*

EXAMPLE ① **Solving Proportions Using Multiplication**

Solve $\dfrac{5}{7} = \dfrac{x}{21}$.

$$\dfrac{5}{7} = \dfrac{x}{21}$$ Write the proportion.

$$21 \cdot \dfrac{5}{7} = 21 \cdot \dfrac{x}{21}$$ Multiplication Property of Equality

$$15 = x$$ Simplify.

⋮· The solution is 15.

On Your Own

Now You're Ready
Exercises 4–9

Use multiplication to solve the proportion.

1. $\dfrac{w}{6} = \dfrac{6}{9}$

2. $\dfrac{12}{10} = \dfrac{a}{15}$

3. $\dfrac{y}{6} = \dfrac{2}{4}$

EXAMPLE ② **Solving Proportions Using the Cross Products Property**

Solve each proportion.

a. $\dfrac{x}{8} = \dfrac{7}{10}$ 　　　　　　　　**b.** $\dfrac{9}{y} = \dfrac{3}{17}$

$x \cdot 10 = 8 \cdot 7$　　Cross Products Property　　$9 \cdot 17 = y \cdot 3$

$10x = 56$　　Multiply.　　$153 = 3y$

$x = 5.6$　　Divide.　　$51 = y$

⋮· The solution is 5.6. 　　　　⋮· The solution is 51.

Laurie's Notes

Lesson Tutorials
Lesson Plans
Answer Presentation Tool

Introduction

Connect
- **Yesterday:** Students solved proportions using the Cross Products Property.
- **Today:** Students will solve proportions using different strategies.

Motivate
- The Cross Products Property can be used to solve any proportion, but you want students to recognize when it is more efficient to use simple mental math or the Multiplication Property of Equality.
- **Make Sense of Problems and Persevere in Solving Them:** As you work through problems with students, share with them the wisdom of analyzing the problem first to decide what method makes the most sense.
- **Common Error:** Students sometimes confuse multiplication of fractions and the Cross Products Property.

Lesson Notes

Example 1
- The Multiplication Property of Equality works because the variable is in the numerator.
- If this same problem had been $\frac{7}{5} = \frac{21}{x}$, you could not solve by multiplying both sides of the equation by $\frac{1}{21}$ because that would simplify to $\frac{1}{15} = \frac{1}{x}$ and you still haven't solved for x. Be sure students understand this.
- Be sure to check for understanding with this idea.
- **?** "Could you use another strategy such as mental math to solve this problem?" Yes. Listen for the idea of equivalent fractions.

On Your Own
- **Think-Pair-Share:** Students should read each question independently and then work in pairs to answer the questions. When they have answered the questions, the pair should compare their answers with another group and discuss any discrepancies.
- **?** Ask students to share their strategies.
- At least one pair of students should solve Question 3 by simplifying $\frac{2}{4} = \frac{1}{2}$, and using mental math to finish.

Example 2
- **?** "How are parts (a) and (b) different?" In part (a), the variable is in the numerator and in part (b), the variable is in the denominator. Part (b) involves one numerator that is a factor of the other numerator.
- **?** "Can you easily use the Multiplication Property of Equality to solve both examples?" Using the Multiplication Property of Equality would be difficult in part (b) because the variable is in the denominator.

Extra Example 1
Solve $\frac{c}{12} = \frac{5}{3}$. 20

On Your Own
1. $w = 4$
2. $a = 18$
3. $y = 3$

Extra Example 2
Solve each proportion.

a. $\frac{3}{4} = \frac{u}{6}$ 4.5

b. $\frac{4}{13} = \frac{12}{h}$ 39

On Your Own

4. $x = 8$

5. $y = 2.5$

6. $z = 15$

Extra Example 3

In Example 3, your toll is $9. How many kilometers did you drive?   about 193 km

On Your Own

7. $\dfrac{7.5}{x} = \dfrac{1}{2.54}$; about 19.05

8. $\dfrac{100}{x} = \dfrac{1}{0.035}$; about 3.5

9. $\dfrac{2}{x} = \dfrac{1}{1.06}$; about 2.12

10. $\dfrac{4}{x} = \dfrac{1}{3.28}$; about 13.12

Differentiated Instruction

Visual

To reinforce the Cross Products Property, write each number and variable of a proportion on a card. Give the set of cards to a student. Have the student set up the proportion.

$$\dfrac{\boxed{3}}{\boxed{8}} = \dfrac{\boxed{k}}{\boxed{4}}$$

Then have the student find the cross products with the cards.

$$\boxed{3} \cdot \boxed{4} = \boxed{8} \cdot \boxed{k}$$

Finish by having the student complete the solution using paper and pencil.

On Your Own

- **Think-Pair-Share:** Students should read each question independently and then work in pairs to answer the questions. When they have answered the questions, the pair should compare their answers with another group and discuss any discrepancies.
- ? Ask students to share their strategies.
- Although the directions say to solve using the Cross Products Property, one pair of students might solve Questions 4 and 5 by using mental math and recognize equivalent fractions.

Example 3

- ? "Which is farther, one mile or one kilometer?" one mile
- Hopefully, students will know that 1 mile ≈ 1.61 kilometers. If not, have them refer to the conversion chart in the back of the book.
- ? "What do the ordered pairs (100, 7.5) and (200, 15) mean on the graph?" If you drive 100 miles the toll is $7.50, and if you drive 200 miles the toll is $15.
- You may need to help students focus on the units for each axis.
- Explain that you want to convert miles to kilometers.
- The first method uses the conversion factor and dimensional analysis. Work through the problem as shown.
- The second method uses a proportion. Although the units for each value are not written, say aloud the units as you write the proportion: "1.61 kilometers is to 1 mile as how many kilometers is to 100 miles?" Each ratio compares kilometers to miles.
- Students may say it is not necessary to set up a proportion because the numbers are so simple. Change the problem by asking what toll would be due for driving 150 miles or 232 miles.

On Your Own

- These exercises provide a helpful review of decimal multiplication and division.
- Students may need to refer to the conversion chart in the back of the book to complete these problems.
- **Attend to Precision:** Remind students that the symbol ≈ means *approximately equal to* and that answers should be rounded to the nearest hundredth, if necessary.

Closure

- Write and solve 3 proportions. One should use mental math to solve, one should use the Multiplication Property of Equality, and one should use the Cross Products Property.

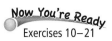
On Your Own

Use the Cross Products Property to solve the proportion.

4. $\dfrac{2}{7} = \dfrac{x}{28}$

5. $\dfrac{12}{5} = \dfrac{6}{y}$

6. $\dfrac{40}{z+1} = \dfrac{15}{6}$

EXAMPLE ③   **Real-Life Application**

TOLL PLAZA
½ MILE
REDUCE SPEED

The graph shows the toll y due on a turnpike for driving x miles. Your toll is $7.50. How many *kilometers* did you drive?

The point (100, 7.5) on the graph shows that the toll is $7.50 for driving 100 miles. Convert 100 miles to kilometers.

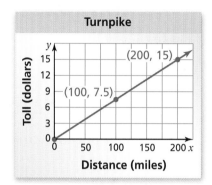

Method 1: Convert using a ratio.

1 mi ≈ 1.61 km

$$100 \ \text{mi} \times \frac{1.61 \ \text{km}}{1 \ \text{mi}} = 161 \ \text{km}$$

∴ So, you drove about 161 kilometers.

Method 2: Convert using a proportion.

Let x be the number of kilometers equivalent to 100 miles.

kilometers → $\dfrac{1.61}{1} = \dfrac{x}{100}$ ← kilometers   Write a proportion. Use 1.61 km ≈ 1 mi.
miles →   ← miles

$1.61 \cdot 100 = 1 \cdot x$ Cross Products Property

$161 = x$ Simplify.

∴ So, you drove about 161 kilometers.

On Your Own

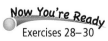
Write and solve a proportion to complete the statement. Round to the nearest hundredth, if necessary.

7. 7.5 in. ≈ ⬜ cm

8. 100 g ≈ ⬜ oz

9. 2 L ≈ ⬜ qt

10. 4 m ≈ ⬜ ft

Check It Out
Help with Homework
BigIdeasMath ✓com

 Vocabulary and Concept Check

1. **WRITING** What are three ways you can solve a proportion?

2. **OPEN-ENDED** Which way would you choose to solve $\dfrac{3}{x} = \dfrac{6}{14}$?
   Explain your reasoning.

3. **NUMBER SENSE** Does $\dfrac{x}{4} = \dfrac{15}{3}$ have the same solution as $\dfrac{x}{15} = \dfrac{4}{3}$?
   Use the Cross Products Property to explain your answer.

 Practice and Problem Solving

Use multiplication to solve the proportion.

① 4. $\dfrac{9}{5} = \dfrac{z}{20}$

5. $\dfrac{h}{15} = \dfrac{16}{3}$

6. $\dfrac{w}{4} = \dfrac{42}{24}$

7. $\dfrac{35}{28} = \dfrac{n}{12}$

8. $\dfrac{7}{16} = \dfrac{x}{4}$

9. $\dfrac{y}{9} = \dfrac{44}{54}$

Use the Cross Products Property to solve the proportion.

② 10. $\dfrac{a}{6} = \dfrac{15}{2}$

11. $\dfrac{10}{7} = \dfrac{8}{k}$

12. $\dfrac{3}{4} = \dfrac{v}{14}$

13. $\dfrac{5}{n} = \dfrac{16}{32}$

14. $\dfrac{36}{42} = \dfrac{24}{r}$

15. $\dfrac{9}{10} = \dfrac{d}{6.4}$

16. $\dfrac{x}{8} = \dfrac{3}{12}$

17. $\dfrac{8}{m} = \dfrac{6}{15}$

18. $\dfrac{4}{24} = \dfrac{c}{36}$

19. $\dfrac{20}{16} = \dfrac{d}{12}$

20. $\dfrac{30}{20} = \dfrac{w}{14}$

21. $\dfrac{2.4}{1.8} = \dfrac{7.2}{k}$

22. **ERROR ANALYSIS** Describe and correct the error
   in solving the proportion $\dfrac{m}{8} = \dfrac{15}{24}$.

$$✗ \quad \dfrac{m}{8} = \dfrac{15}{24}$$
$$8 \cdot m = 24 \cdot 15$$
$$m = 45$$

23. **PENS** Forty-eight pens are packaged in 4 boxes.
   How many pens are packaged in 9 boxes?

24. **PIZZA PARTY** How much does it cost to buy 10 medium pizzas?

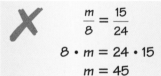
3 Medium Pizzas for $10.50

Solve the proportion.

25. $\dfrac{2x}{5} = \dfrac{9}{15}$

26. $\dfrac{5}{2} = \dfrac{d-2}{4}$

27. $\dfrac{4}{k+3} = \dfrac{8}{14}$

Assignment Guide and Homework Check

Level	Day 1 Activity Assignment	Day 2 Lesson Assignment	Homework Check
Basic	10–13, 39–43	1–3, 5–9 odd, 15–21 odd, 22, 23–27 odd	5, 17, 21, 23, 27
Average	10–13, 39–43	1–3, 5–9 odd, 15–21 odd, 22, 29, 30, 32–35	5, 17, 21, 29, 33
Advanced	10–13, 39–43	1–3, 4–8 even, 14–38 even	18, 26, 30, 32, 36
Accelerated	1–3, 4–8 even, 10–13, 14–38 even, 39–43		18, 26, 30, 32, 36

Common Errors

- **Exercises 4–9** Some students may multiply by the denominator of the fraction without the variable. Remind them that they are trying to get the variable alone, so they want to multiply both sides by the denominator of the fraction with the variable. Give students an example without a fraction on the other side of the equation to remind them of the process.

- **Exercises 10–21** Students may divide instead of multiply when finding the cross products, or they may multiply across the numerators and the denominators as if they were multiplying fractions. Remind students that the ratios have an equal sign between them, not a multiplication sign. Also tell them that when they use the Cross Products Property, it produces an "X," which means multiplication.

5.4 Record and Practice Journal

Use multiplication to solve the proportion.

1. $\frac{a}{40} = \frac{3}{10}$ 2. $\frac{6}{11} = \frac{c}{77}$ 3. $\frac{b}{65} = \frac{7}{13}$

   $a = 12$ $c = 42$ $b = 35$

Use the Cross Products Property to solve the proportion.

4. $\frac{k}{6} = \frac{8}{16}$ 5. $\frac{5.4}{7} = \frac{27}{h}$ 6. $\frac{8}{11} = \frac{4}{y+2}$

   $k = 3$ $h = 35$ $y = 3.5$

Write and solve a proportion to complete the statement.

7. 42 in. = _____ cm   8. 12.6 kg = _____ lb   9. 3 oz = _____ g

   106.68   28   84

10. A cell phone company charges $5 for 250 text messages. How much does the company charge for 300 text messages?

   $6

Vocabulary and Concept Check

1. mental math; Multiplication Property of Equality; Cross Products Property

2. *Sample answer:* mental math; Because $3 \cdot 2 = 6$, the product of x and 2 is 14. So, $x = 7$.

3. yes; Both cross products give the equation $3x = 60$.

Practice and Problem Solving

4. $z = 36$ 5. $h = 80$

6. $w = 7$ 7. $n = 15$

8. $x = 1\frac{3}{4}$ 9. $y = 7\frac{1}{3}$

10. $a = 45$ 11. $k = 5.6$

12. $v = 10.5$ 13. $n = 10$

14. $r = 28$

15. $d = 5.76$

16. $x = 2$

17. $m = 20$

18. $c = 6$

19. $d = 15$

20. $w = 21$

21. $k = 5.4$

22. They did not perform the cross multiplication properly.

    $$\frac{m}{8} = \frac{15}{24}$$

    $$m \cdot 24 = 8 \cdot 15$$

    $$m = 5$$

23. 108 pens

24. $35

25. $x = 1.5$

26. $d = 12$

27. $k = 4$

28. $\dfrac{6}{x} = \dfrac{1}{0.62}$; about 3.7

29. $\dfrac{2.5}{x} = \dfrac{1}{0.26}$; about 0.65

30. $\dfrac{90}{x} = \dfrac{1}{0.45}$; about 40.5

31. true; Both cross products give the equation $3a = 2b$.

32. $769.50   33. 15.5 lb

34. a. about 7.62 cm

    b. 16 mo

    c. 40 mo

35. no; The relationship is not proportional. It should take more people less time to build the swing set.

36. See Taking Math Deeper.

37. 4 bags

38. $\dfrac{1}{5}$; $\dfrac{m}{k} = \dfrac{\frac{n}{2}}{\frac{5n}{2}} = \dfrac{n}{2} \cdot \dfrac{2}{5n} = \dfrac{1}{5}$

Fair Game Review

39–42.

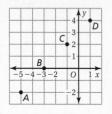

43. D

Mini-Assessment

Solve the proportion.

1. $\dfrac{x}{12} = \dfrac{3}{8}$ $x = 4.5$

2. $\dfrac{6}{11} = \dfrac{9}{m}$ $m = 16.5$

3. $\dfrac{6}{12} = \dfrac{c}{36}$ $c = 18$

4. $\dfrac{18}{3} = \dfrac{24}{b}$ $b = 4$

5. Thirty-six pencils are packed in three boxes. How many pencils are packed in five boxes?   60 pencils

Taking Math Deeper

Exercise 36

Sometimes it is a good suggestion to "forget about algebra and just answer the question." After answering the question, we might "relax" and try to look for an efficient or clever way to answer the question.

 Start with a ratio table. Adults : children is 5 : 3.

Adults	Children	Total
5	3	8
10	6	16
15	9	24

② I can see this will take a while. I'm going to jump ahead in the ratio table.

Adults	Children	Total
100	60	160 too many
80	48	128 too few
90	54	144 just right

③ The answer is 90 adults. I wonder if I can get the answer algebraically.

Adults = $5x$ Children = $3x$

$$5x + 3x = 144$$
$$8x = 144$$
$$x = 18$$

Adults = $5 \cdot 18 = 90$

Cool

Reteaching and Enrichment Strategies

If students need help. . .	If students got it. . .
Resources by Chapter • Practice A and Practice B • Puzzle Time Record and Practice Journal Practice Differentiating the Lesson Lesson Tutorials Skills Review Handbook	Resources by Chapter • Enrichment and Extension • Technology Connection Start the next section

Write and solve a proportion to complete the statement. Round to the nearest hundredth if necessary.

3 **28.** 6 km ≈ [   ] mi

29. 2.5 L ≈ [   ] gal

30. 90 lb ≈ [   ] kg

31. TRUE OR FALSE? Tell whether the statement is *true* or *false*. Explain.

If $\dfrac{a}{b} = \dfrac{2}{3}$, then $\dfrac{3}{2} = \dfrac{b}{a}$.

32. CLASS TRIP It costs $95 for 20 students to visit an aquarium. How much does it cost for 162 students?

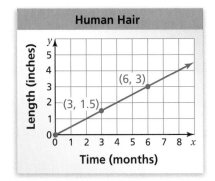

Human Hair

Length (inches) vs *Time (months)*, points (3, 1.5) and (6, 3)

33. GRAVITY A person who weighs 120 pounds on Earth weighs 20 pounds on the Moon. How much does a 93-pound person weigh on the Moon?

34. HAIR The length of human hair is proportional to the number of months it has grown.

    a. What is the hair length in *centimeters* after 6 months?

    b. How long does it take hair to grow 8 inches?

    c. Use a different method than the one in part (b) to find how long it takes hair to grow 20 inches.

35. SWING SET It takes 6 hours for 2 people to build a swing set. Can you use the proportion $\dfrac{2}{6} = \dfrac{5}{h}$ to determine the number of hours h it will take 5 people to build the swing set? Explain.

36. REASONING There are 144 people in an audience. The ratio of adults to children is 5 to 3. How many are adults?

37. PROBLEM SOLVING Three pounds of lawn seed covers 1800 square feet. How many bags are needed to cover 8400 square feet?

38. **Critical Thinking** Consider the proportions $\dfrac{m}{n} = \dfrac{1}{2}$ and $\dfrac{n}{k} = \dfrac{2}{5}$.

What is the ratio $\dfrac{m}{k}$? Explain your reasoning.

 Fair Game Review What you learned in previous grades & lessons

Plot the ordered pair in a coordinate plane. *(Skills Review Handbook)*

39. $A(-5, -2)$

40. $B(-3, 0)$

41. $C(-1, 2)$

42. $D(1, 4)$

43. MULTIPLE CHOICE Which expression is equivalent to $(3w - 8) - 4(2w + 3)$? *(Section 3.2)*

    A $11w + 4$ **B** $-5w - 5$ **C** $-5w + 4$ **D** $-5w - 20$

5.5 Slope

Essential Question How can you compare two rates graphically?

1 ACTIVITY: Comparing Unit Rates

Work with a partner. The table shows the maximum speeds of several animals.

a. Find the missing speeds. Round your answers to the nearest tenth.

b. Which animal is fastest? Which animal is slowest?

c. Explain how you convert between the two units of speed.

Animal	Speed (miles per hour)	Speed (feet per second)
Antelope	61.0	
Black mamba snake		29.3
Cheetah		102.6
Chicken		13.2
Coyote	43.0	
Domestic pig		16.0
Elephant		36.6
Elk		66.0
Giant tortoise	0.2	
Giraffe	32.0	
Gray fox		61.6
Greyhound	39.4	
Grizzly bear		44.0
Human		41.0
Hyena	40.0	
Jackal	35.0	
Lion		73.3
Peregrine falcon	200.0	
Quarter horse	47.5	
Spider		1.76
Squirrel	12.0	
Thomson's gazelle	50.0	
Three-toed sloth		0.2
Tuna	47.0	

Slope

In this lesson, you will
- find the slopes of lines.
- interpret the slopes of lines as rates.

Laurie's Notes

What Your Students Will Learn

- Find the slopes of lines using graphs.
- Interpret the slopes of lines as rates by using tables.

Previous Learning

Students should know how to write and simplify ratios, how to convert measures, and how to graph proportional relationships.

Introduction

Applying Mathematical Practices

- **Model with Mathematics:** Mathematically proficient students use a graph as a tool to show the relationship between two quantities.

Motivate

- If students are not comfortable with dimensional analysis from the previous lesson, you will need to help them get started with this activity.
- **Big Idea:** You begin with a speed in certain units (i.e., miles per hour) and you want a speed with different units (i.e., feet per second).
- You need to set up the factors so that the unwanted units divide out. The units you want to divide out should always appear diagonally from one another. In your final answer, one of the units should end up in the numerator and the other should end up in the denominator.
- Example: $\dfrac{61 \text{ miles}}{\text{hour}} \times \dfrac{1 \text{ hour}}{3600 \text{ seconds}} \times \dfrac{5280 \text{ feet}}{1 \text{ mile}}$
- Because 1 hour = 3600 seconds, multiplying by $\dfrac{1 \text{ hour}}{3600 \text{ seconds}}$ or $\dfrac{3600 \text{ seconds}}{1 \text{ hour}}$ is equivalent to multiplying by 1.
- Divide out by a common factor of 240, and the answer is $89\dfrac{7}{15}$ feet per second.
- Converting from feet per second to miles per hour will require students to multiply by the reciprocal of each of the conversion factors.

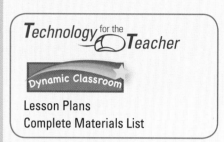

Technology for the Teacher

Dynamic Classroom

Lesson Plans
Complete Materials List

Activity Notes

Activity 1

- It would be appropriate for students to work in pairs and use a calculator to complete the activity.
- **?** "Look through the list and predict the fastest animal. Mark it with the letter F. Predict the slowest animal. Mark it with the letter S."
- Did students select the Peregrine falcon or cheetah as the fastest?
- Did students select the giant tortoise or three-toed sloth as the slowest?
- **Look for and Express Regularity in Repeated Reasoning:** You may want to help students see how to convert from miles per hour to feet per second and vice versa.

$$\underbrace{\dfrac{\text{miles}}{\text{hour}} \times \dfrac{1 \text{ hour}}{3600 \text{ seconds}}}_{\text{equals } 1} \times \underbrace{\dfrac{5280 \text{ feet}}{1 \text{ mile}}}_{\text{equals } 1}$$

- When students have finished, discuss the results, the answer to part (c), and their predictions.

5.5 Record and Practice Journal

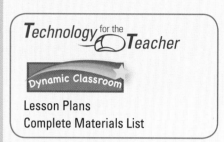

Essential Question How can you compare two rates graphically?

1 ACTIVITY: Comparing Unit Rates

Work with a partner. The table shows the maximum speeds of several animals.

a. Find the missing speeds. Round your answers to the nearest tenth.

b. Which animal is fastest? Which animal is slowest? peregrine falcon; three-toed sloth

c. Explain how you convert between the two units of speed.
See Additional Answers.

Animal	Speed (miles per hour)	Speed (feet per second)
Antelope	61.0	89.5
Black mamba snake	20.0	29.3
Cheetah	70.0	102.6
Chicken	9.0	13.2
Coyote	43.0	63.1
Domestic pig	10.9	16.0
Elephant	25.0	36.6
Elk	45.0	66.0
Giant tortoise	0.2	0.3
Giraffe	32.0	46.9
Gray fox	42.0	61.6
Greyhound	39.4	57.8
Grizzly bear	30.0	44.0
Human	28.0	41.0
Hyena	40.0	58.7
Jackal	35.0	51.3
Lion	50.0	73.3
Peregrine falcon	200.0	293.3
Quarter horse	47.5	69.7
Spider	1.2	1.76
Squirrel	12.0	17.6
Thomson's gazelle	50.0	73.3
Three-Toed sloth	0.1	0.2
Tuna	47.0	68.9

English Language Learners

Labels

English learners may recognize the fraction bar as division, but may not be familiar with its use in the concept of rate. The following unit rates are equivalent.

$\frac{3 \text{ m}}{1 \text{ h}}$, 3 m/h, 3 meters per hour

Each of these rates can be read as "three meters *for every* hour."

5.5 Record and Practice Journal

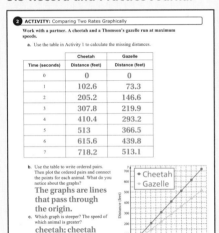

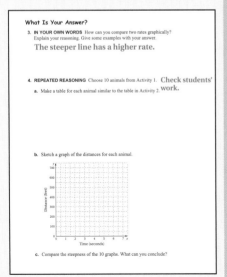

Laurie's Notes

Activity 2

- *Note:* The formal definition of **slope** will come in the lesson. The idea in this activity is to get students to understand that a steeper line translates to a greater speed.
- Discuss the concept of a constant speed. Talk about a car on cruise control or items traveling on an assembly line in a factory. Over short periods of time it is possible for animals to run at a constant speed.
- **?** "If a cheetah is running at a constant speed and has traveled a distance of 102.6 feet in 1 second, how far will it run in two seconds?" 205.2 ft
- Have students complete the table. This is a good review of decimal multiplication or decimal addition (depending on how the student completes the work). You may find it helpful to review the rules for multiplying and adding decimals beforehand.
- **?** **Discuss:** "After 3 seconds, how far has each animal run? after 7 seconds?"
- **?** "Is there ever a time when the gazelle has run farther than the cheetah?"
- Students might find it helpful to use a straightedge or ruler to connect their points.
- **?** "If the cheetah was not running at a constant speed, would your graph look different? Explain." Yes. The points would not be in a line. The graph would go up and down accordingly.
- **Model with Mathematics:** Discuss the relationship between the speed of animals and the steepness of the two graphs. Students should be able to tell from the graph which animal has the greater speed.
- **Model with Mathematics:** Some students may recognize which graph will be steeper after graphing the first two points for each animal. They may also remember from earlier in the chapter that in the graph of a proportional relationship, you can find the unit rate y from the point $(1, y)$.

What Is Your Answer?

- Question 4 can be done as homework.

Closure

- An airplane is traveling at a constant speed of 6 miles per minute. Make a table to show the distance traveled each minute for 8 minutes. Graph your data and connect the points. Then describe the graph.

Minutes, x	0	1	2	3	4	5	6	7	8
Miles, y	0	6	12	18	24	30	36	42	48

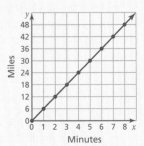

a line that passes through the origin

ACTIVITY: Comparing Two Rates Graphically

Math Practice

Apply Mathematics

How can you use the graph to determine which animal has the greater speed?

Work with a partner. A cheetah and a Thomson's gazelle run at maximum speed.

a. Use the table in Activity 1 to calculate the missing distances.

Time (seconds)	Cheetah Distance (feet)	Gazelle Distance (feet)
0		
1		
2		
3		
4		
5		
6		
7		

b. Use the table to write ordered pairs. Then plot the ordered pairs and connect the points for each animal. What do you notice about the graphs?

c. Which graph is steeper? The speed of which animal is greater?

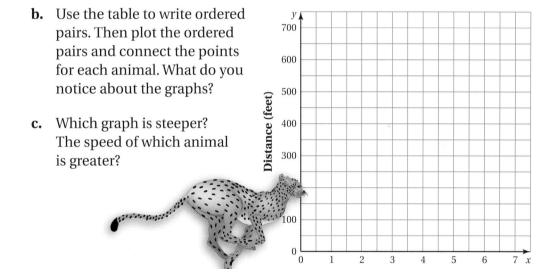

What Is Your Answer?

3. **IN YOUR OWN WORDS** How can you compare two rates graphically? Explain your reasoning. Give some examples with your answer.

4. **REPEATED REASONING** Choose 10 animals from Activity 1.

   a. Make a table for each animal similar to the table in Activity 2.

   b. Sketch a graph of the distances for each animal.

   c. Compare the steepness of the 10 graphs. What can you conclude?

Key Vocabulary ◀))
slope, *p. 194*

Study Tip

The slope of a line is the same between any two points on the line because lines have a *constant* rate of change.

🔑 Key Idea

Slope

Slope is the rate of change between any two points on a line. It is a measure of the *steepness* of a line.

To find the slope of a line, find the ratio of the change in y (vertical change) to the change in x (horizontal change).

$$\text{slope} = \frac{\text{change in } y}{\text{change in } x}$$

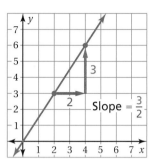

$$\text{Slope} = \frac{3}{2}$$

EXAMPLE 1 **Finding Slopes**

Find the slope of each line.

a.

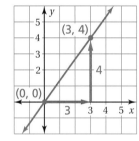

$$\text{slope} = \frac{\text{change in } y}{\text{change in } x}$$

$$= \frac{4}{3}$$

∴ The slope of the line is $\frac{4}{3}$.

b.

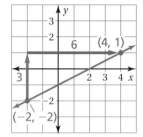

$$\text{slope} = \frac{\text{change in } y}{\text{change in } x}$$

$$= \frac{3}{6} = \frac{1}{2}$$

∴ The slope of the line is $\frac{1}{2}$.

🔵 On Your Own

Now You're Ready
Exercises 4–9

Find the slope of the line.

1.

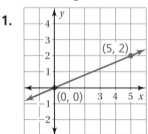

2.

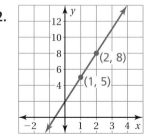

◀)) Multi-Language Glossary at BigIdeasMath✓com

Laurie's Notes

Introduction

Connect

- **Yesterday:** Students explored constant rates of speed in a table and on a graph.
- **Today:** Students will define slope and determine the slope of a line from its graph.

Motivate

- Ask students about the slope of a half-pipe at a skateboard park, the slope of a local hill that they are familiar with, or the slope of a wheelchair ramp at school. (Chances are they will all have different slopes.)
- **?** "What does the word slope mean when talking about a wheelchair ramp?" Listen for words such as "steepness" or "incline." Students often use their hands to demonstrate slope.

Lesson Notes

Key Idea

- Write the definition of slope on the board.
- Remind students that a rate is a ratio. Slopes are often thought of as rates.
- **FYI:** You may remember learning that the formula for the slope of a line passing through the points (x_1, y_1) and (x_2, y_2) is $\dfrac{y_2 - y_1}{x_2 - x_1}$. Students are not formally introduced to this formula until future courses. Try to enforce the concept rather than teach the formula.

Example 1

- **Use Appropriate Tools Strategically:** The arrows on the graphs are good visual aids to help students think about how much each variable changes.
- Writing the amount of change on the graph is good reinforcement.
- At this stage, have them think about reading the graph left-to-right.
- **?** "For the second graph, what would the slope be if you used the points $(0, -1)$ and $(4, 1)$?" the same, $\dfrac{2}{4} = \dfrac{1}{2}$
- **Common Error:** A very common error is for students to find slope by writing the change in x over the change in y.

On Your Own

- **Neighbor Check:** Have students work independently and then have their neighbors check their work. Have students discuss any discrepancies.

Goal Today's lesson is determining the **slope** of a line, when given a graph.

Lesson Tutorials
Lesson Plans
Answer Presentation Tool

Extra Example 1

Find the slope of each line.

a. 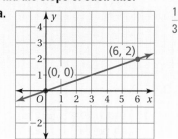 $\dfrac{1}{3}$

b. 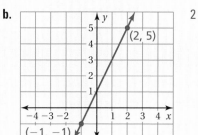 2

On Your Own

1. $\dfrac{2}{5}$

2. 3

Extra Example 2

The table shows your earnings for mowing lawns.

Lawns Mowed, x	3	6	9	12
Earnings, y (dollars)	45	90	135	180

a. Graph the data.

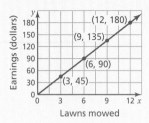

b. Find and interpret the slope of the line through the points. 15; You earn $15 per lawn mowed.

On Your Own

3. 5; No

4. a. Your friend's line is steeper. Your friend's pay rate is greater than yours.

   b. 7; Your friend earns $7 per hour babysitting.

Differentiated Instruction

Auditory

Emphasize that slope relates to rate problems. The slope formula expresses how much the y-coordinate changes for a given change in the x-coordinate. For example, weighing out 3 pounds of apples and paying $4.05 is equivalent to the unit rate of $1.35 per pound. So, the change in the cost is $1.35 for every one pound increase of apples.

Example 2

- Students must first read and understand the data (table of ordered pairs), and then plot the data.
- Start to use language that will prepare students for future courses.
- The number of hours worked is *x*, and the amount of dollars earned is *y*. "The amount of money earned depends upon how many hours you worked."
- **Construct Viable Arguments and Critique the Reasoning of Others:** Slopes are often rates. Here, the rate is *dollars per hour*. Be sure to have students explain the meaning of this concept. Listen for an understanding of the connection between *slope* and *unit rate*.
- **Big Idea:** Students will often be asked to interpret a slope. This means to look at the context of the problem and decide what a slope means with regard to the two variables.

On Your Own

- Share answers as a class.
- **Extension:** "Read information from the graph. How many hours did you work to earn $30?" 6

Closure

- The cost of admission to the local museum is given in the table. Graph the data and determine the slope. Interpret the slope in the context of the problem.

Number of People, x	2	3	4
Total Cost of Admission, y	$4.50	$6.75	$9.00

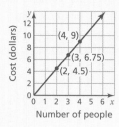

slope: 2.25; The slope represents the cost per person.

EXAMPLE 2   Interpreting a Slope

The table shows your earnings for babysitting.

a. **Graph the data.**

b. **Find and interpret the slope of the line through the points.**

Hours, *x*	0	2	4	6	8	10
Earnings, *y* (dollars)	0	10	20	30	40	50

a. Graph the data. Draw a line through the points.

b. Choose any two points to find the slope of the line.

$$\text{slope} = \frac{\text{change in } y}{\text{change in } x}$$

$$= \frac{20}{4} \quad \leftarrow \boxed{\text{dollars}}$$
$$\qquad\quad \leftarrow \boxed{\text{hours}}$$

$$= 5$$

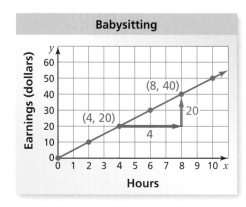

Babysitting

The slope of the line represents the unit rate. The slope is 5. So, you earn $5 per hour babysitting.

On Your Own

Now You're Ready
Exercises 10 and 11

3. In Example 2, use two other points to find the slope. Does the slope change?

4. The graph shows the amounts you and your friend earn babysitting.

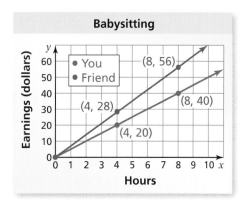

Babysitting

a. Compare the steepness of the lines. What does this mean in the context of the problem?

b. Find and interpret the slope of the blue line.

 Vocabulary and Concept Check

1. **VOCABULARY** Is there a connection between rate and slope? Explain.

2. **REASONING** Which line has the greatest slope?

3. **REASONING** Is it more difficult to run up a ramp with a slope of $\frac{1}{5}$ or a ramp with a slope of 5? Explain.

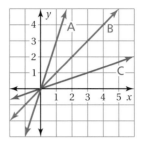

 Practice and Problem Solving

Find the slope of the line.

① 4.
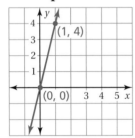

5.
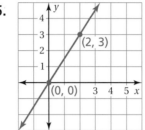

6.
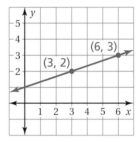

7.
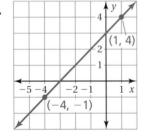

8.
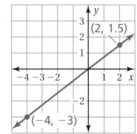

9.
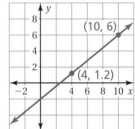

Graph the data. Then find and interpret the slope of the line through the points.

② 10.

Minutes, x	3	5	7	9
Words, y	135	225	315	405

11.

Gallons, x	5	10	15	20
Miles, y	162.5	325	487.5	650

12. **ERROR ANALYSIS** Describe and correct the error in finding the slope of the line passing through (0, 0) and (4, 5).

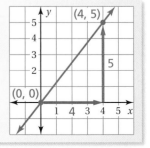

Assignment Guide and Homework Check

Level	Day 1 Activity Assignment	Day 2 Lesson Assignment	Homework Check
Basic	20–23	1–3, 5–11 odd, 12, 13–17 odd	2, 5, 11, 17
Average	20–23	1–3, 5–11 odd, 12, 14–17	2, 5, 11, 16
Advanced	20–23	1–3, 4–18 even	4, 10, 16, 18
Accelerated	1–3, 4–18 even, 20–23		4, 10, 16, 18

Common Errors

- **Exercises 4–11** Students may put the change in *x* over the change in *y*. Remind them that the vertical change is written over the horizontal change and that slope is a rate. Tell the students to label the axes with units that represent a common rate to help them remember which change goes on top. For example, label the *y*-axis "miles" and the *x*-axis "gallons." This should help students identify that the change in *y* goes on top because miles is first in the rate.
- **Exercises 13–15** Students may not remember how to plot ordered pairs with negative numbers. Remind students of Quadrants II, III, and IV, and the signs of the coordinates in each quadrant.

5.5 Record and Practice Journal

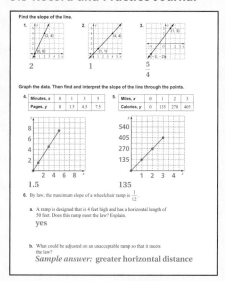

 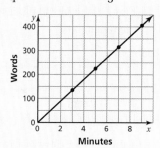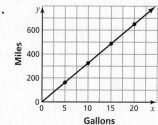

Practice and Problem Solving

13–15. See Additional Answers.

16. See *Taking Math Deeper*.

17. See Additional Answers.

18. 0; The change in *y* is 0 because the *y*-values do not change. So, the slope is 0.

19. $y = 6$

 Fair Game Review

20. $-\dfrac{4}{5}$ **21.** $-\dfrac{3}{5}$

22. 3   **23.** C

Mini-Assessment

Find the slope of the line that passes through the two points.

1. (0, 0), (3, 2) $\dfrac{2}{3}$

2. (−2, −2), (5, 5) 1

3. (−3, −4), (6, 8) $\dfrac{4}{3}$

4. (−4, −2), (−2, −1) $\dfrac{1}{2}$

5. The graph shows the amount of money you are saving for a computer.

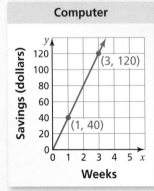

Computer

a. Find the slope of the line. 40

b. Interpret the slope of the line in the context of the problem. You are saving $40 per week.

Taking Math Deeper

Exercise 16

This is a classic type of problem that uses linear models to predict future events. Each person is saving money at a constant rate (constant slope). The fact that the rate is constant is what makes the graph a line. The prediction of when $165 will be saved assumes that the constant rate continues into the future.

 Interpret the slope in context.

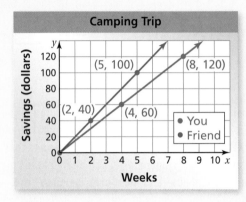

Camping Trip

a. Your friend's graph is steeper than yours. So, your friend's saving rate (in dollars per week) is greater than yours.

 Find the slope of each line.

b.
	Slopes	Rates

You: $\dfrac{60}{4}$ = $15 per week

Friend: $\dfrac{60}{3}$ = $20 per week

c. Your friend saves $20 − $15 = $5 more per week.

Slope = rate

 How long will it take for you to save $165?

d. At $15 per week, it will take $\dfrac{165}{15}$ = 11 weeks.

Reteaching and Enrichment Strategies

If students need help. . .	If students got it. . .
Resources by Chapter • Practice A and Practice B • Puzzle Time Record and Practice Journal Practice Differentiating the Lesson Lesson Tutorials Skills Review Handbook	Resources by Chapter • Enrichment and Extension • Technology Connection Start the next section

Graph the line that passes through the two points. Then find the slope of the line.

13. $(0, 0), \left(\dfrac{1}{3}, \dfrac{7}{3}\right)$

14. $\left(-\dfrac{3}{2}, -\dfrac{3}{2}\right), \left(\dfrac{3}{2}, \dfrac{3}{2}\right)$

15. $\left(1, \dfrac{5}{2}\right), \left(-\dfrac{1}{2}, -\dfrac{1}{4}\right)$

16. CAMPING The graph shows the amount of money you and a friend are saving for a camping trip.

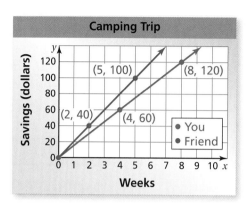

Camping Trip

   a. Compare the steepness of the lines. What does this mean in the context of the problem?

   b. Find the slope of each line.

   c. How much more money does your friend save each week than you?

   d. The camping trip costs $165. How long will it take you to save enough money?

17. MAPS An atlas contains a map of Ohio. The table shows data from the key on the map.

Distance on Map (mm), x	10	20	30	40
Actual Distance (mi), y	25	50	75	100

   a. Graph the data.

   b. Find the slope of the line. What does this mean in the context of the problem?

   c. The map distance between Toledo and Columbus is 48 millimeters. What is the actual distance?

   d. Cincinnati is about 225 miles from Cleveland. What is the distance between these cities on the map?

18. CRITICAL THINKING What is the slope of a line that passes through the points $(2, 0)$ and $(5, 0)$? Explain.

19. **Number Sense** A line has a slope of 2. It passes through the points $(1, 2)$ and $(3, y)$. What is the value of y?

Fair Game Review What you learned in previous grades & lessons

Multiply. *(Section 2.4)*

20. $-\dfrac{3}{5} \times \dfrac{8}{6}$

21. $1\dfrac{1}{2} \times \left(-\dfrac{6}{15}\right)$

22. $-2\dfrac{1}{4} \times \left(-1\dfrac{1}{3}\right)$

23. MULTIPLE CHOICE You have 18 stamps from Mexico in your stamp collection. These stamps represent $\dfrac{3}{8}$ of your collection. The rest of the stamps are from the United States. How many stamps are from the United States? *(Section 3.4)*

   Ⓐ 12   Ⓑ 24   Ⓒ 30   Ⓓ 48

5.6 Direct Variation

Essential Question

How can you use a graph to show the relationship between two quantities that vary directly? How can you use an equation?

1 ACTIVITY: Math in Literature

Gulliver's Travels was written by Jonathan Swift and published in 1726. Gulliver was shipwrecked on the island Lilliput, where the people were only 6 inches tall. When the Lilliputians decided to make a shirt for Gulliver, a Lilliputian tailor stated that he could determine Gulliver's measurements by simply measuring the distance around Gulliver's thumb. He said "Twice around the thumb equals once around the wrist. Twice around the wrist is once around the neck. Twice around the neck is once around the waist."

Direct Variation

In this lesson, you will

- identify direct variation from graphs or equations.
- use direct variation models to solve problems.

Work with a partner. Use the tailor's statement to complete the table.

Thumb, t	Wrist, w	Neck, n	Waist, x
0 in.			
1 in.			
	4 in.		
		12 in.	
			32 in.
	10 in.		

Laurie's Notes

Introduction

Applying Mathematical Practices

- **Model with Mathematics:** In this lesson, students will use a graph as a tool to show that two quantities vary directly.

Motivate

- Write the following table.

	Paper	Pencil	$1 Bill	Stick of Gum
Length (inches)	11	8	6	?
Length (centimeters)	27.94	20.32	?	6

- ❓ "Does anyone have an idea of how to find the missing values in the table without measuring?" number of inches × 2.54 = number of centimeters
- This question is checking to see if students remember the relationship between inches and centimeters. It is also a perfect example for direct variation.

Activity Notes

Discuss

- **Big Idea:** When two quantities *vary directly*, the ratio of one quantity to another is a *constant*. The term *direct variation* will be introduced in tomorrow's lesson.
- The ordered pair (0, 0) is always a solution of an equation describing two quantities that vary directly.
- When solutions of the equation are plotted, the *constant ratio* is the slope of the line.
- When two quantities vary directly, it can also be said that they are directly proportional.

Activity 1

- Ask a student to read the introduction.
- ❓ "Have any of you read *Gulliver's Travels*?" Give students an opportunity to share information about the story if they have read it.
- Students should work with partners to complete the table.
- ❓ "Are there any patterns in the table? Describe the patterns." **Construct Viable Arguments and Critique the Reasoning of Others:** There are many patterns! Listen for patterns in a single column, between adjacent columns, and between non-adjacent columns. Students might also mention the only odd numbers in the table appear in the first column. Ask students whether they can explain each pattern observed. In many cases, they will refer to "twice around the thumb equals twice around the wrist . . . ."

What Your Students Will Learn

- Identify direct variation from tables, graphs, or equations.
- Use direct variation models to solve problems.

Previous Learning

Students have used a variable to write an expression and have plotted points in the coordinate plane. Students have also written equations in two variables.

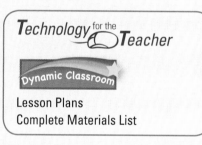

Technology for the *Teacher*

Dynamic Classroom

Lesson Plans
Complete Materials List

5.6 Record and Practice Journal

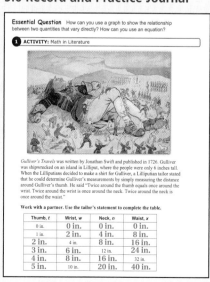

Essential Question How can you use a graph to show the relationship between two quantities that vary directly? How can you use an equation?

1 ACTIVITY: Math in Literature

Gulliver's Travels was written by Jonathan Swift and published in 1726. Gulliver was shipwrecked on an island in Lilliput, where the people were only 6 inches tall. When the Lilliputians decided to make a shirt for Gulliver, a Lilliputian tailor stated that he could determine Gulliver's measurements by simply measuring the distance around Gulliver's thumb. He said "Twice around the thumb equals once around the wrist. Twice around the wrist is once around the neck. Twice around the neck is once around the waist."

Work with a partner. Use the tailor's statement to complete the table.

Thumb, t	Wrist, w	Neck, n	Waist, x
0 in.	0 in.	0 in.	0 in.
1 in.	2 in.	4 in.	8 in.
2 in.	4 in.	8 in.	16 in.
3 in.	6 in.	12 in.	24 in.
4 in.	8 in.	16 in.	32 in.
5 in.	10 in.	20 in.	40 in.

Auditory

A common mistake when plotting points is to confuse the order of the coordinates in the ordered pair. The simple phrase "over and up" will assist the students in moving over the *x*-axis with the first number, and then moving up with the second number.

5.6 Record and Practice Journal

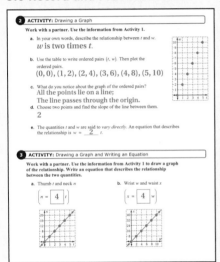

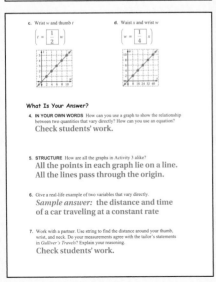

Laurie's Notes

Activity 2

- Many skills and concepts from this chapter are integrated in this activity.
- When asked to describe the relationship between *t* and *w*, students may say "twice around the thumb equals once around the wrist." **Attend to Precision:** Be sure students understand that what is being referred to is the *distance around* the thumb and the wrist.
- ? "What does the ordered pair (0, 0) mean?" In the context of the problem, it does not make sense to talk about distances around the thumb and the wrist that are equal to 0. Each is getting closer to 0.
- Because the line passes through (0, 0) and (1, 2), some students might quickly note that the slope is 2, remembering that in the graph of a proportional relationship, you can find the unit rate *y* from the point (1, *y*).
- ? What is the relationship between the coefficient of *t* in part (e) and the slope? They are the same.

Activity 3

- **Scaffolding:** If time is short, you may need to scaffold this activity. Have different groups do a different problem and record a sample of each on the board.
- **Common Error:** Students may reverse the ordered pairs (wrist length, waist length) versus (waist length, wrist length).
- Remind students that as the problems are written, the first quantity mentioned is the first coordinate.
- They should do a test point to see if it satisfies the equation they wrote.
- **Connection:** There are two pairs of equations that share a pattern: $w = 2t$ and $t = \frac{1}{2}w$; $x = 4w$ and $w = \frac{1}{4}x$. The slopes are reciprocals.

What Is Your Answer?

- Question 6: Students are sometimes stumped by this question, yet it is very common. For instance, the cost of an item varies directly with how many items are purchased (1 newspaper, $1.25), (2 newspapers, $2.50), and so on.
- Question 7: Leave sufficient time for this problem.

Closure

- **Exit Ticket:** Write an equation that describes the relationship between the length of objects measured in inches *x* and measured in centimeters *y*.
  $y = 2.54x$

2 ACTIVITY: Drawing a Graph

Work with a partner. Use the information from Activity 1.

a. In your own words, describe the relationship between t and w.

b. Use the table to write the ordered pairs (t, w). Then plot the ordered pairs.

c. What do you notice about the graph of the ordered pairs?

d. Choose two points and find the slope of the line between them.

e. The quantities t and w are said to *vary directly*. An equation that describes the relationship is

$$w = \boxed{}\, t.$$

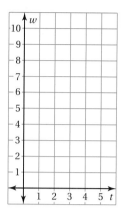

3 ACTIVITY: Drawing a Graph and Writing an Equation

Math Practice

Label Axes

How do you know which labels to use for the axes? Explain.

Work with a partner. Use the information from Activity 1 to draw a graph of the relationship. Write an equation that describes the relationship between the two quantities.

a. Thumb t and neck n $(n = \boxed{}\, t)$

b. Wrist w and waist x $(x = \boxed{}\, w)$

c. Wrist w and thumb t $(t = \boxed{}\, w)$

d. Waist x and wrist w $(w = \boxed{}\, x)$

What Is Your Answer?

4. **IN YOUR OWN WORDS** How can you use a graph to show the relationship between two quantities that vary directly? How can you use an equation?

5. **STRUCTURE** How are all the graphs in Activity 3 alike?

6. Give a real-life example of two variables that vary directly.

7. Work with a partner. Use string to find the distance around your thumb, wrist, and neck. Do your measurements agree with the tailor's statement in *Gulliver's Travels*? Explain your reasoning.

Practice

Use what you learned about quantities that vary directly to complete Exercises 4 and 5 on page 202.

Key Vocabulary 🔊

direct variation, *p. 200*

constant of proportionality, *p. 200*

 **Key Idea**

Direct Variation

Words Two quantities x and y show **direct variation** when $y = kx$, where k is a number and $k \neq 0$. The number k is called the **constant of proportionality**.

Graph The graph of $y = kx$ is a line with a slope of k that passes through the origin. So, two quantities that show direct variation are in a proportional relationship.

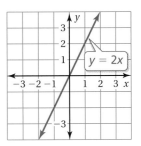

EXAMPLE ① **Identifying Direct Variation**

Tell whether x and y show direct variation. Explain your reasoning.

a.

x	1	2	3	4
y	−2	0	2	4

b.

x	0	2	4	6
y	0	2	4	6

Plot the points. Draw a line through the points.

Plot the points. Draw a line through the points.

Study Tip

Other ways to say that x and y show direct variation are "y varies directly with x" and "x and y are directly proportional."

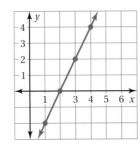

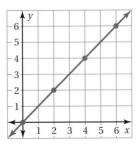

⋮⋮ The line does *not* pass through the origin. So, x and y do *not* show direct variation.

⋮⋮ The line passes through the origin. So, x and y show direct variation.

EXAMPLE ② **Identifying Direct Variation**

Tell whether x and y show direct variation. Explain your reasoning.

a. $y + 1 = 2x$

$y = 2x - 1$ Solve for y.

b. $\dfrac{1}{2}y = x$

$y = 2x$ Solve for y.

⋮⋮ The equation *cannot* be written as $y = kx$. So, x and y do *not* show direct variation.

⋮⋮ The equation can be written as $y = kx$. So, x and y show direct variation.

🔊 Multi-Language Glossary at BigIdeasMath ✓ com

Laurie's Notes

Introduction

Connect

- **Yesterday:** Students completed a table of values, plotted ordered pairs, and wrote equations as they developed an understanding of quantities that vary directly.
- **Today:** Students will use a formal definition of direct variation.

Motivate

- **?** Some states have returnable bottle laws. "If you receive $0.05 for each bottle, how much money do you receive for 4 bottles? 10 bottles?" $0.20; $0.50
- Have students make a table to show the relationship between the number *x* of bottles collected and the amount *y* of money received.
- Then have students make a quick sketch of the ordered pairs.
- Observe that (0, 0) is on the graph and the ordered pairs lie on a line.

Lesson Notes

Key Idea

- Write the Key Idea on the board. Remind students throughout the lesson that *k* is the **constant of proportionality**, and it is also the slope of the line.
- The equation $y = kx$ can be confusing to students. They see three variables. Remind them that this is the *general form*. The variables *y* and *x* will remain in the final equation but *k* will be replaced by a number.
- Examples of equations in general form include $y = 2x$ and $y = 1.25x$. Point out to students that *k* has been replaced with a number.
- **?** "Why should $k \neq 0$?" If *k* did equal 0, the resulting equation would be $y = 0 \cdot x$ or $y = 0$. While *k* has been replaced with a number and *y* still appears in the final equation, *x* no longer does. Because there is no *x*, the equation is not in general form and does not show direct variation.
- Mention a key feature of this graph: it passes through the origin.

Example 1

- **Common Error:** Students may look at the table of values and believe that the first example is not direct variation simply because *x* and *y* are increasing by different rates in the table. *x* is increasing by 1 and *y* is increasing by 2.
- **Connection:** When two variables vary directly, you can also say that they vary proportionally.
- **Extension:** Ask students to find the slope of the line in part (b).

Example 2

- This example requires students to recall equations in two variables.
- Students need to think about how they solved equations using the Properties of Equality.
- **Extension:** Ask students whether the point (0, 0) satisfies either equation.

Goal Today's lesson is using a formal definition of **direct variation**.

Lesson Tutorials
Lesson Plans
Answer Presentation Tool

Extra Example 1

Tell whether *x* and *y* show direct variation. Explain your reasoning.

a.

x	1	2	3	4
y	−1	0	1	2

The line does not pass through the origin. So, *x* and *y* do not show direct variation.

b.

x	0	3	6	9
y	0	1	2	3

The line passes through the origin. So, *x* and *y* show direct variation.

Extra Example 2

Tell whether *x* and *y* show direct variation. Explain your reasoning.

a. $y - 6 = 3x$

The equation cannot be written as $y = kx$. So, *x* and *y* do not show direct variation.

b. $x = 4y$

The equation can be written as $y = kx$. So, *x* and *y* show direct variation.

T-200

On Your Own

1. no; The line does not pass through the origin.

2. yes; The line passes through the origin.

3. no; The points do not lie on a line.

4–6. See Additional Answers.

Extra Example 3

The table shows the area y (in square feet) that a power paint sprayer can paint in x minutes.

x	$\frac{1}{2}$	1	$\frac{3}{2}$	2
y	10	20	30	40

a. Graph the data. Tell whether x and y are directly proportional.

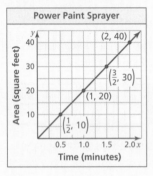

yes

b. Write an equation that represents the line. $y = 20x$

c. Use the equation to find the area painted in 10 minutes.
200 square feet

On Your Own

7. no; There is not a constant rate of change.

English Language Learners

Vocabulary

Students may confuse the words *variation* and *variable*. A variable is a number that changes and is represented by a letter. Stress that variation refers to how the variable y *varies* in relation to the variable x.

On Your Own

- **Think-Pair-Share:** Students should read each question independently and then work in pairs to answer the questions. When they have answered the questions, the pair should compare their answers with another group and discuss any discrepancies.

- If students have difficulty with Questions 4–6, they could make a quick table of values and plot the ordered pairs.

Example 3

- Ask students to interpret an ordered pair in the table. For instance, $\left(\frac{1}{2}, 8\right)$ means that in $\frac{1}{2}$ minute, the robotic vacuum can clean 8 square feet.

- **Model with Mathematics:** Ask students to explain why the table of values is a ratio table, and to identify the unit rate. Plotting the ordered pairs confirms that x and y are directly proportional.

? "What is the constant of proportionality?" 16

? "What is the slope of the line?" 16

? "What is the equation of the line?" $y = 16x$

- Students can use the equation to find the area cleaned for any amount of time.

On Your Own

- **Neighbor Check:** Have students work independently and then have their neighbors check their work. Have students discuss any discrepancies.

Closure

- Tell whether x and y show direct variation. Explain your reasoning.

x	y
1	0
3	2
5	4
7	6

no; The line passes through (1, 0), so it does not pass through the origin.

On Your Own

Now You're Ready
Exercises 6–17

Tell whether x and y show direct variation. Explain your reasoning.

1.

x	y
0	−2
1	1
2	4
3	7

2.

x	y
1	4
2	8
3	12
4	16

3.

x	y
−2	4
−1	2
0	0
1	2

4. $xy = 3$

5. $x = \dfrac{1}{3}y$

6. $y + 1 = x$

EXAMPLE ③ **Real-Life Application**

x	y
$\dfrac{1}{2}$	8
1	16
$\dfrac{3}{2}$	24
2	32

The table shows the area y (in square feet) that a robotic vacuum cleans in x minutes.

a. Graph the data. Tell whether x and y are directly proportional.

Graph the data. Draw a line through the points.

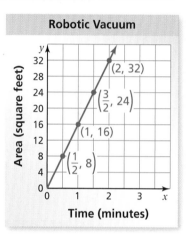
Robotic Vacuum

∴ The graph is a line through the origin. So, x and y are directly proportional.

b. Write an equation that represents the line.

Choose any two points to find the slope of the line.

$$\text{slope} = \frac{\text{change in } y}{\text{change in } x} = \frac{16}{1} = 16$$

∴ The slope of the line is the constant of proportionality, k. So, an equation of the line is $y = 16x$.

c. Use the equation to find the area cleaned in 10 minutes.

$y = 16x$ — Write the equation.

$= 16(10)$ — Substitute 10 for x.

$= 160$ — Multiply.

∴ So, the vacuum cleans 160 square feet in 10 minutes.

On Your Own

Now You're Ready
Exercise 19

7. WHAT IF? The battery weakens and the robot begins cleaning less and less area each minute. Do x and y show direct variation? Explain.

Vocabulary and Concept Check

1. **VOCABULARY** What does it mean for x and y to vary directly?

2. **WRITING** What point is on the graph of every direct variation equation?

3. **DIFFERENT WORDS, SAME QUESTION** Which is different? Find "both" answers.

Do x and y show direct variation?

Are x and y in a proportional relationship?

Is the graph of the relationship a line?

Does y vary directly with x?

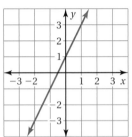

Practice and Problem Solving

Graph the ordered pairs in a coordinate plane. Do you think that graph shows that the quantities vary directly? Explain your reasoning.

4. $(-1, -1)$, $(0, 0)$, $(1, 1)$, $(2, 2)$

5. $(-4, -2)$, $(-2, 0)$, $(0, 2)$, $(2, 4)$

Tell whether x and y show direct variation. Explain your reasoning. If so, find k.

① 6.

x	1	2	3	4
y	2	4	6	8

7.

x	−2	−1	0	1
y	0	2	4	6

8.

x	−1	0	1	2
y	−2	−1	0	1

9.

x	3	6	9	12
y	2	4	6	8

② 10. $y - x = 4$ 11. $x = \frac{2}{5}y$ 12. $y + 3 = x + 6$ 13. $y - 5 = 2x$

14. $x - y = 0$ 15. $\frac{x}{y} = 2$ 16. $8 = xy$ 17. $x^2 = y$

18. **ERROR ANALYSIS** Describe and correct the error in telling whether x and y show direct variation.

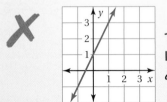

The graph is a line, so it shows direct variation.

③ 19. **RECYCLING** The table shows the profit y for recycling x pounds of aluminum. Graph the data. Tell whether x and y show direct variation. If so, write an equation that represents the line.

Aluminum (lb), x	10	20	30	40
Profit, y	$4.50	$9.00	$13.50	$18.00

Assignment Guide and Homework Check

Level	Day 1 Activity Assignment	Day 2 Lesson Assignment	Homework Check
Basic	4, 5, 30–34	1–3, 7–17 odd, 18, 19–25 odd	2, 9, 15, 23
Average	4, 5, 30–34	1–3, 7–17 odd, 18–28 even	2, 9, 24, 26, 28
Advanced	4, 5, 30–34	1–3, 8–28 even	14, 20, 22, 24, 28
Accelerated	1–5, 8–28 even, 30–34		14, 20, 22, 24, 28

Common Errors

- **Exercises 6–9** Students may immediately state that the table does not show direct variation because (0, 0) is not listed. Encourage them to find the change in x and change in y. Then use that knowledge to go back to x = 0 and determine if the table satisfies both requirements for direct variation.
- **Exercises 10–17** Students may try to identify the direct variation equations without solving for y. Remind them to solve for y first.
- **Exercise 17** Students may say that it shows direct variation because the graph goes through (0, 0). Ask them if the graph will be linear.
- **Exercises 20–22** Students may not grasp how to write the equation. Remind them of slope to determine the coefficient of x.

5.6 Record and Practice Journal

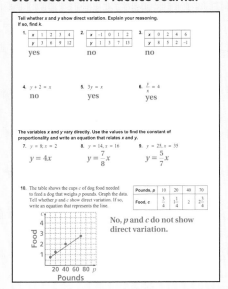

1. $y = kx$, where k is a number and $k \neq 0$.

2. (0, 0)

3. Is the graph of the relationship a line?; yes; no

 Practice and Problem Solving

4.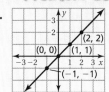

   yes; All the points lie on a line and the line passes through the origin.

5.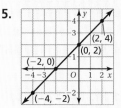

   no; The line does not pass through the origin.

6. yes; The line passes through the origin; $k = 2$

7. no; The line does not pass through the origin.

8. no; The line does not pass through the origin.

9. yes; The line passes through the origin; $k = \dfrac{2}{3}$

10. no; The equation cannot be written as $y = kx$.

11. yes; The equation can be written as $y = kx$; $k = \dfrac{5}{2}$

12. no; The equation cannot be written as $y = kx$.

13. no; The equation cannot be written as $y = kx$.

14–19. See Additional Answers.

20. $k = 24; y = 24x$

21. $k = \frac{5}{3}; y = \frac{5}{3}x$

22. $k = \frac{9}{8}; y = \frac{9}{8}x$

23. $y = 2.54x$

24. See *Taking Math Deeper*.

25. When $x = 0$, $y = 0$. So, the graph of a proportional relationship always passes through the origin.

26. yes; $k = 13$; The cost of 1 ticket is \$13; $y = 13x$; \$182

27. no

28. 76,000 mg

29. Every graph of direct variation is a line; however, not all lines show direct variation because the line must pass through the origin.

Fair Game Review

30. 0.65   31. 0.5625

32. 0.525   33. 0.96

34. D

Mini-Assessment

Tell whether *x* and *y* show direct variation.

1. $y - 3 = 4x$ no

2. $\frac{1}{4}y = x$ yes

3. $x - y = 2$ no

4. $6y + 3 = 12x + 3$ yes

5. One mile is approximately equal to 1.61 kilometers. Write a direct variation equation that relates *x* miles to *y* kilometers. $y = 1.61x$

Taking Math Deeper

Exercise 24

The problem shows students how the coordinate plane can help draw a blueprint.

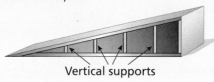

Vertical supports

To design the waterskiing ramp, locate the beginning of the ramp at the origin. The horizontal distances are the *x*-values. The vertical distances are the *y*-values.

① Design the ramp.

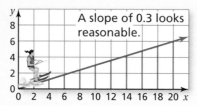

A slope of 0.3 looks reasonable.

The water skier leaves this ramp at a height of 6 feet. If this seems too high, then redesign the ramp with a slope of 0.2. Then, the final height will be 4 feet.

② Write a direct variation equation.

For the slope in the graph, $y = 0.3x$.

③ Plan 10 vertical support heights.

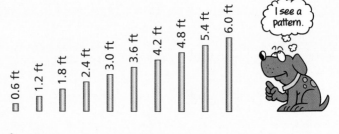

0.6 ft   1.2 ft   1.8 ft   2.4 ft   3.0 ft   3.6 ft   4.2 ft   4.8 ft   5.4 ft   6.0 ft

I see a pattern.

Project

Write a report comparing jumping ramps that are used in a variety of sports. Include your opinion as to why some ramps are higher and others are lower.

Reteaching and Enrichment Strategies

If students need help. . .	If students got it. . .
Resources by Chapter • Practice A and Practice B • Puzzle Time Record and Practice Journal Practice Differentiating the Lesson Lesson Tutorials Skills Review Handbook	Resources by Chapter • Enrichment and Extension • Technology Connection Start the next section

The variables x and y vary directly. Use the values to find the constant of proportionality. Then write an equation that relates x and y.

20. $y = 72; x = 3$

21. $y = 20; x = 12$

22. $y = 45; x = 40$

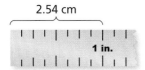

2.54 cm

1 in.

23. **MEASUREMENT** Write a direct variation equation that relates x inches to y centimeters.

24. **MODELING** Design a waterskiing ramp. Show how you can use direct variation to plan the heights of the vertical supports.

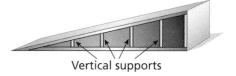

Vertical supports

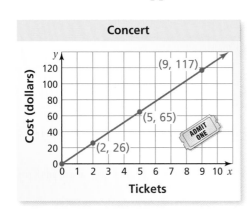

Concert

(9, 117)

(5, 65)

(2, 26)

Cost (dollars)

Tickets

ADMIT ONE

25. **REASONING** Use $y = kx$ to show why the graph of a proportional relationship always passes through the origin.

26. **TICKETS** The graph shows the cost of buying concert tickets. Tell whether x and y show direct variation. If so, find and interpret the constant of proportionality. Then write an equation and find the cost of 14 tickets.

27. **CELL PHONE PLANS** Tell whether x and y show direct variation. If so, write an equation of direct variation.

Minutes, x	500	700	900	1200
Cost, y	$40	$50	$60	$75

28. **CHLORINE** The amount of chlorine in a swimming pool varies directly with the volume of water. The pool has 2.5 milligrams of chlorine per liter of water. How much chlorine is in the pool?

29. Is the graph of every direct variation equation a line? Does the graph of every line represent a direct variation equation? Explain your reasoning.

8000 gallons

![pencil icon] **Fair Game Review** What you learned in previous grades & lessons

Write the fraction as a decimal. *(Section 2.1)*

30. $\dfrac{13}{20}$

31. $\dfrac{9}{16}$

32. $\dfrac{21}{40}$

33. $\dfrac{24}{25}$

34. **MULTIPLE CHOICE** Which rate is *not* equivalent to 180 feet per 8 seconds? *(Section 5.1)*

Ⓐ $\dfrac{225 \text{ ft}}{10 \text{ sec}}$

Ⓑ $\dfrac{45 \text{ ft}}{2 \text{ sec}}$

Ⓒ $\dfrac{135 \text{ ft}}{6 \text{ sec}}$

Ⓓ $\dfrac{180 \text{ ft}}{1 \text{ sec}}$

Check It Out
Progress Check
BigIdeasMath ✓com

Solve the proportion. *(Section 5.4)*

1. $\dfrac{7}{n} = \dfrac{42}{48}$

2. $\dfrac{x}{2} = \dfrac{40}{16}$

3. $\dfrac{3}{11} = \dfrac{27}{z}$

Find the slope of the line. *(Section 5.5)*

4.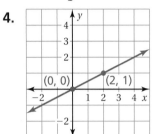

5.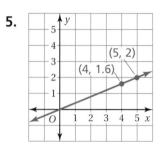

Graph the data. Then find and interpret the slope of the line through the points.
(Section 5.5)

6.

Hours, x	2	4	6	8
Miles, y	10	20	30	40

7.

Packages, x	6	10	14	18
Servings, y	9	15	21	27

Tell whether x and y show direct variation. Explain your reasoning. *(Section 5.6)*

8. $y - 9 = 6 + x$

9. $x = \dfrac{5}{8}y$

10. **CONCERT** A benefit concert with three performers lasts 8 hours. At this rate, how many hours is a concert with four performers? *(Section 5.4)*

11. **LAWN MOWING** The graph shows how much you and your friend each earn mowing lawns. *(Section 5.5)*

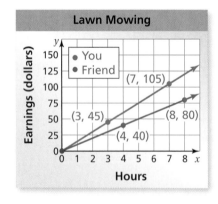

    a. Compare the steepness of the lines. What does this mean in the context of the problem?

    b. Find and interpret the slope of each line.

    c. How much more money do you earn per hour than your friend?

12. **PIE SALE** The table shows the profits of a pie sale. Tell whether x and y show direct variation. If so, write the equation of direct variation. *(Section 5.6)*

Pies Sold, x	10	12	14	16
Profit, y	$79.50	$95.40	$111.30	$127.20

Alternative Assessment Options

| Math Chat | Student Reflective Focus Question |
| Structured Interview | Writing Prompt |

Structured Interview

Interviews can occur formally or informally. Ask a student to perform a task and explain it, describing his or her thought process throughout the task. Probe the student for more information. Do not ask leading questions. Keep a rubric or notes.

Teacher Prompts	Student Answers	Teacher Notes
Tell me a story about taking a vacation. Include this sentence. Three tickets cost $420, so 5 tickets cost ? dollars.	Five friends go to Florida. All of their plane tickets are the same price. Three friends buy their tickets together. Three tickets cost $420, so 5 tickets cost $700.	Student can solve a proportion.
Add to your story using this phrase. 37.5 miles in ? minutes, or 450 miles in 1 hour	While at its cruising altitude, the plane flies 37.5 miles in 5 minutes, or 450 miles in 1 hour.	Student can solve a proportion.

Study Help Sample Answers

Remind students to complete Graphic Organizers for the rest of the chapter.

6.

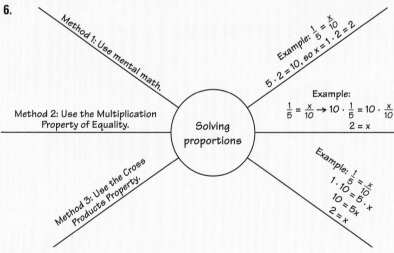

7–8. Available at *BigIdeasMath.com*.

Reteaching and Enrichment Strategies

If students need help. . .	If students got it. . .
Resources by Chapter • Practice A and Practice B • Puzzle Time Lesson Tutorials *BigIdeasMath.com*	Resources by Chapter • Enrichment and Extension • Technology Connection Game Closet at *BigIdeasMath.com* Start the Chapter Review

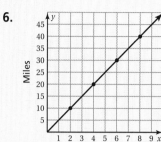
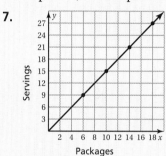

For the Teacher
Additional Review Options
- *BigIdeasMath.com*
- Online Assessment
- Game Closet at *BigIdeasMath.com*
- Vocabulary Help
- Resources by Chapter

Answers

1. 28.9 miles per gallon

2. 2.4 revolutions per second

3. 120 calories per serving

Review of Common Errors

Exercises 1–3
- Students may find the unit rate but forget to include the units. Remind them that the units are necessary for understanding a unit rate, or any rate.

Exercises 4–7
- Students may have difficulty understanding why you can write the ratio in simplest form. Tell students to compare the ratio to a fraction. Simplifying ratios is the same as writing equivalent fractions.

Exercise 8
- Students may immediately state that x and y are not in a proportional relationship because (0, 0) is not listed. Encourage students to use a graph to solve the problem, as stated in the directions.

Exercises 9 and 10
- Students may write half of the proportion using rows and the other half using columns. They will have forgotten to include one of the values. Remind students that they need to pick a method for writing proportions with tables and be consistent throughout the problem.

Exercises 11–14
- Students may divide instead of multiply when finding cross products, or they may multiply across the numerators and the denominators as if they were multiplying fractions. Remind students that the ratios have an equal sign between them, not a multiplication sign. Also tell them that when they use the Cross Products Property, it produces an "X," which means multiplication.

Exercises 15–17
- Students may put the change in x over the change in y. Remind them that the vertical change is written over the horizontal change and that slope is a rate.

Exercises 18–21
- Students may try to identify the direct variation equations without solving for y. Remind them to solve for y first.

5 Chapter Review

Check It Out
Vocabulary Help
BigIdeasMath ✓com

Review Key Vocabulary

ratio, *p. 164*
rate, *p. 164*
unit rate, *p. 164*
complex fraction, *p. 165*

proportion, *p. 172*
proportional, *p. 172*
cross products, *p. 173*
slope, *p. 194*

direct variation, *p. 200*
constant of proportionality,
p. 200

Review Examples and Exercises

5.1 Ratios and Rates *(pp. 162–169)*

There are 15 orangutans and 25 gorillas in a nature preserve.
One of the orangutans swings 75 feet in 15 seconds on a rope.

a. Find the ratio of orangutans to gorillas.

b. How fast is the orangutan swinging?

a. $\dfrac{\text{orangutans}}{\text{gorillas}} = \dfrac{15}{25} = \dfrac{3}{5}$

∴ The ratio of orangutans to gorillas is $\dfrac{3}{5}$.

b. 75 feet in 15 seconds $= \dfrac{75 \text{ ft}}{15 \text{ sec}}$

$= \dfrac{75 \text{ ft} \div 15}{15 \text{ sec} \div 15}$

$= \dfrac{5 \text{ ft}}{1 \text{ sec}}$

∴ The orangutan is swinging 5 feet per second.

Exercises

Find the unit rate.

1. 289 miles on 10 gallons

2. $6\dfrac{2}{5}$ revolutions in $2\dfrac{2}{3}$ seconds

3. calories per serving

Servings	2	4	6	8
Calories	240	480	720	960

5.2 Proportions *(pp. 170–177)*

Tell whether the ratios $\dfrac{9}{12}$ and $\dfrac{6}{8}$ form a proportion.

$\dfrac{9}{12} = \dfrac{9 \div 3}{12 \div 3} = \dfrac{3}{4}$

$\dfrac{6}{8} = \dfrac{6 \div 2}{8 \div 2} = \dfrac{3}{4}$

The ratios are equivalent.

∴ So, $\dfrac{9}{12}$ and $\dfrac{6}{8}$ form a proportion.

Exercises

Tell whether the ratios form a proportion.

4. $\dfrac{4}{9}, \dfrac{2}{3}$

5. $\dfrac{12}{22}, \dfrac{18}{33}$

6. $\dfrac{8}{50}, \dfrac{4}{10}$

7. $\dfrac{32}{40}, \dfrac{12}{15}$

8. Use a graph to determine whether x and y are in a proportional relationship.

x	1	3	6	8
y	4	12	24	32

5.3 ## Writing Proportions *(pp. 178–183)*

Write a proportion that gives the number r of returns on Saturday.

	Friday	Saturday
Sales	40	85
Returns	32	r

One proportion is $\dfrac{40 \text{ sales}}{32 \text{ returns}} = \dfrac{85 \text{ sales}}{r \text{ returns}}$.

Exercises

Use the table to write a proportion.

9.

	Game 1	Game 2
Penalties	6	8
Minutes	16	m

10.

	Concert 1	Concert 2
Songs	15	18
Hours	2.5	h

5.4 ## Solving Proportions *(pp. 186–191)*

Solve $\dfrac{15}{2} = \dfrac{30}{y}$.

$15 \cdot y = 2 \cdot 30$	Cross Products Property
$15y = 60$	Multiply.
$y = 4$	Divide.

The solution is 4.

Exercises

Solve the proportion.

11. $\dfrac{x}{4} = \dfrac{2}{5}$

12. $\dfrac{5}{12} = \dfrac{y}{15}$

13. $\dfrac{8}{20} = \dfrac{6}{w}$

14. $\dfrac{s+1}{4} = \dfrac{4}{8}$

Review Game

Proportions

Materials
- 1 deck of cards for each group
- paper for each student
- pencil for each student

Directions

Play in groups of 4 to 6 people. Each player is dealt four cards. Two are dealt face up. The other two are dealt face down and are held in the player's hand. The remainder of the deck is placed face down between the players.

The face up cards represent the denominators of two fractions. The object is to use the face down cards and other cards to form a proportion. Several cards can be added together in the numerator and denominator of both fractions until a proportion is obtained.

When it is a player's turn, he or she must do one of three things: lay a card down in either fraction's numerator or denominator, ask another player for a specific card, or draw from the pile. A student who draws from the pile or obtains a card from another student must wait until their next turn to lay a card down.

Card values are as follows:

2 through 10: face value

Jack: -1

Queen: -2

King: -3

Ace: -4

Points are awarded as follows:

First person to form a proportion: 10 points

Second: 9 points

Third: 8 points

Fourth: 7 points, and so on.

Who Wins?

After a set amount of time, the player with the most points wins.

For the Student
Additional Practice
- Lesson Tutorials
- Multi-Language Glossary
- Self-Grading Progress Check
- *BigIdeasMath.com*
  Dynamic Student Edition
  Student Resources

Answers

4. no   **5.** yes

6. no   **7.** yes

8.

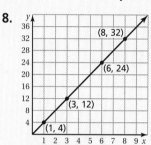

x and y are in a proportional relationship.

9. *Sample answer:*

$$\frac{8 \text{ penalties}}{6 \text{ penalties}} = \frac{m \text{ minutes}}{16 \text{ minutes}}$$

10. *Sample answer:*

$$\frac{15 \text{ songs}}{2.5 \text{ hours}} = \frac{18 \text{ songs}}{h \text{ hours}}$$

11. $x = 1.6$

12. $y = 6.25$

13. $w = 15$

14. $s = 1$

15. slope $= 1$

16. slope $= \dfrac{2}{3}$

17. slope $= 2$

18. no; The equation cannot be written as $y = kx$.

19. yes; The equation can be written as $y = kx$.

20. yes; The equation can be written as $y = kx$.

21. no; The equation cannot be written as $y = kx$.

My Thoughts on the Chapter

What worked. . .

What did not work. . .

What I would do differently. . .

5.5 Slope (pp. 192–197)

The graph shows the number of visits your website received over the past 6 months. Find and interpret the slope.

Choose any two points to find the slope of the line.

$$\text{slope} = \frac{\text{change in } y}{\text{change in } x}$$

$$= \frac{50}{1} \leftarrow \boxed{\text{visits}} \\ \leftarrow \boxed{\text{months}}$$

$$= 50$$

Website Visits

The slope of the line represents the unit rate. The slope is 50. So, the number of visits increased by 50 each month.

Exercises

Find the slope of the line.

15.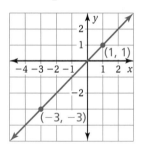

16.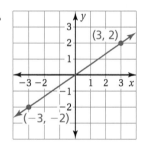

17.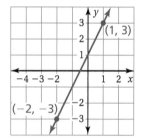

5.6 Direct Variation (pp. 198–203)

Tell whether x and y show direct variation. Explain your reasoning.

a. $x + y - 1 = 3$

$\quad y = 4 - x$ Solve for y.

The equation *cannot* be written as $y = kx$. So, x and y do *not* show direct variation.

b. $x = 8y$

$\quad \dfrac{1}{8}x = y$ Solve for y.

The equation can be written as $y = kx$. So, x and y show direct variation.

Exercises

Tell whether x and y show direct variation. Explain your reasoning.

18. $x + y = 6$

19. $y - x = 0$

20. $\dfrac{x}{y} = 20$

21. $x = y + 2$

Check It Out
Test Practice
BigIdeasMath ✓com

Find the unit rate.

1. 84 miles in 12 days

2. $2\frac{2}{5}$ kilometers in $3\frac{3}{4}$ minutes

Tell whether the ratios form a proportion.

3. $\frac{1}{9}, \frac{6}{54}$

4. $\frac{9}{12}, \frac{8}{72}$

Use a graph to tell whether x and y are in a proportional relationship.

5.

x	2	4	6	8
y	10	20	30	40

6.

x	1	3	5	7
y	3	7	11	15

Use the table to write a proportion.

7.

	Monday	Tuesday
Gallons	6	8
Miles	180	m

8.

	Thursday	Friday
Classes	6	c
Hours	8	4

Solve the proportion.

9. $\frac{x}{8} = \frac{9}{4}$

10. $\frac{17}{3} = \frac{y}{6}$

Graph the line that passes through the two points. Then find the slope of the line.

11. $(15, 9), (-5, -3)$

12. $(2, 9), (4, 18)$

Tell whether x and y show direct variation. Explain your reasoning.

13. $xy - 11 = 5$

14. $x = \frac{3}{y}$

15. $\frac{y}{x} = 8$

16. MOVIE TICKETS Five movie tickets cost $36.25.
What is the cost of 8 movie tickets?

17. CROSSWALK The graph shows the
number of cycles of a crosswalk signal
during the day and during the night.

Don't Walk

Walk

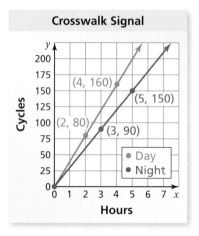

a. Compare the steepness of the
lines. What does this mean in
the context of the problem?

b. Find and interpret the slope of
each line.

18. GLAZE A specific shade of green glaze requires 5 parts blue to 3 parts yellow.
A glaze mixture contains 25 quarts of blue and 9 quarts of yellow. How can
you fix the mixture to make the specific shade of green glaze?

Test Item References

Chapter Test Questions	Section to Review
1, 2	5.1
3–6	5.2
7, 8	5.3
9, 10, 16, 18	5.4
11, 12, 17	5.5
13–15	5.6

Test-Taking Strategies

Remind students to quickly look over the entire test before they start so that they can budget their time. Some students may have difficulty distinguishing between such concepts as ratios, rates, unit rates, proportions, and slopes, so encourage students to jot down definitions on the back of the test before they start. Students should use **Stop** and **Think** strategies to ensure that they understand what is being asked before they write an answer.

Common Errors

- **Exercises 1 and 2** Students may find the unit rate but forget to include the units. Remind them that the units are necessary for understanding a unit rate.
- **Exercises 3 and 4** Students may have difficulty understanding why you can write the ratio in simplest form. Tell students to compare the ratio to a fraction. Simplifying ratios is the same as writing equivalent fractions.
- **Exercises 5 and 6** Students may immediately state that x and y are not in a proportional relationship because (0, 0) is not listed. Encourage students to use a graph to solve the problem, as stated in the directions.
- **Exercises 7 and 8** Students may write half of the proportion using rows and the other half using columns. They will have forgotten to include one of the values. Remind students that they need to pick a method for writing proportions with tables and be consistent throughout the problem.
- **Exercises 13–15** Students may try to identify the direct variation equations without solving for y. Remind them to solve for y first.

Reteaching and Enrichment Strategies

If students need help. . .	If students got it. . .
Resources by Chapter • Practice A and Practice B • Puzzle Time Record and Practice Journal Practice Differentiating the Lesson Lesson Tutorials *BigIdeasMath.com* Skills Review Handbook	Resources by Chapter • Enrichment and Extension • Technology Connection Game Closet at *BigIdeasMath.com* Start Cumulative Assessment

Answers

1. 7 miles per day

2. $\dfrac{16}{25}$ kilometer per minute

3. yes   **4.** no

5. 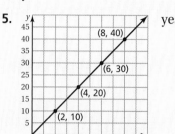 yes

6. 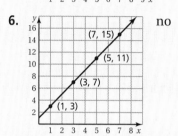 no

7. *Sample answer:*
$$\dfrac{8 \text{ gallons}}{6 \text{ gallons}} = \dfrac{m \text{ miles}}{180 \text{ miles}}$$

8. *Sample answer:*
$$\dfrac{6 \text{ classes}}{8 \text{ hours}} = \dfrac{c \text{ classes}}{4 \text{ hours}}$$

9. $x = 18$ **10.** $y = 34$

11.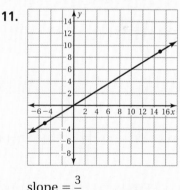

slope $= \dfrac{3}{5}$

12–17. See Additional Answers.

18. Add 6 quarts of yellow

Online Assessment
Assessment Book
ExamView® Assessment Suite

After Answering Easy Questions, Relax

Answer Easy Questions First

Estimate the Answer

Read All Choices before Answering

Read Question before Answering

Solve Directly or Eliminate Choices

Solve Problem before Looking at Choices

Use Intelligent Guessing

Work Backwards

About this Strategy

When taking a multiple choice test, be sure to read each question carefully and thoroughly. It is also very important to read each answer choice carefully. Do not pick the first answer that you think is correct!

Answers

1. A
2. F
3. 29
4. B
5. H

Item Analysis

1. **A.** Correct answer

   B. The student thinks the price of the 4 pencils is the unit price.

   C. The student multiplies 0.80 and 4.

   D. The student divides 4 by 0.80.

2. **F.** Correct answer

   G. The student applies operations incorrectly.

   H. The student applies operations incorrectly.

   I. The student applies operations incorrectly.

3. **Gridded Response:** Correct answer: 29

   Common Error: The student makes a sign error when multiplying and gets an answer of -19.

4. **A.** The student finds the change in x over the change in y.

   B. Correct answer

   C. The student finds the change in x.

   D. The student finds the change in y.

5. **F.** The student does not reverse the inequality sign when dividing by a negative number.

   G. The student reverses the sign when subtracting and does not realize the inequality symbol calls for a closed circle.

   H. Correct answer

   I. The student does not realize that the inequality sign calls for a closed circle.

1. The school store sells 4 pencils for $0.80. What is the unit cost of a pencil?

   A. $0.20   **C.** $3.20

   B. $0.80   **D.** $5.00

2. Which expressions do *not* have a value of 3?

   I. $2 + (-1)$ II. $2 - (-1)$

   III. $-3 \times (-1)$ IV. $-3 \div (-1)$

   F. I only   **H.** II only

   G. III and IV   **I.** I, III, and IV

3. What is the value of the expression below?

   $$-4 \times (-6) - (-5)$$

4. What is the slope of the line shown?

   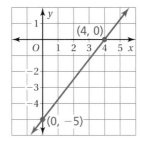

   A. $\frac{4}{5}$ **C.** 4

   B. $\frac{5}{4}$ **D.** 5

5. The graph below represents which inequality?

   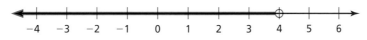

   F. $-3 - 6x < -27$ **H.** $5 - 3x > -7$

   G. $2x + 6 \geq 14$ **I.** $2x + 3 \leq 11$

6. The quantities x and y are proportional. What is the missing value in the table?

x	y
$\dfrac{2}{3}$	6
$\dfrac{4}{3}$	12
$\dfrac{8}{3}$	24
5	

A. 30

B. 36

C. 45

D. 48

7. You are selling tomatoes. You have already earned $16 today. How many additional pounds of tomatoes do you need to sell to earn a total of $60?

F. 4

G. 11

H. 15

I. 19

$4 per pound

8. The distance traveled by the a high-speed train is proportional to the number of hours traveled. Which of the following is *not* a valid interpretation of the graph below?

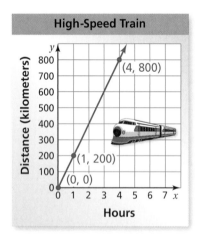

A. The train travels 0 kilometers in 0 hours.

B. The unit rate is 200 kilometers per hour.

C. After 4 hours, the train is traveling 800 kilometers per hour.

D. The train travels 800 kilometers in 4 hours.

Item Analysis (continued)

6. **A.** The student adds 6 to 24 because $12 - 6 = 6$.

   B. The student adds 12 to 24 because $24 - 12 = 12$.

   C. Correct answer

   D. The student doubles 24 because each value in the *y* column is twice the previous value.

7. **F.** The student finds the number of pounds already sold.

   G. Correct answer

   H. The student finds the total number of pounds instead of the additional number of pounds.

   I. The student adds the total number of pounds and the number of pounds already sold.

8. **A.** The student misinterprets the information in the graph.

   B. The student misinterprets the information in the graph.

   C. Correct answer

   D. The student misinterprets the information in the graph.

9. G

10. 3

11. *Part A Sample answer:*

$$\frac{800}{15} = \frac{6000}{m}$$

Part B 112.5 min

12. D

Item Analysis (continued)

9. **F.** The student takes the reciprocal of the dividend.

  G. Correct answer

  H. The student switches denominators instead of taking the reciprocal of the divisor.

  I. The student makes a sign error.

10. **Gridded Response:** Correct answer: 3

  Common Error: The student makes a sign error when subtracting 3 from both sides of the inequality and gets an answer of 2.

11. **2 points** The student demonstrates a thorough understanding of writing and solving proportions. For Part A, the student correctly writes a proportion such as $\frac{800}{15} = \frac{6000}{m}$ and provides an appropriate explanation. For Part B, the student correctly gets a value of 112.5 for m, shows appropriate work, and states that it would take 112.5 minutes.

  1 point The student demonstrates a partial understanding of writing and solving proportions. The student writes a correct proportion but does not solve it successfully, or the student does not write a correct proportion but demonstrates the ability to solve a proportion.

  0 points The student demonstrates insufficient understanding of writing and solving proportions. The student does not write a correct proportion and shows little or no evidence of being able to solve proportions.

12. **A.** The student makes a sign error when dividing each side of the equation by -2.

  B. The student subtracts 6 from the left side of the equation but adds 6 to the right side of the equation, and makes a sign error when dividing each side of the equation by -2.

  C. The student subtracts 6 from the left side of the equation but adds 6 to the right side of the equation.

  D. Correct answer

9. Regina was evaluating the expression below. What should Regina do to correct the error she made?

$$-\frac{3}{2} \div \left(-\frac{8}{7}\right) = -\frac{2}{3} \times \left(-\frac{7}{8}\right)$$

$$= \frac{2 \times 7}{3 \times 8}$$

$$= \frac{14}{24}$$

$$= \frac{7}{12}$$

F. Rewrite $-\frac{3}{2} \div \left(-\frac{8}{7}\right)$ as $-\frac{2}{3} \times \left(-\frac{8}{7}\right)$.

G. Rewrite $-\frac{3}{2} \div \left(-\frac{8}{7}\right)$ as $-\frac{3}{2} \times \left(-\frac{7}{8}\right)$.

H. Rewrite $-\frac{3}{2} \div \left(-\frac{8}{7}\right)$ as $-\frac{3}{7} \times \left(-\frac{8}{2}\right)$.

I. Rewrite $-\frac{2}{3} \times \left(-\frac{7}{8}\right)$ as $-\frac{2 \times 7}{3 \times 8}$.

10. What is the least value of t for which the inequality is true?

$$3 - 6t \le -15$$

11. You can mow 800 square feet of lawn in 15 minutes. At this rate, how many minutes will you take to mow a lawn that measures 6000 square feet?

Part A Write a proportion to represent the problem. Use m to represent the number of minutes. Explain your reasoning.

Part B Solve the proportion you wrote in Part A. Then use it to answer the problem. Show your work.

12. What value of p makes the equation below true?

$$6 - 2p = -48$$

A. -27

B. -21

C. 21

D. 27

6 Percents

6.1 Percents and Decimals

6.2 Comparing and Ordering Fractions, Decimals, and Percents

6.3 The Percent Proportion

6.4 The Percent Equation

6.5 Percents of Increase and Decrease

6.6 Discounts and Markups

6.7 Simple Interest

"Here's my sales strategy. I buy each dog bone for $0.05."

"Then I mark each one up to $1. Then, I have a 75% off sale. Cool, huh?"

"At 4 a day, I have chewed 17,536 dog biscuits. At only 99.9% pure, that means that..."

"I have swallowed seventeen and a half contaminated dog biscuits during the past twelve years."

What Your Students Have Learned

- Multiply and divide by powers of 10 and explain the placement of the decimal point.
- Find equivalent fractions.
- Compare decimals to the thousandths place.
- Use and interpret simple equations.
- Understand ratios and describe ratio relationships.
- Understand and find percent as a rate per 100.
- Find the part and the whole of ratio relationships.
- Identify equivalent expressions.
- Solve one-step equations.

What Your Students Will Learn

- Write percents as decimals and decimals as percents.
- Compare fractions, decimals, and percents.
- Use proportionality to solve percent problems.
- Use the percent equation.
- Solve percent problems involving percents of increase and decrease, and simple interest.

Pacing Guide for Chapter 6

Chapter Opener Regular Accelerated	1 Day 1 Day
Section 1 Regular Accelerated	2 Days 1 Day
Section 2 Regular Accelerated	2 Days 1 Day
Section 3 Regular Accelerated	2 Days 1 Day
Section 4 Regular Accelerated	2 Days 1 Day
Study Help / Quiz Regular Accelerated	1 Day 1 Day
Section 5 Regular Accelerated	2 Days 1 Day
Section 6 Regular Accelerated	2 Days 1 Day
Section 7 Regular Accelerated	2 Days 1 Day
Chapter Review/ Chapter Tests Regular Accelerated	2 Days 2 Days
Total Chapter 6 Regular Accelerated	18 Days 11 Days
Year-to-Date Regular Accelerated	89 Days 51 Days

Technology for the *Teacher*

BigIdeasMath.com
Chapter at a Glance
Complete Materials List
Parent Letters: English and Spanish

- Write percents as fractions in simplest form.
- Write fractions and mixed numbers as percents.

Additional Topics for Review
- Compare and Order Integers
- Writing Decimals as Fractions and Percents
- Writing Fractions as Decimals
- Ratios
- Solving Proportions

Try It Yourself

1. $\dfrac{4}{25}$ 2. $\dfrac{2}{5}$

3. $\dfrac{17}{25}$ 4. $\dfrac{17}{20}$

5. $1\dfrac{12}{25}$ 6. $1\dfrac{1}{2}$

7. $1\dfrac{1}{20}$ 8. $2\dfrac{19}{25}$

9. 36%   10. 86%

11. 55%   12. 60%

13. 125%   14. 148%

15. 180%   16. 230%

Record and Practice Journal
Fair Game Review

1. $\dfrac{1}{4}$ 2. $\dfrac{13}{20}$

3. $1\dfrac{1}{10}$ 4. $2\dfrac{1}{2}$

5. $\dfrac{3}{20}$ 6. $\dfrac{3}{50}$

7. $\dfrac{3}{10}$ 8. 20%

9. 25%   10. 84%

11. 140%   12. 265%

13. 150%   14. 60%

Math Background Notes

Vocabulary Review
- Percent
- Numerator
- Denominator
- Equivalent Fractions

Writing Percents as Fractions
- Students learned how to convert between percents and fractions.
- A 10-by-10 grid has 100 small squares making it a convenient model for percents as well as decimals.
- Be sure students understand that when modeling, for instance 45%, any 45 squares can be shaded. However, for ease of being able to read the model quickly, we generally shade four strips of 10 and one strip of 5.
- ❓ "What is 100% as a fraction?" 1
- ❓ "If the percent is greater than 100%, what do you know about the equivalent fraction?" It will be greater than 1.

Writing Fractions as Percents
- Students have learned how to write equivalent fractions.
- ❓ "What are equivalent fractions?" two fractions that represent the same amount
- Review which fractions can be written as equivalent fractions with a denominator of 100.
- ❓ "How do you write equivalent fractions with a denominator of 100?" Find a number you can multiply the denominator by so that it equals 100. Multiply the numerator by the same number.
- ❓ "If the number is greater than 1, what do you know about the percent?" It will be greater than 100%.

Reteaching and Enrichment Strategies

If students need help. . .	If students got it. . .
Record and Practice Journal • Fair Game Review Skills Review Handbook Lesson Tutorials	Game Closet at *BigIdeasMath.com* Start the next section

What You Learned Before

"The fact that these two percents do not total 100 is a sad commentary on humans."

Writing Percents as Fractions

Example 1   Write 45% as a fraction in simplest form.

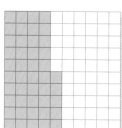

$$45\% = \frac{45}{100}$$ Write as a fraction with a denominator of 100.

$$= \frac{9}{20}$$ Simplify.

So, $45\% = \frac{9}{20}$.

Try It Yourself

Write the percent as a fraction or mixed number in simplest form.

1. 16%   **2.** 40%   **3.** 68%   **4.** 85%

5. 148%   **6.** 150%   **7.** 105%   **8.** 276%

Writing Fractions as Percents

Example 2   Write $\frac{3}{25}$ as a percent.

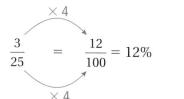

$$\frac{3}{25} = \frac{12}{100} = 12\%$$

Because $25 \times 4 = 100$, multiply the numerator and denominator by 4. Write the numerator with a percent symbol.

Try It Yourself

Write the fraction or mixed number as a percent.

9. $\frac{9}{25}$ **10.** $\frac{43}{50}$ **11.** $\frac{11}{20}$ **12.** $\frac{3}{5}$

13. $1\frac{1}{4}$ **14.** $1\frac{12}{25}$ **15.** $1\frac{4}{5}$ **16.** $2\frac{3}{10}$

Essential Question How does the decimal point move when you rewrite a percent as a decimal and when you rewrite a decimal as a percent?

1   ACTIVITY: Writing Percents as Decimals

Work with a partner. Write the percent shown by the model. Write the percent as a decimal.

a.

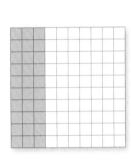

$$\boxed{}\% = \dfrac{\boxed{}}{\boxed{}}$$ ← per   ← cent

$$= \dfrac{\boxed{}}{\boxed{}}$$ Simplify.

$$= \boxed{}$$ Write fraction as a decimal.

b.

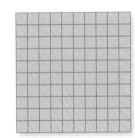

c.

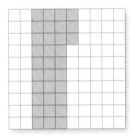

d.

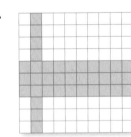

e.

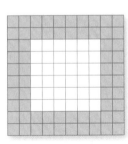

Percents and Decimals

In this lesson, you will
- write percents as decimals.
- write decimals as percents.
- solve real-life problems.

f.

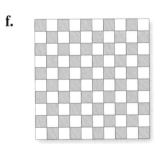

g.

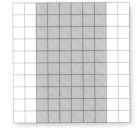

Laurie's Notes

Introduction

Applying Mathematical Practices

- **Model with Mathematics:** Mathematically proficient students use models to help make sense of different representations of numbers. In this lesson, a 100-grid is used to model percents, making the connection to the equivalent fraction and decimal forms.

Motivate

- Share a fictional story about collecting student homework on a USB drive and the need to purchase a new drive with greater capacity—for all of their homework! Work the following nonfictional facts into the story.
  - Most common USB drives hold 4 GB (gigabytes) or 8 GB.
  - A 16 GB USB drive holds 4 times or 400% as much data as the 4 GB, and 2 times or 200% as much data as the 8 GB.

For the Teacher

- Students have converted between fractions and decimals and between fractions and percents. This lesson completes the triangle by converting between decimals and percents.

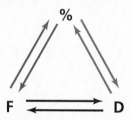

- Relating percents to money (part of a dollar) is often a helpful technique.

Activity Notes

Activity 1

- Check in with students as they work through part (a). There are 3 steps: write the percent; write the percent as a fraction; write the fraction as a decimal.
- A common question from students is whether they can leave the answer as 0.30 versus 0.3. The two decimals are equivalent. However, explain that you generally write the simplified version 0.3 for the same reason you simplify fractions.
- Students should work with partners to answer the remaining parts.
- Probe what strategies were used to count the shaded squares.
- **Common Error:** In part (b), students often write "100% = 100" instead of "100% = 1."

What Your Students Will Learn
- Write percents as decimals and decimals as percents.
- Write fractions as percents and decimals.

Previous Learning
Students should know how to write halves, fourths, tenths, and hundredths as percents. Students also need to be able to convert between fractions and decimals, and multiply and divide decimals by powers of 10.

Lesson Plans
Complete Materials List

6.1 Record and Practice Journal

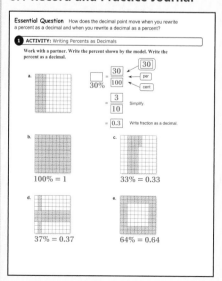

6.1 Record and Practice Journal

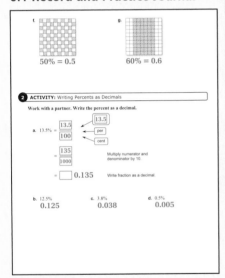

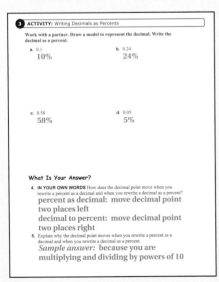

Laurie's Notes

Activity 2

- **Discuss:** Percents do not need to involve whole numbers, and in fact often do not.
- ❓ "What is different about the visual model from what you have seen previously?" *One-half of a square is shaded.*
- **FYI:** Students initially find it odd to have a decimal as part of a fraction. Multiplying the numerator and denominator by a power of 10 makes sense to them.
- **Model with Mathematics:** Sketching models of parts (b)–(d) helps students focus on the fractional portion of 1%. You could also "magnify" one small square and draw it off to the side to shade various portions of 1%, as shown.

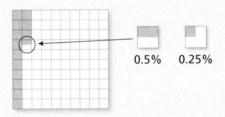

0.5%   0.25%

- **Common Error:** When percents are less than 1%, students often just remove the percent symbol and leave the decimal. In part (d), 0.5% does not equal 0.5.

Activity 3

- This example reverses the process students have just completed. After they sketch a model of the decimal, they write the decimal as a fraction with a denominator of 100, and then write the percent.

What Is Your Answer?

- Students should recognize that the decimal point has moved two decimal places left or right, although they may not have a good understanding of why.
- Listen for explanations about the percent symbol and its meaning.
- ❓ Probe about the location of the decimal point for percents involving whole numbers, such as 25%. "Where is the decimal point located?" *after the 5*
- **Construct Viable Arguments and Critique the Reasoning of Others:** Students may not be able to articulate clearly an explanation for Question 5. Listen to students' explanations and references to place value, definition of percent, and multiplication and division of powers of 10.

Closure

- **Exit Ticket:** Write each percent as a decimal. 60%, 6%, and 0.6% *0.6, 0.06, 0.006*

ACTIVITY: Writing Percents as Decimals

Work with a partner. Write the percent as a decimal.

a. 13.5%

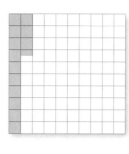

$\% = \dfrac{}{}$ ← per ← cent

$= \dfrac{}{}$ Multiply numerator and denominator by 10.

$= $ Write fraction as a decimal.

b. 12.5%   **c.** 3.8%   **d.** 0.5%

---

③ **ACTIVITY: Writing Decimals as Percents**

Work with a partner. Draw a model to represent the decimal. Write the decimal as a percent.

a. 0.1

$$0.1 \quad = \quad 0.10 = \dfrac{}{} \quad = \quad \%$$

One ▢   Ten ▢   ▢ Percent

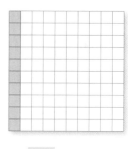

b. 0.24   **c.** 0.58   **d.** 0.05

What Is Your Answer?

4. IN YOUR OWN WORDS How does the decimal point move when you rewrite a percent as a decimal and when you rewrite a decimal as a percent?

5. Explain why the decimal point moves when you rewrite a percent as a decimal and when you rewrite a decimal as a percent.

Practice → Use what you learned about percents and decimals to complete Exercises 7–12 and 19–24 on page 218.

 **Key Idea**

Writing Percents as Decimals

Words Remove the percent symbol. Then divide by 100, or just move the decimal point two places to the left.

Numbers $23\% = 23.\% = 0.23$

EXAMPLE 1 Writing Percents as Decimals

Study Tip

When moving the decimal point, you may need to place one or more zeros in the number.

a. Write 52% as a decimal.

$52\% = 52.\% = 0.52$

Check

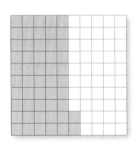

b. Write 7% as a decimal.

$7\% = 07.\% = 0.07$

Check

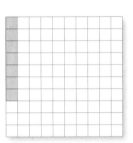

● **On Your Own**

Now You're Ready
Exercises 7–18

Write the percent as a decimal. Use a model to check your answer.

1. 24%   **2.** 3%   **3.** 107%   **4.** 92.7%

 **Key Idea**

Writing Decimals as Percents

Words Multiply by 100, or just move the decimal point two places to the right. Then add a percent symbol.

Numbers $0.36 = 0.36 = 36\%$

EXAMPLE 2 Writing Decimals as Percents

a. Write 0.47 as a percent.

$0.47 = 0.47 = 47\%$

b. Write 0.663 as a percent.

$0.663 = 0.663 = 66.3\%$

c. Write 1.8 as a percent.

$1.8 = 1.80 = 180\%$

d. Write 0.009 as a percent.

$0.009 = 0.009 = 0.9\%$

Laurie's Notes

Introduction

Connect

- **Yesterday:** Students used a visual model to represent percents, then wrote the percents as decimals.
- **Today:** Students will convert between percents and decimals.

Motivate

- Divide the class into 4 groups by any method. Each group has a different rule to follow.

  Group A: Multiply by 100. Group B: Multiply by 0.01.
  Group C: Divide by 100. Group D: Divide by 0.01.

- Use small white boards, an electronic polling system, or scrap paper to record answers.
- Use the numbers 60, 44, and 2.5. Check results from each group after each problem.
- Students should recognize the pattern. Groups A and D have the same answers as do Groups B and C. Discuss the results.

Technology for the Teacher

Dynamic Classroom

Lesson Tutorials
Lesson Plans
Answer Presentation Tool

Lesson Notes

Key Idea

- To reinforce the meaning behind moving the decimal point, say "23 percent, 23 per one hundred, or 23 hundredths."

Example 1

- Work through both parts of the example.

? "52% is close to which benchmark fraction?" $\frac{1}{2}$

- The visual model shows an amount just more than 50% or $\frac{1}{2}$.
- Point out the *Study Tip*. In part (b), 7% means 7 per one hundred or 7 hundredths. To write that, you have to place a zero to the left of the seven, 0.07.

On Your Own

- **Think-Pair-Share:** Students should read each question independently and then work in pairs to answer the questions. When they have answered the questions, the pair should compare their answers with another group and discuss any discrepancies.

Key Idea

- **Connection:** Dividing by 100 is equivalent to multiplying by 0.01. The decimal point moves 2 places to the left. Multiplying by 100 is equivalent to dividing by 0.01. The decimal point moves 2 places to the right.

Example 2

- Work through the examples.

Extra Example 1

a. Write 66% as a decimal. 0.66
b. Write 2% as a decimal. 0.02

On Your Own

1. 0.24   2. 0.03
3. 1.07   4. 0.927

Extra Example 2

a. Write 0.29 as a percent. 29%
b. Write 0.775 as a percent. 77.5%
c. Write 2.5 as a percent. 250%
d. Write 0.0075 as a percent. 0.75%

On Your Own

5. 94%   **6.** 120%

7. 31.6%   **8.** 0.5%

Extra Example 3

On a science test, you get 85 out of a possible 100 points. Write your score as a percent, a fraction, and a decimal.

85%, $\frac{17}{20}$, 0.85

Extra Example 4

In Example 4, how many times more UV rays are reflected by water than by grass? 28 times more

On Your Own

9. $18\% = \frac{9}{50} = 0.18$

10. 5.6

Differentiated Instruction

Kinesthetic

Some students may struggle with writing decimals for very small percents (i.e., 0.0075% as 0.000075) or writing percents for very small decimals (i.e., 0.0075 as 0.75%). To improve their number sense, have students draw a 10-by-10 grid with sides of 20 centimeters. Divide one of the squares to make a smaller 10-by-10 grid. Have students compare 75% and 0.75% by shading squares to represent each percent. Then have students write the decimal equivalents of the percents.

On Your Own

- **Think-Pair-Share:** Students should read each question independently and then work in pairs to answer the questions. When they have answered the questions, the pair should compare their answers with another group and discuss any discrepancies.

Words of Wisdom

- Students quickly recognize that the decimal point moves 2 places left or right depending on the type of conversion.
- Students generally do well with 2-digit problems: 24%, or 0.24.
- Students have difficulty with percents greater than 100% or less than 1% (i.e., 250%, 0.025%). They want to quickly move the decimal point without thinking first.
- Reinforce the meaning behind moving the decimal point.
    - 25% is 25 per one hundred or 0.25.
    - 250% is 250 per one hundred or 2.5.
    - 0.25% is 25 hundredths per one hundred or 0.0025.
    - 0.025% is 25 thousandths per one hundred or 0.00025.
- When you feel that students are comfortable with percents between 1 and 100, practice converting with greater and lesser percents.

Example 3

- Note the test-taking strategy of eliminating choices.

Example 4

- **FYI:** UV rays are necessary for our bodies to produce vitamin D, a substance that helps strengthen bones. The downside: reflection of UV rays off of snow and sand are enough to cause photokeratitis, sunburn of the cornea.
- This example reviews all three forms: percents, decimals, and fractions, as well as division of decimals.
- **?** "What is the key question being asked?" This is a language connection. "How many times more. . . ."

On Your Own

- **Construct Viable Arguments and Critique the Reasoning of Others:** There are several ways in which students may justify their answers. Take time to hear a variety of approaches.

Closure

- **Writing Prompts:**
  To write a percent as a decimal, . . .
  To write a decimal as a percent, . . .

 **On Your Own**

Now You're Ready
Exercises 19–30

Write the decimal as a percent. Use a model to check your answer.

5. 0.94   **6.** 1.2   **7.** 0.316   **8.** 0.005

EXAMPLE 3   **Writing a Fraction as a Percent and a Decimal**

On a math test, you get 92 out of a possible 100 points. Which of the following is *not* another way of expressing 92 out of 100?

   (A) $\dfrac{23}{25}$ **(B)** 92%   **(C)** $\dfrac{17}{20}$ **(D)** 0.92

$$92 \text{ out of } 100 = \frac{92}{100}$$

    $= 92\%$ Eliminate Choice B.

    $= \dfrac{23}{25}$ Eliminate Choice A.

    $= 0.92$ Eliminate Choice D.

   So, the correct answer is **(C)**.

EXAMPLE 4   **Real-Life Application**

The figure shows the portions of ultraviolet (UV) rays reflected by four different surfaces. How many times more UV rays are reflected by water than by sea foam?

Write 25% and $\dfrac{21}{25}$ as decimals.

    Sea foam: $25\% = 25.\% = 0.25$ **Water:** $\dfrac{21}{25} = \dfrac{84}{100} = 0.84$

Divide 0.84 by 0.25: $0.25\overline{)0.84}$ $\longrightarrow$ $25\overline{)84.00}$ with quotient 3.36

   So, water reflects about 3.4 times more UV rays than sea foam.

 **On Your Own**

9. Write "18 out of 100" as a percent, a fraction, and a decimal.

10. In Example 4, how many times more UV rays are reflected by water than by sand?

Check It Out
Help with Homework
BigIdeasMath ✓com

 Vocabulary and Concept Check

MATCHING Match the decimal with its equivalent percent.

1. 0.42   **2.** 4.02   **3.** 0.042   **4.** 0.0402

   A. 4.02%   **B.** 42%   **C.** 4.2%   **D.** 402%

5. OPEN-ENDED Write three different decimals that are between 10% and 20%.

6. WHICH ONE DOESN'T BELONG? Which one does *not* belong with the other three? Explain your reasoning.

| 70% | 0.7 | $\frac{7}{10}$ | 0.07 |

 Practice and Problem Solving

Write the percent as a decimal.

7. 78%   **8.** 55%   **9.** 18.5%

10. 57.4%   **11.** 33%   **12.** 9%

13. 47.63%   **14.** 91.25%   **15.** 166%

16. 217%   **17.** 0.06%   **18.** 0.034%

Write the decimal as a percent.

19. 0.74   **20.** 0.52   **21.** 0.89

22. 0.768   **23.** 0.99   **24.** 0.49

25. 0.487   **26.** 0.128   **27.** 3.68

28. 5.12   **29.** 0.0371   **30.** 0.0046

31. ERROR ANALYSIS Describe and correct the error in writing 0.86 as a percent.

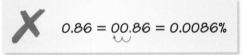

32. MUSIC Thirty-six percent of the songs on your MP3 player are pop songs. Write this percent as a decimal.

33. CAT About 0.34 of the length of a cat is its tail. Write this decimal as a percent.

34. COMPUTER Write the percent of free space on the computer as a decimal.

Volume	Capacity	Free Space	% Free Space
🖫 (C:)	149 GB	133 GB	89 %

Write the percent as a fraction in simplest form and as a decimal.

35. 36%   **36.** 23.5%   **37.** 16.24%

Assignment Guide and Homework Check

Level	Day 1 Activity Assignment	Day 2 Lesson Assignment	Homework Check
Basic	7–12, 19–24, 42–50	1–6, 13, 15, 17, 25–37 odd	13, 25, 33, 35
Average	7–12, 19–24, 42–50	1–6, 16–18, 28–31, 34–38 even, 39	18, 28, 34, 36
Advanced	7–12, 19–24, 42–50	1–6, 18, 30, 31, 32–40 even, 41	18, 30, 34, 36
Accelerated	1–12, 16–30 even, 31, 32–40 even, 41–50		18, 30, 34, 36

Common Errors

- **Exercises 7–30** Students may move the decimal point the wrong way, forget to place zeros as placeholders, or move the decimal point too many places (especially when the percent is greater than 100). As a class, gather some helpful information for remembering how to convert decimals and percents. For example, when the percent is greater than 100%, the decimal equivalent will be greater than 1. Because there are two zeros in 100, you need to move the decimal point two places. When converting from percents to decimals, you move the decimal point to the left because D is to the left of P in the alphabet. When converting from decimals to percents, you move the decimal point to the right because P is to the right of D in the alphabet.
- **Exercises 35–37** Students may move the decimal point to the right instead of to the left. Remind them how to get rid of the decimal point in the percent when writing a fraction.

6.1 Record and Practice Journal

Write the percent as a decimal.
1. 35%   2. 160%   3. 74.8%   4. 0.3%
   0.35   1.6   0.748   0.003

Write the decimal as a percent.
5. 1.23   6. 0.49   7. 0.024   8. 0.881
   123%   49%   2.4%   88.1%

Write the percent as a fraction in simplest form and as a decimal.
9. 48%   10. 15.5%   11. 84.95%
$\frac{12}{25}$; 0.48 $\frac{31}{200}$; 0.155 $\frac{1699}{2000}$; 0.8495

12. People with severe hearing loss were given a sentence and word recognition test six months after they got implants in their ears. The patients scored an average of 82% on the test. Write this percent as a decimal.
0.82

Vocabulary and Concept Check

1. B   2. D

3. C   4. A

5. *Sample answer:* 0.11, 0.13, 0.19

6. 0.07 because it represents 7% instead of 70%.

Practice and Problem Solving

7. 0.78   8. 0.55

9. 0.185   10. 0.574

11. 0.33   12. 0.09

13. 0.4763   14. 0.9125

15. 1.66   16. 2.17

17. 0.0006   18. 0.00034

19. 74%   20. 52%

21. 89%   22. 76.8%

23. 99%

24. 49%

25. 48.7%

26. 12.8%

27. 368%

28. 512%

29. 3.71%

30. 0.46%

31. The decimal point was moved in the wrong direction. $0.86 = 0.86 = 86\%$

32. 0.36

33. 34%

34. 0.89

35. $\frac{9}{25} = 0.36$

36. $\frac{47}{200} = 0.235$

37. $\frac{203}{1250} = 0.1624$

Practice and Problem Solving

38. a. car: 0.2, school bus: 0.48, bicycle: 0.08

b. car: $\frac{1}{5}$, school bus: $\frac{12}{25}$, bicycle: $\frac{2}{25}$

c. 24%

d. *Answer should include, but is not limited to:* A bar graph showing either the *number* of students or the *portion* of students in the class that get to school in various ways.

39. 40%

40. See *Taking Math Deeper.*

41. a. $16.\overline{6}\%$, or $16\frac{2}{3}\%$

b. $\frac{5}{6}$

Fair Game Review

42. $\frac{23}{50}$ **43.** $\frac{31}{100}$

44. $2\frac{1}{5}$ **45.** $4\frac{8}{25}$

46. $-5x + 3$

47. $-1.6n - 1$

48. $-3y + 15$

49. $-b - \frac{3}{2}$

50. B

Mini-Assessment

Write the percent as a decimal.

1. 12% 0.12   **2.** 130% 1.3

Write the decimal as a percent.

3. 0.98 98%   **4.** 2.25 225%

5. Forty-two percent of your cell phone ringtones are pop songs. Write this percent as a decimal. 0.42

Taking Math Deeper

Exercise 40

This is a good review that all regions of a circle graph must add up to 1 or 100%. Begin by making a table to summarize the given information.

Color	Portion Who Said this Color	
	Percent	**Decimal**
●	26%	0.26
●	40%	0.40
●	4%	0.04
●		
●	14%	0.14

Tables help me.

① What % said red, blue, or yellow? "Or" means to add the percents.

a. 26% + 40% + 4% = 70%

② "How many times more" means to divide.

b. 0.26 ÷ 0.04 = 6.5 times more.

③ One way is to use decimals. Add the decimal numbers and write the decimal sum as a percent.

0.26 + 0.40 + 0.04 + 0.14 = 0.84 or 84%

Another way is to add the percent form of the numbers.

26% + 40% + 4% + 14% = 84%

To find the percent of students who said green, subtract from 100%.

c. 100% − 84% = 16% = 0.16

Reteaching and Enrichment Strategies

If students need help. . .	If students got it. . .
Resources by Chapter • Practice A and Practice B • Puzzle Time Record and Practice Journal Practice Differentiating the Lesson Lesson Tutorials Skills Review Handbook	Resources by Chapter • Enrichment and Extension • Technology Connection Start the next section

38. SCHOOL The percents of students who travel to school by car, bus, and bicycle are shown for a school of 825 students.

Car: 20%

School bus: 48%

Bicycle: 8%

   a. Write the percents as decimals.

   b. Write the percents as fractions.

   c. What percent of students use another method to travel to school?

   d. RESEARCH Make a bar graph that represents how the students in your class travel to school.

39. ELECTIONS In an election, the winning candidate receives 60% of the votes. What percent of the votes does the other candidate receive?

40. COLORS Students in a class were asked to tell their favorite color.

   a. What percent said red, blue, or yellow?

   b. How many times more students said red than yellow?

   c. Use two methods to find the percent of students who said green. Which method do you prefer?

41. **Problem Solving** In the first 42 Super Bowls, $0.1\overline{6}$ of the MVPs (most valuable players) were running backs.

   a. What percent of the MVPs were running backs?

   b. What fraction of the MVPs were *not* running backs?

Favorite Color

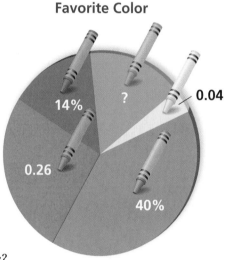

14% ? 0.04 0.26 40%

 Fair Game Review *What you learned in previous grades & lessons*

Write the decimal as a fraction or mixed number in simplest form.
(Skills Review Handbook)

42. 0.46   **43.** 0.31   **44.** 2.2   **45.** 4.32

Simplify the expression. *(Section 3.1)*

46. $4x + 3 - 9x$ **47.** $5 + 3.2n - 6 - 4.8n$

48. $2y - 5(y - 3)$ **49.** $-\frac{1}{2}(8b + 3) + 3b$

50. MULTIPLE CHOICE Ham costs $4.48 per pound. Cheese costs $6.36 per pound. You buy 1.5 pounds of ham and 0.75 pound of cheese. How much more do you pay for the ham? *(Skills Review Handbook)*

   (A) $1.41   **(B)** $1.95   **(C)** $4.77   **(D)** $6.18

Comparing and Ordering Fractions, Decimals, and Percents

Essential Question How can you order numbers that are written as fractions, decimals, and percents?

1 ACTIVITY: Using Fractions, Decimals, and Percents

Work with a partner. Decide which number form (fraction, decimal, or percent) is more common. Then find which is greater.

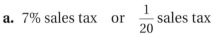

a. 7% sales tax   or $\frac{1}{20}$ sales tax

b. 0.37 cup of flour   or $\frac{1}{3}$ cup of flour

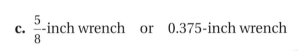

c. $\frac{5}{8}$-inch wrench   or   0.375-inch wrench

d. $12\frac{3}{5}$ dollars   or   12.56 dollars

e. 93% test score   or $\frac{7}{8}$ test score

f. $5\frac{5}{6}$ fluid ounces   or   5.6 fluid ounces

Fractions, Decimals, and Percents

In this lesson, you will

- compare and order fractions, decimals, and percents.
- solve real-life problems.

2 ACTIVITY: Ordering Numbers

Work with a partner to order the following numbers.

$$\frac{1}{8} \qquad 11\% \qquad \frac{3}{20} \qquad 0.172 \qquad 0.32 \qquad 43\% \qquad 7\% \qquad 0.7 \qquad \frac{5}{6}$$

a. Decide on a strategy for ordering the numbers. Will you write them all as fractions, decimals, or percents?

b. Use your strategy and a number line to order the numbers from least to greatest. (Note: Label the number line appropriately.)

Laurie's Notes

Introduction

Applying Mathematical Practices

- **Reason Abstractly and Quantitatively:** In this lesson, students will be converting between different representations of a number so that numbers may be compared. Mathematically proficient students would reason that $\frac{6}{11} > 0.48$ because $\frac{6}{11} > \frac{1}{2}$ and $\frac{1}{2} > 0.48$.

Motivate

- Write each number on an index card: $\frac{1}{10}, \frac{13}{20}, \frac{4}{5}$, 0.18, 0.45, 0.5, 35%, 60%, 85%.
- Hand out the 9 cards. Ask the 3 students holding fraction cards to stand in order (least to greatest). Then ask the 3 students holding decimal cards and the 3 students holding percent cards to do the same.
- Have each group of students describe the strategies for ordering themselves.
- Now ask all 9 students to order themselves.
- Discuss strategies used when all 3 forms are used in the same problem.

Activity Notes

Activity 1

- Remind students of the conversion triangle. Today they will work with all 3 forms.
- **Reason Abstractly and Quantitatively:** Remind students that there are two parts to each question.
- Discuss the results in class.
- **?** "Which fractions were repeating decimals?" Listen for knowledge that thirds and sixths are repeating decimals.
- **Common Error:** Some students who do not have a good understanding of how to convert a fraction to a decimal will take simple fractions such as $\frac{1}{20}$ and write 1.20 or 0.120. For $\frac{1}{3}$, they write 1.3 or 0.13.

Activity 2

- **Reason Abstractly and Quantitatively:** Have students work in groups, with each student sharing suggestions for how to order the numbers. Listen for key understanding, such as 0.7 and $\frac{5}{6}$ are the only two numbers greater than $\frac{1}{2}$.
- Note that the number line has scaling for tenths and only one number will be graphed in a tenth scale mark.
- In completing this activity, many skills are reviewed. This is an opportunity for informal assessment to guide instruction in the sections ahead.
- **Communication:** One group should present their answers to the class. They should discuss their results and the strategies they used.

What Your Students Will Learn

- Compare and order fractions, decimals, and percents by using the same number form.

Previous Learning

Students should know how to write halves, fourths, tenths, and hundredths as percents. Students also need to be able to convert between fractions and decimals, and multiply and divide decimals by powers of 10.

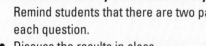

Lesson Plans
Complete Materials List

6.2 Record and Practice Journal

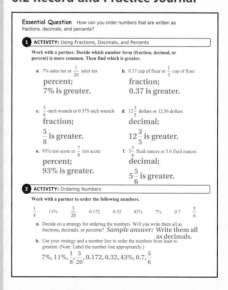

Essential Question How can you order numbers that are written as fractions, decimals, and percents?

1 ACTIVITY: Using Fractions, Decimals, and Percents

Work with a partner. Decide which number form (fraction, decimal, or percent) is more common. Then find which is greater.

a. 7% sales tax or $\frac{1}{20}$ sales tax
percent; 7% is greater.

b. 0.37 cup of flour or $\frac{1}{3}$ cup of flour
fraction; 0.37 is greater.

c. $\frac{5}{8}$-inch wrench or 0.375-inch wrench
fraction; $\frac{5}{8}$ is greater.

d. $12\frac{3}{5}$ dollars or 12.56 dollars
decimal; $12\frac{3}{5}$ is greater.

e. 93% test score or $\frac{7}{8}$ test score
percent; 93% is greater.

f. $5\frac{5}{6}$ fluid ounces or 5.6 fluid ounces
decimal; $5\frac{5}{6}$ is greater.

2 ACTIVITY: Ordering Numbers

Work with a partner to order the following numbers.

$\frac{1}{8}$ 11% $\frac{3}{20}$ 0.172   0.32   43%   7%   0.7 $\frac{5}{6}$

a. Decide on a strategy for ordering the numbers. Will you write them all as fractions, decimals, or percents? *Sample answer:* Write them all as decimals.

b. Use your strategy and a number line to order the numbers from least to greatest. (Note: Label the number line appropriately.)

7%, 11%, $\frac{1}{8}, \frac{3}{20}$, 0.172, 0.32, 43%, 0.7, $\frac{5}{6}$

T-220

6.2 Record and Practice Journal

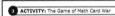

3 ACTIVITY: The Game of Math Card War

Preparation:

- Cut index cards to make 40 playing cards.*
- Write each number in the table onto a card.

75%	$\frac{3}{4}$	$\frac{1}{3}$	$\frac{3}{10}$	0.3	25%	0.4	0.25	100%	0.27
0.75	$66\frac{2}{3}$%	12.5%	40%	$\frac{1}{4}$	4%	0.5%	0.04	$\frac{1}{100}$	$\frac{2}{3}$
0	30%	5%	$\frac{27}{100}$	0.05	$33\frac{1}{3}$%	$\frac{2}{5}$	0.333...	27%	1%
1	0.01	$\frac{1}{20}$	$\frac{1}{8}$	0.125	$\frac{1}{25}$	$\frac{1}{200}$	0.005	0.666...	0%

To Play:

- Play with a partner.
- Deal 20 cards to each player facedown.
- Each player turns one card faceup. The player with the greater number wins. The winner collects both cards and places them at the bottom of his or her cards.
- Suppose there is a tie. Each player lays three cards facedown, then a new card faceup. The player with the greater of these new cards wins. The winner collects all 10 cards and places them at the bottom of his or her cards.
- Continue playing until one player has all the cards. This player wins the game.

  Check students' work.

*Cut-outs are available in the back of the Record and Practice Journal.

What Is Your Answer?

4. IN YOUR OWN WORDS How can you order numbers that are written as fractions, decimals, and percents? Give an example with your answer.

Convert all numbers to one form then order them.

5. All but one of the U.S. coins shown has a name that is related to its value. Which one is it? How are the names of the others related to their values?

Nickel; the name nickel comes from the metal the coin was originally made from; 1 cent = $\frac{1}{100}$; dime comes from the French word "disme" meaning tenth-part; quarter = $\frac{1}{4}$ dollar; half dollar = $\frac{1}{2}$ dollar

Laurie's Notes

Activity 3

- To preserve cards for multiple uses, make cards on colored card stock and store individual sets in plastic zipper bags.
- The card game *War* is common to most students. The question asked for each play is, "Which number is greater?" The player with the greater value collects both cards.
- If the cards have equivalent values $\left(\text{i.e., } 75\% \text{ and } \frac{3}{4}\right)$, there is a tie. As stated in the text, each player lays 3 cards face down and 1 card face up. The player with the card of greater value collects all of the cards in play.
- **Comparing Cards:** The learning component of this activity is when students actually compare two numbers. Listen to students discuss how they are comparing the numbers.
- To start the play, give students the opportunity to preview the cards. Explain the rules and let students begin.
- If one group finishes early, they should shuffle the cards and play again.
- **Extension:** You can use the same set of cards to do a matching activity with the cards either face up (easier) or down. If they make a 3-card match, such as 75%, $\frac{3}{4}$, and 0.75, they get a certain number of points.

  A 2-card match, such as 0 and 0%, is worth fewer points.
- **?** What cards have no match?

What Is Your Answer?

- Listen for the big idea, namely that the farther to the right the number is on the number line, the greater the number.

Closure

- **Exit Ticket:**
  - Write a percent between 25% and 40%. Write the equivalent fraction and decimal. *Sample answer:* 30%, $\frac{3}{10}$, 0.3
  - Write a percent between 75% and 90%. Write the equivalent fraction and decimal. *Sample answer:* 85%, $\frac{17}{20}$, 0.85

Math Practice

Make Sense of Quantities

What strategies can you use to determine which number is greater?

Preparation:

- Cut index cards to make 40 playing cards.
- Write each number in the table onto a card.

To Play:

- Play with a partner.
- Deal 20 cards facedown to each player.
- Each player turns one card faceup. The player with the greater number wins. The winner collects both cards and places them at the bottom of his or her cards.
- Suppose there is a tie. Each player lays three cards facedown, then a new card faceup. The player with the greater of these new cards wins. The winner collects all 10 cards and places them at the bottom of his or her cards.
- Continue playing until one player has all the cards. This player wins the game.

75%	$\frac{3}{4}$	$\frac{1}{3}$	$\frac{3}{10}$	0.3	25%	0.4	0.25	100%	0.27
0.75	$66\frac{2}{3}\%$	12.5%	40%	$\frac{1}{4}$	4%	0.5%	0.04	$\frac{1}{100}$	$\frac{2}{3}$
0	30%	5%	$\frac{27}{100}$	0.05	$33\frac{1}{3}\%$	$\frac{2}{5}$	0.333...	27%	1%
1	0.01	$\frac{1}{20}$	$\frac{1}{8}$	0.125	$\frac{1}{25}$	$\frac{1}{200}$	0.005	0.666...	0%

What Is Your Answer?

4. **IN YOUR OWN WORDS** How can you order numbers that are written as fractions, decimals, and percents? Give an example with your answer.

5. All but one of the U.S. coins shown has a name that is related to its value. Which one is it? How are the names of the others related to their values?

Practice Use what you learned about ordering numbers to complete Exercises 4–7, 16, and 17 on page 224.

When comparing and ordering fractions, decimals, and percents, write the numbers as all fractions, all decimals, or all percents.

EXAMPLE ① **Comparing Fractions, Decimals, and Percents**

a. **Which is greater, $\frac{3}{20}$ or 16%?**

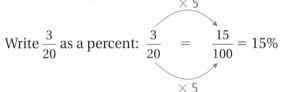

Write $\frac{3}{20}$ as a percent: $\frac{3}{20} = \frac{15}{100} = 15\%$

Study Tip

It is usually easier to order decimals or percents than to order fractions.

⋮∙ 15% is less than 16%. So, 16% is the greater number.

b. **Which is greater, 79% or 0.08?**

Write 79% as a decimal: $79\% = 79.\% = 0.79$

⋮∙ 0.79 is greater than 0.08. So, 79% is the greater number.

On Your Own

Now You're Ready
Exercises 4–15

1. Which is greater, 25% or $\frac{7}{25}$?   2. Which is greater, 0.49 or 94%?

EXAMPLE ② **Real-Life Application**

You, your sister, and a friend each take the same number of shots at a soccer goal. You make 72% of your shots, your sister makes $\frac{19}{25}$ of her shots, and your friend makes 0.67 of his shots. Who made the fewest shots?

Remember

To order numbers from least to greatest, write them as they appear on a number line from left to right.

Write 72% and $\frac{19}{25}$ as decimals.

You: $72\% = 72.\% = 0.72$ **Sister:** $\frac{19}{25} = \frac{76}{100} = 0.76$

Graph the decimals on a number line.

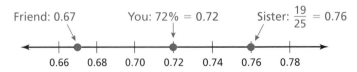

Friend: 0.67   You: 72% = 0.72   Sister: $\frac{19}{25}$ = 0.76

0.66   0.68   0.70   0.72   0.74   0.76   0.78

⋮∙ 0.67 is the least number. So, your friend made the fewest shots.

Laurie's Notes

Introduction

Connect
- **Yesterday:** Students compared and ordered common percents, decimals, and fractions.
- **Today:** Students will compare and order less common percents, decimals, and fractions.

Motivate
- Read the 5 statistics about the United States and have students order the percents from least to greatest.
  1) 90% of the states joined the United States before 1900.
  2) The U.S. population represents about 4.5% of the world population.
  3) About 20% of the U.S. population is under 15 years old.
  4) 62% of the states are entirely east of the Mississippi River.
  5) Water area represents about 6.9% of the total area of the United States.
- ? "Why is it usually easier to order decimals or percents than fractions?" Listen for an understanding of unlike denominators being harder to order than place value.

Lesson Notes

Example 1
- ? "To compare $\frac{3}{20}$ and 16%, should you write $\frac{3}{20}$ as a percent or 16% as a fraction and why?" Discuss both options and perhaps work through both options, if time permits.
- ? "Could you compare $\frac{3}{20}$ and 16% by writing each as a decimal?" yes; Depending on time, show this option as well.
- ? Work through part (b). "What is 0.08 as a percent?" 8%
- **Common Error:** Students might say 80% because they are comparing it to 79%.
- **Reason Abstractly and Quantitatively:** Take time to review and analyze the efficiency of the different strategies. For instance, if the fraction in part (a) had been $\frac{3}{19}$, it would have been difficult to write an equivalent fraction with a denominator of 100. Changing the fraction and the percent to decimals would have been the preferred strategy.

On Your Own
- Discuss methods used by students to make comparisons.

Example 2
- Note the scale on the number line. A common misconception is that the scale must be in units of 1, 5, or 10. Even digits are not as common.
- ? "Why is scaling by even digits helpful in this problem? Could we use a scale of 10?" Even digits show the distance between the 3 numbers in an accurate display. You could use 10, but it would be less accurate.

Goal Today's lesson is comparing and ordering less common fractions, decimals, and percents.

Lesson Tutorials
Lesson Plans
Answer Presentation Tool

Extra Example 1

a. Which is greater, $\frac{17}{20}$ or 80%? $\frac{17}{20}$

b. Which is greater, 28% or 0.29? 0.29

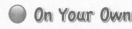

On Your Own

1. $\frac{7}{25}$ 2. 94%

Extra Example 2

You, your sister, and your friend each take the same number of shots at a soccer goal. You make 0.67 of your shots, your sister makes 68% of her shots, and your friend makes $\frac{17}{20}$ of her shots. Who made the fewest shots? You made the fewest shots.

On Your Own

3. You made the most shots.

Extra Example 3

The table shows the portions of the population of Rhode Island that live in each county. List the counties in order by population from least to greatest.

Counties	Fraction	Decimal	Percent
Bristol	$\frac{1}{20}$		
Kent		0.16	
Newport	$\frac{2}{25}$		
Providence			59%
Washington	$\frac{3}{25}$		

Bristol, Newport, Washington, Kent, Providence

On Your Own

4. Washington, Michigan, Ohio, Illinois, New York, Texas, and California

Differentiated Instruction

Kinesthetic

Some students have difficulty seeing that 60% is not only equal to "60 out of 100," but is also equal to "6 out of 10" and "3 out of 5." Give students three 100-grid squares. In the first grid, have students shade 60 out of 100 squares. With the second grid, have students draw thick lines around groups of 10 squares and then shade 6 of the groups. With the third grid, have students draw thick lines around groups of 20 squares and then shade 3 of the groups. When students compare all three grids, they should see that the same amount of 60 squares have been shaded on each of the grids. This can be seen easily if the students have grouped the squares by rows or columns.

Laurie's Notes

On Your Own

- **Think-Pair-Share:** Students should read the question independently and then work in pairs to answer the question. When they have answered the question, the pair should compare their answer with another group and discuss any discrepancies.
- Have students graph the three numbers on a scaled number line.

Example 3

- Give students time to read the information in the problem.
- **?** "Interpret what it means for $\frac{1}{50}$ of the U.S. population to live in Washington." Listen for an understanding that one in every 50 U.S. residents lives in Washington.
- **?** **Look for and Express Regularity in Repeated Reasoning:** "If there were 100 U.S. residents, how many would live in Washington?" 2 "If there were 1000 U.S. residents, how many would live in Washington?" 20
- Discuss the columns of the table. Note that the decimals and percents are easier to compare than the fractions.

On Your Own

- **Construct Viable Arguments and Critique the Reasoning of Others:** There are several ways in which students may justify their answers. Take time to hear a variety of approaches.

Closure

- Complete this table of common fractions, decimals, and percents.

Fraction	$\frac{1}{4}$	$\frac{3}{10}$	$\frac{1}{2}$	$\frac{2}{5}$	$\frac{1}{20}$	$\frac{3}{20}$	$\frac{7}{10}$	$\frac{3}{4}$	$\frac{1}{5}$	1
Decimal	0.25	0.3	0.5	0.4	0.05	0.15	0.7	0.75	0.2	1
Percent	25%	30%	50%	40%	5%	15%	70%	75%	20%	100%

Now You're Ready
Exercises 16–21

On Your Own

3. You make 75% of your shots, your sister makes $\frac{13}{20}$ of her shots, and your friend makes 0.7 of his shots. Who made the most shots?

EXAMPLE 3 **Real-Life Application**

Washington: $\frac{1}{50}$ Michigan: 0.03

New York: 6%

California: 0.12

Ohio: $\frac{1}{25}$

The map shows the portions of the U.S. population that live in five states.

List the five states in order by population from least to greatest.

Begin by writing each portion as a fraction, a decimal, and a percent.

State	Fraction	Decimal	Percent
Michigan	$\frac{3}{100}$	0.03	3%
New York	$\frac{6}{100}$	0.06	6%
Washington	$\frac{1}{50}$	0.02	2%
California	$\frac{12}{100}$	0.12	12%
Ohio	$\frac{1}{25}$	0.04	4%

Graph the percent for each state on a number line.

Michigan: 3%   New York: 6%
Washington: 2%   Ohio: 4%   California: 12%

0% 2% 4% 6% 8% 10% 12% 14%

⋮ The states in order by population from least to greatest are Washington, Michigan, Ohio, New York, and California.

On Your Own

4. The portion of the U.S. population that lives in Texas is $\frac{2}{25}$. The portion that lives in Illinois is 0.042. Reorder the states in Example 3 including Texas and Illinois.

 ## Vocabulary and Concept Check

1. **NUMBER SENSE** Copy and complete the table.

2. **NUMBER SENSE** How would you decide whether $\frac{3}{5}$ or 59% is greater? Explain.

3. **WHICH ONE DOESN'T BELONG?** Which one does *not* belong with the other three? Explain your reasoning.

40%	$\frac{2}{5}$
0.4	0.04

Fraction	Decimal	Percent
$\frac{18}{25}$	0.72	
$\frac{17}{20}$		85%
$\frac{13}{50}$		
	0.62	
		45%

 ## Practice and Problem Solving

Tell which number is greater.

① **4.** 0.9, 95%   **5.** 20%, 0.02   **6.** $\frac{37}{50}$, 37%   **7.** 50%, $\frac{13}{25}$

8. 0.086, 86%   **9.** 76%, 0.67   **10.** 60%, $\frac{5}{8}$ **11.** 0.12, 1.2%

12. 17%, $\frac{4}{25}$ **13.** 140%, 0.14   **14.** $\frac{1}{3}$, 30%   **15.** 80%, $\frac{7}{9}$

Use a number line to order the numbers from least to greatest.

② **16.** 38%, $\frac{8}{25}$, 0.41   **17.** 68%, 0.63, $\frac{13}{20}$

18. $\frac{43}{50}$, 0.91, $\frac{7}{8}$, 84%   **19.** 0.15%, $\frac{3}{20}$, 0.015

20. 2.62, $2\frac{2}{5}$, 26.8%, 2.26, 271%   **21.** $\frac{87}{200}$, 0.44, 43.7%, $\frac{21}{50}$

22. **TEST** You answered 21 out of 25 questions correctly on a test. Did you reach your goal of getting at least 80%?

23. **POPULATION** The table shows the portions of the world population that live in four countries. Order the countries by population from least to greatest.

Country	Brazil	India	Russia	United States
Portion of World Population	2.8%	$\frac{7}{40}$	$\frac{1}{50}$	0.044

Assignment Guide and Homework Check

Level	Day 1 Activity Assignment	Day 2 Lesson Assignment	Homework Check
Basic	4–7, 16, 17, 33–36	1–3, 9–15 odd, 19–29 odd	9, 13, 19, 25
Average	4–7, 16, 17, 33–36	1–3, 12–15, 20–22, 25–29 odd, 30, 31	13, 20, 22, 25, 27
Advanced	4–7, 16, 17, 33–36	1–3, 8–14 even, 18–30 even, 31, 32	18, 22, 24, 26
Accelerated	1–7, 8–30 even, 31–36		18, 22, 24, 26

Common Errors

- **Exercises 4–21** Students may try to order the numbers without converting them or will only convert them mentally and do so incorrectly. Tell them that it is necessary to convert all the numbers to one form and that they should write out the steps to make sure that they are converting correctly.
- **Exercises 24 and 25** Students may try to round the numbers that have repeating decimals and incorrectly order the numbers. Remind them that even though you often round repeating decimals, the decimal is actually less than or greater than the rounded decimal. For example, $0.\overline{6}$ is less than 0.667, but greater than 0.666.

6.2 Record and Practice Journal

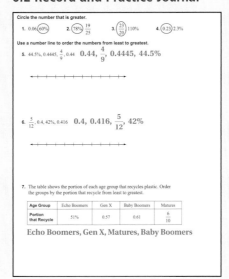

Vocabulary and Concept Check

1.

$\frac{18}{25}$	0.72	72%
$\frac{17}{20}$	0.85	85%
$\frac{13}{50}$	0.26	26%
$\frac{31}{50}$	0.62	62%
$\frac{9}{20}$	0.45	45%

2. *Sample answer*: Write $\frac{3}{5}$ as a percent, then compare 59% to determine which is greater.

3. 0.04; $0.04 = 4\%$, but 40%, $\frac{2}{5}$, and 0.4 are all equal to 40%.

Practice and Problem Solving

4. 95%   **5.** 20%

6. $\frac{37}{50}$ **7.** $\frac{13}{25}$

8. 86%   **9.** 76%

10. $\frac{5}{8}$ **11.** 0.12

12. 17%   **13.** 140%

14. $\frac{1}{3}$ **15.** 80%

16.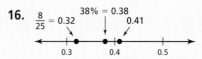

17.

18.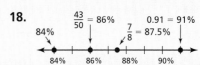

19.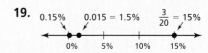

20. See Additional Answers.

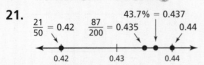

21.

$\frac{21}{50} = 0.42$ $\frac{87}{200} = 0.435$ 43.7% = 0.437   0.44

0.42   0.43   0.44

22. yes

23. Russia, Brazil, United States, India

24. $0.66, 66.1\%, \frac{2}{3}, 0.667$

25. $21\%, 0.2\overline{1}, \frac{11}{50}, \frac{2}{9}$

26. *A*   **27.** *D*

28. *B*   **29.** *C*

30. Stage 21, Stage 7, Stage 8, Stage 1, Stage 17

31. See *Taking Math Deeper*.

32. a. 7

    b. There is none. $\frac{1}{a}$ is less than 33% when a is greater than 3, but when a is greater than 3, $\frac{a}{8}$ is greater than 33%.

33. yes   **34.** no

35. yes   **36.** D

Mini-Assessment

Tell which number is greater.

1. 27%, 0.48 0.48

2. $\frac{2}{5}$, 0.125 $\frac{2}{5}$

3. $\frac{3}{4}$, 76% 76%

4. $\frac{5}{8}$, 60% $\frac{5}{8}$

5. On a quiz, you answer 7 out of 10 questions correctly. Did you reach your goal of getting 80% or better? no

Taking Math Deeper

Exercise 31

The problem doesn't tell students which form to use to order the numbers. It is up to the students to decide when percents or decimals are easier to use than fractions.

1   Use a table to organize the information.

Animal	Portion of Day Sleeping		
	Percent	**Decimal**	**Fraction**
Dolphin	43.3%	0.433	
Lion	56.3%	0.563	
Rabbit	47.5%	0.475	$\frac{19}{40}$
Squirrel	62.0%	0.620	$\frac{31}{50}$
Tiger	65.8%	0.658	

2   Use a number line to order the decimals.

a.

Dolphin   Rabbit   Lion   Squirrel   Tiger

0.4   0.5   0.6   0.7

3   Using an estimate of 8 hours a day, the portion of the day a teen sleeps is

    b. $\frac{8}{24} = \frac{1}{3} = 33.\overline{3}\%$.

    c. On the ordered list, the teen would be first!

Project

Use a chart to keep track of the number of hours you sleep each day for a week. Compare your chart with three of your classmates. Find the average number of hours each of you sleep each day. Which day seems to be the sleepiest day of the week?

Reteaching and Enrichment Strategies

If students need help. . .	If students got it. . .
Resources by Chapter • Practice A and Practice B • Puzzle Time Record and Practice Journal Practice Differentiating the Lesson Lesson Tutorials Skills Review Handbook	Resources by Chapter • Enrichment and Extension • Technology Connection Start the next section

PRECISION Order the numbers from least to greatest.

24. 66.1%, 0.66, $\frac{2}{3}$, 0.667

25. $\frac{2}{9}$, 21%, $0.2\overline{1}$, $\frac{11}{50}$

Tell which letter shows the graph of the number.

26. $\frac{2}{5}$ **27.** 45.2%   **28.** 0.435   **29.** $\frac{4}{9}$

30. TOUR DE FRANCE The Tour de France is a bicycle road race. The whole race is made up of 21 small races called *stages*. The table shows how several stages compare to the whole Tour de France in a recent year. Order the stages from shortest to longest.

Stage	1	7	8	17	21
Portion of Total Distance	$\frac{11}{200}$	0.044	$\frac{6}{125}$	0.06	4%

31. SLEEP The table shows the portions of the day that several animals sleep.

  a. Order the animals by sleep time from least to greatest.

  b. Estimate the portion of the day that you sleep.

  c. Where do you fit on the ordered list?

Animal	Portion of Day Sleeping
Dolphin	0.433
Lion	56.3%
Rabbit	$\frac{19}{40}$
Squirrel	$\frac{31}{50}$
Tiger	65.8%

32. **Number Sense** Tell what whole number you can substitute for a in each list so the numbers are ordered from least to greatest. If there is none, explain why.

  a. $\frac{2}{a}$, $\frac{a}{22}$, 33%

  b. $\frac{1}{a}$, $\frac{a}{8}$, 33%

 Fair Game Review What you learned in previous grades & lessons

Tell whether the ratios form a proportion. *(Section 5.2)*

33. $\frac{6}{10}$, $\frac{9}{15}$

34. $\frac{7}{16}$, $\frac{28}{80}$

35. $\frac{20}{12}$, $\frac{35}{21}$

36. MULTIPLE CHOICE What is the solution of $2n - 4 > -12$? *(Section 4.4)*

  (A) $n < -10$ **(B)** $n < -4$ **(C)** $n > -2$ **(D)** $n > -4$

Essential Question How can you use models to estimate percent questions?

The statement "25% of 12 is 3" has three numbers. In real-life problems, any one of these numbers can be unknown.

Question	Which number is missing?	Type of Question
What is 25% of 12?	3	Find a part of a number.
3 is what percent of 12?	25%	Find a percent.
3 is 25% of what?	12	Find the whole.

1 ACTIVITY: Estimating a Part

Work with a partner. Use a model to estimate the answer to each question.

a. What number is 50% of 30?

⋮ So, from the model, ⬚ is 50% of 30.

b. What number is 75% of 30? **c.** What number is 40% of 30?

d. What number is 6% of 30? **e.** What number is 65% of 30?

2 ACTIVITY: Estimating a Percent

Percent Proportion

In this lesson, you will

• use the percent proportion to find parts, wholes, and percents.

Work with a partner. Use a model to estimate the answer to each question.

a. 15 is what percent of 75?

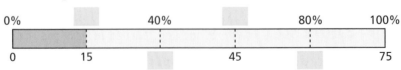

⋮ So, from the model, 15 is ⬚ of 75.

b. 5 is what percent of 20? **c.** 18 is what percent of 40?

d. 50 is what percent of 80? **e.** 75 is what percent of 50?

Laurie's Notes

What Your Students Will Learn
- Use the percent proportion to find parts, wholes, and percents.

Previous Learning
Students should know how to solve simple percent problems and how to use ratio tables.

Introduction

Applying Mathematical Practices
- **Reason Abstractly and Quantitatively** and **Model with Mathematics:** In this lesson, students use the concept of a proportion to solve different types of percent problems. The percent bar model and ratio tables help student reasoning from a visual and numeric display.

Motivate
- Share with students that sometimes their thinking can get *scrambled up* while solving percent problems, so an egg model would be a good way to introduce the chapter!
- **?** Use an egg carton to help visualize a few simple percent problems.
  - "What is 75% of 12?" 9
  - "3 is what percent of 12?" 25%
  - "12 is 50% of what number?" 24

Technology for the **Teacher**

Dynamic Classroom

Lesson Plans
Complete Materials List

Activity Notes

Activity 1
- **FYI:** You may want to begin with a quick review of fractional equivalents of the following common percents:
  10%, 20%, 30%, 40%, 60%, 70%, 80%, 90%, 25%, 50%, 75%, $33\frac{1}{3}$%, $66\frac{2}{3}$%

- **Use Appropriate Tools Strategically:** The percent bar model is an effective tool for estimating an answer, or judging the reasonableness of an answer if students have an understanding of fractional parts of a whole.

- The length of the bar is 100%, the whole. Percents near 50% are about $\frac{1}{2}$ of the whole.
- Students should be able to judge percents near 25% $\left(\frac{1}{4}\right)$ and 75% $\left(\frac{3}{4}\right)$.

- Students should locate the percents on the same model.
- When students have finished, draw a percent bar model on the board. Have volunteers share their answers.
- Remind students that these are approximations. Check for reasonableness in their approximations. For example, 40% is closer to 50% than 25%.

Activity 2
- **FYI:** Some students may find it helpful to use a long strip of paper that they can fold or write on when answering questions.
- **Reason Abstractly and Quantitatively:** Students may wonder why the percent bar model was divided into 5 parts versus 2 parts or 4 parts. Students should reason that if you divide into quarters, half of 75 is about 37, and half of 37 is about 18. Because you want to know what percent 15 is of 75, a percent bar model divided into 4 parts would allow you to estimate that it is less than 25%. To get a closer estimate you would try smaller parts, such as fifths, as shown.
- Students should be able to use mental math to find 10% of any number. Knowing 10%, it is easy to find 20% (double 10%).

6.3 Record and Practice Journal

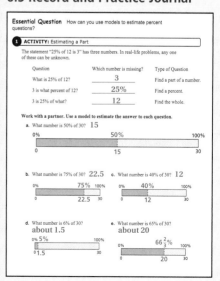

Visual

Have students use colored pencils to write the percent proportion in words in their notebooks. Then use the colored pencils to underline or circle the corresponding numbers in the problem statement.

$$\frac{\text{part}}{\text{whole}} = \frac{\text{percent}}{100}$$

(18) is what (percent) of (40)?

$$\frac{18}{40} = \frac{p}{100}$$

6.3 Record and Practice Journal

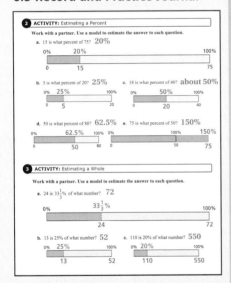

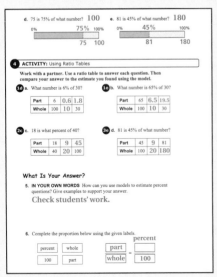

Laurie's Notes

Activity 3

- Encourage students to draw a percent bar model for each problem. You could also provide strips of paper that students can fold or write on.
- It is important that students be able to approximate the part in the whole model. Ask questions such as, "Is it greater than or less than one-half? Is it greater than or less than one-quarter?"
- **Model with Mathematics:** Estimating the whole is generally easier for students. You can draw a model of what you have, as a part and as a percent. Because you know the percent, you know how the percent bar model should be scaled to find the whole.

Activity 4

- Percent bar models help you estimate answers. Ratio tables can be used to find the exact answers. Remind students how they found the percent of a number in a previous course (multiplication) and how they found the whole (division). There are numerous ways to use ratio tables to find these values.
- Break down the ratios shown in each of the tables and compare. Discuss the "common proportion" to help lead into Question 6 and the Key Idea in tomorrow's lesson.

What Is Your Answer?

- **Think-Pair-Share:** Students should read each question independently and then work in pairs to answer the questions. When they have answered the questions, the pair should compare their answers with another group and discuss any discrepancies.

Closure

- Use the model shown. What 3 questions could be asked? "24 is what percent of 60?"; "What is 40% of 60?"; "24 is 40% of what number?"

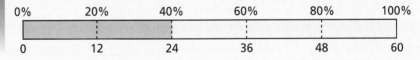

3 ACTIVITY: Estimating a Whole

Math Practice

Use a Model

What quantities are given? How can you use the model to find the unknown quantity?

Work with a partner. Use a model to estimate the answer to each question.

a. 24 is $33\frac{1}{3}\%$ of what number?

∴ So, from the model, 24 is $33\frac{1}{3}\%$ of [ ].

b. 13 is 25% of what number?   **c.** 110 is 20% of what number?

d. 75 is 75% of what number?   **e.** 81 is 45% of what number?

4 ACTIVITY: Using Ratio Tables

Work with a partner. Use a ratio table to answer each question. Then compare your answer to the estimate you found using the model.

1d a. What number is 6% of 30?

Part	6		
Whole	100		30

1e b. What number is 65% of 30?

Part	65		
Whole	100		30

2c c. 18 is what percent of 40?

Part	18		
Whole	40		100

3e d. 81 is 45% of what number?

Part	45		81
Whole	100		

What Is Your Answer?

5. IN YOUR OWN WORDS How can you use models to estimate percent questions? Give examples to support your answer.

6. Complete the proportion below using the given labels.

percent

whole

100

part

$$\frac{}{} = \frac{}{}$$

Practice ▶ Use what you learned about estimating percent questions to complete Exercises 5–10 on page 230.

Key Idea

The Percent Proportion

Words You can represent "a is p percent of w" with the proportion

$$\frac{a}{w} = \frac{p}{100}$$

where a is part of the whole w, and $p\%$, or $\dfrac{p}{100}$, is the percent.

Numbers   3 out of 4 is 75%.

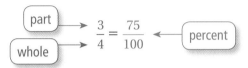

Study Tip

In percent problems, the word *of* is usually followed by the whole.

EXAMPLE ❶ **Finding a Percent**

What percent of 15 is 12?

$$\frac{a}{w} = \frac{p}{100}$$ Write the percent proportion.

$$\frac{12}{15} = \frac{p}{100}$$ Substitute 12 for a and 15 for w.

$$100 \cdot \frac{12}{15} = 100 \cdot \frac{p}{100}$$ Multiplication Property of Equality

$$80 = p$$ Simplify.

∴ So, 80% of 15 is 12.

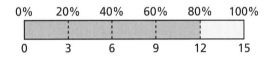

EXAMPLE ❷ **Finding a Part**

What number is 36% of 50?

$$\frac{a}{w} = \frac{p}{100}$$ Write the percent proportion.

$$\frac{a}{50} = \frac{36}{100}$$ Substitute 50 for w and 36 for p.

$$50 \cdot \frac{a}{50} = 50 \cdot \frac{36}{100}$$ Multiplication Property of Equality

$$a = 18$$ Simplify.

∴ So, 18 is 36% of 50.

Laurie's Notes

Introduction

Connect

- **Yesterday:** Students used the percent bar model to explore three types of percent problems.
- **Today:** Students will use the percent proportion to solve three types of percent problems.

Motivate

- Share information about the Enhanced Fujita Scale (EF Scale) used to rate the strength of a tornado based on estimated wind speeds and related damage.
- April through July are the four months with the highest frequency of tornadoes.
- Briefly discuss any experiences students have had with tornadoes and explain that they will come back to tornadoes at the end of class.

Enhanced Fujita Scale	
EF Number	3 Second Gust (mph)
0	65–85
1	86–110
2	111–135
3	136–165
4	166–200
5	Over 200

Lesson Notes

Key Idea

- Students should be familiar with the vocabulary *part*, *whole*, and *percent* from yesterday's activity.

Words of Wisdom

- **Reason Abstractly and Quantitatively** and **Model with Mathematics:** In all of the examples today, draw a percent bar model off to the side to help student reasoning. The model helps students estimate and consider the reasonableness of the answer.

Example 1

- ❓ "What is the whole? 15 What is the part?" 12.
- ❓ "Is 12 more or less than 50% of 15?" more
- Set up the percent proportion, substitute the known quantities, and solve.
- ❓ "What other strategies could be used to solve the proportion?" Simplify $\frac{12}{15} = \frac{4}{5}$, then write $\frac{4}{5}$ as a percent.

Example 2

- ❓ "Are you looking for a whole or a part?" looking for the part
- Draw a percent bar model with 50 as the whole. Divide into 5 parts and label the corresponding percents and amounts. Because 36% is closest to 40%, ask students to estimate an answer, which would be close to 20.
- Set up the percent proportion, substitute the known quantities, and solve.
- Refer back to the model to confirm that the answer makes sense.

Goal Today's lesson is finding percents using the percent proportion.

Lesson Tutorials
Lesson Plans
Answer Presentation Tool

English Language Learners

Vocabulary

English language learners need to be able to distinguish between the *whole* and the *part of the whole* when setting up a percent proportion. Have students write the percent proportion $\frac{a}{w} = \frac{p}{100}$ and work with the following two statements.

    35 is 25% of 140

    175% of 36 is 63

Students should easily be able to substitute the number for the percent *p*. The whole *w* is the number after the word *of*. The remaining number is the part of the whole *a*. So, the percent proportions are

$$\frac{35}{140} = \frac{25}{100} \text{ and } \frac{63}{36} = \frac{175}{100}.$$

Extra Example 1

What percent of 20 is 8?
40%

Extra Example 2

What number is 45% of 80?
36

Extra Example 3

210% of what number is 84?
40

On Your Own

1. $\frac{3}{5} = \frac{p}{100}; p = 60$

2. $\frac{25}{20} = \frac{p}{100}; p = 125$

3. $\frac{a}{60} = \frac{80}{100}; a = 48$

4. $\frac{a}{40.5} = \frac{10}{100}; a = 4.05$

5. $\frac{4}{w} = \frac{0.1}{100}; w = 4000$

6. $\frac{\frac{1}{2}}{w} = \frac{25}{100}; w = 2$

Extra Example 4

Using the bar graph from Example 4, what percent of the tornadoes were EF2s?
20%

On Your Own

7. 29 tornadoes

Example 3

? "Are you looking for a whole or a part?" 24 is the part, so you are looking for the whole.

• Draw a percent bar model. The whole is the number associated with the 100%, which is not obvious to all students. Working with percents greater than 100% is more challenging.

• Set up the percent proportion, substitute the known quantities, and solve.

• Refer back to the percent bar model to confirm that the answer makes sense.

On Your Own

• The decimals and fractions included in these exercises can present problems for some students. Encourage students to set up the percent proportion and take time to estimate an answer. For instance, Question 5 says you take a very small percent of a number, and you get 4. You have to start with a fairly large number because 1% of 100 is 4.

• **Neighbor Check:** Have students work independently and then have their neighbors check their work. Have students discuss any discrepancies.

Example 4

• Have students read the bar graph and discuss the information displayed.

? "How can you find the percent of tornadoes that were EF1s?" The whole would be the total number of tornadoes (145) and the part would be the number that were EF1s (58).

? "Will the answer be more or less than 50%? Explain." Less, because half of 145 is about 72 and 58 < 72.

• Set up the percent proportion, substitute the known quantities, and solve.

On Your Own

• **Neighbor Check:** Have students work independently and then have their neighbors check their work. Have students discuss any discrepancies.

Closure

• An average of 1253 tornadoes occur in the U.S. each year, and about 12.5% of them are in Texas. About how many tornadoes does Texas have each year? about 157

150% of what number is 24?

$$\frac{a}{w} = \frac{p}{100}$$ Write the percent proportion.

$$\frac{24}{w} = \frac{150}{100}$$ Substitute 24 for a and 150 for p.

$$24 \cdot 100 = w \cdot 150$$ Cross Products Property

$$2400 = 150w$$ Multiply.

$$16 = w$$ Divide each side by 150.

∴ So, 150% of 16 is 24.

On Your Own

Now You're Ready
Exercises 11–18

Write and solve a proportion to answer the question.

1. What percent of 5 is 3?
2. 25 is what percent of 20?
3. What number is 80% of 60?
4. 10% of 40.5 is what number?
5. 0.1% of what number is 4?
6. $\frac{1}{2}$ is 25% of what number?

EXAMPLE ④ Real-Life Application

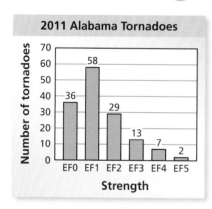

2011 Alabama Tornadoes

The bar graph shows the strengths of tornadoes that occurred in Alabama in 2011. What percent of the tornadoes were EF1s?

The total number of tornadoes, 145, is the *whole*, and the number of EF1 tornadoes, 58, is the *part*.

$$\frac{a}{w} = \frac{p}{100}$$ Write the percent proportion.

$$\frac{58}{145} = \frac{p}{100}$$ Substitute 58 for a and 145 for w.

$$100 \cdot \frac{58}{145} = 100 \cdot \frac{p}{100}$$ Multiplication Property of Equality

$$40 = p$$ Simplify.

∴ So, 40% of the tornadoes were EF1s.

On Your Own

7. Twenty percent of the tornadoes occurred in central Alabama on April 27. How many tornadoes does this represent?

Vocabulary and Concept Check

1. **VOCABULARY** Write the percent proportion in words.

2. **WRITING** Explain how to use a proportion to find 30% of a number.

3. **NUMBER SENSE** Write and solve the percent proportion represented by the model.

0%   20%   40%   60%   80%   100%

0   40

4. **WHICH ONE DOESN'T BELONG?** Which proportion does *not* belong with the other three? Explain your reasoning.

$$\frac{15}{w} = \frac{50}{100}$$ $$\frac{12}{15} = \frac{40}{n}$$ $$\frac{15}{25} = \frac{p}{100}$$ $$\frac{a}{20} = \frac{35}{100}$$

Practice and Problem Solving

Use a model to estimate the answer to the question. Use a ratio table to check your answer.

5. What number is 24% of 80?

6. 15 is what percent of 40?

7. 15 is 30% of what number?

8. What number is 120% of 70?

9. 20 is what percent of 52?

10. 48 is 75% of what number?

Write and solve a proportion to answer the question.

① 11. What percent of 25 is 12?

12. 14 is what percent of 56?

② 13. 25% of what number is 9?

14. 36 is 0.9% of what number?

③ 15. 75% of 124 is what number?

16. 110% of 90 is what number?

17. What number is 0.4% of 40?

18. 72 is what percent of 45?

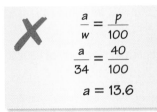

$$\frac{a}{w} = \frac{p}{100}$$
$$\frac{a}{34} = \frac{40}{100}$$
$$a = 13.6$$

19. **ERROR ANALYSIS** Describe and correct the error in using the percent proportion to answer the question below.

"40% of what number is 34?"

20. **FITNESS** Of 140 seventh-grade students, 15% earn the Presidential Physical Fitness Award. How many students earn the award?

21. **COMMISSION** A salesperson receives a 3% commission on sales. The salesperson receives $180 in commission. What is the amount of sales?

Assignment Guide and Homework Check

Level	Day 1 Activity Assignment	Day 2 Lesson Assignment	Homework Check
Basic	5–10, 32–35	1–4, 11–27 odd	11, 13, 15, 21
Average	5–10, 32–35	1–4, 11–21 odd, 22–30 even	11, 13, 15, 26
Advanced	5–10, 32–35	1–4, 12–18 even, 19, 20–30 even, 31	12, 14, 16, 24, 28
Accelerated	1–10, 12–18 even, 19, 20–30 even, 31–35		12, 14, 16, 24, 28

For Your Information

- **Exercise 21** Students may get confused by the word *commission*. Tell students that commission is a fee or percentage allowed to a sales representative or an agent for services rendered.

Common Errors

- **Exercises 5–18** Students may not know what number to substitute for each variable. Walk through each type of question with the students. Emphasize that the word *is* means *equals*, and *of* means *to multiply*. Tell students to write the question and then write the meaning of each word or group of words underneath.
- **Exercises 20 and 21** Students will mix up the whole and the part when trying to write the percent proportion for the word problems. Ask them to identify each part of the proportion before writing it in the proportion format.
- **Exercise 29** Students may struggle with this exercise because there is no vertical scale. Tell students to think of the bars as a model. This will get them heading in the right direction.

6.3 Record and Practice Journal

Write and solve a proportion to answer the question.

1. 40% of 60 is what number?
$a = 0.4 \cdot 60; 24$

2. 17 is what percent of 50?
$17 = p \cdot 50; 34\%$

3. 38% of what number is 57?
$57 = 0.38 \cdot w; 150$

4. 44% of 25 is what number?
$a \cdot 0.44 \cdot 25; 11$

5. 52 is what percent of 50?
$52 = p \cdot 50; 104\%$

6. 150% of what number is 18?
$18 = 1.5 \cdot w; 12$

7. You put 60% of your paycheck into your savings account. Your paycheck is $235. How much money do you put in your savings account?
$141

Vocabulary and Concept Check

1. The percent proportion is $\frac{a}{w} = \frac{p}{100}$ where a is part of the whole w, and $p\%$, or $\frac{p}{100}$, is the percent.

2. 30% is $\frac{30}{100}$ and the number w is the whole. Set up the percent proportion as $\frac{a}{w} = \frac{30}{100}$ and solve for a.

3. $\frac{a}{40} = \frac{60}{100}$; $a = 24$

4. $\frac{12}{15} = \frac{40}{n}$; This proportion is not a percent proportion.

Practice and Problem Solving

5. 20

6. 37.5%

7. 50

8. 84

9. about 37.5%

10. 64

11. $\frac{12}{25} = \frac{p}{100}$; $p = 48$

12. $\frac{14}{56} = \frac{p}{100}$; $p = 25$

13. $\frac{9}{w} = \frac{25}{100}$; $w = 36$

14. $\frac{36}{w} = \frac{0.9}{100}$; $w = 4000$

15. $\frac{a}{124} = \frac{75}{100}$; $a = 93$

16. $\frac{a}{90} = \frac{110}{100}$; $a = 99$

17. $\frac{a}{40} = \frac{0.4}{100}$; $a = 0.16$

18. $\frac{72}{45} = \frac{p}{100}$; $p = 160$

19. See Additional Answers.

20. 21 students

21. $6000

22. $\dfrac{0.5}{20} = \dfrac{p}{100}$; $p = 2.5$

23. $\dfrac{14.2}{w} = \dfrac{35.5}{100}$; $w = 40$

24. $\dfrac{\frac{3}{4}}{w} = \dfrac{60}{100}$; $w = 1\frac{1}{4}$

25. $\dfrac{a}{\frac{7}{8}} = \dfrac{25}{100}$; $a = \dfrac{7}{32}$

26. 4 left   **27.** $8.40

28. $66\frac{2}{3}\%$

29. See Additional Answers.

30. See *Taking Math Deeper*.

31. a. 62.5%

   b. $52x$

Fair Game Review

32. 3   **33.** -0.6

34. -2.5 **35.** B

Mini-Assessment

Write and solve a proportion to answer the question.

1. What percent of 35 is 28?

   $\dfrac{29}{35} = \dfrac{p}{100}$; $p = 80$

2. What number is 28% of 50?

   $\dfrac{a}{50} = \dfrac{28}{100}$; $a = 14$

3. 160% of what number is 144?

   $\dfrac{144}{w} = \dfrac{160}{100}$; $w = 90$

4. 0.15% of what number is 10.5?

   $\dfrac{10.5}{w} = \dfrac{0.15}{100}$; $w = 7000$

5. You score an 80% on your test. You answer 44 questions correctly. How many questions were on the test?

   $\dfrac{44}{w} = \dfrac{80}{100}$; $w = 55$

Taking Math Deeper

Exercise 30

You can write and solve an equation to find the answer to this problem. You can also use other ways, such as models, to solve this problem.

 Use a circular model.

Let the circle represent 100% of the number.

You know that 20% of the number is *x*. Draw a section that represents 20% of the circle and label it *x*.

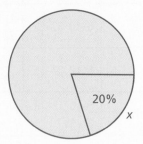

 Notice that you can draw four more 20% sections in the circle, each representing *x*. The circle contains five of these sections.

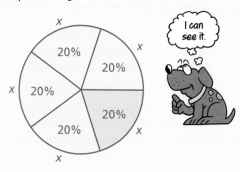

So, 100% of the number is 5*x*.

 You can also use a rectangular model.

You know that 20% of the number is *x*. So, for each 20% you add on the model, add another *x*. Stop when you reach 100%.

0%	20%	40%	60%	80%	100%
0	*x*	2*x*	3*x*	4*x*	5*x*

So, 100% of the number is 5*x*.

Reteaching and Enrichment Strategies

If students need help. . .	If students got it. . .
Resources by Chapter • Practice A and Practice B • Puzzle Time Record and Practice Journal Practice Differentiating the Lesson Lesson Tutorials Skills Review Handbook	Resources by Chapter • Enrichment and Extension • Technology Connection Start the next section

Write and solve a proportion to answer the question.

22. 0.5 is what percent of 20?

23. 14.2 is 35.5% of what number?

24. $\frac{3}{4}$ is 60% of what number?

25. What number is 25% of $\frac{7}{8}$?

26. HOMEWORK You are assigned 32 math exercises for homework. You complete 87.5% of these before dinner. How many do you have left to do after dinner?

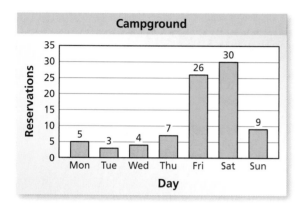

27. HOURLY WAGE Your friend earns $10.50 per hour. This is 125% of her hourly wage last year. How much did your friend earn per hour last year?

28. CAMPSITE The bar graph shows the numbers of reserved campsites at a campground for one week. What percent of the reservations were for Friday or Saturday?

29. PROBLEM SOLVING A classmate displays the results of a class president election in the bar graph shown.

   a. What is missing from the bar graph?

   b. What percent of the votes does the last-place candidate receive? Explain your reasoning.

   c. There are 124 votes total. How many votes does Chloe receive?

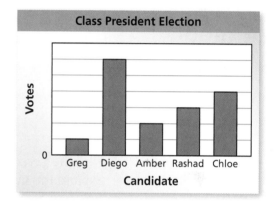

30. REASONING 20% of a number is x. What is 100% of the number? Assume $x > 0$.

31. **Structure** Answer each question. Assume $x > 0$.

   a. What percent of $8x$ is $5x$?

   b. What is 65% of $80x$?

 Fair Game Review What you learned in previous grades & lessons

Evaluate the expression when $a = -15$ and $b = -5$. *(Section 1.5)*

32. $a \div b$

33. $\dfrac{b + 14}{a}$

34. $\dfrac{b^2}{a + 5}$

35. MULTIPLE CHOICE What is the solution of $9x = -1.8$? *(Section 3.4)*

   Ⓐ $x = -5$ **Ⓑ** $x = -0.2$ **Ⓒ** $x = 0.2$ **Ⓓ** $x = 5$

6.4 The Percent Equation

Essential Question
How can you use an equivalent form of the percent proportion to solve a percent problem?

1 ACTIVITY: Solving Percent Problems Using Different Methods

Work with a partner. The circle graph shows the number of votes received by each candidate during a school election. So far, only half the students have voted.

Votes Received by Each Candidate

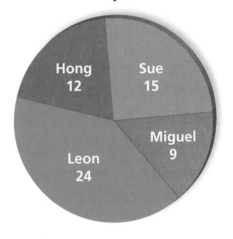

a. Complete the table.

Candidate	Number of votes received / Total number of votes
Sue	
Miguel	
Leon	
Hong	

b. Find the percent of students who voted for each candidate. Explain the method you used to find your answers.

c. Compare the method you used in part (b) with the methods used by other students in your class. Which method do you prefer? Explain.

2 ACTIVITY: Finding Parts Using Different Methods

Percent Equation

In this lesson, you will
- use the percent equation to find parts, wholes, and percents.
- solve real-life problems.

Work with a partner. The circle graph shows the final results of the election.

Final Results

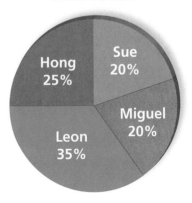

a. Find the number of students who voted for each candidate. Explain the method you used to find your answers.

b. Compare the method you used in part (a) with the methods used by other students in your class. Which method do you prefer? Explain.

Laurie's Notes

Introduction

Applying Mathematical Practices

- **Construct Viable Arguments and Critique the Reasoning of Others:** Mathematically proficient students are able to explain their reasoning in a way in which others can understand. They are also able to compare different solution methods and analyze benefits of each, or why different methods might be used for certain types of problems.

Motivate

- Do a quick review of benchmark percents.
- Write the different forms of each benchmark on index cards.

    Example: | 50% | | 0.5 | | ½ |

- Distribute the cards so that each student has one card. If the number of students is not a multiple of 3, make one or two duplicate cards.
- Without speaking, students should walk around and find the other forms equivalent to their numbers.
- Debrief by having each group display their three representations.

Activity Notes

FYI

- In the first two activities, students are solving percent problems using different methods. It is important to make time for students to share different strategies versus having only one method presented.

Activity 1

- There are several ways in which students could find the percent of votes received by each of the four candidates. Methods include:
  - Simplify the fraction and then write the percent.
  - Double a previous answer. 24 votes is twice as many as 12 votes.
  - Write a proportion and solve.
  - Change the fraction to a decimal and then to a percent.
- **Construct Viable Arguments and Critique the Reasoning of Others:** When students have finished part (b), listen to several students explain their method(s), then give students time to answer part (c).

Activity 2

❓ "How many students voted? Explain." 120; In Activity 1, half of the students had voted, and there were 60. Therefore 100% would be 120 students.

- There are several ways in which students could find the number of votes received by each of the four candidates. Methods include:
  - Mental math
  - Write a proportion and solve.
  - Multiply.
- **Extension:** Have students explore whether voting patterns changed from the first 60 voters to the last 60 voters.

What Your Students Will Learn

- Use the percent equation to find parts, percents, and wholes.

Previous Learning

Students should know how to solve simple percent problems.

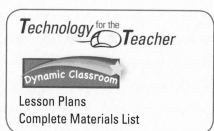

Lesson Plans
Complete Materials List

6.4 Record and Practice Journal

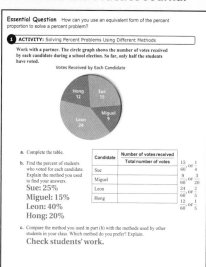

Essential Question How can you use an equivalent form of the percent proportion to solve a percent problem?

1 ACTIVITY: Solving Percent Problems Using Different Methods

Work with a partner. The circle graph shows the number of votes received by each candidate during a school election. So far, only half the students have voted.

Votes Received by Each Candidate

a. Complete the table.

b. Find the percent of students who voted for each candidate. Explain the method you used to find your answers.
Sue: 25%
Miguel: 15%
Leon: 40%
Hong: 20%

Candidate	Number of votes received / Total number of votes	
Sue		$\frac{15}{60}$ or $\frac{1}{4}$
Miguel		$\frac{9}{60}$ or $\frac{3}{20}$
Leon		$\frac{24}{60}$ or $\frac{2}{5}$
Hong		$\frac{12}{60}$ or $\frac{1}{5}$

c. Compare the method you used in part (b) with the methods used by other students in your class. Which method do you prefer? Explain.
Check students' work.

Differentiated Instruction

Visual

Some students will benefit from seeing how fractions and percents relate. Draw a circle on the board. Write 100% on top of the circle and 1 underneath. Explain that both of these values describe the area of the circle. Draw one-half of a circle and one-fourth of a circle and ask students to give you two representations.

$\frac{1}{2}$ and 50%, $\frac{1}{4}$ and 25%

6.4 Record and Practice Journal

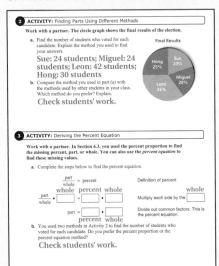

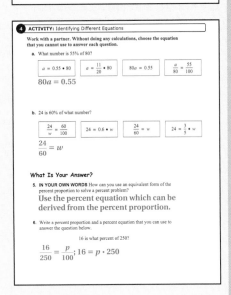

Laurie's Notes

Activity 3

? "How did we solve percent equations in the last lesson?" Students should describe the percent proportion. In particular, they should mention that percents were written as a fraction out of 100.

- In this activity, students derive the percent equation.
- **Note:** In Section 6.3, p is the actual percent. In Section 6.4, p is the percent written as a decimal or fraction.
- When the percent proportion is multiplied by the whole on both sides of the equation, the percent equation is found.
- In a previous course, students multiplied to find the percent of a number so the percent equation should not feel very new.
- **Connection:** This activity shows the connection between the percent proportion and percent equation.

Activity 4

- **Construct Viable Arguments and Critique the Reasoning of Others:** Listen for student justification for their answers.
- **Extension:** Have students answer each question.

What Is Your Answer?

- **Neighbor Check:** Have students work independently and then have their neighbors check their work. Have students discuss any discrepancies.

Closure

- What percent problem is suggested by the following? $12 = \frac{1}{4}w$

  12 is 25% of what number?; 48

3 ACTIVITY: Deriving the Percent Equation

Work with a partner. In Section 6.3, you used the percent proportion to find the missing percent, part, or whole. You can also use the *percent equation* to find these missing values.

a. Complete the steps below to find the percent equation.

$$\frac{\text{part}}{\text{whole}} = \text{percent}$$ Definition of percent

$$\frac{\text{part}}{\text{whole}} \cdot \boxed{} = \boxed{} \cdot \boxed{}$$ Multiply each side by the $\boxed{}$.

$$\text{part} = \boxed{} \cdot \boxed{}$$ Divide out common factors. This is the percent equation.

b. Use the percent equation to find the number of students who voted for each candidate in Activity 2. How does this method compare to the percent proportion?

4 ACTIVITY: Identifying Different Equations

Work with a partner. Without doing any calculations, choose the equation that you cannot use to answer each question.

a. What number is 55% of 80?

$$a = 0.55 \cdot 80 \qquad a = \frac{11}{20} \cdot 80 \qquad 80a = 0.55 \qquad \frac{a}{80} = \frac{55}{100}$$

b. 24 is 60% of what number?

$$\frac{24}{w} = \frac{60}{100} \qquad 24 = 0.6 \cdot w \qquad \frac{24}{60} = w \qquad 24 = \frac{3}{5} \cdot w$$

What Is Your Answer?

5. **IN YOUR OWN WORDS** How can you use an equivalent form of the percent proportion to solve a percent problem?

6. Write a percent proportion and a percent equation that you can use to answer the question below.

16 is what percent of 250?

Practice Use what you learned about solving percent problems to complete Exercises 4–9 on page 236.

Key Idea

The Percent Equation

Words To represent "a is p percent of w," use an equation.

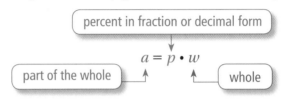

percent in fraction or decimal form

$$a = p \cdot w$$

part of the whole   whole

Numbers $15 = 0.5 \cdot 30$

EXAMPLE **1** **Finding a Part of a Number**

What number is 24% of 50?

Estimate 0%   25%   100%

0   12.5   50

Common Error

Remember to convert a percent to a fraction or a decimal before using the percent equation. For Example 1, write 24% as $\frac{24}{100}$.

$a = p \cdot w$	Write percent equation.
$= \frac{24}{100} \cdot 50$	Substitute $\frac{24}{100}$ for p and 50 for w.
$= 12$	Simplify.

So, 12 is 24% of 50.   **Reasonable?** $12 \approx 12.5$ ✓

EXAMPLE **2** **Finding a Percent**

9.5 is what percent of 25?

Estimate 0%   40%   100%

0   10   25

$a = p \cdot w$	Write percent equation.
$9.5 = p \cdot 25$	Substitute 9.5 for a and 25 for w.
$\frac{9.5}{25} = \frac{p \cdot 25}{25}$	Division Property of Equality
$0.38 = p$	Simplify.

Because 0.38 equals 38%,   **Reasonable?** $38\% \approx 40\%$ ✓
9.5 is 38% of 25.

Laurie's Notes

Introduction

Connect

- **Yesterday:** Students explored the connection between the percent proportion and the percent equation.
- **Today:** Students will use the percent equation to solve three types of percent problems.

Motivate

- The 2010 population of the United States was approximately 309 million (*Source:* U.S. Census Bureau) with about 24% being under 18 years old. About how many people in the U.S. are under the age of 18? about 74,160,000

Lesson Notes

Key Idea

- **Connection:** Students should know how to find a percent of a number by multiplying. The percent equation builds upon this idea to find the missing percent or the unknown whole. When you know two of the three quantities in this equation, you can solve for the third.
- **FYI:** Students often get lost in the language of these problems. It is important to help students translate the problems and make sense of the information that is given.

Words of Wisdom

- **Reason Abstractly and Quantitatively** and **Model with Mathematics:** In all of the examples today, draw a percent bar model off to the side to help student reasoning. The model helps students estimate and consider the reasonableness of the answer.

Example 1

- Another way to phrase this question is "24% of 50 is what number?"
- **Estimate:** 24% is close to 25%, and 25% is $\frac{1}{4}$.
- ❓ "What is $\frac{1}{4}$ of 50?" 12.5
- If time permits, write 24% as a decimal and work the problem again.

Example 2

- Read the example as "9.5 is a part of 25."
- ❓ "Is 9.5 more or less than half of 25?" less
- ❓ Draw the percent bar model explaining that it represents 25. Draw the half mark (50%) and ask how much that would represent. 12.5
- ❓ Now draw the quarter mark (25%) and ask how much that would represent. 6.25 Through this process, students should recognize that 9.5 is between 25% and 50% of 25.
- **Common Error:** Students may forget that the decimal answer to the division problem needs to be rewritten as a percent.

Lesson Tutorials
Lesson Plans
Answer Presentation Tool

Extra Example 1

What number is 73% of 200? 146

Extra Example 2

36.4 is what percent of 40? 91%

Laurie's Notes

Extra Example 3

18 is 15% of what number? 120

On Your Own

1. $a = 0.1 \cdot 20$; 2

2. $a = 1.5 \cdot 40$; 60

3. $3 = p \cdot 600$; 0.5%

4. $18 = p \cdot 20$; 90%

5. $8 = 0.8 \cdot w$; 10

6. $90 = 0.18 \cdot w$; 500

Extra Example 4

Your total cost for lunch is $18.50 for food and $1.48 for tax.

a. Find the percent of sales tax on the food total. 8%

b. Find the amount of an 18% tip on the food total. $3.33

On Your Own

7. $5.50

English Language Learners

Vocabulary

English learners may have trouble identifying which is the *whole* and which is the *part of the whole* in a percent equation. Have students write percent equations for the statements "20% of 300 is 60" and "125% of 50 is 62.5." Suggest that they start by substituting the percent *p* into the equation. Next, substitute the whole *w*. In most cases, this is the number after the word *of*. The remaining number is the part of the whole *a*.

Example 3

- This type of problem, finding a whole, is a bit harder. Knowing fractional equivalents is extremely helpful in developing a sense about the size of the answer.
- ❓ "What is the part?" 39 "So, 39 is a part of something."
- ❓ "How big of a part is it, approximately?" 52%, about half
- **Construct Viable Arguments and Critique the Reasoning of Others:** Help students reason that if 39 is half of something, the whole must be about 80. Only at this point does it make sense to translate what is known into an equation. 39 is 52% of some number.
- **Common Error:** Students may divide 39 by 52 and ignore the decimals completely.

On Your Own

- Have students work with partners on these problems. Encourage students to sketch the percent bar model and record the information they know. Then write the percent equation.
- Have students put their work on the board.

Example 4

- Have students read the bar graph and discuss the information displayed.
- ❓ "In addition to paying for what you ordered (food and drink), what other costs are there when you eat at a restaurant?" sales tax and tip
- Review decimal operations as you work through each part.

On Your Own

- Model finding 10%, and then double for 20%.

Closure

- **Exit Ticket:** Use the percent equation to answer the question, 12 is what percent of 48? 25%

EXAMPLE 3 **Finding a Whole**

39 is 52% of what number? **Estimate**

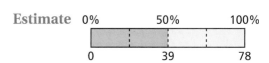

$$a = p \cdot w$$ Write percent equation.

$$39 = 0.52 \cdot w$$ Substitute 39 for a and 0.52 for p.

$$75 = w$$ Divide each side by 0.52.

So, 39 is 52% of 75. **Reasonable?** $75 \approx 78$ ✓

On Your Own

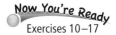
Now You're Ready
Exercises 10–17

Write and solve an equation to answer the question.

1. What number is 10% of 20?
2. What number is 150% of 40?
3. 3 is what percent of 600?
4. 18 is what percent of 20?
5. 8 is 80% of what number?
6. 90 is 18% of what number?

EXAMPLE 4 **Real-Life Application**

8th Street Cafe

DATE: MAY04'13 05:45PM
TABLE: 29
SERVER: JANE

Food Total	27.50
Tax	1.65
Subtotal	29.15

TIP: _____

TOTAL: _____

Thank You

a. **Find the percent of sales tax on the food total.**

Answer the question: $1.65 is what percent of $27.50?

$$a = p \cdot w$$ Write percent equation.

$$1.65 = p \cdot 27.50$$ Substitute 1.65 for a and 27.50 for w.

$$0.06 = p$$ Divide each side by 27.50.

Because 0.06 equals 6%, the percent of sales tax is 6%.

b. **Find the amount of a 16% tip on the food total.**

Answer the question: What tip amount is 16% of $27.50?

$$a = p \cdot w$$ Write percent equation.

$$= 0.16 \cdot 27.50$$ Substitute 0.16 for p and 27.50 for w.

$$= 4.40$$ Multiply.

So, the amount of the tip is $4.40.

On Your Own

7. **WHAT IF?** Find the amount of a 20% tip on the food total.

Check It Out
Help with Homework
BigIdeasMath √com

 Vocabulary and Concept Check

1. **VOCABULARY** Write the percent equation in words.

2. **REASONING** A number *n* is 150% of number *m*. Is *n greater than*, *less than*, or *equal to m*? Explain your reasoning.

3. **DIFFERENT WORDS, SAME QUESTION** Which is different? Find "both" answers.

What number is 20% of 55?	55 is 20% of what number?
20% of 55 is what number?	0.2 • 55 is what number?

 Practice and Problem Solving

Answer the question. Explain the method you chose.

4. What number is 24% of 80?

5. 15 is what percent of 40?

6. 15 is 30% of what number?

7. What number is 120% of 70?

8. 20 is what percent of 52?

9. 48 is 75% of what number?

Write and solve an equation to answer the question.

① 10. 20% of 150 is what number?

11. 45 is what percent of 60?

② 12. 35% of what number is 35?

13. 0.8% of 150 is what number?

③ 14. 29 is what percent of 20?

15. 0.5% of what number is 12?

16. What percent of 300 is 51?

17. 120% of what number is 102?

ERROR ANALYSIS Describe and correct the error in using the percent equation.

18. What number is 35% of 20?

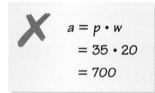

$a = p \cdot w$
$= 35 \cdot 20$
$= 700$

19. 30 is 60% of what number?

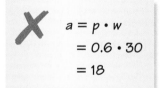

$a = p \cdot w$
$= 0.6 \cdot 30$
$= 18$

20. **COMMISSION** A salesperson receives a 2.5% commission on sales. What commission does the salesperson receive for $8000 in sales?

21. **FUNDRAISING** Your school raised 125% of its fundraising goal. The school raised $6750. What was the goal?

22. **SURFBOARD** The sales tax on a surfboard is $12. What is the percent of sales tax?

SALE $240

Assignment Guide and Homework Check

Level	Day 1 Activity Assignment	Day 2 Lesson Assignment	Homework Check
Basic	4–9, 32–36	1–3, 11–27 odd	11, 15, 21, 27
Average	4–9, 32–36	1–3, 11–19 odd, 22–26, 28, 29	11, 15, 23, 28
Advanced	4–9, 32–36	1–3, 10–30 even, 31	24, 26, 28, 30
Accelerated	1–9, 10–30 even, 31–36		24, 26, 28, 30

Common Errors

- **Exercises 4–17** Students may not know what number to substitute for each variable. Walk through each type of question with the students. Emphasize that the word *is* means *equals*, and *of* means *to multiply*. Tell students to write the question and then write the meaning of each word or group of words underneath.
- **Exercises 20–22** Students may mix up the whole and the part when trying to write the percent equation for the word problems. Ask them to identify each part of the equation before writing it in the equation format. For example, in Exercise 20, ask, "What is the salesperson's total sales, in dollars? 8000 "Which variable in the percent equation does this number represent?" The whole Continue to ask questions for each of the variables.
- **Exercise 28** Students may not realize that the sum of the parts of a circle graph equals 100%.

6.4 Record and Practice Journal

Write and solve an equation to answer the question.

1. What number is 35% of 80?
$a = 0.35 \cdot 80;\ 28$

2. 8 is what percent of 5?
$8 = p \cdot 5;\ 160\%$

3. What percent of 125 is 50?
$50 = p \cdot 125;\ 40\%$

4. 12% of what number is 48?
$48 = 0.12 \cdot w;\ 400$

5. 12 is what percent of 50?
$12 = p \cdot 50;\ 24\%$

6. What percent of 12 is 3?
$3 = p \cdot 12;\ 25\%$

7. You receive 15% of the profit from a car wash. How much money do you receive from a profit of $300?
$45

Vocabulary and Concept Check

1. A part of the whole is equal to a percent times the whole.

2. greater than; Because $150\% = 1.5$, $n = 1.5 \cdot m$.

3. 55 is 20% of what number?; 275; 11

Practice and Problem Solving

4. 19.2

5. 37.5%

6. 50

7. 84

8. about 38.5%

9. 64

10. $a = 0.2 \cdot 150$; 30

11. $45 = p \cdot 60$; 75%

12. $35 = 0.35 \cdot w$; 100

13. $a = 0.008 \cdot 150$; 1.2

14. $29 = p \cdot 20$; 145%

15. $12 = 0.005 \cdot w$; 2400

16. $51 = p \cdot 300$; 17%

17. $102 = 1.2 \cdot w$; 85

18. The percent was not converted to a decimal or fraction.
$$a = p \cdot w$$
$$= 0.35 \cdot 20$$
$$= 7$$

19. 30 represents the part of the whole.
$$30 = 0.6 \cdot w$$
$$50 = w$$

20. $200

21. $5400

22. 5%

Practice and Problem Solving

23. 26 years old

24. 70 years old

25. 56 signers

26. 70%

27. If the percent is less than 100%, the percent of a number is less than the number; 50% of 80 is 40; If the percent is equal to 100%, the percent of a number is equal to the number; 100% of 80 is 80; If the percent is greater than 100%, the percent of a number is greater than the number; 150% of 80 is 120.

28. a. 80 students

   b. 30 students

29. See *Taking Math Deeper*.

30. false; If *W* is 25% of *Z*, then *Z* : *W* is 100 : 25, because *Z* represents the whole.

31. 92%

Fair Game Review

32. 0.6   **33.** 0.88

34. 0.25   **35.** 0.36

36. A

Mini-Assessment

Write and solve an equation to answer the question.

1. 52 is what percent of 80? 65%

2. 28 is 35% of what number? 80

3. What number is 25% of 92? 23

4. What percent of 250 is 60? 24%

5. A new laptop computer costs $800. The sales tax on the computer is $48. What is the percent of sales tax? 6%

Taking Math Deeper

Exercise 29

Any problem that has this much given information is difficult for students. Encourage students to begin by organizing the information with a table or a diagram. When organizing the information, it is a good idea to add as much other information as you can find... *before looking at the questions*.

 Organize the given information.

 Add other information.

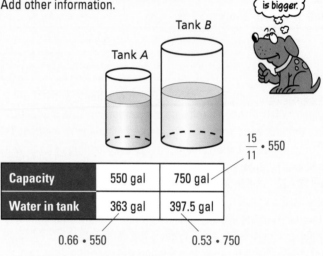

Capacity	550 gal	750 gal
Water in tank	363 gal	397.5 gal

0.66 • 550   0.53 • 750

$\frac{15}{11}$ • 550

③ Now the questions are easy.

   a. Tank *A* has 363 gallons of water.

   b. The capacity of tank *B* is 750 gallons.

   c. Tank *B* has 397.5 gallons of water.

Project

Use your school library or the Internet to research how a water tower works. How does the water get into the tower? How long does it take for the water to drain out? How often is the water completely exchanged; in other words, if a gallon goes in today when will that gallon be draining out? What other interesting things did you discover?

Reteaching and Enrichment Strategies

If students need help. . .	If students got it. . .
Resources by Chapter • Practice A and Practice B • Puzzle Time Record and Practice Journal Practice Differentiating the Lesson Lesson Tutorials Skills Review Handbook	Resources by Chapter • Enrichment and Extension • Technology Connection Start the next section

PUZZLE There were *n* signers of the Declaration of Independence. The youngest was Edward Rutledge, who was *x* years old. The oldest was Benjamin Franklin, who was *y* years old.

23. *x* is 25% of 104. What was Rutledge's age?

24. 7 is 10% of *y*. What was Franklin's age?

25. *n* is 80% of *y*. How many signers were there?

26. *y* is what percent of $(n + y - x)$?

Favorite Sport

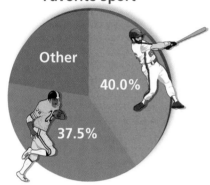

Other

40.0%

37.5%

27. **LOGIC** How can you tell whether the percent of a number will be *greater than*, *less than*, or *equal to* the number? Give examples to support your answer.

28. **SURVEY** In a survey, a group of students were asked their favorite sport. Eighteen students chose "other" sports.

    a. How many students participated?

    b. How many chose football?

29. **WATER TANK** Water tank *A* has a capacity of 550 gallons and is 66% full. Water tank *B* is 53% full. The ratio of the capacity of Tank *A* to Tank *B* is 11 : 15.

    a. How much water is in Tank *A*?

    b. What is the capacity of Tank *B*?

    c. How much water is in Tank *B*?

30. **TRUE OR FALSE?** Tell whether the statement is *true* or *false*. Explain your reasoning.

    If *W* is 25% of *Z*, then *Z* : *W* is 75 : 25.

31. **Reasoning** The table shows your test results for math class. What test score do you need on the last exam to earn 90% of the total points?

Test Score	Point Value
83%	100
91.6%	250
88%	150
?	300

 Fair Game Review *What you learned in previous grades & lessons*

Simplify. Write the answer as a decimal. *(Skills Review Handbook)*

32. $\dfrac{10 - 4}{10}$

33. $\dfrac{25 - 3}{25}$

34. $\dfrac{105 - 84}{84}$

35. $\dfrac{170 - 125}{125}$

36. **MULTIPLE CHOICE** There are 160 people in a grade. The ratio of boys to girls is 3 to 5. Which proportion can you use to find the number *x* of boys? *(Section 5.3)*

    Ⓐ $\dfrac{3}{8} = \dfrac{x}{160}$ Ⓑ $\dfrac{3}{5} = \dfrac{x}{160}$ Ⓒ $\dfrac{5}{8} = \dfrac{x}{160}$ Ⓓ $\dfrac{3}{5} = \dfrac{160}{x}$

You can use a **summary triangle** to explain a concept. Here is an example of a summary triangle for writing a percent as a decimal.

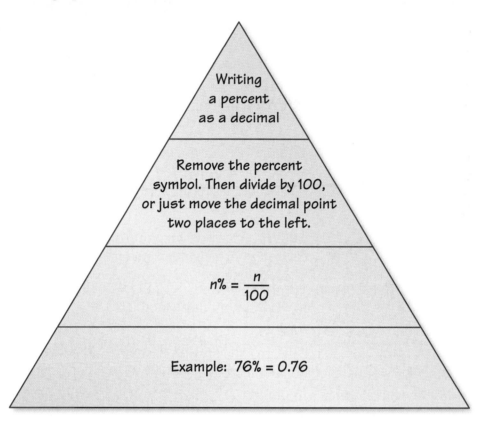

Writing
a percent
as a decimal

Remove the percent symbol. Then divide by 100, or just move the decimal point two places to the left.

$n\% = \dfrac{n}{100}$

Example: $76\% = 0.76$

On Your Own

Make summary triangles to help you study these topics.

1. writing a decimal as a percent

2. comparing and ordering fractions, decimals, and percents

3. the percent proportion

4. the percent equation

After you complete this chapter, make summary triangles for the following topics.

5. percent of change

6. discount

7. markup

8. simple interest

"**I found this great summary triangle in my** *Beautiful Beagle Magazine.*"

Sample Answers

1.

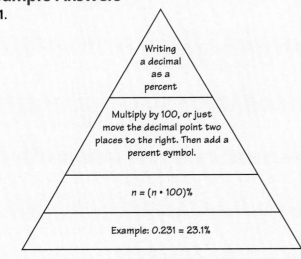

Writing a decimal as a percent

Multiply by 100, or just move the decimal point two places to the right. Then add a percent symbol.

$n = (n \cdot 100)\%$

Example: $0.231 = 23.1\%$

2.

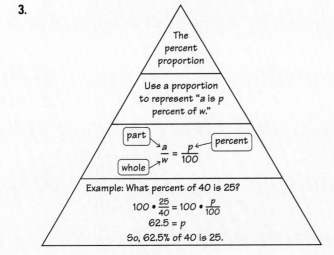

Comparing and ordering fractions, decimals, and percents

Write the numbers as all fractions, all decimals, or all percents. Then compare. Use a number line if necessary.

Examples:
Which is greater, 5.1% or $\frac{1}{20}$?
$\frac{1}{20} = \frac{5}{100} = 5\%$. So, 5.1% is greater.

Which is greater, 3.3% or 0.03?
$3.3\% = 0.033$. So, 3.3% is greater.

3.

The percent proportion

Use a proportion to represent "a is p percent of w."

part $\quad$ percent
$\dfrac{a}{w} = \dfrac{p}{100}$
whole

Example: What percent of 40 is 25?
$100 \cdot \dfrac{25}{40} = 100 \cdot \dfrac{p}{100}$
$62.5 = p$
So, 62.5% of 40 is 25.

4. Available at *BigIdeasMath.com*.

List of Organizers
Available at *BigIdeasMath.com*

Comparison Chart
Concept Circle
Definition (Idea) and Example Chart
Example and Non-Example Chart
Formula Triangle
Four Square
Information Frame
Information Wheel
Notetaking Organizer
Process Diagram
Summary Triangle
Word Magnet
Y Chart

About this Organizer

A **Summary Triangle** can be used to explain a concept. Typically, the summary triangle is divided into 3 or 4 parts. In the top part, students write the concept being explained. In the middle part(s), students write any procedure, explanation, description, definition, theorem, and/or formula(s). In the bottom part, students write an example to illustrate the concept. A summary triangle can be used as an assessment tool, in which blanks are left for students to complete. Also, students can place their summary triangles on note cards to use as a quick study reference.

Technology for the *Teacher*

Editable Graphic Organizer

Answers

1. 0.34
2. 0.0012
3. 0.625
4. 67%
5. 535%
6. 68.5%
7. 74%
8. 0.3

9.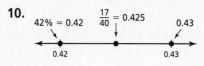

   $\frac{6}{5} = 1.2$ 1.22   125% = 1.25

   1.2 ——•——•———————•—— 1.3
        1.2   1.3

10.
   42% = 0.42 $\frac{17}{40} = 0.425$ 0.43

   ——•———————•———————•——
     0.42   0.43

11. $\frac{6}{15} = \frac{p}{100}; p = 40$

12. $\frac{35}{25} = \frac{p}{100}; p = 140$

13. $\frac{a}{50} = \frac{40}{100}; a = 20$

14. $\frac{5}{w} = \frac{0.5}{100}; w = 1000$

15. $a = 0.28 \cdot 75; 21$

16. $42 = 0.21 \cdot w; 200$

17. 0.38

18. Team 4; Team 5

19. 17 passes

20. 50 messages

Alternative Quiz Ideas

100% Quiz   Math Log
Error Notebook   Notebook Quiz
Group Quiz   Partner Quiz
Homework Quiz   Pass the Paper

Partner Quiz

- Students should work in pairs. Each pair should have a small white board.
- The teacher selects certain problems from the quiz and writes one on the board.
- The pairs work together to solve the problem and write their answer on the white board.
- Students show their answers and, as a class, discuss any differences.
- Repeat for as many problems as the teacher chooses.
- For the word problems, teachers may choose to have students read them out of the book.

Reteaching and Enrichment Strategies

If students need help. . .	If students got it. . .
Resources by Chapter • Practice A and Practice B • Puzzle Time Lesson Tutorials *BigIdeasMath.com*	Resources by Chapter • Enrichment and Extension • Technology Connection Game Closet at *BigIdeasMath.com* Start the next section

6.1–6.4   Quiz

Write the percent as a decimal. *(Section 6.1)*

1. 34%

2. 0.12%

3. 62.5%

Write the decimal as a percent. *(Section 6.1)*

4. 0.67

5. 5.35

6. 0.685

Tell which number is greater. *(Section 6.2)*

7. $\dfrac{11}{15}$, 74%

8. 3%, 0.3

Use a number line to order the numbers from least to greatest. *(Section 6.2)*

9. 125%, $\dfrac{6}{5}$, 1.22

10. 42%, 0.43, $\dfrac{17}{40}$

Write and solve a proportion to answer the question. *(Section 6.3)*

11. What percent of 15 is 6?

12. 35 is what percent of 25?

13. What number is 40% of 50?

14. 0.5% of what number is 5?

Write and solve an equation to answer the question. *(Section 6.4)*

15. What number is 28% of 75?

16. 42 is 21% of what number?

17. FISHING On a fishing trip, 38% of the fish that you catch are perch. Write this percent as a decimal. *(Section 6.1)*

18. SCAVENGER HUNT The table shows the results of 8 teams competing in a scavenger hunt. Which team collected the most items? Which team collected the fewest items? *(Section 6.2)*

Team	1	2	3	4	5	6	7	8
Portion Collected	$\dfrac{3}{4}$	0.8	77.5%	0.825	$\dfrac{29}{40}$	76.25%	$\dfrac{63}{80}$	81.25%

19. COMPLETIONS A quarterback completed 68% of his passes in a game. He threw 25 passes. How many passes did the quarterback complete? *(Section 6.3)*

20. TEXT MESSAGES You have 44 text messages in your inbox. How many messages can your cell phone hold? *(Section 6.4)*

6.5 Percents of Increase and Decrease

Essential Question What is a percent of decrease? What is a percent of increase?

1 ACTIVITY: Percent of Decrease

Work with a partner.

Each year in the Columbia River Basin, adult salmon swim upriver to streams to lay eggs and hatch their young.

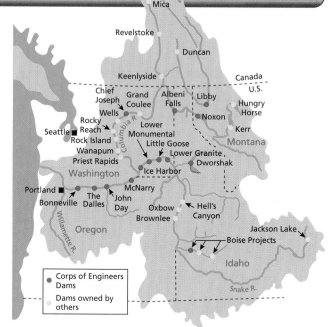

To go up the river, the adult salmon use fish ladders. But to go down the river, the young salmon must pass through several dams.

At one time, there were electric turbines at each of the eight dams on the main stem of the Columbia and Snake Rivers. About 88% of the young salmon passed through these turbines unharmed.

a. Copy and complete the table to show the number of young salmon that made it through the dams.

Dam	0	1	2	3	4	5	6	7	8
Salmon	1000	880	774						

$$88\% \text{ of } 1000 = 0.88 \cdot 1000 \qquad 88\% \text{ of } 880 = 0.88 \cdot 880$$
$$= 880 \qquad\qquad\qquad = 774.4$$
$$\approx 774$$

b. Display the data in a bar graph.

c. By what percent did the number of young salmon decrease when passing through each dam?

Percents

In this lesson, you will
- find percents of increase.
- find percents of decrease.

Laurie's Notes

Introduction

Applying Mathematical Practices

- **Use Appropriate Tools Strategically** and **Look for and Express Regularity in Repeated Reasoning:** Use of calculators allows students to draw important conclusions. When there is a repeated percent change (increase or decrease), the percent remains constant while the amount changes.

Motivate

- Talk about compact fluorescent light bulbs. Fluorescent light bulbs use 75% less energy than incandescent light bulbs (percent decrease) and last up to 900% longer (percent increase).
- If your school has replaced incandescent light bulbs with fluorescent light bulbs, discuss the potential savings.

Activity Notes

Activity 1

- **Representation:** Although the difference between the decimal point and the multiplication dot are clear in the textbook, it may not be as clear when you write it on the board. You may consider using parentheses to show the multiplication: (0.88)(1000).
- Have a student read the story information.
- ? "What percent of salmon makes it through each dam?" 88%
- ? "What percent of salmon does not make it through each dam?" 12%
- Discuss the general concept of fewer salmon at dam 2 than dam 1.
- **FYI:** Electric turbines in the dams generate electricity. These turbines are what affect the survival rate of the young salmon.
- Students should follow the two calculations shown. Remind students to round their answers to a whole number of salmon at each dam.
- **Use Appropriate Tools Strategically:** Use of a calculator will help facilitate the computation so that students can focus on how the numbers are changing.
- Check students' results before completing the graph.
- **Look for and Express Regularity in Repeated Reasoning** and **Big Idea:** Each entry is 12% less than the previous entry. The *amount* of decrease is changing, but the *percent* is not.
- Have students describe patterns they observe in the numbers and the bar graph.

What Your Students Will Learn

- Find percents of increase or percents of decrease.
- Find percent error.

Previous Learning

Students should be able to find a percent of a number, round decimal values, and convert between fractions, decimals, and percents.

Lesson Plans
Complete Materials List

6.5 Record and Practice Journal

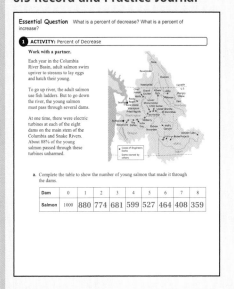

Essential Question What is a percent of decrease? What is a percent of increase?

1 ACTIVITY: Percent of Decrease

Work with a partner.

Each year in the Columbia River Basin, adult salmon swim upriver to streams to lay eggs and hatch their young.

To go up river, the adult salmon use fish ladders. But to go down the river, the young salmon must pass through several dams.

At one time, there were electric turbines at each of the eight dams on the main stem of the Columbia and Snake Rivers. About 88% of the young salmon passed through these turbines unharmed.

a. Complete the table to show the number of young salmon that made it through the dams.

Dam	0	1	2	3	4	5	6	7	8
Salmon	1000	880	774	681	599	527	464	408	359

English Language Learners

Visual Aid

Demonstrate writing a percent as a decimal. Locate the decimal point in the 7%.

7.%

Draw two arrows to show that the decimal point moves two places *left*.

7.%

Write zeros to the left of the number if needed.

007.%

Rewrite as a decimal with the decimal point two places to the left and without the percent sign.

0.07

6.5 Record and Practice Journal

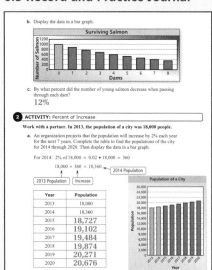

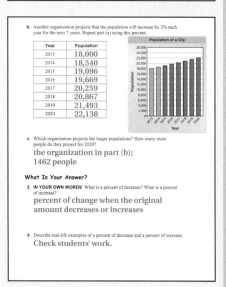

Activity 2

- Ask a student to read the problem.
- Work through the first year with the students. There are two steps involved: 1) find the amount the population has increased, and 2) add this amount to the current population.
- ? "How much did the population increase in 2014?" 360
- ? "What percent did the population increase in 2014?" 2%
- Remind students to round their answers to a whole number and add this number to the current population.
- **Use Appropriate Tools Strategically:** Use of a calculator will help facilitate the computation so that students can focus on how the numbers are changing.
- Check students' results before completing the graph.
- **Look for and Express Regularity in Repeated Reasoning** and **Big Idea:** Each entry is 2% more than the previous entry. The *amount* of increase is changing, but the *percent* is not.
- **Extension:** Discuss how projections are made based upon current trends.

What Is Your Answer?

- For Question 4, students could discuss this at home and bring ideas to class.

Closure

- "You scored 80 points on your first test. If your score increased 10% on the next test, what is your score?" 88 points

Work with a partner. In 2013, the population of a city was 18,000 people.

a. An organization projects that the population will increase by 2% each year for the next 7 years. Copy and complete the table to find the populations of the city for 2014 through 2020. Then display the data in a bar graph.

For 2014:

$$2\% \text{ of } 18{,}000 = 0.02 \cdot 18{,}000$$
$$= 360$$

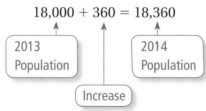

$$18{,}000 + 360 = 18{,}360$$

2013 Population

2014 Population

Increase

Year	Population
2013	18,000
2014	18,360
2015	
2016	
2017	
2018	
2019	
2020	

b. Another organization projects that the population will increase by 3% each year for the next 7 years. Repeat part (a) using this percent.

c. Which organization projects the larger populations? How many more people do they project for 2020?

What Is Your Answer?

3. **IN YOUR OWN WORDS** What is a percent of decrease? What is a percent of increase?

4. Describe real-life examples of a percent of decrease and a percent of increase.

Practice

Use what you learned about percent of increase and percent of decrease to complete Exercises 4–7 on page 244.

Check It Out
Lesson Tutorials
BigIdeasMath.com

Key Vocabulary 🔊
percent of change,
  p. 242
percent of increase,
  p. 242
percent of decrease,
  p. 242
percent error, p. 243

A **percent of change** is the percent that a quantity changes from the original amount.

$$\text{percent of change} = \frac{\text{amount of change}}{\text{original amount}}$$

 **Key Idea**

Percents of Increase and Decrease

When the original amount increases, the percent of change is called a **percent of increase**.

$$\text{percent of increase} = \frac{\text{new amount} - \text{original amount}}{\text{original amount}}$$

When the original amount decreases, the percent of change is called a **percent of decrease**.

$$\text{percent of decrease} = \frac{\text{original amount} - \text{new amount}}{\text{original amount}}$$

EXAMPLE (1) **Finding a Percent of Increase**

The table shows the numbers of hours you spent online last weekend. What is the percent of change in your online time from Saturday to Sunday?

Day	Hours Online
Saturday	2
Sunday	4.5

The number of hours on Sunday is greater than the number of hours on Saturday. So, the percent of change is a percent of increase.

$$\text{percent of increase} = \frac{\text{new amount} - \text{original amount}}{\text{original amount}}$$

$$= \frac{4.5 - 2}{2} \qquad \text{Substitute.}$$

$$= \frac{2.5}{2} \qquad \text{Subtract.}$$

$$= 1.25, \text{ or } 125\% \qquad \text{Write as a percent.}$$

∴ So, your online time increased 125% from Saturday to Sunday.

⬤ **On Your Own**

Find the percent of change. Round to the nearest tenth of a percent if necessary.

1. 10 inches to 25 inches   **2.** 57 people to 65 people

Laurie's Notes

Introduction

Connect

- **Yesterday:** Students explored two real-life problems with quantities that decreased or increased by a percent.
- **Today:** Students will use a percent of change formula to solve problems.

Motivate

- Pose a question such as: "If 400 people in your neighborhood had a cell phone last year and one year later 500 people had a cell phone, what percent has cell phone ownership increased?"

Lesson Notes

Key Idea

- Explain the difference between *amount* of change and *percent* of change. Refer to the salmon and population activities.
- Use the cell phone example to help identify vocabulary:

  original amount $= 400$,

  amount of change $= 100$,

  percent of change $= \dfrac{100}{400} = 25\%$.

Example 1

- ❓ "Did the online use increase or decrease from Saturday to Sunday?" increase
- ❓ "How much did the online use increase from Saturday to Sunday?" 2.5 h
- Have students write the equation, substitute the values, and then simplify. The original amount is 2. The new amount is 4.5. Because the number of hours increased, you are finding a **percent of increase**. Percent of

  increase $= \dfrac{4.5 - 2}{2} = 1.25 = 125\%$.

- **Common Error:** Students think the answer is 1.25. This decimal must still be converted to a percent. This often happens when the percent answer is greater than 100%.
- **Connection:** Draw a percent bar model of this problem.

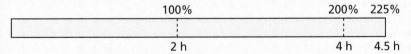

The percent of increase is 125% beyond the 100%.

- **Model with Mathematics** and **Big Idea:** From the percent bar model you can see that a 100% increase doubles the number. Another way of saying this is that when a number doubles, it has increased 100%.

On Your Own

- In Question 1, the length has more than doubled, so the percent of increase is greater than 100%.
- In Question 2, the number of people has not doubled, so the percent of increase is less than 100%.

Goal Today's lesson is using a **percent of change** formula to solve problems.

Lesson Tutorials
Lesson Plans
Answer Presentation Tool

Extra Example 1

Find the percent of change from 40 hours to 50 hours. increase of 25%

On Your Own

1. 150% increase
2. about 14.0% increase

Example 2

- "How much did the number of home runs change each year?" decrease of 8, increase of 18, decrease of 8
- **?** "What is the original amount?" 28
- **?** "What is the new amount?" 20
- Students should now use the **percent of decrease** formula.
- This problem involves a number of skills: reading a bar graph, using the percent of decrease formula, converting a fraction to a decimal, and converting a decimal to a percent.
- The answer is rounded to the nearest tenth of a percent.
- **?** "Is the percent change from 2011–2012 more or less than 100%? How do you know?" The number of home runs more than doubled, so the increase is greater than 100%.

Key Idea

- *Percent error* may be a new topic in your curriculum, but it is really an application of percent change.
- Write the Key Idea. The percent error compares the amount of error to the actual amount.
- An estimate can be too high or too low when compared to the actual, but generally **percent error** is referred to as a positive amount.

Example 3

- Before doing this example, you could ask students to write down their estimates for the length of the classroom. These estimates could be used after doing this example, with students computing their percent error.
- **Use Appropriate Tools Strategically:** Calculators would be appropriate for this problem.

On Your Own

- If you have small white boards available, have each student solve Question 3 on a white board. Have students hold up their white boards and then have students determine their mistakes.

On Your Own

3. decrease of about 44.4%

4. you; Your percent error is about 23.8% and your friend's percent error is about 9.5%.

Closure

- **Writing Prompt:** To find the percent of change...

Differentiated Instruction

Vocabulary

Make sure students understand the difference between *increased by 150%* and *increased to 150%*. The percent of increase in the first case is 150%. The percent of increase in the second case is 50%, because the increase does not include the original amount. By the same token, there is a difference between *decreased by 25%* and *decreased to 25%*. In the first case, it means taking 25% and leaving 75%. In the second case, it means to leave 25% and take away 75%.

EXAMPLE 2 **Finding a Percent of Decrease**

The bar graph shows a softball player's home run totals. What was the percent of change from 2012 to 2013?

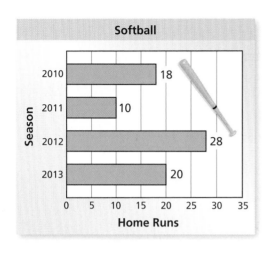

The number of home runs decreased from 2012 to 2013. So, the percent of change is a percent of decrease.

$$\text{percent of decrease} = \frac{\text{original amount} - \text{new amount}}{\text{original amount}}$$

$$= \frac{28 - 20}{28} \qquad \text{Substitute.}$$

$$= \frac{8}{28} \qquad \text{Subtract.}$$

$$\approx 0.286, \text{ or } 28.6\% \qquad \text{Write as a percent.}$$

So, the number of home runs decreased about 28.6%.

Percent Error

A **percent error** is the percent that an estimated quantity differs from the actual amount.

Study Tip

The amount of error is always positive.

$$\text{percent error} = \frac{\text{amount of error}}{\text{actual amount}}$$

EXAMPLE 3 **Finding a Percent Error**

You estimate that the length of your classroom is 16 feet. The actual length is 21 feet. Find the percent error.

The amount of error is $21 - 16 = 5$ feet.

$$\text{percent error} = \frac{\text{amount of error}}{\text{actual amount}} \qquad \text{Write percent error equation.}$$

$$= \frac{5}{21} \qquad \text{Substitute.}$$

$$\approx 0.238, \text{ or } 23.8\% \qquad \text{Write as a percent.}$$

The percent error is about 23.8%.

On Your Own

Now You're Ready
Exercises 8–15 and 18

3. In Example 2, what was the percent of change from 2010 to 2011?

4. **WHAT IF?** In Example 3, your friend estimates that the length of the classroom is 23 feet. Who has the greater percent error? Explain.

 ## Vocabulary and Concept Check

1. **VOCABULARY** How do you know whether a percent of change is a *percent of increase* or a *percent of decrease*?

2. **NUMBER SENSE** Without calculating, which has a greater percent of increase?
   - 5 bonus points on a 50-point exam
   - 5 bonus points on a 100-point exam

3. **WRITING** What does it mean to have a 100% decrease?

 ## Practice and Problem Solving

Find the new amount.

4. 8 meters increased by 25%

5. 15 liters increased by 60%

6. 50 points decreased by 26%

7. 25 penalties decreased by 32%

Identify the percent of change as an *increase* or a *decrease*. Then find the percent of change. Round to the nearest tenth of a percent if necessary.

① ② 8. 12 inches to 36 inches

9. 75 people to 25 people

10. 50 pounds to 35 pounds

11. 24 songs to 78 songs

12. 10 gallons to 24 gallons

13. 72 paper clips to 63 paper clips

14. 16 centimeters to 44.2 centimeters

15. 68 miles to 42.5 miles

16. **ERROR ANALYSIS** Describe and correct the error in finding the percent increase from 18 to 26.

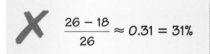

$$\frac{26 - 18}{26} \approx 0.31 = 31\%$$

17. **VIDEO GAME** Last week, you finished Level 2 of a video game in 32 minutes. Today, you finish Level 2 in 28 minutes. What is your percent of change?

③ 18. **PIG** You estimate that a baby pig weighs 20 pounds. The actual weight of the baby pig is 16 pounds. Find the percent error.

19. **CONCERT** You estimate that 200 people attended a school concert. The actual attendance was 240 people.

   a. Find the percent error.

   b. What other estimate gives the same percent error? Explain your reasoning.

Assignment Guide and Homework Check

Level	Day 1 Activity Assignment	Day 2 Lesson Assignment	Homework Check
Basic	4–7, 32–36	1–3, 9–23 odd, 16	9, 13, 17, 19, 21
Average	4–7, 32–36	1–3, 8–18 even, 19–31 odd	10, 12, 18, 21, 25
Advanced	4–7, 32–36	1–3, 8–30 even, 31	12, 18, 24, 28, 30
Accelerated	1–7, 8–30 even, 31–36		12, 18, 24, 28, 30

Common Errors

- **Exercises 4–7** Students may find the percent of the number and forget to add or subtract from the original amount. Remind them that these are two-step problems. Before evaluating, tell students to write down what needs to be done for each step.
- **Exercises 8–15** Students may mix up where to place the numbers in the equation to find percent of change. When they do not put the numbers in the right place, they might find a negative number in the numerator. First, emphasize that students must know if it is increasing or decreasing before they start the problem. Next, tell students that the number in the denominator is going to be the original or starting number given for both increasing and decreasing percents of change. Finally, the numerator should never have a negative answer. If students get a negative number, it is because they found the wrong difference. The numerator is always the greater number minus the lesser number.

6.5 Record and Practice Journal

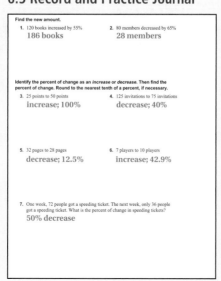

Find the new amount.

1. 120 books increased by 55%
   186 books
2. 80 members decreased by 65%
   28 members

Identify the percent of change as an *increase* or *decrease*. Then find the percent of change. Round to the nearest tenth of a percent, if necessary.

3. 25 points to 50 points
   increase; 100%
4. 125 invitations to 75 invitations
   decrease; 40%

5. 32 pages to 28 pages
   decrease; 12.5%
6. 7 players to 10 players
   increase; 42.9%

7. One week, 72 people got a speeding ticket. The next week, only 36 people got a speeding ticket. What is the percent of change in speeding tickets?
   50% decrease

Vocabulary and Concept Check

1. If the original amount decreases, the percent of change is a percent of decrease. If the original amount increases, the percent of change is a percent of increase.

2. 5 bonus points on a 50-point exam

3. The new amount is now 0.

Practice and Problem Solving

4. 10 m
5. 24 L
6. 37 points
7. 17 penalties
8. increase; 200%
9. decrease; 66.7%
10. decrease; 30%
11. increase; 225%
12. increase; 140%
13. decrease; 12.5%
14. increase; 176.3%
15. decrease; 37.5%
16. The denominator should be 18, which is the original amount.
    $$\frac{26 - 18}{18} \approx 0.44 = 44\%$$
17. 12.5% decrease
18. 25%
19. **a.** about 16.7%

    b. 280 people; To get the same percent error, the amount of error needs to be the same. Because your estimate was 40 people below the actual attendance, an estimate of 40 people above the actual attendance will give the same percent error.

Practice and Problem Solving

20. increase; 100%

21. decrease; 25%

22. increase; 133.3%

23. decrease; 70%

24–25. See Additional Answers.

26. **a.** 100% increase

 b. 300% increase

27. 15.6 ounces; 16.4 ounces

28. about 24.52% decrease

29. See Additional Answers.

30. See *Taking Math Deeper*.

31. 10 girls

Fair Game Review

32. $a = 0.25 \cdot 64$; 16

33. $39.2 = p \cdot 112$; 35%

34. $5 = 0.05 \cdot w$; 100

35. $18 = 0.32 \cdot w$; 56.25

36. B

Mini-Assessment

Identify the percent of change as an *increase* or *decrease*. Then find the percent of change.

1. 15 meters to 36 meters increase; 140%

2. 20 songs to 70 songs increase; 250%

3. 90 people to 45 people decrease; 50%

4. Yesterday, it took 40 minutes to drive to school. Today, it took 32 minutes to drive to school. What is your percent of change? The number of minutes it took to get to school decreased by 20%.

5. You estimate that a box contains 141 envelopes. The actual number of envelopes is 150. Find the percent error. 6%

Taking Math Deeper

Exercise 30

This exercise is difficult because the percent of increase is given with the *new amount*, rather than the *original amount*. A good way to start is to use a table to organize the given information.

Tables help me.

 Organize given information.

	Donation	Increase over previous year
This year	$10,120	15%
1 year ago	x	10%
2 years ago	y	

 Find last year's donation.

$$x + 0.15x = 10{,}120 \qquad \text{Write an equation.}$$
$$1.15x = 10{,}120 \qquad \text{Combine like terms.}$$
$$x = \$8800 \qquad \text{Divide each side by 1.15.}$$

 Find donation from 2 years ago.

$$y + 0.1y = 8800 \qquad \text{Write an equation.}$$
$$1.1y = 8800 \qquad \text{Combine like terms.}$$
$$y = \$8000 \qquad \text{Divide each side by 1.1.}$$

Project

Plan a fundraiser for your school. Write a proposal that includes the purpose of the fundraiser, the type of activity, the length of time, and the amount of money you would like to raise. Be prepared to present your proposal to the class.

Reteaching and Enrichment Strategies

If students need help. . .	If students got it. . .
Resources by Chapter • Practice A and Practice B • Puzzle Time Record and Practice Journal Practice Differentiating the Lesson Lesson Tutorials Skills Review Handbook	Resources by Chapter • Enrichment and Extension • Technology Connection Start the next section

Identify the percent of change as an *increase* or a *decrease*. Then find the percent of change. Round to the nearest tenth of a percent if necessary.

20. $\frac{1}{4}$ to $\frac{1}{2}$ **21.** $\frac{4}{5}$ to $\frac{3}{5}$ **22.** $\frac{3}{8}$ to $\frac{7}{8}$ **23.** $\frac{5}{4}$ to $\frac{3}{8}$

24. CRITICAL THINKING Explain why a change from 20 to 40 is a 100% increase, but a change from 40 to 20 is a 50% decrease.

25. POPULATION The table shows population data for a community.

Year	Population
2007	118,000
2013	138,000

    a. What is the percent of change from 2007 to 2013?

    b. Use this percent of change to predict the population in 2019.

26. GEOMETRY Suppose the length and the width of the sandbox are doubled.

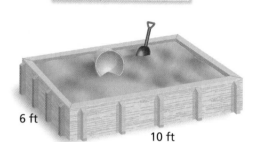

6 ft
10 ft

    a. Find the percent of change in the perimeter.

    b. Find the percent of change in the area.

27. CEREAL A cereal company fills boxes with 16 ounces of cereal. The acceptable percent error in filling a box is 2.5%. Find the least and the greatest acceptable weights.

7:45   June
5:51   September

28. PRECISION Find the percent of change from June to September in the time to run a mile.

29. CRITICAL THINKING A number increases by 10%, and then decreases by 10%. Will the result be *greater than*, *less than*, or *equal to* the original number? Explain.

30. DONATIONS Donations to an annual fundraiser are 15% greater this year than last year. Last year, donations were 10% greater than the year before. The amount raised this year is $10,120. How much was raised 2 years ago?

31. **Reasoning** Forty students are in the science club. Of those, 45% are girls. This percent increases to 56% after new girls join the club. How many new girls join?

Fair Game Review What you learned in previous grades & lessons

Write and solve an equation to answer the question. *(Section 6.4)*

32. What number is 25% of 64?

33. 39.2 is what percent of 112?

34. 5 is 5% of what number?

35. 18 is 32% of what number?

36. MULTIPLE CHOICE Which set of ratios does *not* form a proportion? *(Section 5.2)*

    Ⓐ $\frac{1}{4}, \frac{6}{24}$ **Ⓑ** $\frac{4}{7}, \frac{7}{10}$ **Ⓒ** $\frac{16}{24}, \frac{2}{3}$ **Ⓓ** $\frac{36}{10}, \frac{18}{5}$

Essential Question How can you find discounts and selling prices?

1 ACTIVITY: Comparing Discounts

Work with a partner. The same pair of sneakers is on sale at three stores. Which one is the best buy? Explain.

a. Regular Price: $45   b. Regular Price: $49   c. Regular Price: $39

a.

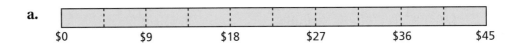

$0 $9 $18 $27 $36 $45

b.

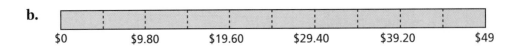

$0 $9.80 $19.60 $29.40 $39.20 $49

c.

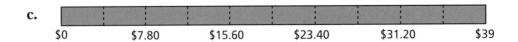

$0 $7.80 $15.60 $23.40 $31.20 $39

2 ACTIVITY: Finding the Original Price

Work with a partner.

Percents

In this lesson, you will
- use percent of discounts to find prices of items.
- use percent of markups to find selling prices of items.

a. You buy a shirt that is on sale for 30% off. You pay $22.40. Your friend wants to know the original price of the shirt. Show how you can use the model below to find the original price.

b. Explain how you can use the percent proportion to find the original price.

$0 $22.40   Original Price

Laurie's Notes

Introduction

Applying Mathematical Practices

- **Model with Mathematics:** Percent applications are abundant. The percent bar model helps students visualize the problem and check on the reasonableness of the answer.

Motivate

- Show a newspaper circular that advertises a discount (sale).

Activity Notes

Activity 1

- Explain that sale items involve a *percent* discount and the *amount* of discount. If possible, use the newspaper circular to make this distinction.
- The percent bar models are divided into 10 equal parts. Dollar amounts for items are shown on the bars.
- Discuss how the dollar amounts can be computed. Students can use mental math to find 10% and multiply by the correct amount.
- **Big Idea:** When you *save* 40% ($18), you *pay* 60% ($27). Starting at $45, move to the left 40%, that is the savings.
- **Extension:** Determine the amount you save *and* the price paid. This will not be possible for the last example because the amount of discount varies.
- **?** "How do you decide the best buy?" Listen for the lowest final price instead of the greatest savings because the original prices may vary.
- **?** "What does the phrase, "up to 70% off" mean?" Percent off will vary from 0% up to 70%.

Activity 2

- **Connection:** Finding the original price is the same as finding the whole. $22.40 is the part.
- **?** "What percent does $22.40 represent of the original price?" 70%
- **?** "How does the percent bar model help you think about the original price?"

  Students might describe the $22.40 as 70% or about $\frac{2}{3}$ of the original price.

  So, another $\frac{1}{3}$ has to be added on to find the original price.

- Use the percent equation: $22.40 is 70% of what number?
- **?** "Why is 70% used instead of 30%?" Because $22.40 is the part and it is 70% of the whole, or original price.
- **Struggling Students:** Students sometimes struggle with this concept. Reinforce by constantly telling students "30% off the original price is the same as paying 70% of the original price."
- **Construct Viable Arguments and Critique the Reasoning of Others:** Ask volunteers to explain each method—the percent equation and the percent proportion. Discuss with students the use of each method. Some students may prefer one method over the other.

What Your Students Will Learn

- Use percents of discounts to find sale prices or original prices of items.
- Use percents of markups to find selling prices of items.

Previous Learning

Students should be able to find a percent of a number, round decimal values, and convert between fractions, decimals, and percents.

Technology for the Teacher

Dynamic Classroom

Lesson Plans
Complete Materials List

6.6 Record and Practice Journal

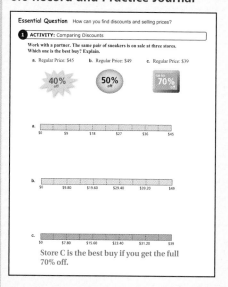

Essential Question How can you find discounts and selling prices?

1 ACTIVITY: Comparing Discounts

Work with a partner. The same pair of sneakers is on sale at three stores. Which one is the best buy? Explain.

a. Regular Price: $45   b. Regular Price: $49   c. Regular Price: $39

40% off   50% off   Up to 70% off

a.
$0 $9 $18 $27 $36 $45

b.
$0 $9.80 $19.60 $29.40 $39.20 $49

c.
$0 $7.80 $15.60 $23.40 $31.20 $39

Store C is the best buy if you get the full 70% off.

Vocabulary

English learners may not be familiar with terms used in business, such as *discount, markup, purchase price,* and *selling price*. Take time to explain these terms.

6.6 Record and Practice Journal

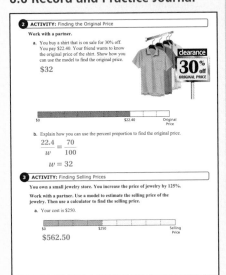

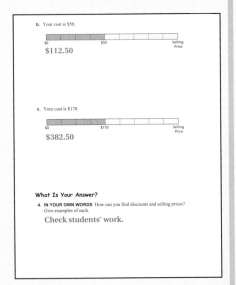

Laurie's Notes

Activity 3

- **Discuss:** A store purchases an item for *x* dollars. The store needs to sell this item for more than *x* dollars to cover operating costs and to make a profit.
- **Explain:** A store purchases an item for $2 and sells it for $4. This represents a 100% increase ($2 + 100% of $2).
- **Tip:** Have students put 100% above the $250 store cost in part (a).
- ? "How does the selling price compare to the price the store paid for the item?" *125% greater, or 225% of the store's cost*
- **Reason Abstractly and Quantitatively:** If a $10 item sells for $25, the $10 item was *increased by 150%* and the selling price is *250% of $10*. One way to show this to students is to write: (100% of $10) + (150% of $10) = (250% of $10) = $25.

Words of Wisdom

- Be careful with language. This is not an obvious concept for students. Try to use consistent language with every example; store purchase price, store selling price, original price, increased amount, discount amount, and sale price.

What Is Your Answer?

- You want students to discover that they can find the selling price after a 25% discount by multiplying by 0.75 (one step) *or* by multiplying by 0.25 and then subtracting the result from the original price (two steps).
- Similarly, for markups, you can multiply the cost by 1.75 (one step) for a 75% markup *or* multiply by 0.75 and then add to the cost (two steps).

Closure

- You purchased an item marked 25% off. What percent of the original price did you pay? *75%*

3 ACTIVITY: Finding Selling Prices

Math Practice

Make Sense of Quantities

What do the quantities represent? What is the relationship between the quantities?

You own a small jewelry store. You increase the price of the jewelry by 125%.

Work with a partner. Use a model to estimate the selling price of the jewelry. Then use a calculator to find the selling price.

a. Your cost is $250.

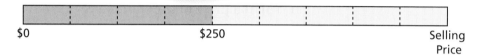

$0 $250   Selling Price

b. Your cost is $50.

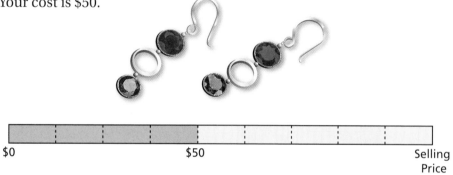

$0 $50   Selling Price

c. Your cost is $170.

$0 $170   Selling Price

What Is Your Answer?

4. IN YOUR OWN WORDS How can you find discounts and selling prices? Give examples of each.

Practice Use what you learned about discounts to complete Exercises 4, 9, and 14 on page 250.

Check It Out
Lesson Tutorials
BigIdeasMath Vcom

Key Vocabulary
discount, *p. 248*
markup, *p. 248*

 Key Ideas

Discounts

A **discount** is a decrease in the original price of an item.

Markups

To make a profit, stores charge more than what they pay. The increase from what the store pays to the selling price is called a **markup**.

EXAMPLE 1 **Finding a Sale Price**

The original price of the shorts is $35. What is the sale price?

Method 1: First, find the discount. The discount is 25% of $35.

$a = p \cdot w$	Write percent equation.
$= 0.25 \cdot 35$	Substitute 0.25 for p and 35 for w.
$= 8.75$	Multiply.

Next, find the sale price.

sale price	=	original price	−	discount
	=	35	−	8.75
	= 26.25			

So, the sale price is $26.25.

Method 2: First, find the percent of the original price.

$$100\% - 25\% = 75\%$$

Next, find the sale price.

$$\text{sale price} = 75\% \text{ of } \$35$$
$$= 0.75 \cdot 35$$
$$= 26.25$$

Study Tip

A 25% discount is the same as paying 75% of the original price.

So, the sale price is $26.25.

Check

0%	25%		75%	100%

| 0 | 8.75 | 26.25 | 35 |

✓

 On Your Own

Now You're Ready
Exercises 4−8

1. The original price of a skateboard is $50. The sale price includes a 20% discount. What is the sale price?

Multi-Language Glossary at BigIdeasMath Vcom

Laurie's Notes

Introduction

Connect

- **Yesterday:** Students explored discounts and selling prices using a percent bar model.
- **Today:** Students will use the percent equation to find discounts and markups of items.

Motivate

? **Story Time:** "A store buys an MP3 player for $100 and marks it up 50%. The store has a 50% off sale. You purchase the MP3 player. What do you pay?" $75 "Did the store lose money?" yes

Lesson Notes

Key Ideas

- Discuss each concept using examples from the previous day's activity.
- Use the following to help students understand the vocabulary.

$$\text{wholesale price} + \text{markup} = \text{retail price}$$
$$\text{(or selling price)}$$

what a store pays	increase in price	price you pay

- **FYI:** Some students may choose to use the percent proportion instead of the percent equation. Remind them of these *equivalent* methods if desired. Examples in this lesson are solved using the percent equation.

Example 1

- Two methods are shown, work through each method. Both methods require two steps. In the first method, you multiply to find the amount of discount, then you subtract to find the sale price. In the second method, you subtract first to find the percent of the original price you will pay, then you use the percent equation to find the sale price.
- **Connection:** The amount of discount is a *part* of the *whole* original price. The percent equation is used to find the amount of the discount.
- **Common Error:** Students find the discount or the amount saved ($8.75) instead of the sale price ($26.25).
- Discuss the *Study Tip.* Try other discounts (i.e., 30%) and ask what percent you are paying (70%).

? "Why is the percent bar model divided into 4 parts?"

Because the discount is 25% or $\frac{1}{4}$.

On Your Own

? "How should the percent bar model be divided and why?" 5 parts

because $20\% = \frac{1}{5}$

Goal Today's lesson is using the percent equation to find **discounts** and **markups**.

Lesson Tutorials
Lesson Plans
Answer Presentation Tool

Extra Example 1

The original price of a T-shirt is $15. The sale price includes a 35% discount. What is the sale price? $9.75

On Your Own

1. $40

Extra Example 2

The discount on a package of athletic socks is 15%. It is on sale for $17. What is the original price of the package of athletic socks? $20

Extra Example 3

A store pays $15 for a baseball cap. The percent markup is 60%. What is the selling price? $24

 On Your Own

2. $20

3. $90

Differentiated Instruction

Visual

Some students may have a hard time remembering the relationships between *sale price*, *selling price*, *discount*, and *markup*. Have them copy the verbal models into their notebooks.

Discount

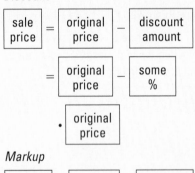

Markup

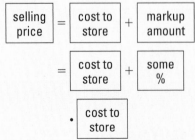

Example 2

- "What is the percent equation?" $a = p \cdot w$
- **?** "What do you know in this problem?" 33 is the part and 60% is the percent.
- **Common Error:** Students multiply 33 by 60% (or 40%). Students need to remember that 33 is a *part* of the original price, it's not the *whole*.

Example 3

- Work through the problem as shown. Encourage students to use mental math to find 20% of $70. 10% of 70 is 7. So, 20% of 70 is 2(7), or 14.
- Two steps were used to answer the question: 1) Find 20% of $70 and 2) add this amount to the original amount of $70.
- **?** "Could this problem be done in one step? Explain." yes; 120% of $70 = $84
- **?** "Explain why the 120% makes sense." You pay 100% of the store's cost plus an additional 20% markup for a total of 120%.
- Method 2 uses the fact that the selling price is 120% of what the store paid. Make sure students understand this. Show students how to use a proportion, $\frac{a}{70} = \frac{120}{100}$ and a one-step percent equation, $a = 1.2(70)$. The ratio table shows the division and multiplication to get to 120% and that this is also the sum of the two rows, which is comparable to the procedure in Method 1.
- **Reason Abstractly and Quantitatively** and **Construct Viable Arguments and Critique the Reasoning of Others:** Now go back and ask students what other ways they can solve the previous examples. Deepening student understanding results from considering different ways in which to solve problems.
- **Common Error:** Students find the markup ($14) instead of the selling price ($84).
- **Extension:** Have students draw a percent bar model for this problem. The model will be divided into fifths.

On Your Own

- **Neighbor Check:** Have students work independently and then have their neighbors check their work. Have students discuss any discrepancies.

Closure

- **Writing Prompt:** Explain two ways to find the sale price for an item marked 30% off. 1) Find the amount of discount and subtract from the original price. 2) Find the percent of the original price and multiply the percent by the original price.
- **Extension:** "If an item is marked up and then discounted the same percent, will the store make a profit? Explain." no; the *amount of markup* will be less than the *amount of discount*, so the store will sell the item for less than what it paid.
- **Extension:** "Is a 25% discount followed by a 10% discount the same as a 35% discount? Explain." no; The sale price for an item discounted 25% followed by a 10% discount would be 0.75(0.9) = 0.675, or 67.5% of the original price. The sale price for an item discounted 35% would be 65% of the original price.

EXAMPLE **2** **Finding an Original Price**

What is the original price of the shoes?

The sale price is
$100\% - 40\% = 60\%$
of the original price.

Answer the question: 33 is 60% of what number?

$a = p \cdot w$ Write percent equation.

$33 = 0.6 \cdot w$ Substitute 33 for a and 0.6 for p.

$55 = w$ Divide each side by 0.6.

So, the original price of the shoes is $55.

Check

0%		60%	100%

| 0 | 33 | 55 | ✓ |

EXAMPLE **3** **Finding a Selling Price**

A store pays $70 for a bicycle. The percent of markup is 20%. What is the selling price?

Method 1: First, find the markup. The markup is 20% of $70.

$a = p \cdot w$

$= 0.20 \cdot 70$

$= 14$

Next, find the selling price.

selling price	=	cost to store	+	markup
	=	70	+	14

$= 84$

So, the selling price is $84.

Method 2: Use a ratio table. The selling price is 120% of the cost to the store.

Percent	Dollars
100%	$70
20%	$14
120%	$84

÷5   ×6   ÷5   ×6

So, the selling price is $84.

Check

0%	40%	80%	120%

| 0 | 14 | 28 | 42 | 56 | 70 | 84 | ✓ |

On Your Own

Now You're Ready
Exercises 9–13 and 17–19

2. The discount on a DVD is 50%. It is on sale for $10. What is the original price of the DVD?

3. A store pays $75 for an aquarium. The markup is 20%. What is the selling price?

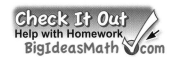
 Vocabulary and Concept Check

1. **WRITING** Describe how to find the sale price of an item that has been discounted 25%.

2. **WRITING** Describe how to find the selling price of an item that has been marked up 110%.

3. **REASONING** Which would you rather pay? Explain your reasoning.

   a. | 6% tax on a discounted price | or | 6% tax on the original price |

   b. | 30% markup on a $30 shirt | or | $30 markup on a $30 shirt |

 Practice and Problem Solving

Copy and complete the table.

		Original Price	Percent of Discount	Sale Price
①	4.	$80	20%	
	5.	$42	15%	
	6.	$120	80%	
	7.	$112	32%	
	8.	$69.80	60%	
②	9.		25%	$40
	10.		5%	$57
	11.		80%	$90
	12.		64%	$72
	13.		15%	$146.54
	14.	$60		$45
	15.	$82		$65.60
	16.	$95		$61.75

Find the selling price.

③ 17. Cost to store: $50
   Markup: 10%

18. Cost to store: $80
   Markup: 60%

19. Cost to store: $140
   Markup: 25%

Assignment Guide and Homework Check

Level	Day 1 Activity Assignment	Day 2 Lesson Assignment	Homework Check
Basic	4, 9, 14, 26–29	1–3, 5, 7, 11–17 odd, 21	5, 11, 15, 17
Average	4, 9, 14, 26–29	1–3, 5, 7, 11–17 odd, 21, 22	5, 11, 15, 17
Advanced	4, 9, 14, 26–29	1–3, 8, 12, 16–24 even, 25	8, 12, 16, 20, 24
Accelerated	1–4, 8, 9, 12–24 even, 25–29		8, 12, 16, 20, 24

Common Errors

- **Exercises 4–8** Students may write the discount amount as the sale price instead of subtracting it from the original amount. When students copy the table, ask them to add another column titled "Discount Amount." Remind them to subtract the discount amount from the original price.
- **Exercises 9–16** Remind students that there is an extra step in the problem. They should subtract the percent of discount from 100% to find the percent of the original price of the item.
- **Exercises 17–19** Students may find the markup and not the selling price. Remind them that they must add the markup to the cost to obtain the selling price.

Vocabulary and Concept Check

1. *Sample answer:* Multiply the original price by 100% − 25% = 75% to find the sale price.

2. Find the markup by taking 110% of the amount. Then add the amount and the markup to find the selling price.

3. **a.** 6% tax on a discounted price; The discounted price is less, so the tax is less.

   b. 30% markup on a $30 shirt; 30% of $30 is less than $30.

Practice and Problem Solving

4. $64
5. $35.70
6. $24
7. $76.16
8. $27.92
9. $53.33
10. $60
11. $450
12. $200
13. $172.40
14. 25%
15. 20%
16. 35%
17. $55
18. $128
19. $175

6.6 Record and Practice Journal

Complete the table.

	Original Price	Percent of Discount	Sale Price
1.	$20	20%	$16
2.	$95	35%	$61.75
3.	$222	75%	$55.50
4.	$130	40%	$78

Find the selling price.

5. Cost to store: $20
   Markup: 15%
   $23

6. Cost to store: $56
   Markup: 80%
   $100.80

7. Cost to store: $110
   Markup: 140%
   $264

8. A store buys an item for $10. To earn a profit of $25, what percent does the store need to markup the item?
   250%

20. no; Only the amount of markup should be in the numerator, $\dfrac{105 - 60}{60} = 0.75$. So, the percent of markup is 75%.

21. "Multiply $45.85 by 0.1" and "Multiply $45.85 by 0.9, then subtract from $45.85." Both will give the sale price of $4.59. The first method is easier because it is only one step.

22. a. Store C

   b. at least 11.82%

23. no; $31.08

24. See *Taking Math Deeper.*

25. $30

 Fair Game Review

26. 170   **27.** 180

28. 1152   **29.** C

Mini-Assessment

Find the price, discount, markup, or cost to store.

1. Original price: $50
   Discount: 15%
   Sale price: ? $42.50

2. Original price: $35
   Discount: ?
   Sale price: $31.50 10%

3. Cost to store: $75
   Markup: ?
   Selling price: $112.50 50%

4. Cost to store: ?
   Markup: 15%
   Selling price: $85.10 $74

5. The sale price for a bicycle is $89.90. The sale price includes a discount of 20%. What is the original price of the bicycle? $112.38

Taking Math Deeper

Exercise 24

A good way to approach this problem is to take things one step at a time. Also, in problems like this, it is much easier to round up to $40 and $30 for easier calculations.

 Find the percent of discount.
   a. 10 is 25% of 40.

   It is easier to round $39.99 to $40 before doing the calculations.

Jeans	$40 (39.99)
Discount	(-10.00)
Subtotal	29.99
Sales Tax	1.95
Total	31.94

 Find the percent of sales tax.

   1.95 is what % of 30?

   $$1.95 = p \cdot 30$$
   $$0.065 = p$$

   b. Sales tax = 6.5%.

Jeans	39.99
Discount	-10.00
Subtotal $30	(29.99)
Sales Tax	(1.95)
Total	31.94

③ Find the actual markup.

   $$x + 0.6x = 40$$
   $$1.6x = 40$$
   $$x = \$25 \quad \text{Wholesale}$$

The $40 jeans cost the store $25. After the discount of $10, the markup is $5. Find the percent of markup by answering "5 is what % of 25?"

   c. $5 is a 20% markup on $25.

$5 markup

Project

Check the newspaper or local advertisements for a store near you. Select five items that are on sale. Prepare a chart that shows the original price, the percent of discount, and the sale price. How much would you save if you purchased all five items at the sale price?

Reteaching and Enrichment Strategies

If students need help. . .	If students got it. . .
Resources by Chapter • Practice A and Practice B • Puzzle Time Record and Practice Journal Practice Differentiating the Lesson Lesson Tutorials Skills Review Handbook	Resources by Chapter • Enrichment and Extension • Technology Connection Start the next section

20. **YOU BE THE TEACHER** The cost to a store for an MP3 player is $60. The selling price is $105. A classmate says that the markup is 175% because $\frac{\$105}{\$60} = 1.75$. Is your classmate correct? If not, explain how to find the correct percent of markup.

21. **SCOOTER** The scooter is on sale for 90% off the original price. Which of the methods can you use to find the sale price? Which method do you prefer? Explain.

| Multiply $45.85 by 0.9. | Multiply $45.85 by 0.1. |

| Multiply $45.85 by 0.9, then add to $45.85. | Multiply $45.85 by 0.9, then subtract from $45.85. |

22. **GAMING** You are shopping for a video game system.

Store	Cost to Store	Markup
A	$162	40%
B	$155	30%
C	$160	25%

   a. At which store should you buy the system?

   b. Store A has a weekend sale. What discount must Store A offer for you to buy the system there?

23. **STEREO** A $129.50 stereo is discounted 40%. The next month, the sale price is discounted 60%. Is the stereo now "free"? If not, what is the sale price?

24. **CLOTHING** You buy a pair of jeans at a department store.

   a. What is the percent of discount to the nearest percent?

   b. What is the percent of sales tax to the nearest tenth of a percent?

   c. The price of the jeans includes a 60% markup. After the discount, what is the percent of markup to the nearest percent?

Department Store

Jeans	39.99
Discount	-10.00
Subtotal	29.99
Sales Tax	1.95
Total	31.94

Thank You

25. **Critical Thinking** You buy a bicycle helmet for $22.26, which includes 6% sales tax. The helmet is discounted 30% off the selling price. What is the original price?

Fair Game Review What you learned in previous grades & lessons

Evaluate. *(Skills Review Handbook)*

26. 2000(0.085)   27. 1500(0.04)(3)   28. 3200(0.045)(8)

29. **MULTIPLE CHOICE** Which measurement is greater than 1 meter? *(Skills Review Handbook)*

   Ⓐ 38 inches   Ⓑ 1 yard   Ⓒ 3.4 feet   Ⓓ 98 centimeters

Essential Question How can you find the amount of simple interest earned on a savings account? How can you find the amount of interest owed on a loan?

Simple interest is money earned on a savings account or an investment. It can also be money you pay for borrowing money.

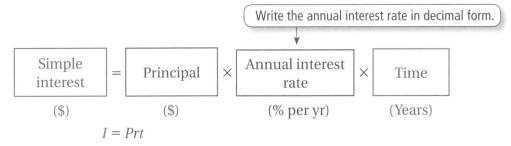

Write the annual interest rate in decimal form.

Simple interest	=	Principal	×	Annual interest rate	×	Time
($)		($)		(% per yr)		(Years)

$$I = Prt$$

1 ACTIVITY: Finding Simple Interest

Work with a partner. You put $100 in a savings account. The account earns 6% simple interest per year. (a) Find the interest earned and the balance at the end of 6 months. (b) Copy and complete the table. Then make a bar graph that shows how the balance grows in 6 months.

a. $I = Prt$ Write simple interest formula.

  = ▢   Substitute values.

  = ▢   Multiply.

∴ At the end of 6 months, you earn $▢ in interest. So, your balance is $▢ .

Percents

In this lesson, you will

- use the simple interest formula to find interest earned or paid, annual interest rates, and amounts paid on loans.

b.

Time	Interest	Balance
0 month	$0	$100
1 month		
2 months		
3 months		
4 months		
5 months		
6 months		

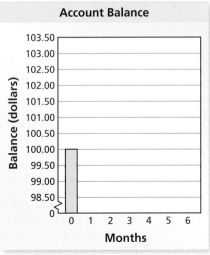

Account Balance

Laurie's Notes

Introduction

Applying Mathematical Practices

- **Model with Mathematics:** Percent applications are abundant in the financial world. Interest rates are stated as percents and students need to understand how interest is paid or charged to consumers.

Motivate

- **Story Time:** Tell students that you just saw an ad for the latest smart phone and you really want to buy it. "It's only $400, but unfortunately I have a few other bills this month and can't afford $400 all at once."
- **?** "What can I do?" Students may suggest that you go to a bank and borrow the money.
- **?** "Will a bank just give me $400?" Hopefully students will know that there is a fee you have to pay to borrow the money.

Activity Notes

Discuss

- Today's investigation involves three activities. Given time constraints and your own students, you may not complete all three.
- **Financial Literacy:** You want students to have some understanding of the cost of borrowing money or the ability to earn money when it is deposited in a bank, not to become trained loan officers.
- **Discuss:** When you *deposit* money, you should *earn* money. When you *borrow* money, you should *pay* money.
- Define *simple interest formula.*
- **Discuss:** Interest earned/owed is influenced by how much money is involved (principal), the rate you pay/earn, and the amount of time.
- Make clear that it is an *annual* interest rate and the time is in *years.*
- Students should assume that deposits are made at the beginning of the interest period in all banking problems, unless otherwise stated.

Activity 1

- This activity uses the simple interest formula. The principal stays the same for each month's calculation. Interest paid is *not* being compounded.
- **Demonstrate:** After one month, you earn $100(0.06)\left(\dfrac{1}{12}\right) = \$0.50$. This $0.50 is added to the principal.
- Get students started on month 2. Students should use $100 for the principal and $\dfrac{2}{12}$ for the time. Interest earned $= \$100(0.06)\left(\dfrac{2}{12}\right) = \$1.00$.
- Students should work with partners to complete the table and the graph.
- **Attend to Precision:** Note the use of the broken axis on the bar graph. In order to show the constant growth of $0.50 per month, the vertical scale has to be quite small.

What Your Students Will Learn

- Use the simple interest formula to find interest earned or paid, annual interest rates, length of loan, or amounts paid on loans.

Previous Learning

Students should be familiar with finding a percent of a number, rounding decimal values, and converting between fractions, decimals, and percents.

Technology for the *Teacher*

Dynamic Classroom

Lesson Plans
Complete Materials List

6.7 Record and Practice Journal

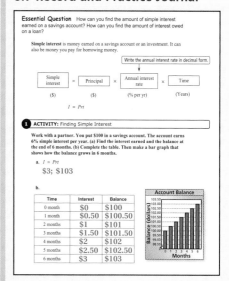

Essential Question How can you find the amount of simple interest earned on a savings account? How can you find the amount of interest owed on a loan?

Simple interest is money earned on a savings account or an investment. It can also be money you pay for borrowing money.

Write the annual interest rate in decimal form.

| Simple interest | = | Principal | × | Annual interest rate | × | Time |
| ($) | | ($) | | (% per yr) | | (Years) |

$I = Prt$

1 ACTIVITY: Finding Simple Interest

Work with a partner. You put $100 in a savings account. The account earns 6% simple interest per year. (a) Find the interest earned and the balance at the end of 6 months. (b) Complete the table. Then make a bar graph that shows how the balance grows in 6 months.

a. $I = Prt$

$3; $103

b.

Time	Interest	Balance
0 month	$0	$100
1 month	$0.50	$100.50
2 months	$1	$101
3 months	$1.50	$101.50
4 months	$2	$102
5 months	$2.50	$102.50
6 months	$3	$103

Account Balance

Auditory

Discuss the meaning of the word *interest*. An interest rate is often expressed as an annual percentage of the principal.

Laurie's Notes

Activity 2

- You may wish to use the information to demonstrate the impact of carrying a large credit card debt.
- **Discuss:** How a credit card operates, how you can get one, and how it works (consumer, store, bank).
- **Community:** If your local bank has an education or outreach coordinator, consider having them come in as a guest speaker.
- Read through the information given. Calculate the interest owed for one month, $\$16,000(0.14)\left(\dfrac{1}{12}\right) \approx \$186.67$, or at the higher interest rate, $\$16,000(0.16)\left(\dfrac{1}{12}\right) \approx \$213.33$.
- **Reason Abstractly and Quantitatively:** Why should you shop around for the lowest interest rates? Why should you keep your principal as small as possible?
- Remind the students that the consumer needs to pay the interest ($\$186.67$ or $\$213.33$) *plus* they need to be paying off the principal.

Activity 3

- The national debt is a complicated concept. The intent of this activity is to raise awareness and to use the simple interest formula with a really large number.
- **Caution:** If you use a calculator for this problem, the debt and simple interest will appear in scientific notation.
- **Representation:** It will be helpful to write out the simple interest formula using the decimal numbers so that students can see all of the zeros: $I = (16,000,000,000,000)(0.01)(1)$. This has a greater impact than scientific notation.
- **Extension:** Many local newspapers print the national debt and the approximate per person debt each day. Record this information once a week for about 2 months to get a sense for how the numbers are changing. You can even do this for the entire school year.

Closure

- **Exit Ticket:** What do you need to know in order to compute simple interest? Principal, annual interest rate, and time

6.7 Record and Practice Journal

2 ACTIVITY: Financial Literacy

Work with a partner. Use the following information to write a report about credit cards. In the report, describe how a credit card works. Include examples that show the amount of interest paid each month on a credit card.

U.S. Credit Card Data

- A typical household with credit card debt in the United States owes about $16,000 to credit card companies.
- A typical credit card interest rate is 14% to 16% per year. This is called the annual percentage rate.

Check students' work.

3 ACTIVITY: The National Debt

Work with a partner. In 2012, the United States owed about $16 trillion in debt. The interest rate on the national debt is about 1% per year.

a. Write $16 trillion in decimal form. How many zeros does this number have?
$16,000,000,000,000; 12 zeros

b. How much interest does the United States pay each year on its national debt?
$160 billion

c. How much interest does the United States pay each day on its national debt?
$438,356,164.40

d. The United States has a population of about 314 million people. Estimate the amount of interest that each person pays per year toward interest on the national debt.
about $509.55

What Is Your Answer?

4. IN YOUR OWN WORDS How can you find the amount of simple interest earned on a savings account? How can you find the amount of interest owed on a loan? Give examples with your answer.
Use $I = Prt$.

2 ACTIVITY: Financial Literacy

Work with a partner. Use the following information to write a report about credit cards. In the report, describe how a credit card works. Include examples that show the amount of interest paid each month on a credit card.

Math Practice

Use Other Resources

What resources can you use to find more information about credit cards?

U.S. Credit Card Data

- A typical household with credit card debt in the United States owes about $16,000 to credit card companies.

- A typical credit card interest rate is 14% to 16% per year. This is called the annual percentage rate.

3 ACTIVITY: The National Debt

Work with a partner. In 2012, the United States owed about $16 trillion in debt. The interest rate on the national debt is about 1% per year.

a. Write $16 trillion in decimal form. How many zeros does this number have?

b. How much interest does the United States pay each year on its national debt?

c. How much interest does the United States pay each day on its national debt?

d. The United States has a population of about 314 million people. Estimate the amount of interest that each person pays per year toward interest on the national debt.

$16 Trillion in Debt

What Is Your Answer?

4. IN YOUR OWN WORDS How can you find the amount of simple interest earned on a savings account? How can you find the amount of interest owed on a loan? Give examples with your answer.

Practice

Use what you learned about simple interest to complete Exercises 4–7 on page 256.

6.7 Lesson

Key Vocabulary
interest, *p. 254*
principal, *p. 254*
simple interest,
   p. 254

Interest is money paid or earned for the use of money. The **principal** is the amount of money borrowed or deposited.

 **Key Idea**

Simple Interest

Words **Simple interest** is money paid or earned only on the principal.

Algebra

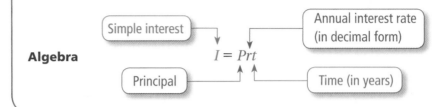

EXAMPLE (1) **Finding Interest Earned**

You put $500 in a savings account. The account earns 3% simple interest per year. (a) What is the interest earned after 3 years? (b) What is the balance after 3 years?

a. $I = Prt$ Write simple interest formula.

    $= 500(0.03)(3)$ Substitute 500 for *P*, 0.03 for *r*, and 3 for *t*.

    $= 45$ Multiply.

     So, the interest earned is $45 after 3 years.

b. To find the balance, add the interest to the principal.

     So, the balance is $500 + $45 = $545 after 3 years.

EXAMPLE (2) **Finding an Annual Interest Rate**

You put $1000 in an account. The account earns $100 simple interest in 4 years. What is the annual interest rate?

         $I = Prt$ Write simple interest formula.

     $100 = 1000(r)(4)$ Substitute 100 for *I*, 1000 for *P*, and 4 for *t*.

     $100 = 4000r$ Simplify.

   $0.025 = r$ Divide each side by 4000.

     So, the annual interest rate of the account is 0.025, or 2.5%.

      ◀)) Multi-Language Glossary at BigIdeasMath✓.com

Laurie's Notes

Introduction

Connect

- **Yesterday:** Students explored the simple interest formula, applying it to several consumer applications.
- **Today:** Students will use the simple interest formula and knowledge of equation solving to solve for different variables in the formula.

Motivate

- Just imagine that when you are older, you win a $5 million lottery. If you deposit that money for 10 years at 6% simple interest, how much will you have at the end of 10 years? $8,000,000

Lesson Notes

Key Idea

- **Vocabulary:** interest, money paid or earned, principal, amount of money borrowed or deposited, balance
- **Representation:** Write the formula in words first.
  Simple Interest = (Principal)(Annual interest rate)(Time)
- **Explain:** Simple interest is only one type of interest. There are also compound and exponential interest calculations. The interest rate is written as a decimal. Time is written in terms of years. When time is given in months, remember to express it as a fraction of a year or as a decimal. For example, 9 months = $\frac{9}{12}$ or 0.75 year.
- **Reason Abstractly and Quantitatively:** This formula is similar to the volume formula for a rectangular prism; three variables are multiplied together. Knowing 3 of the 4 variables, you can solve for the fourth.

Example 1

- There are two parts to the problem: Calculate the interest earned and then determine the amount (balance) in the account.
- **?** "What operation is performed in writing *Prt*?" multiplication
- **?** "In calculating 500(0.03)(3), what order is the multiplication performed?" Order doesn't matter, multiplication is commutative.
- **Explain:** Your balance is the original principal *plus* the interest earned.
- **Reason Abstractly and Quantitatively:** If time permits, "What would your balance be if the interest rate had been 6% instead of 3%?" $590 Doubling the interest rate doubles the amount earned. This can be shown in the equation $I = 500(0.06)(3) = 500(0.03)(2)(3)$.

Example 2

- This example uses the Division Property of Equality to solve for the interest rate.
- **?** "Why does 1000(*r*)(4) = 4000*r*?" Commutative Property of Multiplication
- **Common Error:** Students divide 4000 by 100 instead of 100 by 4000.
- **?** "How do you write a decimal as a percent?" Move the decimal point two places to the right. (Multiply by 100.) Then add a percent symbol.

Extra Example 1

You put $200 in a savings account. The account earns 2% simple interest per year.

a. What is the interest earned after 5 years? $20

b. What is the balance after 5 years? $220

Extra Example 2

You put $700 in an account. The account earns $224 simple interest in 8 years. What is the annual interest rate? 4%

On Your Own

1. $511.25

2. 2%

Extra Example 3

Using the pictograph in Example 3, how long does it take an account with a principal of $400 to earn $36 interest?
6 years

Extra Example 4

You borrow $300 to buy a guitar. The simple interest rate is 12%. You pay off the loan after 4 years. How much do you pay for the loan? $444

On Your Own

3. 2.5 yr

4. $270

English Language Learners

Vocabulary

Review with English learners the mathematical meanings of principal, interest, and balance because these words have multiple meanings in the English language. They should understand that interest is paid to customers when they deposit money into an account. When a person borrows money, the person pays interest to the bank.

On Your Own

- **Neighbor Check**: Have students work independently and then have their neighbors check their work. Have students discuss any discrepancies.
- Check accuracy of decimals in these problems.

Example 3

- Discuss the diagram.
- ? "Why would a bank offer different interest rates for different principals?" Students may not understand that banks are using deposited money to loan to other people.
- Work through the problem.
- ? "What is 6.25 as a mixed number?" $6\frac{1}{4}$
- **Connection:** Students may wonder why anyone would want to know how long it takes to earn $100 in interest. Use an example of depositing money for a future purchase (car, house, college education).

Example 4

- Remind students that the simple interest formula is used to calculate interest *earned* when you *deposit* money and to calculate interest *owed* when you *borrow* money.
- **Discuss:** There are two parts to the problem: 1) Calculate the interest owed and 2) determine the total cost you must pay back for the loan.
- **Extension:** Have students find the monthly payment. $1050 \div 60 = $17.50

On Your Own

- **Neighbor Check:** Have students work independently and then have their neighbors check their work. Have students discuss any discrepancies.

Closure

- **Exit Ticket:** Assume $1000 was deposited at 5% simple interest when you were born. Approximately how much is the account worth today?
  age 11: $1550, age 12: $1600, age 13: $1650, age 14: $1700

Now You're Ready
Exercises 4–16

On Your Own

1. In Example 1, what is the balance of the account after 9 months?

2. You put $350 in an account. The account earns $17.50 simple interest in 2.5 years. What is the annual interest rate?

EXAMPLE ③ **Finding an Amount of Time**

A bank offers three savings accounts. The simple interest rate is determined by the principal. How long does it take an account with a principal of $800 to earn $100 in interest?

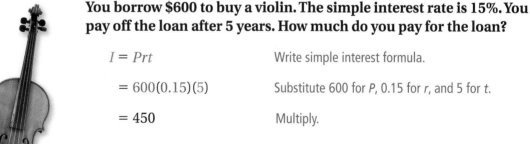

The pictogram shows that the interest rate for a principal of $800 is 2%.

$I = Prt$ Write simple interest formula.

$100 = 800(0.02)(t)$ Substitute 100 for *I*, 800 for *P*, and 0.02 for *r*.

$100 = 16t$ Simplify.

$6.25 = t$ Divide each side by 16.

⋮ So, the account earns $100 in interest in 6.25 years.

EXAMPLE ④ **Finding an Amount Paid on a Loan**

You borrow $600 to buy a violin. The simple interest rate is 15%. You pay off the loan after 5 years. How much do you pay for the loan?

$I = Prt$ Write simple interest formula.

$= 600(0.15)(5)$ Substitute 600 for *P*, 0.15 for *r*, and 5 for *t*.

$= 450$ Multiply.

To find the amount you pay, add the interest to the loan amount.

⋮ So, you pay $600 + $450 = $1050 for the loan.

On Your Own

Now You're Ready
Exercises 17–20
and 24–27

3. In Example 3, how long does it take an account with a principal of $10,000 to earn $750 in interest?

4. **WHAT IF?** In Example 4, you pay off the loan after 2 years. How much money do you save?

 Vocabulary and Concept Check

1. **VOCABULARY** Define each variable in $I = Prt$.

2. **WRITING** In each situation, tell whether you would want a *higher* or *lower* interest rate. Explain your reasoning.

   a. you borrow money   **b.** you open a savings account

3. **REASONING** An account earns 6% simple interest. You want to find the interest earned on $200 after 8 months. What conversions do you need to make before you can use the formula $I = Prt$?

 Practice and Problem Solving

An account earns simple interest. (a) Find the interest earned. (b) Find the balance of the account.

① **4.** $600 at 5% for 2 years   **5.** $1500 at 4% for 5 years

6. $350 at 3% for 10 years   **7.** $1800 at 6.5% for 30 months

8. $700 at 8% for 6 years   **9.** $1675 at 4.6% for 4 years

10. $925 at 2% for 2.4 years   **11.** $5200 at 7.36% for 54 months

12. **ERROR ANALYSIS** Describe and correct the error in finding the simple interest earned on $500 at 6% for 18 months.

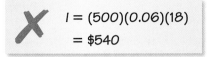

$$I = (500)(0.06)(18)$$
$$= \$540$$

Find the annual interest rate.

② **13.** $I = \$24$, $P = \$400$, $t = 2$ years   **14.** $I = \$562.50$, $P = \$1500$, $t = 5$ years

15. $I = \$54$, $P = \$900$, $t = 18$ months   **16.** $I = \$160.67$, $P = \$2000$, $t = 8$ months

Find the amount of time.

③ **17.** $I = \$30$, $P = \$500$, $r = 3\%$ **18.** $I = \$720$, $P = \$1000$, $r = 9\%$

19. $I = \$54$, $P = \$800$, $r = 4.5\%$ **20.** $I = \$450$, $P = \$2400$, $r = 7.5\%$

21. **BANKING** A savings account earns 5% simple interest per year. The principal is $1200. What is the balance after 4 years?

22. **SAVINGS** You put $400 in an account. The account earns $18 simple interest in 9 months. What is the annual interest rate?

23. **CD** You put $3000 in a CD (certificate of deposit) at the promotional rate. How long will it take to earn $336 in interest?

Certificate of Deposit

This certificate is the original Specimen and valid document from the treasury and Securities department of state have trust financial group & associates. The agreement herein construed are thorough, correct and binding on the parties. Alterations made are also equally valid.

Promotional Rate 5.6% Simple Interest

DIRECTOR'S SIGNATURE

Assignment Guide and Homework Check

Level	Day 1 Activity Assignment	Day 2 Lesson Assignment	Homework Check
Basic	4–7, 38–41	1–3, 9, 11, 12, 13–27 odd	9, 13, 17, 23, 25
Average	4–7, 38–41	1–3, 8–12 even, 13–19 odd, 23–27 odd, 32–34	8, 13, 17, 25, 32
Advanced	4–7, 38–41	1–3, 12–36 even, 37	24, 30, 34, 36
Accelerated	1–7, 12–36 even, 37–41		24, 30, 34, 36

Common Errors

- **Exercises 4–11** Students may forget to change the percent to a decimal. Remind them that before they can put the percent into the equation, they must change the percent to a fraction or a decimal.
- **Exercises 7 and 11** Students may not change months into years and calculate a much greater interest amount. Remind them that the simple interest formula is for *years* and that the time must be changed to years.
- **Exercises 15 and 16** Students may not change the time from months to years. Remind them that the time is in years.
- **Exercises 24–27** Students may only find the amount of interest paid for the loan. Remind them that the total amount paid on a loan is the original principal plus the interest.

6.7 Record and Practice Journal

An account earns simple interest. (a) Find the interest earned. (b) Find the balance of the account.

1. $400 at 7% for 3 years
   a. $84
   b. $484

2. $1200 at 5.6% for 4 years
   a. $268.80
   b. $1468.80

Find the annual interest rate.

3. $I = \$18$, $P = \$200$, $t = 18$ months
   6%

4. $I = \$310$, $P = \$1000$, $t = 5$ years
   6.2%

Find the amount of time.

5. $I = \$60$, $P = \$750$, $r = 4\%$
   2 years

6. $I = \$825$, $P = \$2500$, $r = 5.5\%$
   6 years

7. You put $500 in a savings account. The account earns $15.75 simple interest in 6 months. What is the annual interest rate?
   6.3%

Vocabulary and Concept Check

1. I = simple interest, P = principal, r = annual interest rate (in decimal form), t = time (in years)

2. **a.** lower interest rate because you would pay less

   b. higher interest rate because you would receive more

3. You have to change 6% to a decimal and 8 months to a fraction of a year.

Practice and Problem Solving

4. **a.** $60   **b.** $660

5. **a.** $300   **b.** $1800

6. **a.** $105   **b.** $455

7. **a.** $292.50 **b.** $2092.50

8. **a.** $336   **b.** $1036

9. **a.** $308.20 **b.** $1983.20

10. **a.** $44.40 **b.** $969.40

11. **a.** $1722.24 **b.** $6922.24

12. They did not convert 18 months to years.
$$I = 500(0.06)\left(\frac{18}{12}\right)$$
$$= \$45$$

13. 3%   14. 7.5%

15. 4%   16. 12.05%

17. 2 yr

18. 8 yr

19. 1.5 yr

20. 2.5 yr

21. $1440

22. 6%

23. 2 yr

Practice and Problem Solving

24. $1770   **25.** $2720

26. $3660   **27.** $6700.80

28. $2550   **29.** $8500

30. 4 yr   **31.** 5.25%

32. See *Taking Math Deeper*.

33. 4 yr

34. $77.25

35. 12.5 yr; Substitute $2000 for *P* and *I*, 0.08 for *r*, and solve for *t*.

36. $300

37. Year 1 = $520
Year 2 = $540.80
Year 3 = $562.43

Fair Game Review

38. $x < -3$;
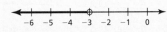

39. $b \geq 1$;
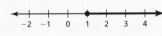

40. $w \leq -9$;

41. A

Mini-Assessment

Find the annual simple interest rate.

1. $I = \$60$, $P = \$500$, $t = 3$ years 4%

2. $I = \$45$, $P = \$600$, $t = 2$ years 3.75%

Find the amount of time.

3. $I = \$117$, $P = \$1300$, $r = 3\%$ 3 yr

4. $I = \$71.50$, $P = \$1100$, $r = 3.25\%$ 2 yr

5. A savings account earns 4.5% annual simple interest. The principal is $1300. What is the balance after 3 years? $1475.50

Taking Math Deeper

Exercise 32

This problem isn't particularly difficult. However, it is a good opportunity for students to pick up some **financial literacy**. That is, when you pay for items with a credit card, you often have to pay interest. In other words, you are taking out a loan.

(1) Find the amount spent.

Total = $175.54

Zoo Trip	
Tickets	67.70
Food	62.34
Gas	45.50
Total Cost	175.54

(2) Find the interest paid.

$I = Prt$ Write the formula.

$= 175.54 \cdot 0.12 \cdot \dfrac{3}{12}$ Substitute amounts.

$\approx \$5.27$ Simplify.

Wow

(3) Find the total cost of the trip.

Total = 175.54 + 5.27
     = $180.81

How much interest would I pay if I didn't pay the charge for 1 year? for 2 years?

Project

Many credit cards charge different rates of interest. Use the school library or the Internet to research the amount of interest charged by three different credit card companies. Compare the cost of the trip to the zoo based on the different interest rates. Why should you be careful when selecting a credit card and charging items to the card?

Reteaching and Enrichment Strategies

If students need help. . .	If students got it. . .
Resources by Chapter • Practice A and Practice B • Puzzle Time Record and Practice Journal Practice Differentiating the Lesson Lesson Tutorials Skills Review Handbook	Resources by Chapter • Enrichment and Extension • Technology Connection Start the next section

Find the amount paid for the loan.

④ **24.** $1500 at 9% for 2 years

25. $2000 at 12% for 3 years

26. $2400 at 10.5% for 5 years

27. $4800 at 9.9% for 4 years

Copy and complete the table.

	Principal	Interest Rate	Time	Simple Interest
28.	$12,000	4.25%	5 years	
29.		6.5%	18 months	$828.75
30.	$15,500	8.75%		$5425.00
31.	$18,000		54 months	$4252.50

32. ZOO A family charges a trip to the zoo on a credit card. The simple interest rate is 12%. The charges are paid after 3 months. What is the total amount paid for the trip?

Zoo Trip

Tickets	67.70
Food	62.34
Gas	45.50
Total Cost	?

33. MONEY MARKET You deposit $5000 in an account earning 7.5% simple interest. How long will it take for the balance of the account to be $6500?

11.8% Simple Interest
Equal monthly
payments for 2 years

34. LOANS A music company offers a loan to buy a drum set for $1500. What is the monthly payment?

35. REASONING How many years will it take for $2000 to double at a simple interest rate of 8%? Explain how you found your answer.

36. PROBLEM SOLVING You have two loans, for 2 years each. The total interest for the two loans is $138. On the first loan, you pay 7.5% simple interest on a principal of $800. On the second loan, you pay 3% simple interest. What is the principal for the second loan?

37. Critical Thinking You put $500 in an account that earns 4% annual interest. The interest earned each year is added to the principal to create a new principal. Find the total amount in your account after each year for 3 years.

Fair Game Review What you learned in previous grades & lessons

Solve the inequality. Graph the solution. *(Section 4.2)*

38. $x + 5 < 2$

39. $b - 2 \geq -1$

40. $w + 6 \leq -3$

41. MULTIPLE CHOICE What is the solution of $4x + 5 = -11$? *(Section 3.5)*

Ⓐ $x = -4$ Ⓑ $x = -1.5$ Ⓒ $x = 1.5$ Ⓓ $x = 4$

Identify the percent of change as an *increase* or a *decrease*. Then find the percent of change. Round to the nearest tenth of a percent if necessary.   *(Section 6.5)*

1. 8 inches to 24 inches

2. 300 miles to 210 miles

Find the original price, discount, sale price, or selling price.   *(Section 6.6)*

3. Original price: $30
   Discount: 10%
   Sale price: ?

4. Original price: $55
   Discount: ?
   Sale price: $46.75

5. Original price: ?
   Discount: 75%
   Sale price: $74.75

6. Cost to store: $152
   Markup: 50%
   Selling price: ?

An account earns simple interest. Find the interest earned, principal, interest rate, or time.   *(Section 6.7)*

7. Interest earned: ?
   Principal: $1200
   Interest rate: 2%
   Time: 5 years

8. Interest earned: $25
   Principal: $500
   Interest rate: 5%
   Time: ?

9. Interest earned: $76
   Principal: $800
   Interest rate: ?
   Time: 2 years

10. Interest earned: $119.88
    Principal: ?
    Interest rate: 3.6%
    Time: 3 years

11. **HEIGHT** You estimate that your friend is 50 inches tall. The actual height of your friend is 54 inches. Find the percent error.   *(Section 6.5)*

12. **DIGITAL CAMERA** A digital camera costs $230. The camera is on sale for 30% off, and you have a coupon for an additional 15% off the sale price. What is the final price?   *(Section 6.6)*

13. **WATER SKIS** The original price of the water skis was $200. What is the percent of discount?   *(Section 6.6)*

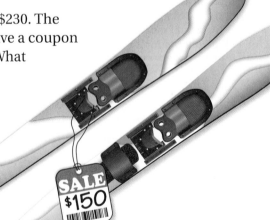

SALE $150

2 Ways to Own:
1. $75 cash back with 3.5% simple interest
2. No interest for 2 years

14. **SAXOPHONE** A saxophone costs $1200. A store offers two loan options. Which option saves more money if you pay the loan in 2 years?   *(Section 6.7)*

15. **LOAN** You borrow $200. The simple interest rate is 12%. You pay off the loan after 2 years. How much do you pay for the loan?   *(Section 6.7)*

Alternative Assessment Options

Math Chat   Student Reflective Focus Question

Structured Interview   **Writing Prompt**

Writing Prompt

Ask students to write a story about making purchases and saving money. The students should include discounts and markups in the story. If they have money left over from their purchases, they should place it in a savings account. The students should include simple interest in the story. Then have students share their stories with the class.

Study Help Sample Answers

Remind students to complete Graphic Organizers for the rest of the chapter.

5.

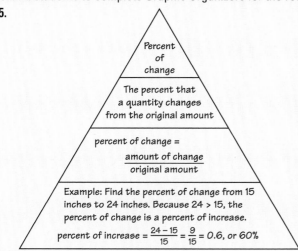

6–8. Available at *BigIdeasMath.com*.

Reteaching and Enrichment Strategies

If students need help. . .	If students got it. . .
Resources by Chapter • Practice A and Practice B • Puzzle Time Lesson Tutorials *BigIdeasMath.com*	Resources by Chapter • Enrichment and Extension • Technology Connection Game Closet at *BigIdeasMath.com* Start the Chapter Review

Technology for the *Teacher*

Online Assessment

Assessment Book

ExamView® Assessment Suite

Review of Common Errors

Exercises 1–6

- Students may move the decimal point the wrong way, forget to insert zeros as placeholders, or move the decimal too many places (especially when the percent is greater than 100).

Exercises 7–14

- Students may try to order the numbers without converting them, or will only convert them mentally and do so incorrectly.

Answers

1. 0.76
2. 0.06
3. 3.34
4. 15%
5. 124%
6. 9.7%
7. 52%
8. 245%
9. 0.46
10. 22%
11.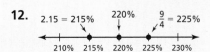
12.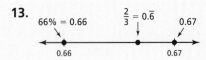
13.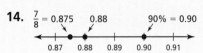
14.
$\frac{7}{8} = 0.875$ 0.88   90% = 0.90

0.87   0.88   0.89   0.90   0.91

6 Chapter Review

Check It Out
Vocabulary Help
BigIdeasMath ✓.com

Review Key Vocabulary

percent of change, *p. 242*   percent error, *p. 243*   interest, *p. 254*
percent of increase, *p. 242*   discount, *p. 248*   principal, *p. 254*
percent of decrease, *p. 242*   markup, *p. 248*   simple interest, *p. 254*

Review Examples and Exercises

6.1 Percents and Decimals *(pp. 214–219)*

a. Write 64% as a decimal.

$64\% = 64.\% = 0.64$

b. Write 0.023 as a percent.

$0.023 = 0.023 = 2.3\%$

Exercises

Write the percent as a decimal. Use a model to check your answer.

1. 76%   **2.** 6%   **3.** 334%

Write the decimal as a percent. Use a model to check your answer.

4. 0.15   **5.** 1.24   **6.** 0.097

6.2 Comparing and Ordering Fractions, Decimals, and Percents *(pp. 220–225)*

Which is greater, $\frac{9}{10}$ or 88%?

Write $\frac{9}{10}$ as a percent: $\frac{9}{10} = \frac{90}{100} = 90\%$

∴ 88% is less than 90%. So, $\frac{9}{10}$ is the greater number.

Exercises

Tell which number is greater.

7. $\frac{1}{2}$, 52%   **8.** $\frac{12}{5}$, 245%

9. 0.46, 43%   **10.** 0.023, 22%

Use a number line to order the numbers from least to greatest.

11. $\frac{41}{50}$, 0.83, 80%   **12.** $\frac{9}{4}$, 220%, 2.15

13. 0.67, 66%, $\frac{2}{3}$ **14.** 0.88, $\frac{7}{8}$, 90%

6.3 **The Percent Proportion** *(pp. 226–231)*

a. **What percent of 24 is 9?**

$$\frac{a}{w} = \frac{p}{100}$$ Write the percent proportion.

$$\frac{9}{24} = \frac{p}{100}$$ Substitute 9 for *a* and 24 for *w*.

$$100 \cdot \frac{9}{24} = 100 \cdot \frac{p}{100}$$ Multiplication Property of Equality

$$37.5 = p$$ Simplify.

So, 37.5% of 24 is 9.

b. **What number is 15% of 80?**

$$\frac{a}{w} = \frac{p}{100}$$ Write the percent proportion.

$$\frac{a}{80} = \frac{15}{100}$$ Substitute 80 for *w* and 15 for *p*.

$$80 \cdot \frac{a}{80} = 80 \cdot \frac{15}{100}$$ Multiplication Property of Equality

$$a = 12$$ Simplify.

So, 12 is 15% of 80.

c. **120% of what number is 54?**

$$\frac{a}{w} = \frac{p}{100}$$ Write the percent proportion.

$$\frac{54}{w} = \frac{120}{100}$$ Substitute 54 for *a* and 120 for *p*.

$$54 \cdot 100 = w \cdot 120$$ Cross Products Property

$$5400 = 120w$$ Multiply.

$$45 = w$$ Divide each side by 120.

So, 120% of 45 is 54.

Exercises

Write and solve a proportion to answer the question.

15. What percent of 60 is 18?

16. 40 is what percent of 32?

17. What number is 70% of 70?

18. $\frac{3}{4}$ is 75% of what number?

Review of Common Errors (continued)

Exercises 15–18

- Students may not know what number to substitute for each variable. Walk through each type of question with the students. Emphasize that the word "is" means "equals," and "of" means "multiplied by."

19. $a = 0.24 \cdot 25$; 6

20. $9 = p \cdot 20$; 45%

21. $60.8 = p \cdot 32$; 190%

22. $91 = 1.3 \cdot w$; 70

23. $10.2 = 0.85 \cdot w$; 12

24. $a = 0.83 \cdot 20$; 16.6

25. 120 parking spaces

26. 64%

Review of Common Errors (continued)

Exercises 19–26

- Students may not know what number to substitute for each variable. Walk through each type of question with the students. Emphasize that the word "is" means "equals," and "of" means "multiplied by."
- Students may mix up the whole and the part when trying to write the percent equation for the word problems. Ask students to identify each part of the equation before writing it in the equation format.

Exercises 27–29

- Students may mix up where to place the numbers in the equation to find percent of change. When students do not put the numbers in the right place, they might find a negative number in the numerator. Emphasize that students must know if it is increasing or decreasing before they can do anything else. The numerator should never have a negative answer. If students get a negative number, then they need to switch the order of the numbers in the problem and then subtract.

Exercises 30 and 31

- Students may just find the markup and not the selling price. Remind them that they must add the markup to the cost to store.
- Remind students that the sale price is not the percent of discount multiplied by the original price.

Exercises 32–38

- Students may forget to change the percent to a decimal. Remind them that before they can put the percent into the equation, they must change the percent to a fraction or a decimal.

6.4 The Percent Equation (pp. 232–237)

a. What number is 72% of 25?

$$a = p \cdot w \qquad \text{Write percent equation.}$$
$$= 0.72 \cdot 25 \qquad \text{Substitute 0.72 for } p \text{ and 25 for } w.$$
$$= 18 \qquad \text{Multiply.}$$

So, 72% of 25 is 18.

b. 28 is what percent of 70?

$$a = p \cdot w \qquad \text{Write percent equation.}$$
$$28 = p \cdot 70 \qquad \text{Substitute 28 for } a \text{ and 70 for } w.$$
$$\frac{28}{70} = \frac{p \cdot 70}{70} \qquad \text{Division Property of Equality}$$
$$0.4 = p \qquad \text{Simplify.}$$

Because 0.4 equals 40%, 28 is 40% of 70.

c. 22.1 is 26% of what number?

$$a = p \cdot w \qquad \text{Write percent equation.}$$
$$22.1 = 0.26 \cdot w \qquad \text{Substitute 22.1 for } a \text{ and 0.26 for } p.$$
$$85 = w \qquad \text{Divide each side by 0.26.}$$

So, 22.1 is 26% of 85.

Exercises

Write and solve an equation to answer the question.

19. What number is 24% of 25?

20. 9 is what percent of 20?

21. 60.8 is what percent of 32?

22. 91 is 130% of what number?

23. 85% of what number is 10.2?

24. 83% of 20 is what number?

25. PARKING 15% of the school parking spaces are handicap spaces. The school has 18 handicap spaces. How many parking spaces are there?

26. FIELD TRIP Of the 25 students on a field trip, 16 students bring cameras. What percent of the students bring cameras?

6.5 Percents of Increase and Decrease (pp. 240–245)

The table shows the numbers of skim boarders at a beach on Saturday and Sunday. What was the percent of change in boarders from Saturday to Sunday?

The number of skim boarders on Sunday is less than the number of skim boarders on Saturday. So, the percent of change is a percent of decrease.

$$\text{percent of decrease} = \frac{\text{original amount} - \text{new amount}}{\text{original amount}}$$

Day	Number of Skim Boarders
Saturday	12
Sunday	9

$= \dfrac{12 - 9}{12}$ Substitute.

$= \dfrac{3}{12}$ Subtract.

$= 0.25 = 25\%$ Write as a percent.

So, the number of skim boarders decreased by 25% from Saturday to Sunday.

Exercises

Identify the percent of change as an *increase* or a *decrease*. Then find the percent of change. Round to the nearest tenth of a percent if necessary.

27. 6 yards to 36 yards

28. 120 meals to 52 meals

29. MARBLES You estimate that a jar contains 68 marbles. The actual number of marbles is 60. Find the percent error.

6.6 Discounts and Markups (pp. 246–251)

What is the original price of the tennis racquet?

The sale price is 100% − 30% = 70% of the original price.

Answer the question: 21 is 70% of what number?

$a = p \cdot w$ Write percent equation.

$21 = 0.7 \cdot w$ Substitute 21 for a and 0.7 for p.

$30 = w$ Divide each side by 0.7.

So, the original price of the tennis racquet is $30.

SALE 30% off Now $21

Exercises

Find the sale price or original price.

30. Original price: $50
Discount: 15%
Sale price: ?

31. Original price: ?
Discount: 20%
Sale price: $75

Review Game

Percents of Increase and Decrease

Materials per Group
- 1 deck of cards with the jacks, queens, kings, and aces removed
- paper
- pencil
- calculator

Directions

Each group starts with 108 points. The cards are placed face down in the middle of the group. One member of the group turns a card over. If the card is red, the face value of the card is subtracted from the number of points. If the card is black, the face value of the card is added to the number of points. Group members take turns calculating the percent increase or decrease and turning cards over. The starting number of points at each player's turn is the same as the ending number of points at the previous player's turn. The group should be back to 108 points after going through all of the cards.

Who Wins?

The group with the highest mean percent increase wins. To find the mean percent increase, add the percent increases and divide the sum by 18.

For the Student
Additional Practice
- Lesson Tutorials
- Multi-Language Glossary
- Self-Grading Progress Check
- *BigIdeasMath.com*
    Dynamic Student Edition
    Student Resources

Answers

27. increase; 500%
28. decrease; 56.7%
29. about 13.3%
30. $42.50
31. $93.75
32. **a.** $36
    b. $336
33. **a.** $280
    b. $2280
34. 1.7%
35. 7.1%
36. 3 years
37. 6 years
38. 4%

My Thoughts on the Chapter

What worked. . .

What did not work. . .

What I would do differently. . .

6.7 Simple Interest (pp. 252–257)

You put $200 in a savings account. The account earns 2% simple interest per year.

a. What is the interest earned after 4 years?

b. What is the balance after 4 years?

a.
$I = Prt$	Write simple interest formula.
$= 200(0.02)(4)$	Substitute 200 for P, 0.02 for r, and 4 for t.
$= 16$	Multiply.

So, the interest earned is $16 after 4 years.

b. To find the balance, add the interest to the principal.

So, the balance is $200 + $16 = $216 after 4 years.

You put $500 in an account. The account earns $55 simple interest in 5 years. What is the annual interest rate?

$I = Prt$	Write simple interest formula.
$55 = 500(r)(5)$	Substitute 55 for I, 500 for P, and 5 for t.
$55 = 2500r$	Simplify.
$0.022 = r$	Divide each side by 2500.

So, the annual interest rate of the account is 0.022, or 2.2%.

Exercises

An account earns simple interest.

a. Find the interest earned.

b. Find the balance of the account.

32. $300 at 4% for 3 years

33. $2000 at 3.5% for 4 years

Find the annual simple interest rate.

34. $I = $17, P = $500, t = 2$ years

35. $I = $426, P = $1200, t = 5$ years

Find the amount of time.

36. $I = $60, P = $400, r = 5\%$

37. $I = $237.90, P = $1525, r = 2.6\%$

38. **SAVINGS** You put $100 in an account. The account earns $2 simple interest in 6 months. What is the annual interest rate?

Check It Out
Test Practice
BigIdeasMath ✓com

Write the percent as a decimal.

1. 0.96%

2. 65%

3. 25.7%

Write the decimal as a percent.

4. 0.42

5. 7.88

6. 0.5854

Tell which number is greater.

7. $\frac{16}{25}$, 65%

8. 56%, 5.6

Use a number line to order the numbers from least to greatest.

9. 85%, $\frac{15}{18}$, 0.84

10. 58.3%, 0.58, $\frac{7}{12}$

Answer the question.

11. What percent of 28 is 21?

12. 64 is what percent of 40?

13. What number is 80% of 45?

14. 0.8% of what number is 6?

Identify the percent of change as an *increase* or a *decrease*. Then find the percent of change. Round to the nearest tenth of a percent if necessary.

15. 4 strikeouts to 10 strikeouts

16. $24 to $18

Find the sale price or selling price.

17. Original price: $15
Discount: 5%
Sale price: ?

18. Cost to store: $5.50
Markup: 75%
Selling price: ?

An account earns simple interest. Find the interest earned or the principal.

19. Interest earned: ?
Principal: $450
Interest rate: 6%
Time: 8 years

20. Interest earned: $27
Principal: ?
Interest rate: 1.5%
Time: 2 years

21. BASKETBALL You, your cousin, and a friend each take the same number of free throws at a basketball hoop. Who made the most free throws?

22. PARKING LOT You estimate that there are 66 cars in a parking lot. The actual number of cars is 75.

a. Find the percent error.

b. What other estimate gives the same percent error? Explain your reasoning.

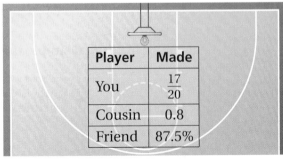

Player	Made
You	$\frac{17}{20}$
Cousin	0.8
Friend	87.5%

23. INVESTMENT You put $800 in an account that earns 4% simple interest. Find the total amount in your account after each year for 3 years.

Test Item References

Chapter Test Questions	Section to Review
1–6	6.1
7–10, 21	6.2
11–14	6.3
11–14	6.4
15, 16, 22	6.5
17, 18	6.6
19, 20, 23	6.7

Test-Taking Strategies

Remind students to quickly look over the entire test before they start so they can budget their time. Remind them that the test is on fractions, decimals, *and* percents and that they need to read the problems carefully. Students need to use the **Stop** and **Think** strategy before they answer a question. Students need to remember to think of the different representations of each number as they work through the test, such as 0.5, $\frac{1}{2}$, and 50%.

Common Errors

- **Exercises 7–10** Students may try to order the numbers without converting them or will only convert them mentally and do so incorrectly. Tell students that it is necessary to convert all the numbers to one form.
- **Exercises 11–14** Students may not know what numbers to substitute for the variables. Review each type of question with students. Emphasize that the word "is" means "equals" and "of" means "multiplied by." Ask students to identify the whole, the part of the whole, and the percent.
- **Exercises 15 and 16** Students might place the numbers in the percent of change formulas incorrectly. Remind them that they should have the difference between the greater amount and the lesser amount in the numerator, so the numerator should never be negative. Also point out that the original amount should always be in the denominator.

Reteaching and Enrichment Strategies

If students need help. . .	If students got it. . .
Resources by Chapter • Practice A and Practice B • Puzzle Time Record and Practice Journal Practice Differentiating the Lesson Lesson Tutorials *BigIdeasMath.com* Skills Review Handbook	Resources by Chapter • Enrichment and Extension • Technology Connection Game Closet at *BigIdeasMath.com* Start Cumulative Assessment

Answers

1. 0.0096
2. 0.65
3. 0.257
4. 42%
5. 788%
6. 58.54%
7. 65%
8. 5.6
9.
   $\frac{15}{18} = 0.8\overline{3}$ 0.84   85% = 0.85
10.
    0.58   58.3% = 0.583 $\frac{7}{12} = 0.58\overline{3}$
11. 75%
12. 160%
13. 36
14. 750
15. increase; 150%
16. decrease; 25%
17. $14.25
18. $9.63
19. $216
20. $900
21. Your friend
22. **a.** 12%

    b. 84 cars; To get the same percent error, the amount of error needs to be the same. Because your estimate was 9 cars below the actual number, an estimate of 9 cars above the actual number will give the same percent error.

23. Year 1: $832
    Year 2: $864
    Year 3: $896

Technology for the *Teacher*

Online Assessment
Assessment Book
ExamView® Assessment Suite

Test-Taking Strategies

Available at *BigIdeasMath.com*

After Answering Easy Questions, Relax
Answer Easy Questions First
Estimate the Answer
Read All Choices before Answering
Read Question before Answering
Solve Directly or Eliminate Choices
Solve Problem before Looking at Choices
Use Intelligent Guessing
Work Backwards

About this Strategy

When taking a multiple choice test, be sure to read each question carefully and thoroughly. It is also very important to read each answer choice carefully. Do not pick the first answer you think is correct! If two answer choices are the same, eliminate them both. There can only be one correct answer.

Answers

1. C
2. G
3. 6
4. D

Item Analysis

1. **A.** The student finds 30% of $8.50 but does not subtract this amount from $8.50.

   B. The student thinks that 30% is equivalent to $3.00 and subtracts this amount from $8.50.

   C. Correct answer

   D. The student thinks that 30% is equivalent to $0.30 and subtracts this amount from $8.50.

2. **F.** The student divides incorrectly or converts measures incorrectly to choose an incorrect box.

   G. Correct answer

   H. The student divides incorrectly or converts measures incorrectly to choose an incorrect box.

   I. The student divides incorrectly or converts measures incorrectly to choose an incorrect box.

3. **Gridded Response:** Correct answer: 6

   Common Error: The student makes a sign error when dividing and gets an answer of $x = -6$.

4. **A.** The student chooses a proportion that will find what percent 17 is of 43.

   B. The student chooses a proportion that will find 43% of 17.

   C. The student chooses a proportion that will find 17% of 43.

   D. Correct answer

Technology for the *Teacher*

Performance Tasks
Online Assessment
Assessment Book
ExamView® Assessment Suite

1. A movie theater offers 30% off the price of a movie ticket to students from your school. The regular price of a movie ticket is $8.50. What is the discounted price that you would pay for a ticket?

   A. $2.55

   B. $5.50

   C. $5.95

   D. $8.20

Which amount of increase in your catnip allowance do you want?

Ⓐ 50%　Ⓑ 75%　Ⓒ 98%　Ⓓ 10%

I get it. C for catnip.

"Reading all choices before answering can really pay off!"

2. You are comparing the prices of four boxes of cereal. Two of the boxes contain free extra cereal.

   - Box F costs $3.59 and contains 16 ounces.

   - Box G costs $3.79 and contains 16 ounces, plus an additional 10% for free.

   - Box H costs $4.00 and contains 500 grams.

   - Box I costs $4.69 and contains 500 grams, plus an additional 20% for free.

   Which box has the least unit cost? (1 ounce = 28.35 grams)

   F. Box F

   G. Box G

   H. Box H

   I. Box I

3. What value makes the equation $11 - 3x = -7$ true?

4. Which proportion represents the problem below?

   "17% of a number is 43. What is the number?"

   A. $\dfrac{17}{43} = \dfrac{n}{100}$

   B. $\dfrac{n}{17} = \dfrac{43}{100}$

   C. $\dfrac{n}{43} = \dfrac{17}{100}$

   D. $\dfrac{43}{n} = \dfrac{17}{100}$

5. Which list of numbers is in order from least to greatest?

F. $0.8, \frac{5}{8}, 70\%, 0.09$

H. $\frac{5}{8}, 70\%, 0.8, 0.09$

G. $0.09, \frac{5}{8}, 0.8, 70\%$

I. $0.09, \frac{5}{8}, 70\%, 0.8$

6. What is the value of $\frac{9}{8} \div \left(-\frac{11}{4}\right)$?

7. A pair of running shoes is on sale for 25% off the original price.

ORIGINAL PRICE $123.75

Which price is closest to the sale price of the running shoes?

A. $93

C. $124

B. $99

D. $149

8. What is the slope of the line?

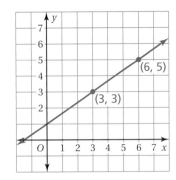

F. $\frac{2}{3}$

H. 2

G. $\frac{3}{2}$

I. 3

Item Analysis (continued)

5. **F.** The student thinks that 0.8 is less than 0.09 and does not know how to compare decimal numbers with fractions or percents.

   G. The student orders the decimal numbers and fractions correctly but thinks that 70% equals 70.

   H. The student orders the numbers using either the numerator or the leading digit.

   I. Correct answer

6. **Gridded Response:** Correct answer: $-\dfrac{9}{22}$

   Common Error: The student takes the reciprocal of the wrong fraction and gets an answer of $-\dfrac{22}{9}$.

7. **A.** Correct answer

   B. The student thinks that 25% = $25 and subtracts $25 from $123.75.

   C. The student thinks that 25% = $0.25, and then either adds $0.25 to or subtracts $0.25 from $123.75.

   D. The student thinks that 25% = $25 and adds $25 to $123.75.

8. **F.** Correct answer

   G. The student finds the change in x over the change in y.

   H. The student finds the change in y.

   I. The student finds the change in x.

Answers

5. I

6. $-\dfrac{9}{22}$

7. A

8. F

Answers

9. D

10. *Part A* $371 at the hardware store; $355.20 at the online store

   Part B The hardware store offers the better final cost by $5.30.

11. F

Item Analysis (continued)

9. **A.** The student does not perform the correct operation.

   B. The student does not perform the correct operation.

   C. The student does not perform the correct operation.

   D. Correct answer

10. **4 points** The student demonstrates a thorough understanding of solving problems involving finding percents of numbers. In Part A, the student correctly calculates the cost of the ladder at each store, getting $371 at the hardware store and $355.20 at the online store. In Part B, the student correctly subtracts 10% of the cost of the ladder at the hardware store before adding the tax. The student also correctly recalculates the cost of the ladder at the online store without the shipping and handling charge. The student then compares the final costs, showing that the hardware store offers the better final cost by $5.30. The student shows accurate, complete work for both parts and provides clear and complete explanations.

   3 points The student demonstrates an understanding of solving problems involving finding percents of numbers, but the student's work and explanations demonstrate an essential but less than thorough understanding.

   2 points The student demonstrates a partial understanding of solving problems involving finding percents of numbers. The student's work and explanations demonstrate a lack of essential understanding.

   1 point The student demonstrates very limited understanding of solving problems involving finding percents of numbers. The student's response is incomplete and exhibits many flaws.

   0 points The student provided no response, a completely incorrect or incomprehensible response, or a response that demonstrates insufficient understanding of solving problems involving finding percents of numbers.

11. **F.** Correct answer

   G. The student does not reverse the inequality.

   H. The student makes a sign error.

   I. The student makes a sign error and does not reverse the inequality.

9. Brad solved the equation in the box shown.

What should Brad do to correct the error that he made?

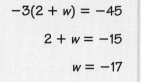

$$-3(2 + w) = -45$$
$$2 + w = -15$$
$$w = -17$$

A. Multiply -45 by -3 to get $2 + w = 135$.

B. Add 3 to -45 to get $2 + w = -42$.

C. Add 2 to -15 to get $w = -13$.

D. Divide -45 by -3 to get 15.

10. You are comparing the costs of a certain model of ladder at a hardware store and at an online store.

Part A What is the cost of the ladder at each of the stores? Show your work and explain your reasoning.

Part B Suppose that the hardware store is offering 10% off the price of the ladder and that the online store is offering free shipping and handling. Which store offers the better final cost? by how much? Show your work and explain your reasoning.

11. Which graph represents the inequality below?

$$-5 - 3x \geq -11$$

F.
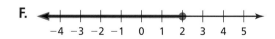

H.

G.
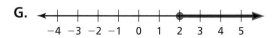

I.

7 Constructions and Scale Drawings

7.1 Adjacent and Vertical Angles

7.2 Complementary and Supplementary Angles

7.3 Triangles

7.4 Quadrilaterals

7.5 Scale Drawings

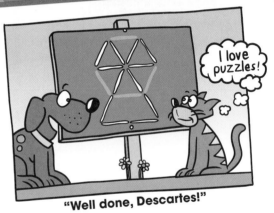

"Move 4 of the lines to make 3 equilateral triangles."

"Well done, Descartes!"

"I'm at 3rd base. You are running to 1st base, and Fluffy is running to 2nd base."

"Should I throw the ball to 2nd to get Fluffy out or throw it to 1st to get you out?"

What Your Students Have Learned

- Classify two dimensional figures into categories based on properties.
- Interpret multiplication as scaling.
- Convert standard measurement units within a measurement system.
- Draw polygons in the coordinate plane given vertices and find lengths of sides.
- Understand ratios and describe ratio relationships.
- Use ratio reasoning to convert measurement units.

What Your Students Will Learn

- Use supplementary, complementary, vertical, and adjacent angles.
- Draw geometric shapes with given conditions, focusing on triangles and quadrilaterals.
- Reproduce a scale drawing at a different scale.
- Represent proportional relationships with equations.
- Use proportionality to solve ratio problems.
- Use scale drawings to compute actual lengths and areas.
- Find scale factors using scale drawings and models.

Pacing Guide for Chapter 7

Chapter Opener	
Regular	1 Day
Accelerated	1 Day
Section 1	
Regular	2 Days
Accelerated	1 Day
Section 2	
Regular	2 Days
Accelerated	1 Day
Section 3	
Regular	4 Days
Accelerated	3 Days
Study Help / Quiz	
Regular	1 Days
Accelerated	0 Days
Section 4	
Regular	3 Days
Accelerated	2 Days
Section 5	
Regular	2 Days
Accelerated	1 Day
Chapter Review/ Chapter Tests	
Regular	2 Days
Accelerated	2 Days
Total Chapter 7	
Regular	17 Days
Accelerated	11 Days
Year-to-Date	
Regular	106 Days
Accelerated	62 Days

Technology for the *Teacher*

BigIdeasMath.com
Chapter at a Glance
Complete Materials List
Parent Letters: English and Spanish

What Your Students Have Learned

- Measure angles and draw angles using a protractor.

Additional Topics for Review

- Lines
- Intersection
- Coordinate Plane
- Graphing Ordered Pairs
- Polygons
- Triangles
- Ratios
- Proportions
- Perimeter
- Area

Try It Yourself

1. 70°; acute

2. 90°; right

3. 115°; obtuse

4.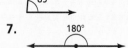
   55°

5. 160°

6. 85°

7. 180°

Record and Practice Journal
Fair Game Review

1. 60°; acute

2. 90°; right

3. 120°; obtuse

4. 65°; acute

5. 180°; straight

6. 165°; obtuse

7–10. See Additional Answers.

Math Background Notes

Vocabulary Review

- Angle
- Degrees
- Protractor
- Acute
- Obtuse
- Right
- Straight
- Vertex
- Endpoint
- Ray

Measuring Angles

- Students should know how to measure angles using a protractor.
- Remind students that an angle can be classified by its measure. A right angle is 90°, an acute angle is less than 90°, an obtuse angle is between 90° and 180°, and a straight angle is 180°.
- **Common Error:** Students may use the wrong set of angles on a protractor. Encourage them to decide which set to use by comparing the angle measure to 90°.
- **Teaching Tip:** Ask students to find real-life examples of angles, such as the hands of a clock.
- **Representation:** Show students how to measure an angle when one of the rays does not pass through the 0° mark on a protractor. Suppose in Example 1(a), the center of the protractor is aligned at the angle's vertex but the rays pass through the 40° mark and the 60° mark. Students can find the angle measure by subtracting 40 from 60.

Drawing Angles

- Students should know how to draw angles of specified measure.
- **Common Error:** Again, students may use the wrong set of angles on a protractor. Encourage them to decide which set to use by comparing the angle measure to 90°.

Reteaching and Enrichment Strategies

If students need help. . .	If students got it. . .
Record and Practice Journal • Fair Game Review Skills Review Handbook Lesson Tutorials	Game Closet at *BigIdeasMath.com* Start the next section

What You Learned Before

"Look at this baby crocodile! Isn't it cute?"

Yes, it's very acute.

Measuring Angles

Example 1 Use a protractor to find the measure of each angle. Then classify the angle as *acute*, *obtuse*, *right*, or *straight*.

a.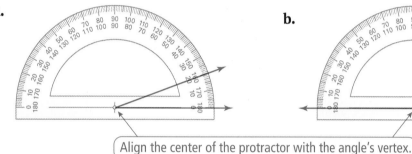

b.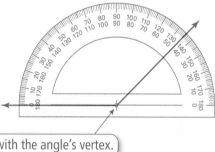

Align the center of the protractor with the angle's vertex.

⋮• The angle measure is 20°. So, the angle is acute.

⋮• The angle measure is 135°. So, the angle is obtuse.

Drawing Angles

Example 2 Use a protractor to draw a 45° angle.

Draw a ray. Place the center of the protractor on the endpoint of the ray and align the protractor so the ray passes through the 0° mark. Make a mark at 45°. Then draw a ray from the endpoint at the center of the protractor through the mark at 45°.

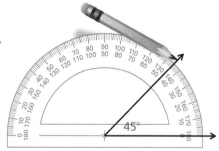

Try It Yourself

Use a protractor to find the measure of the angle. Then classify the angle as *acute*, *obtuse*, *right*, or *straight*.

1.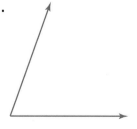

2.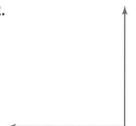

3.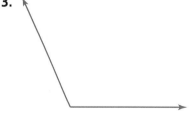

Use a protractor to draw an angle with the given measure.

4. 55°   **5.** 160°   **6.** 85°   **7.** 180°

Essential Question What can you conclude about the angles formed by two intersecting lines?

Classification of Angles

Acute:
Less than 90°

Right:
Equal to 90°

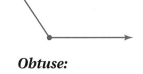

Obtuse:
Greater than 90° and
less than 180°

Straight:
Equal to 180°

1 ACTIVITY: Drawing Angles

Work with a partner.

a. Draw the hands of the clock to represent the given type of angle.

Acute   Straight   Right   Obtuse

b. What is the measure of the angle formed by the hands of the clock at the given time?

9:00   6:00   12:00

The Meaning of a Word ● Adjacent

Geometry

In this lesson, you will

● identify adjacent and vertical angles.
● find angle measures using adjacent and vertical angles.

When two states are **adjacent**,   they are next to each other and they share a common border.

Laurie's Notes

Introduction

Applying Mathematical Practices

- **Construct Viable Arguments and Critique the Reasoning of Others:** Throughout this chapter, students will investigate and make conjectures about geometric properties. Students need practice giving supporting evidence for their conjectures.

Motivate

- **Preparation:** Make a model to practice estimation skills with angle measures. Cut two circles (6-inch diameter) out of file folders. Cut a slit in each. On one circle, label every 10°. The second circle is shaded. Insert one circle into the other so the angle measure faces you and the shaded angle faces the students.
- Ask students to estimate the measure of the shaded angle. You can read the answer from your side of the model. Repeat several times.

| Labeled | Shaded | Your view | Students' view |

Activity Notes

Discuss

- ❓ "What names do you use to classify angles, and what does each mean?" acute: less than 90°, right: 90°, obtuse: greater than 90° and less than 180°, straight: 180°
- **Caution:** Do not draw every angle in this chapter with the initial ray horizontal and extending rightward. Use varied orientation to gauge students' understanding of reading angle measures.

Activity 1

- In part (a), students could share what time they drew on each clock face.
- **FYI:** Students may ask about angles greater than 180°. If so, tell them that they will study such angles in future math classes. Snowboarders and skateboarders may be familiar with such angles as 360°, 540°, and 720°.
- **Extension:** A *reflex angle* is greater than 180° and less than 360°. For instance, the clockwise angle formed by 2 and 12 is a reflex angle.

The Meaning of the Word

- Discuss the meaning of *adjacent*, using the given map to illustrate.
- ❓ "Can you give other examples of objects that are adjacent?" Answers will vary.

What Your Students Will Learn

- Identify adjacent and vertical angles.
- Find angle measures using adjacent and vertical angles.
- Construct angles using a protractor.

Previous Learning

Students should know basic vocabulary associated with angles.

7.1 Record and Practice Journal

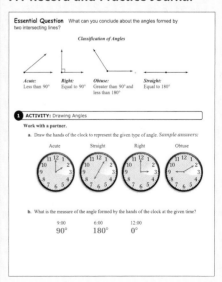

Explain to English learners that the name *right angle* does not come from the orientation of the angle opening to the right, as shown in the activity. Students might think that if the angle opens to the left, it is called a *left angle*. Point out that any angle that measures 90° is a *right angle*.

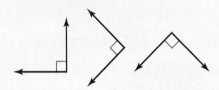

7.1 Record and Practice Journal

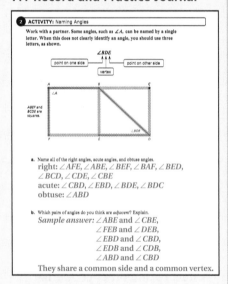

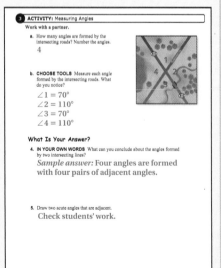

Laurie's Notes

Activity 2

• This activity reviews how angles are named. Discuss when there is a need for three letters instead of one. Also discuss that ∠ *EBD* and ∠ *DBE* name the same angle (because the vertex position is the same).

? "What other angles could be named using just one letter?" ∠ *C*, ∠ *F*

• Although the right angles are not labeled in this diagram, it is assumed that those which appear to be right angles are right angles.

Activity 3

• Do not assume that all students will recall how to measure an angle. It might be helpful to review how to use a protractor before students begin part (b).

• **Attend to Precision:** Discuss the need for precise measurement. It is important for students to take their time and carefully align the center of the protractor with the vertex of the angle.

• Walk around and observe to make sure students are measuring the angles correctly.

• If students ask, tell them that it is not necessary to measure the two straight angles.

? "What do you notice about the angles you measured?" Depending on how carefully students measured, they should have two pairs of corresponding angles that have the same measure.

What Is Your Answer?

• **Think-Pair-Share:** Students should read each question independently and then work in pairs to answer the questions. When they have answered the questions, the pair should compare their answers with another group and discuss any discrepancies.

Closure

• Look around the room. Name angles that appear to be acute, right, obtuse, or straight. (You could also include reflex angles, if you discussed these earlier.)

Work with a partner. Some angles, such as ∠A, can be named by a single letter. When this does not clearly identify an angle, you should use three letters, as shown.

Math Practice

Justify Conclusions

When you name an angle, does the order in which you write the letters matter? Explain.

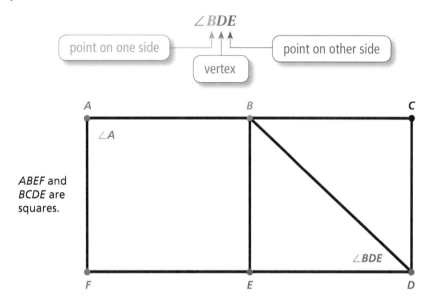

∠BDE

point on one side — vertex — point on other side

ABEF and *BCDE* are squares.

a. Name all the right angles, acute angles, and obtuse angles.

b. Which pairs of angles do you think are *adjacent*? Explain.

3 **ACTIVITY: Measuring Angles**

Work with a partner.

a. How many angles are formed by the intersecting roads? Number the angles.

b. **CHOOSE TOOLS** Measure each angle formed by the intersecting roads. What do you notice?

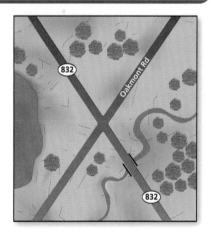

What Is Your Answer?

4. **IN YOUR OWN WORDS** What can you conclude about the angles formed by two intersecting lines?

5. Draw two acute angles that are adjacent.

Practice

Use what you learned about angles and intersecting lines to complete Exercises 3 and 4 on page 274.

Check It Out
Lesson Tutorials
BigIdeasMath ✓com

Key Vocabulary
adjacent angles,
  p. 272
vertical angles, p. 272
congruent angles,
  p. 272

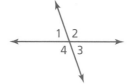 Key Ideas

Adjacent Angles

Words Two angles are **adjacent angles** when they share a common side and have the same vertex.

Examples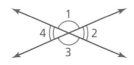

∠1 and ∠2 are adjacent.

∠2 and ∠4 are not adjacent.

Vertical Angles

Words Two angles are **vertical angles** when they are opposite angles formed by the intersection of two lines. Vertical angles are **congruent angles**, meaning they have the same measure.

Examples

∠1 and ∠3 are vertical angles.

∠2 and ∠4 are vertical angles.

EXAMPLE **1** **Naming Angles**

Use the figure shown.

a. Name a pair of adjacent angles.

∠ABC and ∠ABF share a common side and have the same vertex B.

⋮⋮ So, ∠ABC and ∠ABF are adjacent angles.

b. Name a pair of vertical angles.

∠ABF and ∠CBD are opposite angles formed by the intersection of two lines.

⋮⋮ So, ∠ABF and ∠CBD are vertical angles.

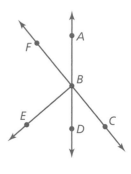

On Your Own

Now You're Ready
Exercises 5 and 6

Name two pairs of adjacent angles and two pairs of vertical angles in the figure.

1.

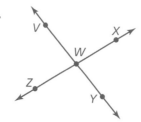

2.

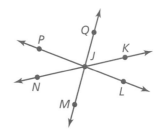

Laurie's Notes

Introduction

Connect

- **Yesterday:** Students explored drawing, naming, and measuring angles, and investigated relationships between angles.
- **Today:** Students will identify adjacent or vertical angles.

Motivate

- Because this chapter will be focusing on geometry, students should know we credit Euclid for the study of geometry. He is often called the Father of Geometry. Euclid was a Greek mathematician best known for his 13 books on geometry known as *The Elements*. This work influenced the development of Western mathematics for more than 2000 years.

Lesson Notes

Key Ideas

- ❓ "Does anyone know what the term *congruent angles* means?" two angles that have the same measure
- Draw two pairs of congruent angles and show students the marks used to indicate that the angles are congruent.
- ❓ "What does the word *adjacent* mean?" side-by-side
- Write the Key Ideas.
- **Model:** When two lines intersect, two pairs of vertical angles are formed. Vertical angles are congruent. Demonstrate this with a pair of scissors that have straight blades.

Example 1

- Draw the figure and ask students to identify the lines and rays. There are 5 rays, and two pairs of rays are also collinear (on the same line). Only $\overrightarrow{BE}$ is not part of a line.
- **FYI:** $\angle ABE$ and $\angle CBE$ are adjacent angles. Some students may give this response to part (a).
- ❓ "Is $\angle ABE$ adjacent to $\angle FBE$? Explain." Students may say yes because the definition is satisfied; however, adjacent angles do not overlap. In a geometry class, this will be included in the definition.
- ❓ "Name a pair of vertical angles." Listen for the two pairs of vertical angles. One pair is acute and the other pair is obtuse.
- ❓ **Construct Viable Arguments and Critique the Reasoning of Others:** "When two lines intersect, will there always be one pair of obtuse angles and one pair of acute angles? Explain." no; The two lines could form four right angles.

On Your Own

- **Construct Viable Arguments and Critique the Reasoning of Others:** Question 2 has many correct answers. Students should be able to explain why their answers are correct.

Goal Today's lesson is identifying adjacent or vertical angles.

Lesson Tutorials
Lesson Plans
Answer Presentation Tool

Extra Example 1

Use the figure shown.

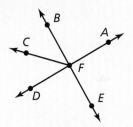

a. Name a pair of adjacent angles.
   Sample answer: $\angle AFB$ and $\angle CFB$
b. Name a pair of vertical angles.
   Sample answer: $\angle AFB$ and $\angle EFD$

On Your Own

1. *Sample answers:* adjacent: $\angle XWY$ and $\angle ZWY$, $\angle XWY$ and $\angle XWV$; vertical: $\angle VWX$ and $\angle ZWY$, $\angle YWX$ and $\angle ZWV$

2. *Sample answers:* adjacent: $\angle LJM$ and $\angle LJK$, $\angle LJM$ and $\angle NJM$; vertical: $\angle KJL$ and $\angle PJN$, $\angle PJQ$ and $\angle MJL$

Extra Example 2

Tell whether the angles are *adjacent* or *vertical*. Then find the value of *x*.

a.

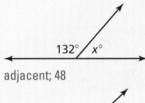

adjacent; 48

b.

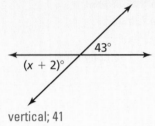

vertical; 41

Extra Example 3

Draw a pair of vertical angles with a measure of 55°.

 On Your Own

3. adjacent; 95

4. vertical; 90

5. adjacent; 11

6.

Differentiated Instruction

Visual

Help students visualize vertical angles. Draw vertical angles on the board or overhead.

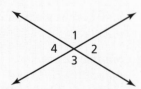

Point out that the lines creating vertical angles form an "X" and that vertical angles do *not* share sides.

Example 2

- Work through each part as shown.
- In part (b), remind students of the corner mark used to designate a right angle.
- Also in part (b), point out the information in the *remember* box. If the angles overlap, then their measures would not sum to the measure of the larger angle.
- **?** **Construct Viable Arguments and Critique the Reasoning of Others: Extension:** "In part (a), what are the measures of the two remaining angles? How do you know?" 110°; Any two adjacent angles in the figure form a straight angle, which measures 180°, and 180° − 70° = 110°.

Example 3

- Ask students to use their protractors to draw a 40° angle. The first step is not obvious to many students! Draw a ray. Its endpoint will be the vertex of the angle. Then use the protractor to place a mark at 40°, and draw another ray from the vertex through the mark.
- A second method is to draw a line and place a point on the line to represent the vertex. Then place the protractor on the point and place a mark at 40°, and draw a line through the mark and the vertex.
- **Common Error:** When the protractor has two scales (clockwise and counterclockwise), students may draw an angle of 140°. If this happens, then ask whether a 40° angle is acute or obtuse.

On Your Own

- **Think-Pair-Share:** Students should read each question independently and then work in pairs to answer the questions. When they have answered the questions, the pair should compare their answers with another group and discuss any discrepancies.

Closure

- True or False?
  1. Vertical angles are always acute. false
  2. Adjacent angles could be acute. true
  3. Adjacent angles could be obtuse. true
  4. Vertical angles are congruent. true
  5. Adjacent angles could be congruent. true

EXAMPLE 2 **Using Adjacent and Vertical Angles**

Tell whether the angles are *adjacent* or *vertical*. Then find the value of x.

a.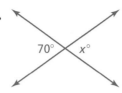

The angles are vertical angles. Because vertical angles are congruent, the angles have the same measure.

⋮⋮ So, the value of x is 70.

Remember

You can add angle measures. When two or more adjacent angles form a larger angle, the sum of the measures of the smaller angles is equal to the measure of the larger angle.

b.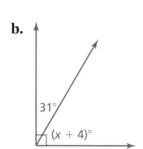

The angles are adjacent angles. Because the angles make up a right angle, the sum of their measures is 90°.

$(x + 4) + 31 = 90$ Write equation.

$x + 35 = 90$ Combine like terms.

$x = 55$ Subtract 35 from each side.

⋮⋮ So, the value of x is 55.

EXAMPLE 3 **Constructing Angles**

Draw a pair of vertical angles with a measure of 40°.

Step 1: Use a protractor to draw a 40° angle.

Step 2: Use a straightedge to extend the sides to form two intersecting lines.

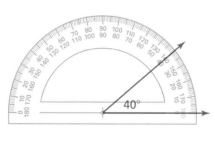

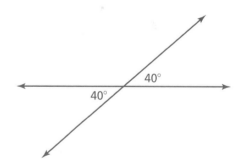

On Your Own

Now You're Ready
Exercises 8–17

Tell whether the angles are *adjacent* or *vertical*. Then find the value of x.

3.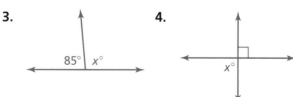

4.

5.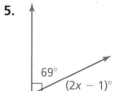

6. Draw a pair of vertical angles with a measure of 75°.

Vocabulary and Concept Check

1. **VOCABULARY** When two lines intersect, how many pairs of vertical angles are formed? How many pairs of adjacent angles are formed?

2. **REASONING** Identify the congruent angles in the figure. Explain your reasoning.

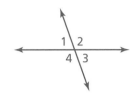

Practice and Problem Solving

Use the figure at the right.

3. Measure each angle formed by the intersecting lines.

4. Name two angles that are adjacent to ∠ABC.

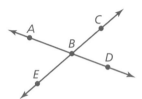

Name two pairs of adjacent angles and two pairs of vertical angles in the figure.

① 5.

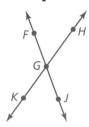

6.

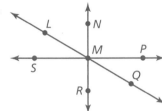

7. **ERROR ANALYSIS** Describe and correct the error in naming a pair of vertical angles.

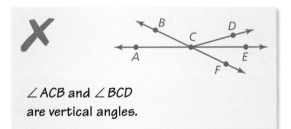

∠ACB and ∠BCD
are vertical angles.

Tell whether the angles are *adjacent* or *vertical*. Then find the value of *x*.

② 8.

$x°$

$35°$

9.

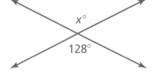

$x°$

$128°$

10.

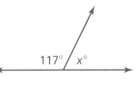

$117°$ $x°$

11.

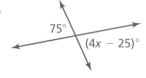

$75°$

$(4x - 25)°$

12.

$4x°$

$2x°$

13.

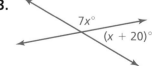

$7x°$

$(x + 20)°$

Assignment Guide and Homework Check

Level	Day 1 Activity Assignment	Day 2 Lesson Assignment	Homework Check
Basic	3, 4, 27–30	1, 2, 5–21 odd	5, 9, 11, 19
Average	3, 4, 27–30	1, 2, 5–13 odd, 14–24 even	5, 9, 11, 20
Advanced	3, 4, 27–30	1, 2, 6, 7, 8–24 even, 25, 26	10, 12, 18, 20, 24
Accelerated	1–4, 6, 7, 8–24 even, 25–30		10, 12, 18, 20, 24

Common Errors

- **Exercises 8–13** Students may think that there is not enough information to determine the value of x. Ask them to think about the information given in each figure. For instance, Exercise 8 shows two angles making up a right angle. So, the sum of the two angle measures must be 90°. Students can use this information to set up and solve a simple equation to find the value of x.
- **Exercise 26** Students may guess at an answer because they are unsure of how to solve this problem. To get them on the right track, you may need to give them the hint that the sum of the angle measures of a triangle is 180°. Students have not formally learned this yet, but they have seen it in earlier material, such as activities.

7.1 Record and Practice Journal

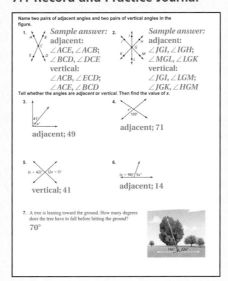

14. 25° 25°

15–25. See Additional Answers.

26. See *Taking Math Deeper*.

Fair Game Review

27. $n < -9$;
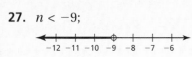

28. $x \geq -34$;
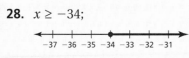

29. $m < 4$;

30. B

Mini-Assessment

Use the figure shown.

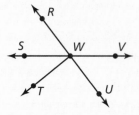

1. Name a pair of adjacent angles.
  Sample answer: $\angle RWS$ and $\angle TWS$

2. Name a pair of vertical angles.
  Sample answer: $\angle RWS$ and $\angle VWU$

3. Tell whether the angles are *adjacent* or *vertical*. Then find the value of *x*.

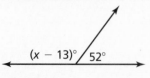
$(x - 13)°$ 52°

adjacent; 141

4. Draw a pair of vertical angles with a measure of 62°.

62°   62°

Taking Math Deeper

Exercise 26

This exercise may be difficult for students. They may not be aware that the sum of the interior angle measures of a triangle is 180°, which is formally presented later in this chapter. Students can solve the problem by carefully constructing the triangle that is created by the ground, wall, and ladder.

1 Use the diagram to find as many angle measures as possible.

- You can assume that the angle created by the ground and the wall is 90°.

- Because a straight angle is 180°, the acute angle created by the ladder and the ground is $180° - 120° = 60°$.

2 Construct the triangle.

Draw a "backwards L" to represent the right angle. Then use a protractor to draw the 60° angle.

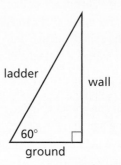
ladder   wall
60°
ground

3 No matter how large or small you draw the "backwards L," when you measure the remaining angle, you see that it is 30°.

So, the ladder shown is not leaning at a safe angle.

Project

Have students lean straight objects, such as a yardstick, up against a wall and observe the angles formed. Tell them to move the base of the object closer to and farther from the wall, noting its affect on the two acute angles.

Reteaching and Enrichment Strategies

If students need help...	If students got it...
Resources by Chapter • Practice A and Practice B • Puzzle Time Record and Practice Journal Practice Differentiating the Lesson Lesson Tutorials Skills Review Handbook	Resources by Chapter • Enrichment and Extension • Technology Connection Start the next section

Draw a pair of vertical angles with the given measure.

③ 14. 25°   **15.** 85°   **16.** 110°   **17.** 135°

18. IRON CROSS The iron cross is a skiing trick in which the tips of the skis are crossed while the skier is airborne. Find the value of x in the iron cross shown.

19. OPEN-ENDED Draw a pair of adjacent angles with the given description.

    a. Both angles are acute.

    b. One angle is acute, and one is obtuse.

    c. The sum of the angle measures is 135°.

127°

$(2x + 41)°$

20. PRECISION Explain two procedures that you can use to draw adjacent angles with given measures.

Determine whether the statement is *always, sometimes,* or *never* true.

21. When the measure of $\angle 1$ is 70°, the measure of $\angle 3$ is 110°.

22. When the measure of $\angle 4$ is 120°, the measure of $\angle 1$ is 60°.

23. $\angle 2$ and $\angle 3$ are congruent.

24. The measure of $\angle 1$ plus the measure of $\angle 2$ equals the measure of $\angle 3$ plus the measure of $\angle 4$.

25. REASONING Draw a figure in which $\angle 1$ and $\angle 2$ are acute vertical angles, $\angle 3$ is a right angle adjacent to $\angle 2$, and the sum of the measure of $\angle 1$ and the measure of $\angle 4$ is 180°.

26. ✦Structure✦ For safety reasons, a ladder should make a 15° angle with a wall. Is the ladder shown leaning at a safe angle? Explain.

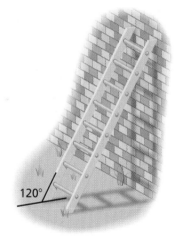

120°

Fair Game Review What you learned in previous grades & lessons

Solve the inequality. Graph the solution. *(Section 4.3)*

27. $-6n > 54$ **28.** $-\dfrac{1}{2}x \le 17$ **29.** $-1.6 < \dfrac{m}{-2.5}$

30. MULTIPLE CHOICE What is the slope of the line that passes through the points (2, 3) and (6, 8)? *(Section 5.5)*

  Ⓐ $\dfrac{4}{5}$ **Ⓑ** $\dfrac{5}{4}$ **Ⓒ** $\dfrac{4}{3}$ **Ⓓ** $\dfrac{3}{2}$

Complementary and Supplementary Angles

Essential Question How can you classify two angles as complementary or supplementary?

---

1 ACTIVITY: Complementary and Supplementary Angles

Work with a partner.

a. The graph represents the measures of *complementary angles.* Use the graph to complete the table.

x		20°		30°	45°		75°
y	80°		65°	60°		40°	

b. How do you know when two angles are complementary? Explain.

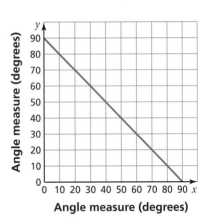

Angle measure (degrees)

c. The graph represents the measures of *supplementary angles.* Use the graph to complete the table.

x	20°		60°	90°		140°	
y		150°		90°	50°		30°

d. How do you know when two angles are supplementary? Explain.

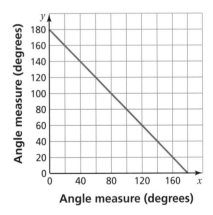

Angle measure (degrees)

---

2 ACTIVITY: Exploring Rules About Angles

Geometry

In this lesson, you will

- classify pairs of angles as complementary, supplementary, or neither.
- find angle measures using complementary and supplementary angles.

Work with a partner. Copy and complete each sentence with *always*, *sometimes*, or *never*.

a. If x and y are complementary angles, then both x and y are _____ acute.

b. If x and y are supplementary angles, then x is _____ acute.

c. If x is a right angle, then x is _____ acute.

d. If x and y are complementary angles, then x and y are _____ adjacent.

e. If x and y are supplementary angles, then x and y are _____ vertical.

Laurie's Notes

Introduction

Applying Mathematical Practices

- **Construct Viable Arguments and Critique the Reasoning of Others:**
  Students will make conjectures about the relationships between two angles that are complementary and two angles that are supplementary.

Motivate

- When students arrive, start making complimentary remarks such as, "Your sweater is nice. I like your glasses. Your shoelaces are cool." Make enough remarks so students catch on that you are giving compliments.
- Next, make a few remarks such as, "Your necklace complements your outfit. We have a full complement of faculty members. The great dessert complements a nice meal."
- You want students to recognize the difference between the homonyms *compliment* and *complement*. Today they will investigate complementary angles.

Activity Notes

Activity 1

- Complementary and supplementary angles are not defined in this activity. Instead, students will read ordered pairs from the graphs to complete the tables. From the tables, students should be able to recognize the relationship between two angles that are complementary and the relationship between two angles that are supplementary.
- This activity is a good review of reading ordered pairs from a graph.
- **Construct Viable Arguments and Critique the Reasoning of Others:** In each part, students need to explain the relationship between the two angles.
- **?** "Are *x* and *y* proportional in part (a) or part (b)? Explain." no; In both parts, neither graph passes through the origin.

Activity 2

- Give time for partners to discuss the problems. The graphs and tables of values from Activity 1 should help students think through their answers.
- **Construct Viable Arguments and Critique the Reasoning of Others:**
  Teaching Tip: When answers are *sometimes* true, it is important to give students a sample of when the statement is true and when the statement is false. For example, in part (b), $x = 75°$ and $y = 105°$ which makes x acute, or $x = 105°$ and $y = 75°$ which makes x obtuse.

What Your Students Will Learn

- Classify pairs of angles as complementary, supplementary, or neither.
- Find angle measures using complementary and supplementary angles.
- Construct complementary and supplementary angles using a protractor.

Previous Learning

Students should know basic vocabulary associated with angles.

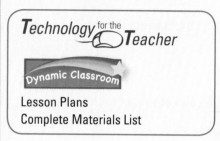

Lesson Plans
Complete Materials List

7.2 Record and Practice Journal

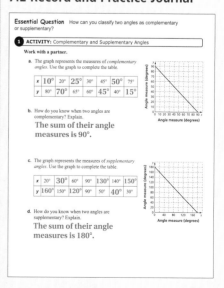

English Language Learners

Vocabulary

For English language learners, discuss the meanings of the words *complement* and *compliment*. As mentioned in the Motivate section, the words are homonyms. They sound the same when they are pronounced, but they have different meanings.

Complement means either of two parts needed to complete the whole.

Compliment is an expression of praise, commendation, or admiration.

7.2 Record and Practice Journal

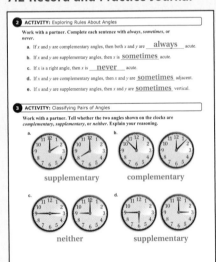

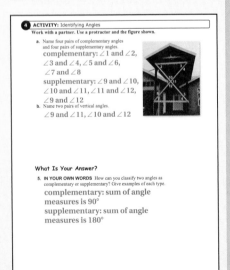

Laurie's Notes

Activity 3

- Many students think that complementary or supplementary angles must always be adjacent because teachers often draw them only that way! There is nothing in either definition (given in the lesson) that requires the angles to be adjacent.
- This activity focuses on the relationships between the angles and not on how they are drawn.
- After students have finished, discuss the answers.

Activity 4

- **Use Appropriate Tools Strategically:** Students will use protractors to justify their answers for this activity. Also, you should discuss the use of numbers instead of three letters to name the angles. When a figure contains many angles, it is much easier to number them.
- Have students measure the angles to the nearest whole degree.

What Is Your Answer?

- **Think-Pair-Share:** Students should read each question independently and then work in pairs to answer the question. When they have answered the question, the pair should compare their answer with another group and discuss any discrepancies.

Closure

- Look around the room. Name angles that appear to be complementary or supplementary.

3 ACTIVITY: Classifying Pairs of Angles

Work with a partner. Tell whether the two angles shown on the clocks are *complementary*, *supplementary*, or *neither*. Explain your reasoning.

a.

b.

c.

d.

4 ACTIVITY: Identifying Angles

Work with a partner. Use a protractor and the figure shown.

a. Name four pairs of complementary angles and four pairs of supplementary angles.

b. Name two pairs of vertical angles.

What Is Your Answer?

5. **IN YOUR OWN WORDS** How can you classify two angles as complementary or supplementary? Give examples of each type.

Math Practice

Use Definitions

How can you use the definitions of *complementary*, *supplementary*, and *vertical angles* to answer the questions?

Practice ▶ Use what you learned about complementary and supplementary angles to complete Exercises 3–5 on page 280.

Key Vocabulary 🔊
complementary
  angles, p. 278
supplementary
  angles, p. 278

 **Key Ideas**

Complementary Angles

Words Two angles are **complementary angles** when the sum of their measures is 90°.

Examples

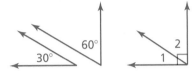

∠1 and ∠2 are complementary angles.

Supplementary Angles

Words Two angles are **supplementary angles** when the sum of their measures is 180°.

Examples

∠3 and ∠4 are supplementary angles.

EXAMPLE ① **Classifying Pairs of Angles**

Tell whether the angles are *complementary*, *supplementary*, or *neither*.

a. [70° 110°] $70° + 110° = 180°$

⋮ So, the angles are supplementary.

b. [49° 41°] $41° + 49° = 90°$

⋮ So, the angles are complementary.

c. [128° 62°] $128° + 62° = 190°$

⋮ So, the angles are *neither* complementary nor supplementary.

🌑 **On Your Own**

Now You're Ready
Exercises 6–11

Tell whether the angles are *complementary*, *supplementary*, or *neither*.

1.

2.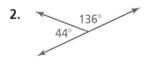

3.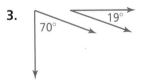

🔊 Multi-Language Glossary at BigIdeasMathcom

Laurie's Notes

Goal Today's lesson is classifying angles as **complementary** or **supplementary**.

Introduction

Connect
- **Yesterday:** Students explored complementary and supplementary angles.
- **Today:** Students will classify several pairs of angles as complementary or supplementary.

Motivate
- Yesterday, the homonyms *compliment* and *complement* were used.
- Ask students to give different meanings for the word *supplement*. Student answers may include some or all of the following.
  - a part added to a book or document
  - a part added to a newspaper that might be a special feature
  - something added to complete a deficiency (such as a dietary supplement)
  - something added to support (such as a learning supplement or tutor)

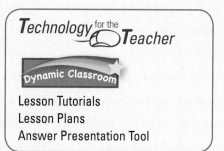

Technology for the **Teacher**

Dynamic Classroom

Lesson Tutorials
Lesson Plans
Answer Presentation Tool

Lesson Notes

Key Ideas
- Write the Key Ideas. Define and sketch complementary angles and supplementary angles.
- In the figures, the angles are drawn with an orientation to suggest that the sum is 90° (complementary) or 180° (supplementary), but they do not need to have this orientation. For example, $\angle A$ and $\angle B$ below are complementary, however, it is not immediately obvious because of their orientation.

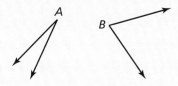

Example 1
- **Construct Viable Arguments and Critique the Reasoning of Others:** In this example, students sum the angle measures and determine if they add to 90°, 180°, or neither. Make sure students do not rely on their eyesight. They should actually add the angle measures.

On Your Own
- **Think-Pair-Share:** Students should read each question independently and then work in pairs to answer the questions. When they have answered the questions, the pair should compare their answers with another group and discuss any discrepancies.

Extra Example 1

Tell whether the angles are *complementary*, *supplementary*, or *neither*.

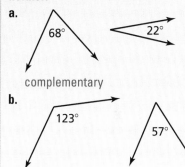

a.

68°   22°

complementary

b.

123°   57°

supplementary

c.

65°   24°

neither

On Your Own

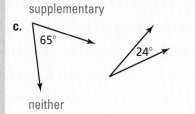

1. complementary
2. supplementary
3. neither

Extra Example 2

Tell whether the angles are *complementary* or *supplementary*. Then find the value of *x*.

a.

complementary; 61

b.

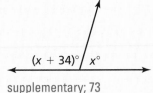

supplementary; 73

Extra Example 3

Draw a pair of adjacent supplementary angles so that one angle has a measure of 70°.

On Your Own

4. supplementary; 33.5°

5. complementary; 31°

6.

Differentiated Instruction

Inclusion

Students may have problems remembering the measures of *complementary* and *supplementary* angles. Point out that *c* comes before *s* in the alphabet and 90 comes before 180 numerically.

Laurie's Notes

Example 2

- In this example, students practice writing and solving equations.
- **?** "What do you know about the two angles in part (a)? Explain." The sum of their measures is 90° because they make up a right angle.
- Work through part (a) as shown.
- **?** "What do you know about the two angles in part (b)? Explain." The sum of their measures is 180° because they make up a straight angle.
- Work through part (b) as shown.
- Have students check their answers by substituting each value of *x* in the corresponding figure.

Example 3

- **Use Appropriate Tools Strategically:** Ask students to use their protractors to draw the supplementary angles. Draw a line and place a point on the line to represent the vertex. Then place the protractor on the point and place a mark at 60°, and draw a ray from the vertex through the mark.

On Your Own

- **Think-Pair-Share:** Students should read each question independently and then work in pairs to answer the questions. When they have answered the questions, the pair should compare their answers with another group and discuss any discrepancies.

Closure

- True or False?
  1. Supplementary angles could both be acute. false
  2. Supplementary angles could be congruent. true
  3. Complementary angles sum to 180°. false
  4. Complementary angles could be obtuse. false
  5. Every angle has a complement and a supplement. false

EXAMPLE **2**

Using Complementary and Supplementary Angles

Tell whether the angles are *complementary* or *supplementary*. Then find the value of *x*.

a.

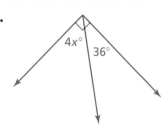

The two angles make up a right angle. So, the angles are complementary angles, and the sum of their measures is 90°.

$4x + 36 = 90$	Write equation.
$4x = 54$	Subtract 36 from each side.
$x = 13.5$	Divide each side by 4.

b.

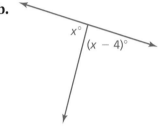

The two angles make up a straight angle. So, the angles are supplementary angles, and the sum of their measures is 180°.

$x + (x - 4) = 180$	Write equation.
$2x - 4 = 180$	Combine like terms.
$2x = 184$	Add 4 to each side.
$x = 92$	Divide each side by 2.

EXAMPLE **3**

Constructing Angles

Draw a pair of adjacent supplementary angles so that one angle has a measure of 60°.

Step 1: Use a protractor to draw a 60° angle.

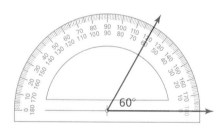

Step 2: Extend one of the sides to form a line.

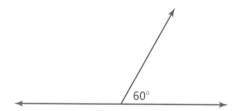

 On Your Own

Now You're Ready
Exercises 12–14
and 17–20

Tell whether the angles are *complementary* or *supplementary*. Then find the value of *x*.

4.

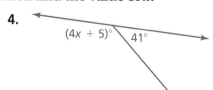

5.

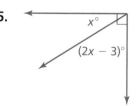

6. Draw a pair of adjacent supplementary angles so that one angle has a measure of 15°.

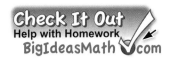

 ## Vocabulary and Concept Check

1. **VOCABULARY** Explain how complementary angles and supplementary angles are different.

2. **REASONING** Can adjacent angles be supplementary? complementary? neither? Explain.

 ## Practice and Problem Solving

Tell whether the statement is *always*, *sometimes*, or *never* true. Explain.

3. If *x* and *y* are supplementary angles, then *x* is obtuse.

4. If *x* and *y* are right angles, then *x* and *y* are supplementary angles.

5. If *x* and *y* are complementary angles, then *y* is a right angle.

Tell whether the angles are *complementary*, *supplementary*, or *neither*.

① 6.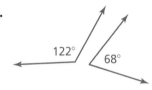

7.

8.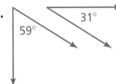

9.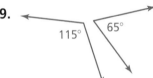

10.

11.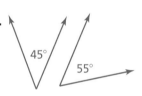

Tell whether the angles are *complementary* or *supplementary*. Then find the value of *x*.

② 12.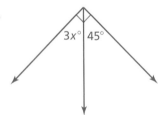

13.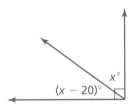

14.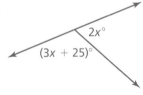

15. **INTERSECTION** What are the measures of the other three angles formed by the intersection?

16. **TRIBUTARY** A tributary joins a river at an angle. Find the value of *x*.

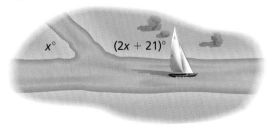

Assignment Guide and Homework Check

Level	Day 1 Activity Assignment	Day 2 Lesson Assignment	Homework Check
Basic	3–5, 28–31	1, 2, 7–23 odd	2, 7, 9, 13, 15
Average	3–5, 28–31	1, 2, 7–15 odd, 16–26 even	7, 9, 13, 16, 24
Advanced	3–5, 28–31	1, 2, 6–26 even, 27	6, 14, 22, 24
Accelerated	1–5, 6–26 even, 27–31		6, 14, 22, 24

For Your Information

- **Exercise 23** Students may not understand what a *vanishing point* is. A vanishing point is a point in a perspective drawing to which parallel lines appear to converge.

Common Errors

- **Exercises 6–11** Students may mix up the terms *supplementary* and *complementary*. Remind them of the definitions and use the alliteration that complementary angles are corners and supplementary angles are straight.
- **Exercises 12–14** Students may think that there is not enough information to determine the value of x. Remind them of the definitions they have learned in the lesson and ask whether either could apply. For instance, Exercise 12 shows two angles making up a right angle, so a student can use the definition of complementary angles to find x.

7.2 Record and Practice Journal

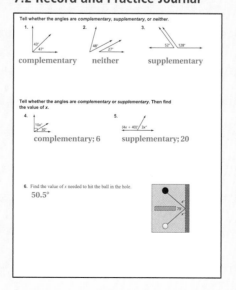

Tell whether the angles are *complementary, supplementary,* or *neither.*

1. complementary
2. neither
3. supplementary

Tell whether the angles are *complementary* or *supplementary.* Then find the value of x.

4. complementary; 6
5. supplementary; 20

6. Find the value of x needed to hit the ball in the hole.
   50.5°

Vocabulary and Concept Check

1. The sum of the measures of two complementary angles is 90°. The sum of the measures of two supplementary angles is 180°.

2. Adjacent angles are not defined by their measure, so they can be complementary, supplementary, or neither.

Practice and Problem Solving

3. sometimes; Either x or y may be obtuse.

4. always; $90° + 90° = 180°$

5. never; Because x and y must both be less than 90° and greater than 0°.

6. neither

7. complementary

8. complementary

9. supplementary

10. supplementary

11. neither

12. complementary; 15

13. complementary; 55

14. supplementary; 31

15. $\angle 1 = 130°$, $\angle 2 = 50°$, $\angle 3 = 130°$

16. 53

17–23. See Additional Answers.

24. yes; *Sample answer:* ∠LMQ is a straight angle. By removing ∠NMP, the remaining two angles (∠LMN and ∠PMQ) have a sum of 90°.

25. 54°

26. See *Taking Math Deeper.*

27. $x = 10$; $y = 20$

Fair Game Review

28. $x = -15$ **29.** $n = -\dfrac{5}{12}$

30. $y = -9.3$ **31.** B

Mini-Assessment

Tell whether the angles are *complementary*, *supplementary*, **or** *neither*.

1.

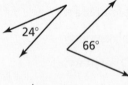

complementary

2.

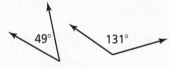

supplementary

3.

neither

4. Tell whether the angles are *complementary* or *supplementary*. Then find the value of *x*.

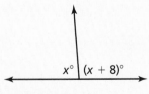

supplementary; 86

Taking Math Deeper

Exercise 26

This exercise is a good lesson for students. The definition of vertical angles is related to the *position* of the angles. However, the definitions of complementary angles and supplementary angles are only based on the *measures* of the angles and not on the position of the angles.

 Draw the angles.

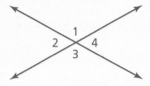

 ∠2 and ∠4 are complementary.

$$\angle 2 = \angle 4 \qquad \text{Vertical angles}$$
$$\angle 2 + \angle 4 = 90 \qquad \text{Complementary angles}$$

Solving this implies that ∠2 = 45° and ∠4 = 45°.

3 ∠2 and ∠4 are supplementary.

$$\angle 2 = \angle 4 \qquad \text{Vertical angles}$$
$$\angle 2 + \angle 4 = 180 \qquad \text{Supplementary angles}$$

Solving this implies that ∠2 = 90° and ∠4 = 90°.

Project

Look around your classroom, school, home, or anywhere you go. Find examples of complementary and supplementary angles. How do you know they are complementary or supplementary? What is the most common angle you find?

Reteaching and Enrichment Strategies

If students need help. . .	If students got it. . .
Resources by Chapter • Practice A and Practice B • Puzzle Time Record and Practice Journal Practice Differentiating the Lesson Lesson Tutorials Skills Review Handbook	Resources by Chapter • Enrichment and Extension • Technology Connection Start the next section

Draw a pair of adjacent supplementary angles so that one angle has the given measure.

③ **17.** 20°   **18.** 35°   **19.** 80°   **20.** 130°

21. PRECISION Explain two procedures that you can use to draw two adjacent complementary angles. Then draw a pair of adjacent complementary angles so that one angle has a measure of 30°.

22. OPEN-ENDED Give an example of an angle that can be a supplementary angle but cannot be a complementary angle. Explain.

23. VANISHING POINT The vanishing point of the picture is represented by point B.

     a. The measure of ∠ABD is 6.2 times greater than the measure of ∠CBD. Find the measure of ∠CBD.

     b. ∠FBE and ∠EBD are congruent. Find the measure of ∠FBE.

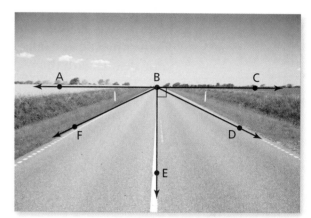

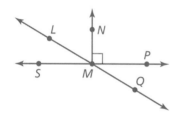

24. LOGIC Your friend says that ∠LMN and ∠PMQ are complementary angles. Is she correct? Explain.

25. RATIO The measures of two complementary angles have a ratio of 3 : 2. What is the measure of the larger angle?

26. REASONING Two angles are vertical angles. What are their measures if they are also complementary angles? supplementary angles?

27. *Problem Solving* Find the values of *x* and *y*.

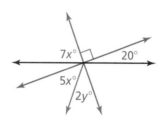

Fair Game Review *What you learned in previous grades & lessons*

Solve the equation. Check your solution. *(Section 3.3)*

28. $x + 7 = -8$ **29.** $\frac{1}{3} = n + \frac{3}{4}$ **30.** $-12.7 = y - 3.4$

31. MULTIPLE CHOICE Which decimal is equal to 3.7%? *(Section 6.1)*

     Ⓐ 0.0037   **Ⓑ** 0.037   **Ⓒ** 0.37   **Ⓓ** 3.7

7.3 Triangles

Essential Question How can you construct triangles?

1 ACTIVITY: Constructing Triangles Using Side Lengths

Work with a partner. Cut different-colored straws to the lengths shown. Then construct a triangle with the specified straws if possible. Compare your results with those of others in your class.

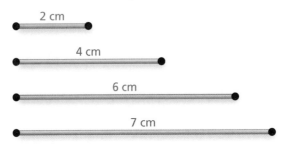

2 cm

4 cm

6 cm

7 cm

a. blue, green, purple

b. red, green, purple

c. red, blue, purple

d. red, blue, green

2 ACTIVITY: Using Technology to Draw Triangles (Side Lengths)

Work with a partner. Use geometry software to draw a triangle with the two given side lengths. What is the length of the third side of your triangle? Compare your results with those of others in your class.

a. 4 units, 7 units

Geometry

In this lesson, you will
- construct triangles with given angle measures.
- construct triangles with given side lengths.

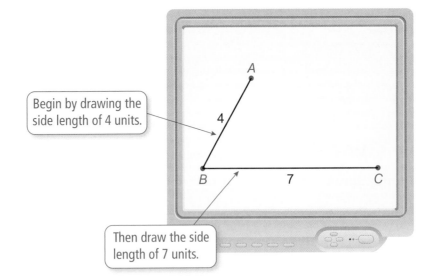

Begin by drawing the side length of 4 units.

A

4

B   7   C

Then draw the side length of 7 units.

b. 3 units, 5 units

c. 2 units, 8 units

d. 1 unit, 1 unit

Laurie's Notes

Introduction

Applying Mathematical Practices

- **Use Appropriate Tools Strategically:** Students will investigate four activities, including two with technology and two without technology. It is important for students to select tools strategically as they develop understanding of mathematical concepts. Discussion of different approaches is essential.

Motivate

- Play a quick game that will help students remember vocabulary relating to triangles. Divide the class into two groups. Give a vocabulary word and each group must write the definition on a piece of paper and hand it to you. Definitions must be written in complete sentences. The first team with a correct definition gets a point. The team with the most points at the end wins.
- Some examples: obtuse angle, acute angle, right angle, scalene triangle, isosceles triangle, right triangle, equilateral triangle, equiangular triangle

Activity Notes

Words of Wisdom

- If geometry software is available, then let students experience exploring with it. The activities may take more than a day to complete, especially when students are not familiar with the software.

Activity 1

- **Teacher Tip:** To save time, you could pre-cut the straws and prepare reclosable bags with the pieces necessary for each pair of students ahead of time. In place of straws, you could use colored pipe cleaners or uncooked linguine.
- In this activity, students investigate whether different combinations of three side lengths form a triangle.
- Tell students to place the straws end-to-end and pivot the outer two straws to form a triangle, if possible.
- **?** "What conclusions can you make?" In parts (a) and (b), it is possible to form a triangle, but in parts (c) and (d), it is not.

Activity 2

- **Construct Viable Arguments and Critique the Reasoning of Others:** Each part of the activity has a range of correct answers. Encourage students to find these ranges and explain their reasoning.
- **FYI:** With some geometry software, you can round the side lengths from 0 to 15 decimal places. When rounding to 0 decimal places, it is possible to obtain erroneous results, such as a triangle with side lengths of 3, 4, and 7 units. To be safe, use a setting of 1 or 2 decimal places.
- **Use Appropriate Tools Strategically:** Discuss any discoveries students make using the software.

What Your Students Will Learn

- Classify triangles when given angle measures or side lengths.
- Construct triangles with given angle measures or side lengths.

Previous Learning

Students should know how to classify two-dimensional figures based on properties, draw polygons, and draw angles.

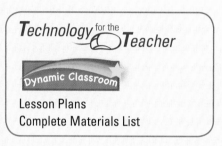

Technology for the Teacher

Dynamic Classroom

Lesson Plans
Complete Materials List

7.3 Record and Practice Journal

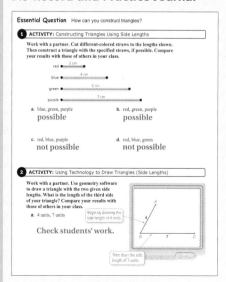

Essential Question How can you construct triangles?

1 ACTIVITY: Constructing Triangles Using Side Lengths

Work with a partner. Cut different-colored straws to the lengths shown. Then construct a triangle with the specified straws, if possible. Compare your results with those of others in your class.

red ● —— 2 cm
blue ● —— 4 cm
green ● —— 6 cm
purple ● —— 7 cm

a. blue, green, purple
possible

b. red, green, purple
possible

c. red, blue, purple
not possible

d. red, blue, green
not possible

2 ACTIVITY: Using Technology to Draw Triangles (Side Lengths)

Work with a partner. Use geometry software to draw a triangle with the two given side lengths. What is the length of the third side of your triangle? Compare your results with those of others in your class.

a. 4 units, 7 units
Begin by drawing the side length of 4 units.

Check students' work.

Then draw the side length of 7 units.

Kinesthetic

When talking about right, acute, and obtuse angles of a triangle, ask students if it is possible to draw a triangle with 2 right angles. Students should see by drawing the two right angles with a common side that the remaining two sides of the right angles will never meet. So, no triangle can be formed with 2 right angles. Ask students if it is possible for a triangle to have 2 obtuse angles. Students should reach the same conclusion. No triangle can be formed with 2 obtuse angles.

7.3 Record and Practice Journal

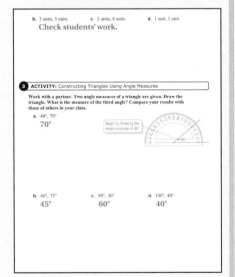

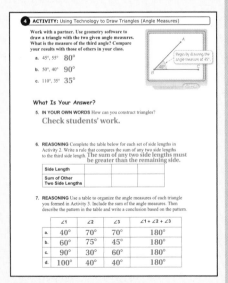

Laurie's Notes

Activity 3

- In this activity, students construct triangles using two given angles, and then find the measure of the third angle. No constraints are given for the side lengths. One way to accomplish this is to draw the first given angle at one end of a segment, draw the second given angle at the other end, and extend the rays for the two angles until they intersect. Then, use a protractor to measure the third angle formed by the intersection.
- After students finish this activity, discuss the results.
- **?** "In part (a), what is the measure of the third angle?" 70°, if students construct the triangle and measure the angle accurately
- **Extension:** Show students two different triangles that satisfy the requirements of part (a), where one triangle is clearly larger than the other. Probe students about the different corresponding side lengths but the same corresponding angle measures. This is a preview of similar triangles, which are covered in a future course.

Activity 4

- If necessary, show students how to use the software to construct a triangle with two given angle measures.
- Set the software to round angle measures to the nearest whole degree.
- After students finish this activity, discuss the results.
- **?** "In part (a), what is the measure of the third angle?" 80°, if students construct the triangle accurately
- **Use Appropriate Tools Strategically :** Discuss any further discoveries students make using the software.

What Is Your Answer?

- Have students work in pairs.

Closure

- **Exit Ticket:**
  - If the lengths of two sides of a triangle are 3 cm and 5 cm, then what are some possible lengths for the third side? *Sample answers:* 4 cm, 2.3 cm, 7 cm
  - If the measures of two angles of a triangle are 45° and 65°, then what is the measure of the third angle? 70°

3 ACTIVITY: Constructing Triangles Using Angle Measures

Work with a partner. Two angle measures of a triangle are given. Draw the triangle. What is the measure of the third angle? Compare your results with those of others in your class.

a. 40°, 70°

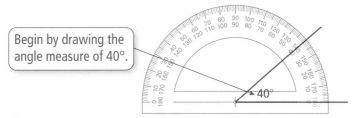

Begin by drawing the angle measure of 40°.

40°

b. 60°, 75°   **c.** 90°, 30°   **d.** 100°, 40°

4 ACTIVITY: Using Technology to Draw Triangles (Angle Measures)

Math Practice

Recognize Usefulness of Tools

What are some advantages and disadvantages of using geometry software to draw a triangle?

Work with a partner. Use geometry software to draw a triangle with the two given angle measures. What is the measure of the third angle? Compare your results with those of others in your class.

a. 45°, 55°

b. 50°, 40°

c. 110°, 35°

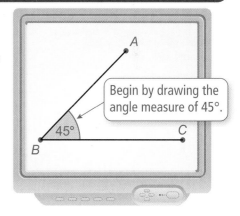

Begin by drawing the angle measure of 45°.

45°

What Is Your Answer?

5. **IN YOUR OWN WORDS** How can you construct triangles?

6. **REASONING** Complete the table below for each set of side lengths in Activity 2. Write a rule that compares the sum of any two side lengths to the third side length.

Side Length			
Sum of Other Two Side Lengths			

7. **REASONING** Use a table to organize the angle measures of each triangle you formed in Activity 3. Include the sum of the angle measures. Then describe the pattern in the table and write a conclusion based on the pattern.

Practice

Use what you learned about constructing triangles to complete Exercises 3–5 on page 286.

Key Vocabulary 🔊
congruent sides,
p. 284

You can use side lengths and angle measures to classify triangles.

 **Key Ideas**

Classifying Triangles Using Angles

| *acute* triangle | *obtuse* triangle | *right* triangle | *equiangular* triangle |

| all acute angles | 1 obtuse angle | 1 right angle | 3 congruent angles |

Classifying Triangles Using Sides

Congruent sides have the same length.

| *scalene* triangle | *isosceles* triangle | *equilateral* triangle |

| no congruent sides | at least 2 congruent sides | 3 congruent sides |

Reading

Red arcs indicate congruent angles.
Red tick marks indicate congruent sides.

EXAMPLE **1** **Classifying Triangles**

Classify each triangle.

a.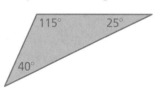

b.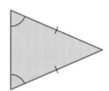

The triangle has one obtuse angle and no congruent sides.

⋮• So, the triangle is an obtuse scalene triangle.

The triangle has all acute angles and two congruent sides.

⋮• So, the triangle is an acute isosceles triangle.

● **On Your Own**

Now You're Ready
Exercises 6–11

Classify the triangle.

1.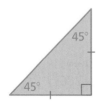

2.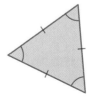

🔊 Multi-Language Glossary at BigIdeasMath✓com

Laurie's Notes

Introduction

Connect

- **Yesterday:** Students investigated constructing triangles using side lengths and angle measures.
- **Today:** Students will classify and further investigate constructing triangles.

Motivate

- On the overhead or interactive board, display a collection of polygons such as the following.

- Ask students how they could sort or classify these polygons. They may suggest sorting by color, number of sides, whether the polygon has a right angle, and so on.

Technology for the *Teacher*

Dynamic Classroom

Lesson Tutorials
Lesson Plans
Answer Presentation Tool

Lesson Notes

Key Ideas

- Terminology used in the Key Ideas should be familiar to students. Some of this terminology is used in a previous course.
- **?** "What does *congruent* mean?" the same length or measure
- Recall that sides as well as angles can be congruent.
- **?** "How can triangles be classified using angles, that is, what names for triangles refer to the angles?" Listen for the four different names for triangles classified using angles.
- **?** "How can triangles be classified using sides, that is, what names for triangles refer to the sides?" Listen for the three different names for triangles classified using sides.
- Sketch and identify each type of triangle shown in the Key Ideas, being sure to mark the right angle and the congruent angles and sides.
- **?** "Is it possible to classify a triangle using angles *and* sides?" yes; Listen for examples such as an acute isosceles triangle.

Example 1

- You could have students explore the figures with protractors and rulers.
- **Attend to Precision:** Caution students against making assumptions. When angles or sides are congruent, they will be marked as such. When an angle is a right angle, it will be marked as such.

On Your Own

- **Neighbor Check:** Have students work independently and then have their neighbors check their work. Have students discuss any discrepancies.

Extra Example 1

Classify each triangle.

a.

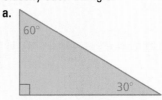

right scalene triangle

b.

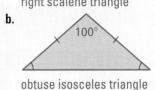

obtuse isosceles triangle

On Your Own

1. right isosceles triangle

2. equilateral equiangular triangle

Extra Example 2

Draw a triangle with angle measures of 35°, 45°, and 100°. Then classify the triangle.

obtuse scalene triangle

Extra Example 3

Draw a triangle with a 3-centimeter side and a 5-centimeter side that meet at a 75° angle. Then classify the triangle.

5 cm

75°

3 cm

Not actual size

acute scalene triangle

On Your Own

3.

    45°   45°

    right isosceles triangle

4. See Additional Answers.

English Language Learners

Illustrate

Have students copy the empty table into their notebooks and then complete it with triangles that represent both attributes.

	Acute	Right	Obtuse
Scalene	△	◿	◺
Isosceles	△	◿	△
Equilateral	△	not possible	not possible

Laurie's Notes

Example 2

- Work through the example as shown.
- ❓ **Construct Viable Arguments and Critique the Reasoning of Others:** "Does the order in which you draw the angles matter? Explain." no; After you draw two angles, the third angle should have the measure of the remaining angle.
- **Attend to Precision:** Students should measure to verify that the third angle has the desired measure.
- **Teaching Tip:** Encourage students not to make tiny drawings. It can be difficult to measure angles when the side lengths are shorter than the radius of the protractor.
- The triangle is classified as right scalene. Students should compare their triangles to those of their neighbors. They should realize that many different sized triangles can have angles of 30°, 60°, and 90°. They should understand that every neighbor's triangle should be right scalene. Point this out to students if they do not make this conclusion on their own.

Example 3

- Explain that students are going to construct and classify a triangle using three given pieces of information: two sides and the angle between the sides.
- Follow the steps shown. If you wish, have students draw the sides in inches instead of centimeters in order to make the triangle slightly larger and easier to draw.
- Measure the third side. It should be approximately 1.56 centimeters (or inches).
- The triangle is classified as obtuse scalene. Students should compare their triangles to those of their neighbors. This time, each student should have the same sized triangle.
- **Extension:** Repeat the construction but with the 4 centimeter leg horizontal (instead of the 3 centimeter leg). Ask students whether they think this triangle is different from the one in the example.

On Your Own

- **Think-Pair-Share:** Students should read each question independently and then work in pairs to answer the questions. When they have answered the questions, the pair should compare their answers with another group and discuss any discrepancies.

Closure

- Sketch a triangle that is (a) right isosceles and (b) obtuse scalene. Label the sides and angles.

EXAMPLE 2 **Constructing a Triangle Using Angle Measures**

Draw a triangle with angle measures of 30°, 60°, and 90°. Then classify the triangle.

Step 1: Use a protractor to draw the 30° angle.

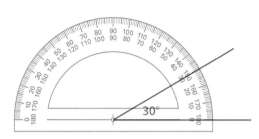

Step 2: Use a protractor to draw the 60° angle.

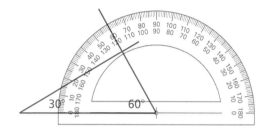

Step 3: The protractor shows that the measure of the remaining angle is 90°.

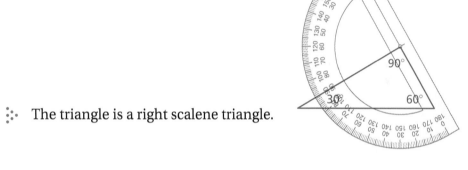

Study Tip

After drawing the first two angles, make sure you check the remaining angle.

∴ The triangle is a right scalene triangle.

EXAMPLE 3 **Constructing a Triangle Using Side Lengths**

Draw a triangle with a 3-centimeter side and a 4-centimeter side that meet at a 20° angle. Then classify the triangle.

Step 1: Use a protractor to draw a 20° angle.

Step 2: Use a ruler to mark 3 centimeters on one ray and 4 centimeters on the other ray.

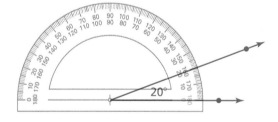

Step 3: Draw the third side to form the triangle.

∴ The triangle is an obtuse scalene triangle.

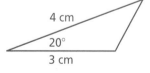

On Your Own

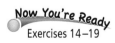

Now You're Ready
Exercises 14–19

3. Draw a triangle with angle measures of 45°, 45°, and 90°. Then classify the triangle.

4. Draw a triangle with a 1-inch side and a 2-inch side that meet at a 60° angle. Then classify the triangle.

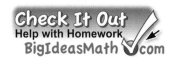
 Vocabulary and Concept Check

1. **WRITING** How can you classify triangles using angles? using sides?

2. **DIFFERENT WORDS, SAME QUESTION** Which is different? Find "both" answers.

Construct an equilateral triangle.	Construct a triangle with 3 congruent sides.
Construct an equiangular triangle.	Construct a triangle with no congruent sides.

 Practice and Problem Solving

Construct a triangle with the given description.

3. side lengths: 4 cm, 6 cm
4. side lengths: 5 cm, 12 cm
5. angles: 65°, 55°

Classify the triangle.

① 6.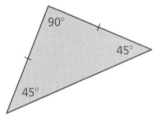

7.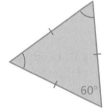

8.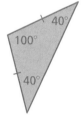

9.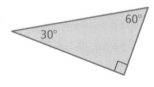

10.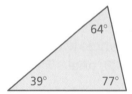

11.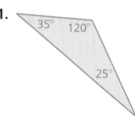

12. **ERROR ANALYSIS** Describe and correct the error in classifying the triangle.

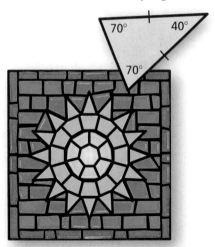

✗ 98° 43° 39°

The triangle is acute and scalene because it has two acute angles and no congruent sides.

13. **MOSAIC TILE** A mosaic is a pattern or picture made of small pieces of colored material. Classify the yellow triangle used in the mosaic.

Assignment Guide and Homework Check

Level	Day 1 Activity Assignment	Day 2 Lesson Assignment	Homework Check
Basic	3–5, 28–31	1, 2, 7–11 odd, 12, 13–23 odd	9, 13, 15, 21
Average	3–5, 28–31	1, 2, 6–12 even, 13–27 odd	10, 13, 15, 23
Advanced	3–5, 28–31	1, 2, 6–26 even, 27	10, 16, 20, 24, 27
Accelerated	1–5, 6–26 even, 27–31		10, 16, 20, 24, 27

Common Errors

- **Exercises 6–11** Students may classify the triangle using sides only and forget to classify it using angles (or vice versa). Remind students to classify the triangles using sides *and* angles.

7.3 Record and Practice Journal

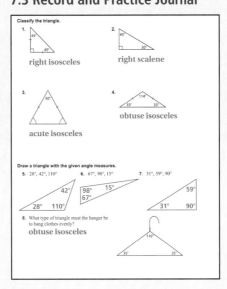

Vocabulary and Concept Check

1. *Angles:* When a triangle has 3 acute angles, it is an acute triangle. When a triangle has 1 obtuse angle, it is an obtuse triangle. When a triangle has 1 right angle, it is a right triangle. When a triangle has 3 congruent angles, it is an equiangular triangle.

   Sides: When a triangle has no congruent sides, it is a scalene triangle. When a triangle has 2 congruent sides, it is an isosceles triangle. When a triangle has 3 congruent sides, it is an equilateral triangle.

2. Construct a triangle with no congruent sides;

   (no congruent sides)

   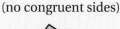

   (equilateral, equiangular, 3 congruent sides)

Practice and Problem Solving

3–5. See Additional Answers.

6. right isosceles

7. equilateral equiangular

8. obtuse isosceles

9. right scalene

10. acute scalene

11. obtuse scalene

12–13. See Additional Answers.

Practice and Problem Solving

14–20. See Additional Answers.

21. no; The sum of the angle measures must be 180°.

22. See *Taking Math Deeper*.

23–27. See Additional Answers.

Fair Game Review

28. yes; The equation can be written as $y = kx$ where $k = \frac{1}{2}$.

29. no; The equation cannot be written as $y = kx$.

30. no; The equation cannot be written as $y = kx$.

31. B

Mini-Assessment

Classify the triangle.

1.

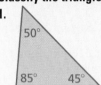

acute scalene triangle

2.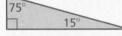

right scalene triangle

3. Draw a triangle with angle measures of 20°, 70°, and 90°. Then classify the triangle.

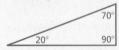

right scalene triangle

4. Draw a triangle with a 3-centimeter side and a 4-centimeter side that meet at a 15° angle. Then classify the triangle.

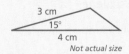

Not actual size

obtuse scalene triangle

Taking Math Deeper

Exercise 22

First, see whether it is possible to construct a triangle with one angle measure of 60° and one 4-centimeter side. If it is, see whether you can construct another triangle that is different from the first one.

① Use a protractor to draw a 60° angle. Make one of the rays 4 centimeters long. Choose a length for the second ray and draw the third side.

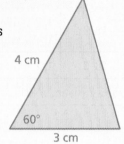

② Now that you have drawn one triangle, stop and think about what you might be able to change to create a different triangle that still satisfies the given description.

In the triangle above, you chose the length of the second ray to be 3 centimeters. Notice that you can change this length without changing the 60° angle or the 4-centimeter side. However, it changes the other side length and angles.

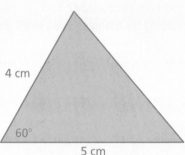

③ Answer the question.

So, you can construct *many* triangles with one angle measure of 60° and one 4-centimeter side.

Project

Tell students to create three problems that give descriptions of triangles. One description should lead to *no* possible triangle, another should lead to *one* possible triangle, and the other should lead to *many* possible triangles. Students can exchange problems and determine how many triangles are possible.

Reteaching and Enrichment Strategies

If students need help...	If students got it...
Resources by Chapter • Practice A and Practice B • Puzzle Time Record and Practice Journal Practice Differentiating the Lesson Lesson Tutorials Skills Review Handbook	**Resources by Chapter** • Enrichment and Extension • Technology Connection Start the next section

Draw a triangle with the given angle measures. Then classify the triangle.

2 **14.** 15°, 75°, 90°   **15.** 20°, 60°, 100°   **16.** 30°, 30°, 120°

Draw a triangle with the given description.

3 **17.** a triangle with a 2-inch side and a 3-inch side that meet at a 40° angle

18. a triangle with a 45° angle connected to a 60° angle by an 8-centimeter side

19. an acute scalene triangle

20. **LOGIC** You are constructing a triangle. You draw the first angle, as shown. Your friend says that you must be constructing an acute triangle. Is your friend correct? Explain your reasoning.

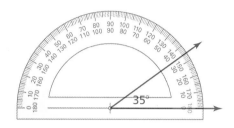

Determine whether you can construct *many, one,* or *no* triangle(s) with the given description. Explain your reasoning.

21. a triangle with angle measures of 50°, 70°, and 100°

22. a triangle with one angle measure of 60° and one 4-centimeter side

23. a scalene triangle with a 3-centimeter side and a 7-centimeter side

24. an isosceles triangle with two 4-inch sides that meet at an 80° angle

25. an isosceles triangle with two 2-inch sides and one 5-inch side

26. a right triangle with three congruent sides

27. Consider the three isosceles triangles.

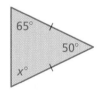 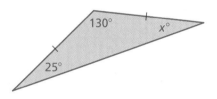

    a. Find the value of *x* for each triangle.

    b. What do you notice about the angle measures of each triangle?

    c. Write a rule about the angle measures of an isosceles triangle.

Fair Game Review What you learned in previous grades & lessons

Tell whether *x* and *y* show direct variation. Explain your reasoning. If so, find the constant of proportionality. *(Section 5.6)*

28. $x = 2y$ **29.** $y - x = 6$ **30.** $xy = 5$

31. **MULTIPLE CHOICE** A savings account earns 6% simple interest per year. The principal is $800. What is the balance after 18 months? *(Section 6.7)*

    A $864   **B** $872   **C** $1664   **D** $7200

 **Key Idea**

Sum of the Angle Measures of a Triangle

Words The sum of the angle measures of a triangle is 180°.

Algebra $x + y + z = 180$

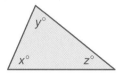

EXAMPLE ① **Finding Angle Measures**

Find each value of x. Then classify each triangle.

Geometry

In this extension, you will

- understand that the sum of the angle measures of any triangle is 180°.
- find missing angle measures in triangles.

a.

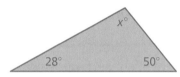

$$x + 28 + 50 = 180$$
$$x + 78 = 180$$
$$x = 102$$

⋮ The value of x is 102. The triangle has one obtuse angle and no congruent sides. So, it is an obtuse scalene triangle.

b.

$$x + 45 + 90 = 180$$
$$x + 135 = 180$$
$$x = 45$$

⋮ The value of x is 45. The triangle has a right angle and two congruent sides. So, it is a right isosceles triangle.

● **Practice**

Find the value of x. Then classify the triangle.

1.

2.

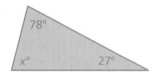

3.

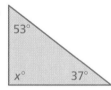

4.

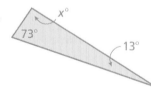

5.

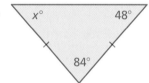

6.

Tell whether a triangle can have the given angle measures. If not, change the first angle measure so that the angle measures form a triangle.

7. 76.2°, 81.7°, 22.1°

8. 115.1°, 47.5°, 93°

9. $5\frac{2}{3}°$, $64\frac{1}{3}°$, 87°

10. $31\frac{3}{4}°$, $53\frac{1}{2}°$, $94\frac{3}{4}°$

Laurie's Notes

Introduction

Connect
- **Yesterday:** Students classified and constructed triangles.
- **Today:** Students will find the missing angle measure of a triangle and classify the triangle.

Motivate
? Discuss the Ohio State flag.

The blue triangle represents hills and valleys. The red and white stripes represent roads and waterways. The 13 leftmost stars represent the 13 original colonies. The 4 stars on the right bring the total to 17, representing that Ohio was the 17th state admitted to the Union.

Lesson Notes

Key Idea
- The property is written with variables to suggest that you can solve an equation to find the third angle when you know the other two angles. This is also called the *Triangle Sum Theorem*.
- ? "What type of angles are the remaining angles of a right triangle? a triangle with an obtuse angle?" Both are acute.
- ? "Do you think an obtuse triangle could have a right angle? Explain." no; The sum of the angle measures would be greater than 180°.

Example 1
- Some students may argue that all they need to do is add the angle measures and subtract from 180. Remind them that they are practicing a *process*, one that works when the three angle measures are given as algebraic expressions, such as $(x + 10)°$, $(x + 20)°$, and $(x + 30)°$.

Practice
- **Construct Viable Arguments and Critique the Reasoning of Others:** Students may observe that the isosceles triangles in Exercises 5 and 6 have a pair of congruent angles. This is revisited in Exercises 11–13.
- Exercises 7–10 provide practice with fractions and decimals.

What Your Students Will Learn
- Understand that the sum of the angle measures of any triangle is 180°.
- Find missing angle measures in triangles using the Triangle Sum Theorem.

Goal Today's lesson is finding the missing angle measure of a triangle and classifying the triangle.

Extra Example 1
Find each value of x. Then classify each triangle.

a.
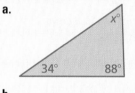
58; acute scalene triangle

b.

25; right scalene triangle

Practice
1. 91; obtuse scalene triangle
2. 75; acute scalene triangle
3. 90; right scalene triangle
4. 94; obtuse scalene triangle
5. 48; acute isosceles triangle
6. 60; equiangular equilateral triangle
7. yes
8. no; 39.5°
9. no; $28\frac{2}{3}$
10. yes

Record and Practice Journal
Extension 7.3 Practice
1–11. See Additional Answers.

Extra Example 2

Find each value of *x*. Then classify each triangle.

a.

54; acute isosceles triangle

b.

60; equiangular equilateral triangle

 Practice

11. 67.5; acute isosceles triangle

12. 60; equiangular equilateral triangle

13. 24; obtuse isosceles triangle

14. 25; right scalene triangle

15. 35; obtuse scalene triangle

16–17. See Additional Answers.

Mini-Assessment

Find the value of *x*. Then classify the triangle.

1.

48; acute scalene triangle

2.

18; obtuse isosceles triangle

3.

20; obtuse isosceles triangle

T-289

Example 2

- Share the symbolism of each flag.
  - **Jamaica:** The yellow divides the flag into four triangles and represents sunshine and natural resources. Black represents the burdens overcome by the people and the hardships in the future. Green represents the land and hope for the future.
  - **Cuba:** The blue stripes refer to the three old divisions of the island and the two white stripes represent the strength of the independent ideal. The red triangle symbolizes equality, fraternity and freedom, and the blood shed in the struggle for independence. The white star symbolizes the absolute freedom among the Cuban people.
- Set up and solve the equations as shown.
- **?** "How would you classify the green triangle on the Jamaican flag?" obtuse isosceles

Practice

- In both examples, students set up and solved simple equations to find the value of *x*. This experience should help them with Exercises 11–15 and 17.
- **Construct Viable Arguments and Critique the Reasoning of Others:** For Exercise 16, ask a volunteer to explain his or her reasoning.

Closure

- **Exit Ticket:** Find the value of *x*. Then classify the triangle.

a.

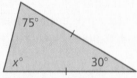

75; acute isosceles triangle

b.

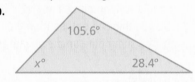

46; obtuse scalene triangle

EXAMPLE (2) **Finding Angle Measures**

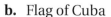

Math Practice

Analyze Givens
What information is given in the problem? How can you use this information to answer the question?

Find each value of *x*. Then classify each triangle.

a. Flag of Jamaica

$x + x + 128 = 180$

$2x + 128 = 180$

$2x = 52$

$x = 26$

⁝∙ The value of *x* is 26. The triangle has one obtuse angle and two congruent sides. So, it is an obtuse isosceles triangle.

b. Flag of Cuba

$x + x + 60 = 180$

$2x + 60 = 180$

$2x = 120$

$x = 60$

⁝∙ The value of *x* is 60. All three angles are congruent. So, it is an equilateral and equiangular triangle.

Practice

Find the value of *x*. Then classify the triangle.

11.

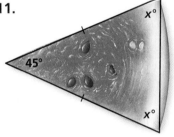

45°

12.

60°

13.

132°

14.

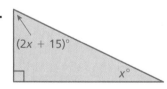

$(2x + 15)°$

15.

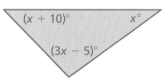

$(x + 10)°$ $x°$

$(3x - 5)°$

16. REASONING Explain why all triangles have at least two acute angles.

17. CARDS One method of stacking cards is shown.

a. Find the value of *x*.

b. Describe how to stack the cards with different angles. Is the value of *x* limited? If so, what are the limitations? Explain your reasoning.

36°

You can use an **example and non-example chart** to list examples and non-examples of a vocabulary word or item. Here is an example and non-example chart for complementary angles.

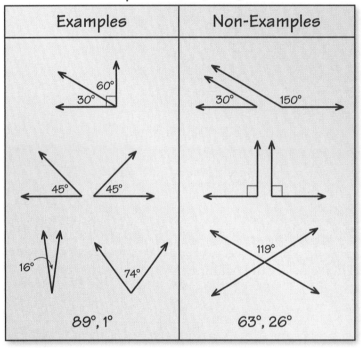

On Your Own

Make example and non-example charts to help you study these topics.

1. adjacent angles

2. vertical angles

3. supplementary angles

After you complete this chapter, make example and non-example charts for the following topics.

4. quadrilaterals

5. scale factor

"What do you think of my example & non-example chart for popular cat toys?"

Sample Answers

1.

Adjacent angles

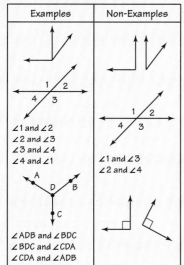

2.

Vertical angles

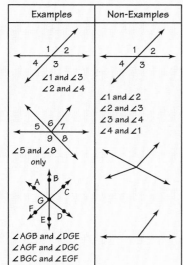

3.

Supplementary angles

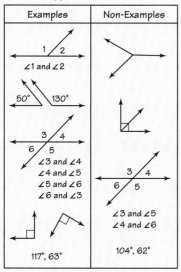

List of Organizers
Available at *BigIdeasMath.com*

Comparison Chart
Concept Circle
Definition (Idea) and Example Chart
Example and Non-Example Chart
Formula Triangle
Four Square
Information Frame
Information Wheel
Notetaking Organizer
Process Diagram
Summary Triangle
Word Magnet
Y Chart

About this Organizer

An **Example and Non-Example Chart** can be used to list examples and non-examples of a vocabulary word or term. Students write examples of the word or term in the left column and non-examples in the right column. This type of organizer serves as a good tool for assessing students' knowledge of pairs of topics that have subtle but important differences, such as complementary and supplementary angles. Blank example and non-example charts can be included on tests or quizzes for this purpose.

Technology for the *Teacher*

Editable Graphic Organizer

Answers

1. *Sample answer:*
   adjacent: $\angle PQR$ and $\angle RQS$,
   $\angle PQT$ and $\angle TQS$;
   vertical: $\angle PQR$ and $\angle TQS$,
   $\angle PQT$ and $\angle RQS$

2. *Sample answer:*
   adjacent: $\angle YUZ$ and $\angle ZUV$,
   $\angle ZUV$ and $\angle VUW$;
   vertical: $\angle YUX$ and $\angle VUW$,
   $\angle YUV$ and $\angle XUW$

3. adjacent; 146

4. adjacent; 16

5. vertical; 49

6. supplementary; 50

7. complementary; 24

8.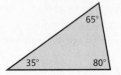

9. See Additional Answers.

10.

11. 115; obtuse scalene

12. 45; right isosceles

13. 80; equilateral equiangular

14. Use vertical angles to find that
    the measure of $\angle 2$ is 115°.
    Use supplementary angles to
    find that the measure of $\angle 3$ is
    65°. Then use supplementary
    angles to find that the
    measure of $\angle 2$ is 115°.

Alternative Quiz Ideas

100% Quiz	Math Log
Error Notebook	Notebook Quiz
Group Quiz	Partner Quiz
Homework Quiz	Pass the Paper

Notebook Quiz

A notebook quiz is used to check students' notebooks. Students should be told at the beginning of the course what the expectations are for their notebooks: notes, class work, homework, date, problem number, goals, definitions, or anything else that you feel is important for your class. They also need to know that it is their responsibility to obtain the notes when they miss class.

1. On a certain day, how was this vocabulary term defined?
2. For Section 7.1, what is the answer to On Your Own Question 5?
3. For Section 7.3 Extension, what is the answer to Practice Question 1?
4. For Section 7.4, what is the answer to the Essential Question?
5. On a certain day, what was the homework assignment?

Give the students 5 minutes to answer these questions.

Reteaching and Enrichment Strategies

If students need help. . .	If students got it. . .
Resources by Chapter	Resources by Chapter
• Practice A and Practice B	• Enrichment and Extension
• Puzzle Time	• Technology Connection
Lesson Tutorials	Game Closet at *BigIdeasMath.com*
BigIdeasMath.com	Start the next section

Name two pairs of adjacent angles and two pairs of vertical angles in the figure. *(Section 7.1)*

1.

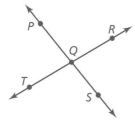

2.

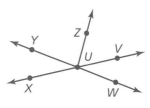

Tell whether the angles are *adjacent* or *vertical*. Then find the value of *x*. *(Section 7.1)*

3.

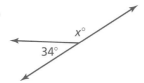

4.

5.

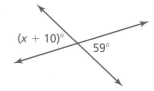

Tell whether the angles are *complementary* or *supplementary*. Then find the value of *x*. *(Section 7.2)*

6.

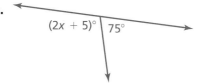

7.

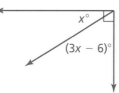

Draw a triangle with the given description. *(Section 7.3)*

8. a triangle with angle measures of 35°, 65°, and 80°

9. a triangle with a 5-centimeter side and a 7-centimeter side that meet at a 70° angle

10. an obtuse scalene triangle

Find the value of *x*. Then classify the triangle. *(Section 7.3)*

11.

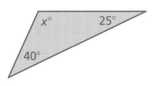

12.

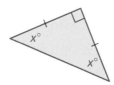

13.

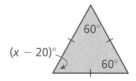

14. RAILROAD CROSSING Describe two ways to find the measure of ∠2. *(Section 7.1 and Section 7.2)*

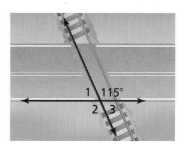

7.4 Quadrilaterals

Essential Question How can you classify quadrilaterals?

Quad means *four* and *lateral* means *side*. So, *quadrilateral* means a polygon with *four sides*.

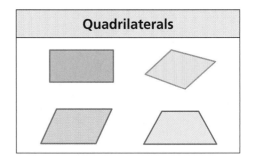

Quadrilaterals

1 ACTIVITY: Using Descriptions to Form Quadrilaterals

Work with a partner. Use a geoboard to form a quadrilateral that fits the given description. Record your results on geoboard dot paper.

a. Form a quadrilateral with exactly one pair of parallel sides.

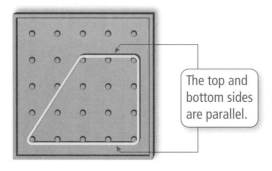

The top and bottom sides are parallel.

b. Form a quadrilateral with four congruent sides and four right angles.

c. Form a quadrilateral with four right angles that is *not* a square.

d. Form a quadrilateral with four congruent sides that is *not* a square.

e. Form a quadrilateral with two pairs of congruent adjacent sides and whose opposite sides are *not* congruent.

f. Form a quadrilateral with congruent and parallel opposite sides that is *not* a rectangle.

Geometry

In this lesson, you will

- understand that the sum of the angle measures of any quadrilateral is 360°.
- find missing angle measures in quadrilaterals.
- construct quadrilaterals.

2 ACTIVITY: Naming Quadrilaterals

Work with a partner. Match the names *square, rectangle, rhombus, parallelogram, trapezoid,* and *kite* with your 6 drawings in Activity 1.

Laurie's Notes

Introduction

Applying Mathematical Practices

- **Construct Viable Arguments and Critique the Reasoning of Others:** Students will be making observations and statements about the attributes of quadrilaterals. Listen carefully to their reasoning.

Motivate

- **?** "Have any of you created designs on grid or dot paper?"
- Quilters often sketch the design they are going to use before they begin to make a quilt. Share an example of a quilt design that contains different geometric shapes. The one shown is called *Symmetry in Motion*.
- **?** "What geometric shapes do you see in this design?" triangles, a square, trapezoids, and parallelograms
- You want to gain a sense of what pre-knowledge students have. Do not dwell on the attributes of the different quadrilaterals at this time.

Activity Notes

Write

- Write "quad" on the board, and ask students to think about different words that begin with this prefix. Some examples include *quadrilateral, quadrant, quadruple, quadruplets, quadruped,* and *quadrillion.*

Activity 1

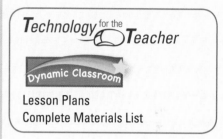

- Observe the orientations of students' quadrilaterals as they form them. For part (a), ask if the parallel sides need to be horizontal or vertical. Do some students form quadrilaterals with a diagonal orientation such as the one shown?
- Students are sometimes challenged to make quadrilaterals that have a particular attribute. For instance, part (e) may not be obvious to all students. Without giving away the answer, assure students that it is possible and have them explore different quadrilaterals on their geoboards.
- For part (e), explain to students that *adjacent sides* mean that the sides share a common vertex.

Activity 2

- **Construct Viable Arguments and Critique the Reasoning of Others:** After students have completed this activity, ask them if it is possible for a quadrilateral to have more than one name. For instance, a rectangle might also be called a parallelogram. However, if students are not yet ready to recognize that all rectangles are parallelograms, but not all parallelograms are rectangles, do not dwell on this.
- Look around the room and identify different types of quadrilaterals.

7.4 Record and Practice Journal

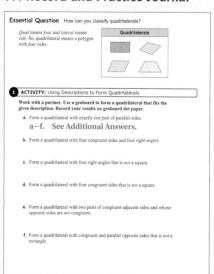

Essential Question How can you classify quadrilaterals?

Quad means *four* and *lateral* means *side.* So, quadrilateral means a polygon with *four* sides.

Quadrilaterals

1 ACTIVITY: Using Descriptions to Form Quadrilaterals

Work with a partner. Use a geoboard to form a quadrilateral that fits the given description. Record your results on geoboard dot paper.

a. Form a quadrilateral with exactly one pair of parallel sides.

    a–f. See Additional Answers.

b. Form a quadrilateral with four congruent sides and four right angles.

c. Form a quadrilateral with four right angles that is *not* a square.

d. Form a quadrilateral with four congruent sides that is *not* a square.

e. Form a quadrilateral with two pairs of congruent adjacent sides and whose opposite sides are *not* congruent.

f. Form a quadrilateral with congruent and parallel opposite sides that is *not* a rectangle.

English learners may have difficulty pronouncing words such as *parallelogram* and *quadrilateral*. Pronounce these words slowly and have students repeat after you. You may find it helpful to write these words on the board broken down phonetically:

pa-ruh-**lel**-uh-gram

kwa-druh-**la**-tuh-rul

7.4 Record and Practice Journal

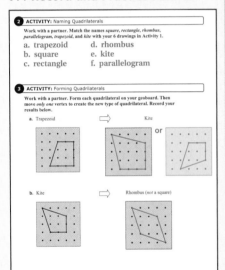

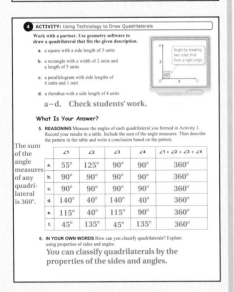

Laurie's Notes

Activity 3

- In this activity, students are asked to keep three vertices fixed and change one vertex by moving the rubber band. Two side lengths will change and two side lengths will remain the same.
- If the desired quadrilateral is not formed on their first attempt at moving the rubber band, students should start from the beginning.
- **?** "Do you think there is more than one way to do each part of the activity?" Part (a) has two answers. Part (b) has one answer. Have students share how they answered each part.

Activity 4

- Depending upon the software used, there may or may not be enough time to complete this activity. If the software does not have a menu for selecting a particular type of quadrilateral, then it will take longer.
- **Use Appropriate Tools Strategically** and **Construct Viable Arguments and Critique the Reasoning of Others:** There is more than one way to draw these quadrilaterals. Ask students to share their methods.

What Is Your Answer?

- For Question 6, ask students to include sketches with their explanations.

Closure

- Distribute to each student an index card on which a quadrilateral has been drawn. There should be quadrilaterals from this investigation as well as scalene quadrilaterals. Give a series of quadrilateral attributes and as you do so, students stand up if the quadrilateral on their card has the attribute.

3 ACTIVITY: Forming Quadrilaterals

Work with a partner. Form each quadrilateral on your geoboard. Then move *only one* vertex to create the new type of quadrilateral. Record your results on geoboard dot paper.

a. Trapezoid ⟹ Kite

b. Kite ⟹ Rhombus (*not* a square)

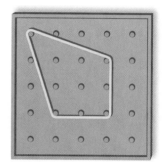

4 ACTIVITY: Using Technology to Draw Quadrilaterals

Math Practice

Use Technology to Explore

How does geometry software help you learn about the characteristics of a quadrilateral?

Work with a partner. Use geometry software to draw a quadrilateral that fits the given description.

a. a square with a side length of 3 units

b. a rectangle with a width of 2 units and a length of 5 units

c. a parallelogram with side lengths of 6 units and 1 unit

d. a rhombus with a side length of 4 units

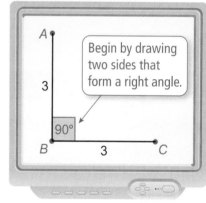

Begin by drawing two sides that form a right angle.

What Is Your Answer?

5. REASONING Measure the angles of each quadrilateral you formed in Activity 1. Record your results in a table. Include the sum of the angle measures. Then describe the pattern in the table and write a conclusion based on the pattern.

6. IN YOUR OWN WORDS How can you classify quadrilaterals? Explain using properties of sides and angles.

Practice

Use what you learned about quadrilaterals to complete Exercises 4–6 on page 296.

Key Vocabulary
kite, p. 294

A quadrilateral is a polygon with four sides. The diagram shows properties of different types of quadrilaterals and how they are related. When identifying a quadrilateral, use the name that is most specific.

Reading

Red arrows indicate parallel sides.

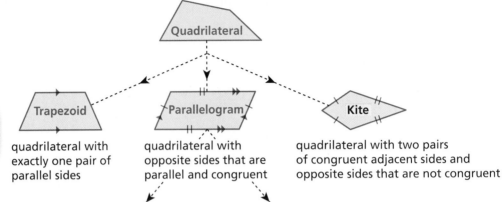

Quadrilateral

Trapezoid
quadrilateral with exactly one pair of parallel sides

Parallelogram
quadrilateral with opposite sides that are parallel and congruent

Kite
quadrilateral with two pairs of congruent adjacent sides and opposite sides that are not congruent

Rectangle
parallelogram with four right angles

Rhombus
parallelogram with four congruent sides

Square
parallelogram with four congruent sides and four right angles

EXAMPLE 1 **Classifying Quadrilaterals**

Study Tip

In Example 1(a), the square is also a parallelogram, a rectangle, and a rhombus. Square is the most specific name.

Classify the quadrilateral.

a.

The quadrilateral has four congruent sides and four right angles.

∴ So, the quadrilateral is a square.

b.

The quadrilateral has two pairs of congruent adjacent sides and opposite sides that are not congruent.

∴ So, the quadrilateral is a kite.

● **On Your Own**

Now You're Ready
Exercises 4–9

Classify the quadrilateral.

1.

2.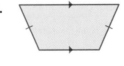

3.

🔊 Multi-Language Glossary at BigIdeasMath✓com

Laurie's Notes

Introduction

Connect
- **Yesterday:** Students explored properties of quadrilaterals.
- **Today:** Students will classify quadrilaterals by the attributes they possess.

Motivate
- **Preparation:** Make several quadrilaterals out of bendable drinking straws. You can make a slit in one end of a straw and insert it into another straw.
- To make a square, use four full-length straws. To make a rectangle or a kite, cut off 1 to 2 inches from two straws and use two full-length straws.

- The square can be flexed to make a rhombus. The rectangle can be flexed to make a parallelogram.
- Use the models to help students develop a sense of the relationship between the various quadrilaterals.

Lesson Notes

Discuss
- Discuss the quadrilateral classifications of trapezoids, parallelograms, and kites. These groups are disjoint, that is, a quadrilateral cannot belong to more than one of these classifications. Discuss the marks on the quadrilaterals.
- ❓ "What do the arrows mean?" parallel sides "What do the tick marks mean?" congruent sides
- Discuss the difference between *opposite* sides and *adjacent* sides.
- The dashed line from the parallelogram means that rectangles and rhombuses (or rhombi) are parallelograms with additional attributes. A rectangle is a parallelogram with four right angles, and a rhombus is a parallelogram with four congruent sides. A square is a parallelogram with *both* of these attributes.

Example 1
- ❓ "What properties does the quadrilateral in Example 1(a) have?" four congruent sides and four right angles "What is it?" square
- Although students could say that the quadrilateral in Example 1(a) is a parallelogram, which is correct, they should use the most specific name when identifying a quadrilateral.

On Your Own
- **Neighbor Check:** Have students work independently and then have their neighbors check their work. Have students discuss any discrepancies.

Technology for the **Teacher**

Dynamic Classroom

Lesson Tutorials
Lesson Plans
Answer Presentation Tool

Extra Example 1

Classify the quadrilateral.

a.

trapezoid

b.

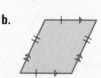

parallelogram

On Your Own
1. rhombus
2. trapezoid
3. rectangle

Visual

Bring in examples of figures that have the same size and shape (congruent) and figures that have the same shape but not necessarily the same size (similar). Ask students to identify the figures that are congruent. Then ask students to identify figures that are not congruent.

Extra Example 2

Find the value of x. $x = 80$

Extra Example 3

Draw a parallelogram with a 50° angle and a 130° angle.

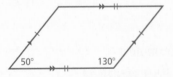

On Your Own

4. $x = 80$

5. $x = 65$

6. *Sample answer:*

Not actual size

Laurie's Notes

Key Idea

? "What observation did you have yesterday regarding the sum of the angle measures of a quadrilateral?" The sum is 360°.

- **Model with Mathematics:** Without measuring, the same conclusion can be reached in the following manner. Cut out a paper quadrilateral. On the overhead projector, tear off the four angles and arrange them about a point. Students should make the observation that the angles complete one revolution or 360°.

? **Extension:** "Does this work for any quadrilateral?" yes

? "If you know three of the four angle measures of a quadrilateral, could you find the missing angle measure?" yes "How?" Let x represent the missing angle measure. Then write an expression for the sum of the four angle measures, set it equal to 360, and solve.

Example 2

- Although students may want to solve this problem in their heads, work through the example to review equation-solving skills.

Example 3

- You could begin the construction by drawing a pair of parallel lines. Trace lines along both edges of a ruler, and then draw the angles from one of the lines through the other. Ask students whether they prefer this method or the method shown in the example.

- Students may realize that they do not need to "force" corresponding sides to be parallel or congruent. Ask them why they think this is so and whether this would work for any pair of angles. Try to lead them to see that the angles need to be supplementary to construct a parallelogram. This is a good preview of the work in a future course with transversals and parallel lines.

- **Extension:** Investigate the relationships between opposite angles.

- If time allows, repeat the construction with 120° on the bottom left and 60° on the bottom right. Ask students whether they think this parallelogram is different from the one in the example.

On Your Own

- After students finish Question 4, ask if they have an observation about the angles of a parallelogram. Opposite angles are congruent and adjacent angle measures sum to 180°.

Closure

- **Exit Ticket:** Find the value of x.

  a.

  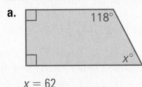

  $x = 62$

  b.

  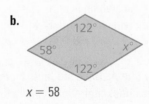

  $x = 58$

 **Key Idea**

Sum of the Angle Measures of a Quadrilateral

Words   The sum of the angle measures
of a quadrilateral is 360°.

Algebra $w + x + y + z = 360$

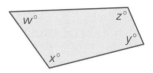

EXAMPLE 2   **Finding an Angle Measure of a Quadrilateral**

Find the value of x.

$70 + 75 + 115 + x =$	360	Write an equation.
$260 + x =$	360	Combine like terms.
-260	-260	Subtraction Property of Equality
$x =$	100	Simplify.

∴ The value of x is 100.

EXAMPLE 3   **Constructing a Quadrilateral**

Draw a parallelogram with a 60° angle and a 120° angle.

Step 1: Draw a line.

Step 2: Draw a 60° angle
and a 120° angle
that each have
one side on
the line.

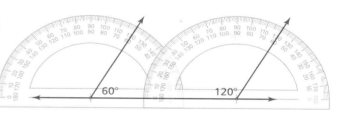

Step 3: Draw the remaining side. Make sure
that both pairs of opposite sides are
parallel and congruent.

● **On Your Own**

Now You're Ready
Exercises 10–12
and 14–17

Find the value of x.

4.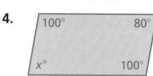

5.

6. Draw a right trapezoid whose parallel sides have lengths
of 3 centimeters and 5 centimeters.

 Vocabulary and Concept Check

1. **VOCABULARY** Which statements are true?

   a. All squares are rectangles.   b. All squares are parallelograms.

   c. All rectangles are parallelograms.   d. All squares are rhombuses.

   e. All rhombuses are parallelograms.

2. **REASONING** Name two types of quadrilaterals with four right angles.

3. **WHICH ONE DOESN'T BELONG?** Which type of quadrilateral does *not* belong with the other three? Explain your reasoning.

   | rectangle | parallelogram | square | kite |

 Practice and Problem Solving

Classify the quadrilateral.

 4.

5.

6.
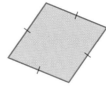

7.

8.
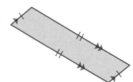

9.

Find the value of x.

 10.

11.
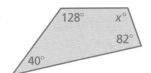

12.

13. **KITE MAKING** What is the measure of the angle at the tail end of the kite?

Assignment Guide and Homework Check

Level	Day 1 Activity Assignment	Day 2 Lesson Assignment	Homework Check
Basic	4–6, 27–30	1–3, 7–19 odd	3, 9, 13, 17
Average	4–6, 27–30	1–3, 7–13 odd, 14–24 even	3, 9, 13, 20
Advanced	4–6, 27–30	1–3, 8–24 even, 25, 26	1, 8, 12, 24
Accelerated	1–6, 8–24 even, 25–30		1, 8, 12, 24

Common Errors

- **Exercises 1 and 18–23** Students may have difficulty answering these questions correctly. Encourage them to learn the diagram at the top of page 294 to help them answer these questions correctly.
- **Exercises 4–9** Students may not use the most specific name when identifying the quadrilaterals. For instance, they may identify the quadrilateral in Exercise 9 as a parallelogram instead of a rectangle. Remind students to identify the quadrilaterals with the most specific name.
- **Exercises 10–13** Students may forget that the sum of the angle measures of a quadrilateral is 360° and that a right angle measures 90°. Remind students of these facts.

7.4 Record and Practice Journal

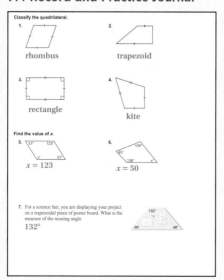

1. all of them

2. rectangle, square

3. kite; It is the only type of quadrilateral listed that does not have opposite sides that are parallel and congruent.

 Practice and Problem Solving

4. square   5. trapezoid

6. rhombus   7. kite

8. parallelogram

9. rectangle

10. 65   11. 110

12. 128   13. 58°

14.

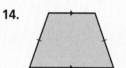

15.

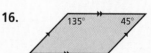

16.

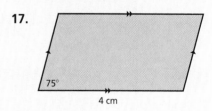

17.

18. always   19. always

20. sometimes

21. never   22. never

23. sometimes

24. See *Taking Math Deeper.*

25–26. See Additional Answers.

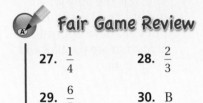
Fair Game Review

27. $\frac{1}{4}$ **28.** $\frac{2}{3}$

29. $\frac{6}{5}$ **30.** B

Mini-Assessment

1. Classify the quadrilateral.

a.

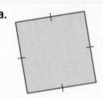

rhombus

b.

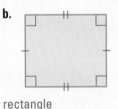

rectangle

2. Find the value of x.

a.

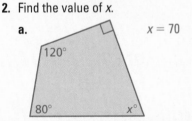

$x = 70$

b.

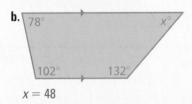

$x = 48$

3. What is the measure of the top right angle of the gold bar? 102°

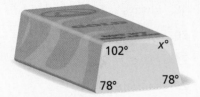

Taking Math Deeper

Exercise 24

One way to begin the problem is to make a diagram of the old door, the new door, and the piece you remove.

① Draw a diagram.

Removed piece

a. The new door is a quadrilateral with exactly one pair of parallel sides. So, it is a trapezoid.

② Write and solve an equation to find the value of x.

There are 360° in a trapezoid. However, rather than setting the sum of the four angle measures equal to 360° and solving, you can write and solve a simpler equation.

Notice that the two angles at the top of the door are supplementary. So, the angles at the bottom of the door must also be supplementary in order for the four angle measures to add up to 360°.

$$x + 91.5 = 180$$
$$x = 88.5$$

③ Answer the question.

b. So, the new angle at the bottom left side of the door is 88.5°.

Reteaching and Enrichment Strategies

If students need help. . .	If students got it. . .
Resources by Chapter • Practice A and Practice B • Puzzle Time Record and Practice Journal Practice Differentiating the Lesson Lesson Tutorials Skills Review Handbook	Resources by Chapter • Enrichment and Extension • Technology Connection Start the next section

Draw a quadrilateral with the given description.

③ 14. a trapezoid with a pair of congruent, nonparallel sides

15. a rhombus with 3-centimeter sides and two 100° angles

16. a parallelogram with a 45° angle and a 135° angle

17. a parallelogram with a 75° angle and a 4-centimeter side

Copy and complete using *always*, *sometimes*, or *never*.

18. A square is ⎯?⎯ a rectangle.

19. A square is ⎯?⎯ a rhombus.

20. A rhombus is ⎯?⎯ a square.

21. A parallelogram is ⎯?⎯ a trapezoid.

22. A trapezoid is ⎯?⎯ a kite.

23. A rhombus is ⎯?⎯ a rectangle.

24. DOOR The dashed line shows how you cut the bottom of a rectangular door so it opens more easily.

    a. Identify the new shape of the door. Explain.

    b. What is the new angle at the bottom left side of the door? Explain.

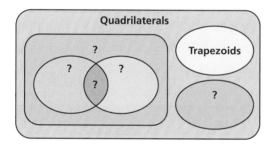

25. VENN DIAGRAM The diagram shows that some quadrilaterals are trapezoids, and all trapezoids are quadrilaterals. Copy the diagram. Fill in the names of the types of quadrilaterals to show their relationships.

26. **Structure** Consider the parallelogram.

    a. Find the values of x and y.

    b. Make a conjecture about opposite angles in a parallelogram.

    c. In polygons, consecutive interior angles share a common side. Make a conjecture about consecutive interior angles in a parallelogram.

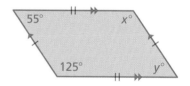

 Fair Game Review *What you learned in previous grades & lessons*

Write the ratio as a fraction in simplest form. *(Section 5.1)*

27. 3 turnovers : 12 assists

28. 18 girls to 27 boys

29. 42 pens : 35 pencils

30. MULTIPLE CHOICE Computer sales decreased from 40 to 32. What is the percent of decrease? *(Section 6.5)*

    Ⓐ 8%   **Ⓑ** 20%   **Ⓒ** 25%   **Ⓓ** 80%

7.5 Scale Drawings

Essential Question How can you enlarge or reduce a drawing proportionally?

1 ACTIVITY: Comparing Measurements

Work with a partner. The diagram shows a food court at a shopping mall. Each centimeter in the diagram represents 40 meters.

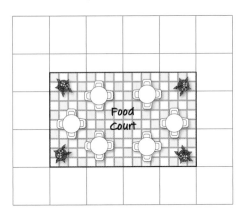

a. Find the length and the width of the drawing of the food court.

length: ▢ cm   width: ▢ cm

b. Find the actual length and width of the food court. Explain how you found your answers.

length: ▢ m   width: ▢ m

c. Find the ratios $\dfrac{\text{drawing length}}{\text{actual length}}$ and $\dfrac{\text{drawing width}}{\text{actual width}}$. What do you notice?

2 ACTIVITY: Recreating a Drawing

Work with a partner. Draw the food court in Activity 1 on the grid paper so that each centimeter represents 20 meters.

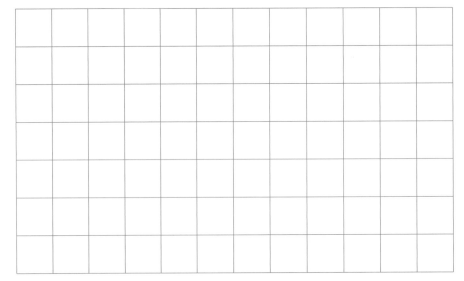

a. What happens to the size of the drawing?

b. Find the length and the width of your drawing. Compare these dimensions to the dimensions of the original drawing in Activity 1.

Geometry

In this lesson, you will
- use scale drawings to find actual distances.
- find scale factors.
- use scale drawings to find actual perimeters and areas.
- recreate scale drawings at a different scale.

Laurie's Notes

Introduction

Applying Mathematical Practices

- **Attend to Precision:** Mathematically proficient students understand that precision in measurement and in labeling units is important.

Motivate

- For each activity in this section, students should work with partners.
- Give each pair of students a map of the United States. Be sure the map has a scale on it.
- Tell students you are ready to start planning your summer vacation and you would like their input.
- Have students mark the map with a dot to represent your current location. Then tell them you would like to take a road trip this summer and not travel more than 1500 miles from home. As a benchmark, tell them that the distance between New York to Los Angeles is about 2800 miles. Give them a few minutes to decide where you should take your road trip.
- Make a list of the various locations students selected. Return to this list at the end of the period.

Activity Notes

Activity 1

- Distribute centimeter rulers to students. Check to see that all students know how to measure and read lengths correctly.
- **FYI:** Not all rulers start with 0 flush at the end of the ruler. It could be 0.1 to 0.2 centimeters from the edge.
- **Attend to Precision:** Have students measure accurately to the nearest 0.1 centimeter.
- After they have finished measuring, compare results.
- **?** "To answer part (b), did anyone set up a ratio table or write a proportion?" Answers will vary.
- **?** "What do you notice about the ratios you found in part (c)?" They are both equal to 1 cm : 40 m.
- **Connection:** The units are important when writing the ratios, so they should be included. It is given that each centimeter in the diagram represents 40 meters.

Activity 2

- Point out to students that each centimeter on the grid represents 20 meters.
- After students have finished the new drawing, discuss part (a).
- **?** **Construct Viable Arguments and Critique the Reasoning of Others:** "Why did the dimensions double?" 1 centimeter represents only half of what it did in Activity 1.

What Your Students Will Learn

- Use scale drawings to find actual distances, perimeters, and areas.
- Use actual measurements to find distances in models.
- Find scale factors.
- Recreate scale drawings at a different scale.

Previous Learning

Students should know common units of length, both U.S. customary and metric systems. Students should also know how to solve proportions.

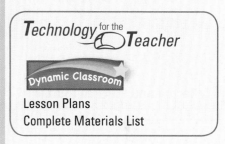

Technology for the **Teacher**

Dynamic Classroom

Lesson Plans
Complete Materials List

7.5 Record and Practice Journal

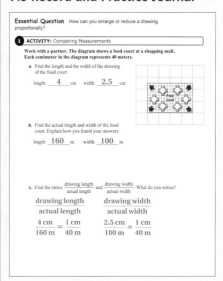

Essential Question How can you enlarge or reduce a drawing proportionally?

1 ACTIVITY: Comparing Measurements

Work with a partner. The diagram shows a food court at a shopping mall. Each centimeter in the diagram represents 40 meters.

a. Find the length and the width of the drawing of the food court.

length: __4__ cm   width: __2.5__ cm

b. Find the actual length and width of the food court. Explain how you found your answers.

length: __160__ m   width: __100__ m

c. Find the ratios $\frac{\text{drawing length}}{\text{actual length}}$ and $\frac{\text{drawing width}}{\text{actual width}}$. What do you notice?

$$\frac{\text{drawing length}}{\text{actual length}}$$

$$\frac{4 \text{ cm}}{160 \text{ m}} = \frac{1 \text{ cm}}{40 \text{ m}}$$

$$\frac{\text{drawing width}}{\text{actual width}}$$

$$\frac{2.5 \text{ cm}}{100 \text{ m}} = \frac{1 \text{ cm}}{40 \text{ m}}$$

Differentiated Instruction

Kinesthetic

Materials needed: poster board, ruler, colored markers or pencils, copies of the design below.

Students are to recreate the design in a larger size. They will need to choose a ratio to enlarge the design, measure the design, and set up proportions to find the new dimensions.

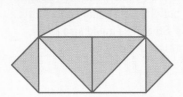

7.5 Record and Practice Journal

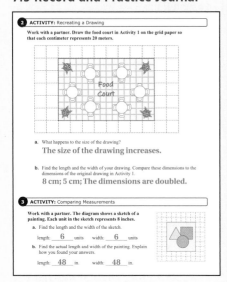

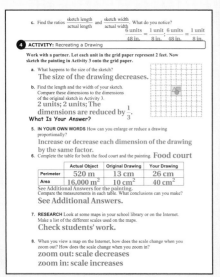

Laurie's Notes

Activity 3

- This activity is similar to Activity 1. Point out that each unit in the sketch represents 8 inches.
- **?** "To answer part (b), did anyone set up a ratio table or write a proportion?" Answers will vary.
- **?** "What do you notice about the ratios you found in part (c)?" They are both equal to 1 unit: 8 in.

Activity 4

- Point out to students that each unit on the grid represents 2 feet.
- After students have finished the new sketch, discuss part (a).
- **?** **Construct Viable Arguments and Critique the Reasoning of Others:** "Why did the dimensions decrease by a factor of $\frac{1}{3}$?" Because 2 ft = 24 in. = 3 × 8 in., 1 unit represents 3 times what it did in Activity 3.

What Is Your Answer?

- Question 6 explores a big idea in mathematics. When the dimensions change by a factor of k, the perimeter also changes by a factor of k, but the area changes by a factor of k^2. This idea is presented at the end of the lesson.

Closure

- Refer to the map of the United States given to students at the beginning of class. Point out the scale on the map. Have students determine the actual distance from your home to the location they selected.

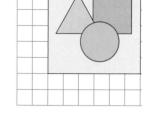

3 ACTIVITY: Comparing Measurements

Work with a partner. The diagram shows a sketch of a painting. Each unit in the sketch represents 8 inches.

a. Find the length and the width of the sketch.

length: ▭ units   width: ▭ units

b. Find the actual length and width of the painting. Explain how you found your answers.

length: ▭ in.   width: ▭ in.

c. Find the ratios $\dfrac{\text{sketch length}}{\text{actual length}}$ and $\dfrac{\text{sketch width}}{\text{actual width}}$.
What do you notice?

4 ACTIVITY: Recreating a Drawing

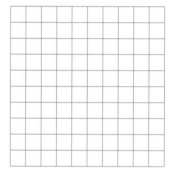

Work with a partner. Let each unit in the grid paper represent 2 feet. Now sketch the painting in Activity 3 onto the grid paper.

a. What happens to the size of the sketch?

b. Find the length and the width of your sketch. Compare these dimensions to the dimensions of the original sketch in Activity 3.

Math Practice

Specify Units

How do you know whether to use feet or units for each measurement?

What Is Your Answer?

5. **IN YOUR OWN WORDS** How can you enlarge or reduce a drawing proportionally?

6. Complete the table for both the food court and the painting.

	Actual Object	Original Drawing	Your Drawing
Perimeter			
Area			

Compare the measurements in each table. What conclusions can you make?

7. **RESEARCH** Look at some maps in your school library or on the Internet. Make a list of the different scales used on the maps.

8. When you view a map on the Internet, how does the scale change when you zoom out? How does the scale change when you zoom in?

Practice
Use what you learned about enlarging or reducing drawings to complete Exercises 4–7 on page 303.

Check It Out
Lesson Tutorials
BigIdeasMath ✓.com

Key Ideas

Scale Drawings and Models

A **scale drawing** is a proportional, two-dimensional drawing of an object. A **scale model** is a proportional, three-dimensional model of an object.

Scale

The measurements in scale drawings and models are proportional to the measurements of the actual object. The **scale** gives the ratio that compares the measurements of the drawing or model with the actual measurements.

$$\frac{1 \text{ in.}}{10 \text{ mi}}$$ ← drawing distance
  ← actual distance

1 in. : 10 mi
 drawing   actual

Study Tip

Scales are written so that the drawing distance comes first in the ratio.

EXAMPLE 1 Finding an Actual Distance

What is the actual distance *d* between Cadillac and Detroit?

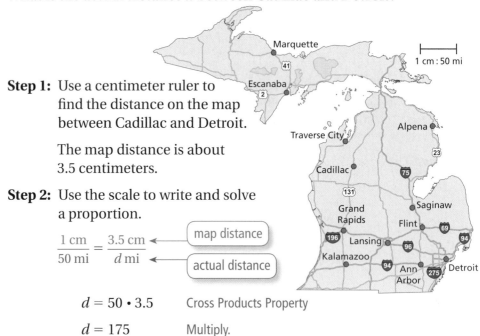

Step 1: Use a centimeter ruler to find the distance on the map between Cadillac and Detroit.

The map distance is about 3.5 centimeters.

Step 2: Use the scale to write and solve a proportion.

$$\frac{1 \text{ cm}}{50 \text{ mi}} = \frac{3.5 \text{ cm}}{d \text{ mi}}$$ ← map distance
  ← actual distance

$$d = 50 \cdot 3.5$$ Cross Products Property

$$d = 175$$ Multiply.

⋮ So, the distance between Cadillac and Detroit is about 175 miles.

On Your Own

Now You're Ready
Exercises 8–11

1. What is the actual distance between Traverse City and Marquette?

🔊 Multi-Language Glossary at BigIdeasMath✓.com

Laurie's Notes

Introduction

Connect
- **Yesterday:** Students compared measurements and recreated drawings.
- **Today:** Students will use a scale drawing to find a missing measure.

Motivate
- Show the class some items that have scales written on them: map (print one from the Internet), matchbook car, blueprint, or floor plan. Ask about the meaning of the scale for each item.
- **Trivia:** The world's biggest baseball bat is 120 feet and leans against the Louisville Slugger Museum & Factory in Kentucky. It is an exact-scale replica of Babe Ruth's 34-inch Louisville Slugger bat.

Lesson Notes

Key Ideas
- Be careful and consistent with your language today. Continually refer to the ratio of the scale drawing (or scale model) to the actual object.
- Review the *Study Tip*.

Example 1
- Have students explore the map. Ask if they see the scale. You can use the width of your baby finger to approximate 1 centimeter. Ask students to use their baby fingers to approximate the distance across the bottom of Michigan. about 160 mi
- Use centimeter rulers to measure the distance from Cadillac to Detroit.
- **Common Error:** Students might measure in inches instead of in centimeters, or they may confuse centimeters with millimeters.
- **?** "What is the map distance from Cadillac to Detroit?" 3.5 cm
- Set up the proportion using the language, "1 centimeter is to 50 miles as 3.5 centimeters is to what?"
- **Common Question:** Students will often ask how you can use the Cross Products Property because you are multiplying two different units together (50 mi and 3.5 cm). Because both numerators are the same unit (cm) and both denominators are the same unit (mi), it is okay to multiply. If the units were not the same in the numerators and were the same in the denominators, this could not be done. A quick way to explain this to students is to use a simple problem: If the scale is 1 cm : 5 ft, then 2 cm would be what actual distance? 10 ft
- **Attend to Precision:** Encourage students to write the units when they write the initial proportion. This ensures that the proportion has been set up correctly. When the numeric answer has been found, label with the correct units of measure.
- **Check for Reasonableness:** If the map distance is 3.5 centimeters, the actual distance should be 3.5 times the scale distance of 50 miles. Three times 50 miles is 150 miles, so an answer of 175 miles seems reasonable.

Goal Today's lesson is using **scale drawings** to find missing measurements.

Lesson Tutorials
Lesson Plans
Answer Presentation Tool

Extra Example 1
Using the map from Example 1, what is the distance *d* between Detroit and Marquette? about 350 mi

On Your Own
1. about 150 mi

Extra Example 2

The Earth's crust has a thickness of 80 kilometers on some of the continents. Using the scale model from Example 2, how thick is the crust of the model? 0.16 in.

 On Your Own

2. 5.8 in.

Extra Example 3

A sketch of a fashion designer's shirt is 9 centimeters long. The actual shirt is 1 meter long.

a. What is the scale of the drawing?
9 cm : 1 m

b. What is the scale factor of the drawing? 9 : 100

 On Your Own

3. 1 : 200

English Language Learners

Illustrate

Students have seen maps in classrooms and perhaps on road trips with their families. Hand out road maps to students in small groups. Have students find distances between cities on the map. Ask students how they can find the distances between cities using proportions.

Example 2

- This question is looking for a dimension of the scale model, not for an actual distance as in Example 1. Be careful with the language: scale to actual.
- Explain that this is an example of a scale model (3-Dimensional), not a scale drawing (2-Dimensional).
- Note that the units are written in the original proportion.
- **Check for Reasonableness:** Because the scale is 1 in. : 500 km, 2 inches would represent 1000 kilometers and 4 inches would represent 2000 kilometers. An answer of 4.6 inches is reasonable.

On Your Own

- **Think-Pair-Share:** Students should read each question independently and then work in pairs to answer the questions. When they have answered the questions, the pair should compare their answers with another group and discuss any discrepancies.

Write

- Write and discuss the definition of a scale factor.
- Connect the definition to yesterday's activities. For instance, in Activity 1, the scale used was 1 cm : 40 m. Because 40 meters is equal to 4000 centimeters, the scale factor would be 1 cm : 4000 cm = 1 : 4000.

Example 3

- **FYI:** The Sergeant Floyd Monument is located in Sioux City, Iowa. It memorializes Sgt. Charles Floyd, the only man who died on the Lewis and Clark Expedition. The monument is the nation's first nationally registered historical landmark.
- Read the problem and ask students to write the ratio of the model height to the actual height and then find the scale.
- ❓ "How can you find the scale factor when you know the scale?" Write the scale with the same units and simplify.
- ❓ "What conversion factor can you use?" 1 foot = 12 inches
- Work through the rest of the problem as shown.
- **Extension:** Use $\frac{1}{12}$: 10 to show that the answer is the same when converting inches to feet.

On Your Own

❓ "Is the scale model larger or smaller than the actual item? Explain." The scale model is smaller because 1 millimeter is shorter than 20 centimeters.

EXAMPLE **2** **Finding a Distance in a Model**

The liquid outer core of Earth is 2300 kilometers thick. A scale model of the layers of Earth has a scale of 1 in. : 500 km. How thick is the liquid outer core of the model?

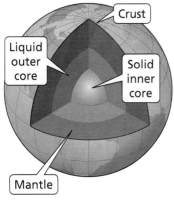

Crust
Liquid outer core
Solid inner core
Mantle

Ⓐ 0.2 in. Ⓑ 4.6 in. Ⓒ 0.2 km Ⓓ 4.6 km

$$\frac{1 \text{ in.}}{500 \text{ km}} = \frac{x \text{ in.}}{2300 \text{ km}}$$ ← model thickness
← actual thickness

$$\frac{1 \text{ in.}}{500 \text{ km}} \cdot 2300 \text{ km} = \frac{x \text{ in.}}{2300 \text{ km}} \cdot 2300 \text{ km}$$ Multiplication Property of Equality

$$4.6 = x$$ Simplify.

So, the liquid outer core of the model is 4.6 inches thick. The correct answer is Ⓑ.

On Your Own

2. The mantle of Earth is 2900 kilometers thick. How thick is the mantle of the model?

A scale can be written without units when the units are the same. A scale without units is called a **scale factor**.

EXAMPLE **3** **Finding a Scale Factor**

A scale model of the Sergeant Floyd Monument is 10 inches tall. The actual monument is 100 feet tall.

a. What is the scale of the model?

$$\frac{\text{model height}}{\text{actual height}} = \frac{10 \text{ in.}}{100 \text{ ft}} = \frac{1 \text{ in.}}{10 \text{ ft}}$$

The scale is 1 in. : 10 ft.

b. What is the scale factor of the model?

Write the scale with the same units. Use the fact that 1 ft = 12 in.

$$\text{scale factor} = \frac{1 \text{ in.}}{10 \text{ ft}} = \frac{1 \text{ in.}}{120 \text{ in.}} = \frac{1}{120}$$

The scale factor is 1 : 120.

On Your Own

3. A drawing has a scale of 1 mm : 20 cm. What is the scale factor of the drawing?

Now You're Ready
Exercises 12–16

1 cm : 2 mm

The scale drawing of a computer chip helps you see the individual components on the chip.

a. **Find the perimeter and the area of the computer chip in the scale drawing.**

When measured using a centimeter ruler, the scale drawing of the computer chip has a side length of 4 centimeters.

⋮⋅ So, the perimeter of the computer chip in the scale drawing is $4(4) = 16$ centimeters, and the area is $4^2 = 16$ square centimeters.

b. **Find the actual perimeter and area of the computer chip.**

$$\frac{1\ cm}{2\ mm} = \frac{4\ cm}{s\ mm}$$ ← drawing distance
← actual distance

$s = 2 \cdot 4$ Cross Products Property

$s = 8$ Multiply.

The side length of the actual computer chip is 8 millimeters.

⋮⋅ So, the actual perimeter of the computer chip is $4(8) = 32$ millimeters, and the actual area is $8^2 = 64$ square millimeters.

c. **Compare the ratios** $\dfrac{\text{drawing perimeter}}{\text{actual perimeter}}$ **and** $\dfrac{\text{drawing area}}{\text{actual area}}$ **to the scale factor.**

Use the fact that 1 cm = 10 mm.

$$\text{scale factor} = \frac{1\ cm}{2\ mm} = \frac{10\ \cancel{mm}}{2\ \cancel{mm}} = \frac{5}{1}$$

$$\frac{\text{drawing perimeter}}{\text{actual perimeter}} = \frac{16\ cm}{32\ mm} = \frac{1\ cm}{2\ mm} = \frac{5}{1}$$

$$\frac{\text{drawing area}}{\text{actual area}} = \frac{16\ cm^2}{64\ mm^2} = \frac{1\ cm^2}{4\ mm^2} = \left(\frac{1\ cm}{2\ mm}\right)^2 = \left(\frac{5}{1}\right)^2$$

⋮⋅ So, the ratio of the perimeters is equal to the scale factor, and the ratio of the areas is equal to the square of the scale factor.

Study Tip

The ratios tell you that the perimeter of the drawing is 5 times the actual perimeter, and the area of the drawing is $5^2 = 25$ times the actual area.

● **On Your Own**

Now You're Ready
Exercises 22 and 23

4. **WHAT IF?** The scale of the drawing of the computer chip is 1 cm : 3 mm. How do the answers in parts (a)–(c) change? Justify your answer.

Laurie's Notes

Example 4

- Read the problem and point out the scale drawing.
- **? Construct Viable Arguments and Critique the Reasoning of Others:** "Is the actual chip larger or smaller than the chip in the scale drawing? Explain." smaller; According to the scale, 1 centimeter represents only 2 millimeters.
- Measure to find the side length of the chip in the scale drawing.
- **?** "What is the perimeter of the chip in the scale drawing?" $4(4) = 16$ cm
- **?** "What is the area of the chip in the scale drawing?" $4^2 = 16$ cm^2
- To find the actual perimeter and area of the chip, you can begin by setting up and solving a proportion to find the side length of the actual chip. Be careful to label units and use precise language. As shown in the solution, the numerator of each ratio is a drawing distance and the denominator of each ratio is an actual distance.
- Some students will want to bypass the step of writing the proportion and use mental math. Remind them that they are practicing a *process* that will enable them to solve more difficult problems that they may not be able to solve using mental math.
- Knowing the side length of the actual chip enables you to find the actual perimeter and area of the chip.
- **Attend to Precision:** Work slowly through part (c). Remind students that when finding the scale factor, it is necessary to have the same units in the numerator and in the denominator.
- Make sure students understand the manipulation of the square units in the step $\dfrac{1 \text{ cm}^2}{4 \text{ mm}^2} = \left(\dfrac{1 \text{ cm}}{2 \text{ mm}}\right)^2$.

Closure

- **Exit Ticket:** A common model train scale is called the HO Scale, where the scale factor is 1 : 87. If the diameter of a wheel on a model train is 0.3 inch, what is the diameter of the actual wheel? 26.1 in.

Extra Example 4

The scale drawing of a miniature glass mosaic tile helps you see the detail on the tile.

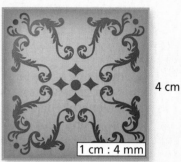

4 cm

1 cm : 4 mm

4 cm

a. Find the perimeter and area of the tile in the scale drawing. 16 cm, 16 cm^2

b. Find the actual perimeter and area of the tile. 64 mm, 256 mm^2

c. Compare the ratios $\dfrac{\text{drawing perimeter}}{\text{actual perimeter}}$ and $\dfrac{\text{drawing area}}{\text{actual area}}$ to the scale factor. The ratio of the perimeters is equal to the scale factor, $\dfrac{5}{2}$, and the ratio of the areas is equal to the square of the scale factor, $\left(\dfrac{5}{2}\right)^2$.

On Your Own

4. **a.** does not change; The size of the drawing does not change, just the scale.

b. actual perimeter = 48 mm, actual area = 144 mm^2; The side length of the actual computer chip increases to 12 millimeters, so the actual perimeter and area increase accordingly.

c. See Additional Answers.

1. A scale is the ratio that compares the measurements of the drawing or model with the actual measurements. A scale factor is a scale without any units.

2. larger; because 2 cm > 1 mm

3. Convert one of the lengths into the same units as the other length. Then, form the scale and simplify.

Practice and Problem Solving

4. 25 ft

5. 10 ft by 10 ft

6. 50 ft; 35 ft

7. 112.5%

8. 100 mi

9. 50 mi

10. 200 mi

11. 110 mi

12. 75 in.

13. 15 in.

14. 3.84 m

15. 21.6 yd

16. 17.5 mm

17. The 5 cm should be in the numerator.

$$\frac{1 \text{ cm}}{20 \text{ m}} = \frac{5 \text{ cm}}{x \text{ m}}$$

$$x = 100 \text{ m}$$

Assignment Guide and Homework Check

Level	Day 1 Activity Assignment	Day 2 Lesson Assignment	Homework Check
Basic	4–7, 32–36	1–3, 9–21 odd	2, 9, 15, 19
Average	4–7, 32–36	1–3, 9–17 odd, 18–30	2, 9, 13, 18
Advanced	4–7, 32–36	1–3, 10–16 even, 17, 18–30 even, 31	10, 14, 18, 22
Accelerated	1–7, 10–16 even, 17, 18–30 even, 31–36		10, 14, 18, 22

Common Errors

- **Exercises 8–11** When measuring with a centimeter ruler, students may not start at zero on the ruler. As a result, they will get an incorrect distance. Ask students to estimate the distance before measuring so they can check the reasonableness of their measurement.
- **Exercises 12–16** Students may mix up the proportion values when solving for the missing dimension. Remind them that the model dimension is in the numerator and the actual dimension is in the denominator in both ratios.

7.5 Record and Practice Journal

Find the missing dimension. Use the scale factor 1 : 8.

Item	Model	Actual
1. Statue	Height: 168 in.	Height: __112__ ft
2. Painting	Width: __2500__ cm	Width: 200 m
3. Alligator	Height: __9.6__ in.	Height: 6.4 ft
4. Train	Length: 36.5 in.	Length: _____ ft $24\frac{1}{3}$

5. The diameter of the moon is 2160 miles. A model has a scale of 1 in. : 150 mi. What is the diameter of the model?

   14.4 in.

6. A map has a scale of 1 in. : 4 mi.

   a. You measure 3 inches between your house and the movie theater. How many miles is it from your house to the movie theater?

   12 mi

   b. It is 17 miles to the mall. How many inches is that on the map?

   4.25 in.

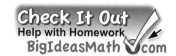
Check It Out
Help with Homework
BigIdeasMath.com

Vocabulary and Concept Check

1. **VOCABULARY** Compare and contrast the terms *scale* and *scale factor*.

2. **CRITICAL THINKING** The scale of a drawing is 2 cm : 1 mm. Is the scale drawing *larger* or *smaller* than the actual object? Explain.

3. **REASONING** How would you find the scale factor of a drawing that shows a length of 4 inches when the actual object is 8 feet long?

Practice and Problem Solving

Use the drawing and a centimeter ruler. Each centimeter in the drawing represents 5 feet.

4. What is the actual length of the flower garden?

5. What are the actual dimensions of the rose bed?

6. What are the actual perimeters of the perennial beds?

7. The area of the tulip bed is what percent of the area of the rose bed?

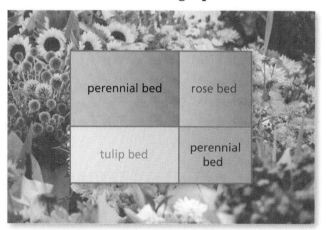

Use the map in Example 1 to find the actual distance between the cities.

8. Kalamazoo and Ann Arbor

9. Lansing and Flint

10. Grand Rapids and Escanaba

11. Saginaw and Alpena

Find the missing dimension. Use the scale factor 1 : 12.

Item	Model	Actual
12. Mattress	Length: 6.25 in.	Length: ___ in.
13. Corvette	Length: ___ in.	Length: 15 ft
14. Water tower	Depth: 32 cm	Depth: ___ m
15. Wingspan	Width: 5.4 ft	Width: ___ yd
16. Football helmet	Diameter: ___ mm	Diameter: 21 cm

17. **ERROR ANALYSIS** A scale is 1 cm : 20 m. Describe and correct the error in finding the actual distance that corresponds to 5 centimeters.

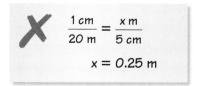

$$\frac{1\ cm}{20\ m} = \frac{x\ m}{5\ cm}$$

$$x = 0.25\ m$$

Use a centimeter ruler to measure the segment shown. Find the scale of the drawing.

18. |——— 120 m ———|

19.

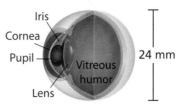

Iris
Cornea
Pupil
Vitreous humor
Lens
24 mm

20. REASONING You know the length and the width of a scale model. What additional information do you need to know to find the scale of the model?

21. OPEN-ENDED You are in charge of creating a billboard advertisement with the dimensions shown.

   a. Choose a product. Then design the billboard using words and a picture.

   b. What is the scale factor of your design?

|——— 16 ft ———|

8 ft

YOUR AD HERE

④ **22. CENTRAL PARK** Central Park is a rectangular park in New York City.

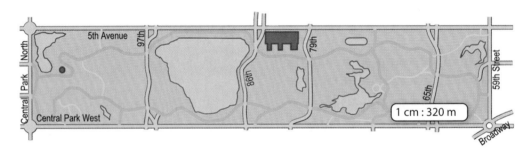

Central Park North
5th Avenue
97th
79th
86th
65th
59th Street
Central Park West
Broadway
1 cm : 320 m

   a. Find the perimeter and the area of Central Park in the scale drawing.

   b. Find the actual perimeter and area of Central Park.

23. ICON You are designing an icon for a mobile app.

   a. Find the perimeter and the area of the icon in the scale drawing.

   b. Find the actual perimeter and area of the icon.

1 cm : 2.5 mm

24. CRITICAL THINKING Use the results of Exercises 22 and 23 to make a conjecture about the relationship between the scale factor of a drawing and the ratios $\dfrac{\text{drawing perimeter}}{\text{actual perimeter}}$ and $\dfrac{\text{drawing area}}{\text{actual area}}$.

Common Errors

- **Exercise 30** Students may count the squares in the blueprint of the bathroom and use that as the area of the bathroom. Remind them that they need to find the actual length and width of each room and then find the area.

18. 4 cm; 1 cm : 30 m

19. 2.4 cm; 1 cm : 10 mm

20. The length or width of the actual item

21. a. *Answer should include, but is not limited to:* Make sure words and picture match the product.

b. Answers will vary.

22. a. 30 cm; 31.25 cm^2

b. 9600 m; 3,200,000 m^2

23. a. 16 cm; 16 cm^2

b. 40 mm; 100 mm^2

24. The ratio of the perimeters is the scale factor and the ratio of the areas is the square of the scale factor.

Differentiated Instruction

Kinesthetic

Materials needed: large piece of construction paper, ruler, and protractor. Have students work in pairs to draw a large right triangle on the paper. Record the lengths of the sides and the measures of the angles in a table. Connect the midpoints of each side of the large triangle and record the side lengths and angle measures of the second triangle in the table. Connect the midpoints of the sides of the second triangle to form a third triangle. Record the side lengths and angle measures of the third triangle in the table. Have students determine if the triangles are scale drawings of each other. If they are, then have students find the ratios of the perimeters and areas of each pair of triangles.

Practice and Problem Solving

25. See Additional Answers.

26.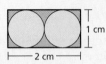

27. 15 ft^2

28. 4.5 ft^2

29. 3 ft^2

30. a. $480

    b. $1536

    c. tile; Because $5 per square foot is greater than $2 per square foot, the tile has a higher unit cost.

31. See *Taking Math Deeper*.

Fair Game Review

32–35.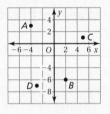

36. D

Mini-Assessment

Find the missing dimension. Use the scale factor 1 : 6.

	Model	Actual
1.	12 in.	72 in.
2.	3 ft	18 ft
3.	20 cm	120 cm
4.	2 yd	12 yd

5. A fish in an aquarium is 4 feet long. A scale model of the fish is 2 inches long. What is the scale factor?

    1 : 24

Taking Math Deeper

Exercise 31

This is a short, but nice problem. It requires that students make a reasonable estimate for the radius or diameter of a baseball. Then it requires that students use this information to determine the reasonableness of making a scale model.

1 Draw and label a diagram.

Model diameter (baseball): 3 in.   Model diameter: *x* in.

Actual radius (Earth): 6378 km

Actual radius (Sun): 695,500 km

2 Write and solve a proportion.

$$\frac{x}{695,500} = \frac{3}{6378}$$

$$695,500 \cdot \frac{x}{695,500} = 695,500 \cdot \frac{3}{6378}$$

$$x \approx 327 \text{ in.}$$

3 Answer the question.

327 inches is equal to 27.25 feet. The model for the Sun would have to be about the width of a classroom. So, it is not reasonable to use a baseball for Earth. A reasonable model for Earth would have a diameter of one-quarter inch and the diameter for the model of the Sun would be 27.26 inches.

Project

Make a scale drawing of the solar system. Make sure students include the scale they used.

Reteaching and Enrichment Strategies

If students need help. . .	If students got it. . .
Resources by Chapter • Practice A and Practice B • Puzzle Time Record and Practice Journal Practice Differentiating the Lesson Lesson Tutorials Skills Review Handbook	Resources by Chapter • Enrichment and Extension • Technology Connection Start the next section

Recreate the scale drawing so that it has a scale of 1 cm : 4 m.

25.
1 cm : 8 m

26.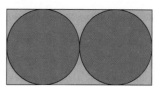
1 cm : 2 m

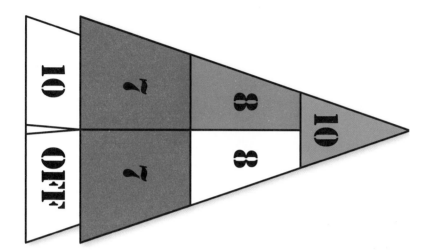

The shuffleboard diagram has a scale of 1 cm : 1 ft. Find the actual area of the region.

27. red region

28. blue region

29. green region

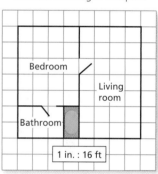

Reduced Drawing of Blueprint

30. **BLUEPRINT** In a blueprint, each square has a side length of $\frac{1}{4}$ inch.

    a. Ceramic tile costs $5 per square foot. How much would it cost to tile the bathroom?

    b. Carpet costs $18 per square yard. How much would it cost to carpet the bedroom and living room?

    c. Which has a greater unit cost, the tile or the carpet? Explain.

31. **Modeling** You are making a scale model of the solar system. The radius of Earth is 6378 kilometers. The radius of the Sun is 695,500 kilometers. Is it reasonable to choose a baseball as a model of Earth? Explain your reasoning.

 Fair Game Review What you learned in previous grades & lessons

Plot and label the ordered pair in a coordinate plane. *(Skills Review Handbook)*

32. $A(-4, 3)$ 33. $B(2, -6)$ 34. $C(5, 1)$ 35. $D(-3, -7)$

36. **MULTIPLE CHOICE** Which set of numbers is ordered from least to greatest? *(Section 6.2)*

    (A) $\frac{7}{20}$, 32%, 0.45   (B) 17%, 0.21, $\frac{3}{25}$ (C) 0.88, $\frac{7}{8}$, 93%   (D) 57%, $\frac{11}{16}$, 5.7

Classify the quadrilateral. *(Section 7.4)*

1.

2.

Find the value of x. *(Section 7.4)*

3.

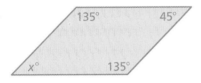

4.

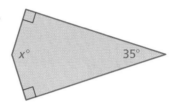

Draw a quadrilateral with the given description. *(Section 7.4)*

5. a rhombus with 2-centimeter sides and two 50° angles

6. a parallelogram with a 65° angle and a 5-centimeter side

Find the missing dimension. Use the scale factor 1 : 20. *(Section 7.5)*

	Item	Model	Actual
7.	Basketball player	Height: in.	Height: 90 in.
8.	Dinosaur	Length: 3.75 ft	Length: ft

9. SHED The side of the storage shed is in the shape of a trapezoid. Find the value of *x*. *(Section 7.4)*

10. DOLPHIN A dolphin in an aquarium is 12 feet long. A scale model of the dolphin is $3\frac{1}{2}$ inches long. What is the scale factor of the model? *(Section 7.5)*

11. SOCCER A scale drawing of a soccer field is shown. The actual soccer field is 300 feet long. *(Section 7.5)*

  a. What is the scale of the drawing?

  b. What is the scale factor of the drawing?

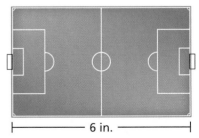

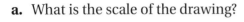

Alternative Assessment Options

Math Chat   Student Reflective Focus Question
Structured Interview Writing Prompt

Math Chat
- Put students in pairs to complete and discuss the exercises from the quiz. The discussion should include classifying quadrilaterals and using scale drawings.
- The teacher should walk around the classroom listening to the pairs and ask questions to ensure understanding.

Study Help Sample Answers

Remind students to complete Graphic Organizers for the rest of the chapter.

4.

Quadrilaterals

Examples	Non-Examples

5.

Scale factor

Examples	Non-Examples
5 : 1	1 cm : 2 mm
1 : 200	1 mm : 20 cm
1 : 1	12 in. : 1 ft
3 : 2	3 mi : 2 in.

1. rhombus

2. kite

3. 45

4. 145

5.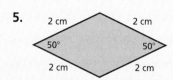

6. See Additional Answers.

7. 4.5 in.

8. 75 ft

9. 70

10. 7 : 288

11. **a.** 1 in. : 50 ft

  b. 1 : 600

Reteaching and Enrichment Strategies

If students need help...	If students got it...
Resources by Chapter • Practice A and Practice B • Puzzle Time Lesson Tutorials *BigIdeasMath.com*	Resources by Chapter • Enrichment and Extension • Technology Connection Game Closet at *BigIdeasMath.com* Start the Chapter Review

Technology for the *Teacher*

Online Assessment
Assessment Book
ExamView® Assessment Suite

For the Teacher
Additional Review Options

- *BigIdeasMath.com*
- Online Assessment
- Game Closet at *BigIdeasMath.com*
- Vocabulary Help
- Resources by Chapter

Answers

1. adjacent; 21
2. vertical; 81

Review of Common Errors

Exercises 1 and 2

- Students may think that there is not enough information to determine the value of x. Ask them to think about the information given in each figure. For instance, Exercise 1 shows two angles making up a right angle. So, the sum of the two angle measures must be 90°. Students can use this information to set up and solve a simple equation to find the value of x.

Exercises 3 and 4

- Students may think that there is not enough information to determine the value of x. Remind them of the definitions they have learned in the lesson and ask whether either could apply. For instance, Exercise 3 shows two angles making up a right angle, so a student can use the definition of complementary angles to find x.

Exercises 5 and 6

- Students may have difficulty using a protractor to construct the triangles. A quick review of the methods presented in Section 7.3 may be helpful.

Exercises 7 and 8

- Students may think that there is not enough information to find the value of x. Remind them that the sum of the angle measures of a triangle is 180°, and (in the case of Exercise 7), a box in a vertex signifies a right angle, which has a measure of 90°.
- Students may classify the triangle using sides only and forget to classify it using angles (or vice versa). Remind students to classify the triangles using sides *and* angles.

Exercises 9 and 10

- Students may forget that the sum of the angle measures of a quadrilateral is 360°. Remind them of this fact.

Exercise 11

- Students may forget what a rhombus is. If so, refer them to the diagram at the top of page 294.

Exercises 12 and 13

- When measuring with a centimeter ruler, students may not start at zero on the ruler. As a result, they will get an incorrect measurement and scale. Remind students to check their answers for reasonableness.

Check It Out
Vocabulary Help
BigIdeasMath com

Review Key Vocabulary

adjacent angles, *p. 272*
vertical angles, *p. 272*
congruent angles, *p. 272*
complementary angles,
  p. 278

supplementary angles,
  p. 278
congruent sides, *p. 284*
kite, *p. 294*
scale drawing, *p. 300*

scale model, *p. 300*
scale, *p. 300*
scale factor, *p. 301*

Review Examples and Exercises

7.1 Adjacent and Vertical Angles *(pp. 270–275)*

Tell whether the angles are *adjacent* or *vertical*. Then find the value of *x*.

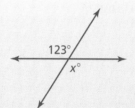

The angles are vertical angles. Because vertical angles are congruent, the angles have the same measure.

∴ So, the value of x is 123.

Exercises

Tell whether the angles are *adjacent* or *vertical*. Then find the value of *x*.

1.

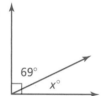

2.

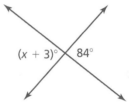

7.2 Complementary and Supplementary Angles *(pp. 276–281)*

Tell whether the angles are *complementary* or *supplementary*. Then find the value of *x*.

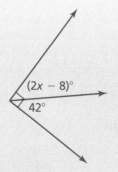

The two angles make up a right angle. So, the angles are complementary angles, and the sum of their measures is 90°.

$(2x - 8) + 42 = 90$	Write equation.
$2x + 34 = 90$	Combine like terms.
$2x = 56$	Subtract 34 from each side.
$x = 28$	Divide each side by 2.

∴ So, the value of x is 28.

Exercises

Tell whether the angles are *complementary* or *supplementary*. Then find the value of x.

3.

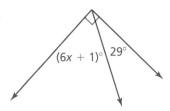

(6x + 1)° 29°

4.

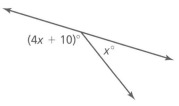

(4x + 10)° x°

Triangles *(pp. 282–289)*

Draw a triangle with a 3.5-centimeter side and a 4-centimeter side that meet at a 25° angle. Then classify the triangle.

Step 1: Use a protractor to draw a 25° angle.

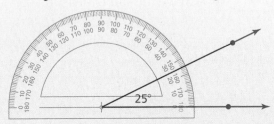

25°

Step 2: Use a ruler to mark 3.5 centimeters on one ray and 4 centimeters on the other ray.

Step 3: Draw the third side to form the triangle.

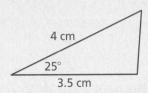

4 cm

25°

3.5 cm

∴ The triangle is an obtuse scalene triangle.

Exercises

Draw a triangle with the given description.

5. a triangle with angle measures of 40°, 50°, and 90°

6. a triangle with a 3-inch side and a 4-inch side that meet at a 30° angle

Find the value of x. Then classify the triangle.

7.

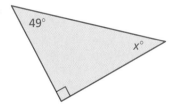

49°

x°

8.

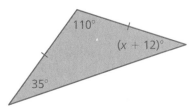

110°

(x + 12)°

35°

Review Game

Constructions and Figures

Materials per group:
- game cards
- pencil
- paper

Directions

Play in groups of two. One student shuffles the cards and places them face down. The other student turns over a card. Both students write as many details as possible about the figure (right angle, equal sides, etc.).

Who wins?

The student with the most correct details wins the round. Play continues until all cards are turned.

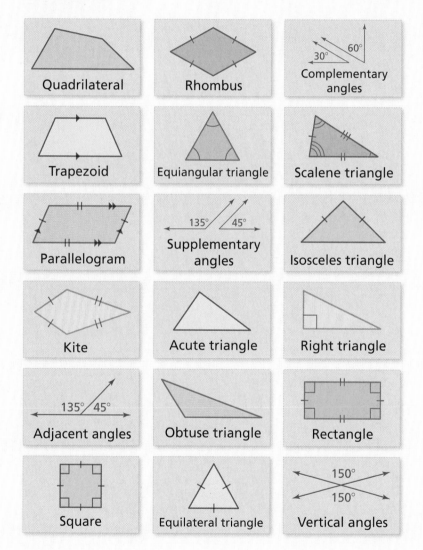

For the Student
Additional Practice
- Lesson Tutorials
- Multi-Language Glossary
- Self-Grading Progress Check
- *BigIdeasMath.com*
    Dynamic Student Edition
    Student Resources

Answers

3. complementary; 10

4. supplementary; 34

5.

6. See Additional Answers.

7. 41; right scalene

8. 23; isosceles obtuse

9. 52

10. 147

11.

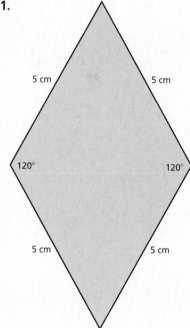

12. 6 cm; 1 cm : 5 in.

13. 2.5 cm; 1 cm : 3 in.

My Thoughts on the Chapter

What worked. . .

What did not work. . .

What I would do differently. . .

7.4 Quadrilaterals *(pp. 292–297)*

Draw a parallelogram with a 50° angle and a 130° angle.

Step 1: Draw a line.

Step 2: Draw a 50° angle and a 130° angle that each have one side on the line.

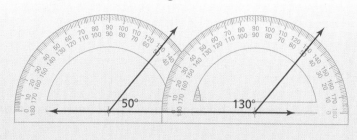

Step 3: Draw the remaining side. Make sure that both pairs of opposite sides are parallel and congruent.

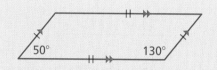

Exercises

Find the value of x.

9.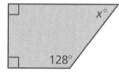

10.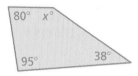

11. Draw a rhombus with 5-centimeter sides and two 120° angles.

7.5 Scale Drawings *(pp. 298–305)*

A lighthouse is 160 feet tall. A scale model of the lighthouse has a scale of 1 in. : 8 ft. How tall is the model of the lighthouse?

$$\frac{1 \text{ in.}}{8 \text{ ft}} = \frac{x \text{ in.}}{160 \text{ ft}}$$

← model height

← actual height

$$\frac{1 \text{ in.}}{8 \text{ ft}} \cdot 160 \text{ ft} = \frac{x \text{ in.}}{160 \text{ ft}} \cdot 160 \text{ ft} \qquad \text{Multiplication Property of Equality}$$

$$20 = x \qquad\qquad\qquad \text{Simplify.}$$

So, the model of the lighthouse is 20 inches tall.

Exercises

Use a centimeter ruler to measure the segment shown. Find the scale of the drawing.

12. |———— 30 in. ————|

13. |— 7.5 in. —|

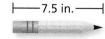

Chapter Review **309**

Check It Out
Test Practice
BigIdeasMath.com

Tell whether the angles are *adjacent* or *vertical*. Then find the value of *x*.

1.

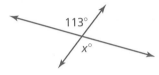

113°

x°

2.

(*x* + 6)°

56°

Tell whether the angles are *complementary* or *supplementary*. Then find the value of *x*.

3.

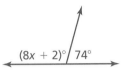

(8*x* + 2)° 74°

4.

15°

(4*x* − 5)°

Draw a triangle with the given angle measures. Then classify the triangle.

5. 10°, 80°, 90°

6. 30°, 40°, 110°

Draw a triangle with the given description.

7. a triangle with a 5-inch side and a 6-inch side that meet at a 50° angle

8. a right isosceles triangle

Find the value of *x*. Then classify the triangle.

9.

23° 129° *x*°

10.

x°
68° *x*°

11.

x°
x° *x*°

Find the value of *x*.

12.

x°

13.

95° *x*°

95°

14.

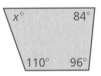

x° 84°

110° 96°

Draw a quadrilateral with the given description.

15. a rhombus with 6-centimeter sides and two 80° angles

16. a parallelogram with a 20° angle and a 160° angle

17. FISH Use a centimeter ruler to measure the fish. Find the scale factor of the drawing.

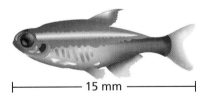

15 mm

18. CAD An engineer is using computer-aided design (CAD) software to design a component for a space shuttle. The scale of the drawing is 1 cm : 60 in. The actual length of the component is 12.5 feet. What is the length of the component in the drawing?

Test Item References

Chapter Test Questions	Section to Review
1, 2	7.1
3, 4	7.2
5–11	7.3
12–16	7.4
17, 18	7.5

Test-Taking Strategies

Remind students to quickly look over the entire test before they start so that they can budget their time. This chapter contains many definitions that some students may find difficult to keep straight. Encourage them to jot down definitions on the back of the test before they start. Students need to **Stop** and **Think** as they work through the test.

Common Errors

- **Exercises 1–4** Students may think that there is not enough information to determine the value of *x*. Remind them of the definitions they have learned and ask whether either could apply. For instance, Exercise 4 shows two angles making up a right angle, so a student can use the definition of complementary angles to find *x*.
- **Exercises 5–8** Students may have difficulty using a protractor to construct the triangles. A quick review of the methods presented in Section 7.3 may be helpful.
- **Exercises 9–14** Students may forget that the sum of the angle measures of a triangle is 180° or that the sum of the angle measures of a quadrilateral is 360°. Remind them of these facts.
- **Exercises 9–11** Students may classify the triangle using sides only and forget to classify it using angles (or vice versa). Remind students to classify the triangles using sides *and* angles.
- **Exercises 15 and 16** Students may forget what a rhombus or a parallelogram is. If so, refer them to the diagram at the top of page 294.

Reteaching and Enrichment Strategies

If students need help. . .	If students got it. . .
Resources by Chapter • Practice A and Practice B • Puzzle Time Record and Practice Journal Practice Differentiating the Lesson Lesson Tutorials *BigIdeasMath.com* Skills Review Handbook	Resources by Chapter • Enrichment and Extension • Technology Connection Game Closet at *BigIdeasMath.com* Start Cumulative Assessment

Answers

1. vertical; 113
2. adjacent; 28
3. supplementary; 13
4. complementary; 20
5.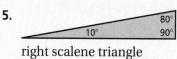
   right scalene triangle
6.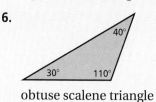
   obtuse scalene triangle
7. See Additional Answers.
8.
9. 28; obtuse scalene triangle
10. 56; acute isosceles triangle
11. 60; equilateral equiangular triangle
12. 90
13. 80
14. 70
15. See Additional Answers.
16.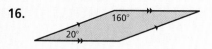
17. 5 cm; 10 : 3
18. 2.5 cm

Online Assessment
Assessment Book
ExamView® Assessment Suite

After Answering Easy Questions, Relax

Answer Easy Questions First

Estimate the Answer

Read All Choices before Answering

Read Question before Answering

Solve Directly or Eliminate Choices

Solve Problem before Looking at Choices

Use Intelligent Guessing

Work Backwards

About this Strategy

When taking a multiple choice test, be sure to read each question carefully and thoroughly. Sometimes it is easier to solve the problem and then look for the answer among the choices.

Answers

1. D

2. 75%

3. H

Item Analysis

1. **A.** The student divides 1 by 9 instead of dividing 9 by 1.

    B. The student confuses the meanings of *x* and *y*.

    C. The student looks from (0, 0) to (1, 9), not recognizing that the line segment between the points represents calories being burned between $x = 0$ and $x = 1$.

    D. Correct answer

2. **Gridded Response:** Correct answer: 75%

    Common Error: The student finds what percent 10 is of 40, and gets an answer of 25%.

3. **F.** The student finds the value of $2 - 6 - 9$.

    G. The student finds the value of $-2 + 6 - 9$.

    H. Correct answer

    I. The student finds the value of $-2 + 6 - (-9)$.

Technology for the *Teacher*

Performance Tasks
Online Assessment
Assessment Book
ExamView® Assessment Suite

T-311

1. The number of calories you burn by playing basketball is proportional to the number of minutes you play. Which of the following is a valid interpretation of the graph below?

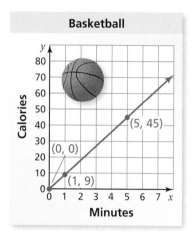

Basketball

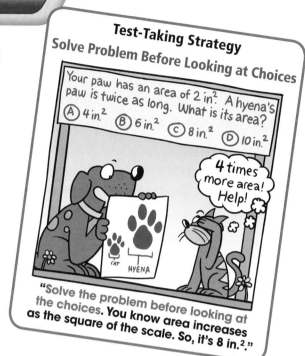
A. The unit rate is $\frac{1}{9}$ calorie per minute.

B. You burn 5 calories by playing basketball for 45 minutes.

C. You do not burn any calories if you do not play basketball for at least 1 minute.

D. You burn an additional 9 calories for each minute of basketball you play.

2. A lighting store is holding a clearance sale. The store is offering discounts on all the lamps it sells. As the sale progresses, the store will increase the percent of discount it is offering.

   You want to buy a lamp that has an original price of $40. You will buy the lamp when its price is marked down to $10. What percent discount will you have received?

3. What is the value of the expression below?

$$2 - 6 - (-9)$$

F. -13

G. -5

H. 5

I. 13

4. What is the solution to the proportion below?

$$\frac{8}{12} = \frac{x}{18}$$

5. Which graph represents the inequality below?

$$-5 - 6x \le -23$$

A.

B.

C.

D.

6. You are building a scale model of a park that is planned for a city. The model uses the scale below.

$$1 \text{ centimeter} = 2 \text{ meters}$$

The park will have a rectangular reflecting pool with a length of 20 meters and a width of 12 meters. In your scale model, what will be the area of the reflecting pool?

F. 60 cm²

G. 120 cm²

H. 480 cm²

I. 960 cm²

7. The quantities x and y are proportional. What is the missing value in the table?

x	y
$\frac{5}{7}$	10
$\frac{9}{7}$	18
$\frac{15}{7}$	30
4	

A. 38

B. 42

C. 46

D. 56

Item Analysis (continued)

Answers

4. 12

5. B

6. F

7. D

4. Gridded Response: Correct answer: 12

Common Error: The student incorrectly uses the Cross Products Property, getting an answer of 27.

5. A. The student does not reverse the inequality symbol.

   B. Correct answer

   C. The student does not reverse the inequality symbol and excludes $x = 3$.

   D. The student excludes $x = 3$.

6. F. Correct answer

   G. The student computes the area in square meters but then uses the given scale factor, which is for length, not area.

   H. The student reverses the relationship between the actual park and the scale model. The student also computes the area in square meters but then uses the given scale factor, which is for length, not area.

   I. The student reverses the relationship between the actual park and the scale model.

7. A. The student adds 8 to 30 because $18 - 10 = 8$.

   B. The student adds 12 to 30 because $30 - 18 = 12$.

   C. The student thinks the amount being added to y is being increased by 4 each time.

   D. Correct answer

Answers

8. H

9. C

10. *Part A* 90 miles

Part B $3\frac{1}{4}$ inches

Item Analysis (continued)

8. **F.** The student confuses a straight angle with a right angle.

G. The student confuses a straight angle with a right angle and adds instead of subtracts to find the measure of $\angle 2$.

H. Correct answer

I. The student adds instead of subtracts to find the measure of $\angle 2$.

9. **A.** The student subtracts incorrectly. The original difference of -50 is correct.

B. The student subtracts incorrectly. The original expression, $\frac{c}{5} + 15$, is correct.

C. Correct answer

D. The student identifies the operation error but then thinks that division by a number is undone by multiplication by the opposite of the number.

10. **2 points** The student demonstrates a thorough understanding of working with scale drawings. In Part A, the student correctly determines that the actual distance is 90 miles. In Part B, the student correctly determines that the distance on the map should be $3\frac{1}{4}$ inches. The student provides clear and complete work and explanations.

1 point The student demonstrates a partial understanding of working with scale drawings. The student provides some correct work and explanation.

0 points The student demonstrates insufficient understanding of working with scale drawings. The student is unable to make any meaningful progress toward a correct answer.

8. ∠1 and ∠2 form a straight angle. ∠1 has a measure of 28°. What is the measure of ∠2?

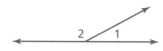

 F. 62°   **H.** 152°

 G. 118°   **I.** 208°

9. Brett solved the equation in the box below.

$$\frac{c}{5} - (-15) = -35$$

$$\frac{c}{5} + 15 = -35$$

$$\frac{c}{5} + 15 - 15 = -35 - 15$$

$$\frac{c}{5} = -50$$

$$\frac{c}{5} = \frac{-50}{5}$$

$$c = -10$$

What should Brett do to correct the error that he made?

A. Subtract 15 from -35 to get -20.

B. Rewrite $\frac{c}{5} - (-15)$ as $\frac{c}{5} - 15$.

C. Multiply each side of the equation by 5 to get $c = -250$.

D. Multiply each side of the equation by -5 to get $c = 250$.

10. A map of the state where Donna lives has the scale shown below.

$$\frac{1}{2} \text{ inch} = 10 \text{ miles}$$

Part A Donna measured the distance between her town and the state capital on the map. Her measurement was $4\frac{1}{2}$ inches. Based on Donna's measurement, what is the actual distance, in miles, between her town and the state capital? Show your work and explain your reasoning.

Part B Donna wants to mark her favorite campsite on the map. She knows that the campsite is 65 miles north of her town. What distance on the map, in inches, represents an actual distance of 65 miles? Show your work and explain your reasoning.

8 Circles and Area

8.1 Circles and Circumference

8.2 Perimeters of Composite Figures

8.3 Areas of Circles

8.4 Areas of Composite Figures

"Think of any number between 1 and 9."

"Okay, now add 4 to the number, multiply by 3, subtract 12, and divide by your original number."

"You end up with 3, don't you?"

"What do you get when you divide the circumference of a jack-o-lantern by its diameter?"

"Pumpkin pi, HE HE HE."

What Your Students Have Learned

- Find areas of rectangles with fractional side lengths.
- Classify two-dimensional figures into categories based on properties.
- Use formulas to find the areas of parallelograms, triangles, and trapezoids.
- Write and evaluate numerical expressions involving whole-number exponents.

What Your Students Will Learn

- Understand pi and its estimates.
- Use values of pi to estimate and calculate the circumference and area of circles.
- Find perimeters and areas of composite two-dimensional figures, including semi-circles.

Pacing Guide for Chapter 8

Chapter Opener Regular Accelerated	1 Day 1 Day
Section 1 Regular Accelerated	2 Days 1 Day
Section 2 Regular Accelerated	2 Days 1 Day
Study Help / Quiz Regular Accelerated	1 Day 0 Days
Section 3 Regular Accelerated	2 Days 1 Day
Section 4 Regular Accelerated	2 Days 1 Day
Chapter Review/ Chapter Tests Regular Accelerated	2 Days 2 Days
Total Chapter 8 Regular Accelerated	12 Days 7 Days
Year-to-Date Regular Accelerated	118 Days 69 Days

Technology for the *Teacher*

BigIdeasMath.com
Chapter at a Glance
Complete Materials List
Parent Letters: English and Spanish

What Your Students Have Learned

- Classify shapes and polygons used in figures.
- Square numbers and use the order of operations to evaluate expressions.

Additional Topics for Review

- Exponents
- Perimeter
- Area

Try It Yourself

1. trapezoids
2. circles
3. trapezoid, triangle
4. triangles
5. rectangle, triangle
6. rectangle, triangles
7. 25   8. 144
9. 12   10. 196
11. 243   12. 188

Record and Practice Journal
Fair Game Review

1. trapezoid and triangle
2. rectangles
3. rectangle and trapezoids
4. trapezoids, rectangles, circles
5. trapezoids, parallelogram
6. 49   7. 121
8. 100   9. 700
10. 144   11. 405
12. 744   13. 566
14. 5000 meters

Math Background Notes

Vocabulary Review

- Evaluate
- Simplify
- Exponent

Classifying Figures

- Students should know how to classify figures.
- **Teaching Tip:** Encourage students to redraw each figure on a sheet of paper. By constructing it themselves, students are more likely to identify the polygons used in the figure.
- **Teaching Tip:** Tactile learners might benefit from doing these examples using polygon templates. Allow them to build the shapes they see using the polygon templates. This could help students to distinguish which polygons they are seeing.

Squaring Numbers and Using Order of Operations

- Students should know the order of operations.
- You may want to review the correct order of operations with students. Many students probably learned the pneumonic device *Please Excuse My Dear Aunt Sally.* Ask a volunteer to explain why this phrase is helpful.
- **Common Error:** Remind students that the exponent describes how many times the base acts as a factor. The exponent itself will not appear as a factor.

Reteaching and Enrichment Strategies

If students need help. . .	If students got it. . .
Record and Practice Journal • Fair Game Review Skills Review Handbook Lesson Tutorials	Game Closet at *BigIdeasMath.com* Start the next section

What You Learned Before

"The area of the circle is pi *r* squared. The area of the triangle is one-half *bh*."

Classifying Figures

Identify the basic shapes in the figure.

Example 1

∴ Rectangle, right triangle

Example 2

Semicircle, square, and triangle

Try It Yourself

Identify the basic shapes in the figure.

1.

2.

3.

4.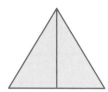

5.

6.

Squaring Numbers and Using Order of Operations

Example 3   Evaluate 4^2.

$$4^2 = 4 \cdot 4 = 16$$

4^2 means to multiply 4 by itself.

Example 4   Evaluate $3 \cdot 6^2$.

$$3 \cdot 6^2 = 3 \cdot (6 \cdot 6) = 3 \cdot 36 = 108$$

Use order of operations. Evaluate the exponent, and then multiply.

Try It Yourself

Evaluate the expression.

7. 5^2

8. 12^2

9. $3 \cdot 2^2$

10. $4 \cdot 7^2$

11. $3(1 + 8)^2$

12. $2(3 + 7)^2 - 3 \cdot 4$

Essential Question How can you find the circumference of a circle?

Archimedes was a Greek mathematician, physicist, engineer, and astronomer.

Archimedes discovered that in any circle the ratio of circumference to diameter is always the same. Archimedes called this ratio pi, or π (a letter from the Greek alphabet).

$$\pi = \frac{\text{circumference}}{\text{diameter}}$$

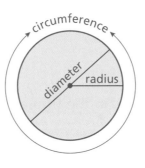

In Activities 1 and 2, you will use the same strategy Archimedes used to approximate π.

1   ACTIVITY: Approximating Pi

Work with a partner. Copy the table. Record your results in the table.

- **Measure the perimeter of the large square in millimeters.**

- **Measure the diameter of the circle in millimeters.**

- **Measure the perimeter of the small square in millimeters.**

- **Calculate the ratios of the two perimeters to the diameter.**

- **The average of these two ratios is an approximation of π.**

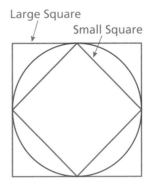

Large Square

Small Square

Geometry

In this lesson, you will
- describe a circle in terms of radius and diameter.
- understand the concept of pi.
- find circumferences of circles and perimeters of semicircles.

Sides	Large Perimeter	Diameter of Circle	Small Perimeter	$\dfrac{\text{Large Perimeter}}{\text{Diameter}}$	$\dfrac{\text{Small Perimeter}}{\text{Diameter}}$	Average of Ratios
4						
6						
8						
10						

Laurie's Notes

Introduction

Applying Mathematical Practices

- **Look for and Express Regularity in Repeated Reasoning:** Today's activity gives students the opportunity to understand that pi is a calculable number. Students will replicate a method used by Archimedes.

Motivate

- Share the history and some of Archimedes' inventions to interest students in the person whose work they will replicate today. (See the next page.)

Discuss

? "Does the diameter have to go through the center of the circle?" yes

? "What is the relationship between the radius and the diameter of a circle?" The diameter is twice the radius.

Activity Notes

Activity 1

- Discuss the first diagram involving the squares and a circle. A circle is inscribed in the large square—inscribed means the circle is entirely within the square and touches the square at exactly 4 points, the middle of each side of the square. The small square is inscribed in the circle—the small square is entirely within the circle and touches the circle at exactly 4 points, the vertices of the square.

? "What does perimeter mean?" distance around a figure "How do you calculate the perimeter of a square?" length of one side × 4

? "How will you measure the diameter of the circle?" measure the diagonal of the small square

? "What is another name for average?" mean "How do you calculate the average in this problem?" Sum the two ratios and divide by 2.

- **Attend to Precision:** Remind students to measure to the nearest millimeter.
- When students have finished the first row in the table, discuss the results and reflect on what the answers mean with reference to the diagram.

? **Look for and Express Regularity in Repeated Reasoning:** "If the process were repeated for a different size square to start, would the perimeter of the larger square always be greater than the perimeter of the smaller square?" yes

? "How does the distance around the circle compare to the distance around the two squares?" (Note: The definition for circumference is presented in the next lesson.) The distance around the circle is a number between the perimeters of the two squares.

What Your Students Will Learn

- Describe a circle in terms of radius and diameter.
- Understand the concept of pi.
- Find circumferences of circles and perimeters of semicircles.
- Estimate diameters of circles.

Previous Learning

Students should know how to find an average (mean).

Lesson Plans
Complete Materials List

8.1 Record and Practice Journal

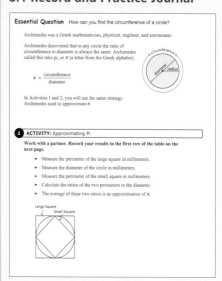

Kinesthetic

Have students bring in circular objects, such as cans or flying discs. Use string or a tape measure to find the diameter and circumference of each object. Have students divide the circumference by the diameter and record the results on the board. Using the classroom data, find the average to see how close the results are to π. Check to see if students make the connection that the ratio circumference : diameter is always the same.

8.1 Record and Practice Journal

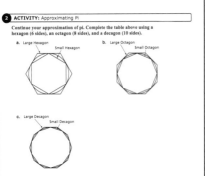

Laurie's Notes

Activity 2

- **Look for and Express Regularity in Repeated Reasoning:** Students should be able to complete this activity on their own after the guidance given for Activity 1.

? When students have finished ask, "Do you have any observations about the perimeters of the polygons in each problem?" The two perimeters are getting closer together and closer to the value of the circumference.

? The activity is called *Approximating Pi*, and there is no reason why a student should know the value of pi, even though they may. Explain that the last column is the approximation of pi. Ask, "What do you observe about the values in the last column?" As the number of sides of the polygons increase, the values get closer to the value of pi.

- **Use Appropriate Tools Strategically:** In discussing part (e), remind students that Archimedes did this investigation more than 2000 years ago without the aid of computer or calculator. Consider how long it might have taken him to construct the polygons with 96 sides!

- **Extension:** Read the short book referenced in the illustration. Your students will enjoy the story.

What Is Your Answer?

- **Big Idea:** Because the two perimeters converge to the circumference and each of the perimeters is divided by the diameter, it should seem reasonable to students that the circumference divided by the diameter approximates pi.

Closure

- Describe Archimedes' method for approximating pi.

History

- **Archimedes** is considered to be one of the greatest mathematicians of all time.
  - born in Syracuse, Italy; 287–212 B.C.
  - mathematician, physicist, engineer, and inventor
  - Archimedes was killed by a Roman soldier. Here are two accounts of his death.
    - He was working a math problem and refused to leave to meet the Roman general. So, the soldier killed him with his sword.
    - He was on his way to surrender and was carrying mathematical instruments. The soldier thought they were valuable and killed Archimedes.
- **His inventions include:**
  - Archimedes (water) screw for irrigating fields
  - Various instruments of war, such as catapults, cranes, and mirrors

A page from *Sir Cumference and the First Round Table* by Cindy Neuschwander

2 ACTIVITY: Approximating Pi

Math Practice

Make Conjectures

How can you use the results of the activity to find an approximation of pi?

Continue your approximation of pi. Complete the table from Activity 1 using a hexagon (6 sides), an octagon (8 sides), and a decagon (10 sides).

a.
Large Hexagon
Small Hexagon

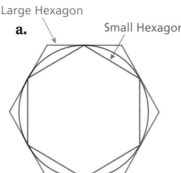

b.
Large Octagon
Small Octagon

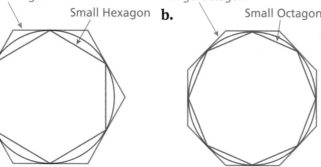

c.
Large Decagon
Small Decagon

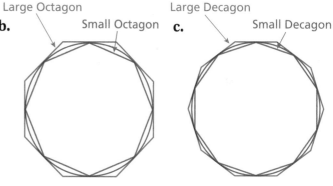

d. From the table, what can you conclude about the value of π? Explain your reasoning.

e. Archimedes calculated the value of π using polygons with 96 sides. Do you think his calculations were more or less accurate than yours?

What Is Your Answer?

3. **IN YOUR OWN WORDS** Now that you know an approximation for pi, explain how you can use it to find the circumference of a circle. Write a formula for the circumference C of a circle whose diameter is d.

4. **CONSTRUCTION** Use a compass to draw three circles. Use your formula from Question 3 to find the circumference of each circle.

Practice

Use what you learned about circles and circumference to complete Exercises 9–11 on page 321.

Check It Out
Lesson Tutorials
BigIdeasMath ✓com

Key Vocabulary 🔊
circle, *p. 318*
center, *p. 318*
radius, *p. 318*
diameter, *p. 318*
circumference, *p. 319*
pi, *p. 319*
semicircle, *p. 320*

A **circle** is the set of all points in a plane that are the same distance from a point called the **center**.

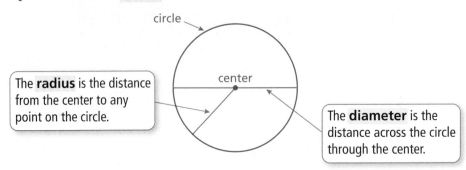

The **radius** is the distance from the center to any point on the circle.

The **diameter** is the distance across the circle through the center.

Key Idea

Radius and Diameter

Words The diameter d of a circle is twice the radius r. The radius r of a circle is one-half the diameter d.

Algebra   **Diameter:** $d = 2r$ **Radius:** $r = \dfrac{d}{2}$

EXAMPLE **1**   **Finding a Radius and a Diameter**

a. **The diameter of a circle is 20 feet. Find the radius.**

b. **The radius of a circle is 7 meters. Find the diameter.**

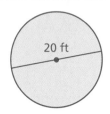

20 ft

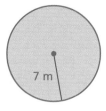

7 m

$r = \dfrac{d}{2}$ Radius of a circle

$= \dfrac{20}{2}$ Substitute 20 for d.

$= 10$ Divide.

⋮ The radius is 10 feet.

$d = 2r$ Diameter of a circle

$= 2(7)$ Substitute 7 for r.

$= 14$ Multiply.

⋮ The diameter is 14 meters.

On Your Own

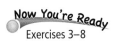

Now You're Ready
Exercises 3–8

1. The diameter of a circle is 16 centimeters. Find the radius.

2. The radius of a circle is 9 yards. Find the diameter.

🔊 Multi-Language Glossary at BigIdeasMath✓com

Laurie's Notes

Introduction

Connect
- **Yesterday:** Students investigated a technique for calculating the value of pi.
- **Today:** Students will use the formula for circumference to solve real-life problems.

Motivate
- **Whole Class Activity:** How observant are your students? Tell them that you are going to give them 1 minute to write a list of objects in the room that have a special characteristic. Give them time to get scrap paper and a pencil.
- Announce that they need to list objects that are circular or have a circle on them. Go!
- Items will vary but expected items include: clock and/or watch face, bottom of coffee cup, pencil's eraser, pupils of your eyes, Person X's glasses, metal feet on the chair, etc.

Lesson Notes

Key Idea
- Draw a circle on the board and label the center, a radius, and a diameter. Discuss each.
- Write the Key Idea.
- Discuss with students that if you know either the diameter or radius of a circle, you can find the other. There is a 2 : 1 relationship between **diameter** and **radius**.

Example 1
- Work through each example.
- Remind students to label answers with the appropriate units.

On Your Own
- It helps some students to draw a sketch of a circle and label the given information. These students need to see the relationship instead of only reading the given information.

Goal Today's lesson is estimating and calculating the **circumference** of **circles** using common estimates for **pi**.

Technology for the *Teacher*

Dynamic Classroom

Lesson Tutorials
Lesson Plans
Answer Presentation Tool

Extra Example 1

a. The diameter of a circle is 4 yards. Find the radius. 2 yd
b. The radius of a circle is 15 millimeters. Find the diameter. 30 mm

On Your Own
1. 8 cm
2. 18 yd

Vocabulary

Use everyday objects, such as a clock, to discuss the meaning of *circle, center, radius, diameter,* and *circumference.* Have students add these terms to their notebooks along with a sketch of a circle with the parts labeled. Students should also include pi with its symbol, π, and its approximations, 3.14 and $\frac{22}{7}$, in their notebooks.

Key Idea

- **Make Sense of Problems and Persevere in Solving Them:** Read the information at the top of the page. Note the two approximations of pi and how each is used in the problems that follow. Note the *Study Tip.*
- Write the Key Idea.
- **Common Misconception:** Students may know that pi is a number; however, when they see it in a formula, they can become confused. Students may ask if π is a variable. Pi is a constant whose value is approximately 3.14 or $\frac{22}{7}$. In each formula, remind students that πd means π times the diameter and $2\pi r$ means 2 times π times the radius.
- **Big Idea:** The ratio of the circumference to the diameter is pi. To get the formula for circumference, multiply both sides of the equation by the diameter.

$$\frac{\text{circumference}}{\text{diameter}} = \text{pi}$$
$$\frac{C}{d} = \pi$$
$$C = \pi d$$

Extra Example 2

a. Find the circumference of a flying disc with a radius of 8 centimeters. Use 3.14 for π. *about 50.24 cm*

b. Find the circumference of a clock with a diameter of 21 inches. Use $\frac{22}{7}$ for π. *about 66 in.*

Example 2

- Work through each example.
- **Make Sense of Problems and Persevere in Solving Them:** Point out to students when to use the different forms of the circumference formula given a radius or a diameter.
- Remind students that the symbol $\approx$ means *approximately equal to.*
- **?** "Why is the equal sign replaced by the approximately equal ($\approx$) sign?" *3.14 is an approximation for π.*
- **?** "Why is $2 \times 3.14 \times 5$ equal to $2 \times 5 \times 3.14$?" *Commutative Property of Multiplication*
- Note that in part (b), the diameter is a multiple of 7, so $\frac{22}{7}$ is used as the approximation for π.

On Your Own

- **Think-Pair-Share:** Students should read each question independently and then work in pairs to answer the questions. When they have answered the questions, the pair should compare their answers with another group and discuss any discrepancies.
- Check which estimate for π students used and why.

On Your Own

3. about 12.56 cm

4. about 44 ft

5. about 28.26 in.

The distance around a circle is called the **circumference**. The ratio $\frac{\text{circumference}}{\text{diameter}}$ is the same for *every* circle and is represented by the Greek letter π, called **pi**. The value of π can be approximated as 3.14 or $\frac{22}{7}$.

Study Tip

When the radius or diameter is a multiple of 7, it is easier to use $\frac{22}{7}$ as the estimate of π.

Key Idea

Circumference of a Circle

Words   The circumference C of a circle is equal to π times the diameter d or π times twice the radius r.

Algebra $C = \pi d$ or $C = 2\pi r$

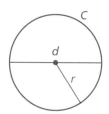

EXAMPLE 2   **Finding Circumferences of Circles**

5 in.

a. **Find the circumference of the flying disc. Use 3.14 for π.**

$C = 2\pi r$ Write formula for circumference.

$\approx 2 \cdot 3.14 \cdot 5$ Substitute 3.14 for π and 5 for r.

$= 31.4$ Multiply.

The circumference is about 31.4 inches.

b. **Find the circumference of the watch face. Use $\frac{22}{7}$ for π.**

28 mm

$C = \pi d$ Write formula for circumference.

$\approx \frac{22}{7} \cdot 28$ Substitute $\frac{22}{7}$ for π and 28 for d.

$= 88$ Multiply.

The circumference is about 88 millimeters.

On Your Own

Now You're Ready
Exercises 9–11

Find the circumference of the object. Use 3.14 or $\frac{22}{7}$ for π.

3.
2 cm

4.
14 ft

5.
9 in.

EXAMPLE 3 **Estimating a Diameter**

C = 31.4 in.

The circumference of the roll of caution tape decreases 10.5 inches after a construction worker uses some of the tape. Which is the best estimate of the diameter of the roll after the decrease?

   (A) 5 inches   (B) 7 inches   (C) 10 inches   (D) 12 inches

After the decrease, the circumference of the roll is
$31.4 - 10.5 = 20.9$ inches.

$C = \pi d$	Write formula for circumference.
$20.9 \approx 3.14 \cdot d$	Substitute 20.9 for C and 3.14 for π.
$21 \approx 3d$	Round 20.9 up to 21. Round 3.14 down to 3.
$7 = d$	Divide each side by 3.

∴ The correct answer is (B).

On Your Own

6. **WHAT IF?** The circumference of the roll of tape decreases 5.25 inches. Estimate the diameter of the roll after the decrease.

EXAMPLE 4 **Finding the Perimeter of a Semicircular Region**

A semicircle is one-half of a circle. Find the perimeter of the semicircular region.

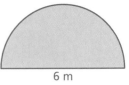

6 m

The straight side is 6 meters long. The distance around the curved part is one-half the circumference of a circle with a diameter of 6 meters.

$\dfrac{C}{2} = \dfrac{\pi d}{2}$	Divide the circumference by 2.
$\approx \dfrac{3.14 \cdot 6}{2}$	Substitute 3.14 for π and 6 for d.
$= 9.42$	Simplify.

∴ So, the perimeter is about $6 + 9.42 = 15.42$ meters.

On Your Own

Now You're Ready
Exercises 15 and 16

Find the perimeter of the semicircular region.

7.
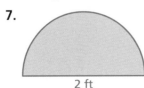
2 ft

8. 7 cm
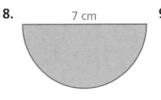

9.
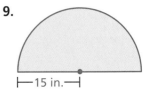
15 in.

Laurie's Notes

Example 3
- Ask a volunteer to read the given information.
- Note that this multiple choice question asks for the best *estimate*. Rounding the circumference and π is sufficient to select the correct answer.

Example 4
- Draw the diagram. It may be helpful to draw the other half of the semicircle as a dotted line.
- Another way to visualize the distance around the curved part is to overlap a circle with a diameter of 6 meters with a semicircle with a straight side of 6 meters.
- **Common Error:** Students may find half the circumference and forget to add the distance across the diameter.

Closure
- **Exit Ticket:** A peso has a diameter of 21 millimeters.
  a. What is the radius of the peso? 10.5 mm
  b. What is the circumference of the peso? about 66 mm

Extra Example 3

In Example 3, the circumference of the roll of tape decreases 7.5 inches. Estimate the diameter of the roll after the decrease. about 8 in.

 On Your Own
  6. about 9 in.

Extra Example 4

Find the perimeter of a semicircular region with a radius of 5 yards. about 25.7 yd

 On Your Own
  7. about 5.14 ft
  8. about 18 cm
  9. about 77.1 in.

1. The radius is one-half the diameter.

2. the distance from the center to any point on the circle; This phrase describes the radius of a circle, whereas the other phrases describe the circumference of a circle.

Practice and Problem Solving

3. 2.5 cm

4. 14 mm

5. $1\frac{3}{4}$ in.

6. 12 cm

7. 4 in.

8. 1.6 ft

9. about 31.4 in.

10. about 44 in.

11. about 56.52 in.

12. *Sample answer:* A lawn game has two circular targets with 28-inch diameters. You lost one. You want to use a length of wire to make a replacement.

$C = \pi d \approx \dfrac{22}{7} \cdot 28 = 88$

You need a piece of wire 88 inches long.

Assignment Guide and Homework Check

Level	Day 1 Activity Assignment	Day 2 Lesson Assignment	Homework Check
Basic	9–11, 26–29	1, 2, 3–7 odd, 13–21 odd	5, 7, 13, 15
Average	9–11, 26–29	1, 2, 3–7 odd, 13–19 odd, 20–24 even	5, 7, 15, 20
Advanced	9–11, 26–29	1, 2, 4–8 even, 12–24 even	16, 18, 20, 24
Accelerated	1, 2, 4–8 even, 9–11, 12–24 even, 26–29		16, 18, 20, 24

Common Errors

- **Exercises 3–8** Students may confuse what they are finding and double the diameter or halve the radius. Remind them that the radius is half the diameter. Encourage students to draw a line representing the radius or diameter for each problem so that they have a visual reference.

- **Exercises 9–11** Students may use the wrong formula for circumference when given a radius or diameter. Remind them of the different equations. If students are struggling with the two equations, tell them to use only one equation and to convert the dimension given to the one in the chosen formula.

- **Exercises 9–11** Students may use $\dfrac{22}{7}$ when it would be easier to use 3.14 and get frustrated. Remind them to use $\dfrac{22}{7}$ when the radius or diameter is a multiple of 7.

8.1 Record and Practice Journal

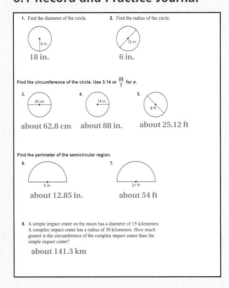

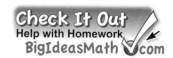
 Vocabulary and Concept Check

1. **VOCABULARY** What is the relationship between the radius and the diameter of a circle?

2. **WHICH ONE DOESN'T BELONG?** Which phrase does *not* belong with the other three? Explain your reasoning.

the distance around a circle	π times twice the radius

π times the diameter	the distance from the center to any point on the circle

 Practice and Problem Solving

Find the radius of the button.

① 3.
5 cm

4.
28 mm

5.
$3\frac{1}{2}$ in.

Find the diameter of the object.

6.
6 cm

7.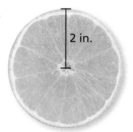
2 in.

8.
0.8 ft

Find the circumference of the pizza. Use 3.14 or $\frac{22}{7}$ for π.

② 9.
10 in.

10.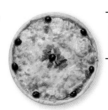
7 in.

11.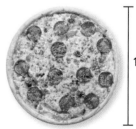
18 in.

12. **CHOOSE TOOLS** Choose a real-life circular object. Explain why you might need to know its circumference. Then find the circumference.

13. **SINKHOLE** A circular sinkhole has a circumference of 75.36 meters. A week later, it has a circumference of 150.42 meters.

   a. Estimate the diameter of the sinkhole each week.

   b. How many times greater is the diameter of the sinkhole now compared to the previous week?

14. **REASONING** Consider the circles A, B, C, and D.

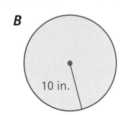
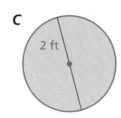
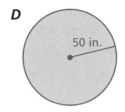

A 8 ft   B 10 in.   C 2 ft   D 50 in.

   a. Without calculating, which circle has the greatest circumference?

   b. Without calculating, which circle has the least circumference?

Find the perimeter of the window.

④ 15.

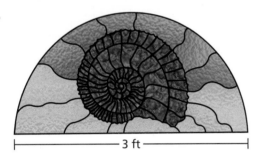

3 ft

16.

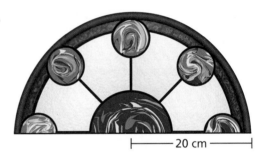

20 cm

Find the circumferences of both circles.

17.

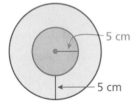

5 cm
5 cm

18.

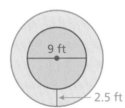

9 ft
2.5 ft

19.

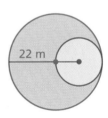

22 m

20. **STRUCTURE** Because the ratio $\dfrac{\text{circumference}}{\text{diameter}}$ is the same for every circle, is the ratio $\dfrac{\text{circumference}}{\text{radius}}$ the same for every circle? Explain.

21. **WIRE** A wire is bent to form four semicircles. How long is the wire?

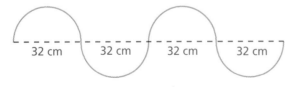

32 cm   32 cm   32 cm   32 cm

22. **CRITICAL THINKING** Explain how to draw a circle with a circumference of π^2 inches. Then draw the circle.

Common Errors

- **Exercise 14** Students may try to compare the radii or diameters without converting to the same units. Remind them that they need to have all the same units before comparing.
- **Exercises 15 and 16** Students may find the circumference and forget to divide it in half. Remind them that because it is not a whole circle, they must divide it in half.
- **Exercises 15 and 16** Students may forget to add the diameter onto the perimeter after they have found the circumference part. Remind them that the perimeter includes all the sides, not just the circular part.
- **Exercises 17–19** Students may use the incorrect radius or diameter for the larger or smaller circle. Remind them that they will need to figure out the radius or diameter of the other circle.

 Practice and Problem Solving

13. **a.** about 25 m; about 50 m

    b. about 2 times greater

14. **a.** D

    b. B

15. about 7.71 ft

16. about 102.8 cm

17. about 31.4 cm;
    about 62.8 cm

18. about 28.26 ft; about 44 ft

19. about 69.08 m;
    about 138.16 m

20. yes; Because

    $$\frac{\text{circumference}}{\text{radius}} = \frac{2\pi r}{r}$$
    $$= \frac{2\pi \cancel{r}}{\cancel{r}}$$
    $$= 2\pi,$$

    the ratio is the same for every circle.

21. about 200.96 cm

22. The circle has a diameter of π inches, so use a diameter of about 3.1 inches.

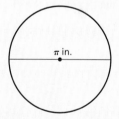

π in.

Differentiated Instruction

Auditory

For students that confuse radius and diameter, emphasize that a **di**ameter **di**vides a circle in half. In an exercise when the radius is needed in a formula and the diameter is given, emphasize that you **di**vide the **di**ameter to find the radius.

Practice and Problem Solving

23. See *Taking Math Deeper.*

24. a. small tire: about 127 rotations; large tire: about 38 rotations

    b. *Sample answer:* A bicycle with large wheels would allow you to travel farther with each rotation of the pedal.

25. a. about 254.34 mm; First find the length of the minute hand. Then find $\frac{3}{4}$ of the circumference of a circle whose radius is the length of the minute hand.

    b. See Additional Answers.

Fair Game Review

26. 22 ft   **27.** 20 m

28. 65 in.   **29.** D

Mini-Assessment

Find the circumference of the circle. Use 3.14 or $\frac{22}{7}$ for π.

1.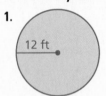
12 ft

about 75.36 ft

2.
28 in.

about 88 in.

3.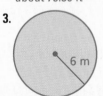
6 m

about 37.68 m

4.
4 cm

about 12.56 cm

5. Find the perimeter of the window.
about 20.56 ft

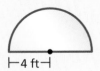
├─4 ft─┤

Taking Math Deeper

Exercise 23

Speed records for flights around the world are kept by Fédération Aéronautique Internationale. See *fai.org*. There are many different categories, depending on the type of plane and other factors.

 a. Find the circumference of the Tropic of Cancer.

$$C = 2\pi r$$
$$= 11,708\,\pi$$
$$\approx 36,763 \text{ km}$$

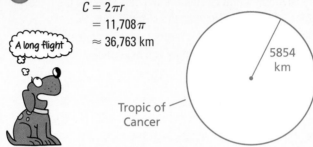

A long flight

5854 km

Tropic of Cancer

 Here is a route for a recent attempt to break the speed record for a flight around the world.

 b. One record was set by Claude Delorme of France. He flew around the world at an average speed of 1231 kilometers per hour (westbound).

At this rate, it took Claude about 30 hours to fly around the world. If your students estimate a time that is less than this, you might point out that they should try for the next world record!

Project

Draw a picture of the path you would take to fly around the world in an attempt to set a new world record.

Reteaching and Enrichment Strategies

If students need help. . .	If students got it. . .

Resources by Chapter
  • Practice A and Practice B
  • Puzzle Time
Record and Practice Journal Practice
Differentiating the Lesson
Lesson Tutorials
Skills Review Handbook
	Resources by Chapter
	  • Enrichment and Extension
	  • Technology Connection
	Start the next section

23. AROUND THE WORLD "Lines" of latitude on Earth are actually circles. The Tropic of Cancer is the northernmost line of latitude at which the Sun appears directly overhead at noon. The Tropic of Cancer has a radius of 5854 kilometers.

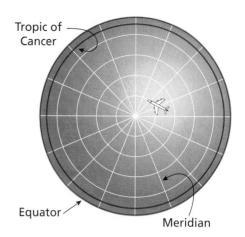

To qualify for an around-the-world speed record, a pilot must cover a distance no less than the circumference of the Tropic of Cancer, cross all meridians, and land on the same airfield where he started.

a. What is the minimum distance that a pilot must fly to qualify for an around-the-world speed record?

b. **RESEARCH** Estimate the time it would take for a pilot to qualify for the speed record.

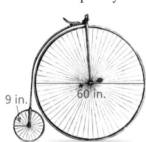

24. PROBLEM SOLVING Bicycles in the late 1800s looked very different than they do today.

a. How many rotations does each tire make after traveling 600 feet? Round your answers to the nearest whole number.

b. Would you rather ride a bicycle made with two large wheels or two small wheels? Explain.

25. **Logic** The length of the minute hand is 150% of the length of the hour hand.

a. What distance will the tip of the minute hand move in 45 minutes? Explain how you found your answer.

b. In 1 hour, how much farther does the tip of the minute hand move than the tip of the hour hand? Explain how you found your answer.

 Fair Game Review What you learned in previous grades & lessons

Find the perimeter of the polygon. (*Skills Review Handbook*)

26.

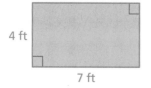

4 ft

7 ft

27.

6 m   5 m

9 m

28.

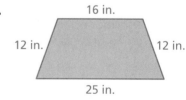

16 in.

12 in.   12 in.

25 in.

29. MULTIPLE CHOICE What is the median of the data set? (*Skills Review Handbook*)

12, 25, 16, 9, 5, 22, 27, 20

Ⓐ 7   Ⓑ 16   Ⓒ 17   Ⓓ 18

8.2 Perimeters of Composite Figures

Essential Question How can you find the perimeter of a composite figure?

1 ACTIVITY: Finding a Pattern

Work with a partner. Describe the pattern of the perimeters. Use your pattern to find the perimeter of the tenth figure in the sequence. (Each small square has a perimeter of 4.)

a.

b.

c.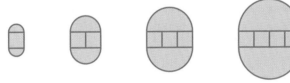

2 ACTIVITY: Combining Figures

Work with a partner.

a. A rancher is constructing a rectangular corral and a trapezoidal corral, as shown. How much fencing does the rancher need to construct both corrals?

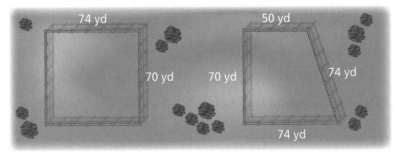

b. Another rancher is constructing one corral by combining the two corrals above, as shown. Does this rancher need more or less fencing? Explain your reasoning.

c. How can the rancher in part (b) combine the two corrals to use even less fencing?

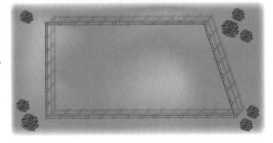

Geometry

In this lesson, you will
- find perimeters of composite figures.

Laurie's Notes

Introduction

Applying Mathematical Practices

- **Construct Viable Arguments and Critique the Reasoning of Others:** In the first activity, students describe and generalize a pattern. The ability to offer evidence for their conjectures is good practice for students.

Motivate

- Draw the following "equations." Ask students how they would find the perimeter of the last figure knowing the dimensions of the figures on the left side of the equation.

- These puzzles should help focus students' thinking on how they will find the perimeter of a composite figure, without defining composite figures!

Activity Notes

Activity 1

- Students should describe the pattern of the perimeter in words. For instance, the figures in part (a) have perimeters which are *increasing by 2* each time.
- **Common Error:** Remind students that perimeter is the distance *around* the figure, and that interior segments are not included.
- **Representation:** For part (c), if students leave their answers in terms of π, they are more likely to see the pattern.
- ? "How is the perimeter changing in part (a)? Can you explain why?" increasing by 2; Explanations will vary.
- ? "How is the perimeter changing in part (b)? Can you explain why?" increasing by 4; Explanations will vary.
- ? "How is the perimeter changing in part (c)? Can you explain why?" increasing by π; Explanations will vary.

Activity 2

- It may not be obvious that the second rancher uses less fencing.
- Only two sides of the rectangular corral are labeled. Be sure that students have computed the perimeter correctly.
- **Model with Mathematics:** For some students it would be helpful to have cut-out pieces of the two shapes that they can manipulate. This helps them to see the perimeter that is lost when the pieces are moved together.
- **Construct Viable Arguments and Critique the Reasoning of Others:** Listen for student reasoning as they describe their solutions for part (c).
- **Extension:** Ask the students about the area enclosed by ranchers. Are they the same or different? Explain. The area is the same for both ranchers. The area does not change when the two polygons are combined.

What Your Students Will Learn

- Estimate perimeters of figures using grid paper.
- Find perimeters of composite figures using formulas.

Previous Learning

Students should know how to find the perimeter and circumference for common shapes.

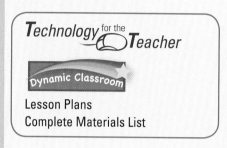

Lesson Plans
Complete Materials List

8.2 Record and Practice Journal

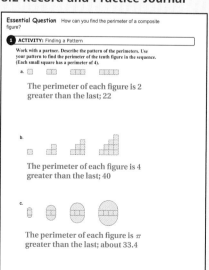

Essential Question How can you find the perimeter of a composite figure?

1 ACTIVITY: Finding a Pattern

Work with a partner. Describe the pattern of the perimeters. Use your pattern to find the perimeter of the tenth figure in the sequence. (Each small square has a perimeter of 4).

a.

The perimeter of each figure is 2 greater than the last; 22

b.

The perimeter of each figure is 4 greater than the last; 40

c.

The perimeter of each figure is π greater than the last; about 33.4

English Language Learners

Visual

Have students find rectangular, trapezoidal, triangular, and circular regions in the school and on the school grounds. Then ask students to calculate the perimeter of these regions.

8.2 Record and Practice Journal

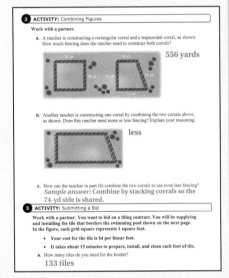

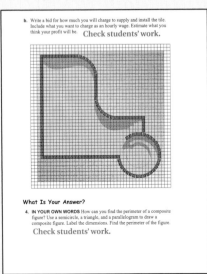

Laurie's Notes

Activity 3

- **Model with Mathematics:** Share with your students that builders and contractors submit bids for work that they want to do. If more than one bid is received, the consumer selects the builder or contractor based upon a number of factors, one of which is the cost that is quoted.
- In this problem, assume that each pair of students is bidding on the job. They could even have a name for their two-person company. Their bid sheet (work done) should be neat and organized and easily understood by the pool's owner—you!
- Explain the term *$4 per linear foot* so that all students understand. In construction, when the width of material (in this case, the tile) is predetermined, the cost is given in terms of the length, not in terms of area (square feet).
- Students should count the number of tiles surrounding the pool and use the scale given on the diagram to determine how much tile is needed.
- Make sure students include a labor charge based on the information given. Students will set their own hourly wage. Is it realistic?
- Discuss general results with the whole class. Compare the quotes, separating out the material and labor.
- **Extension:** List the hourly wages and the number of hours of labor that each company charges. Use this data and find the mean, median, and range of the hourly wages and the total labor charge.

What Is Your Answer?

- **Neighbor Check:** Have students work independently and then have their neighbors check their work. Have students discuss any discrepancies.

Closure

- Find the perimeter of the arrow. 60 cm

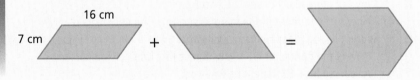

ACTIVITY: Submitting a Bid

Work with a partner. You want to bid on a tiling contract. You will be supplying and installing the brown tile that borders the swimming pool. In the figure, each grid square represents 1 square foot.

- Your cost for the tile is $4 per linear foot.
- It takes about 15 minutes to prepare, install, and clean each foot of tile.

a. How many brown tiles do you need for the border?

b. Write a bid for how much you will charge to supply and install the tile. Include what you want to charge as an hourly wage. Estimate what you think your profit will be.

Math Practice

Communicate Precisely

What do you need to include to create an accurate bid? Explain.

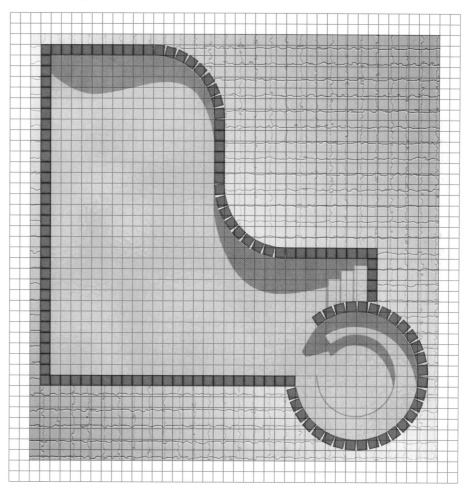

What Is Your Answer?

4. IN YOUR OWN WORDS How can you find the perimeter of a composite figure? Use a semicircle, a triangle, and a parallelogram to draw a composite figure. Label the dimensions. Find the perimeter of the figure.

Practice Use what you learned about perimeters of composite figures to complete Exercises 3–5 on page 328.

Key Vocabulary
composite figure,
p. 326

A **composite figure** is made up of triangles, squares, rectangles, semicircles, and other two-dimensional figures. Here are two examples.

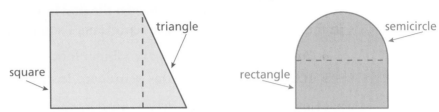

To find the perimeter of a composite figure, find the distance around the figure.

EXAMPLE 1 **Estimating a Perimeter Using Grid Paper**

Estimate the perimeter of the arrow.

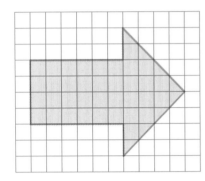

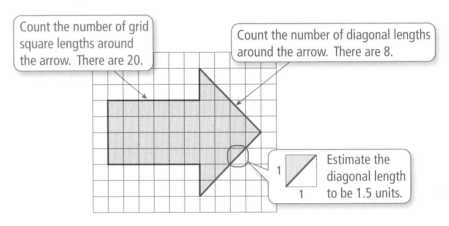

Count the number of grid square lengths around the arrow. There are 20.

Count the number of diagonal lengths around the arrow. There are 8.

Estimate the diagonal length to be 1.5 units.

Length of 20 grid square lengths: $20 \times 1 = 20$ units

Length of 8 diagonal lengths: $8 \times 1.5 = 12$ units

So, the perimeter is about $20 + 12 = 32$ units.

On Your Own

Now You're Ready
Exercises 3–8

Estimate the perimeter of the figure.

1.

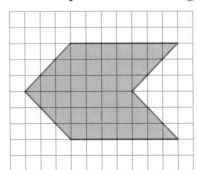

2.

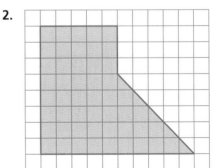

🔊 Multi-Language Glossary at BigIdeasMath✓com

Laurie's Notes

Introduction

Connect

- **Yesterday:** Students used a variety of problem-solving skills to find the perimeter of several composite figures.
- **Today:** Students will use formulas to find the perimeter of composite figures.

Motivate

- If you have a set of tangrams, use them to share a puzzle with students. Place the 7 pieces on the overhead and ask a volunteer to rearrange the pieces into a square. The solution is shown. Then rearrange the pieces to form the bird shown. Ask how the perimeter changes from the square to the bird.

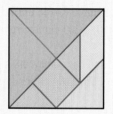 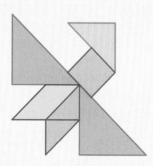

- The goal is not to find the perimeter of the bird, only to recognize that as more segments are exposed (on the perimeter), the perimeter will increase.
- The figures that students work with today are composed of common geometric shapes. Students should be familiar with how to find the perimeter of each.
- Introduce the vocabulary word *composite figure*.

Lesson Notes

Example 1

- The arrow is a shape that students discussed at the beginning of yesterday's lesson.
- Students are asked to estimate the perimeter, because the slanted lengths are irrational numbers.
- **Common Error:** Students may think the diagonal of the square is 1 unit because the side lengths are 1 unit. Have them think about the *shortcut* of walking across the diagonal of a square versus walking the two side lengths.

On Your Own

- Question 1 is very similar to the first example.
- Ask a volunteer to share his or her thinking and solution to the problem.

Technology for the *Teacher*

Dynamic Classroom

Lesson Tutorials
Lesson Plans
Answer Presentation Tool

Extra Example 1

Estimate the perimeter of the arrow.

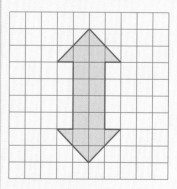

about 24 units

On Your Own

1. about 32 units
2. about 33.5 units

Laurie's Notes

Extra Example 2

The figure is made of a semicircle and a square. Find the perimeter.

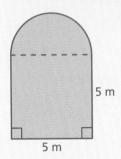

5 m

5 m

about 22.85 m

Extra Example 3

Find the perimeter of the running track in Example 3 when the radius of the semicircle is 20 yards and the length of the rectangle is 50 yards. about 225.6 yd

On Your Own

   3. about 74.82 cm

   4. about 41.12 m

Differentiated Instruction

Auditory

Review the two formulas for circumference, $C = \pi d$ and $C = 2\pi r$. On the board or overhead, derive the formulas for diameter, $d = \dfrac{C}{\pi}$, and radius, $r = \dfrac{C}{2\pi}$. Have students copy the four formulas in their notebooks to use as a quick reference when doing word problems. Remind students that the circumference of a semicircle is half of a full circle.

Example 2

- Draw the diagram for the problem.
- **?** "What is the diameter of the semicircle?" 10 ft
- Explain to students that although the third side of the triangle is shown (which is also the diameter of the circle), it is not part of the perimeter.
- **?** **Construct Viable Arguments and Critique the Reasoning of Others** and **Attend to Precision:** "Could $\dfrac{22}{7}$ be used for π?" yes "Why do you think 3.14 was used?" The diameter is 10, which is not a multiple of 7.

Example 3

- Draw the diagram for the problem.
- **?** "What are the dimensions of the rectangle?" 100 m by 64 m
- **?** "What is the diameter of the semicircles?" 64 m
- Make sure students understand that they have to add both straightaways of the running track. This is where the 100 + 100 comes from in the answer statement.
- Point out to students that in Examples 2 and 3 the answers are stated to be *about* a certain number. The answers are approximations because 3.14 is only an approximation for π.

On Your Own

- Students may think there is not enough information to solve Question 4. The quadrilateral is a square, so the diameter of the semicircles is 8 meters.
- **Neighbor Check:** Have students work independently and then have their neighbors check their work. Have students discuss any discrepancies.
- When students have finished, ask volunteers to show their work at the board.

Closure

- Find the perimeter of the room shown. 64 ft

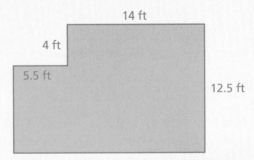

14 ft

4 ft

5.5 ft

12.5 ft

EXAMPLE **2** **Finding a Perimeter**

The figure is made up of a semicircle and a triangle. Find the perimeter.

The distance around the triangular part of the figure is $6 + 8 = 14$ feet.

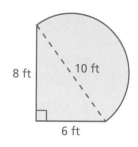

8 ft   10 ft

6 ft

The distance around the semicircle is one-half the circumference of a circle with a diameter of 10 feet.

$$\frac{C}{2} = \frac{\pi d}{2}$$ Divide the circumference by 2.

$$\approx \frac{3.14 \cdot 10}{2}$$ Substitute 3.14 for π and 10 for d.

$$= 15.7$$ Simplify.

So, the perimeter is about $14 + 15.7 = 29.7$ feet.

EXAMPLE **3** **Finding a Perimeter**

The running track is made up of a rectangle and two semicircles. Find the perimeter.

The semicircular ends of the track form a circle with a radius of 32 meters. Find its circumference.

$$C = 2\pi r$$ Write formula for circumference.

$$\approx 2 \cdot 3.14 \cdot 32$$ Substitute 3.14 for π and 32 for r.

$$= 200.96$$ Multiply.

So, the perimeter is about $100 + 100 + 200.96 = 400.96$ meters.

32 m

100 m

On Your Own

Now You're Ready
Exercises 9–11

3. The figure is made up of a semicircle and a triangle. Find the perimeter.

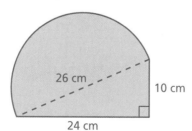

26 cm   10 cm

24 cm

4. The figure is made up of a square and two semicircles. Find the perimeter.

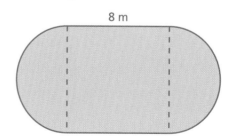

8 m

Vocabulary and Concept Check

1. **REASONING** Is the perimeter of the composite figure equal to the sum of the perimeters of the individual figures? Explain.

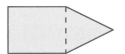

2. **OPEN-ENDED** Draw a composite figure formed by a parallelogram and a trapezoid.

Practice and Problem Solving

Estimate the perimeter of the figure.

3.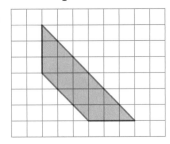

4.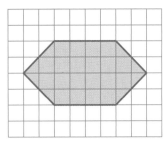

5.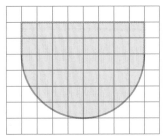

6.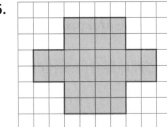

7.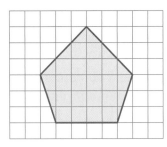

8.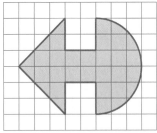

Find the perimeter of the figure.

9.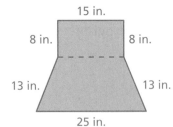

5 m
5 m
11 m
7 m

10.

15 in.
8 in.   8 in.
13 in.   13 in.
25 in.

11.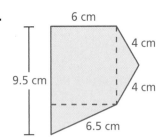

6 cm
4 cm
9.5 cm
4 cm
6.5 cm

12. **ERROR ANALYSIS** Describe and correct the error in finding the perimeter of the figure.

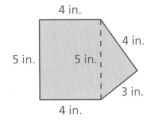

4 in.
4 in.
5 in.   5 in.
3 in.
4 in.

✗ Perimeter = 4 + 3 + 4 + 5 + 4 + 5
= 25 in.

Assignment Guide and Homework Check

Level	Day 1 Activity Assignment	Day 2 Lesson Assignment	Homework Check
Basic	3–5, 20–24	1, 2, 6, 7, 9, 10, 12, 13, 15, 16	6, 9, 12, 13
Average	3–5, 20–24	1, 2, 6, 7, 9, 10, 12, 13, 15–17	6, 9, 12, 13
Advanced	3–5, 20–24	1, 2, 8–14 even, 17–19	8, 10, 12, 14
Accelerated	1–5, 8–14 even, 17–24		8, 10, 12, 14

Common Errors

- **Exercises 1, 9–11, 13–15** Students may include the dotted lines in their calculation of the perimeter. Remind them that the dotted lines are for reference and sometimes give information to find another length. Only the outside lengths are counted.
- **Exercises 3–8** Students may count the sides of the squares inside the figure. Remind them that they are finding the perimeter, so they only need to count the outside lengths.
- **Exercises 5 and 8** Students may have difficulty estimating the curved portions of the figure. Give students tracing paper and tell them to trace the line as straight instead of curved and compare it with the length of the side of a square to help them estimate the length.

8.2 Record and Practice Journal

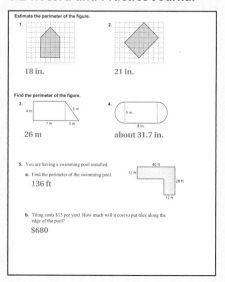

Vocabulary and Concept Check

1. no; The perimeter of the composite figure does not include the measure of the shared side.

2. *Sample answer:*

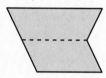

Practice and Problem Solving

3. 19.5 units

4. 20 units

5. 25.5 units

6. 28 units

7. 19 units

8. 30 units

9. 56 m

10. 82 in.

11. 30 cm

12. The length of the rectangle was counted twice.
Perimeter = 4 + 3 + 4 + 5 + 4 = 20 in.

Practice and Problem Solving

13. about 26.85 in.

14. about 50.26 in.

15. about 36.84 ft

16. $16,875

17. See *Taking Math Deeper.*

18. See *Additional Answers.*

19. *Sample answer:* By adding the triangle shown by the dashed line to the L-shaped figure, you *reduce* the perimeter.

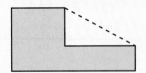

Fair Game Review

20. 19.35   **21.** 279.68

22. 153.86   **23.** 205

24. D

Mini-Assessment

Find the perimeter of the figure.

1.

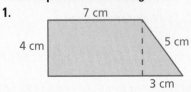

26 cm

2.

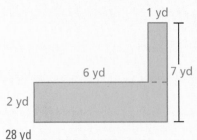

28 yd

3. Find the perimeter of the garden.

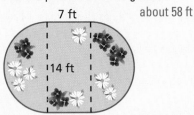

7 ft   about 58 ft

14 ft

T-329

Taking Math Deeper

Exercise 17

Notice that the dimensions of the field and the distance in the unit rate are multiples of 3. So, you can convert feet to yards and use smaller numbers in the calculations.

① Convert the baseball field dimensions to yards and your running rate to yards per second.

Radius: $225 \text{ ft} \times \dfrac{1 \text{ yd}}{3 \text{ ft}} = 75 \text{ yd}$

Sides: $300 \text{ ft} \times \dfrac{1 \text{ yd}}{3 \text{ ft}} = 100 \text{ yd}$

Rate: $\dfrac{9 \text{ ft}}{1 \text{ sec}} \times \dfrac{1 \text{ yd}}{3 \text{ ft}} = \dfrac{3 \text{ yd}}{1 \text{ sec}}$

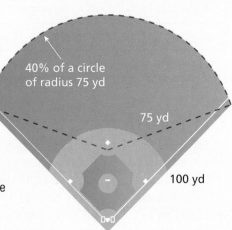

40% of a circle of radius 75 yd

75 yd

100 yd

② Find the distance around the field.

The circumference *C* of a circle with a radius of 75 yards is $C \approx 2 \cdot 3.14 \cdot 75 = 471$ yards.

The distance around the curved part of the field is 40% of this circumference. The other two sides of the field are 100 yards each. So, the perimeter of the field is about $100 + 100 + 0.4(471) = 388.4$ yards.

③ Divide the distance by the rate to find the time.

$$388.4 \text{ yd} \div \dfrac{3 \text{ yd}}{1 \text{ sec}} = 388.4 \text{ yd} \times \dfrac{1 \text{ sec}}{3 \text{ yd}} \approx 129.5 \text{ sec}$$

It takes you about 129.5 seconds, or about 2 minutes and 10 seconds to run around the baseball field.

Play ball!

Project

Research the dimensions of several different baseball fields. Estimate how long it would take you to run around the perimeter of each field.

Reteaching and Enrichment Strategies

If students need help. . .	If students got it. . .
Resources by Chapter • Practice A and Practice B • Puzzle Time Record and Practice Journal Practice Differentiating the Lesson Lesson Tutorials Skills Review Handbook	Resources by Chapter • Enrichment and Extension • Technology Connection Start the next section

Find the perimeter of the figure.

13.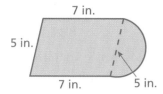
7 in.
5 in.
7 in.
5 in.

14.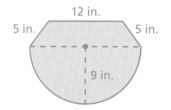
12 in.
5 in.
5 in.
9 in.

15.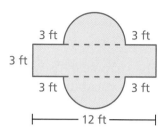
3 ft
3 ft
3 ft
3 ft
3 ft
12 ft

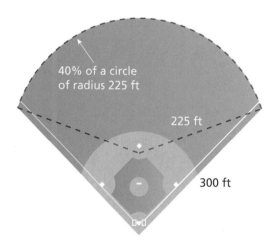
40% of a circle of radius 225 ft
225 ft
300 ft

240 ft 285 ft
450 ft 450 ft
450 ft

16. PASTURE A farmer wants to fence a section of land for a horse pasture. Fencing costs $27 per yard. How much will it cost to fence the pasture?

17. BASEBALL You run around the perimeter of the baseball field at a rate of 9 feet per second. How long does it take you to run around the baseball field?

18. TRACK In Example 3, the running track has six lanes. Explain why the starting points for the six runners are staggered. Draw a diagram as part of your explanation.

19. *Critical Thinking* How can you add a figure to a composite figure without increasing its perimeter? Draw a diagram to support your answer.

Fair Game Review What you learned in previous grades & lessons

Evaluate the expression. (Skills Review Handbook)

20. $2.15(3)^2$ **21.** $4.37(8)^2$ **22.** $3.14(7)^2$ **23.** $8.2(5)^2$

24. MULTIPLE CHOICE Which expression is equivalent to $(5y + 4) - 2(7 - 2y)$? (Section 3.2)

Ⓐ $y - 10$ Ⓑ $9y + 18$ Ⓒ $3y - 10$ Ⓓ $9y - 10$

Check It Out
Graphic Organizer
BigIdeasMath ✓com

You can use a **word magnet** to organize formulas or phrases that are associated with a vocabulary word or term. Here is an example of a word magnet for circle.

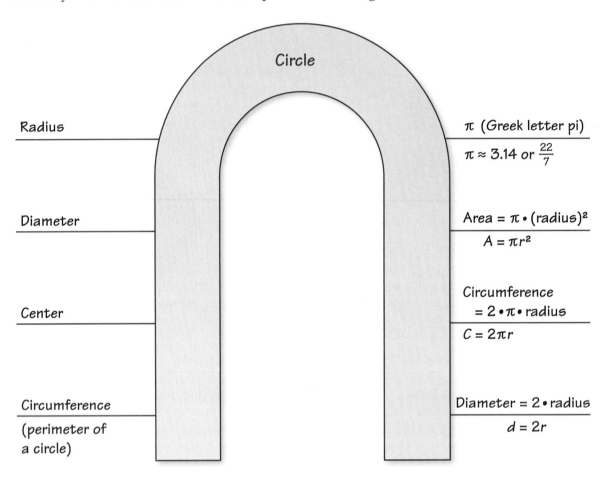

Circle

Radius

Diameter

Center

Circumference
(perimeter of
a circle)

π (Greek letter pi)

$\pi \approx 3.14$ or $\frac{22}{7}$

Area = $\pi \cdot$ (radius)^2

$A = \pi r^2$

Circumference
= $2 \cdot \pi \cdot$ radius

$C = 2\pi r$

Diameter = $2 \cdot$ radius

$d = 2r$

On Your Own

Make word magnets to help you study these topics.

1. semicircle

2. composite figure

3. perimeter

After you complete this chapter, make word magnets for the following topics.

4. area of a circle

5. area of a composite figure

"I'm trying to make a word magnet for happiness, but I can only think of two words."

Sample Answers

1.

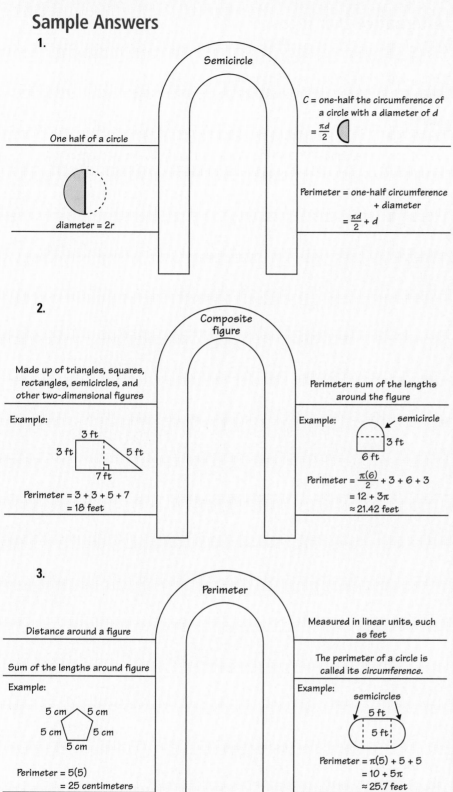

Semicircle

One half of a circle

One half of a circle

diameter = 2r

C = one-half the circumference of a circle with a diameter of d

$= \dfrac{\pi d}{2}$

Perimeter = one-half circumference + diameter

$= \dfrac{\pi d}{2} + d$

2.

Composite figure

Made up of triangles, squares, rectangles, semicircles, and other two-dimensional figures

Example:

3 ft
3 ft   5 ft
7 ft

Perimeter = 3 + 3 + 5 + 7
= 18 feet

Perimeter: sum of the lengths around the figure

Example: semicircle
3 ft
6 ft

Perimeter = $\dfrac{\pi(6)}{2}$ + 3 + 6 + 3
= 12 + 3π
≈ 21.42 feet

3.

Perimeter

Distance around a figure

Sum of the lengths around figure

Example:

5 cm   5 cm
5 cm   5 cm
5 cm

Perimeter = 5(5)
= 25 centimeters

Measured in linear units, such as feet

The perimeter of a circle is called its *circumference*.

Example: semicircles
5 ft
5 ft

Perimeter = π(5) + 5 + 5
= 10 + 5π
≈ 25.7 feet

Technology for the *Teacher*

Editable Graphic Organizer

Answers

1. 18 cm

2. 22 in.

3. 18 units

4. 21.4 units

5. 21 units

6. about 37.68 mm

7. about 4.71 ft

8. about 22 cm

9. 88 in.

10. about 49.12 ft

11. about 7.71 ft

12. about 25.12 mm

13. 60 ft

14. about 15.68 in.

Alternative Quiz Ideas

100% Quiz	**Math Log**
Error Notebook	Notebook Quiz
Group Quiz	Partner Quiz
Homework Quiz	Pass the Paper

Math Log

Ask students to keep a math log for the chapter. Have them include diagrams, definitions, and examples. Everything should be clearly labeled. It might be helpful if they put the information in a chart. Students can add to the log as new topics are introduced.

Reteaching and Enrichment Strategies

If students need help. . .	If students got it. . .
Resources by Chapter • Practice A and Practice B • Puzzle Time Lesson Tutorials *BigIdeasMath.com*	Resources by Chapter • Enrichment and Extension • Technology Connection Game Closet at *BigIdeasMath.com* Start the next section

Check It Out
Progress Check
BigIdeasMath.com

1. The diameter of a circle is 36 centimeters. Find the radius. *(Section 8.1)*

2. The radius of a circle is 11 inches. Find the diameter. *(Section 8.1)*

Estimate the perimeter of the figure. *(Section 8.2)*

3.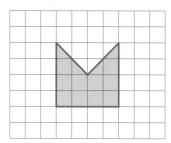

4.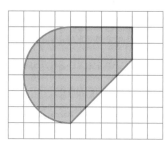

5.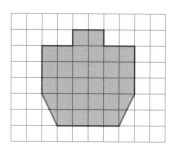

Find the circumference of the circle. Use 3.14 or $\frac{22}{7}$ for π. *(Section 8.1)*

6.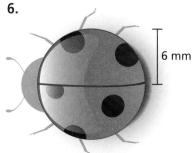
6 mm

7.
1.5 ft

8.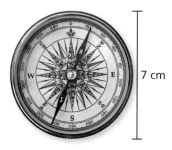
7 cm

Find the perimeter of the figure. *(Section 8.1 and Section 8.2)*

9.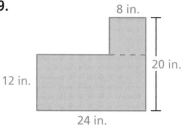
8 in.
20 in.
12 in.
24 in.

10.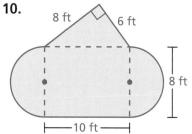
8 ft   6 ft
8 ft
10 ft

11.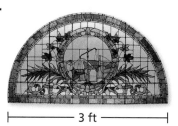
3 ft

12. **BUTTON** What is the circumference of a circular button with a diameter of 8 millimeters? *(Section 8.1)*

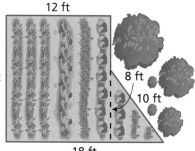
12 ft
14 ft
8 ft
10 ft
18 ft

13. **GARDEN** You want to fence part of a yard to make a vegetable garden. How many feet of fencing do you need to surround the garden? *(Section 8.2)*

14. **BAKING** A baker is using two circular pans. The larger pan has a diameter of 12 inches. The smaller pan has a diameter of 7 inches. How much greater is the circumference of the larger pan than that of the smaller pan? *(Section 8.1)*

Essential Question How can you find the area of a circle?

1 ACTIVITY: Estimating the Area of a Circle

Work with a partner. Each square in the grid is 1 unit by 1 unit.

a. Find the area of the large 10-by-10 square.

b. Copy and complete the table.

Region			
Area (square units)			

c. Use your results to estimate the area of the circle. Explain your reasoning.

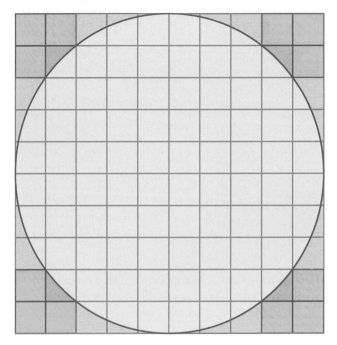

Geometry

In this lesson, you will
- find areas of circles and semicircles.

d. Fill in the blanks. Explain your reasoning.

$$\text{Area of large square} = \boxed{} \cdot 5^2 \text{ square units}$$

$$\text{Area of circle} \approx \boxed{} \cdot 5^2 \text{ square units}$$

e. What dimension of the circle does 5 represent? What can you conclude?

Laurie's Notes

Introduction

Applying Mathematical Practices
- **Model with Mathematics:** Students are able to develop the formula for the area of a circle from working through two different activities.

Motivate
- Sketch this diagram.
- ❓ "The lawn at my house is a square. A rotating sprinkler reaches each side but misses the four corners. What percent of my lawn do you think is getting watered?"
- Students should visually estimate at this stage. They should guess close to 75%. Record their guesses and return to them at the end of the class.

Activity Notes

Activity 1
- **Model with Mathematics:** Using square units to measure the area of a circle is a difficult concept for some students. I have had students ask how the square pieces fit into the curve. This activity helps students visualize how to count the area of a circle and then compute it.
- Students should observe that the circle is inscribed in a large square that has an area of 100 square units.
- As a class, discuss the area of each colored region. The orange square is a unit square. The green triangle is close to half of a 1×2 rectangle. The pink triangle is about half the unit square.
- ❓ "How does knowing the area of these pieces help you approximate the area of the circle?" Ask the question and let students begin their work.
- When groups have finished, have several record their work at the board. They should have Area of circle $= 100 - 4(3) - 8(1) - 4\left(\dfrac{1}{2}\right) = 78$ units^2.
- ❓ "The area of the large square is 100 units^2. Can you write 100 as the product of a number times 5^2?" $100 = 4(5^2)$
- ❓ "Let's do the same with the area of the circle. Can you write 78 as the product of a number times 5^2? Explain." yes; divide 78 by 25; $78 = 3.12(5^2)$
- ❓ "Thinking about the circle, what part of the circle is 5 units long?" radius
- It is very likely that at this point students may guess that 3.12 is supposed to be π and may want to know what they did wrong because they got 3.12 instead of 3.14. It's important to say that they did nothing wrong. The areas of the two triangles are approximations because they aren't even triangles—there is a curved edge.
- The conclusion you are hoping for is that the area of this circle is πr^2.
- ❓ "Do you think this is a formula that works for all circles?" open-ended

What Your Students Will Learn
- Find areas of circles and semicircles using formulas.

Previous Learning
Students should know how to find areas of parallelograms.

Technology for the Teacher
Dynamic Classroom
Lesson Plans
Complete Materials List

8.3 Record and Practice Journal

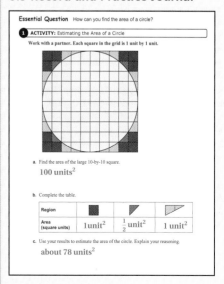

Essential Question How can you find the area of a circle?

1 ACTIVITY: Estimating the Area of a Circle

Work with a partner. Each square in the grid is 1 unit by 1 unit.

a. Find the area of the large 10-by-10 square.
100 units^2

b. Complete the table.

Region			
Area (square units)	1 unit^2	$\frac{1}{2}$ unit^2	1 unit^2

c. Use your results to estimate the area of the circle. Explain your reasoning.
about 78 units^2

Differentiated Instruction

Visual

To help students visualize the relationship between area and radius, have them use a compass and grid paper to draw a circle with a radius of 5 units. Ask them to estimate the area of the circle by counting the squares. They should get an estimate that is slightly greater than $3r^2$. Then have students draw a square on the grid paper that has a vertex at the center of the circle and two perpendicular radii as sides. Students should see that one-quarter of the area is enclosed in the square. Because the area of the square is r^2, the area of the circle is less than $4r^2$. So, the area of the circle is between $3r^2$ and $4r^2$, or more precisely πr^2.

8.3 Record and Practice Journal

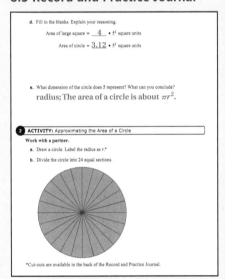

d. Fill in the blanks. Explain your reasoning.

Area of large square = __4__ • 5^2 square units

Area of circle = __3.12__ • 5^2 square units

e. What dimension of the circle does 5 represent? What can you conclude?

radius; The area of a circle is about πr^2.

2 ACTIVITY: Approximating the Area of a Circle

Work with a partner.

a. Draw a circle. Label the radius as *r*.*

b. Divide the circle into 24 equal sections.

*Cut-outs are available in the back of the Record and Practice Journal.

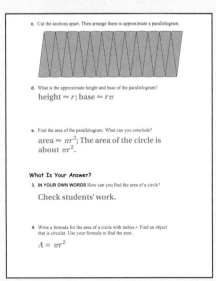

c. Cut the sections apart. Then arrange them to approximate a parallelogram.

d. What is the approximate height and base of the parallelogram?

height ≈ r; base ≈ $r\pi$

e. Find the area of the parallelogram. What can you conclude?

area ≈ πr^2; The area of the circle is about πr^2.

What Is Your Answer?

3. **IN YOUR OWN WORDS** How can you find the area of a circle?

Check students' work.

4. Write a formula for the area of a circle with radius *r*. Find an object that is circular. Use your formula to find the area.

$A = \pi r^2$

Laurie's Notes

Activity 2

- **Preparation:** Trace and cut 2 circles from file folder weight paper. Cut one circle into 4 sectors (pie-shaped pieces) and one circle into 12 sectors.
- In this activity, a different approach is used to develop the area formula.
- To connect the two activities, tell students that you want to think about the circle in the large square.
- ❓ "If you cut the circle into 4 parts and rearranged them, do you think it would make a shape that was familiar?"
- Take out the 4 prepared pieces. Arrange the 4 pieces in a circle and then rearrange them in a fashion similar to the design in the book. The curvature of the circle is still very apparent, so students will not suggest a similarity to any figure they know.
- ❓ "What if the circle were cut into smaller pieces and then rearranged?"
- Take out the 12 prepared pieces. Arrange the 12 pieces in a circle and then rearrange them in a fashion similar to the design in the book. The curvature of the circle is much less apparent.
- **Look for and Express Regularity in Repeated Reasoning** and **Big Idea:** The 24 pieces approximate a parallelogram. The curvature of the circle is not obvious. The area of a parallelogram is base × height.

  The base is half the circumference. (Pieces are alternated so that $\frac{1}{2}$ the circumference is on the top and $\frac{1}{2}$ is on the bottom.) The height is the radius of the original circle.

  Area = base × height
  $$= \frac{1}{2}C \cdot r$$
  $$= \frac{1}{2}(2\pi r)r$$
  $$= \pi r^2$$

What Is Your Answer?

- **Neighbor Check:** Have students work independently and then have their neighbors check their work. Have students discuss any discrepancies.

Closure

- What percent of the lawn is getting watered? 78%
- **Connection:** This is an example of a geometric probability question where the area of the circle is being compared to the area of the square. In Activity 1, students found the area of the square and the circle.

ACTIVITY: Approximating the Area of a Circle

Work with a partner.

a. Draw a circle. Label the radius as *r*.

b. Divide the circle into 24 equal sections.

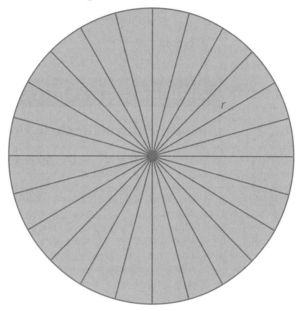

c. Cut the sections apart. Then arrange them to approximate a parallelogram.

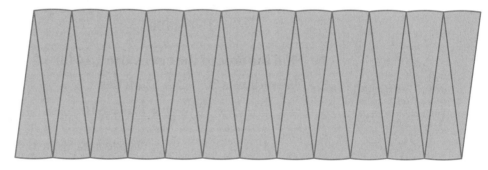

Math Practice

Interpret a Solution

What does the area of the parallelogram represent? Explain.

d. What is the approximate height and base of the parallelogram?

e. Find the area of the parallelogram. What can you conclude?

What Is Your Answer?

3. **IN YOUR OWN WORDS** How can you find the area of a circle?

4. Write a formula for the area of a circle with radius *r*. Find an object that is circular. Use your formula to find the area.

Practice

Use what you learned about areas of circles to complete Exercises 3–5 on page 336.

 **Key Idea**

Area of a Circle

Words The area A of a circle is the product of π and the square of the radius.

Algebra $A = \pi r^2$

EXAMPLE **1**   **Finding Areas of Circles**

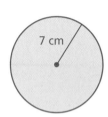

7 cm

a. Find the area of the circle. Use $\dfrac{22}{7}$ for π.

Estimate $3 \times 7^2 \approx 3 \times 50 = 150$

$$A = \pi r^2 \qquad \text{Write formula for area.}$$

$$\approx \frac{22}{7} \cdot 7^2 \qquad \text{Substitute } \frac{22}{7} \text{ for } \pi \text{ and 7 for } r.$$

$$= \frac{22}{\cancel{7}_1} \cdot \cancel{49}^{\,7} \qquad \text{Evaluate } 7^2. \text{ Divide out the common factor.}$$

$$= 154 \qquad \text{Multiply.}$$

⋮⋅ The area is about 154 square centimeters.

Reasonable? $154 \approx 150$ ✓

b. Find the area of the circle. Use 3.14 for π.

The radius is $26 \div 2 = 13$ inches.

Estimate $3 \times 13^2 \approx 3 \times 170 = 510$

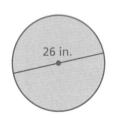

26 in.

$$A = \pi r^2 \qquad \text{Write formula for area.}$$

$$\approx 3.14 \cdot 13^2 \qquad \text{Substitute 3.14 for } \pi \text{ and 13 for } r.$$

$$= 3.14 \cdot 169 \qquad \text{Evaluate } 13^2.$$

$$= 530.66 \qquad \text{Multiply.}$$

⋮⋅ The area is about 530.66 square inches.

Reasonable? $530.66 \approx 510$ ✓

● **On Your Own**

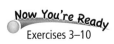
Now You're Ready
Exercises 3–10

1. Find the area of a circle with a radius of 6 feet. Use 3.14 for π.

2. Find the area of a circle with a diameter of 28 meters. Use $\dfrac{22}{7}$ for π.

Laurie's Notes

Introduction

Connect

- **Yesterday:** Students gained an intuitive understanding about how to find the area of a circle.
- **Today:** Students will use the formula for area to solve real-life problems.

Motivate

- **?** "At a pizza restaurant, you have a choice of ordering a 16-inch pizza or a 12-inch pizza for dinner. What measurement is being used to describe the pizza?" the diameter
- **?** "Both pizzas are cut into the same number of pieces. Do both pizzas give you the same amount of pizza?" No, the size of the slices is different.
- **?** "How would you figure out how many times more pizza you get with the 16-inch pizza than with the 12-inch pizza?" Listen for students to mention the idea of comparing the areas of the pizzas.
- Mention to students that after today's lesson, they will be able to answer this question. You may want to return to this question at the end of class.

Lesson Notes

Key Idea

- Discuss the formula, written in words and algebraically.
- **Common Error:** Students square the product of π and r, instead of just the radius.

Example 1

- Work through each example. Remind students that pi can be rounded to 3 when estimating.
- **Common Error:** Make sure students are reading the figures correctly. A line segment from the center to the outside edge indicates a radius. A line segment across the circle through the center is a diameter.
- **?** "What would a reasonable answer be?" Answers will vary, but should be similar to what is shown in the text.
- **?** "In part (a), what estimate should be used for π? Why?" $\frac{22}{7}$; The radius is a multiple of 7.
- Remind students to label answers with the appropriate units.
- **Teaching Tip:** Many students view a diagram differently than a written problem: *Find the area of a circle with a radius of 7 centimeters.* The visual learner is aided by a simple diagram.
- **Common Error:** For part (b), students may forget to divide the diameter by 2 to find the radius before substituting in the area formula.
- If students are not using calculators, ask them to give a reasonable estimate for 3.14×169 before they multiply.

On Your Own

- **Neighbor Check:** Have students work independently and then have their neighbors check their work. Have students discuss any discrepancies.

Goal Today's lesson is finding the area of a circle and a semicircle.

Lesson Tutorials
Lesson Plans
Answer Presentation Tool

Extra Example 1

a. Find the area of a circle with a radius of 21 feet. Use $\frac{22}{7}$ for π. about 1386 ft^2

b. Find the area of a circle with a diameter of 16 meters. Use 3.14 for π. about 200.96 m^2

On Your Own

1. about 113.04 ft^2

2. about 616 m^2

Extra Example 2

The 1893 World's Fair in Chicago boasted the first ever Ferris wheel with a diameter of 250 feet. How far did a person travel in one revolution of the wheel? about 785 ft

 On Your Own

   3. the diameter of the tire

Extra Example 3

Find the area of a semicircle with a radius of 9 inches. about 127.17 in.^{2}

 On Your Own

   4. about 25.12 m^2

   5. about 9.8125 yd^2

   6. about 189.97 cm^2

English Language Learners

Vocabulary

Discuss with students the word *approximation*. Give and ask for examples of times they use approximations in their daily life. (e.g., "I'll meet you in ten minutes.") Discuss when using an approximation for an answer is as useful as an exact answer. Discuss the usefulness of approximations for π, 3.14 and $\frac{22}{7}$.

Laurie's Notes

Example 2

- This monster truck example will grab students' attention!
- **Model with Mathematics:** A circle is a common shape. In this question, students connect vocabulary of a circle to a real-life application of a circle.

On Your Own

- Review this together as a class.

Example 3

- Ask if anyone knows what an orchestra pit is. An orchestra pit is a lowered area in front of the stage where musicians perform.
- **?** "How do you find the area of a semicircle?" Find $\frac{1}{2}$ the area of the circle.
- **?** "What is the radius of the semicircle?" 15 ft
- Work through the problem.
- **Extension:** If time permits and students are using the area formula correctly, try the following problems.
  - **?** "If you make a mistake and use the diameter instead of the radius in the area formula, is the area doubled? Explain." No; It is quadrupled.
  - **?** "Find the area of the doughnut region." $8\pi \approx 25.12$ in.^{2}

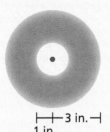

├─┼─3 in.─┤
1 in.

On Your Own

- Check to see that students recognize the difference in how the dimensions are labeled. The radius is marked with a line segment from the center to the outer edge. The diameter is marked with a line segment from one edge through the center to the opposite edge.

Closure

- Find the areas of a 16-inch pizza and a 12-inch pizza. How many times larger is the 16-inch pizza than the 12-inch pizza? about 1.8 times
- Students will find this easier to calculate if the area is kept in terms of π.
- This situation can also be thought of as "how many times more toppings are needed for the larger pizza?"

EXAMPLE (2) **Describing a Distance**

You want to find the distance the monster truck travels when the tires make one 360-degree rotation. Which best describes this distance?

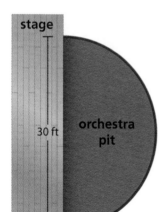

  A   the radius of the tire   **B**   the diameter of the tire

  C   the circumference of the tire   **D**   the area of the tire

The distance the truck travels after one rotation is the same as the distance *around* the tire. So, the circumference of the tire best describes the distance in one rotation.

   The correct answer is **C**.

● **On Your Own**

3. You want to find the height of one of the tires. Which measurement would best describe the height?

EXAMPLE (3) **Finding the Area of a Semicircle**

Find the area of the semicircular orchestra pit.

The area of the orchestra pit is one-half the area of a circle with a diameter of 30 feet.

The radius of the circle is 30 ÷ 2 = 15 feet.

$$\frac{A}{2} = \frac{\pi r^2}{2}$$ Divide the area by 2.

$$\approx \frac{3.14 \cdot 15^2}{2}$$ Substitute 3.14 for π and 15 for *r*.

$$= \frac{3.14 \cdot 225}{2}$$ Evaluate 15^2.

$$= 353.25$$ Simplify.

   So, the area of the orchestra pit is about 353.25 square feet.

● **On Your Own**

Now You're Ready
Exercises 13–15

Find the area of the semicircle.

4.
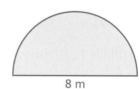
8 m

5. 5 yd
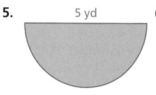

6.
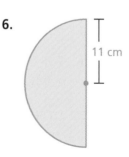
11 cm

Check It Out
Help with Homework
BigIdeasMath.com

 Vocabulary and Concept Check

1. **VOCABULARY** Explain how to find the area of a circle given its diameter.

2. **DIFFERENT WORDS, SAME QUESTION** Which is different? Find "both" answers.

What is the area of a circle with a diameter of 1 m?	What is the area of a circle with a diameter of 100 cm?
What is the area of a circle with a radius of 100 cm?	What is the area of a circle with a radius of 500 mm?

 Practice and Problem Solving

Find the area of the circle. Use 3.14 or $\frac{22}{7}$ for π.

3.

9 mm

4.

14 cm

5.

10 in.

6.

3 in.

7.

2 cm

8.
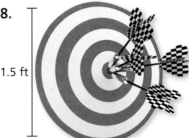
1.5 ft

9. Find the area of a circle with a diameter of 56 millimeters.

10. Find the area of a circle with a radius of 5 feet.

11. **TORTILLA** The diameter of a flour tortilla is 12 inches. What is the area?

12. **LIGHTHOUSE** The Hillsboro Inlet Lighthouse lights up how much more area than the Jupiter Inlet Lighthouse?

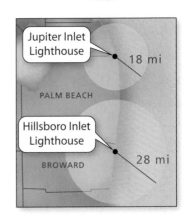

Jupiter Inlet Lighthouse — 18 mi

PALM BEACH

Hillsboro Inlet Lighthouse — 28 mi

BROWARD

Assignment Guide and Homework Check

Level	Day 1 Activity Assignment	Day 2 Lesson Assignment	Homework Check
Basic	3–5, 22–25	1, 2, 6–13, 15	1, 6, 12, 13
Average	3–5, 22–25	1, 2, 7–17 odd, 12, 16, 17, 19	1, 7, 12, 13
Advanced	3–5, 22–25	1, 2, 6–12, 14, 16–21	12, 14, 16, 18, 21
Accelerated	1–12, 14, 16–25		12, 14, 16, 18, 21

For Your Information

- **Exercise 16** Tell students that *radii* is the plural form of radius.
- **Exercise 16** Have students complete the table with their answers in terms of pi. The patterns in the table will be easier for the students to see.

Common Errors

- **Exercises 3–8, 13–15** Students may forget to divide the diameter by 2 to find the area. Remind them that they need the *radius* for the area formula of a circle.
- **Exercises 3–8, 13–15** Students may write the incorrect units for the area. Remind them to carefully check the units and to square the units as well.
- **Exercises 3–8** Students may refer to the formula for circumference and forget to square the radius when finding the area. Remind them that the area formula uses the radius squared.
- **Exercises 13–15** Students may forget to divide the area in half. Remind them that they are finding the area of half of a circle so the area is half of a whole circle.

1. Divide the diameter by 2 to get the radius. Then use the formula $A = \pi r^2$ to find the area.

2. What is the area of a circle with a radius of 100 cm; about 31,400 cm^2; about 7850 cm^2?

Practice and Problem Solving

3. about 254.34 mm^2

4. about 616 cm^2

5. about 314 in.^{2}

6. about 7.065 in.^{2}

7. about 3.14 cm^2

8. about 1.76625 ft^2

9. about 2461.76 mm^2

10. about 78.5 ft^2

11. about 113.04 in.^{2}

12. about 1444.4 mi^2

8.3 Record and Practice Journal

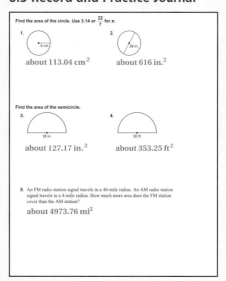

Find the area of the circle. Use 3.14 or $\frac{22}{7}$ for π.

1. 6 cm — about 113.04 cm^2

2. 28 in. — about 616 in.^{2}

Find the area of the semicircle.

3. 18 in. — about 127.17 in.^{2}

4. 30 ft — about 353.25 ft^2

5. An FM radio station signal travels in a 40-mile radius. An AM radio station signal travels in a 4-mile radius. How much more area does the FM station cover than the AM station?

   about 4973.76 mi^2

Practice and Problem Solving

13–16. See Additional Answers.

17. See *Taking Math Deeper*.

18. greater than; The circle's diameter is one-half as long, so it equals the radius of the semicircle. A diagram shows that the area of the semicircle is greater.

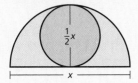

19–21. See Additional Answers.

Fair Game Review

22. 44   **23.** 53

24. 73   **25.** A

Mini-Assessment

Find the area of the circle. Use 3.14 or $\frac{22}{7}$ for π.

1.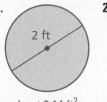

about 3.14 ft²

2.

about 154 m²

3.

about 200.96 in.²

4.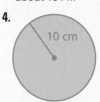

about 314 cm²

5. Find the area of the rug.
about 28.26 ft²

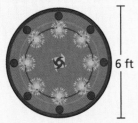

Taking Math Deeper

Exercise 17

This problem is not difficult, once a student realizes that the dog's running area is three-quarters of a circle.

1 Draw and label a diagram.

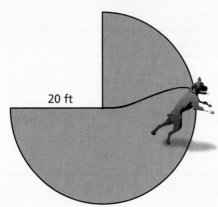

20 ft

2 Find the area.

$$\text{Area} = \frac{3}{4}(\text{Area of circle})$$
$$= \frac{3}{4} \cdot \pi \cdot 20^2$$
$$\approx 942 \text{ ft}^2$$

Almost 1000

3 Ask students to talk about the advantages and disadvantages of this type of outside exercise plan for a dog. What other ways can you allow your dog to be outside and get exercise?
- Build a dog run.
- Fence in a back yard.
- Use an invisible fence.
- Take the dog for a long walk each morning and night.

Project

Design a dog run for your back yard. Do some research to determine the cost of the run.

Reteaching and Enrichment Strategies

If students need help. . .	If students got it. . .

Resources by Chapter
  • Practice A and Practice B
  • Puzzle Time
Record and Practice Journal Practice
Differentiating the Lesson
Lesson Tutorials
Skills Review Handbook
	Resources by Chapter
	  • Enrichment and Extension
	  • Technology Connection
	Start the next section

Find the area of the semicircle.

③ 13.

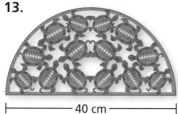

├────── 40 cm ──────┤

14.

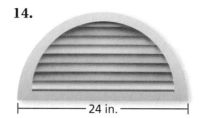

├────── 24 in. ──────┤

15.

├────── 2 ft ──────┤

16. REPEATED REASONING Consider five circles with radii of 1, 2, 4, 8, and 16 inches.

    a. Copy and complete the table. Write your answers in terms of π.

    b. Compare the areas and circumferences. What happens to the circumference of a circle when you double the radius? What happens to the area?

    c. What happens when you triple the radius?

Radius	Circumference	Area
1	2π in.	π in.^{2}
2		
4		
8		
16		

20 ft

17. DOG A dog is leashed to the corner of a house. How much running area does the dog have? Explain how you found your answer.

18. CRITICAL THINKING Is the area of a semicircle with a diameter of x *greater than*, *less than*, or *equal to* the area of a circle with a diameter of $\frac{1}{2}x$? Explain.

Reasoning Find the area of the shaded region. Explain how you found your answer.

19.

5 in.

20.

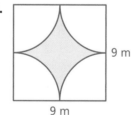

9 m

9 m

21.

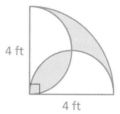

4 ft

4 ft

Fair Game Review *What you learned in previous grades & lessons*

Evaluate the expression. *(Skills Review Handbook)*

22. $\frac{1}{2}(7)(4) + 6(5)$

23. $\frac{1}{2} \cdot 8^2 + 3(7)$

24. $12(6) + \frac{1}{4} \cdot 2^2$

25. MULTIPLE CHOICE What is the product of $-8\frac{1}{3}$ and $3\frac{2}{5}$? *(Section 2.4)*

    Ⓐ $-28\frac{1}{3}$ **Ⓑ** $-24\frac{2}{15}$ **Ⓒ** $24\frac{2}{15}$ **Ⓓ** $28\frac{1}{3}$

8.4 Areas of Composite Figures

Essential Question How can you find the area of a composite figure?

1 ACTIVITY: Estimating Area

Work with a partner.

a. Choose a state. On grid paper, draw a larger outline of the state.

b. Use your drawing to estimate the area (in square miles) of the state.

c. Which state areas are easy to find? Which are difficult? Why?

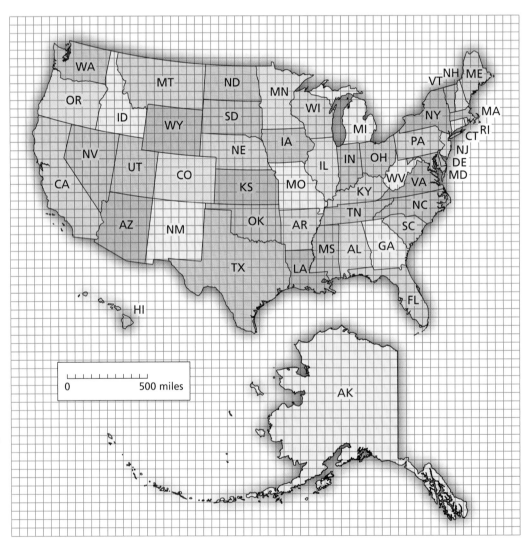

Geometry

In this lesson, you will
- find areas of composite figures by separating them into familiar figures.
- solve real-life problems.

Laurie's Notes

Introduction

Applying Mathematical Practices

- **Make Sense of Problems and Persevere in Solving Them:** Students have worked with a number of area formulas. In making sense of composite figures, they need to view the figures as composed of smaller, familiar figures.

Motivate

- Share U.S. state trivia in question/answer format.
- Largest states: Alaska, Texas, California
- Smallest states: Rhode Island, Delaware, Connecticut
- Most densely populated states: New Jersey, Rhode Island, Massachusetts

Activity Notes

For Your Information

- The three activities may likely take longer than one class period. You may want to do just Activities 1 and 3, or Activities 2 and 3. Students really like Activity 1, and it is interesting to hear their comments about different states or regions of the country.

Activity 1

- ? Ask a variety of questions about the attributes of the states: state with most coastline; states with no coastline; state furthest north, south, east, and west.
- Questions are answered by using eyesight only. The goal is to get students thinking about the states and how each might be described.
- **Make Sense of Problems and Persevere in Solving Them** and **Big Idea:** Each square in the grid is 2500 square miles. Each state is composed of whole squares and parts of whole squares. Students will find the area of the state by finding the area of composite figures.
- Here are a few strategies for implementing this activity.
  - Students are assigned states within one particular region.
  - Students use $\frac{1}{2}$-inch or 1-inch grid paper to draw the larger outline.

    This is *not* a lesson on scale drawings. A larger copy of the state allows students to record their areas for each part on the drawing.
  - Students are given an enlarged photocopy of the state to use.
  - Assign the same state to two students. Have each student make a drawing and then compare their estimates for the non-square parts of the state.
- Discuss part (c). States with straighter borders are easier to draw and to estimate their areas.
- **Extension:** Drawings of adjacent states can be cut out and taped together on a bulletin board. Students can do outside research to gather information about their state: population, population density; or to rank the states by area and find the mean and median of the state areas.

What Your Students Will Learn

- Find areas of composite figures using grid paper.
- Find areas of composite figures by separating them into familiar figures, finding those areas, and then adding up the areas.

Previous Learning

Students should know area formulas.

Lesson Plans
Complete Materials List

8.4 Record and Practice Journal

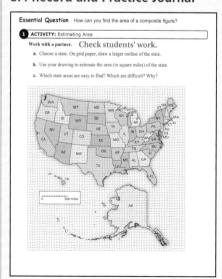

Differentiated Instruction

Kinesthetic

Using grid paper, have students draw and cut out a square 4 units on each side. Next, have students draw and cut out a rectangle that is 3 units by 2 units. Have students place the two figures so that they are touching, but not overlapping. Ask them to find the combined area by finding the area of each piece. Have them draw other shapes that are touching on the grid paper, for example a square with a semicircle on top. Find the total area. Discuss how they could use this idea to find the area of other figures.

8.4 Record and Practice Journal

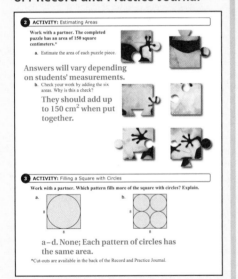

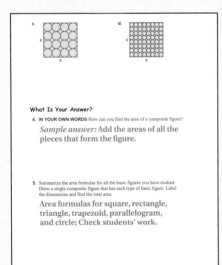

Activity 2

- **Make Sense of Problems and Persevere in Solving Them:** Students will work with actual-size pieces. The interlocking cuts are geometric shapes of which students should know how to find the area. Like Activity 1, to find the area, component parts are used. Unlike Activity 1, a component part may be missing so the area is subtracted off.

- Students will need to make accurate measurements to the nearest tenth of a centimeter.
- Scaled pieces are shown in the textbook.
- This activity reviews measurement, computation with decimals, and area formulas.
- Watch for computation errors.
- The completed puzzle is shown at the right.

Activity 3

- **Big Idea:** Each square has the same amount of green. Students can determine the radius of the circles. Then they can find the sum of the areas of circles in each diagram.
- You may need to suggest a table format to help guide student thinking.

Number of Circles	Radius of One Circle	Total Area
1	4	$1 \times \pi(4)^2 = 16\pi$ square units
4	2	$4 \times \pi(2)^2 = 16\pi$ square units
16	1	$16 \times \pi(1)^2 = 16\pi$ square units
64	$\frac{1}{2}$	$64 \times \pi\left(\frac{1}{2}\right)^2 = 16\pi$ square units

- Students find the results of this activity surprising.

What Is Your Answer?

- Question 5 is a nice summary problem that could become a small project.

Closure

- Does the area of a figure change when it is divided into smaller regions? Explain. No. As each region becomes smaller, the number of regions increases, but the total area remains the same.

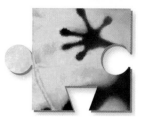

2   ACTIVITY: Estimating Areas

Work with a partner. The completed puzzle has an area of 150 square centimeters.

a. Estimate the area of each puzzle piece.

b. Check your work by adding the six areas. Why is this a check?

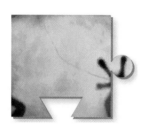

3   ACTIVITY: Filling a Square with Circles

Math Practice

Make a Plan

What steps will you use to solve this problem?

Work with a partner. Which pattern fills more of the square with circles? Explain.

a.
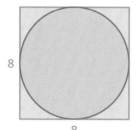

b.
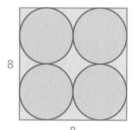

c.
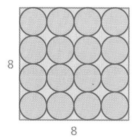

d.
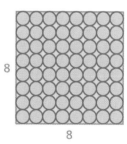

What Is Your Answer?

4. IN YOUR OWN WORDS How can you find the area of a composite figure?

5. Summarize the area formulas for all the basic figures you have studied. Draw a single composite figure that has each type of basic figure. Label the dimensions and find the total area.

 Use what you learned about areas of composite figures to complete Exercises 3–5 on page 342.

To find the area of a composite figure, separate it into figures with areas you know how to find. Then find the sum of the areas of those figures.

EXAMPLE 1   Finding an Area Using Grid Paper

Find the area of the yellow figure.

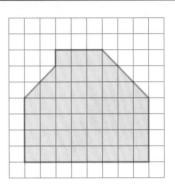

Count the number of squares that lie entirely in the figure. There are 45.

Count the number of half squares in the figure. There are 5.

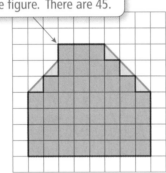

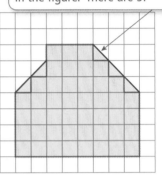

The area of a half square is $1 \div 2 = 0.5$ square unit.

Area of 45 squares: $45 \times 1 = 45$ square units

Area of 5 half squares: $5 \times 0.5 = 2.5$ square units

So, the area is $45 + 2.5 = 47.5$ square units.

On Your Own

Now You're Ready
Exercises 3–8

Find the area of the shaded figure.

1.

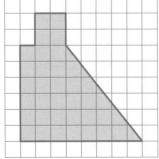

2.

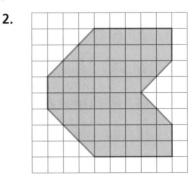

Laurie's Notes

Goal Today's lesson is finding the areas of composite figures.

Technology for the **Teacher**

Dynamic Classroom

Lesson Tutorials
Lesson Plans
Answer Presentation Tool

Introduction

Connect

- **Yesterday:** Students gained an intuitive understanding about how to find the area of composite figures.
- **Today:** Students will divide composite figures into familiar geometric shapes and use known area formulas to find the total area.

Motivate

- Arrange a set of tangram pieces in a square. Tell students that the area of the square is 16 square units.

? "What are the dimensions of the square?" 4 units by 4 units

- Now rearrange all of the tangram pieces to make a new shape.

? "Can you find the area of each of these? Explain" yes, 16 square units; Each new figure is composed of the same 7 pieces that made the square.

Lesson Notes

Example 1

- Work through the example as shown and then try an alternate approach.
- **Construct Viable Arguments and Critique the Reasoning of Others:** There are several ways to find the area of this figure. It is important for students to describe the process. The needed dimensions are unknown, so student explanations are key.

? "What is the name of the yellow polygon?" heptagon

? "Can you think of another way to find the area of the heptagon? Explain." Yes. Listen for a suggestion that divides the heptagon into a rectangle, a triangle, and a trapezoid.

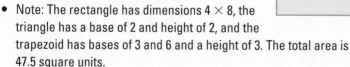

- Note: The rectangle has dimensions 4×8, the triangle has a base of 2 and height of 2, and the trapezoid has bases of 3 and 6 and a height of 3. The total area is 47.5 square units.

? "Are there other strategies for finding the area of the yellow heptagon?" Yes, it could be divided into a rectangle and two trapezoids using two parallel horizontal lines (or two parallel vertical lines).

- **Extension:** Ask students if the area of their classroom is more or less than 48 square meters. Hold two meter sticks perpendicular to each other as a visual clue.

On Your Own

- Encourage students to find more than one method to find the area of each.
- **Construct Viable Arguments and Critique the Reasoning of Others:** Have students share different strategies at the board.

Extra Example 1

Find the area of the figure.

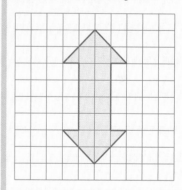

16 square units

On Your Own

1. 37 square units

2. 51 square units

Extra Example 2

Find the area of the pool and the deck. The figure is made up of a right triangle, a square, and a semicircle.

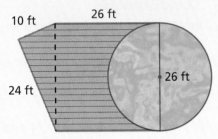

about 1061.33 ft²

Extra Example 3

Find the area of the figure made up of a triangle, a rectangle, and a trapezoid.

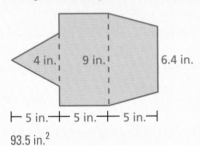

93.5 in.²

🔘 On Your Own

3. 90 m²

4. about 10.28 ft²

English Language Learners

Vocabulary

Formulas use letters to represent specific items or measures. For instance, in the formula for the area of a rectangle, $A = bh$, A represents area, b represents the base, and h represents the height. Review with your English learners the capital and lowercase letters used in formulas.

area (A)	base (b)
circumference (C)	diameter (d)
distance (d)	height (h)
length (ℓ)	perimeter (P)
radius (r)	rate (r)
time (t)	volume (V)
width (w)	

Laurie's Notes

Example 2

- After Example 1, it should be clear to students that this composite figure will be divided into a semicircle and a rectangle.
- **?** "What are the dimensions of the rectangle?" 19 ft by 12 ft
- **?** "What is the radius of the circle?" 6 ft
- In calculating the area of the semicircle, watch for student errors in multiplying and dividing decimals.
- **Extension:** Ask students if the area of a semicircle of radius 6 is the same as the area of a circle of radius 3. The answer is no! 18π compared to 9π

Example 3

- Draw the composite figure without the interior dotted lines. Ask students to think about how the figure could be divided into regions with known area formulas. They may come up with more than three regions.
- Label the diagram with the dimensions given. I find it helpful to draw small arrows from the measurement to the segment it is associated with, especially for interior dimensions.
- Work through the problem, questioning students along the way about the process.
- **?** "How do you find the area of a triangle? How do you multiply $\frac{1}{2}$ by 11.2? What units are used to label area?"
- **Common Misconception:** In a right triangle, the height is a side length. It is not an interior measurement.

On Your Own

- Have students share their strategies for these two problems. Question 4 might be done as 4 semicircles or 2 circles.

🔘 Closure

- **Exit Ticket:** Find the area. 100 units²

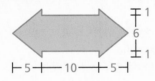

EXAMPLE 2 **Finding an Area**

Find the area of the portion of the basketball court shown.

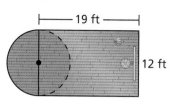

The figure is made up of a rectangle and a semicircle. Find the area of each figure.

Area of Rectangle

$A = \ell w$

$= 19(12)$

$= 228$

Area of Semicircle

$A = \dfrac{\pi r^2}{2}$

$\approx \dfrac{3.14 \cdot 6^2}{2}$

> The semicircle has a radius of $\dfrac{12}{2} = 6$ feet.

$= 56.52$

⁖ So, the area is about $228 + 56.52 = 284.52$ square feet.

EXAMPLE 3 **Finding an Area**

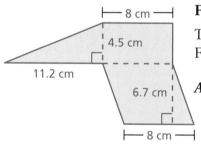

Find the area of the figure.

The figure is made up of a triangle, a rectangle, and a parallelogram. Find the area of each figure.

Area of Triangle

$A = \dfrac{1}{2}bh$

$= \dfrac{1}{2}(11.2)(4.5)$

$= 25.2$

Area of Rectangle

$A = \ell w$

$= 8(4.5)$

$= 36$

Area of Parallelogram

$A = bh$

$= 8(6.7)$

$= 53.6$

⁖ So, the area is $25.2 + 36 + 53.6 = 114.8$ square centimeters.

On Your Own

Now You're Ready
Exercises 9 and 10

Find the area of the figure.

3.

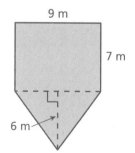

4.

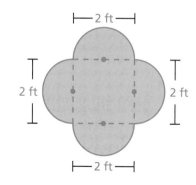

Vocabulary and Concept Check

1. **REASONING** Describe two different ways to find the area of the figure. Name the types of figures you used and the dimensions of each.

2. **REASONING** Draw a trapezoid. Explain how you can think of the trapezoid as a composite figure to find its area.

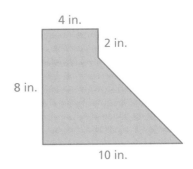

4 in.
2 in.
8 in.
10 in.

Practice and Problem Solving

Find the area of the figure.

① **3.**

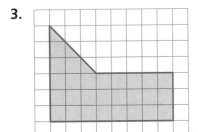

4.

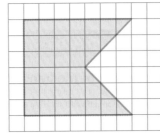

5.

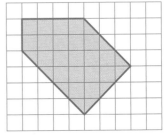

6.

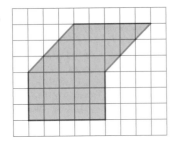

7.

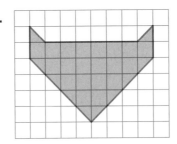

8.

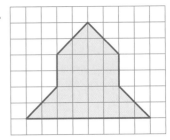

Find the area of the figure.

② ③ **9.**

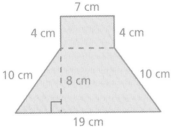

7 cm
4 cm   4 cm
10 cm   10 cm
8 cm
19 cm

10.

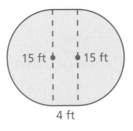

15 ft   15 ft
4 ft

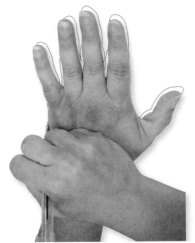

11. **OPEN-ENDED** Trace your hand and your foot on grid paper. Then estimate the area of each. Which one has the greater area?

Assignment Guide and Homework Check

Level	Day 1 Activity Assignment	Day 2 Lesson Assignment	Homework Check
Basic	3–5, 18–22	1, 2, 6–13	2, 6, 9, 12
Average	3–5, 18–22	1, 2, 8–13, 15, 16	8, 9, 12, 15
Advanced	3–5, 18–22	1, 2, 7, 8, 12–17	8, 12, 15, 16
Accelerated	1–5, 8–10, 12–22		8, 12, 15, 16

Common Errors

- **Exercises 3–8** Students may forget to count all the squares inside the figure and just count the ones along the border because this is what they did for perimeter. Remind them that the area includes everything inside the figure.

- **Exercises 9 and 10** Students may forget to include one of the areas of the composite figures or may count one area more than once. Tell them to break apart the figure into several figures. Draw and label each figure and then find the area of each part. Finally add the areas of each part together for the area of the whole figure.

- **Exercises 12–14** Students may forget to subtract the unshaded area from the figure. Remind them that in this situation instead of adding on the area of a figure they must subtract a portion out. Give a real-life example to help students understand, like tiling a bathroom floor but taking out the part where the sink or toilet is.

8.4 Record and Practice Journal

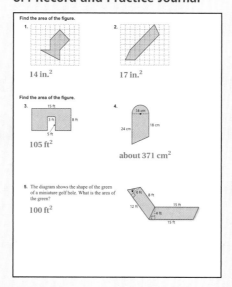

 Vocabulary and Concept Check

1. *Sample answer:* You could add the areas of an 8-inch × 4-inch rectangle and a triangle with a base of 6 inches and a height of 6 inches. Also you could add the area of a 2-inch × 4-inch rectangle to the area of a trapezoid with a height of 6 inches, and base lengths of 4 inches and 10 inches.

2. *Sample answer:* You can think of the trapezoid as a rectangle and two triangles.

 **Practice and Problem Solving**

3. 28.5 units^2

4. 33 units^2

5. 25 units^2

6. 30 units^2

7. 25 units^2

8. 24 units^2

9. 132 cm^2

10. about 236.625 ft^2

11. *Answer should include, but is not limited to:* Tracings of a hand and foot on grid paper, estimates of the areas, and a statement of which is greater.

Practice and Problem Solving

12–15. See Additional Answers.

16. $P =$ about 94.2 ft
$A =$ about 628 ft^2

17. See *Taking Math Deeper.*

Fair Game Review

18. $x - 12$ **19.** $y \div 6$

20. $b + 3$ **21.** $7w$

22. A

Mini-Assessment

Find the area of the figure.

1.

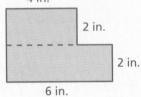

4 in.
2 in.
2 in.
6 in.

20 in.^{2}

2.

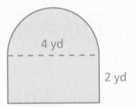

4 yd
2 yd

about 14.28 yd^2

3.

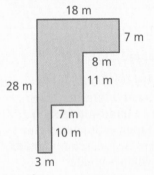

18 m
7 m
8 m
11 m
28 m
7 m
10 m
3 m

266 m^2

4. Find the area of the red region.
about 3.44 in.^{2}

2 in.

Taking Math Deeper

Exercise 17

This problem is not conceptually difficult. However, it does have a lot of calculations and it is difficult to do without a calculator. Before starting the calculations, suggest that students think about which design is more efficient. Each design will fit on an 11-inch × 17-inch sheet of paper. By cutting and folding each pattern, a student can see how much overlap each design has. The design with the greatest overlap is the least efficient.

 1 Find the area of the first design.

$\frac{1}{2} \cdot 5.5 \cdot 2.5 = 6.875$

$\frac{1}{2} \cdot 4.5 \cdot 3 = 6.75$

$4.5 \cdot 5.5 = 24.75$

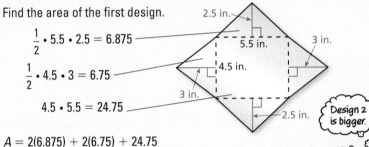

2.5 in.
5.5 in.
3 in.
4.5 in.
3 in.
2.5 in.
Design 2 is bigger.

$A = 2(6.875) + 2(6.75) + 24.75$
$\quad = 52$ in.^{2}

 2 Find the area of the second design.

$2 \cdot 5.5 = 11$

$\frac{1}{2}(4 + 4.5) \cdot 0.75 = 3.1875$

$4.5 \cdot 5.5 = 24.75$

$3.5 \cdot 5.5 = 19.25$

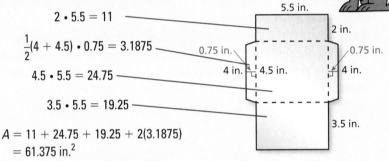

5.5 in.
2 in.
0.75 in.
0.75 in.
4 in.
4.5 in.
4 in.
3.5 in.

$A = 11 + 24.75 + 19.25 + 2(3.1875)$
$\quad = 61.375$ in.^{2}

 3 **a.** So, the second design has the greater area.
Answer the question.

$500 \cdot 61.375 = 30{,}687.5$ in.^{2} (500 envelopes using Design 2)

$\dfrac{30{,}687.5}{52} \approx 590.1$ (number of envelopes using Design 1)

b. You could make 90 more envelopes using Design 1.

Reteaching and Enrichment Strategies

If students need help...	If students got it...
Resources by Chapter • Practice A and Practice B • Puzzle Time Record and Practice Journal Practice Differentiating the Lesson Lesson Tutorials Skills Review Handbook	Resources by Chapter • Enrichment and Extension • Technology Connection Start the next section

Find the area of the figure.

12.

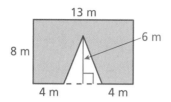

13 m
6 m
8 m
4 m   4 m

13.

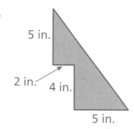

5 in.
2 in.
4 in.
5 in.

14.

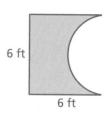

6 ft
6 ft

15. STRUCTURE The figure is made up of a square and a rectangle. Find the area of the shaded region.

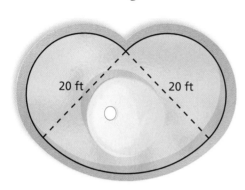

7 m
16 m
3 m

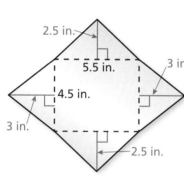

20 ft   20 ft

16. FOUNTAIN The fountain is made up of two semicircles and a quarter circle. Find the perimeter and the area of the fountain.

17. **Critical Thinking** You are deciding on two different designs for envelopes.

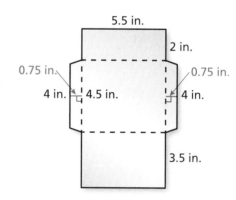

2.5 in.
5.5 in.
3 in.
4.5 in.
3 in.
2.5 in.

5.5 in.
2 in.
0.75 in.
4 in.   4.5 in.   4 in.
0.75 in.
3.5 in.

a. Which design has the greater area?

b. You make 500 envelopes using the design with the greater area. Using the same amount of paper, how many more envelopes can you make with the other design?

Fair Game Review *What you learned in previous grades & lessons*

Write the phrase as an expression. *(Skills Review Handbook)*

18. 12 less than a number x

19. a number y divided by 6

20. a number b increased by 3

21. the product of 7 and a number w

22. MULTIPLE CHOICE What number is 0.02% of 50? *(Section 6.4)*

   Ⓐ 0.01   Ⓑ 0.1   Ⓒ 1   Ⓓ 100

Check It Out
Progress Check
BigIdeasMath ✓ .com

Find the area of the figure. *(Section 8.4)*

1.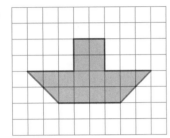

2.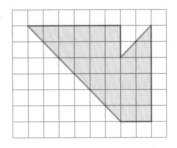

3.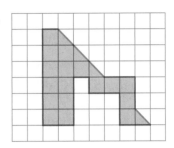

Find the area of the circle. Use 3.14 or $\frac{22}{7}$ for π. *(Section 8.3)*

4.
12 in.

5.
6 cm

6.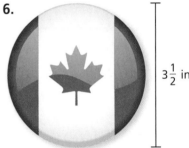
$3\frac{1}{2}$ in.

Find the area of the figure. *(Section 8.4)*

7.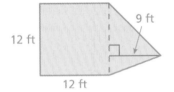
9 ft
12 ft
12 ft

8.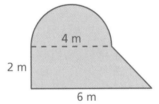
4 m
2 m
6 m

9.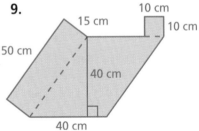
10 cm
15 cm
10 cm
50 cm
40 cm
40 cm

10. **POT HOLDER** A knitted pot holder is shaped like a circle. Its radius is 3.5 inches. What is its area? *(Section 8.3)*

11. **CARD** The heart-shaped card is made up of a square and two semicircles. What is the area of the card? *(Section 8.4)*

8 cm   8 cm

12. **DESK** A desktop is shaped like a semicircle with a diameter of 28 inches. What is the area of the desktop? *(Section 8.3)*

├── 14 ft ──┤

13. **RUG** The circular rug is placed on a square floor. The rug touches all four walls. How much of the floor space is *not* covered by the rug? *(Section 8.4)*

Alternative Assessment Options

Math Chat	Student Reflective Focus Question
Structured Interview	Writing Prompt

Math Chat

- Have students work in pairs. Assign Exercises 10–13 from the Quiz to each pair. Each student works through all four problems. After the students have worked through the problems, they take turns talking through the process that they used to get the answer. Students analyze and evaluate the mathematical thinking and strategies used.
- The teacher should walk around the classroom listening to the pairs and asking questions to ensure understanding.

Study Help Sample Answers

Remind students to complete Graphic Organizers for the rest of the chapter.

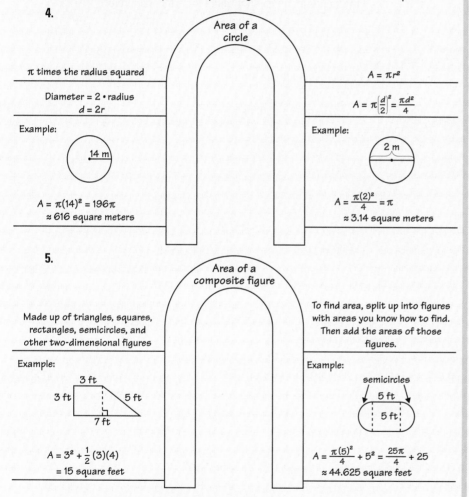

Reteaching and Enrichment Strategies

If students need help. . .	If students got it. . .
Resources by Chapter	Resources by Chapter
• Practice A and Practice B	• Enrichment and Extension
• Puzzle Time	• Technology Connection
Lesson Tutorials	Game Closet at *BigIdeasMath.com*
BigIdeasMath.com	Start the Chapter Review

Answers

1. 16 units^2
2. 28 units^2
3. 19 units^2
4. about 452.16 in.^{2}
5. about 28.26 cm^2
6. about 9.625 in.^{2}
7. 198 ft^2
8. about 16.28 m^2
9. 2450 cm^2
10. about 38.465 in.^{2}
11. about 114.24 cm^2
12. about 308 in.^{2}
13. about 42 ft^2

Online Assessment
Assessment Book
ExamView® Assessment Suite

- *BigIdeasMath.com*
- Online Assessment
- Game Closet at *BigIdeasMath.com*
- Vocabulary Help
- Resources by Chapter

Answers

1. 4 in.
2. 30 mm
3. 50 m
4. 1.5 yd
5. 40 ft
6. 10 m
7. 2 in.
8. 50 mm
9. about 18.84 ft
10. about 66 cm
11. about 132 in.

Review of Common Errors

Exercises 1–8
- Students may confuse what they are finding and double the diameter or halve the radius. Remind them that the radius is half the diameter.

Exercises 9–11
- Students may use the radius in the circumference formula that calls for diameter or the diameter in the circumference formula that calls for radius. Remind them of the different formulas.

Exercises 12–17
- Students may include the dashed lines in their calculation of the perimeter. Remind them that the dashed lines are for reference and sometimes give information to find another length. Only the outside lengths are counted.

Exercises 18–20
- Students may forget to divide the diameter by 2 to find the area. Remind them that they need the *radius* in the given formula for area of a circle.
- Students may use the formula for circumference or forget to square the radius when finding the area.

Exercises 21–23
- Students may forget to include the area of one or more parts of a composite figure or count a part more than once. Tell them to draw and label each part, find the area of each part, and add the areas of the parts to find the area of the composite figure.

Check It Out
Vocabulary Help
BigIdeasMath ✓com

Review Key Vocabulary

circle, *p. 318*

center, *p. 318*

radius, *p. 318*

diameter, *p. 318*

circumference, *p. 319*

pi, *p. 319*

semicircle, *p. 320*

composite figure, *p. 326*

Review Examples and Exercises

8.1 **Circles and Circumference** *(pp. 316–323)*

Find the circumference of the circle. Use 3.14 for π.

The radius is 4 millimeters.

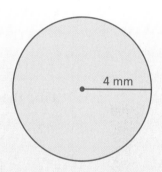

4 mm

$C = 2\pi r$	Write formula for circumference.
$\approx 2 \cdot 3.14 \cdot 4$	Substitute 3.14 for π and 4 for r.
$= 25.12$	Multiply.

⋮ The circumference is about 25.12 millimeters.

Exercises

Find the radius of the circle with the given diameter.

1. 8 inches

2. 60 millimeters

3. 100 meters

4. 3 yards

Find the diameter of the circle with the given radius.

5. 20 feet

6. 5 meters

7. 1 inch

8. 25 millimeters

Find the circumference of the circle. Use 3.14 or $\dfrac{22}{7}$ for π.

9.

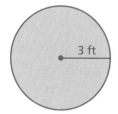

3 ft

10.

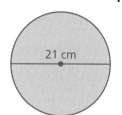

21 cm

11.

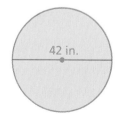

42 in.

8.2 Perimeters of Composite Figures *(pp. 324–329)*

The figure is made up of a semicircle and a square. Find the perimeter.

The distance around the square part is $6 + 6 + 6 = 18$ meters. The distance around the semicircle is one-half the circumference of a circle with $d = 6$ meters.

$$\frac{C}{2} = \frac{\pi d}{2} \qquad \text{Divide the circumference by 2.}$$

$$\approx \frac{3.14 \cdot 6}{2} \qquad \text{Substitute 3.14 for } \pi \text{ and 6 for } d.$$

$$= 9.42 \qquad \text{Simplify.}$$

6 m

∴ So, the perimeter is about $18 + 9.42 = 27.42$ meters.

Exercises

Find the perimeter of the figure.

12.

5 in.
4 in.
3 in.
5 in.
9 in.

13.

9 ft
9 ft
9 ft
9 ft
30 ft

14.

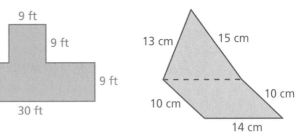

13 cm   15 cm
10 cm   10 cm
14 cm

15.

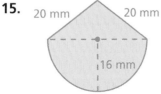

20 mm   20 mm
16 mm

16.

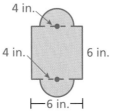

4 in.
4 in.   6 in.
6 in.

17.

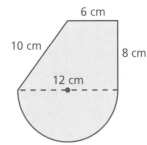

6 cm
10 cm
8 cm
12 cm

8.3 Areas of Circles *(pp. 332–337)*

Find the area of the circle. Use 3.14 for π.

$$A = \pi r^2 \qquad \text{Write formula for area.}$$

$$\approx 3.14 \cdot 20^2 \qquad \text{Substitute 3.14 for } \pi \text{ and 20 for } r.$$

$$= 1256 \qquad \text{Multiply.}$$

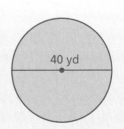

40 yd

∴ The area is about 1256 square yards.

Review Game

Area, Perimeter, and Circumference

Materials per Pair:
- paper
- pencils (colored pencils optional)

Directions:

Each group of four divides into teams of two. Each team competes with the other team. The class is directed to draw and label a composite figure using selected shapes and a predetermined area. For example, the teacher could say, "Using squares, rectangles, and circles, draw and label a composite figure with an area of 50 square units." Each team then draws their figure and calculates the area. Papers are exchanged within the group and teams check each other's work. The team whose correctly-drawn figure has an area that is closest to the area specified receives 1 point. If both teams are equally close, both teams receive 1 point. Play for a set amount of time.

Who Wins?

When time is up, the team with the most points wins.

For the Student
Additional Practice
- Lesson Tutorials
- Multi-Language Glossary
- Self-Grading Progress Check
- *BigIdeasMath.com*
  Dynamic Student Edition
  Student Resources

Answers

12. 24 in.

13. 96 ft

14. 62 cm

15. about 90.24 mm

16. about 28.56 in.

17. about 42.84 cm

18. about 50.24 in.^{2}

19. about 379.94 cm^2

20. about 1386 mm^2

21. about 79.25 in.^{2}

22. 29 in.^{2}

23. about 31.625 ft^2

My Thoughts on the Chapter

What worked. . .

What did not work. . .

What I would do differently. . .

Exercises

Find the area of the circle. Use 3.14 or $\frac{22}{7}$ for π.

18.
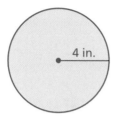
4 in.

19.
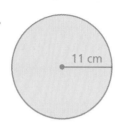
11 cm

20.

42 mm

8.4 **Areas of Composite Figures** *(pp. 338–343)*

Find the area of the figure.

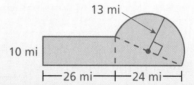
13 mi
10 mi
26 mi · 24 mi

The figure is made up of a rectangle, a triangle and a semicircle. Find the area of each figure.

Area of Rectangle

$A = \ell w$

$= 26(10)$

$= 260$

Area of Triangle

$A = \frac{1}{2}bh$

$= \frac{1}{2}(10)(24)$

$= 120$

Area of Semicircle

$A = \frac{\pi r^2}{2}$

$\approx \frac{3.14 \cdot 13^2}{2}$

$= 265.33$

∴ So, the area is about $260 + 120 + 265.33 = 645.33$ square miles.

Exercises

Find the area of the figure.

21.
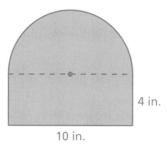
4 in.
10 in.

22.
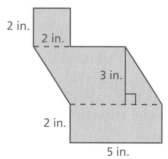
2 in.
2 in.
3 in.
2 in.
5 in.

23.
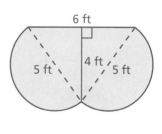
6 ft
5 ft
4 ft
5 ft

8 Chapter Test

Find the radius of the circle with the given diameter.

1. 10 inches

2. 5 yards

Find the diameter of the circle with the given radius.

3. 34 feet

4. 19 meters

Find the circumference and the area of the circle. Use 3.14 or $\frac{22}{7}$ for π.

5.
4 ft

6.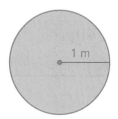
1 m

7.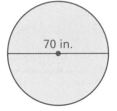
70 in.

8. Estimate the perimeter of the figure. Then find the area.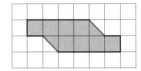

Find the perimeter and the area of the figure. Use 3.14 or $\frac{22}{7}$ for π.

9.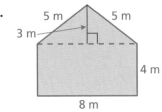

10.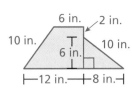

11.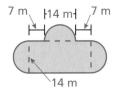

12. MUSEUM A museum plans to rope off the perimeter of the L-shaped exhibit. How much rope does it need?

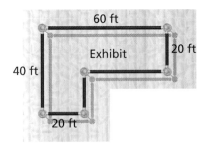

13. ANIMAL PEN You unfold chicken wire to make a circular pen with a diameter of 2.9 meters. How many meters of chicken wire do you need?

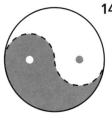

14. YIN AND YANG In the Chinese symbol for yin and yang, the dashed curve shows two semicircles formed by the curve separating the yin (dark) and the yang (light). Is the circumference of the entire yin and yang symbol *less than*, *greater than*, or *equal to* the perimeter of the yin?

Test Item References

Chapter Test Questions	Section to Review
1–7, 13, 14	8.1
8–12, 14	8.2
5–7	8.3
8–11	8.4

Test-Taking Strategies

Remind students to quickly look over the entire test before they start so that they can budget their time. There is a lot of vocabulary in this chapter, so students should have been making flash cards as they worked through the chapter. Words that get mixed up should be jotted on the back of the test before they start. (For instance, they should write down *radius* and *circumference*.) Students need to use the **Stop** and **Think** strategy before they answer the questions.

Common Errors

- **Exercises 1–4** Students may confuse what they are finding and double the diameter or halve the radius. Remind them that the radius is half the diameter. Encourage students to draw a line segment representing each radius or diameter so that they have a visual for reference.
- **Exercises 5–7** Students may use the radius in a formula that calls for diameter, the diameter in a formula that calls for radius, the area formula when calculating circumference, or the circumference formula when calculating area. Again, remind students that the radius is half the diameter. Also, remind them of the circumference and area formulas.
- **Exercise 8** Students may count the sides of the squares inside the figure when finding the perimeter or confuse area and perimeter. Remind students of how to find perimeter and area.
- **Exercises 9–11** When calculating the perimeter of the figure, students may count the dashed lines. Remind them that the dashed lines are for reference and that only the outside lengths are counted. When calculating the area of the figure, students may forget to include the area of one or more parts or count part(s) more than once. Tell them to draw and label each part, find the area of each part, and add the areas of the parts.

Reteaching and Enrichment Strategies

If students need help. . .	If students got it. . .
Resources by Chapter • Practice A and Practice B • Puzzle Time Record and Practice Journal Practice Differentiating the Lesson Lesson Tutorials *BigIdeasMath.com* Skills Review Handbook	Resources by Chapter • Enrichment and Extension • Technology Connection Game Closet at *BigIdeasMath.com* Start Cumulative Assessment

Answers

1. 5 in.
2. 2.5 yd
3. 68 ft
4. 38 m
5. $C \approx 12.56$ ft, $A \approx 12.56$ ft^2
6. $C \approx 6.28$ m, $A \approx 3.14$ m^2
7. $C \approx 220$ in., $A \approx 3850$ in.^{2}
8. $P \approx 15$ units, $A = 9$ units^2
9. $P = 26$ m, $A = 44$ m^2
10. $P = 48$ in., $A = 96$ in.^{2}
11. $P \approx 108$ m, $A \approx 623$ m^2
12. 200 ft
13. about 9.106 m
14. *equal to*

Online Assessment
Assessment Book
ExamView® Assessment Suite

After Answering Easy Questions, Relax

Answer Easy Questions First

Estimate the Answer

Read All Choices before Answering

Read Question before Answering

Solve Directly or Eliminate Choices

Solve Problem before Looking at
    Choices

Use Intelligent Guessing

Work Backwards

About this Strategy

When taking a multiple choice test, be sure to read each question carefully and thoroughly. When taking a timed test, it is often best to skim the test and answer the easy questions first. Be careful that you record your answer in the correct position on the answer sheet.

Answers

1. C

2. 42

3. I

4. C

Item Analysis

1. **A.** The student does not convert to the same units (either 2 quarts to 8 cups or 5 cups to 1.25 quarts) and inverts one of the ratios.

   B. The student does not convert to the same units (either 2 quarts to 8 cups or 5 cups to 1.25 quarts).

   C. Correct answer

   D. The student uses the correct numbers but inverts one of the ratios.

2. **Gridded Response:** Correct answer: 42

   Common Error: The student sets up the equation incorrectly. For instance, $2x + 1 + 85 = 180$ and finds $x = 47$.

3. **F.** On the left side of the equation, the student writes an expression that represents the product of n and a number that is 7 less than 5.

   G. On the left side of the equation, the student writes an expression that represents the product of n and a number that is 5 less than 7.

   H. On the left side of the equation, the student misinterprets the use of "less than" and reverses the order of the subtraction.

   I. Correct answer

4. **A.** The student did not square the radius.

   B. The student multiplied the radius by 2 instead of squaring it.

   C. Correct answer

   D. The student squared the diameter instead of the radius.

Technology for the *Teacher*

Performance Tasks
Online Assessment
Assessment Book
ExamView® Assessment Suite

1. To make 6 servings of soup, you need 5 cups of chicken broth. You want to know how many servings you can make with 2 quarts of chicken broth. Which proportion should you use?

   A. $\dfrac{6}{5} = \dfrac{2}{x}$

   B. $\dfrac{6}{5} = \dfrac{x}{2}$

   C. $\dfrac{6}{5} = \dfrac{x}{8}$

   D. $\dfrac{5}{6} = \dfrac{x}{8}$

Test-Taking Strategy
Answer Easy Questions First

What is the radius of a cat food can that has a diameter of 4 inches?
Ⓐ 1 in. Ⓑ 2 in. Ⓒ 3 in. Ⓓ 4 in.

I love easy questions!

"Scan the test and answer the easy questions first. You know that the radius is half the diameter."

2. What is the value of x?

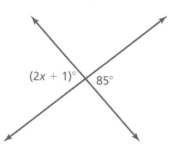

$(2x + 1)°$ $85°$

3. Your mathematics teacher described an equation in words. Her description is in the box below.

   "5 less than the product of 7 and an unknown number is equal to 42."

   Which equation matches your mathematics teacher's description?

   F. $(5 - 7)n = 42$

   G. $(7 - 5)n = 42$

   H. $5 - 7n = 42$

   I. $7n - 5 = 42$

4. What is the area of the circle below? $\left(\text{Use } \dfrac{22}{7} \text{ for } \pi.\right)$

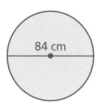

84 cm

   A. 132 cm^2

   B. 264 cm^2

   C. 5544 cm^2

   D. 22,176 cm^2

5. John was finding the area of the figure below.

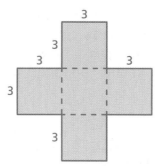

John's work is in the box below.

area of horizontal rectangle

$A = 3 \times (3 + 3 + 3)$

$= 3 \times 9$

$= 27$ square units

area of vertical rectangle

$A = (3 + 3 + 3) \times 3$

$= 9 \times 3$

$= 27$ square units

total area of figure

$A = 27 + 27$

$= 54$ square units

What should John do to correct the error that he made?

F. Add the area of the center square to the 54 square units.

G. Find the area of one square and multiply this number by 4.

H. Subtract the area of the center square from the 54 square units.

I. Subtract 54 from the area of a large square that is 9 units on each side.

6. Which value of x makes the equation below true?

$$5x - 3 = 11$$

A. 1.6

C. 40

B. 2.8

D. 70

Item Analysis (continued)

5. **F.** The student adds the area of the center square that was included twice, instead of subtracting it.

   G. The student does not include the area of the center square at all.

   H. Correct answer

   I. The student employs a plausible strategy but makes the same error as in John's work of counting the center square twice.

6. **A.** The student subtracts 3 instead of adding 3.

   B. Correct answer

   C. The student subtracts 3 instead of adding and then multiplies instead of dividing.

   D. The student multiplies instead of dividing.

Answers

7. 29.42

8. G

9. A

10. *Part A* 942 ft^2

Part B 134.2 ft

Item Analysis (continued)

7. **Gridded Response:** Correct answer: 29.42

   Common Error: The student includes the circumference of the circle instead of just the circumference of the semicircle.

8. **F.** The student does not reverse the inequality sign, leading to $x \geq 1$.

   G. Correct answer

   H. The student does not realize that $x < 5$ does not include 5.

   I. The student does not realize that $x > 5$ does not include 5.

9. **A.** Correct answer

   B. The student chooses the store offering the greatest percent off but ignores the fact that this store also had the greatest original price.

   C. The student chooses the store that had the least original price but ignores the fact that this store also offered the least percent off.

   D. The student chooses the store with the highest sale price.

10. **4 points** The student demonstrates a thorough understanding of solving problems involving area and perimeter of composite figures. In Part A, the student correctly finds the area of the sprayed region to be approximately 942 square feet. In Part B, the student correctly finds the perimeter of the sprayed region to be approximately 134.2 feet. The student shows accurate, complete work for both parts and provides clear and complete explanations.

    3 points The student demonstrates an understanding of solving problems involving area and perimeter of composite figures, but the student's work and explanations demonstrate an essential but less than thorough understanding.

    2 points The student demonstrates a partial understanding of solving problems involving area and perimeter of composite figures. The student's work and explanations demonstrate a lack of essential understanding.

    1 point The student demonstrates very limited understanding of solving problems involving area and perimeter of composite figures. The student's response is incomplete and exhibits many flaws.

    0 points The student provided no response, a completely incorrect or incomprehensible response, or a response that demonstrates insufficient understanding of solving problems involving area and perimeter of composite figures.

7. What is the perimeter of the figure below? (Use 3.14 for π.)

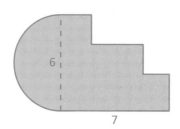

8. Which inequality has 5 in its solution set?

F. $5 - 2x \geq 3$

G. $3x - 4 \geq 8$

H. $8 - 3x > -7$

I. $4 - 2x < -6$

9. Four jewelry stores are selling an identical pair of earrings.

- Store A: original price of $75; 20% off during sale
- Store B: original price of $100; 35% off during sale
- Store C: original price of $70; 10% off during sale
- Store D: original price of $95; 30% off during sale

Which store has the least sale price for the pair of earrings?

A. Store A

B. Store B

C. Store C

D. Store D

10. A lawn sprinkler sprays water onto part of a circular region, as shown below.

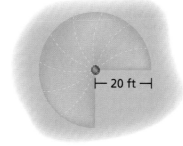

Part A What is the area, in square feet, of the region that the sprinkler sprays with water? Show your work and explain your reasoning. (Use 3.14 for π.)

Part B What is the perimeter, in feet, of the region that the sprinkler sprays with water? Show your work and explain your reasoning. (Use 3.14 for π.)

9 Surface Area and Volume

9.1 Surface Areas of Prisms

9.2 Surface Areas of Pyramids

9.3 Surface Areas of Cylinders

9.4 Volumes of Prisms

9.5 Volumes of Pyramids

"I was thinking that I want the Pagodal roof instead of the Swiss chalet roof for my new doghouse."

"Sometimes you are a true genius."

"Because PAGODAL rearranges to spell 'A DOG PAL.'"

"Take a deep breath and hold it."

"Can I breathe now?"

"Now, do you feel like your surface area or your volume is increasing more?"

What Your Students Have Learned

- Find the areas of rectangles with fractional side lengths.
- Classify two-dimensional figures into categories based on properties.
- Understand volume, and measure it by counting unit cubes.
- Find the volumes of rectangular prisms using a formula.
- Find the areas of triangles, special quadrilaterals, and polygons.
- Use nets made up of rectangles and triangles to find surface areas.
- Find the volumes of prisms with fractional edge lengths.

What Your Students Will Learn

- Use two-dimensional nets to represent three-dimensional solids.
- Solve real-world problems involving surface areas and volumes of objects composed of prisms, pyramids, and cylinders.
- Describe the cross sections that result from slicing three-dimensional figures.

Pacing Guide for Chapter 9

Chapter Opener Regular Accelerated	1 Day 1 Day
Section 1 Regular Accelerated	2 Days 1 Day
Section 2 Regular Accelerated	3 Days 2 Days
Section 3 Regular Accelerated	2 Days 1 Day
Study Help / Quiz Regular Accelerated	1 Day 0 Days
Section 4 Regular Accelerated	2 Days 1 Day
Section 5 Regular Accelerated	3 Days 2 Days
Chapter Review/ Chapter Tests Regular Accelerated	2 Days 2 Days
Total Chapter 9 Regular Accelerated	16 Days 10 Days
Year-to-Date Regular Accelerated	134 Days 79 Days

Technology for the *Teacher*

BigIdeasMath.com
Chapter at a Glance
Complete Materials List
Parent Letters: English and Spanish

What Your Students Have Learned
- Find areas of rectangles, squares, and triangles using formulas.

Additional Topics for Review
- Three-Dimensional Figures
- Using Nets to Find Surface Areas
- Finding the Volumes of Rectangular Prisms

Try It Yourself
1. 99 m^2
2. 35.7 ft^2
3. $\frac{4}{9} \text{ in.}^2$
4. 39 ft^2
5. 140 m^2
6. 225 cm^2

Record and Practice Journal
Fair Game Review
1. 64 cm^2
2. 84 yd^2
3. 58.88 in.^2
4. $\frac{25}{36} \text{ m}^2$
5. $3\frac{1}{9} \text{ mm}^2$
6. 321.63 ft^2
7. 6.25 ft^2
8. 20 cm^2
9. 12 ft^2
10. 21 m^2
11. 30 yd^2
12. 10 in.^2
13. 9 mm^2
14. 24 ft^2

Math Background Notes

Vocabulary Review
- Area
- Square Units
- Base of a Triangle
- Height of a Triangle

Finding Areas of Squares and Rectangles
- Students should be familiar with area formulas for squares and rectangles.
- You may want to review units with students. A complete answer will include a numeric solution and the correct units. Remind students that area is always measured in square units.
- The length and width of a rectangle are interchangeable. Some students may think that the length of the rectangle must always be the side that sits on the ground. Turn the rectangle vertically to demonstrate that the length and width are relative to interpretation.

Finding Areas of Triangles
- Students should be familiar with area formulas for triangles.
- Review how to identify a base and the corresponding height of the triangle.
- **Common Error:** Students may want to take $\frac{1}{2}$ of 6 and $\frac{1}{2}$ of 7 because they think it is the Distributive Property.
- Note that the product of 6 and 7 was found (Associative Property), and then $\frac{1}{2}$ of that product was computed. It would also be correct to multiply from left to right according to the order of operations.

Reteaching and Enrichment Strategies

If students need help. . .	If students got it. . .
Record and Practice Journal • Fair Game Review Skills Review Handbook Lesson Tutorials	Game Closet at *BigIdeasMath.com* Start the next section

What You Learned Before

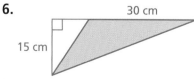

"Descartes, how would you like it if I could double the height of your cat food can?"

● Finding Areas of Squares and Rectangles

Example 1 Find the area of the rectangle.

3 mm

7 mm

$$\text{Area} = \ell w \qquad \text{Write formula for area.}$$
$$= 7(3) \qquad \text{Substitute 7 for } \ell \text{ and 3 for } w.$$
$$= 21 \qquad \text{Multiply.}$$

⋮∙ The area of the rectangle is 21 square millimeters.

Try It Yourself
Find the area of the square or rectangle.

1.

9 m

11 m

2.   4.2 ft
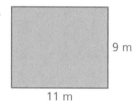
8.5 ft

3. $\frac{2}{3}$ in.
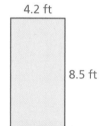
$\frac{2}{3}$ in.

● Finding Areas of Triangles

Example 2 Find the area of the triangle.

$$A = \frac{1}{2}bh \qquad \text{Write formula.}$$
$$= \frac{1}{2}(6)(7) \qquad \text{Substitute 6 for } b \text{ and 7 for } h.$$
$$= \frac{1}{2}(42) \qquad \text{Multiply 6 and 7.}$$
$$= 21 \qquad \text{Multiply } \frac{1}{2} \text{ and 42.}$$

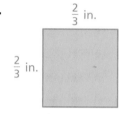

7 in.

6 in.

⋮∙ The area of the triangle is 21 square inches.

Try It Yourself
Find the area of the triangle.

4.
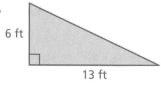
6 ft

13 ft

5.
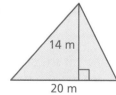
14 m

20 m

6.   30 cm
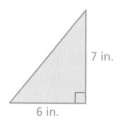
15 cm

Essential Question How can you find the surface area of a prism?

1 ACTIVITY: Surface Area of a Rectangular Prism

Work with a partner. Copy the net for a rectangular prism. Label each side as *h*, *w*, or ℓ. Then use your drawing to write a formula for the surface area of a rectangular prism.

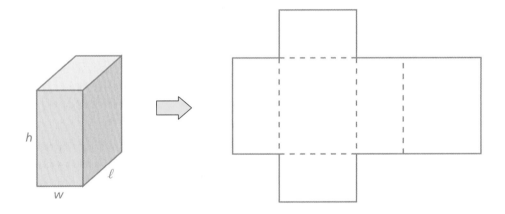

2 ACTIVITY: Surface Area of a Triangular Prism

Work with a partner.

a. Find the surface area of the solid shown by the net. Copy the net, cut it out, and fold it to form a solid. Identify the solid.

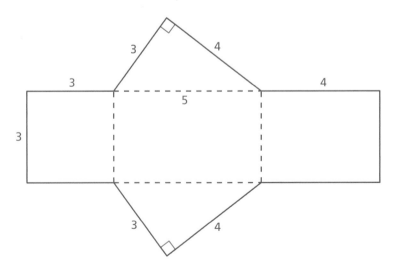

Geometry

In this lesson, you will

- use two-dimensional nets to represent three-dimensional solids.
- find surface areas of rectangular and triangular prisms.
- solve real-life problems.

b. Which of the surfaces of the solid are bases? Why?

Laurie's Notes

Introduction

Applying Mathematical Practices

- **Model with Mathematics:** Finding the surface area of a solid can be challenging when the faces are not all visible. Drawing a net allows students to see all of the faces of the solid.

Motivate

- Begin by standing on a solid box. Students will naturally be curious.
- **?** "I'm standing on a face of this rectangular prism. Is there another face that is congruent to the one I'm standing on? Explain." The opposite face, on the bottom, is congruent to the top.
- Rotate the box and stand on a different face. Repeat the question.
- **?** "How many pairs of congruent faces are there?" three

Write

- Write the definition of surface area. Describe and show examples of nets. Real-life models of nets (donut box, pizza box) are helpful.

Activity Notes

Activity 1

- **Classroom Management:** Collect cardboard samples of prisms such as donut, pizza, and tissue boxes. These can be taken apart to demonstrate nets.
- Explain that any pair of opposite faces of a rectangular prism can be identified as the bases, and the other four faces are the lateral faces.
- **Common Misconception:** The bases do not have to be on the top and bottom, although they are often identified that way.
- **Model with Mathematics:** A net helps students to see how to find the surface area of a prism.
- Students may be challenged to visualize how the prism unfolds as the net.
- Students will likely label the vertical segments to the far left and right as h.
- Encourage students to label all of the segments.
- **Construct Viable Arguments and Critique the Reasoning of Others:** Have students discuss ways to write the formula. Here are a few.
  - $S = \ell h + wh + \ell h + wh + \ell w + \ell w$
  - $S = 2\ell h + 2wh + 2\ell w$
  - $S = 2(\ell h + wh + \ell w)$
- Have students verify suggested formulas and discuss their usability.
- **Extension:** Find how many different nets there are for a cube. Use standard grid paper or investigate at *illuminations.nctm.org*. There are 11.

Activity 2

- **Management Tip:** To save time, you can precut the nets before class.
- **?** "What are the lateral faces of this triangular prism?" The 3 rectangles
- A prism is identified by the type of base it has, such as a triangular prism and a pentagonal prism. The lateral faces are all rectangles.

What Your Students Will Learn

- Use two-dimensional nets to represent three-dimensional solids.
- Find surface areas of rectangular prisms, triangular prisms, and cubes using formulas.

Previous Learning

Students should know how to make a net for a prism and how to find the surface area of a prism by adding unit squares.

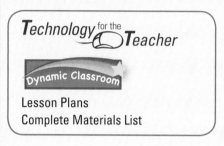

Lesson Plans
Complete Materials List

9.1 Record and Practice Journal

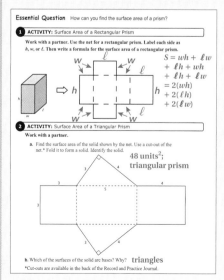

Differentiated Instruction

Kinesthetic

After students draw a net and find the surface area, have them cut out the net, fold it, and tape it together to create the prism.

9.1 Record and Practice Journal

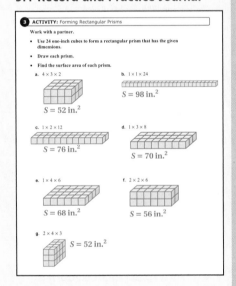

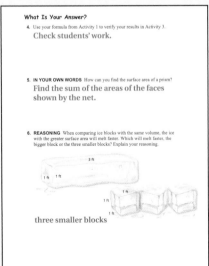

Laurie's Notes

Activity 3

- To help facilitate this activity, you can package 24 one-inch cubes in plastic bags before class.
- **?** "How does the $4 \times 3 \times 2$ prism compare to the $2 \times 4 \times 3$ prism?" The volumes are the same. One has a base of 4×3 and a height of 2. The other has a base of 2×4 and a height of 3.
- **?** "Was the surface area always the same?" no
- **Common Misconception:** Students often think that because the number of wooden cubes stays the same (volume), the surface area should stay the same as well. This activity should help students recognize that the surface areas vary.

What Is Your Answer?

- Review the answers to the questions together as a class. Have volunteers share their answers.

Closure

- Draw the net for a pizza box. Label with approximate dimensions and find the surface area.

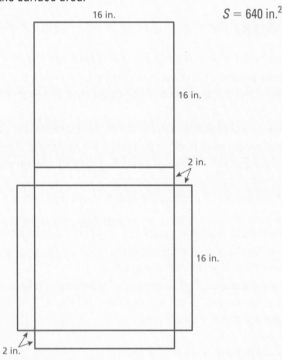

3 ACTIVITY: Forming Rectangular Prisms

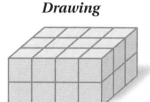

Math Practice

Construct Arguments

What method did you use to find the surface area of the rectangular prism? Explain.

Work with a partner.

- Use 24 one-inch cubes to form a rectangular prism that has the given dimensions.
- Draw each prism.
- Find the surface area of each prism.

a. $4 \times 3 \times 2$ ***Drawing***   ***Surface Area***

[ ] in.^{2}

b. $1 \times 1 \times 24$ **c.** $1 \times 2 \times 12$ **d.** $1 \times 3 \times 8$

e. $1 \times 4 \times 6$ **f.** $2 \times 2 \times 6$ **g.** $2 \times 4 \times 3$

What Is Your Answer?

4. Use your formula from Activity 1 to verify your results in Activity 3.

5. IN YOUR OWN WORDS How can you find the surface area of a prism?

6. REASONING When comparing ice blocks with the same volume, the ice with the greater surface area will melt faster. Which will melt faster, the bigger block or the three smaller blocks? Explain your reasoning.

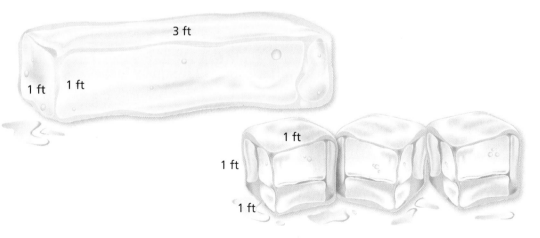

Practice

Use what you learned about the surface areas of rectangular prisms to complete Exercises 4–6 on page 359.

Key Vocabulary
lateral surface area,
   p. 358

 Key Idea

Surface Area of a Rectangular Prism

Words   The surface area S of a rectangular prism is the sum of the areas of the bases and the lateral faces.

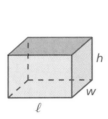

Algebra $S = 2\ell w + 2\ell h + 2wh$

Areas of
bases

Areas of
lateral faces

EXAMPLE 1   Finding the Surface Area of a Rectangular Prism

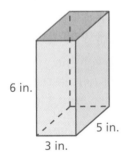

6 in.

5 in.

3 in.

Find the surface area of the prism.

Draw a net.

$$S = 2\ell w + 2\ell h + 2wh$$
$$= 2(3)(5) + 2(3)(6) + 2(5)(6)$$
$$= 30 + 36 + 60$$
$$= 126$$

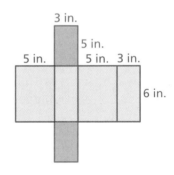

3 in.

5 in.

5 in.   5 in.   3 in.

6 in.

∴ The surface area is 126 square inches.

On Your Own

Now You're Ready
Exercises 7–9

Find the surface area of the prism.

1.

4 ft

3 ft

2 ft

2.

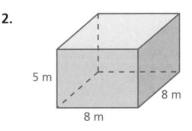

5 m

8 m

8 m

◀)) Multi-Language Glossary at BigIdeasMath.com

Laurie's Notes

Introduction

Connect
- **Yesterday:** Students explored surface area using the nets of two prisms.
- **Today:** Students will work with a formula for the surface area of a prism.

Motivate
- Hold a prism, perhaps a box that has a wrapping on it. It could be wrapping paper or simply clear shrink wrapping.
- ❓ "Why would I need to know the surface area of this prism?" Listen for the concept of wrapping the box in some type of paper.
- Explain to students that when items are mass produced, someone needs to calculate the amount of wrapping needed to cover the prism. The wrapping is the surface area.

Lesson Notes

Key Idea
- Have a net available as a visual, whether it is a cardboard box or a rectangular prism made from the snap together polygon frames.
- Note the color coding of the formula, prism, and net.
- **FYI:** Remind students that any two opposite faces can be called the bases. Once the bases are identified, the remaining 4 faces form the lateral portion.
- ❓ "Can you explain why there are three parts to finding the surface area of the rectangular prism?" Students should recognize that there are 3 pairs of congruent faces.

Example 1
- The challenge for students is not to get bogged down in symbols. Students need to remember that there are 3 pairs of congruent faces. They need to make sure that they calculate the area of each one of the 3 different faces, then double the answer to account for the pair.
- **Common Error:** When students multiply (2)(3)(5), they sometimes multiply $2 \times 3 \times 2 \times 5$, similar to using the Distributive Property. Remind them that multiplication is both commutative and associative, and they can multiply in order ($2 \times 3 \times 5$) or in a different order ($2 \times 5 \times 3$).
- **Teaching Tip:** Write the equation for surface area as:
  $S =$ bases + sides + (front and back). Students follow the words and find the area of each pair without thinking about the variables.

On Your Own
- Students need to record their work neatly so they, and you, can look back and see what corrections are needed.
- **Think-Pair-Share:** Students should read each question independently and then work in pairs to answer the questions. When they have answered the questions, the pair should compare their answers with another group and discuss any discrepancies.

Lesson Tutorials
Lesson Plans
Answer Presentation Tool

Extra Example 1
Find the surface area of a rectangular prism with a length of 6 yards, a width of 4 yards, and a height of 9 yards.
228 yd²

On Your Own
1. 52 ft²
2. 288 m²

Differentiated Instruction

Visual

Use a rectangular box to demonstrate three ways of finding surface area. The first method is to find the area of each face and then add areas. The second method is to open the box into a net and find the area of the net. The third method is to use the formula for finding surface area. Students should see that the three methods have the same result.

Extra Example 2

Find the surface area of the triangular prism.

216 in.^{2}

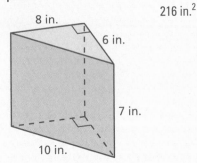

On Your Own

3. 150 m^2

4. 60 cm^2

Key Idea

- This is the general formula for a prism without variables. Most students are comfortable with this form.

? "How many faces are there that make up the bases?" two

? "How many faces are there that make up the lateral faces?" It depends on how many sides the bases have.

Example 2

- Note that the net is a visual reminder of each face whose area must be found. Color coding the faces should help students keep track of their work.

- **Model with Mathematics**: Encourage students to write the formula in words for each new problem: $S =$ area of bases $+$ areas of lateral faces.

On Your Own

- Give students sufficient time to do their work before asking volunteers to share their work *and* sketch at the board.

- Students having difficulty with Question 4 may want to redraw the triangular prism with the base on the bottom.

 **Key Idea**

Surface Area of a Prism

The surface area S of any prism is the sum of the areas of the bases and the lateral faces.

$$S = \text{areas of bases} + \text{areas of lateral faces}$$

EXAMPLE 2 Finding the Surface Area of a Triangular Prism

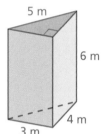

Find the surface area of the prism.

Draw a net.

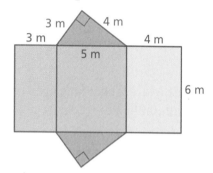

Area of a Base

Red base: $\frac{1}{2} \cdot 3 \cdot 4 = 6$

Areas of Lateral Faces

Green lateral face: $3 \cdot 6 = 18$

Purple lateral face: $5 \cdot 6 = 30$

Blue lateral face: $4 \cdot 6 = 24$

Add the areas of the bases and the lateral faces.

$S = \text{areas of bases} + \text{areas of lateral faces}$

$ = 6 + 6 + 18 + 30 + 24$

There are two identical bases. Count the area twice.

$ = 84$

∴ The surface area is 84 square meters.

On Your Own

Now You're Ready
Exercises 10–12

Find the surface area of the prism.

3.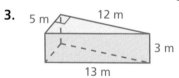

4.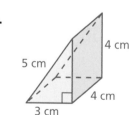

When all the edges of a rectangular prism have the same length *s*, the rectangular prism is a cube. The formula for the surface area of a cube is

$$S = 6s^2.$$ Formula for surface area of a cube

EXAMPLE 3 Finding the Surface Area of a Cube

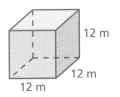

Find the surface area of the cube.

$$S = 6s^2$$ Write formula for surface area of a cube.

$$= 6(12)^2$$ Substitute 12 for *s*.

$$= 864$$ Simplify.

∴ The surface area of the cube is 864 square meters.

The **lateral surface area** of a prism is the sum of the areas of the lateral faces.

EXAMPLE 4 Real-Life Application

The outsides of purple traps are coated with glue to catch emerald ash borers. You make your own trap in the shape of a rectangular prism with an open top and bottom. What is the surface area that you need to coat with glue?

Find the lateral surface area.

$$S = 2\ell h + 2wh$$ ← Do not include the areas of the bases in the formula.

$$= 2(12)(20) + 2(10)(20)$$ Substitute.

$$= 480 + 400$$ Multiply.

$$= 880$$ Add.

∴ So, you need to coat 880 square inches with glue.

On Your Own

5. Which prism has the greater surface area?

6. WHAT IF? In Example 4, both the length and the width of your trap are 12 inches. What is the surface area that you need to coat with glue?

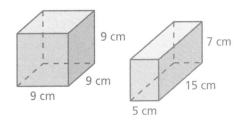

Laurie's Notes

Discuss

- Ask students to discuss how they would find the surface area of a cube. Students should recognize that all six faces are congruent, so finding the area of one face and multiplying by 6 will give you the surface area of the cube.

Example 3

- Write the formula for the surface area of a cube.
- Substitute for the variable and work through the problem.

Example 4

- **FYI:** The emerald ash borer is an exotic beetle that was discovered in southeastern Michigan near Detroit in the summer of 2002. The adult beetles nibble on ash foliage, but cause little damage. The larvae (the immature stage) feed on the inner bark of ash trees, disrupting the tree's ability to transport water and nutrients. The emerald ash borer probably arrived in the United States on solid wood packing material carried in cargo ships or airplanes originating in its native Asia. Since its discovery in Michigan, the emerald ash borer has killed millions of ash trees.
- ❓ "How many faces does the trap have?" four
- ❓ "How will you find the surface area of the trap?" Only find the area of the lateral faces.
- Write the modified surface area formula and work through the problem.
- Encourage students to write the formula in words.

    S = area of lateral faces

On Your Own

- These two problems are a good summary of the lesson. Students must find the surface area of each prism and then compare the answers. The surface areas are not that different so students cannot simply guess.

Closure

- **Writing:** Hold the prism from the Motivate section of today's lesson and ask students to write about how they would find the surface area of the prism. Look for: finding the areas of the bases and lateral faces and adding those together, *or* unwrapping the prism and finding the area of the wrapping.

Extra Example 3

Find the surface area of the cube.

54 yd^2

Extra Example 4

In Example 4, the height of the trap is 30 inches. What is the surface area that you need to coat with glue? 1320 in.^{2}

On Your Own

5. the cube

6. 960 in.^{2}

Vocabulary and Concept Check

1. *Sample answer:* 1) Use a net. 2) Use the formula $S = 2\ell w + 2\ell h + 2wh$.

2. same number of faces, vertices and edges; A cube is made of 6 square faces. A rectangular prism is not.

3. Find the area of the bases of the prism; 24 in.^{2}; 122 in.^{2}

Practice and Problem Solving

4.

22 in.^{2}

5.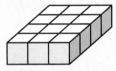

38 in.^{2}

6.

32 in.^{2}

7. 324 m^2

8. 166 mm^2

9. 49.2 yd^2

10. 920 ft^2

11. 136 m^2

12. 382.5 in.^{2}

13. 294 yd^2

14. 1.5 cm^2

15. $2\frac{2}{3}$ ft^2

Assignment Guide and Homework Check

Level	Day 1 Activity Assignment	Day 2 Lesson Assignment	Homework Check
Basic	4–6, 29–32	1–3, 7–15 odd, 16, 17, 19	9, 13, 17, 19
Average	4–6, 29–32	1–3, 9–15 odd, 16, 20–26 even	11, 13, 20, 24
Advanced	4–6, 29–32	1–3, 8–26 even, 27, 28	8, 12, 14, 18, 26
Accelerated	1–6, 8–26 even, 27–32		8, 12, 14, 18, 26

Common Errors

- **Exercises 7–9** Students may find the area of only three of the faces instead of all six. Remind them that each face is paired with another. Show students the net of a rectangular solid to remind them of the six faces.
- **Exercises 7–9** Some students may multiply length by width by height to find the surface area. Show them that the surface area is the sum of the areas of all six faces, so they must multiply and add to find the solution.
- **Exercises 10–12** Students may try to use the formula for a rectangular prism to find the surface area of a triangular prism. Show them that this will not work by focusing on the area of the triangular base. For students who are struggling to identify all the faces, draw a net of the prism and tell them to label the length, width, and height of each part before finding the surface area.

9.1 Record and Practice Journal

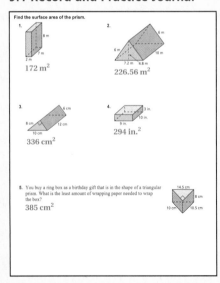

Vocabulary and Concept Check

1. **VOCABULARY** Describe two ways to find the surface area of a rectangular prism.

2. **WRITING** Compare and contrast a rectangular prism to a cube.

3. **DIFFERENT WORDS, SAME QUESTION** Which is different? Find "both" answers.

Find the surface area of the prism.	Find the area of the bases of the prism.

Find the area of the net of the prism.	Find the sum of the areas of the bases and the lateral faces of the prism.

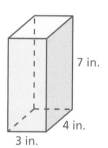

Practice and Problem Solving

Use one-inch cubes to form a rectangular prism that has the given dimensions. Then find the surface area of the prism.

4. $1 \times 2 \times 3$

5. $3 \times 4 \times 1$

6. $2 \times 3 \times 2$

Find the surface area of the prism.

① **7.**
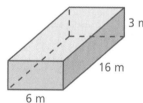

8.
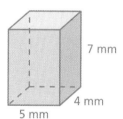

9.
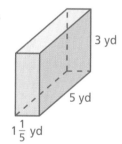

② **10.**
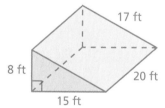

11.
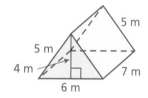

12.
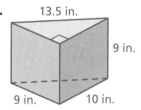

③ **13.**
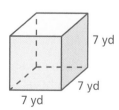

14.
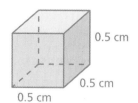

15.

16. ERROR ANALYSIS Describe and correct the error in finding the surface area of the prism.

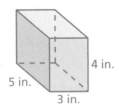

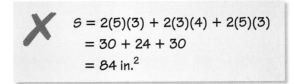

$$S = 2(5)(3) + 2(3)(4) + 2(5)(3)$$
$$= 30 + 24 + 30$$
$$= 84 \text{ in.}^2$$

17. GAME Find the surface area of the tin game case.

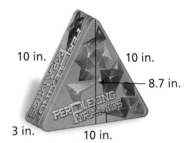

18. WRAPPING PAPER A cube-shaped gift is 11 centimeters long. What is the least amount of wrapping paper you need to wrap the gift?

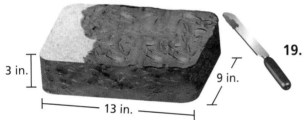

19. FROSTING One can of frosting covers about 280 square inches. Is one can of frosting enough to frost the cake? Explain.

Find the surface area of the prism.

20.

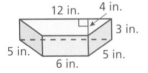

21.

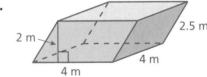

22. OPEN-ENDED Draw and label a rectangular prism that has a surface area of 158 square yards.

23. LABEL A label that wraps around a box of golf balls covers 75% of its lateral surface area. What is the value of x?

24. BREAD Fifty percent of the surface area of the bread is crust. What is the height h?

Common Errors

- **Exercise 19** Students may include the bottom of the cake as part of the area that needs to be frosted. Remind them to pay attention to the context of the problem.
- **Exercises 25 and 26** Students may use the ratio of a side length of the red prism to the corresponding side length of the blue prism. Remind them that the areas are not necessarily in the same proportions as the side lengths.

English Language Learners
Vocabulary
The term *net* means a two-dimensional representation of a solid figure. This definition is very different from its everyday meanings such as a device used to catch birds and fish or a fabric barricade dividing a tennis or volleyball court in half.

Practice and Problem Solving

16. The area of the 3×5 face is used 4 times rather than just twice.

    $S = 2(3)(5) + 2(3)(4) + 2(5)(4)$
    $ = 30 + 24 + 40$
    $ = 94$ in.^{2}

17. 177 in.^{2}

18. 726 cm^2

19. yes; Because you do not need to frost the bottom of the cake, you only need 249 square inches of frosting.

20. 156 in.^{2}

21. 68 m^2

22. Answers will vary.

23. $x = 4$

Practice and Problem Solving

24. See *Taking Math Deeper*.

25. The dimensions of the red prism are three times the dimensions of the blue prism. The surface area of the red prism is 9 times the surface area of the blue prism.

26–28. See Additional Answers.

Fair Game Review

29. 160 ft^2 **30.** 54 m^2

31. 28 ft^2 **32.** B

Mini-Assessment

Find the surface area of the prism.

1.

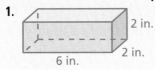

2 in.
6 in.
2 in.

56 in.^2

2.

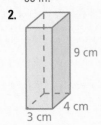

9 cm
4 cm
3 cm

150 cm^2

3. Find the least amount of wrapping paper needed to cover the box. 22 ft^2

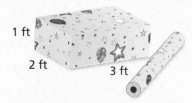

1 ft
2 ft
3 ft

4. Find the least amount of fabric needed to make the tent. 152 ft^2

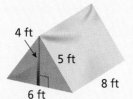

4 ft
5 ft
6 ft
8 ft

Taking Math Deeper

Exercise 24

To complete this problem, students may use a percent. Remind them to write the percent as a decimal before multiplying.

 Find the total surface area.

$$S = 2(10^2) + 2(10h) + 2(10h)$$
$$= 200 + 40h$$

h
10 cm
10 cm

2️⃣ Find the surface area of the crust.

$$C = 2(10h) + 2(10h)$$
$$= 40h$$

3️⃣ In part 1, notice that 200 is the area of the top and bottom, and $40h$ is the area of the crust. These areas are equal, so $40h = 200$, or $h = 5$. Otherwise you can write and solve an equation with *variables on both sides*.

Solve for h.

$$0.5(200 + 40h) = 40h$$
$$100 + 20h = 40h$$
$$100 = 20h$$
$$5 = h$$

So, the height of the bread is 5 centimeters.

Fun Bread Facts:

- It takes 1 second for a combine to harvest enough wheat to make about 8 loaves of bread.

- On average, each American consumes 53 pounds of bread per year.

- An average slice of packaged bread contains only 1 gram of fat and about 70 calories.

Reteaching and Enrichment Strategies

If students need help. . .	If students got it. . .
Resources by Chapter • Practice A and Practice B • Puzzle Time Record and Practice Journal Practice Differentiating the Lesson Lesson Tutorials Skills Review Handbook	Resources by Chapter • Enrichment and Extension • Technology Connection Start the next section

Compare the dimensions of the prisms. How many times greater is the surface area of the red prism than the surface area of the blue prism?

25.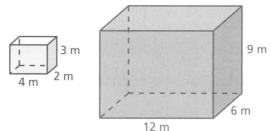

3 m
2 m
4 m
9 m
6 m
12 m

26.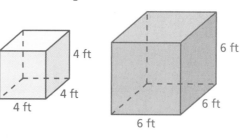

4 ft
4 ft
4 ft
6 ft
6 ft
6 ft

27. STRUCTURE You are painting the prize pedestals shown (including the bottoms). You need 0.5 pint of paint to paint the red pedestal.

  a. The side lengths of the green pedestal are one-half the side lengths of the red pedestal. How much paint do you need to paint the green pedestal?

  b. The side lengths of the blue pedestal are triple the side lengths of the green pedestal. How much paint do you need to paint the blue pedestal?

  c. Compare the ratio of paint amounts to the ratio of side lengths for the green and red pedestals. Repeat for the green and blue pedestals. What do you notice?

 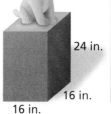 

24 in.
16 in.
16 in.

28. **Number Sense** A keychain-sized Rubik's Cube® is made up of small cubes. Each small cube has a surface area of 1.5 square inches.

  a. What is the side length of each small cube?

  b. What is the surface area of the entire Rubik's Cube®?

Fair Game Review What you learned in previous grades & lessons

Find the area of the triangle *(Skills Review Handbook)*

29.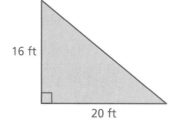

16 ft
20 ft

30.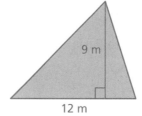

9 m
12 m

31.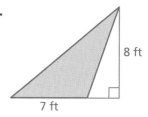

8 ft
7 ft

32. MULTIPLE CHOICE What is the circumference of the basketball? Use 3.14 for π. *(Section 8.1)*

9 in.

  Ⓐ 14.13 in.   **Ⓑ** 28.26 in.   **Ⓒ** 56.52 in.   **Ⓓ** 254.34 in.

Essential Question How can you find the surface area of a pyramid?

Even though many well-known pyramids have square bases, the base of a pyramid can be any polygon.

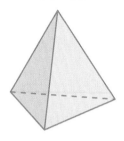

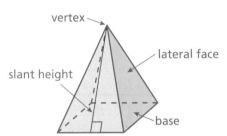

vertex

lateral face

slant height

base

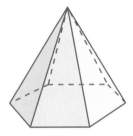

Triangular Base　　　　**Square Base**　　　　**Hexagonal Base**

---

1 ACTIVITY: Making a Scale Model

Work with a partner. Each pyramid has a square base.

- **Draw a net for a scale model of one of the pyramids. Describe your scale.**
- **Cut out the net and fold it to form a pyramid.**
- **Find the lateral surface area of the real-life pyramid.**

a. Cheops Pyramid in Egypt

Side = 230 m, Slant height ≈ 186 m

b. Muttart Conservatory in Edmonton

Side = 26 m, Slant height ≈ 27 m

c. Louvre Pyramid in Paris

Side = 35 m, Slant height ≈ 28 m

d. Pyramid of Caius Cestius in Rome

Side = 22 m, Slant height ≈ 29 m

Geometry

In this lesson, you will
- find surface areas of regular pyramids.
- solve real-life problems.

Laurie's Notes

What Your Students Will Learn
- Find surface areas of square and triangular pyramids using formulas.

Previous Learning
Students should know how to find the area of a triangle and should know the general properties of squares and isosceles triangles.

Introduction

Applying Mathematical Practices
- **Model with Mathematics:** Finding the surface area of a solid can be challenging when the faces are not all visible. Drawing a net allows students to see all the faces of the pyramid.

Motivate
- Share information about the Great Pyramid of Egypt, also known as Cheops Pyramid.
- The Great Pyramid is the largest of the original *Seven Wonders of the World*. It was built in the 5th century B.C. and is estimated to have taken 100,000 men over 20 years to build it.
- The Great Pyramid is a square pyramid. It covers an area of 13 acres. The original height of the Great Pyramid was 485 feet, but due to erosion its height has declined to 450 feet. Each side of the square base is 755.5 feet in length (about 2.5 football field lengths).
- The Great Pyramid consists of approximately 2.5 million blocks that weigh from 2 tons to over 70 tons. The stones are cut so precisely that a credit card cannot fit between them.

Technology for the Teacher

Dynamic Classroom

Lesson Plans
Complete Materials List

Activity Notes

Discuss
- Discuss the vocabulary of pyramids and how they are named according to the base. Make a distinction between the slant height and the height.

Activity 1
- This activity connects scale drawings with the study of pyramids.
- This activity may take up most of the class period, but there may be time to complete Activity 2 or 3 as well.
- To ensure a variety, assign one pyramid to each pair of students and make sure about $\frac{1}{4}$ of the class makes each pyramid.
- Students will need to decide on the scale they will use.
  Example: To make a scale model for pyramid A, assume the scale selected is 1 cm = 20 m.

$$\frac{1 \text{ cm}}{20 \text{ m}} = \frac{x \text{ cm}}{230 \text{ m}} \rightarrow x = 11.5 \qquad \frac{1 \text{ cm}}{20 \text{ m}} = \frac{x \text{ cm}}{186 \text{ m}} \rightarrow x = 9.3$$

- Students will use their eyesight and knowledge of squares and isosceles triangles to construct the square and four isosceles triangles.
- **Attend to Precision:** Have groups discuss the scale they used and how they found the lateral surface area. Listen for how clearly they communicate the process. Multiply the area of one triangular face by 4.
- **Model with Mathematics:** Making a pyramid from a net and finding the lateral surface area helps students remember the process and understand the formula.

9.2 Record and Practice Journal

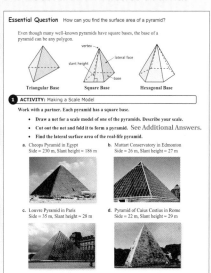

English Language Learners

Vocabulary

English learners may struggle with understanding the *slant height* of a pyramid. Use a skateboard ramp as an example. Ask students to find the length of the ramp. Most likely students will find the length of the slanted portion of the ramp. Compare this length to the slant height of a pyramid.

Activity 2

- Note that students are only asked to find the lateral surface area. This means that they are finding the surface area of 8 congruent isosceles triangles.
- **FYI:** The prefix octa- means eight. An octopus has 8 arms; when October was named in the Roman calendar, it was the 8th month; an octave on the piano has 8 notes; an octad is a group of 8 things.
- **?** "What common road sign is an octagon?" Stop sign
- The net includes the octagonal base, but the surface area of the base is not needed for this problem.
- Ask a volunteer to sketch his or her net at the board. If you have the appropriate snap together polygon frames, make the net.

Activity 3

- Students can reason that the area of a triangle with 8-inch sides must be less than the area of a square with 8-inch sides. By reasoning that the lateral surface area of the triangular pyramid is also less than that of the square pyramid, they can answer part (a) without any computations.
- **?** "What is true about the lateral faces of both pyramids?" All the lateral faces of the two pyramids are congruent triangles.
- **?** "Which pyramid has the greater lateral surface area? Explain." The square pyramid has a greater lateral surface area, because there is one more isosceles triangle.

What Is Your Answer?

- Have students work in pairs to answer the question.

Closure

- **Exit Ticket:** Sketch a net for a hexagonal pyramid and describe how to find the lateral surface area. Find the area of one of the lateral faces and multiply by 6.

9.2 Record and Practice Journal

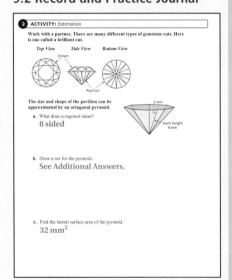

2 ACTIVITY: Estimation

Work with a partner. There are many different types of gemstone cuts. Here is one called a brilliant cut.

The size and shape of the pavilion can be approximated by an octagonal pyramid.

a. What does *octagonal* mean?
   8 sided

b. Draw a net for the pyramid.
   See Additional Answers.

c. Find the lateral surface area of the pyramid.
   32 mm²

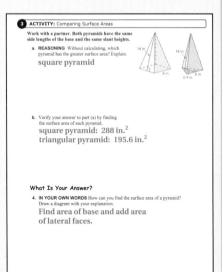

3 ACTIVITY: Comparing Surface Areas

Work with a partner. Both pyramids have the same side lengths of the base and the same slant heights.

a. **REASONING** Without calculating, which pyramid has the greater surface area? Explain.
   square pyramid

b. Verify your answer to part (a) by finding the surface area of each pyramid.
   square pyramid: 288 in.²
   triangular pyramid: 195.6 in.²

What Is Your Answer?

4. **IN YOUR OWN WORDS** How can you find the surface area of a pyramid? Draw a diagram with your explanation.
   Find area of base and add area of lateral faces.

2 **ACTIVITY: Estimation**

Work with a partner. There are many different types of gemstone cuts. Here is one called a brilliant cut.

Top View   *Side View*   *Bottom View*

Crown

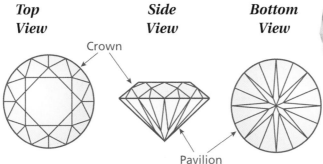

Pavilion

The size and shape of the pavilion can be approximated by an octagonal pyramid.

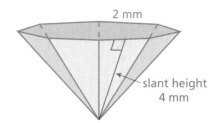

2 mm

slant height 4 mm

a. What does *octagonal* mean?

b. Draw a net for the pyramid.

c. Find the lateral surface area of the pyramid.

3 **ACTIVITY: Comparing Surface Areas**

Work with a partner. Both pyramids have the same side lengths of the base and the same slant heights.

a. **REASONING** Without calculating, which pyramid has the greater surface area? Explain.

b. Verify your answer to part (a) by finding the surface area of each pyramid.

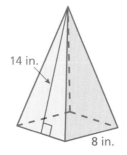

14 in.

8 in.

14 in.

8 in.

6.9 in.

What Is Your Answer?

4. **IN YOUR OWN WORDS** How can you find the surface area of a pyramid? Draw a diagram with your explanation.

Use what you learned about the surface area of a pyramid to complete Exercises 4–6 on page 366.

9.2 Lesson

Key Vocabulary
regular pyramid,
  p. 364
slant height, p. 364

A **regular pyramid** is a pyramid whose base is a regular polygon. The lateral faces are triangles. The height of each triangle is the **slant height** of the pyramid.

 **Key Idea**

Surface Area of a Pyramid

The surface area S of a pyramid is the sum of the areas of the base and the lateral faces.

Remember

In a regular polygon, all the sides are congruent and all the angles are congruent.

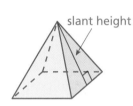

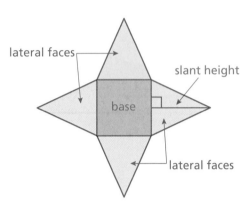

$S = $ area of base + areas of lateral faces

EXAMPLE 1 Finding the Surface Area of a Square Pyramid

Find the surface area of the regular pyramid.

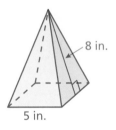

Draw a net.

Area of Base

$5 \cdot 5 = 25$

Area of a Lateral Face

$\dfrac{1}{2} \cdot 5 \cdot 8 = 20$

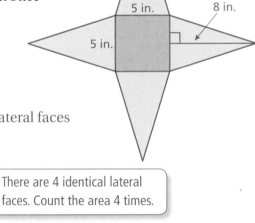

Find the sum of the areas of the base and the lateral faces.

$S = $ area of base + areas of lateral faces

$= 25 + 20 + 20 + 20 + 20$

$= 105$

There are 4 identical lateral faces. Count the area 4 times.

∴ The surface area is 105 square inches.

On Your Own

1. What is the surface area of a square pyramid with a base side length of 9 centimeters and a slant height of 7 centimeters?

Laurie's Notes

Introduction

Connect
- **Yesterday:** Students discovered how to find the surface area of a pyramid by examining the net that makes up a pyramid.
- **Today:** Students will work with a formula for the surface area of a pyramid.

Motivate
- Ask students where they have heard about pyramids or have seen them before. Give groups of students 3–4 minutes to brainstorm a list. They may mention the pyramid on the back of U.S. dollar bills, camping tents, roof designs, tetrahedral dice, and of course, Egyptian pyramids.

Technology for the **Teacher**

Dynamic Classroom

Lesson Tutorials
Lesson Plans
Answer Presentation Tool

Lesson Notes

Key Idea
- Introduce the vocabulary: regular pyramid, regular polygon, slant height.
- **?** "What information does the type of base give you about the lateral faces?" number of sides in the base = number of congruent isosceles triangles for the lateral surface area
- **?** "If you know the length of each side of the base, what else do you know?" the length of the base of the triangular lateral faces

Example 1
- **Model with Mathematics:** Draw the net and label the known information. This should remind students of the work they did yesterday making a scale model of a pyramid.
- Write the formula in words first to model good problem-solving techniques.
- Continue to ask questions as you find the total surface area: "How do you find the area of the base? How many lateral faces are there? What is the area of just one lateral face? How do you find the area of a triangle?"
- **Common Error:** In using the area formula for a triangle, the $\frac{1}{2}$ often produces a computation mistake. In this instance, students must multiply $\frac{1}{2} \times 5 \times 8$. Remind students that it's okay to change the order of the factors (Commutative Property). Rewriting the problem as $\frac{1}{2} \times 8 \times 5$ means that you can work with whole numbers: $\frac{1}{2} \times 8 \times 5 = 4 \times 5 = 20$.

Extra Example 1

What is the surface area of a square pyramid with a base side length of 3 meters and a slant height of 6 meters? 45 m^2

On Your Own
- **Model with Mathematics:** Encourage students to sketch a three-dimensional model of the pyramid and the net for the pyramid. Label the net with the known information.
- Ask a volunteer to share his or her work at the board.

On Your Own
1. 207 cm^2

Extra Example 2

Find the surface area of the regular pyramid.

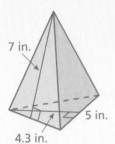

7 in.

5 in.

4.3 in.

63.25 in.²

Extra Example 3

The slant height of the roof in Example 3 is 13 feet. One bundle of shingles covers 30 square feet. How many bundles of shingles should you buy to cover the roof? 16 bundles of shingles

 On Your Own

   2. 105.6 ft²

   3. 17 bundles

Differentiated Instruction

Kinesthetic

Photocopy nets of solids for students to cut out and assemble. Then have students draw their own nets to cut out and assemble.

Laurie's Notes

Example 2

- Remind students of the definition of a regular pyramid. This is important because the base, as drawn, doesn't look like an equilateral triangle. This is the challenge of representing a 3-dimensional figure on a flat 2-dimensional sheet of paper.
- Drawing the net is an important step. It allows the key dimensions to be labeled in a way that can be seen.
- Encourage mental math when multiplying $\frac{1}{2} \times 10 \times 8.7$ and $\frac{1}{2} \times 10 \times 14$. Ask students to share their strategies with other students.

Example 3

? "How does the lateral surface area of the roof relate to the bundles of shingles needed?" The lateral surface area divided by the area covered per bundle gives the number of bundles needed.

- Have students compute the lateral surface area. Some students may need to draw the triangular lateral face first before performing the computation.

? **Reason Abstractly and Quantitatively:** "Suppose you compute the number of bundles needed on another roof, and you get an exact answer of 25.2. Is it okay to round down to 25 bundles? Explain." No; Always round up to the next whole number, so you do not run short of shingles.

- **FYI:** When shingles are placed on a roof, they need to overlap the shingle below. The coverage given per bundle takes into account the overlap.
- **Extension:** Suppose a bundle of shingles sells for $34.75. What will the total cost be for the shingles? $764.50

On Your Own

- Give students sufficient time to do their work for each problem before asking volunteers to share their work at the board.

Closure

- **Exit Ticket:** Sketch a square pyramid with a slant height of 4 centimeters and a base side length of 3 centimeters. Sketch the net and find the surface area. 33 cm²

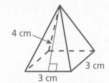
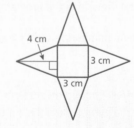

4 cm

3 cm

3 cm

4 cm

3 cm

3 cm

EXAMPLE 2 Finding the Surface Area of a Triangular Pyramid

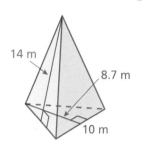

Find the surface area of the regular pyramid.

Draw a net.

Area of Base	**Area of a Lateral Face**
$\frac{1}{2} \cdot 10 \cdot 8.7 = 43.5$	$\frac{1}{2} \cdot 10 \cdot 14 = 70$

Find the sum of the areas of the base and the lateral faces.

$$S = \text{area of base} + \text{areas of lateral faces}$$
$$= 43.5 + \underbrace{70 + 70 + 70}$$
$$= 253.5$$

> There are 3 identical lateral faces. Count the area 3 times.

⋮· The surface area is 253.5 square meters.

EXAMPLE 3 Real-Life Application

A roof is shaped like a square pyramid. One bundle of shingles covers 25 square feet. How many bundles should you buy to cover the roof?

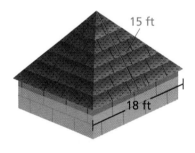

The base of the roof does not need shingles. So, find the sum of the areas of the lateral faces of the pyramid.

Area of a Lateral Face

$$\frac{1}{2} \cdot 18 \cdot 15 = 135$$

There are four identical lateral faces. So, the lateral surface area is

$$135 + 135 + 135 + 135 = 540.$$

Because one bundle of shingles covers 25 square feet, it will take $540 \div 25 = 21.6$ bundles to cover the roof.

⋮· So, you should buy 22 bundles of shingles.

On Your Own

Now You're Ready
Exercises 7–12

2. What is the surface area of the regular pyramid at the right?

3. **WHAT IF?** In Example 3, one bundle of shingles covers 32 square feet. How many bundles should you buy to cover the roof?

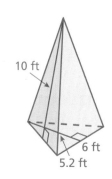

Vocabulary and Concept Check

1. **VOCABULARY** Can a pyramid have rectangles as lateral faces? Explain.

2. **CRITICAL THINKING** Why is it helpful to know the slant height of a pyramid to find its surface area?

3. **WHICH ONE DOESN'T BELONG?** Which description of the solid does *not* belong with the other three? Explain your answer.

square pyramid	regular pyramid
rectangular pyramid	triangular pyramid

5 m
5 m

Practice and Problem Solving

Use the net to find the surface area of the regular pyramid.

4.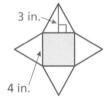
3 in.
4 in.

5.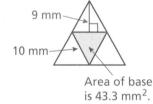
9 mm
10 mm
Area of base is 43.3 mm².

6.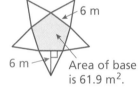
6 m
6 m
Area of base is 61.9 m².

In Exercises 7–11, find the surface area of the regular pyramid.

① ② 7.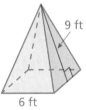
9 ft
6 ft

8.
6 cm
4 cm

9.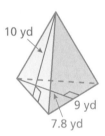
10 yd
9 yd
7.8 yd

10.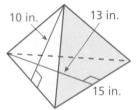
10 in.
13 in.
15 in.

11.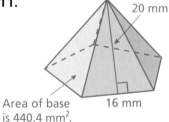
20 mm
Area of base is 440.4 mm².
16 mm

10 in.

③ 12. **LAMPSHADE** The base of the lampshade is a regular hexagon with a side length of 8 inches. Estimate the amount of glass needed to make the lampshade.

13. **GEOMETRY** The surface area of a square pyramid is 85 square meters. The base length is 5 meters. What is the slant height?

Assignment Guide and Homework Check

Level	Day 1 Activity Assignment	Day 2 Lesson Assignment	Homework Check
Basic	4–6, 21–24	1–3, 7–19 odd	2, 9, 13, 15, 17
Average	4–6, 21–24	1–3, 7–13 odd, 14–18 even, 19	2, 9, 13, 14, 18
Advanced	4–6, 21–24	1–3, 8–20 even	10, 12, 16, 18, 20
Accelerated	1–6, 8–20 even, 21–24		10, 12, 16, 18, 20

Common Errors

- **Exercises 7–11** Students may forget to add on the area of the base when finding the surface area. Remind them that when asked to find the surface area, the base is included.
- **Exercises 7–11** Students may add the wrong number of lateral face areas to the area of the base. Examine several different pyramids with different bases and ask if they can find a relationship between the number of sides of the base and the number of lateral faces. (They are the same.) Remind students that the number of sides on the base determines how many triangles make up the lateral surface area.
- **Exercise 12** Students may think that there is not enough information to solve the problem because it is not all labeled in the picture. Tell them to use the information in the word problem to finish labeling the picture. Also ask students to identify how many lateral faces are part of the lamp before they find the area of one face.

Vocabulary and Concept Check

1. no; The lateral faces of a pyramid are triangles.

2. Knowing the slant height helps because it represents the height of the triangle that makes up each lateral face. So, the slant height helps you to find the area of each lateral face.

3. triangular pyramid; The other three are names for the pyramid.

Practice and Problem Solving

4. 40 in.²

5. 178.3 mm²

6. 151.9 m²

7. 144 ft²

8. 64 cm²

9. 170.1 yd²

10. 322.5 in.²

11. 1240.4 mm²

12. 240 in.²

13. 6 m

9.2 Record and Practice Journal

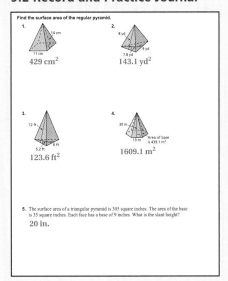

T-366

Practice and Problem Solving

14. 165 ft^2 **15.** 283.5 cm^2

16. 281 ft^2

17. See *Taking Math Deeper.*

18–20. See *Additional Answers.*

Fair Game Review

21. $A \approx 452.16 \text{ units}^2$;
$C \approx 75.36 \text{ units}$

22. $A \approx 200.96 \text{ units}^2$;
$C \approx 50.24 \text{ units}$

23. $A \approx 572.265 \text{ units}^2$;
$C \approx 84.78 \text{ units}$

24. B

Mini-Assessment

Find the surface area of the regular pyramid.

1.

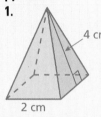

4 cm

2 cm

20 cm^2

2.

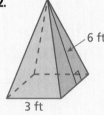

6 ft

3 ft

45 ft^2

3.

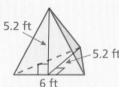

5.2 ft

5.2 ft

6 ft

62.4 ft^2

4. Find the surface area of the roof of the doll house. 480 in.^2

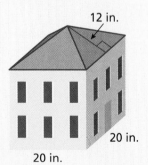

12 in.

20 in.

20 in.

T-367

Taking Math Deeper

Exercise 17

Creating something with fabric takes precision. You need to be precise with your measurements and pattern placement. Once you cut the fabric, there is no way to undo it.

1 **a.** Find the area of the 8 pieces.

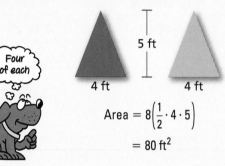

Four of each

5 ft

4 ft   4 ft

$\text{Area} = 8\left(\frac{1}{2} \cdot 4 \cdot 5\right)$

$= 80 \text{ ft}^2$

2 **b.** Draw a diagram.

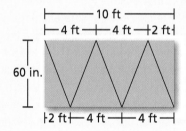

10 ft

4 ft — 4 ft — 2 ft

60 in.

2 ft — 4 ft — 4 ft

3 Answer the question.

For each color, you cut the four pieces from fabric that is 5 feet wide and 10 feet long.

Fabric Area $= 2(5 \cdot 10) = 100 \text{ ft}^2$
Area of 8 pieces $= 80 \text{ ft}^2$

c. Area of waste $= 100 - 80 = 20 \text{ ft}^2$

Project

Use construction paper and a pencil to create an "umbrella" using the least possible amount of paper. The umbrella should be a scale model of the one in the exercise, using a scale of 1 inch to 1 foot.

Reteaching and Enrichment Strategies

If students need help...	If students got it...
Resources by Chapter • Practice A and Practice B • Puzzle Time Record and Practice Journal Practice Differentiating the Lesson Lesson Tutorials Skills Review Handbook	Resources by Chapter • Enrichment and Extension • Technology Connection Start the next section

Find the surface area of the composite solid.

14.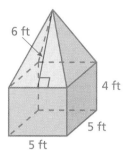

6 ft
4 ft
5 ft
5 ft

15.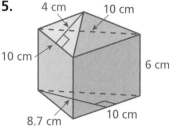

4 cm
10 cm
10 cm
6 cm
8.7 cm
10 cm

16.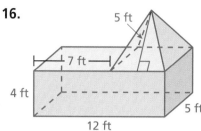

5 ft
7 ft
4 ft
5 ft
12 ft

17. PROBLEM SOLVING You are making an umbrella that is shaped like a regular octagonal pyramid.

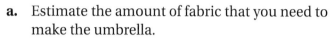

a. Estimate the amount of fabric that you need to make the umbrella.

b. The fabric comes in rolls that are 60 inches wide. Draw a diagram of how you can cut the fabric from a roll that is 10 feet long.

c. How much fabric is wasted?

5 ft

4 ft

18. REASONING The *height* of a pyramid is the perpendicular distance between the base and the top of the pyramid. Which is greater, the height of a pyramid or the slant height? Explain your reasoning.

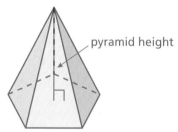

pyramid height

19. TETRAHEDRON A tetrahedron is a triangular pyramid whose four faces are identical equilateral triangles. The total lateral surface area is 93 square centimeters. Find the surface area of the tetrahedron.

20. Reasoning Is the total area of the lateral faces of a pyramid *greater than*, *less than*, or *equal to* the area of the base? Explain.

 Fair Game Review What you learned in previous grades & lessons

Find the area and the circumference of the circle. Use 3.14 for π.
(Section 8.1 and Section 8.3)

21.

12

22.

8

23.

27

24. MULTIPLE CHOICE The distance between bases on a youth baseball field is proportional to the distance between bases on a professional baseball field. The ratio of the youth distance to the professional distance is 2 : 3. Bases on a youth baseball field are 60 feet apart. What is the distance between bases on a professional baseball field? *(Section 5.4)*

Ⓐ 40 ft   Ⓑ 90 ft   Ⓒ 120 ft   Ⓓ 180 ft

Essential Question How can you find the surface area of a cylinder?

A *cylinder* is a solid that has two parallel, identical circular bases.

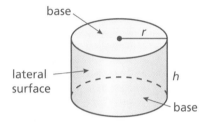

base

r

lateral surface

h

base

1 ACTIVITY: Finding Area

Work with a partner. Use a cardboard cylinder.

- **Talk about how you can find the area of the outside of the roll.**

- **Estimate the area using the methods you discussed.**

- **Use the roll and the scissors to find the actual area of the cardboard.**

- **Compare the actual area to your estimates.**

2 ACTIVITY: Finding Surface Area

Work with a partner.

Geometry

In this lesson, you will
- find surface areas of cylinders.

- **Make a net for the can. Name the shapes in the net.**

- **Find the surface area of the can.**

- **How are the dimensions of the rectangle related to the dimensions of the can?**

Laurie's Notes

Introduction

Applying Mathematical Practices

- **Model with Mathematics:** Drawing, cutting, and measuring the components of a cylinder allow students to make sense of the formula for the surface area of a cylinder.

Motivate

- Use two different cans (cylinders), where the taller can has a lesser radius. A tuna can and a 6-ounce vegetable can work well.
- ❓ Hold both cans. "Which can required more metal to make?" Answers will vary depending on can sizes.
- This question focuses attention on the surface areas of the cans, the need to consider their components, and how they were made.

Activity Notes

Activity 1

- **Big Idea:** Recall the connection between a prism and a cylinder— *structurally they are the same*. Unlike the prism, you don't have cardboard models to unfold. Use cardboard rolls from paper towels or toilet paper, or make rolls from strips of file folder paper.
- By estimating the area before they cut, students are engaging their spatial skills to think about the area of a curved surface.
- ❓ "What shape do you have when the roll is flattened out?" rectangle
- ❓ "How are the dimensions of the rectangle related to the dimensions of the original cylinder?" The height of the rectangle is the height of the cylinder. The length of the rectangle is the circumference of the cylinder.
- ❓ "What units do you use to measure the dimensions of your rectangle?" depending upon the ruler, centimeters or inches
- ❓ "What units do you use to label your answer?" cm^2 or $in.^2$

Activity 2

- Each pair of students will need scrap paper, tape, scissors, and a can. It is more interesting when the cans around the room are of different sizes.
- **Teaching Tip:** Recycle plastic bags and tape to desks for trash disposal.
- Students wrap paper around the can to make the net for the lateral surface. This helps them relate the cylinder and rectangle dimensions.
- **Discuss Results:** Students should describe the parts of a cylinder, how to find the area of each part, and how the dimensions of each part are related to the dimensions of the cylinder.
- **Make Sense of Problems and Persevere in Solving Them:** Making the cylinder and calculating its surface area help students to remember the process and understand the formula.

What Your Students Will Learn

- Find surface areas of cylinders using a formula.

Previous Learning

Students should know the formulas for the area of a rectangle and the area of a circle.

Lesson Plans
Complete Materials List

9.3 Record and Practice Journal

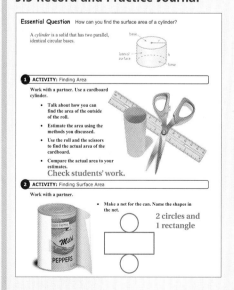

English Language Learners

Vocabulary

Have students work in pairs, one English learner and one English speaker. Have each pair write a problem involving the surface area of a cylinder. On a separate piece of paper, students should solve their own problem. Then have students exchange their problem with another pair of students. Students solve the new problem. After solving the problem, the four students discuss the problems and solutions.

Activity 3

- When estimating the dimensions of the common cylinders, encourage students to use their hands to visualize the size of the cylinder.
- If time permits, set up stations in the room with a different cylinder at each. Provide rulers. In small groups, students move from one station to the next. Make sure you have a good variety of common cylinders: soup can, soft drink can, tuna can, AA battery, etc.

? "What dimensions did you measure for each cylinder?" Students will often say diameter and height.

? "Could the radius be measured?" yes; Take $\frac{1}{2}$ of the diameter to find the radius.

- **Extension:** Gather the results of the activity, and then find the mean for several of the cylinders.

What Is Your Answer?

- Have students work in pairs. Review answers as a class.

Closure

- Hold the two cans from the Motivate section and ask, "Which of the two cans from the beginning of today's lesson required more metal to make?" Answers will vary depending on can sizes.

9.3 Record and Practice Journal

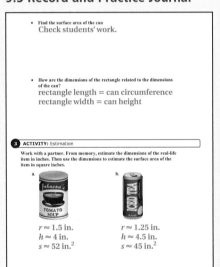

- Find the surface area of the can
  Check students' work.

- How are the dimensions of the rectangle related to the dimensions of the can?
  rectangle length = can circumference
  rectangle width = can height

3 ACTIVITY: Estimation

Work with a partner. From memory, estimate the dimensions of the real-life item in inches. Then use the dimensions to estimate the surface area of the item in square inches.

a.
$r \approx 1.5$ in.
$h \approx 4$ in.
$s \approx 52$ in.^{2}

b.
$r \approx 1.25$ in.
$h \approx 4.5$ in.
$s \approx 45$ in.^{2}

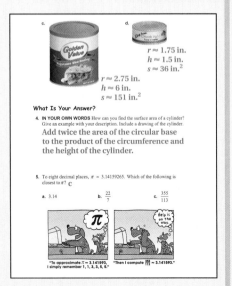

c.
$r \approx 2.75$ in.
$h \approx 6$ in.
$s \approx 151$ in.^{2}

d.
$r \approx 1.75$ in.
$h \approx 1.5$ in.
$s \approx 36$ in.^{2}

What Is Your Answer?

4. **IN YOUR OWN WORDS** How can you find the surface area of a cylinder? Give an example with your description. Include a drawing of the cylinder.
   Add twice the area of the circular base to the product of the circumference and the height of the cylinder.

5. To eight decimal places, $\pi \approx 3.14159265$. Which of the following is closest to π? **c**
   a. 3.14   b. $\frac{22}{7}$ c. $\frac{355}{113}$

Work with a partner. From memory, estimate the dimensions of the real-life item in inches. Then use the dimensions to estimate the surface area of the item in square inches.

a.

b.

c.

d.

What Is Your Answer?

4. **IN YOUR OWN WORDS** How can you find the surface area of a cylinder? Give an example with your description. Include a drawing of the cylinder.

5. To eight decimal places, $\pi \approx 3.14159265$. Which of the following is closest to π?

   a. 3.14 **b.** $\dfrac{22}{7}$ **c.** $\dfrac{355}{113}$

"To approximate $\pi \approx 3.141593$, I simply remember 1, 1, 3, 3, 5, 5."

"Then I compute $\frac{355}{113} \approx 3.141593$."

Practice

Use what you learned about the surface area of a cylinder to complete Exercises 3–5 on page 372.

Key Idea

Surface Area of a Cylinder

Words The surface area S of a cylinder is the sum of the areas of the bases and the lateral surface.

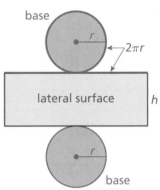

Remember

Pi can be approximated as 3.14 or $\dfrac{22}{7}$.

Algebra $S = 2\pi r^2 + 2\pi rh$

Areas of bases

Area of lateral surface

EXAMPLE 1 **Finding the Surface Area of a Cylinder**

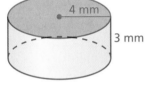

Find the surface area of the cylinder. Round your answer to the nearest tenth.

Draw a net.

$$S = 2\pi r^2 + 2\pi rh$$

$$= 2\pi(4)^2 + 2\pi(4)(3)$$

$$= 32\pi + 24\pi$$

$$= 56\pi$$

$$\approx 175.8$$

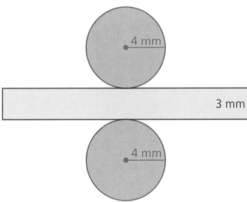

∴ The surface area is about 175.8 square millimeters.

On Your Own

Now You're Ready
Exercises 6–8

Find the surface area of the cylinder. Round your answer to the nearest tenth.

1.
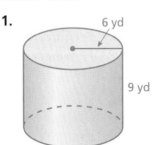

2.
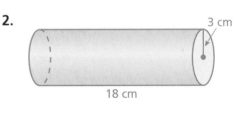

Laurie's Notes

Introduction

Connect

- **Yesterday:** Students discovered how to find the surface area of a cylinder by examining the net that makes up a cylinder.
- **Today:** Students will work with a formula for the surface area of a cylinder.

Motivate

- Find two cans (cylinders) that have volumes in the ratio of 1 : 2 (i.e., 10 fl oz and 20 fl oz).
- ❓ "The larger can has twice the volume of the smaller can. Do you think the surface area is twice as much?" Answers may differ, but most students believe this is true. Return to this question at the end of the lesson.

Lesson Notes

Key Idea

- ❓ "How are cylinders and rectangular prisms alike?" Both have 2 congruent bases and a lateral portion. "Different?" Cylinders have circular bases, while the rectangular prism has a rectangular base.
- Refer to the diagram with the radius marked. Review the formulas for area and circumference.
- Write the formula in words first. Before writing the formula in symbols, ask direct questions to help students make the connection between the words and the symbols.
- ❓ "How do you find the area of the bases?" Find the area of one base, πr^2, and then multiply by 2.
- ❓ "How do you find the area of the lateral portion?" The dimensions of its rectangular net are the height and circumference of the cylinder, so the area of the lateral portion is $2\pi rh$.
- Write the formula in symbols with each part identified (area of bases + lateral surface area).

Example 1

- Write the formula first to model good problem-solving techniques.
- Notice that the values of the variables are substituted, with each term being left in terms of π.
- **Common Misconception:** Students are unsure of how to perform the multiplication with π in the middle of the term. Remind students that π is a number, a factor in this case, just like the other numbers. Because of the Commutative and Associative Properties, the whole numbers can be multiplied first. Then the two like terms, 32π and 24π, are combined. The last step is to substitute 3.14 for π.
- ❓ Review ≈. "What does this symbol mean? Why is it used?" approximately equal to; π is an irrational number and an estimate for pi is used in the calculation.

On Your Own

- Ask volunteers to share their answers.

Goal Today's lesson is finding the surface area of a cylinder using a formula.

Technology for the Teacher

Dynamic Classroom

Lesson Tutorials
Lesson Plans
Answer Presentation Tool

Extra Example 1

Find the surface area of a cylinder with a radius of 3 inches and a height of 4 inches. Round your answer to the nearest tenth. $42\pi \approx 131.9$ in.^{2}

On Your Own

1. $180\pi \approx 565.2$ yd^2
2. $126\pi \approx 395.6$ cm^2

Extra Example 2

Find the lateral surface area of a cylinder with a radius of 2 inches and a height of 6 inches. Round your answer to the nearest tenth. $24\pi \approx 75.4$ in.^{2}

Extra Example 3

You earn $0.07 for recycling a can with a radius of 3 inches and a height of 4 inches (Extra Example 1). How much can you expect to earn for recycling a can with a radius of 2 inches and a height of 6 inches (Extra Example 2)? Assume that the recycle value is proportional to the surface area. $0.05

 On Your Own

3. a. yes

    b. no; Only the lateral surface area doubled. Because the surface area of the can does not double, the recycle value does not double.

Differentiated Instruction

Visual

Encourage students to estimate their answers for reasonableness. For the surface area of a cylinder, a common error is using the diameter in the formula instead of the radius. Have students imagine the cylinder inside of a rectangular prism. By calculating the surface area of the prism, the student has an overestimate of the surface area of the cylinder.

Laurie's Notes

Example 2

- Note that only the lateral surface area is asked for in this example.
- If you have a small can with a label on it, use it as a model for this problem.
- Note again that the answer is left in terms of π until the last step.

Example 3

- **Make Sense of Problems and Persevere in Solving Them:** Ask a volunteer to read the problem. Check to see if students understand what is being asked in this problem.
- **?** "How does the surface area of the can relate to the recycling value?" The value of the recycled can is proportional to the surface area of the can.
- Have students compute the surface area of each can.
- **?** "Approximately how much more metal is there in the larger can?" 24π in.$^2 \approx 75$ in.^{2}
- **?** "How many times more metal is there in the larger can?" 5 times
- This is a good review of proportions. Calculators are helpful.

On Your Own

- Give students sufficient time to do their work before asking volunteers to share their work at the board.

Closure

- Hold the two cans from the Motivate section of today's lesson and ask students to find the surface area of each. Answers will vary depending on can sizes.

EXAMPLE **2** **Finding Surface Area**

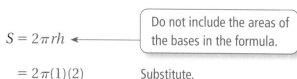

How much paper is used for the label on the can of peas?

1 in.

2 in.

Find the lateral surface area of the cylinder.

$$S = 2\pi rh \longleftarrow$$

Do not include the areas of the bases in the formula.

$$= 2\pi(1)(2) \qquad \text{Substitute.}$$

$$= 4\pi \approx 12.56 \qquad \text{Multiply.}$$

∴ About 12.56 square inches of paper is used for the label.

EXAMPLE **3** **Real-Life Application**

2 in.

5.5 in.

You earn \$0.01 for recycling the can in Example 2. How much can you expect to earn for recycling the tomato can? Assume that the recycle value is proportional to the surface area.

Find the surface area of each can.

Tomatoes

$$S = 2\pi r^2 + 2\pi rh$$

$$= 2\pi(2)^2 + 2\pi(2)(5.5)$$

$$= 8\pi + 22\pi$$

$$= 30\pi$$

Peas

$$S = 2\pi r^2 + 2\pi rh$$

$$= 2\pi(1)^2 + 2\pi(1)(2)$$

$$= 2\pi + 4\pi$$

$$= 6\pi$$

Use a proportion to find the recycle value x of the tomato can.

$$\frac{30\pi\ \text{in.}^2}{x} = \frac{6\pi\ \text{in.}^2}{\$0.01} \longleftarrow \begin{array}{l}\text{surface area}\end{array}$$

$\longleftarrow$ recycle value

$$30\pi \cdot 0.01 = x \cdot 6\pi \qquad \text{Cross Products Property}$$

$$5 \cdot 0.01 = x \qquad \text{Divide each side by } 6\pi.$$

$$0.05 = x \qquad \text{Simplify.}$$

∴ You can expect to earn \$0.05 for recycling the tomato can.

● **On Your Own**

Now You're Ready
Exercises 9–11

3. **WHAT IF?** In Examples 2 and 3, the height of the can of peas is doubled.

   a. Does the amount of paper used in the label double?

   b. Does the recycle value double? Explain.

 ## Vocabulary and Concept Check

1. **CRITICAL THINKING** Which part of the formula $S = 2\pi r^2 + 2\pi rh$ represents the lateral surface area of a cylinder?

2. **CRITICAL THINKING** You are given the height and the circumference of the base of a cylinder. Describe how to find the surface area of the entire cylinder.

 ## Practice and Problem Solving

Make a net for the cylinder. Then find the surface area of the cylinder. Round your answer to the nearest tenth.

3.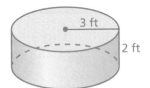
   3 ft
   2 ft

4.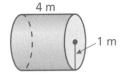
   4 m
   1 m

5.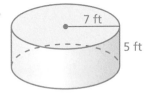
   7 ft
   5 ft

Find the surface area of the cylinder. Round your answer to the nearest tenth.

① 6.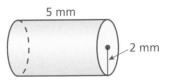
   5 mm
   2 mm

7.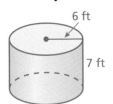
   6 ft
   7 ft

8.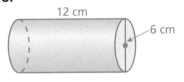
   12 cm
   6 cm

Find the lateral surface area of the cylinder. Round your answer to the nearest tenth.

② 9.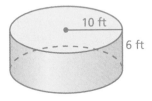
   10 ft
   6 ft

10.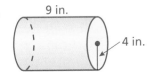
    9 in.
    4 in.

11.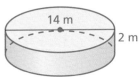
    14 m
    2 m

12. **ERROR ANALYSIS** Describe and correct the error in finding the surface area of the cylinder.

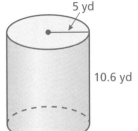
5 yd
10.6 yd

$$
\begin{aligned}
S &= \pi r^2 + 2\pi rh \\
&= \pi(5)^2 + 2\pi(5)(10.6) \\
&= 25\pi + 106\pi \\
&= 131\pi \approx 411.3 \text{ yd}^2
\end{aligned}
$$

13. **TANKER** The truck's tank is a stainless steel cylinder. Find the surface area of the tank.

50 ft

radius = 4 ft

Assignment Guide and Homework Check

Level	Day 1 Activity Assignment	Day 2 Lesson Assignment	Homework Check
Basic	3–5, 19–22	1, 2, 7–11 odd, 12, 13–17 odd	7, 9, 11, 13, 15
Average	3–5, 19–22	1, 2, 6–12 even, 15, 17	8, 10, 15, 17
Advanced	3–5, 19–22	1, 2, 6–18 even	8, 10, 14, 16, 18
Accelerated	1–5, 6–18 even, 19–22		8, 10, 14, 16, 18

Common Errors

- **Exercises 6–8** Students may add the area of only one base. Remind them of the net for a cylinder and that there are two circles as bases.
- **Exercises 6–8** Students may double the radius instead of squaring it. Remind them of the area of a circle and also the order of operations.
- **Exercise 8** Students may use the diameter instead of the radius. Remind them that the radius is in the formula, so they should find the radius before finding the surface area.
- **Exercises 9–11** Students may multiply the height by the area of the circular base instead of the circumference. Review with them how the lateral surface is created to show that the length of the rectangle is the circumference of the circular bases.

9.3 Record and Practice Journal

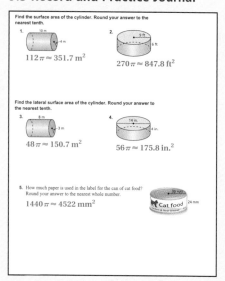

1. $2\pi rh$

2. Use the given circumference to find the radius by solving $C = 2\pi r$ for r. Then use the formula for the surface area of a cylinder.

Practice and Problem Solving

3.

$30\pi \approx 94.2 \text{ ft}^2$

4.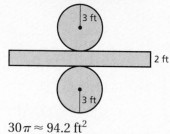

$10\pi \approx 31.4 \text{ m}^2$

5.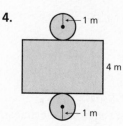

$168\pi \approx 527.5 \text{ ft}^2$

6. $28\pi \approx 87.9 \text{ mm}^2$

7. $156\pi \approx 489.8 \text{ ft}^2$

8. $90\pi \approx 282.6 \text{ cm}^2$

9. $120\pi \approx 376.8 \text{ ft}^2$

10. $72\pi \approx 226.1 \text{ in.}^2$

11. $28\pi \approx 87.9 \text{ m}^2$

12. See Additional Answers.

13. $432\pi \approx 1356.48 \text{ ft}^2$

14. about 36.4%

15. The surface area of the cylinder with the height of 8.5 inches is greater than the surface area of the cylinder with the height of 11 inches.

16. a. $S = 41.125\pi \approx 129.1 \text{ cm}^2$,
$S = 149.875\pi \approx 470.6 \text{ cm}^2$

   b. about 4.0 lb

17. See *Taking Math Deeper.*

18. See *Additional Answers.*

Fair Game Review

19. 10 ft^2 **20.** 16 cm^2

21. 47.5 in.^2 **22.** C

Mini-Assessment

Find the surface area of the cylinder. Round your answer to the nearest tenth.

1.   3 ft   **2.**   2 in.

   8 ft      6 in.

$66\pi \approx 207.2 \text{ ft}^2$ $14\pi \approx 44.0 \text{ in.}^2$

3. Find the surface area of the roll of paper towels.

 11 in.

$67.5\pi \approx 212.0 \text{ in.}^2$

├─5 in.─┤

4. How much paper is used for the label on the can of tuna?

2 in.

1 in.

12.56 in.^2

Taking Math Deeper

Exercise 17

This is a real-life problem. That is, when you leave cheese in the refrigerator without covering it, the amount that dries out is proportional to the surface area.

1   Find the surface area of the uncut cheese.

$$S = 2\pi r^2 + 2\pi rh$$
$$= 2\pi \cdot 3^2 + 2\pi \cdot 3 \cdot 1$$
$$= 24\pi$$
a. $\approx 75.36 \text{ in.}^2$

├── 3 in. ──┤

1 in.

2   Find the surface area of the remaining cheese. One-eighth of the surface area is removed. But, two 3-by-1 rectangular regions are added.

$$S = \frac{7}{8} \cdot 24\pi + 2(3 \cdot 1)$$
$$= 21\pi + 6$$
b. $\approx 71.94 \text{ in.}^2$

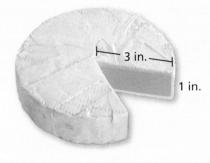
├── 3 in. ──┤

1 in.

1/8 removed

3   Answer the question.
   b. The surface area decreased.

Reteaching and Enrichment Strategies

If students need help. . .	If students got it. . .
Resources by Chapter • Practice A and Practice B • Puzzle Time Record and Practice Journal Practice Differentiating the Lesson Lesson Tutorials Skills Review Handbook	Resources by Chapter • Enrichment and Extension • Technology Connection Start the next section

14. OTTOMAN What percent of the surface area of the ottoman is green (not including the bottom)?

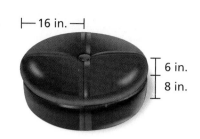
├─ 16 in. ─┤
6 in.
8 in.

15. REASONING You make two cylinders using 8.5-by-11-inch pieces of paper. One has a height of 8.5 inches, and the other has a height of 11 inches. Without calculating, compare the surface areas of the cylinders.

10 cm
3.5 cm
24.5 cm
5.5 cm

16. INSTRUMENT A *ganza* is a percussion instrument used in samba music.

  a. Find the surface area of each of the two labeled ganzas.

  b. The weight of the smaller ganza is 1.1 pounds. Assume that the surface area is proportional to the weight. What is the weight of the larger ganza?

17. BRIE CHEESE The cut wedge represents one-eighth of the cheese.

  a. Find the surface area of the cheese before it is cut.

  b. Find the surface area of the remaining cheese after the wedge is removed. Did the surface area increase, decrease, or remain the same?

├── 3 in. ──┤
1 in.

18. 🔄 *Repeated Reasoning* A cylinder has radius *r* and height *h*.

  a. How many times greater is the surface area of a cylinder when both dimensions are multiplied by a factor of 2? 3? 5? 10?

  b. Describe the pattern in part (a). How many times greater is the surface area of a cylinder when both dimensions are multiplied by a factor of 20?

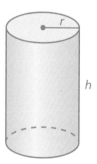
r
h

Fair Game Review What you learned in previous grades & lessons

Find the area. *(Skills Review Handbook)*

19.
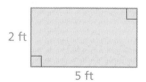
2 ft
5 ft

20.
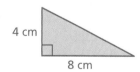
4 cm
8 cm

21.
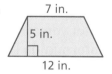
7 in.
5 in.
12 in.

22. MULTIPLE CHOICE 40% of what number is 80? *(Section 6.4)*

    Ⓐ 32   Ⓑ 48   Ⓒ 200   Ⓓ 320

You can use an **information frame** to help you organize and remember concepts.
Here is an example of an information frame for surface areas of rectangular prisms.

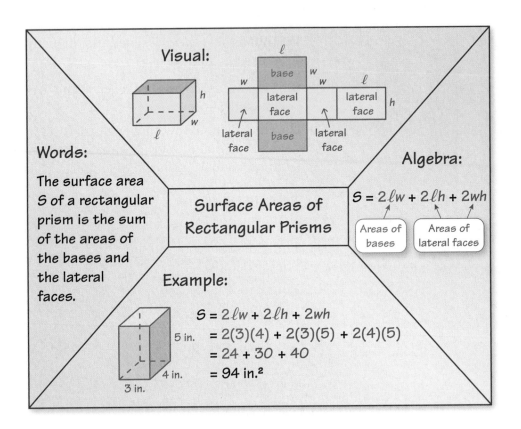

Visual:

Words:

The surface area S of a rectangular prism is the sum of the areas of the bases and the lateral faces.

Surface Areas of Rectangular Prisms

Algebra:

$S = 2\ell w + 2\ell h + 2wh$

Areas of bases

Areas of lateral faces

Example:

$S = 2\ell w + 2\ell h + 2wh$

$= 2(3)(4) + 2(3)(5) + 2(4)(5)$

$= 24 + 30 + 40$

$= 94 \text{ in.}^2$

5 in.

4 in.

3 in.

On Your Own

Make information frames to help you study the topics.

1. surface areas of prisms

2. surface areas of pyramids

3. surface areas of cylinders

After you complete this chapter, make information frames for the following topics.

4. volumes of prisms

5. volumes of pyramids

"I'm having trouble thinking of a good title for my information frame."

Sample Answers

1.

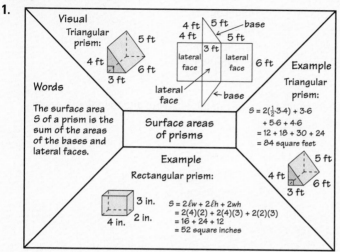

Visual

Triangular prism: 5 ft, 4 ft, 6 ft, 3 ft

4 ft, 5 ft, base, 4 ft, 5 ft, 3 ft, lateral face, 6 ft, lateral face, lateral face, base

Words

The surface area S of a prism is the sum of the areas of the bases and lateral faces.

Surface areas of prisms

Example

Triangular prism:

$S = 2(\frac{1}{2}\cdot 3\cdot 4) + 3\cdot 6$
$ + 5\cdot 6 + 4\cdot 6$
$ = 12 + 18 + 30 + 24$
$ = 84$ square feet

5 ft, 4 ft, 6 ft, 3 ft

Example

Rectangular prism:

3 in., 4 in., 2 in.

$S = 2\ell w + 2\ell h + 2wh$
$ = 2(4)(2) + 2(4)(3) + 2(2)(3)$
$ = 16 + 24 + 12$
$ = 52$ square inches

2.

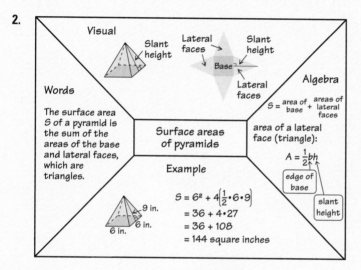

Visual

Slant height, Lateral faces, Slant height, Base, Lateral faces

Words

The surface area S of a pyramid is the sum of the areas of the base and lateral faces, which are triangles.

Surface areas of pyramids

Algebra

$S = $ area of base $+$ areas of lateral faces

area of a lateral face (triangle):

$A = \frac{1}{2}bh$

edge of base, slant height

Example

9 in., 6 in., 6 in.

$S = 6^2 + 4\left(\frac{1}{2}\cdot 6\cdot 9\right)$
$ = 36 + 4\cdot 27$
$ = 36 + 108$
$ = 144$ square inches

3.

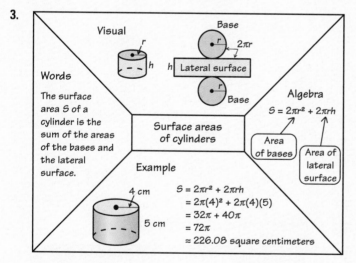

Visual

r, h

Base, r, $2\pi r$, h, Lateral surface, r, Base

Words

The surface area S of a cylinder is the sum of the areas of the bases and the lateral surface.

Surface areas of cylinders

Algebra

$S = 2\pi r^2 + 2\pi rh$

Area of bases, Area of lateral surface

Example

4 cm, 5 cm

$S = 2\pi r^2 + 2\pi rh$
$ = 2\pi(4)^2 + 2\pi(4)(5)$
$ = 32\pi + 40\pi$
$ = 72\pi$
$ \approx 226.08$ square centimeters

List of Organizers

Available at *BigIdeasMath.com*

Comparison Chart
Concept Circle
Example and Non-Example Chart
Formula Triangle
Four Square
Idea (Definition) and Examples Chart
Information Frame
Information Wheel
Notetaking Organizer
Process Diagram
Summary Triangle
Word Magnet
Y Chart

About this Organizer

An **Information Frame** can be used to help students organize and remember concepts. Students write the topic in the middle rectangle. Then students write related concepts in the spaces around the rectangle. Related concepts can include *Words, Numbers, Algebra, Example, Definition, Non-Example, Visual, Procedure, Details,* and *Vocabulary.* Students can place their information frames on note cards to use as a quick study reference.

Technology for the *Teacher*

Editable Graphic Organizer

Answers

1. 132 cm^2

2. 100 mm^2

3. 245 m^2

4. 28 cm^2

5. $78\pi \approx 244.9 \text{ ft}^2$

6. $110\pi \approx 345.4 \text{ m}^2$

7. $126\pi \approx 395.6 \text{ cm}^2$

8. $97.6\pi \approx 306.5 \text{ mm}^2$

9. **a.** 18 ft^2

   b. yes

10. $108\pi \approx 339.12 \text{ in.}^2$

11. $11,520 \text{ in.}^2$

Alternative Quiz Ideas

100% Quiz	Math Log
Error Notebook	Notebook Quiz
Group Quiz	Partner Quiz
Homework Quiz	Pass the Paper

100% Quiz

This is a quiz where students are given the answers and then they have to explain and justify each answer.

Reteaching and Enrichment Strategies

If students need help. . .	If students got it. . .
Resources by Chapter • Practice A and Practice B • Puzzle Time Lesson Tutorials *BigIdeasMath.com*	Resources by Chapter • Enrichment and Extension • Technology Connection Game Closet at *BigIdeasMath.com* Start the next section

Technology for the *Teacher*

Online Assessment
Assessment Book
ExamView® Assessment Suite

Find the surface area of the prism. *(Section 9.1)*

1.
3 cm   4 cm
10 cm
5 cm

2.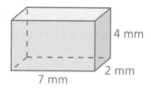
4 mm
2 mm
7 mm

Find the surface area of the regular pyramid. *(Section 9.2)*

3.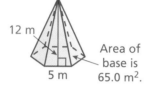
12 m
Area of base is 65.0 m².
5 m

4.
6 cm
2 cm

Find the surface area of the cylinder. Round your answer to the nearest tenth. *(Section 9.3)*

5.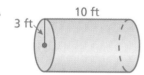
10 ft
3 ft

6.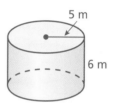
5 m
6 m

Find the lateral surface area of the cylinder. Round your answer to the nearest tenth. *(Section 9.3)*

7.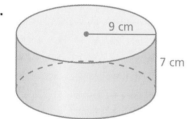
9 cm
7 cm

8.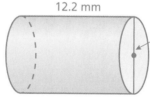
12.2 mm
8 mm

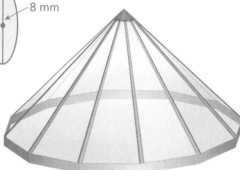

9. **SKYLIGHT** You are making a skylight that has 12 triangular pieces of glass and a slant height of 3 feet. Each triangular piece has a base of 1 foot. *(Section 9.2)*

   a. How much glass will you need to make the skylight?

   b. Can you cut the 12 glass triangles from a sheet of glass that is 4 feet by 8 feet? If so, draw a diagram showing how this can be done.

10. **MAILING TUBE** What is the least amount of material needed to make the mailing tube? *(Section 9.3)*

3 ft
3 in.

11. **WOODEN CHEST** All the faces of the wooden chest will be painted except for the bottom. Find the area to be painted, in *square inches*. *(Section 9.1)*

4 ft
4 ft   4 ft

Essential Question How can you find the volume of a prism?

1 ACTIVITY: Pearls in a Treasure Chest

Work with a partner. A treasure chest is filled with valuable pearls. Each pearl is about 1 centimeter in diameter and is worth about $80.

Use the diagrams below to describe two ways that you can estimate the number of pearls in the treasure chest.

a.

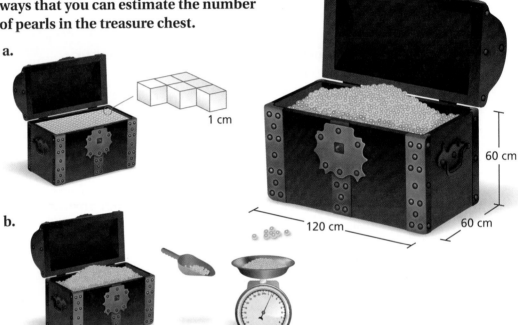

1 cm

60 cm

120 cm

60 cm

b.

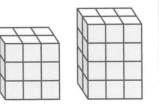

c. Use the method in part (a) to estimate the value of the pearls in the chest.

2 ACTIVITY: Finding a Formula for Volume

Geometry

In this lesson, you will
- find volumes of prisms.
- solve real-life problems.

Work with a partner. You know that the formula for the volume of a rectangular prism is $V = \ell wh$.

a. Write a formula that gives the volume in terms of the area of the base B and the height h.

b. Use both formulas to find the volume of each prism. Do both formulas give you the same volume?

Laurie's Notes

Introduction

Applying Mathematical Practices

- **Look for and Express Regularity in Repeated Reasoning:** The approach used in developing a formula for the volume of a prism is to consider repeated layers with the same base. Mathematically proficient students notice that each layer increases the volume by the number of units of the area of the base.

Motivate

- Hold up a variety of common containers and ask what is commonly found inside. Examples: egg carton (12 eggs); playing cards box (52 cards); crayon box (8 crayons)
- Discuss with students these examples of volume. Each container is filled with objects of the same size. How many eggs fit in the egg carton, or how many crayons fit in the crayon box? Because the units are different (eggs, cards, crayons), you can't compare the volumes.

Activity Notes

Activity 1

- If you have beads (marbles), use them to model this activity. "I've filled this box with beads. How would you estimate the number of beads in the box?"
- **?** "How big is the treasure chest? Compare it to an object in this room." Students should recognize that 120 centimeters is more than 3 feet long.
- **?** "Do you think there is a thousand dollars worth of pearls in the treasure chest? a million dollars? a billion dollars?" Students will likely have only a wild guess about the value of the pearls in the chest at this point.
- In part (a), listen for students to say you can fit a layer of centimeter cubes on the bottom and a total of 60 layers in the chest.
- **?** "How did you estimate the number of pearls using the method in part (a)?" The bottom layer holds about 7200 pearls. Times 60 layers is 432,000 pearls.
- **Attend to Precision:** Discuss how the estimate compares to the actual number of pearls. It will be less because a 1-centimeter pearl has less volume than a cubic centimeter, so more will fit in the chest.
- **?** Explain how the method in part (b) could be used to estimate the number of pearls in the chest. *Sample answer:* Weigh the chest full, and then empty to find the weight of the pearls. Then weigh 10 pearls and use this information to estimate the total number of pearls.

Activity 2

- From the figure, students should see that the bottom layer has 6 cubes, the second layer has 6 cubes, the third layer has 6 cubes, and so on.
- **Look for and Express Regularity in Repeated Reasoning:** If students are not thinking about layers (height), suggest that writing the volume of each prism would be helpful: 6, 12, 18, 24, 30.
- **Big Idea:** The area of the base (denoted *B*) is 6. The height (denoted *h*) is how many layers?

What Your Students Will Learn

- Find volumes of prisms using formulas.

Previous Learning

Students should know how to find areas of two-dimensional figures, surface areas of three-dimensional figures, and volumes of rectangular prisms using $V = \ell w h$ or by counting unit cubes.

Lesson Plans
Complete Materials List

9.4 Record and Practice Journal

Essential Question How can you find the volume of a prism?

1 ACTIVITY: Pearls in a Treasure Chest

Work with a partner. A treasure chest is filled with valuable pearls. Each pearl is about 1 centimeter in diameter and is worth about $80.

Use the diagrams below to describe two ways that you can estimate the number of pearls in the treasure chest.

a.

Because each pearl is about 1 cubic centimeter, the number of pearls in the chest can be estimated by multiplying the length, width, and height of the treasure chest.

b.

Weigh one pearl then weigh all the pearls.

c. Use the method in part (a) to estimate the value of the pearls in the chest.
$34,560,000

Visual

Students may think that prisms with the same volume have the same surface area. Have them work together to find prisms with the same volume, but different dimensions. Then direct the students to find the surface areas of each of the prisms. Ask them to share their results.

9.4 Record and Practice Journal

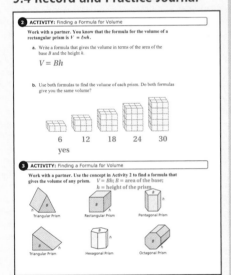

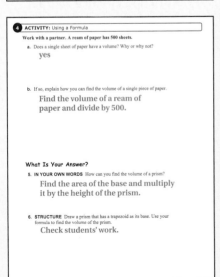

Laurie's Notes

Activity 3

- The formula discovered in Activity 2 is now used in Activity 3.
- **Make Sense of Problems and Persevere in Solving Them:** Students can memorize formulas and have little understanding of why the formula makes sense. It is important throughout this chapter that students see that the formulas are all similar. The volume is found by finding the area of the base (B) and then multiplying by the number of layers (h).
- **Model with Mathematics:** Having models of these prisms is very helpful.
- **Common Misconception:** The height of a prism does not need to be the vertical measure. Demonstrate this by holding a rectangular prism (a tissue box is fine). Ask students to identify the base (a face of the prism) and the height (an edge). Chances are students will identify the (standard) bottom of the box as the base. Now, rotate the tissue box so that the base is vertical. Again ask students to identify the base and height. Students may stick with their first answers or may now switch to the "bottom face" as the base.
- A prism is named by its base. A triangular prism has 2 triangular bases.
- **Attend to Precision:** Give students time to discuss the solids in this activity. If you have physical models of each of these, ask six volunteers to describe how to find the volume of the solid. Expect the student volunteer to point to the base and the height as they explain how to find the volume.

Activity 4

- **Common Misconception:** Students may believe a sheet of paper has no height and so, no volume, only area. It may be difficult to measure the height with tools available to us, but a sheet of paper does have a height.
- **Use Appropriate Tools Strategically:** In part (b), students should deduce that they can measure one sheet of paper indirectly by first measuring a whole ream of paper.
- The ream of copy paper is a good visual model.

What Is Your Answer?

- **Think-Pair-Share:** Students should read each question independently and then work in pairs to answer the questions. When they have answered the questions, the pair should compare their answers with another group and discuss any discrepancies.

Closure

- **Writing Prompt:** To find the volume of a tissue box …

3 ACTIVITY: Finding a Formula for Volume

Math Practice

Use a Formula
What are the given quantities? How can you use the quantities to write a formula?

Work with a partner. Use the concept in Activity 2 to find a formula that gives the volume of any prism.

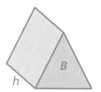

Triangular Prism

Rectangular Prism

Pentagonal Prism

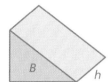

Triangular Prism

Hexagonal Prism

Octagonal Prism

4 ACTIVITY: Using a Formula

Work with a partner. A ream of paper has 500 sheets.

a. Does a single sheet of paper have a volume? Why or why not?

b. If so, explain how you can find the volume of a single sheet of paper.

What Is Your Answer?

5. **IN YOUR OWN WORDS** How can you find the volume of a prism?

6. **STRUCTURE** Draw a prism that has a trapezoid as its base. Use your formula to find the volume of the prism.

Use what you learned about the volumes of prisms to complete Exercises 4–6 on page 380.

Check It Out
Lesson Tutorials
BigIdeasMath ✓com

The *volume* of a three-dimensional figure is a measure of the amount of space that it occupies. Volume is measured in cubic units.

 Key Idea

Volume of a Prism

Words The volume V of a prism is the product of the area of the base and the height of the prism.

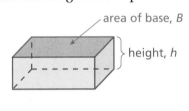

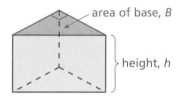

Algebra

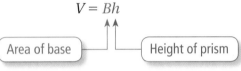

$$V = Bh$$

Area of base ⟶ ⟵ Height of prism

> **Remember**
>
> The volume V of a cube with an edge length of s is $V = s^3$.

EXAMPLE 1 **Finding the Volume of a Prism**

> **Study Tip**
>
> The area of the base of a rectangular prism is the product of the length ℓ and the width w.
>
> You can use $V = \ell wh$ to find the volume of a rectangular prism.

Find the volume of the prism.

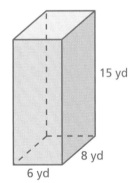

$V = Bh$	Write formula for volume.
$\quad = 6(8) \cdot 15$	Substitute.
$\quad = 48 \cdot 15$	Simplify.
$\quad = 720$	Multiply.

15 yd
8 yd
6 yd

∴ The volume is 720 cubic yards.

EXAMPLE 2 **Finding the Volume of a Prism**

Find the volume of the prism.

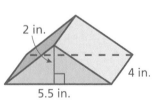

2 in.
4 in.
5.5 in.

$V = Bh$	Write formula for volume.
$\quad = \dfrac{1}{2}(5.5)(2) \cdot 4$	Substitute.
$\quad = 5.5 \cdot 4$	Simplify.
$\quad = 22$	Multiply.

∴ The volume is 22 cubic inches.

Laurie's Notes

Introduction

Connect
- **Yesterday:** Students explored how to find the volume of a prism.
- **Today:** Students will use the formula for the volume of a prism to solve problems.

Motivate
- **True Story:** Baseball legend Ken Griffey Jr. owed teammate Josh Fogg some money and paid him back in pennies. Griffey stacked 60 cartons, each holding $25 worth of pennies, in Fogg's locker.
- ❓ Ask the following questions.
  - "How does this story relate to the volume of a prism?" The volume of the carton is being measured in pennies.
  - "How big is a carton that can hold $25 worth of pennies?" open-ended
  - "How many pennies were in each carton?" 2500
  - "How much did Griffey owe Fogg?" $1500
  - "How much do you think each carton weighed?" A $25 carton of pennies weighs about 16 pounds.

Lesson Notes

Key Idea
- ❓ "What is a prism?" three-dimensional solid with two congruent bases and lateral faces that are rectangles
- ❓ "What are cubic units? Give an example." Cubic units are cubes which fill a space completely without overlapping or leaving gaps. Cubic inches and cubic centimeters are common examples.
- Point out to students that the bases of the prisms are shaded red. The height will be perpendicular to the two congruent bases. The dotted lines are edges that would not be visible through the solid prism.
- **Teaching Tip:** Use words (area of base, height) and symbols (B, h) when writing the formula.
- **Review Vocabulary:** *Product* is the answer to a multiplication problem.

Example 1
- Discuss the *Study Tip* with students.
- ❓ "Could the face measuring 8 yards by 15 yards be the base?" Yes, the height would then be 6 yards.
- **Extension:** Point out to students that all of the measurements are in terms of yards. "What if the 6 yard edge had been labeled 18 feet. Now how would you find the volume?" Convert all 3 dimensions to yards or to feet.

Example 2
- Ask a volunteer to describe the base of this triangular prism.
- ❓ "What property is used to simplify the area of the base?" Commutative Property of Multiplication
- Caution students to distinguish between the height of the base and the height of the prism.

Goal Today's lesson is finding the volumes of prisms.

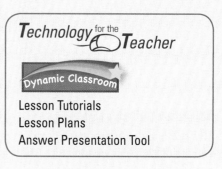

Technology for the Teacher

Dynamic Classroom

Lesson Tutorials
Lesson Plans
Answer Presentation Tool

Extra Example 1

Find the volume of a rectangular prism with a length of 2 meters, a width of 6 meters, and a height of 3 meters. 36 m^3

Extra Example 2

Find the volume of the prism.

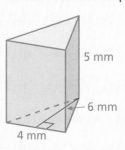

5 mm

6 mm

4 mm

60 mm^3

On Your Own

1. 64 ft^3

2. 270 m^3

Extra Example 3

Two rectangular prisms each have a volume of 120 cubic centimeters. The base of Prism A is 2 centimeters by 4 centimeters. The base of Prism B is 4 centimeters by 6 centimeters.

a. Find the height of each prism.
   Prism A: 15 cm, Prism B: 5 cm

b. Which prism has the lesser surface area? Prism B

On Your Own

3. yes; Because it has the same volume as the other two bags, but its surface area is 107.2 square inches which is less than both Bag A and Bag B.

English Language Learners

Vocabulary

Discuss the meaning of the words *volume* and *cubic units*. Have students add these words to their notebooks.

Laurie's Notes

On Your Own

- Ask volunteers to share their work at the board.

Example 3

- This example connects volume, surface area, and solving equations.
- Ask a student to read the example.
- **Make Sense of Problems and Persevere in Solving Them:** Ask probing questions to make sure that students understand the problem.
- "What type of measurement is 96 cubic inches?" volume
- "What type of measurement is part (b) concerned with?" surface area
- Work through part (a).
- Before beginning part (b), ask students to review the formula for surface area of a rectangular prism. Note that only five of the six faces are considered.
- Work through part (b).
- "Both bags hold the same amount of popcorn. Are there any practical advantages of one bag over the other?" Bag A: can grip it in your hand more easily; Bag B: less likely to tip over and uses less paper

On Your Own

- Given the discussion of the practical features of Bags A and B, students should have a sense of the problems in the design of Bag C.

Closure

- Sketch a rectangular prism. Label the dimensions 4 centimeters, 6 centimeters, and 10 centimeters. Find the volume of the prism.
  $V = 4 \cdot 6 \cdot 10 = 240$ cm^3

Now You're Ready
Exercises 4–12

On Your Own

Find the volume of the prism.

1.
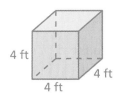
4 ft
4 ft
4 ft

2.
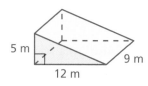
5 m
9 m
12 m

EXAMPLE ③ **Real-Life Application**

A movie theater designs two bags to hold 96 cubic inches of popcorn.
(a) Find the height of each bag. (b) Which bag should the theater
choose to reduce the amount of paper needed? Explain.

Bag A
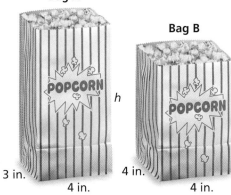
Bag B
POPCORN h
POPCORN h
3 in.   4 in.
4 in.   4 in.

a. Find the height of each bag.

Bag A	**Bag B**
$V = Bh$	$V = Bh$
$96 = 4(3)(h)$	$96 = 4(4)(h)$
$96 = 12h$	$96 = 16h$
$8 = h$	$6 = h$

⋮ The height is 8 inches. ⋮ The height is 6 inches.

b. To determine the amount of paper needed, find the surface
area of each bag. Do not include the top base.

Bag A	**Bag B**
$S = \ell w + 2\ell h + 2wh$	$S = \ell w + 2\ell h + 2wh$
$= 4(3) + 2(4)(8) + 2(3)(8)$	$= 4(4) + 2(4)(6) + 2(4)(6)$
$= 12 + 64 + 48$	$= 16 + 48 + 48$
$= 124 \text{ in.}^2$	$= 112 \text{ in.}^2$

⋮ The surface area of Bag B is less than the surface area of Bag A.
So, the theater should choose Bag B.

On Your Own

Bag C

POPCORN h
4 in.
4.8 in.

3. You design Bag C that has a volume of
96 cubic inches. Should the theater in
Example 3 choose your bag? Explain.

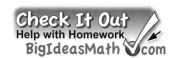
Check It Out
Help with Homework
BigIdeasMath.com

 Vocabulary and Concept Check

1. **VOCABULARY** What types of units are used to describe volume?

2. **VOCABULARY** Explain how to find the volume of a prism.

3. **CRITICAL THINKING** How are volume and surface area different?

 Practice and Problem Solving

Find the volume of the prism.

 4.
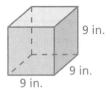
9 in.
9 in.
9 in.

5.
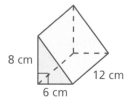
8 cm
12 cm
6 cm

6.

$8\frac{1}{2}$ m
7 m
4 m

7.
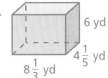
6 yd
$4\frac{1}{5}$ yd
$8\frac{1}{3}$ yd

8.

6 ft
9 ft
4.5 ft

9.
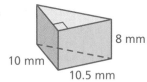
8 mm
10 mm
10.5 mm

10.
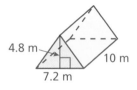
4.8 m
10 m
7.2 m

11.
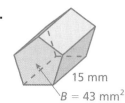
15 mm
$B = 43$ mm^2

12.

20 ft
$B = 166$ ft^2

13. **ERROR ANALYSIS** Describe and correct the error in finding the volume of the triangular prism.

7 cm
10 cm
5 cm

$$\begin{aligned}
V &= Bh \\
&= 10(5)(7) \\
&= 50 \cdot 7 \\
&= 350 \text{ cm}^3
\end{aligned}$$

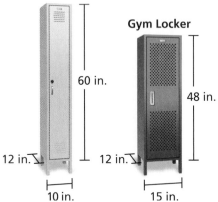
School Locker
60 in.
12 in.
10 in.

Gym Locker
48 in.
12 in.
15 in.

14. **LOCKER** Each locker is shaped like a rectangular prism. Which has more storage space? Explain.

15. **CEREAL BOX** A cereal box is 9 inches by 2.5 inches by 10 inches. What is the volume of the box?

Assignment Guide and Homework Check

Level	Day 1 Activity Assignment	Day 2 Lesson Assignment	Homework Check
Basic	4–6, 25–28	1–3, 7–21 odd	7, 11, 15, 17
Average	4–6, 25–28	1–3, 7–15 odd, 16–22 even	7, 9, 16, 18
Advanced	4–6, 25–28	1–3, 8–12 even, 13, 14–24 even	16, 18, 20, 22
Accelerated	1–6, 8–12 even, 13, 14–24 even, 25–28		16, 18, 20, 22

Common Errors

- **Exercises 4–12** Students may write the units incorrectly, often writing square units instead of cubic units. Remind them that they are working in three dimensions, so the units are cubed. Give an example showing the formula for the base as three units multiplied together. For example, write the volume of Exercise 5 as $V = \frac{1}{2}(6 \text{ cm})(8 \text{ cm})(12 \text{ cm})$.

Vocabulary and Concept Check

1. cubic units

2. Find the area of the base and multiply it by the height.

3. The volume of an object is the amount of space it occupies. The surface area of an object is the sum of the areas of all its faces.

Practice and Problem Solving

4. 729 in.^{3} 5. 288 cm^3

6. 238 m^3 7. 210 yd^3

8. 121.5 ft^3 9. 420 mm^3

10. 172.8 m^3 11. 645 mm^3

12. 3320 ft^3

13. The area of the base is wrong.
$$V = \frac{1}{2}(7)(5) \cdot 10$$
$$= 175 \text{ cm}^3$$

14. The gym locker has more storage space because it has a greater volume.

15. 225 in.^{3}

9.4 Record and Practice Journal

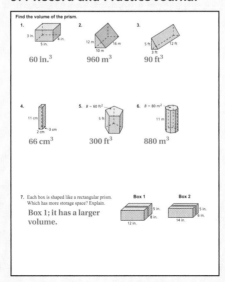

Find the volume of the prism.

1. 60 in.^{3} 2. 960 m^3 3. 90 ft^3

4. 66 cm^3 5. $B \approx 60 \text{ ft}^2$ 300 ft^3 6. $B \approx 80 \text{ m}^2$ 880 m^3

7. Each box is shaped like a rectangular prism. Which has more storage space? Explain.
Box 1; it has a larger volume.

Practice and Problem Solving

16. 1440 in.^{3} **17.** 7200 ft^3

18. sometimes; The prisms in Example 3 have different surface areas, but the same volume. Two prisms that are exactly the same will have the same surface area.

19. See Additional Answers.

20. 48 packets

21. 20 cm

22. *Sample answer:* gas about $3 per gallon; $36

23. See *Taking Math Deeper*.

24. The volume is 2 times greater; The volume is 8 times greater.

Fair Game Review

25. $90   **26.** $144

27. $240.50   **28.** D

Mini-Assessment

Find the volume of the prism.

1.
6 in.
5 in.
4 in.
120 in.^{3}

2.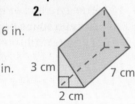
3 cm
7 cm
2 cm
21 cm^3

3.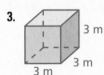
3 m
3 m
3 m
27 m^2

4. Find the volume of the fish tank.

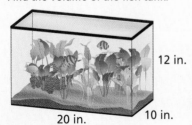
12 in.
20 in.
10 in.
2400 in.^{3}

Taking Math Deeper

Exercise 23

This problem gives students a chance to relate dimensions of a solid with the volume of a solid. It also gives students an opportunity to work with prime factorization. Although aquariums are traditionally the shape of a rectangular prism, remember that other shapes are also possible.

① Find the volume of the aquarium in cubic inches.

$$\text{Volume} = (450 \text{ gal})\left(231 \frac{\text{in.}^3}{\text{gal}}\right) = 103{,}950 \text{ in.}^3$$

② You could choose two of the dimensions and solve for the third, or you can use prime factorization to find whole number dimensions.

Find the prime factorization of 103,950.

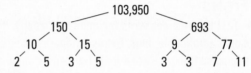

The prime factorization is $2 \times 3 \times 3 \times 3 \times 5 \times 5 \times 7 \times 11$.

③ Rearrange the factors to find one set of possible dimensions.

Length: $2 \times 3 \times 5 \times 5 = 150$ in.
Width: $3 \times 7 = 21$ in.
Height: $3 \times 11 = 33$ in.

A long tank

33 in.
150 in.
21 in.

Project

Research a local aquarium. Select an exhibit and draw a picture of the exhibit you selected. Include a short report about the details of the exhibit.

Reteaching and Enrichment Strategies

If students need help. . .	If students got it. . .
Resources by Chapter • Practice A and Practice B • Puzzle Time Record and Practice Journal Practice Differentiating the Lesson Lesson Tutorials Skills Review Handbook	Resources by Chapter • Enrichment and Extension • Technology Connection Start the next section

Find the volume of the prism.

16.

12 in.

12 in.   10 in.

17.

24 ft

30 ft

20 ft

18. LOGIC Two prisms have the same volume. Do they *always*, *sometimes*, or *never* have the same surface area? Explain.

19. CUBIC UNITS How many cubic inches are in a cubic foot? Use a sketch to explain your reasoning.

20. CAPACITY As a gift, you fill the calendar with packets of chocolate candy. Each packet has a volume of 2 cubic inches. Find the maximum number of packets you can fit inside the calendar.

6 in.

8 in.   4 in.

21. PRECISION Two liters of water are poured into an empty vase shaped like an octagonal prism. The base area is 100 square centimeters. What is the height of the water? ($1 \text{ L} = 1000 \text{ cm}^3$)

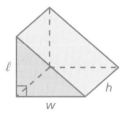

22. GAS TANK The gas tank is 20% full. Use the current price of regular gasoline in your community to find the cost to fill the tank. ($1 \text{ gal} = 231 \text{ in.}^3$)

23. OPEN-ENDED You visit an aquarium. One of the tanks at the aquarium holds 450 gallons of water. Draw a diagram to show one possible set of dimensions of the tank. ($1 \text{ gal} = 231 \text{ in.}^3$)

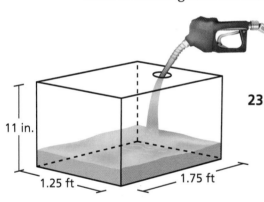

11 in.

1.25 ft   1.75 ft

24. *Critical Thinking* How many times greater is the volume of a triangular prism when one of its dimensions is doubled? when all three dimensions are doubled?

ℓ

h

w

Fair Game Review *What you learned in previous grades & lessons*

Find the selling price. *(Section 6.6)*

25. Cost to store: $75
Markup: 20%

26. Cost to store: $90
Markup: 60%

27. Cost to store: $130
Markup: 85%

28. MULTIPLE CHOICE What is the approximate surface area of a cylinder with a radius of 3 inches and a height of 10 inches? *(Section 9.3)*

A 30 in.^{2} **B** 87 in.^{2} **C** 217 in.^{2} **D** 245 in.^{2}

Essential Question How can you find the volume of a pyramid?

1 ACTIVITY: Finding a Formula Experimentally

Work with a partner.

- Draw the two nets on cardboard and cut them out.

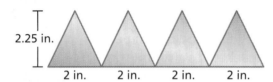

2.25 in.

2 in.   2 in.   2 in.   2 in.

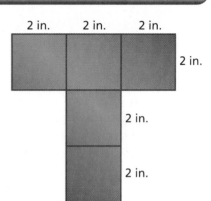

2 in.   2 in.   2 in.

2 in.

2 in.

2 in.

- Fold and tape the nets to form an open square box and an open pyramid.

- Both figures should have the same size square base and the same height.

- Fill the pyramid with pebbles. Then pour the pebbles into the box. Repeat this until the box is full. How many pyramids does it take to fill the box?

- Use your result to find a formula for the volume of a pyramid.

2 ACTIVITY: Comparing Volumes

Work with a partner. You are an archaeologist studying two ancient pyramids. What factors would affect how long it took to build each pyramid? Given similar conditions, which pyramid took longer to build? Explain your reasoning.

Geometry

In this lesson, you will
- find volumes of pyramids.
- solve real-life problems.

The Sun Pyramid in Mexico
Height: about 246 ft
Base: about 738 ft by 738 ft

Cheops Pyramid in Egypt
Height: about 480 ft
Base: about 755 ft by 755 ft

Laurie's Notes

Introduction

Applying Mathematical Practices

- **Reason Abstractly and Quantitatively** and **Model with Mathematics**: In constructing physical models of a prism and a pyramid that have the same base area and height and then comparing their volumes, students make sense of the formula for the volume of a pyramid.

Motivate

- Share information about two well known pyramid-shaped buildings.
- The Luxor Resort and Casino in Las Vegas reaches 350 feet into the sky and is 36 stories tall. Luxor has over 4400 guest rooms, making it one of the top ten largest hotels in the world.
- The Rock and Roll Hall of Fame in Cleveland was designed by architect I. M. Pei. The design of the glass-faced main building uses a pyramid shape to invoke the image of a guitar neck rising to the sky.

Activity Notes

Activity 1

- ❓ "How are the two shapes alike?" same base, same height "How are they different?" One is a square pyramid. The other is a square prism.
- ❓ "How do you think their volumes compare?" Most students guess that the prism has twice the volume as the pyramid.
- ❓ "How can you test your hunch about the volumes?" If students have looked at the activity, they'll want to fill the pyramid.
- After the first pour, students should start to suspect that their guess might be off.
- After the second pour, students are pretty sure the relationship is 3 to 1.
- **Reason Abstractly and Quantitatively** and **Model with Mathematics**: This hands-on experience of making and filling the prism will help students remember the factor of $\frac{1}{3}$. The formula should now make sense to them. The volume of a pyramid should be $\frac{1}{3}$ the volume of a prism with the same base and height as the pyramid.

Activity 2

- ❓ **Attend to Precision:** "What do you know about the pyramids from looking only at the pictures?" They look like square pyramids.
- ❓ "What do you know about the pyramids from looking at their dimensions?" Cheops has a larger base and is nearly twice as tall.
- Give time for students to calculate the volume. From the first activity, they should feel comfortable finding the area of the base and multiplying by the height (this would be the prism's volume), and then taking $\frac{1}{3}$ of this answer.
- ❓ "Which pyramid has the greater volume, and about how many times greater is its volume than the other pyramid?" Cheops Pyramid has about twice the volume of The Sun Pyramid.

What Your Students Will Learn

- Find volumes of pyramids using formulas.

Previous Learning

Students should know how to find the volumes of prisms and how to perform operations on rational numbers.

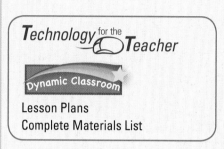

Lesson Plans
Complete Materials List

9.5 Record and Practice Journal

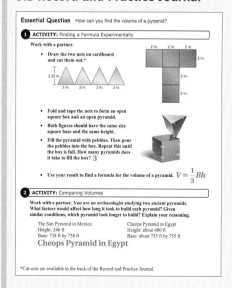

Differentiated Instruction

Visual

Students may confuse pyramids and triangular prisms. Show the students models and point out the following characteristics. A pyramid has one base, which can be any polygon. The remaining faces are triangles. A triangular prism has two bases, which are triangles. The remaining faces are rectangles.

9.5 Record and Practice Journal

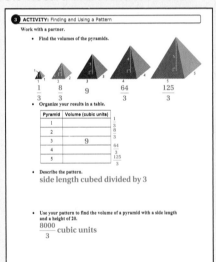

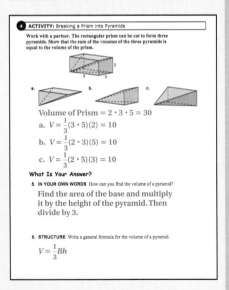

Laurie's Notes

Activity 3

- **Look for and Express Regularity in Repeated Reasoning:** Ask a student to describe the five pyramids shown. You want to make sure that students recognize that the bases are all squares, and the height of the pyramid is the same as the length of the base edge.
- **Use Appropriate Tools Strategically:** Reinforce good problem solving by having students organize their data in a table.
- Allow time for students to record the volume of each pyramid.
- ? "What was the volume of the smallest pyramid and how did you find it?"

$$V = \frac{1}{3} \cdot 1^3 = \frac{1}{3}$$

- ? "What was the volume of the next pyramid and how did you find it?"

$$V = \frac{1}{3} \cdot 2^3 = \frac{8}{3}$$

- **Look for and Express Regularity in Repeated Reasoning:** Repeat the question for the remaining pyramids, and then ask a volunteer to clearly summarize the pattern of the pyramids and their volumes. The height of each pyramid is equal to its side length *s*. The volume of each pyramid is given by $V = \frac{1}{3} \cdot s^3$.

Activity 4

- Discuss with students how to follow the color coding so that correct dimensions can be matched up.

What Is Your Answer?

- **Neighbor Check:** Have students work independently and then have their neighbors check their work. Have students discuss any discrepancies.

Closure

- Does the volume formula you wrote for Question 5 need to have a square base? Explain your thinking. No; Students should try to sketch or make pyramids with other polygonal bases.

ACTIVITY: Finding and Using a Pattern

Math Practice

Look for Patterns

As the height and the base lengths increase, how does this pattern affect the volume? Explain.

Work with a partner.

- Find the volumes of the pyramids.

- Organize your results in a table.

- Describe the pattern.

- Use your pattern to find the volume of a pyramid with a base length and a height of 20.

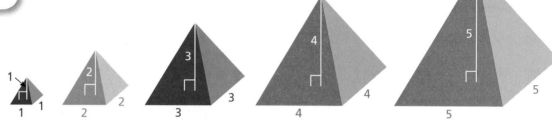

4 **ACTIVITY: Breaking a Prism into Pyramids**

Work with a partner. The rectangular prism can be cut to form three pyramids. Show that the sum of the volumes of the three pyramids is equal to the volume of the prism.

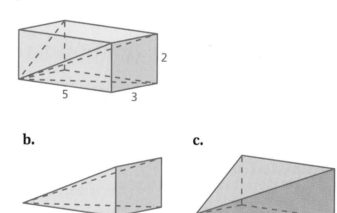

a.   **b.**   **c.**

What Is Your Answer?

5. **IN YOUR OWN WORDS** How can you find the volume of a pyramid?

6. **STRUCTURE** Write a general formula for the volume of a pyramid.

 Use what you learned about the volumes of pyramids to complete Exercises 4–6 on page 386.

Key Idea

Volume of a Pyramid

Words The volume V of a pyramid is one-third the product of the area of the base and the height of the pyramid.

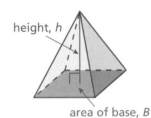
height, h

area of base, B

> **Study Tip**
>
> The *height* of a pyramid is the perpendicular distance from the base to the vertex.

Area of base

Algebra $V = \dfrac{1}{3}Bh$

Height of pyramid

EXAMPLE 1 **Finding the Volume of a Pyramid**

Find the volume of the pyramid.

$$V = \frac{1}{3}Bh \qquad \text{Write formula for volume.}$$

$$= \frac{1}{3}(48)(9) \qquad \text{Substitute.}$$

$$= 144 \qquad \text{Multiply.}$$

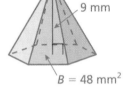
9 mm

$B = 48 \text{ mm}^2$

∴ The volume is 144 cubic millimeters.

EXAMPLE 2 **Finding the Volume of a Pyramid**

Find the volume of the pyramid.

> **Study Tip**
>
> The area of the base of a rectangular pyramid is the product of the length ℓ and the width w.
>
> You can use $V = \dfrac{1}{3}\ell wh$ to find the volume of a rectangular pyramid.

a.

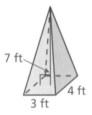
7 ft
4 ft
3 ft

$$V = \frac{1}{3}Bh$$

$$= \frac{1}{3}(4)(3)(7)$$

$$= 28$$

∴ The volume is 28 cubic feet.

b.

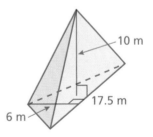
10 m
17.5 m
6 m

$$V = \frac{1}{3}Bh$$

$$= \frac{1}{3}\left(\frac{1}{2}\right)(17.5)(6)(10)$$

$$= 175$$

∴ The volume is 175 cubic meters.

Laurie's Notes

Introduction

Connect

- **Yesterday:** Students discovered how to find the volume of a pyramid by comparing it to the volume of a prism with the same base area and same height.
- **Today:** Students will work with a formula for the volume of a pyramid.

Motivate

- Show and discuss a picture of the Transamerica Pyramid building.
- It is the tallest building in San Francisco.
- The tapered design reduces the building's shadow to let more light reach the streets below.
- A San Francisco regulation limits the ratio of surface area to height for a building.

Lesson Notes

Key Idea

- Write the formula in words, and then in symbols.
- **?** "How will you find the area of the base?" It depends on what type of polygon the base is.
- Discuss the dotted lines and the shaded base. Pyramids are difficult to draw in two dimensions. Have students sketch a triangular pyramid and a square pyramid to practice.

Example 1

- Model good problem solving. Write the formula in words. Write the symbols underneath the words. Substitute the values for the symbols.
- **Common Error:** In using the volume formula, students often find $\frac{1}{3}$ of both B and h $\left(\frac{1}{3} \text{ of } 48 \text{ and } \frac{1}{3} \text{ of } 9\right)$ as though they are using the Distributive Property. Remind them that the Distributive Property is used when there is addition or subtraction involved. The correct steps for this problem are to multiply from left to right.

Example 2

- **?** "Describe the base of each pyramid." part (a): rectangular base; part (b): triangular base
- **?** "How do you find the area of the base in part (b)?" $\frac{1}{2} \times 17.5 \times 6$
- **FYI:** A statement like "one-half of the base times the height" will confuse students in part (b). Specify when you are speaking of parts of the base triangle and when you are speaking of parts of the pyramid.
- Students may need help in multiplying the fractions in each problem. They can apply the Commutative Property to get products of reciprocals in each problem. In part (a), $\frac{1}{3} \times 3 = 1$ and in part (b), $\frac{1}{3} \times \frac{1}{2} \times 6 = 1$.

Goal Today's lesson is finding the volumes of pyramids.

Technology for the *Teacher*

Dynamic Classroom

Lesson Tutorials
Lesson Plans
Answer Presentation Tool

Extra Example 1

Find the volume of a pentagonal pyramid with a base area of 24 square feet and a height of 8 feet. 64 ft³

Extra Example 2

a. Find the volume of a rectangular pyramid with a base of 2 meters by 6 meters and a height of 3 meters. 12 m³

b. Find the volume of a triangular pyramid with a height of 8 inches and where the triangular base has a width of 4 inches and an altitude of 9 inches. 48 in.³

On Your Own

1. 42 ft^3

2. $186\frac{2}{3} \text{ in.}^3$

3. 231 cm^3

Extra Example 3

a. The volume of lotion in Bottle B is how many times the volume in Bottle A? $1\frac{2}{3}$

b. Which is the better buy? Bottle B

Bottle A
$6.60

Bottle B
$10.00

15 cm

10 cm

4 cm   3 cm

6 cm   5 cm

On Your Own

4. yes; Bottles B and C have the same volume, but Bottle C has a unit cost of $2.20.

English Language Learners

Forming Answers

Encourage English learners to form complete sentences in their responses. Students can use the question to help them form the answer.

Question: If you know the area of the base of a pyramid, what else do you need to know to find the volume?

Response: If you know the area of the base of a pyramid, you need to know the height of the pyramid to find the volume.

Laurie's Notes

On Your Own

- Have students name each pyramid and describe what they know about each base. Note that for the pentagonal pyramid, the area of the base has already been computed.
- In Question 2, none of the dimensions contain factors of 3. In computing the volume, $V = \frac{1}{3} \times 10 \times 8 \times 7$, suggest to students that they multiply the whole numbers for a product of 560 and then multiply by $\frac{1}{3}$. Remind students how to rewrite the improper fraction $\frac{560}{3}$ as a mixed number, $186\frac{2}{3}$.

Example 3

- If you have any lotion or shampoo that is in a pyramidal bottle, use it as a model.
- Work through the computation of volume for each bottle.
- **Reason Abstractly and Quantitatively:** Explain different approaches to multiplying the factors in Bottle A: (1) multiply in order from left to right or (2) use the Commutative Property to multiply the whole numbers, and then multiply by $\frac{1}{3}$.
- Discuss the phrase "how many times." Because the volume of Bottle B is not a multiple of the volume of Bottle A, students are uncertain how to compare the volumes.
- ? "How do you decide which bottle is the better buy?" Students will try to describe how to find the cost for one cubic inch. This is the unit price.
- Use the language, cost per volume or cost per cubic inch.

On Your Own

- Give students time to complete this problem. Ask volunteers to share their work at the board.

Closure

- **Exit Ticket:** Sketch a rectangular pyramid with base 3 units by 4 units, and a height of 5 units. What is the volume? 20 units^3

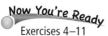

Find the volume of the pyramid.

1.   2.   3.

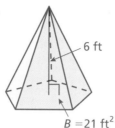 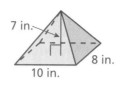 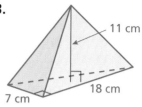

EXAMPLE ③ **Real-Life Application**

a. The volume of sunscreen in Bottle B is about how many times
the volume in Bottle A?

b. Which is the better buy?

a. Use the formula for the volume of a pyramid to estimate the
amount of sunscreen in each bottle.

Bottle A	*Bottle B*
$V = \dfrac{1}{3}Bh$	$V = \dfrac{1}{3}Bh$
$= \dfrac{1}{3}(2)(1)(6)$	$= \dfrac{1}{3}(3)(1.5)(4)$
$= 4 \text{ in.}^3$	$= 6 \text{ in.}^3$

⋮ So, the volume of sunscreen in Bottle B is about $\dfrac{6}{4} = 1.5$ times
the volume in Bottle A.

b. Find the unit cost for each bottle.

Bottle A	*Bottle B*
$\dfrac{\text{cost}}{\text{volume}} = \dfrac{\$9.96}{4 \text{ in.}^3}$	$\dfrac{\text{cost}}{\text{volume}} = \dfrac{\$14.40}{6 \text{ in.}^3}$
$= \dfrac{\$2.49}{1 \text{ in.}^3}$	$= \dfrac{\$2.40}{1 \text{ in.}^3}$

⋮ The unit cost of Bottle B is less than the unit cost of Bottle A.
So, Bottle B is the better buy.

On Your Own

4. Bottle C is on sale for $13.20. Is Bottle C
a better buy than Bottle B in Example 3?
Explain.

Bottle C

Check It Out
Help with Homework
BigIdeasMath✓com

 Vocabulary and Concept Check

1. **WRITING** How is the formula for the volume of a pyramid different from the formula for the volume of a prism?

2. **OPEN-ENDED** Describe a real-life situation that involves finding the volume of a pyramid.

3. **REASONING** A triangular pyramid and a triangular prism have the same base and height. The volume of the prism is how many times the volume of the pyramid?

 Practice and Problem Solving

Find the volume of the pyramid.

① ② **4.**

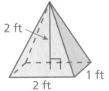
2 ft
1 ft
2 ft

5.

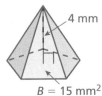
4 mm
$B = 15$ mm²

6.

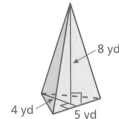
8 yd
4 yd
5 yd

7.

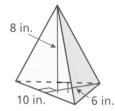
8 in.
10 in.
6 in.

8.

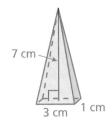
7 cm
3 cm
1 cm

9.

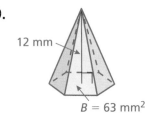
12 mm
$B = 63$ mm²

10.

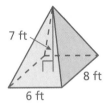
7 ft
8 ft
6 ft

11.

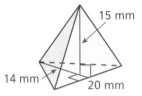
15 mm
14 mm
20 mm

12. **PARACHUTE** In 1483, Leonardo da Vinci designed a parachute. It is believed that this was the first parachute ever designed. In a notebook, he wrote, "If a man is provided with a length of gummed linen cloth with a length of 12 yards on each side and 12 yards high, he can jump from any great height whatsoever without injury." Find the volume of air inside Leonardo's parachute.

Not drawn to scale

Assignment Guide and Homework Check

Level	Day 1 Activity Assignment	Day 2 Lesson Assignment	Homework Check
Basic	4–6, 21–24	1–3, 7–17 odd	7, 11, 13, 17
Average	4–6, 21–24	1–3, 7–11 odd, 14–20 even	7, 11, 14, 16, 18
Advanced	4–6, 21–24	1–3, 7, 8–20 even	7, 10, 12, 18, 20
Accelerated	1–7, 8–20 even, 21–24		7, 10, 12, 18, 20

For Your Information

- **Exercise 12** Skydiver Adrian Nicholas tested the design, jumping from a hot-air balloon at 3000 meters. The parachute weighed over 90 kilograms.

Common Errors

- **Exercises 4–11** Students may write the units incorrectly, often writing square units instead of cubic units. This is especially true when the area of the base is given. Remind them that the units are cubed because there are three dimensions.
- **Exercises 4–11** Students may forget to multiply by one of the measurements, especially when finding the area of the base. Encourage them to find the area of the base separately and then substitute it into the equation. Using colored pencils for each part can also assist students. Tell them to write the formula using different colors for the base and height, as in the lesson. When they substitute values into the equation for volume, they will be able to clearly see that they have accounted for all of the dimensions.

9.5 Record and Practice Journal

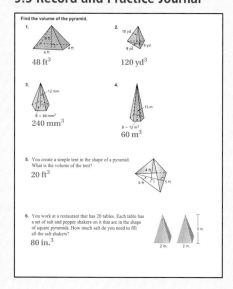

Practice and Problem Solving

13. 156 ft^3

14. 240 m^3

15. 340.4 in.^3

16. Spire B; 4 in.^3

17. See Additional Answers.

18. See *Taking Math Deeper.*

19. *Sample answer:* 5 ft by 4 ft

20. yes; Prism: $V = xyz$

   Pyramid: $V = \frac{1}{3}(xy)(3z) = xyz$

Fair Game Review

21. $153°; 63°$ 22. $98°; 8°$

23. $60°; \text{none}$ 24. C

Mini-Assessment

Find the volume of the pyramid.

1.

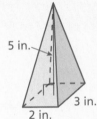

5 in.

3 in.

2 in.

10 in.^3

2.

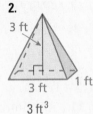

3 ft

3 ft

3 ft

1 ft

3 ft^3

3.

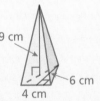

9 cm

6 cm

4 cm

36 cm^3

4. Find the volume of the paper weight.

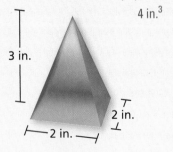

4 in.^3

3 in.

2 in.

2 in.

Taking Math Deeper

Exercise 18

Students have to think a bit about this question. At first it seems like you cannot tell the shape of the base. However, you can count the number of support sticks to find the shape of the base.

1 Count the supports.
   a. There are 12. So, the base is a dodecagon (a 12-sided polygon).

2 Using a ruler, the base of the teepee appears to be about the same as its height. So, estimate the width of the base to be 10 feet.

3 Use a 10-by-10 grid to estimate the area of the base. It appears to have an area of about 80 square feet.

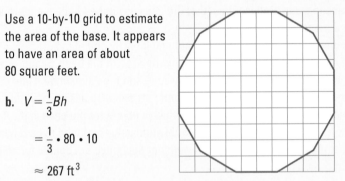

   b. $V = \frac{1}{3}Bh$

      $= \frac{1}{3} \cdot 80 \cdot 10$

      $\approx 267 \text{ ft}^3$

You need to give some leeway in the answers. Anything from 250 cubic feet to 300 cubic feet is a reasonable answer.

Reteaching and Enrichment Strategies

If students need help...	If students got it...
Resources by Chapter • Practice A and Practice B • Puzzle Time Record and Practice Journal Practice Differentiating the Lesson Lesson Tutorials Skills Review Handbook	Resources by Chapter • Enrichment and Extension • Technology Connection Start the next section

Find the volume of the composite solid.

13.

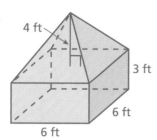

4 ft
3 ft
6 ft
6 ft

14.

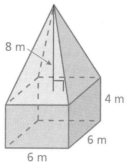

8 m
4 m
6 m
6 m

15.

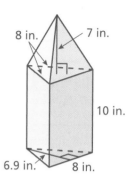

8 in.   7 in.
10 in.
6.9 in.   8 in.

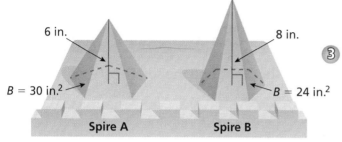

6 in.   8 in.
B = 30 in.^{2} B = 24 in.^{2}
Spire A   Spire B

③ 16. SPIRE Which sand-castle spire has a greater volume? How much more sand do you need to make the spire with the greater volume?

17. PAPERWEIGHT How much glass is needed to manufacture 1000 paperweights? Explain your reasoning.

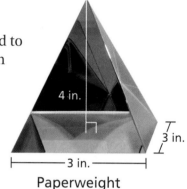

4 in.
3 in.
3 in.
Paperweight

18. PROBLEM SOLVING Use the photo of the tepee.

   a. What is the shape of the base? How can you tell?

   b. The tepee's height is about 10 feet. Estimate the volume of the tepee.

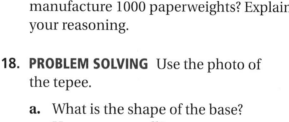

19. OPEN-ENDED A pyramid has a volume of 40 cubic feet and a height of 6 feet. Find one possible set of dimensions of the rectangular base.

20. ⟪Reasoning⟫ Do the two solids have the same volume? Explain.

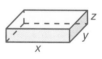
z
y
x

3z
y
x

For the given angle measure, find the measure of a supplementary angle and the measure of a complementary angle, if possible. *(Section 7.2)*

21. 27°

22. 82°

23. 120°

24. MULTIPLE CHOICE The circumference of a circle is 44 inches. Which estimate is closest to the area of the circle? *(Section 8.3)*

  (A) 7 in.^{2} **(B)** 14 in.^{2} **(C)** 154 in.^{2} **(D)** 484 in.^{2}

Check It Out
Lesson Tutorials
BigIdeasMath⚡com

Key Vocabulary
cross section, *p. 388*

Consider a plane "slicing" through a solid. The intersection of the plane and the solid is a two-dimensional shape called a **cross section**. For example, the diagram shows that the intersection of the plane and the rectangular prism is a rectangle.

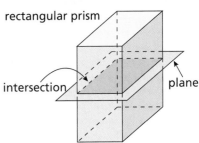

rectangular prism

intersection

plane

EXAMPLE **1** **Describing the Intersection of a Plane and a Solid**

Describe the intersection of the plane and the solid.

Geometry
In this extension, you will
• describe the intersections of planes and solids.

a.
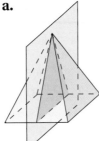

b.
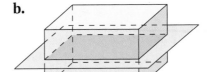

c.
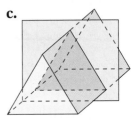

a. The intersection is a triangle.

b. The intersection is a rectangle.

c. The intersection is a triangle.

Practice

Describe the intersection of the plane and the solid.

1.
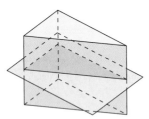

2.
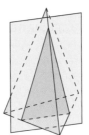

3.
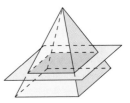

4.
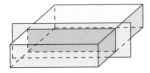

5.
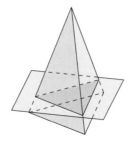

6.
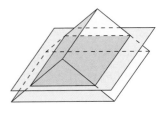

7. **REASONING** A plane that intersects a prism is parallel to the bases of the prism. Describe the intersection of the plane and the prism.

◀ Multi-Language Glossary at BigIdeasMath⚡com

Laurie's Notes

Introduction

Connect
- **Yesterday:** Students found the volumes of pyramids.
- **Today:** Students will describe cross sections of three-dimensional figures.

Motivate
- Use a flashlight, overhead projector, or document camera to show two-dimensional projections of three-dimensional objects. For instance, hold a rectangular prism in different orientations to show different projections. Do the same for a triangular prism (a door stop works well) and a cylinder.
- The goal is for students to visualize different cross sections of solids.
- **Model with Mathematics:** If cross sections are difficult for students to visualize, tell them to consider a cheese slicer cutting through a wedge of cheese. The face exposed when the cheese is cut is a cross section.

Lesson Notes

Example 1
- It is helpful for students to see and hold models of solids when thinking about how a plane intersects a solid. You can stretch a rubber band around the model of a solid to help students visualize the intersection of a plane with the solid.
- Consider using technology that displays the animation of a plane cutting through a solid.

Practice
- **Common Error:** Students may incorrectly describe the intersection because of the orientation of the drawing.

Goal Today's lesson is describing **cross sections** of three-dimensional figures.

Lesson Tutorials
Lesson Plans
Answer Presentation Tool

Extra Example 1

Describe the intersection of the plane and the solid.

a.

b.

a triangle   a square

c.

a rectangle

Practice
1. triangle   2. triangle
3. rectangle   4. rectangle
5. triangle   6. rectangle
7. The intersection is the shape of the base.

Record and Practice Journal
Extension 9.5 Practice

1. triangle   2. rectangle
3. rectangle   4. rectangle
5. triangle   6. rectangle
7. triangle   8. circle
9. circle   10. rectangle
11. circle   12. circle

Extra Example 2

Describe the intersection of the plane and the solid.

a.

a circle

b.

a circle

Practice

8. rectangle

9. circle

10. line segment

11. circle

12. circle

13. rectangle

14. circle

15. The intersection occurs at the vertex of the cone.

Mini-Assessment

Describe the intersection of the plane and the solid.

1.

a triangle

2.

a point

3.

a rectangle

Discuss

• Describe the features of a cone. Structurally it is the same as a pyramid. There is one base and a lateral surface with one vertex not on the base.

Example 2

• You can investigate these cross sections for a cylinder and cone using a flashlight as described in the Motivate activity. A rubber band stretched around the solid will also reveal the perimeter or circumference of the cross section.

Practice

• The intersections are shaded and should be clear to students.
• **Extension:** Ask students how the size of the cross section changes as the plane moves through the cylinder in Question 8.

Closure

• What are the possible cross sections of a cube? point, line, triangle, square, rectangle, parallelogram, quadrilateral, trapezoid, pentagon, hexagon

Example 1 shows how a plane intersects a polyhedron. Now consider the intersection of a plane and a solid having a curved surface, such as a cylinder or cone. As shown, a *cone* is a solid that has one circular base and one vertex.

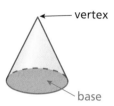
vertex

base

EXAMPLE **Describing the Intersection of a Plane and a Solid**

Math Practice

Analyze Givens
What solid is shown? What are you trying to find? Explain.

Describe the intersection of the plane and the solid.

a.

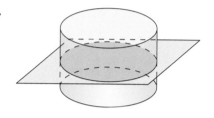

b.

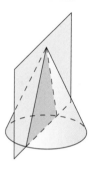

a. The intersection is a circle.

b. The intersection is a triangle.

Practice

Describe the intersection of the plane and the solid.

8.

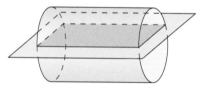

9.

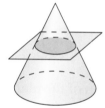

10.

11.

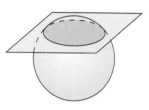

Describe the shape that is formed by the cut made in the food shown.

12.

13.

14.

15. REASONING Explain how a plane can be parallel to the base of a cone and intersect the cone at exactly one point.

Find the volume of the prism. *(Section 9.4)*

1.

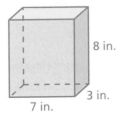

8 in.

3 in.

7 in.

2.

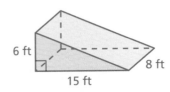

6 ft

8 ft

15 ft

3.

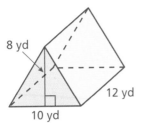

8 yd

12 yd

10 yd

4.

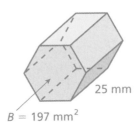

25 mm

$B = 197 \text{ mm}^2$

Find the volume of the solid. Round your answer to the nearest tenth if necessary. *(Section 9.5)*

5.

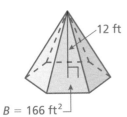

12 ft

$B = 166 \text{ ft}^2$

6.

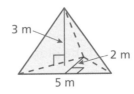

3 m

2 m

5 m

Describe the intersection of the plane and the solid. *(Section 9.5)*

7.

8.

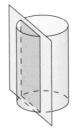

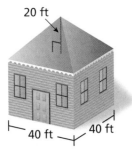

20 ft

40 ft

40 ft

9. ROOF A pyramid hip roof is a good choice for a house in a hurricane area. What is the volume of the roof to the nearest tenth? *(Section 9.5)*

10. CUBIC UNITS How many cubic feet are in a cubic yard? Use a sketch to explain your reasoning. *(Section 9.4)*

Alternative Assessment Options

Math Chat   **Student Reflective Focus Question**

Structured Interview   Writing Prompt

Student Reflective Focus Question

Ask students to summarize the similarities and differences between volumes and cross sections of prisms and pyramids. Be sure that they include examples. Select students at random to present their summaries to the class.

Study Help Sample Answers

Remind students to complete Graphic Organizers for the rest of the chapter.

4.

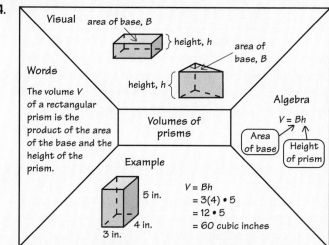

5.

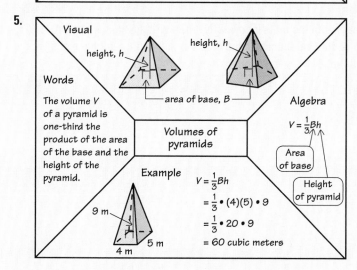

Reteaching and Enrichment Strategies

If students need help. . .	If students got it. . .

Resources by Chapter
  • Practice A and Practice B
  • Puzzle Time
Lesson Tutorials
BigIdeasMath.com
	Resources by Chapter
	  • Enrichment and Extension
	  • Technology Connection
	Game Closet at *BigIdeasMath.com*
	Start the Chapter Review

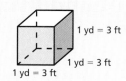
Technology for the *Teacher*

Online Assessment
Assessment Book
ExamView® Assessment Suite

Answers

1. 158 in.^{2}

2. 400 cm^2

3. 108 m^2

Review of Common Errors

Exercises 1–3
- Students may sum the areas of only 3 of the faces of the rectangular prism instead of all 6. Remind them that a rectangular prism has 6 faces.
- Students may try to use the formula for the surface area of a rectangular prism to find the surface area of a triangular prism. Show them that this will not work by comparing the nets of the two types of prisms.

Exercises 4–6
- Students may forget to add the area of the base when finding the surface area. Remind them that when asked to find the surface area, the base is included.

Exercises 4–6
- Students may add the wrong number of lateral face areas to the area of the base. Examine several different pyramids with different bases and ask if they can find a relationship between the number of sides of the base and the number of lateral faces. (They are the same.) Remind students that the number of sides on the base determines how many triangles make up the lateral surface area.

Exercises 7 and 8
- Students may add the area of only one base. Remind them that a cylinder has *two* bases.
- Students may double the radius instead of squaring it, or forget the correct order of operations when using the formula for the surface area of a cylinder. Remind them of the formula and remind them of the order of operations.

Exercises 10–15
- Students may write the units incorrectly, often writing square units instead of cubic units. Remind them that they are working in three dimensions, so the units are cubed.

9 Chapter Review

Check It Out
Vocabulary Help
BigIdeasMath ✓com

Review Key Vocabulary

lateral surface area, *p. 358*

regular pyramid, *p. 364*

slant height, *p. 364*

cross section, *p. 388*

Review Examples and Exercises

9.1 Surface Areas of Prisms *(pp. 354–361)*

Find the surface area of the prism.

Draw a net.

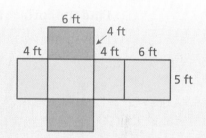

$$S = 2\ell w + 2\ell h + 2wh$$
$$= 2(6)(4) + 2(6)(5) + 2(4)(5)$$
$$= 48 + 60 + 40$$
$$= 148$$

⋮· The surface area is 148 square feet.

Exercises

Find the surface area of the prism.

1.
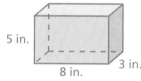
5 in. 8 in. 3 in.

2.
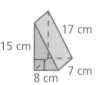
17 cm 15 cm 7 cm 8 cm

3.
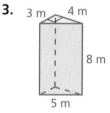
3 m 4 m 8 m 5 m

9.2 Surface Areas of Pyramids *(pp. 362–367)*

Find the surface area of the regular pyramid.

Draw a net.

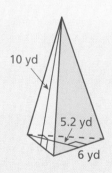
10 yd 5.2 yd 6 yd

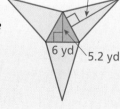
10 yd 6 yd 5.2 yd

Area of Base
$$\frac{1}{2} \cdot 6 \cdot 5.2 = 15.6$$

Area of a Lateral Face
$$\frac{1}{2} \cdot 6 \cdot 10 = 30$$

Find the sum of the areas of the base and all three lateral faces.

$$S = 15.6 + \underbrace{30 + 30 + 30}$$
$$= 105.6$$

> There are 3 identical lateral faces. Count the area 3 times.

⋮· The surface area is 105.6 square yards.

Exercises

Find the surface area of the regular pyramid.

4.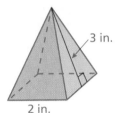

3 in.

2 in.

5.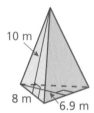

10 m

8 m   6.9 m

6.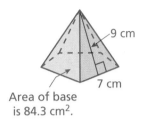

9 cm

7 cm

Area of base
is 84.3 cm².

9.3 **Surface Areas of Cylinders**   *(pp. 368–373)*

Find the surface area of the cylinder. Round your answer to the nearest tenth.

Draw a net.

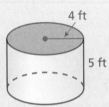

4 ft

5 ft

$$S = 2\pi r^2 + 2\pi r h$$
$$= 2\pi(4)^2 + 2\pi(4)(5)$$
$$= 32\pi + 40\pi$$
$$= 72\pi \approx 226.1$$

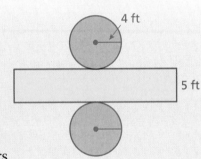

4 ft

5 ft

:• The surface area is about 226.1 square millimeters.

Exercises

Find the surface area of the cylinder. Round your answer to the nearest tenth.

7.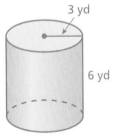

3 yd

6 yd

8. 0.8 cm

6 cm

9. ORANGES Find the lateral surface area of the can of
mandarin oranges.

4 cm

11 cm

9.4 **Volumes of Prisms**   *(pp. 376–381)*

Find the volume of the prism.

$$V = Bh$$ Write formula for volume.

$$= \frac{1}{2}(7)(3) \cdot 5$$ Substitute.

$$= 52.5$$ Multiply.

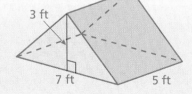

3 ft

7 ft   5 ft

:• The volume is 52.5 cubic feet.

Review Game

3-Dingo

Materials per Group:
- one 3-Dingo card*
- objects to cover the 3-Dingo card squares, such as bingo chips

Directions:

This activity is played like bingo. Divide the class into groups. Call out a three-dimensional figure and all of the dimensions necessary to calculate the surface area or volume. The group calculates the indicated surface area or volume. If that value is in a square on their card under the correct figure, they cover it. Keep calling out figures and their dimensions until a group wins.

Who Wins?

Just like bingo, the first group to get a row, column, or diagonal of covered squares wins. The winning team yells 3-Dingo!

*A 3-Dingo card has 5 columns of 5 squares and a three-dimensional figure (rectangular prism, triangular prism, square pyramid, triangular pyramid, and cylinder) with variable dimensions shown at the top of each column of squares. Different values for the surface area or volume of each figure are shown in each of the 5 squares below the figure. Different cards show the values in different orders, so no two 3-Dingo cards in a set are the same. A 3-Dingo card set is available at *BigIdeasMath.com*.

For the Student
Additional Practice
- Lesson Tutorials
- Multi-Language Glossary
- Self-Grading Progress Check
- *BigIdeasMath.com*
    Dynamic Student Edition
    Student Resources

Answers

4. 16 in.^2

5. 147.6 m^2

6. 241.8 cm^2

7. $54\pi \approx 169.6 \text{ yd}^2$

8. $10.88\pi \approx 34.2 \text{ cm}^2$

9. $88\pi \approx 276.3 \text{ cm}^2$

10. 96 in.^3

11. 120 m^3

12. 607.5 mm^3

13. 850 ft^3

14. 2100 in.^3

15. 192 mm^3

16. rectangle

17. triangle

My Thoughts on the Chapter

What worked. . .

What did not work. . .

What I would do differently. . .

Exercises

Find the volume of the prism.

10.

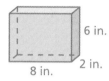

6 in.
2 in.
8 in.

11.

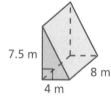

7.5 m
8 m
4 m

12.

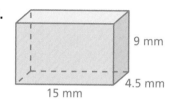

9 mm
15 mm
4.5 mm

9.5 **Volumes of Pyramids** *(pp. 382–389)*

a. Find the volume of the pyramid.

$$V = \frac{1}{3}Bh \qquad \text{Write formula for volume.}$$

$$= \frac{1}{3}(6)(5)(10) \qquad \text{Substitute.}$$

$$= 100 \qquad \text{Multiply.}$$

∴ The volume is 100 cubic yards.

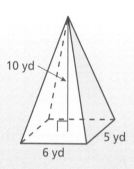

10 yd
5 yd
6 yd

b. Describe the intersection of the plane and the solid.

i.

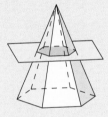

The intersection is a hexagon.

ii.

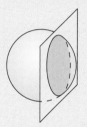

The intersection is a circle.

Exercises

Find the volume of the pyramid.

13.

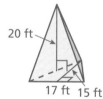

20 ft
17 ft   15 ft

14.

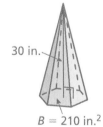

30 in.

$B = 210$ in.^{2}

15.

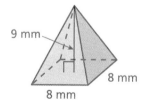

9 mm
8 mm
8 mm

Describe the intersection of the plane and the solid.

16.

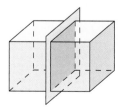

17.

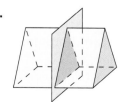

Find the surface area of the prism or regular pyramid.

1.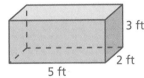

3 ft
2 ft
5 ft

2.

2 in.
1 in.

3.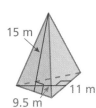

15 m
11 m
9.5 m

Find the surface area of the cylinder. Round your answer to the nearest tenth.

4.

2 cm
3 cm

5.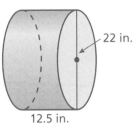

22 in.
12.5 in.

Find the volume of the solid.

6.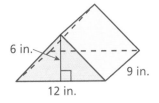

6 in.
9 in.
12 in.

7.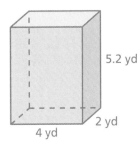

5.2 yd
2 yd
4 yd

8.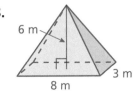

6 m
3 m
8 m

9. SKATEBOARD RAMP A quart of paint covers 80 square feet. How many quarts should you buy to paint the ramp with two coats? (Assume you will not paint the bottom of the ramp.)

15.2 ft
6 ft
19.5 ft
14 ft

h = 9 in.
ℓ = 6 in.
w = 2 in.

10. GRAHAM CRACKERS A manufacturer wants to double the volume of the graham cracker box. The manufacturer will either double the height or double the width.

  a. Which option uses less cardboard? Justify your answer.

  b. What is the volume of the new graham cracker box?

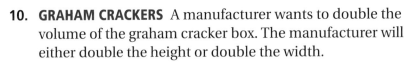

11. SOUP The label on the can of soup covers about 354.2 square centimeters. What is the height of the can? Round your answer to the nearest whole number.

4.7 cm

Test Item References

Chapter Test Questions	Section to Review
1, 9, 10	9.1
2, 3	9.2
4, 5, 11	9.3
6, 7, 10	9.4
8	9.5

Test-Taking Strategies

Remind students to quickly look over the entire test before they start so that they can budget their time. This test is very visual and requires that students remember many terms. It might be helpful for them to jot down some of the terms on the back of the test before they start.

Common Errors

- **Exercise 1** Students may sum the areas of only 3 of the faces of the rectangular prism instead of all 6. Remind them that a rectangular prism has 6 faces.
- **Exercises 2 and 3** Students may add the wrong number of lateral face areas to the area of the base. Remind students that the number of sides on the base determines how many triangles make up the lateral surface area.
- **Exercises 4 and 5** Students may double the radius instead of squaring it or forget the correct order of operations when using the formula for the surface area of a cylinder. Remind them of the formula, and remind them of the order of operations.
- **Exercises 6–8** Students may write the units incorrectly, often writing square units instead of cubic units. Remind them that they are working in three dimensions, so the units are cubed.

Reteaching and Enrichment Strategies

If students need help...	If students got it...
Resources by Chapter • Practice A and Practice B • Puzzle Time Record and Practice Journal Practice Differentiating the Lesson Lesson Tutorials *BigIdeasMath.com* Skills Review Handbook	Resources by Chapter • Enrichment and Extension • Technology Connection Game Closet at *BigIdeasMath.com* Start Cumulative Assessment

Answers

1. 62 ft^2
2. 5 in.^2
3. 299.8 m^2
4. 62.8 cm^2
5. 1623.4 in.^2
6. 324 in.^3
7. 41.6 yd^3
8. 48 m^3
9. 13 quarts of paint
10. **a.** doubling the width
    b. 216 in.^3
11. 12 cm

Technology for the *Teacher*

Online Assessment
Assessment Book
ExamView® Assessment Suite

Answers
1. D
2. G
3. B

Item Analysis

1. **A.** The student multiplies the length and height and adds the width.

   B. The student adds the areas of only three unique faces.

   C. The student multiplies the length, width, and height.

   D. Correct answer

2. **F.** The student finds what percent 60 is of 660.

   G. Correct answer

   H. The student finds 60% of 660 and misplaces the decimal point.

   I. The student thinks that the difference of the scores is equivalent to the percent.

3. **A.** The student incorrectly applies the Cross Products Property to get $3(x - 3) = 8(24)$

   B. Correct answer

   C. The student performs an order of operations error by not multiplying each side by 24 first.

   D. The student thinks that dividing both sides by 24 is the first step to isolating the variable on one side. The correct first step is to multiply both sides by 24.

1. A gift box and its dimensions are shown below.

2 in.

8 in.   4 in.

What is the least amount of wrapping paper that you could have used to wrap the box?

A. 20 in.^{2}

C. 64 in.^{2}

B. 56 in.^{2}

D. 112 in.^{2}

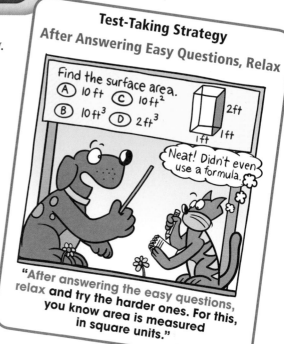

2. A student scored 600 the first time she took the mathematics portion of her college entrance exam. The next time she took the exam, she scored 660. Her second score represents what percent increase over her first score?

F. 9.1%

H. 39.6%

G. 10%

I. 60%

3. Raj was solving the proportion in the box below.

$$\frac{3}{8} = \frac{x - 3}{24}$$

$$3 \cdot 24 = (x - 3) \cdot 8$$

$$72 = x - 24$$

$$96 = x$$

What should Raj do to correct the error that he made?

A. Set the product of the numerators equal to the product of the denominators.

B. Distribute 8 to get $8x - 24$.

C. Add 3 to each side to get $\frac{3}{8} + 3 = \frac{x}{24}$.

D. Divide both sides by 24 to get $\frac{3}{8} \div 24 = x - 3$.

4. A line contains the two points plotted in the coordinate plane below.

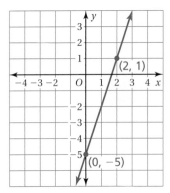

What is the slope of the line?

F. $\dfrac{1}{3}$

H. 3

G. 2

I. 6

5. James is getting ready for wrestling season. As part of his preparation, he plans to lose 5% of his body weight. James currently weighs 160 pounds. How much will he weigh, in pounds, after he loses 5% of his weight?

6. How much material is needed to make the popcorn container?

4 in.

9.5 in.

A. 76π in.^{2}

C. 92π in.^{2}

B. 84π in.^{2}

D. 108π in.^{2}

7. To make 10 servings of soup you need 4 cups of broth. You want to know how many servings you can make with 8 pints of broth. Which proportion should you use?

F. $\dfrac{10}{4} = \dfrac{x}{8}$

H. $\dfrac{10}{4} = \dfrac{8}{x}$

G. $\dfrac{4}{10} = \dfrac{x}{16}$

I. $\dfrac{10}{4} = \dfrac{x}{16}$

Item Analysis (continued)

4. **F.** The student finds the change in *x* divided by the change in *y*.

  G. The student makes an error in subtracting the integers to find the change in *y*.

  H. Correct answer

  I. The student finds the change in *y* but forgets to divide by the change in *x*.

5. **Gridded Response:** Correct answer: 152 lb

Common Error: The student finds only the loss, getting an answer of 8.

6. **A.** The student only calculates the lateral surface area. They do not include the area of the base.

  B. The student doubles the radius instead of squaring it when finding the area of the base.

  C. Correct answer

  D. The student includes the area of both bases instead of just one base.

7. **F.** The student does not convert to the same units (either 8 pints to 16 cups or 4 cups to 2 pints).

  G. The student uses the correct numbers but inverts one of the ratios.

  H. The student does not convert to the same units (either 8 pints to 16 cups or 4 cups to 2 pints) and inverts one of the ratios.

  I. Correct answer

Answers

4. H

5. 152 lb

6. C

7. I

Answers

8. 648 in.^{3}

9. A

10. H

11. *Part A*

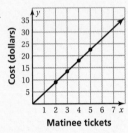

Part B 4.5; the cost of 1 movie ticket is $4.50.

Part C $36

Item Analysis (continued)

8. Gridded Response: Correct answer: 648 in.^{3}

Common Error: The student does not read the entire question and gets an answer of 24 cubic inches.

9. A. Correct answer

B. The student solves $2x + 4 = 90$.

C. The student subtracts 46 from 90.

D. The student solves $2x + 4 + 46 = 180$.

10. F. The student does not realize that the sum of the angle measures cannot be less than 180°.

G. The student does not realize that the sum of the angle measures cannot be greater than 180°.

H. Correct answer

I. The student does not realize that a triangle cannot have an angle with a measure of 0°.

11. 2 points The student demonstrates a thorough understanding of graphing data points and finding the slope. In Part A, the student correctly plots the points and graphs the function $y = 4.5x$, for $x \geq 0$. In Part B, the student correctly determines that the slope is 4.5 and that it costs $4.50 per movie ticket. In Part C, the student correctly determines that it costs $36 to buy 8 movie tickets. The student provides clear and complete work and explanations.

1 point The student demonstrates a partial understanding of graphing data points and finding the slope. The student provides some correct work and explanation.

0 points The student demonstrates insufficient understanding of graphing data points and finding the slope. The student is unable to make any meaningful progress toward a correct answer.

8. A rectangular prism and its dimensions are shown below.

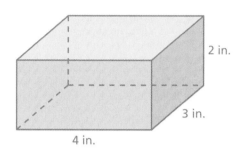

2 in.

3 in.

4 in.

What is the volume, in cubic inches, of a rectangular prism whose dimensions are three times greater?

9. What is the value of x?

A. 20   **C.** 44

B. 43   **D.** 65

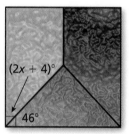

$(2x + 4)°$

$46°$

10. Which of the following could be the angle measures of a triangle?

F. 60°, 50°, 20°   **H.** 30°, 60°, 90°

G. 40°, 80°, 90°   **I.** 0°, 90°, 90°

11. The table below shows the costs of buying matinee movie tickets.

Matinee Tickets, x	2	3	4	5
Cost, y	$9	$13.50	$18	$22.50

Part A   Graph the data.

Part B   Find and interpret the slope of the line through the points.

Part C   How much does it cost to buy 8 matinee movie tickets?

10 Probability and Statistics

10.1 **Outcomes and Events**

10.2 **Probability**

10.3 **Experimental and Theoretical Probability**

10.4 **Compound Events**

10.5 **Independent and Dependent Events**

10.6 **Samples and Populations**

10.7 **Comparing Populations**

"If there are 7 cats in a sack and I draw one at random,..."

"... what is the probability that I will draw you?"

"I'm just about finished making my two number cubes."

"Now, here's how the game works. You toss the two cubes."

"If the sum is even, I win. If it's odd, you win."

What Your Students Have Learned

- Make a line plot with data in fractions of a unit.
- Understand that a measure of center summarizes all of the values in a data set with a single number.
- Understand that a measure of variation summarizes how all of the values in a data set vary with a single number.
- Display data on a number line in dot plots and box-and-whisker plots.
- Choose measures of center and variation based on shape.

What Your Students Will Learn

- Understand representative samples (random sampling) and populations.
- Use samples to draw inferences about populations.
- Compare two populations from random samples using measures of center and variability.
- Understand that probability is the likelihood of an event occurring, expressed as a number from 0 to 1.
- Use simulations to find experimental probabilities.
- Develop probability models and use them to find probabilities.
- Find the probabilities of compound events.

Pacing Guide for Chapter 10

Chapter Opener Regular Accelerated	1 Day 1 Day
Section 1 Regular Accelerated	2 Days 1 Day
Section 2 Regular Accelerated	2 Days 1 Day
Section 3 Regular Accelerated	2 Days 1 Day
Section 4 Regular Accelerated	2 Days 1 Day
Section 5 Regular Accelerated	3 Days 2 Days
Study Help / Quiz Regular Accelerated	1 Day 1 Day
Section 6 Regular Accelerated	3 Days 2 Day
Section 7 Regular Accelerated	2 Days 1 Day
Chapter Review/ Chapter Tests Regular Accelerated	2 Days 2 Days
Total Chapter 10 Regular Accelerated	20 Days 13 Days
Year-to-Date Regular Accelerated	154 Days 92 Days

Technology for the *Teacher*

BigIdeasMath.com
Chapter at a Glance
Complete Materials List
Parent Letters: English and Spanish

What Your Students Have Learned
- Write ratios in simplest form.

Additional Topics for Review
- Multiplying Fractions
- Converting Between Fractions, Decimals, and Percents
- Box-and-Whisker Plots
- Dot Plots
- Stem-and-Leaf Plots

Try It Yourself

1. $1:3$ 2. $3:4$
3. $1:2$ 4. $1:3$
5. $1:2$ 6. $4:3$
7. $1:5$

**Record and Practice Journal
Fair Game Review**

1. $2:3$ 2. $5:3$
3. $1:2$ 4. $1:4$
5. $3:20$ 6. $2:3$
7. $1:5$ 8. $7:17$
9. $2:3$ 10. $3:7$

Math Background Notes

Vocabulary Review
- Fraction
- Simplest Form
- Ratio

Writing Ratios
- Students should know how to write ratios.
- Remind students that a ratio is a comparison between two quantities.
- Ratios can be written in three different ways. They can be expressed as a fraction, using a colon, or using the word "to."
  Example: $\frac{3}{4}$, $3:4$, 3 to 4
- **Common Error:** Order matters when writing ratios. A ratio of $3:4$ carries a different meaning than a ratio of $4:3$. Remind students to write the ratio in the same order that the problem asks for it.
- It is best to express the final ratio in simplest form. Writing a ratio in simplest form is similar to simplifying fractions.

Reteaching and Enrichment Strategies

If students need help. . .	If students got it. . .
Record and Practice Journal • Fair Game Review Skills Review Handbook Lesson Tutorials	Game Closet at *BigIdeasMath.com* Start the next section

What You Learned Before

"Let's spin to decide what we will have for lunch."

M = Mouse
B = Dog Biscuit

Why do we always have to use this spinner?

● Writing Ratios

Example 1   There are 32 football players and 16 cheerleaders at your school. Write the ratio of cheerleaders to football players.

cheerleaders ⟶ $\dfrac{16}{32} = \dfrac{1}{2}$ Write in simplest form.

football players ⟶

⋮∴ So, the ratio of cheerleaders to football players is $\dfrac{1}{2}$.

Example 2

a. Write the ratio of girls to boys in Classroom A.

$$\dfrac{\text{Girls in Classroom A}}{\text{Boys in Classroom A}} = \dfrac{11}{14}$$

	Boys	Girls
Classroom A	14	11
Classroom B	12	8

⋮∴ So, the ratio of girls to boys in Classroom A is $\dfrac{11}{14}$.

b. Write the ratio of boys in Classroom B to the total number of students in both classes.

$$\dfrac{\text{Boys in Classroom B}}{\text{Total number of students}} = \dfrac{12}{14 + 11 + 12 + 8} = \dfrac{12}{45} = \dfrac{4}{15}$$ Write in simplest form.

⋮∴ So, the ratio of boys in Classroom B to the total number of students is $\dfrac{4}{15}$.

Try It Yourself

Write the ratio in simplest form.

1. baseballs to footballs

2. footballs to total pieces of equipment

3. sneakers to ballet slippers

4. sneakers to total number of shoes

5. green beads to blue beads

6. red beads : green beads

7. green beads : total number of beads

Essential Question In an experiment, how can you determine the number of possible results?

An *experiment* is an investigation or a procedure that has varying results. Flipping a coin, rolling a number cube, and spinning a spinner are all examples of experiments.

1 ACTIVITY: Conducting Experiments

Work with a partner.

a. You flip a dime.

There are _____ possible results.

Out of 20 flips, you think you will flip heads _____ times.

Flip a dime 20 times. Tally your results in a table. How close was your guess?

b. You spin the spinner shown.

There are _____ possible results.

Out of 20 spins, you think you will spin orange _____ times.

Spin the spinner 20 times. Tally your results in a table. How close was your guess?

c. You spin the spinner shown.

There are _____ possible results.

Out of 20 spins, you think you will spin a 4 _____ times.

Spin the spinner 20 times. Tally your results in a table. How close was your guess?

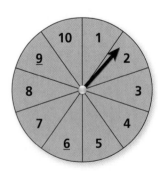

Probability and Statistics

In this lesson, you will

● identify and count the outcomes of experiments.

2 ACTIVITY: Comparing Different Results

Work with a partner. Use the spinner in Activity 1(c).

a. Do you have a better chance of spinning an even number or a multiple of 4? Explain your reasoning.

b. Do you have a better chance of spinning an even number or an odd number? Explain your reasoning.

Laurie's Notes

Introduction

Applying Mathematical Practices

- **Model with Mathematics:** Students gain a conceptual sense of probability by performing these activities. The concept of *possible outcomes* is also developed.

Motivate

? "How many of you have ever played Rock Paper Scissors?

- Different versions of Rock Paper Scissors are played in different countries. In Japan it is called *Jan-ken-pon*. In Indonesia the objects are elephant (one thumb up out of a clapped hand), person (showing one index finger), and ant (showing one little finger). The elephant beats the person. The person beats the ant. The ant beats the elephant, because if an ant gets into the elephant's ear, the elephant cannot do anything about the itchiness!

- Students will be eager to play with their partners. They will play the game as part of today's activity.

Activity Notes

Activity 1

- It is important to have manipulatives for this activity. If dimes are not available, substitute pennies or two-colored counters. You can make spinners using a paper clip.

- Remind students to make a prediction about the outcomes before they perform the experiment.

- If time is short, one partner can record while the other performs the experiment, or have enough materials so that each partner can do the experiment at the same time.

- When students are finished, try to gather class data quickly for one outcome in each part of the activity. Below is an example.

Result	Total number of repetitions	Total number of occurrences
Flipping heads	20 × Number of students	?
Spinning red	20 × Number of students	?
Spinning 4	20 × Number of students	?

- Ask for volunteers to explain how they came up with their guesses in each part. This will help students prepare for Activity 2.

Activity 2

- **Construct Viable Arguments and Critique the Reasoning of Others:** Ask for volunteers to share their explanations with the class.

- **Extension:** Redraw the spinner in part (c) so the numbers are not consecutive. For example, the even numbers are consecutive and the odd numbers are consecutive. Ask students if this new spinner changes how they think about the problem. Some students may say that a particular result is more likely because the numbers are clustered together.

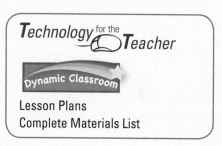

Technology for the Teacher

Dynamic Classroom

Lesson Plans
Complete Materials List

10.1 Record and Practice Journal

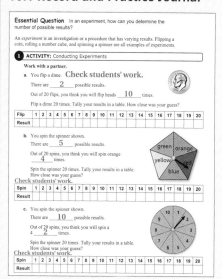

Differentiated Instruction

Auditory

Discuss with students what it means to say that something happens by chance. Ask students to describe situations when a result is determined by chance, such as rolling a number cube or playing Rock Paper Scissors. Then ask students to describe an event that is impossible and an event that is certain.

10.1 Record and Practice Journal

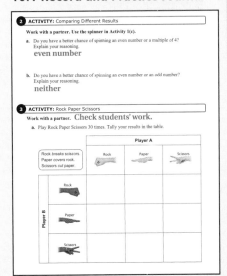

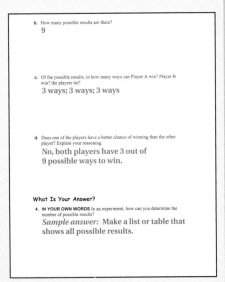

Laurie's Notes

Activity 3

- Review with students the rules for playing Rock Paper Scissors.
- Recording the results means making a tally mark for each game played.
- Students will get caught up in playing and may forget to record the results. You may want to modify the activity and have groups of 3 where the third person is the recorder.
- When students have finished, discuss student results as a whole class.

What Is Your Answer?

- **Neighbor Check:** Have students work independently and then have their neighbors check their work. Have students discuss any discrepancies.

Closure

- **Exit Ticket:** If you played a game similar to Rock Paper Scissors that had a fourth object such as a glove, how many possible results do you think there would be? Explain. 16; Each person has 4 possible results. So, two people have $4 \times 4 = 16$ possible results.

3 ACTIVITY: Rock Paper Scissors

Work with a partner.

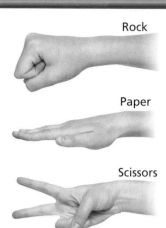

Rock

Paper

Scissors

Math Practice

Interpret a Solution

How do your results compare to the possible results? Explain.

a. Play Rock Paper Scissors 30 times. Tally your results in the table.

b. How many possible results are there?

c. Of the possible results, in how many ways can Player A win? Player B win? the players tie?

d. Does one of the players have a better chance of winning than the other player? Explain your reasoning.

Rock *breaks* scissors.
Paper *covers* rock.
Scissors *cut* paper.

	Player A		
Player B			

What Is Your Answer?

4. IN YOUR OWN WORDS In an experiment, how can you determine the number of possible results?

Use what you learned about experiments to complete Exercises 3 and 4 on page 404.

Check It Out
Lesson Tutorials
BigIdeasMath.com

Key Vocabulary ◀))
experiment, *p. 402*
outcomes, *p. 402*
event, *p. 402*
favorable outcomes, *p. 402*

 **Key Ideas**

Outcomes and Events

An **experiment** is an investigation or a procedure that has varying results. The possible results of an experiment are called **outcomes**. A collection of one or more outcomes is an **event**. The outcomes of a specific event are called **favorable outcomes**.

For example, randomly selecting a marble from a group of marbles is an experiment. Each marble in the group is an outcome. Selecting a green marble from the group is an event.

Reading

When an experiment is performed *at random* or *randomly*, all of the possible outcomes are equally likely.

Possible outcomes

Event: Choosing a green marble
Number of favorable outcomes: 2

EXAMPLE ⓵ **Identifying Outcomes**

You roll the number cube.

a. What are the possible outcomes?

∴ The six possible outcomes are rolling a 1, 2, 3, 4, 5, and 6.

b. What are the favorable outcomes of rolling an even number?

even	*not* even
2, 4, 6	1, 3, 5

∴ The favorable outcomes of the event are rolling a 2, 4, and 6.

c. What are the favorable outcomes of rolling a number greater than 5?

greater than 5	*not* greater than 5
6	1, 2, 3, 4, 5

∴ The favorable outcome of the event is rolling a 6.

◀)) Multi-Language Glossary at BigIdeasMath.com

Laurie's Notes

Introduction

Connect

- **Yesterday:** Students counted the number of possible outcomes of an experiment.
- **Today:** Students will describe the outcomes of an experiment.

Motivate

- Give students a chance to have no homework. Hold a standard deck of cards and tell them one student is going to select a card from the deck. If it is a 7, there is no homework. If it is not a 7, there is homework.
- Students immediately say it is not fair. Act indignant, and suggest that if they do not draw a card, then there definitely will be homework. This will change their attitude.
- If you are willing to accept the outcome, let one student draw a card. The *odds* are in your favor!

Lesson Notes

Key Idea

- Discuss the vocabulary words: experiment, outcomes, event, and favorable outcomes. You can relate the vocabulary to the various activities yesterday and to tossing two number cubes.
- **?** Ask students to identify the favorable outcomes for the events of choosing each color of marble. green (2), blue (1), red (1), yellow (1), purple (1)
- Discuss the Study Tip. This tip provides the statistical meaning. Go over the "meaning of the word" in everyday life; *having no specific pattern, purpose, or objective.*

Example 1

- Make sure that students understand that there can be more than one favorable outcome.
- **?** "What are some other examples of experiments and events? What are the favorable outcomes for these events?" *Sample answer:* An experiment is spinning a spinner with the numbers 1–12. An event is spinning a number greater than 10, with the favorable outcomes 11 and 12.
- **?** "What are the favorable outcomes of rolling a prime number?" 2, 3, and 5
- **?** "What are the favorable outcomes of rolling a number divisible by 3?" 3 and 6
- **?** "What are the favorable outcomes of rolling a number that is a multiple of 8?" There are none.

Lesson Tutorials
Lesson Plans
Answer Presentation Tool

Extra Example 1

You roll a number cube.

a. What are the possible outcomes?
   1, 2, 3, 4, 5, 6

b. What are the favorable outcomes of rolling an odd number? 1, 3, 5

c. What are the favorable outcomes of rolling a number greater than 4?
   5, 6

Laurie's Notes

On Your Own

1. **a.** A, B, C, D, E, F, G, H, I, J, K

    b. A, E, I

Extra Example 2

You spin the spinner shown in Example 2.

a. How many ways can spinning blue occur? 1 way

b. How many ways can spinning *not* green occur? 5 ways

c. What are the favorable outcomes of spinning *not* green? red, red, red, purple, blue

On Your Own

2. **a.** 8 outcomes

    b. 2 ways

    c. 5 ways; blue, blue, red, green, purple

English Language Learners

Vocabulary

Some English learners may confuse the words *outcome* and *event*. Help students see how the two words are related using Example 2 parts (a) and (b).

Possible Outcomes			*Favorable*
red	red	red	← *Outcomes*
purple	blue	green	*of event choosing red*

On Your Own

- **Think-Pair-Share:** Students should read each question independently and then work in pairs to answer the questions. When they have answered the questions, the pair should compare their answers with another group and discuss any discrepancies.

Example 2

- **Make Sense of Problems and Persevere in Solving Them:** Discuss the difference between outcomes and favorable outcomes. Refer back to the marbles in the Key Ideas.
- **Common Error:** When answering part (a), many students will say that there are only 4 possible outcomes, not 6, because there are only four colors in the spinner. Explain to students that they need to count every occurrence of a color as a possible outcome. Students may be able to understand this concept better after part (c).

On Your Own

- Discuss answers as a class.

Closure

- **Exit Ticket:** What are the favorable outcomes of drawing a face card from a deck of cards? There are three face cards (jack, queen, and king) and four suits, so there are 12 favorable outcomes.

On Your Own

Now You're Ready
Exercises 5–11

1. You randomly choose a letter from a hat that contains the letters A through K.

   a. What are the possible outcomes?

   b. What are the favorable outcomes of choosing a vowel?

EXAMPLE 2 **Counting Outcomes**

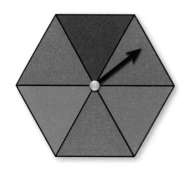

You spin the spinner.

a. How many possible outcomes are there?

The spinner has 6 sections. So, there are 6 possible outcomes.

b. In how many ways can spinning red occur?

The spinner has 3 red sections. So, spinning red can occur in 3 ways.

c. In how many ways can spinning *not* purple occur? What are the favorable outcomes of spinning *not* purple?

The spinner has 5 sections that are *not* purple. So, spinning *not* purple can occur in 5 ways.

purple	*not* purple
purple	red, red, red, green, blue

The favorable outcomes of the event are red, red, red, green, and blue.

On Your Own

Now You're Ready
Exercises 12–17

2. You randomly choose a marble.

   a. How many possible outcomes are there?

   b. In how many ways can choosing blue occur?

   c. In how many ways can choosing *not* yellow occur? What are the favorable outcomes of choosing *not* yellow?

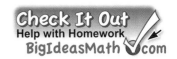
 Vocabulary and Concept Check

1. **VOCABULARY** Is rolling an even number on a number cube an *outcome* or an *event*? Explain.

2. **WRITING** Describe how an outcome and a favorable outcome are different.

 Practice and Problem Solving

You spin the spinner shown.

3. How many possible results are there?

4. Of the possible results, in how many ways can you spin an even number? an odd number?

① 5. **TILES** What are the possible outcomes of randomly choosing one of the tiles shown?

You randomly choose one of the tiles shown above. Find the favorable outcomes of the event.

6. Choosing a 6

7. Choosing an odd number

8. Choosing a number greater than 5

9. Choosing an odd number less than 5

10. Choosing a number less than 3

11. Choosing a number divisible by 3

You randomly choose one marble from the bag. (a) Find the number of ways the event can occur. (b) Find the favorable outcomes of the event.

② 12. Choosing blue

13. Choosing green

14. Choosing purple

15. Choosing yellow

16. Choosing *not* red

17. Choosing *not* blue

18. **ERROR ANALYSIS** Describe and correct the error in finding the number of ways that choosing *not* purple can occur.

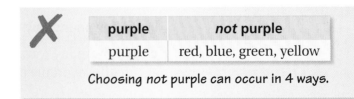

X

purple	*not* purple
purple	red, blue, green, yellow

Choosing not purple can occur in 4 ways.

Assignment Guide and Homework Check

Level	Day 1 Activity Assignment	Day 2 Lesson Assignment	Homework Check
Basic	3, 4, 27–31	1, 2, 5–17 odd, 18, 19–23 odd	7, 13, 19, 21
Average	3, 4, 27–31	1, 2, 5–17 odd, 18–26 even	7, 13, 20, 26
Advanced	3, 4, 27–31	1, 2, 6–26 even	10, 16, 24, 26
Accelerated	1–4, 6–26 even, 27–31		10, 16, 24, 26

Common Errors

- **Exercises 6–11** Students may forget to include, or include too many, favorable outcomes. Encourage them to write out all of the possible outcomes and then circle the favorable outcomes for the given event.
- **Exercises 12–17** Students may forget to include the repeats of a color when describing the favorable outcomes and how many ways the event can occur. Ask students to describe how many times you could pull out a specific color from the bag if you happened to pull out the same color each time (without replacing). For example, if the event is choosing a red marble, you can pull out a red marble three times before there are no more red marbles in the bag. So, the event can occur three times.

10.1 Record and Practice Journal

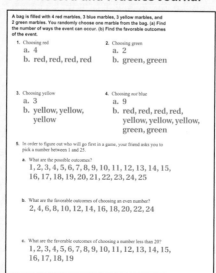

A bag is filled with 4 red marbles, 3 blue marbles, 3 yellow marbles, and 2 green marbles. You randomly choose one marble from the bag. (a) Find the number of ways the event can occur. (b) Find the favorable outcomes of the event.

1. Choosing red
   a. 4
   b. red, red, red, red

2. Choosing green
   a. 2
   b. green, green

3. Choosing yellow
   a. 3
   b. yellow, yellow, yellow

4. Choosing *not* blue
   a. 9
   b. red, red, red, red, yellow, yellow, yellow, green, green

5. In order to figure out who will go first in a game, your friend asks you to pick a number between 1 and 25.
   a. What are the possible outcomes?
      1, 2, 3, 4, 5, 6, 7, 8, 9, 10, 11, 12, 13, 14, 15, 16, 17, 18, 19, 20, 21, 22, 23, 24, 25
   b. What are the favorable outcomes of choosing an even number?
      2, 4, 6, 8, 10, 12, 14, 16, 18, 20, 22, 24
   c. What are the favorable outcomes of choosing a number less than 20?
      1, 2, 3, 4, 5, 6, 7, 8, 9, 10, 11, 12, 13, 14, 15, 16, 17, 18, 19

Vocabulary and Concept Check

1. event; It is a collection of several outcomes.

2. An outcome is one possible result of an experiment. A favorable outcome is an outcome of a specific event.

Practice and Problem Solving

3. 8

4. 4 ways; 4 ways

5. 1, 2, 3, 4, 5, 6, 7, 8, 9

6. 6   7. 1, 3, 5, 7, 9

8. 6, 7, 8, 9   9. 1, 3

10. 1, 2   11. 3, 6, 9

12. a. 2 ways   b. blue, blue

13. a. 1 way   b. green

14. a. 2 ways
    b. purple, purple

15. a. 1 way   b. yellow

16. a. 6 ways
    b. yellow, green, blue, blue, purple, purple

17. a. 7 ways
    b. red, red, red, purple, purple, green, yellow

18. There are 7 marbles that are *not* purple, even though there are only 4 colors. Choosing *not* purple could be red, red, red, blue, blue, green, or yellow.

T-404

Practice and Problem Solving

19. 7 ways

20. false; red

21. true

22. false; five

23. true

24. false; eight

25. 30 rock CDs

26. See *Taking Math Deeper*.

Fair Game Review

27. $x = 2$

28. $n = 21$

29. $w = 12$

30. $b = 68$

31. C

Mini-Assessment

You randomly choose one number below. Find the favorable outcomes of the event.

10, 11, 12, 13, 14, 15, 16, 17, 18, 19

1. Choosing a 14 14

2. Choosing an even number 10, 12, 14, 16, 18

3. Choosing an odd number less than 15 11, 13

4. Choosing a number greater than 16 17, 18, 19

5. Choosing a number divisible by 2 10, 12, 14, 16, 18

Taking Math Deeper

Exercise 26

This problem previews the concept of dependent events.

 With all five cards available, the number of possible outcomes is 5.

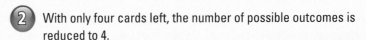

Choose 1 at random.

 With only four cards left, the number of possible outcomes is reduced to 4.

Choose 1 at random.

③ Answer the question.

After the baker is chosen, the number of possible outcomes decreases.

Project

Create a game that uses picture cards and the changing probabilities indicated in the problem. Create the cards. Play the game with a partner.

Reteaching and Enrichment Strategies

If students need help. . .	If students got it. . .
Resources by Chapter • Practice A and Practice B • Puzzle Time Record and Practice Journal Practice Differentiating the Lesson Lesson Tutorials Skills Review Handbook	Resources by Chapter • Enrichment and Extension • Technology Connection Start the next section

19. COINS You have 10 coins in your pocket. Five are Susan B. Anthony dollars, two are Kennedy half-dollars, and three are presidential dollars. You randomly choose a coin. In how many ways can choosing *not* a presidential dollar occur?

Kennedy half-dollar

Presidential dollar

Susan B. Anthony dollar

Spinner A

Tell whether the statement is *true* or *false*. If it is false, change the italicized word to make the statement true.

20. Spinning blue and spinning *green* have the same number of favorable outcomes on Spinner A.

21. Spinning blue has one *more* favorable outcome than spinning green on Spinner B.

22. There are *three* possible outcomes of spinning Spinner A.

23. Spinning *red* can occur in four ways on Spinner B.

24. Spinning not green can occur in *three* ways on Spinner B.

Spinner B

Dancer

Firefighter

Baker

Pirate

Bellhop

25. MUSIC A bargain bin contains classical and rock CDs. There are 60 CDs in the bin. Choosing a rock CD and *not* choosing a rock CD have the same number of favorable outcomes. How many rock CDs are in the bin?

26. **Precision** You randomly choose one of the cards and set it aside. Then you randomly choose a second card. Describe how the number of possible outcomes changes after the first card is chosen.

 Fair Game Review What you learned in previous grades & lessons

Solve the proportion. *(Section 5.4)*

27. $\dfrac{x}{10} = \dfrac{1}{5}$

28. $\dfrac{60}{n} = \dfrac{20}{7}$

29. $\dfrac{1}{3} = \dfrac{w}{36}$

30. $\dfrac{25}{17} = \dfrac{100}{b}$

31. MULTIPLE CHOICE What is the surface area of the rectangular prism? *(Section 9.1)*

Ⓐ 162 in.²   Ⓑ 264 in.²

Ⓒ 324 in.²   Ⓓ 360 in.²

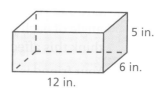

5 in.
6 in.
12 in.

Auditory

Ask students what it means when something happens by chance. Discuss with students situations in which the outcomes happen by chance, such as flipping a coin, rolling a number cube, or spinning a spinner. Ask students what it means for an event to be impossible.

10.2 Record and Practice Journal

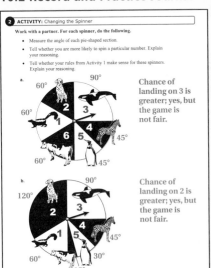

Laurie's Notes

Activity 2

- In this activity, students must consider spinners with different central angle measures. The Black-and-White Spinner Game students created for Activity 1 will be played using these new spinners. First, they need to measure the central angles.

- **?** "How can you tell whether you are more likely to spin a particular number?" Listen for comparisons of the central angle measures.

- **Big Idea:** The outcomes are not equally likely. Students should recognize that not all of the central angle measures are the same, so the likelihood of spinning a particular number varies.

- Ask students to compare and contrast the spinners. Note that for the second spinner, spinning black and spinning white are equally likely, which is a key observation in Activity 3.

Activity 3

- **?** "What would make the game fair?" if each player has an equal chance of winning

- Give time for students to play 10 games on each spinner. Have each pair of students gather data in a table.

- Collect class data for wins and losses on each spinner.

- **?** "What does the data tell you about the fairness of the three spinners?" The data suggests that using these rules with the spinners in Activity 1 and Activity 2(b) results in fair games.

What Is Your Answer?

- Discuss Question 5 as a class. Try to lead them to describe the likelihood as *impossible* by first asking them if it is possible to spin an 8, and then having them complete the statement "There is not an 8 on the spinner, so spinning an 8 is _____."

Closure

- Draw the spinners below on the board. You might offer students an opportunity to spin for no homework if they can come to a consensus as to which of the spinners to use (N = no homework and Y = homework).

② ACTIVITY: Changing the Spinner

Work with a partner. For each spinner, do the following.

- Measure the angle of each pie-shaped section.
- Tell whether you are more likely to spin a particular number. Explain your reasoning.
- Tell whether your rules from Activity 1 make sense for these spinners. Explain your reasoning.

a.   b.

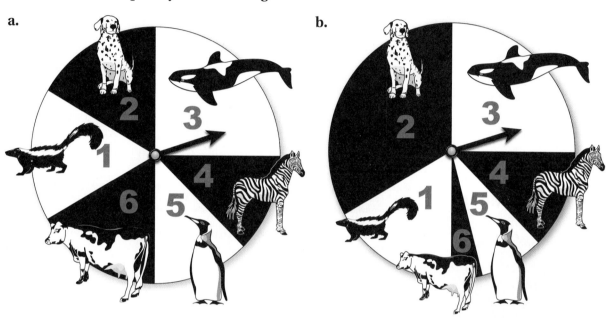

③ ACTIVITY: Is This Game Fair?

Math Practice

Use Prior Results

How can you use the results of the previous activities to determine whether the game is fair?

Work with a partner. Apply the following rules to each spinner in Activities 1 and 2. Is the game fair? Why or why not? If not, who has the better chance of winning?

- Take turns spinning the spinner.
- If you spin an odd number, Player 1 wins.
- If you spin an even number, Player 2 wins.

What Is Your Answer?

4. **IN YOUR OWN WORDS** How can you describe the likelihood of an event?

5. Describe the likelihood of spinning an 8 in Activity 1.

6. Describe a career in which it is important to know the likelihood of an event.

Practice Use what you learned about the likelihood of an event to complete Exercises 4 and 5 on page 410.

Key Vocabulary
probability, *p. 408*

Key Idea

Probability

The **probability** of an event is a number that measures the likelihood that the event will occur. Probabilities are between 0 and 1, including 0 and 1. The diagram relates likelihoods (above the diagram) and probabilities (below the diagram).

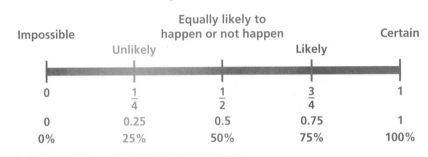

Study Tip

Probabilities can be written as fractions, decimals, or percents.

EXAMPLE 1 **Describing the Likelihood of an Event**

There is an 80% chance of thunderstorms tomorrow. Describe the likelihood of the event.

80% chance

The probability of thunderstorms tomorrow is 80%.

⋮ Because 80% is close to 75%, it is *likely* that there will be thunderstorms tomorrow.

On Your Own

Now You're Ready
Exercises 6–9

Describe the likelihood of the event given its probability.

1. The probability that you land a jump on a snowboard is $\frac{1}{2}$.

2. There is a 100% chance that the temperature will be less than 120°F tomorrow.

Key Idea

Finding the Probability of an Event

When all possible outcomes are equally likely, the probability of an event is the ratio of the number of favorable outcomes to the number of possible outcomes. The probability of an event is written as P(event).

$$P(\text{event}) = \frac{\text{number of favorable outcomes}}{\text{number of possible outcomes}}$$

◀) Multi-Language Glossary at BigIdeasMath☑com

Laurie's Notes

Introduction

Connect
- **Yesterday:** Students developed an intuitive understanding of how to predict the likelihood of the results of a spinner.
- **Today:** Students will compute the probability of an event.

Motivate
- The name of a county in a U.S. state is often named after a person, such as a president. In fact, there are 31 states in the U.S. with a county named Washington. Sometimes the county is named after a historical figure. Madison is a county name in 20 states and Calhoun is a county name in 11 states.
- Draw a spinner representing this information.

> Key:
> C – Calhoun   M – Madison   W – Washington

❓ "If you spin the spinner 100 times, how many times would you expect it to land on Washington? Explain." 50; There are 62 different counties represented and half of them are named Washington.

Lesson Notes

Key Idea
- Discuss possible events which have probabilities near each benchmark. Make them personal for your situation, if possible. Examples: the sun rising tomorrow = 1, math homework = 0.75, winning the softball game = 0.50, skipping breakfast = 0.25, a winter in Vermont with no snow = 0
- Spend time discussing what *equally likely* means. Give examples of events that are equally likely and not equally likely. Equally likely: number cube; Not equally likely: spinner with sections that are not all the same size

Example 1
❓ "Has anyone heard a weather report for tomorrow?" Try to turn student responses into a percent or fraction. For example, you could translate "it's supposed to be nice tomorrow" into "there's a 90% chance of sunshine."
- Be sure that students understand that probabilities can be represented as fractions, decimals, or percents.

On Your Own
- Discuss answers as a class.

Key Idea
- Write the Key Idea.
- Discuss probability and give several examples with which students would be familiar, such as cards, number cubes, and marbles in a bag. Stress that the outcomes must be equally likely to use this ratio.

Goal Today's lesson is finding the **probability** of an event.

Lesson Tutorials
Lesson Plans
Answer Presentation Tool

Extra Example 1
There is a 20% chance of snow flurries tomorrow. Describe the likelihood of the event. Because 20% is close to 25%, it is *unlikely* that there will be snow flurries tomorrow.

 On Your Own

1. equally likely to happen or not happen

2. certain

English Language Learners

Comprehension

This chapter has more word problems and fewer skill problems than most chapters. It may be more difficult for some students. You can present problems in a predictable format so the students will not get stuck on reading and be able to focus on the mathematics.

Extra Example 2

In Example 2, what is the probability of rolling a number greater than 4? $\frac{1}{3}$

Extra Example 3

The probability that you randomly draw a short straw from a group of 50 straws is $\frac{9}{25}$. How many are short straws? 18

 On Your Own

3. $\frac{2}{3}$

4. 0

5. 5

Laurie's Notes

Example 2

? "How many possible outcomes are there?" 6 "How many are favorable outcomes?" 3

? "Can you use the ratio in the previous Key Idea to find the probability of rolling an odd number? Explain." Yes, because the outcomes are equally likely.

- **Attend to Precision:** Students should communicate precisely, explaining why the outcomes are equally likely when providing their explanations.
- **Extension:** Ask about the probability of rolling an even number, rolling a prime number, or rolling a number greater than 8. $\frac{1}{2}, \frac{1}{2}, 0$

Example 3

- Work through the example by writing and solving the proportion.
- **Alternate Solution:** Because $\frac{3}{20} = \frac{15}{100} = 15\%$, this problem could be solved by finding the percent of a number.

$$15\% \text{ of } 40 = 0.15 \times 40$$
$$= 6$$

It is important for students to see this connection.

On Your Own

- **Think-Pair-Share:** Students should read each question independently and then work in pairs to answer the questions. When they have answered the questions, the pair should compare their answers with another group and discuss any discrepancies.

Closure

- Write an example of an event that has the following probabilities.
  a. close to 1
  b. exactly $\frac{1}{2}$
  c. close to 0
  d. exactly $\frac{2}{5}$

EXAMPLE **2** **Finding a Probability**

You roll the number cube. What is the probability of rolling an odd number?

$$P(\text{event}) = \frac{\text{number of favorable outcomes}}{\text{number of possible outcomes}}$$

$$P(\text{odd}) = \frac{3}{6}$$ ← There are 3 odd numbers (1, 3, and 5).
← There is a total of 6 numbers.

$$= \frac{1}{2}$$ Simplify.

∴ The probability of rolling an odd number is $\frac{1}{2}$, or 50%.

EXAMPLE **3** **Using a Probability**

The probability that you randomly draw a short straw from a group of 40 straws is $\frac{3}{20}$. How many are short straws?

Ⓐ 4   Ⓑ 6

Ⓒ 15   Ⓓ 34

$$P(\text{short}) = \frac{\text{number of short straws}}{\text{total number of straws}}$$

$$\frac{3}{20} = \frac{n}{40}$$ Substitute. Let n be the number of short straws.

$$6 = n$$ Solve for n.

There are 6 short straws.

∴ So, the correct answer is Ⓑ.

● **On Your Own**

Exercises 11–15

3. In Example 2, what is the probability of rolling a number greater than 2?

4. In Example 2, what is the probability of rolling a 7?

5. The probability that you randomly draw a short straw from a group of 75 straws is $\frac{1}{15}$. How many are short straws?

Vocabulary and Concept Check

1. **VOCABULARY** Explain how to find the probability of an event.

2. **REASONING** Can the probability of an event be 1.5? Explain.

3. **OPEN-ENDED** Give a real-life example of an event that is impossible. Give a real-life example of an event that is certain.

Practice and Problem Solving

You are playing a game using the spinners shown.

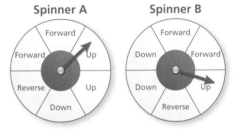

4. You want to move down. On which spinner are you more likely to spin "Down"? Explain.

5. You want to move forward. Which spinner would you spin? Explain.

Describe the likelihood of the event given its probability.

6. Your soccer team wins $\frac{3}{4}$ of the time.

7. There is a 0% chance that you will grow 12 more feet.

8. The probability that the sun rises tomorrow is 1.

9. It rains on $\frac{1}{5}$ of the days in July.

10. **VIOLIN** You have a 50% chance of playing the correct note on a violin. Describe the likelihood of playing the correct note.

You randomly choose one shirt from the shelves. Find the probability of the event.

11. Choosing a red shirt

12. Choosing a green shirt

13. *Not* choosing a white shirt

14. *Not* choosing a black shirt

15. Choosing an orange shirt

16. **ERROR ANALYSIS** Describe and correct the error in finding the probability of *not* choosing a blue shirt from the shelves above.

$P(not\ blue) = \frac{4}{10} = \frac{2}{5}$

Assignment Guide and Homework Check

Level	Day 1 Activity Assignment	Day 2 Lesson Assignment	Homework Check
Basic	4, 5, 23–27	1–3, 7–15 odd, 16, 17, 19	9, 11, 15, 17, 19
Average	4, 5, 23–27	1–3, 6–10 even, 11–15 odd, 16, 18, 19	8, 13, 18, 19
Advanced	4, 5, 23–27	1–3, 6–22 even	8, 14, 18, 20, 22
Accelerated	1–5, 6–22 even, 23–27		8, 14, 18, 20, 22

Common Errors

- **Exercises 11–15** Students may write the probability as the ratio of the number of favorable outcomes to the number of unfavorable outcomes. Remind them that the probability of an event is the ratio of the number of favorable outcomes to the number of possible outcomes.
- **Exercise 18** In part (a), students may write and solve the proportion to find the number of winning ducks correctly. But then they forget to subtract their result from 25 to find the number of *not* winning ducks.

10.2 Record and Practice Journal

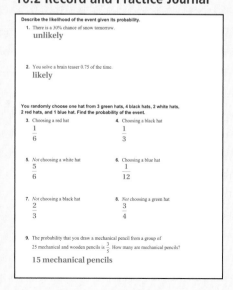

Vocabulary and Concept Check

1. The probability of an event is the ratio of the number of favorable outcomes to the number of possible outcomes.

2. no; Probabilities are between 0 and 1, including 0 and 1.

3. *Sample answer:* You will not have any homework this week.; You will fall asleep tonight.

Practice and Problem Solving

4. Spinner B; There are more chances to land on "Down" with Spinner B.

5. either; Both spinners have the same number of chances to land on "Forward."

6. likely   7. impossible

8. certain   9. unlikely

10. equally likely to happen or not happen

11. $\dfrac{1}{10}$ 12. $\dfrac{1}{5}$

13. $\dfrac{9}{10}$ 14. $\dfrac{4}{5}$

15. 0

16. The student found the probability of choosing a blue shirt.;

$P(not \text{ blue}) = \dfrac{6}{10} = \dfrac{3}{5}$

17. 20

18 See *Taking Math Deeper*.

19. a. $\dfrac{2}{3}$; likely

   b. $\dfrac{1}{3}$; unlikely

   c. $\dfrac{1}{2}$; equally likely to happen or not happen

20.

Mother's Genes

	X	X
X	XX	XX
Y	XY	XY

Father's Genes

21. There are 2 combinations for each.

22. a.

Parent 1

$\frac{1}{4}$, or 25%

	C	s
C	CC	Cs
s	Cs	ss

Parent 2

b. $\frac{3}{4}$, or 75%

Fair Game Review

23. $x < 4$;

24. $b \geq -5$;

25. $w > -3$;

26. $g \leq -3$;

27. C

Mini-Assessment

You randomly choose one number below. Find the probability of the event.

2, 5, 6, 9, 13, 16, 22, 25, 27, 31

1. Choosing an even number $\frac{2}{5}$

2. Choosing an odd number $\frac{3}{5}$

3. Choosing a prime number $\frac{2}{5}$

4. Choosing a number greater than 30 $\frac{1}{10}$

5. Choosing a number less than 2 0

Taking Math Deeper

Exercise 18

Students have not been formally introduced to the *complement* of an event, but they can still use this concept to solve the problem.

 Find the probability of choosing a rubber duck that is *not* a winner.

The probability that you choose a winning rubber duck is 0.24, or 24%. So, 100% − 24% = 76% of the rubber ducks are *not* winners.

24%
Win

76%
Lose

76%

 Find 76% of 25.

$0.76 \cdot 25 = 19$

a. There are 19 rubber ducks that are *not* winners.

Describe the likelihood of *not* choosing a winning duck.

b. There is a 76% chance of *not* choosing a winning rubber duck. So, it is *likely* that you will choose a rubber duck that is *not* a winner.

Project

Many games at carnivals can be modified so that it is harder for you to win a prize. Research a few carnival games and explain how they can be modified to decrease your probability of winning.

Reteaching and Enrichment Strategies

If students need help. . .	If students got it. . .
Resources by Chapter • Practice A and Practice B • Puzzle Time Record and Practice Journal Practice Differentiating the Lesson Lesson Tutorials Skills Review Handbook	Resources by Chapter • Enrichment and Extension • Technology Connection Start the next section

17. CONTEST The rules of a contest say that there is a 5% chance of winning a prize. Four hundred people enter the contest. Predict how many people will win a prize.

18. RUBBER DUCKS At a carnival, the probability that you choose a winning rubber duck from 25 ducks is 0.24.

   a. How many are *not* winning ducks?

   b. Describe the likelihood of *not* choosing a winning duck.

19. DODECAHEDRON A dodecahedron has twelve sides numbered 1 through 12. Find the probability and describe the likelihood of each event.

   a. Rolling a number less than 9

   b. Rolling a multiple of 3

   c. Rolling a number greater than 6

A Punnett square is a grid used to show possible gene combinations for the offspring of two parents. In the Punnett square shown, a boy is represented by *XY*. A girl is represented by *XX*.

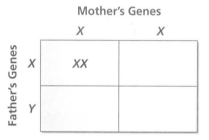

20. Complete the Punnett square.

21. Explain why the probability of two parents having a boy or having a girl is equally likely.

22. *Critical Thinking* Two parents each have the gene combination *Cs*. The gene *C* is for curly hair. The gene *s* is for straight hair.

   a. Make a Punnett square for the two parents. When all outcomes are equally likely, what is the probability of a child having the gene combination *CC*?

   b. Any gene combination that includes a *C* results in curly hair. When all outcomes are equally likely, what is the probability of a child having curly hair?

 Fair Game Review What you learned in previous grades & lessons

Solve the inequality. Graph the solution. *(Section 4.2 and Section 4.3)*

23. $x + 5 < 9$ **24.** $b - 2 \geq -7$ **25.** $1 > -\dfrac{w}{3}$ **26.** $6 \leq -2g$

27. MULTIPLE CHOICE Find the value of *x*. *(Section 7.4)*

   Ⓐ 85   Ⓑ 90

   Ⓒ 93   Ⓓ 102

Essential Question How can you use relative frequencies to find probabilities?

When you conduct an experiment, the **relative frequency** of an event is the fraction or percent of the time that the event occurs.

$$\text{relative frequency} = \frac{\text{number of times the event occurs}}{\text{total number of times you conduct the experiment}}$$

1 ACTIVITY: Finding Relative Frequencies

Work with a partner.

a. Flip a quarter 20 times and record your results. Then complete the table. Are the relative frequencies the same as the probability of flipping heads or tails? Explain.

	Flipping Heads	Flipping Tails
Relative Frequency		

b. Compare your results with those of other students in your class. Are the relative frequencies the same? If not, why do you think they differ?

c. Combine all of the results in your class. Then complete the table again. Did the relative frequencies change? What do you notice? Explain.

d. Suppose everyone in your school conducts this experiment and you combine the results. How do you think the relative frequencies will change?

2 ACTIVITY: Using Relative Frequencies

Probability and Statistics

In this lesson, you will
- find relative frequencies.
- use experimental probabilities to make predictions.
- use theoretical probabilities to find quantities.
- compare experimental and theoretical probabilities.

Work with a partner. You have a bag of colored chips. You randomly select a chip from the bag and replace it. The table shows the number of times you select each color.

Red	Blue	Green	Yellow
24	12	15	9

a. There are 20 chips in the bag. Can you use the table to find the exact number of each color in the bag? Explain.

b. You randomly select a chip from the bag and replace it. You do this 50 times, then 100 times, and you calculate the relative frequencies after each experiment. Which experiment do you think gives a better approximation of the exact number of each color in the bag? Explain.

Laurie's Notes

Introduction

Applying Mathematical Practices

- **Attend to Precision:** Mathematically proficient students try to communicate precisely to others. In these activities, students should pay close attention to the equally likely aspect of the outcomes based upon the relative frequencies of the experiments.

Motivate

- Ask ten students to stand at the front of the room. Hand each one a penny.
- **?** "If [student 1] flips his or her penny, what is the probability it will land on heads?" $\frac{1}{2}$ "If [student 4] flips his or her penny, what is the probability it will land on heads?" $\frac{1}{2}$
- **?** "If all of the students toss their pennies, what is the probability they will all land on heads?" Students may say $\frac{1}{2}$.
- **?** Have all 10 students toss their coins, then ask, "Did everyone's penny land on heads?" It is unlikely that all 10 students' pennies will land on heads.

Discuss

- Discuss the definition of *relative frequency* and relate it to the Motivate.
- Make sure that students understand that tossing a coin 10 times and having it land on heads about half the time is a different experiment than tossing 10 coins and asking how likely it is that all 10 coins land on heads.

Activity Notes

Activity 1

- **?** "What is the probability of flipping heads?" $\frac{1}{2}$ "What is the probability of flipping tails?" $\frac{1}{2}$ "Is each outcome equally likely?" yes
- **Use Appropriate Tools Strategically:** As data are gathered and recorded, several students with calculators can be summarizing the results.
- **?** "Did the relative frequencies change as the results were combined?" Answers will vary. The additional data should smooth the results so that the relative frequencies approach 50% for each outcome.
- Online simulators are available to model this experiment. Be sure students understand as the number of trials increases, the experimental probability gets closer to the theoretical probability.

Activity 2

- **?** "Why is it important to replace the chip each time?" The probability changes for each trial if the chip is not replaced.
- **Construct Viable Arguments and Critique the Reasoning of Others** and **Attend to Precision:** Listen to students' explanations of how the increase in the number of trials improves the accuracy of the approximation.

What Your Students Will Learn

- Find relative frequencies and experimental probabilities of events.
- Use experimental probabilities to make predictions of future outcomes.
- Find theoretical probabilities and use them to find quantities.
- Compare experimental and theoretical probabilities.

Previous Learning

Students should know how to find the probability of an event.

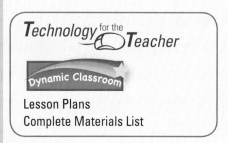

Lesson Plans
Complete Materials List

10.3 Record and Practice Journal

Essential Question How can you use relative frequencies to find probabilities?

When you conduct an experiment, the **relative frequency** of an event is the fraction or percent of the time that the event occurs.

$$\text{relative frequency} = \frac{\text{number of times the event occurs}}{\text{total number of times you conduct the experiment}}$$

1 ACTIVITY: Finding Relative Frequencies

Work with a partner. Check students' work.

a. Flip a quarter 20 times and record your results. Then complete the table. Are the relative frequencies the same as the probability of flipping heads or tails? Explain.

	Flipping Heads	Flipping Tails
Relative Frequency		

b. Compare your results with those of other students in your class. Are the relative frequencies the same? If not, why do you think they differ?

c. Combine all of the results in your class. Then complete the table again. Did the relative frequencies change? What do you notice? Explain.

d. Suppose everyone in your school conducts this experiment and you combine the results. How do you think the relative frequencies will change? The relative frequencies should be close to the probabilities of $\frac{1}{2}$.

Differentiated Instruction

Kinesthetic

Set up 6 groups and assign each group a number from 1 to 6. Each group will predict whether a number less than, greater than, or equal to their assigned number will occur when rolling a number cube 20 times. Have each group track their results in a frequency table. Each group then determines what fraction of the results is less than, equal to, or greater than their assigned number and presents their findings to the class.

10.3 Record and Practice Journal

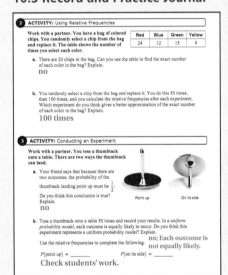

Laurie's Notes

Activity 3

- Be aware of student safety for this activity. Instruct students to toss the thumbtacks carefully so that they stay on the table and no one is hurt.
- ❓ "When you toss a thumbtack, what are the possible outcomes?" It can land with the point up or land on its side. Some students may say it can land with the point down, but if it is tossed onto a hard surface, it is not possible.
- Similar to flipping a coin, tossing a tack also has two outcomes.
- Have students perform part (b) and discuss the results.
- ❓ "What is a *uniform probability model*?" An experiment where the outcomes are equally likely to occur.
- ❓ "Does this experiment represent a uniform probability model?" No; the two outcomes are not equally likely to occur.
- ❓ **Attend to Precision:** "How are the experiments in Activities 1 and 3 alike? How are they different?" Each activity has two possible outcomes. The two outcomes in Activity 1 are equally likely, and the two outcomes in Activity 3 are not equally likely.

What Is Your Answer?

- Ask volunteers to share their reasoning in Question 5. Are students able to distinguish between the relative frequency and the theoretical probability?
- Students often have difficulty thinking of experiments on their own that represent uniform or non-uniform probability models. If students are able to think of some, share these as they might trigger the thinking of other students in the class.
- In Question 8, these spinners were used in the previous section. Students should make a connection between the central angles that they measured and the relative frequencies that they tallied today. Each represents a method for understanding when outcomes of an experiment are equally likely.

Closure

- **Exit Ticket:**
  a. You flip a coin 100 times. How many times do you expect to flip heads? about 50 times
  b. A bag contains 3 red chips and 9 blue chips. You draw a chip and replace it 100 times. How many times do you expect to draw a blue chip? about 75 times

3 ACTIVITY: Conducting an Experiment

Work with a partner. You toss a thumbtack onto a table. There are two ways the thumbtack can land.

Point up   On its side

Math Practice

Analyze Relationships

How can you use the results of your experiment to determine whether this is a uniform probability model?

a. Your friend says that because there are two outcomes, the probability of the thumbtack landing point up must be $\frac{1}{2}$. Do you think this conclusion is true? Explain.

b. Toss a thumbtack onto a table 50 times and record your results. In a *uniform probability model*, each outcome is equally likely to occur. Do you think this experiment represents a uniform probability model? Explain.

Use the relative frequencies to complete the following.

$$P(\text{point up}) = \boxed{} \qquad P(\text{on its side}) = \boxed{}$$

What Is Your Answer?

4. **IN YOUR OWN WORDS** How can you use relative frequencies to find probabilities? Give an example.

5. Your friend rolls a number cube 500 times. How many times do you think your friend will roll an odd number? Explain your reasoning.

6. In Activity 2, your friend says, "There are no orange-colored chips in the bag." Do you think this conclusion is true? Explain.

7. Give an example of an experiment that represents a uniform probability model.

8. Tell whether you can use each spinner to represent a uniform probability model. Explain your reasoning.

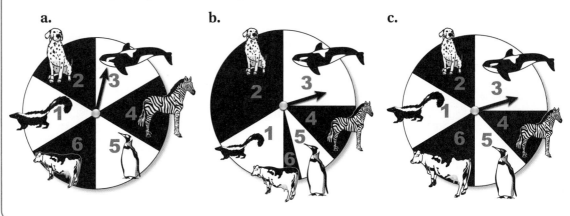

a.   b.   c.

Practice

Use what you learned about relative frequencies to complete Exercises 6 and 7 on page 417.

Check It Out
Lesson Tutorials
BigIdeasMath ✓.com

Key Vocabulary 🔊
relative frequency,
  p. 412
experimental
  probability, p. 414
theoretical
  probability, p. 415

 **Key Idea**

Experimental Probability

Probability that is based on repeated trials of an experiment is called **experimental probability**.

$$P(\text{event}) = \frac{\text{number of times the event occurs}}{\text{total number of trials}}$$

EXAMPLE ① **Finding an Experimental Probability**

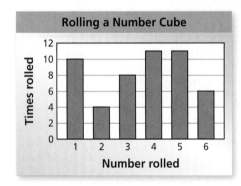

The bar graph shows the results of rolling a number cube 50 times. What is the experimental probability of rolling an odd number?

The bar graph shows 10 ones, 8 threes, and 11 fives. So, an odd number was rolled $10 + 8 + 11 = 29$ times in a total of 50 rolls.

$$P(\text{event}) = \frac{\text{number of times the event occurs}}{\text{total number of trials}}$$

$$P(\text{odd}) = \frac{29}{50}$$

> An odd number was rolled 29 times.

> There was a total of 50 rolls.

⋮ The experimental probability is $\frac{29}{50}$, 0.58, or 58%.

EXAMPLE ② **Making a Prediction**

"April showers bring May flowers." Old Proverb, 1557

It rains 2 out of the last 12 days in March. If this trend continues, how many rainy days would you expect in April?

Find the experimental probability of a rainy day.

$$P(\text{event}) = \frac{\text{number of times the event occurs}}{\text{total number of trials}}$$

$$P(\text{rain}) = \frac{2}{12} = \frac{1}{6}$$

> It rains 2 days.

> There is a total of 12 days.

To make a prediction, multiply the probability of a rainy day by the number of days in April.

$$\frac{1}{6} \cdot 30 = 5$$

⋮ So, you can predict that there will be 5 rainy days in April.

🔊 Multi-Language Glossary at BigIdeasMath✓.com

Laurie's Notes

Introduction

Connect

- **Yesterday:** Students used relative frequencies to find probabilities.
- **Today:** Students will compute the experimental probability of an event and the theoretical probability of an event.

Motivate

- Play *Mystery Bag*. Before students arrive, place 10 cubes of the same shape and size in a paper bag; five of one color and five of a second color.
- Ask a volunteer to be the detective.
- ❓ "There are 10 cubes in my bag. Can you guess what color they are?" not likely
- Let the student remove a cube and look at its color.
- ❓ "Can you guess what color my cubes are?" not likely
- *Replace the cube.* Let the student pick again and see the color. Repeat your question.
- Try this 5–8 times until the student is ready to guess. The number of trials will depend upon the results and the student. You want students to see that they are collecting data and making a prediction.

Lesson Notes

Key Idea

- Discuss experimental probability and make the connection to the activities and to relative frequencies that students found.

Example 1

- ❓ "What information is given in the bar graph that will help answer the question?" The total number of times an odd number was rolled.
- ❓ "How do you write a fraction as a percent?" *Sample answer:* Write an equivalent fraction with a denominator of 100. Then write the numerator with the percent symbol.

Example 2

- Note the important phrase, *if this trend continues*. Knowing the weather for the last 12 days in March, you make a prediction about the weather in April.
- ❓ "Does it seem reasonable to use information from late March to predict weather in April?" Some students may say the weather in April is different, which is why the problem was phrased *if this trend continues*.
- **Big Idea:** The experimental probability is used to make a prediction when you expect a trend to continue, or you believe the experiment reflects what might be true about a larger population.

Goal   Today's lesson is finding the **experimental probability** of an event and the **theoretical probability** of an event.

Technology for the Teacher

Dynamic Classroom

Lesson Tutorials
Lesson Plans
Answer Presentation Tool

Extra Example 1

Using the bar graph and results from Example 1, what is the experimental probability of rolling a prime number? $\frac{23}{50}$, 0.46, or 46%

Extra Example 2

It rains 3 out of the last 15 days in May. If this trend continues, how many rainy days would you expect in June? 6 days

JUNE						
SUN	MON	TUE	WED	THU	FRI	SAT
			1	2	3	4
5	6	7	8	9	10	11
12	13	14	15	16	17	18
19	20	21	22	23	24	25
26	27	28	29	30		

Laurie's Notes

On Your Own

1. $\frac{21}{50}$, 0.42, or 42%

2. 125

Extra Example 3

The letters in the word JACKSON are placed in a hat. You randomly choose a letter from the hat. What is the theoretical probability of choosing a vowel? $\frac{2}{7}$

Extra Example 4

The theoretical probability that you randomly choose a red marble from a bag is $\frac{5}{8}$. There are 40 marbles in the bag. How many are red? 25

On Your Own

- **Neighbor Check:** Have students work independently and then have their neighbors check their work. Have students discuss any discrepancies.
- ❓ "What do you notice about P(odd) and P(even)?" sum to 1

Key Idea

- Write the Key Idea.
- Discuss theoretical probability and give several examples with which students would be familiar, such as cards, dice, and marbles in a bag. Stress that the outcomes must be equally likely.
- Explain to students that this is the type of probability we found in Section 10.2. Now we are calling it theoretical probability.

Example 3

- Work through the example.
- ❓ "What is the probability of *not* choosing a vowel?" $\frac{4}{7}$
- ❓ "How do you write $\frac{3}{7}$ as a percent?" *Sample answer:* Divide 3 by 7 and write as a decimal. Then write the decimal as a percent.

Example 4

- ❓ "What is the probability of winning a bobblehead?" $\frac{1}{6}$
- ❓ "If the spinner has 6 sections, then how many sections are bobblehead sections?" 1
- ❓ "Because there are 3 bobblehead sections on the prize wheel, what do you know about the total number of sections?" There must be more than 6.
- Write the proportion and solve.

T-415

On Your Own

1. In Example 1, what is the experimental probability of rolling an even number?

2. At a clothing company, an inspector finds 5 defective pairs of jeans in a shipment of 200. If this trend continues, about how many pairs of jeans would you expect to be defective in a shipment of 5000?

 Key Idea

Theoretical Probability

When all possible outcomes are equally likely, the **theoretical probability** of an event is the ratio of the number of favorable outcomes to the number of possible outcomes.

$$P(\text{event}) = \frac{\text{number of favorable outcomes}}{\text{number of possible outcomes}}$$

EXAMPLE 3 **Finding a Theoretical Probability**

You randomly choose one of the letters shown. What is the theoretical probability of choosing a vowel?

$$P(\text{event}) = \frac{\text{number of favorable outcomes}}{\text{number of possible outcomes}}$$

$$P(\text{vowel}) = \frac{3}{7}$$

There are 3 vowels.

There is a total of 7 letters.

The probability of choosing a vowel is $\frac{3}{7}$, or about 43%.

EXAMPLE 4 **Using a Theoretical Probability**

The theoretical probability of winning a bobblehead when spinning a prize wheel is $\frac{1}{6}$. The wheel has 3 bobblehead sections. How many sections are on the wheel?

$$P(\text{bobblehead}) = \frac{\text{number of bobblehead sections}}{\text{total number of sections}}$$

$$\frac{1}{6} = \frac{3}{s}$$ Substitute. Let s be the total number of sections.

$$s = 18$$ Cross Products Property

So, there are 18 sections on the wheel.

Now You're Ready
Exercises 15–23

On Your Own

3. In Example 3, what is the theoretical probability of choosing an X?

4. The theoretical probability of spinning an odd number on a spinner is 0.6. The spinner has 10 sections. How many sections have odd numbers?

5. The prize wheel in Example 4 was spun 540 times at a baseball game. About how many bobbleheads would you expect were won?

EXAMPLE 5 Comparing Experimental and Theoretical Probability

The bar graph shows the results of rolling a number cube 300 times.

a. What is the experimental probability of rolling an odd number?

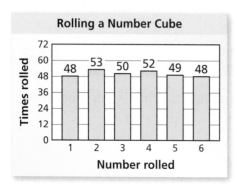

The bar graph shows 48 ones, 50 threes, and 49 fives. So, an odd number was rolled $48 + 50 + 49 = 147$ times in a total of 300 rolls.

$$P(\text{event}) = \frac{\text{number of times the event occurs}}{\text{total number of trials}}$$

$$P(\text{odd}) = \frac{147}{300}$$

An odd number was rolled 147 times.

There was a total of 300 rolls.

$$= \frac{49}{100}, \text{ or } 49\%$$

b. How does the experimental probability compare with the theoretical probability of rolling an odd number?

In Section 10.2, Example 2, you found that the theoretical probability of rolling an odd number is 50%. The experimental probability, 49%, is close to the theoretical probability.

c. Compare the experimental probability in part (a) to the experimental probability in Example 1.

As the number of trials increased from 50 to 300, the experimental probability decreased from 58% to 49%. So, it became closer to the theoretical probability of 50%.

On Your Own

Now You're Ready
Exercises 25–27

6. Use the bar graph in Example 5 to find the experimental probability of rolling a number greater than 1. Compare the experimental probability to the theoretical probability of rolling a number greater than 1.

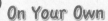

Laurie's Notes

On Your Own

- **Think-Pair-Share:** Students should read each question independently and then work in pairs to answer the questions. When they have answered the questions, the pair should compare their answers with another group and discuss any discrepancies.

Example 5

 "What is different about the bar graph in Example 5 compared to the bar graph in Example 1?" The bar graph in Example 5 shows a greater number of trials; 300 instead of 50.

- Work through each part of the example as shown.
- **Reason Abstractly and Quantitatively:** Revisit the activities and discuss the fact that when the relative frequencies increase, the experimental probability gets closer and closer to the theoretical probability.

On Your Own

- **Neighbor Check:** Have students work independently and then have their neighbors check their work. Have students discuss any discrepancies.

Closure

- Use the colored cubes in the Mystery Bag. Reveal the contents, or use a different color ratio if you wish, and ask the following:
  - "If the contents of the Mystery Bag came from a bag of 1000 colored cubes, how many of each color would you predict in the bag of 1000?" Answers will vary depending upon materials.

 On Your Own

3. $\frac{1}{7}$, or about 14.3%

4. 6 sections

5. 90 bobbleheads

Extra Example 5

Use the bar graph and results from Example 5.

a. What is the experimental probability of rolling an even number?
$\frac{51}{100}$, 0.51, or 51%

b. How does the experimental probability compare with the theoretical probability of rolling an even number? The experimental probability, 51%, is close to the theoretical probability, 50%.

On Your Own

6. 84%; It is close to the theoretical probability of $83\frac{1}{3}$%.

English Language Learners

Vocabulary

This chapter contains many new terms that may cause English learners to struggle. Students should write the key vocabulary in their notebooks along with definitions and diagrams so that they become familiar and comfortable with the vocabulary.

Vocabulary and Concept Check

1. Perform an experiment several times. Count how often the event occurs and divide by the number of trials.

2. yes; You could flip tails 7 out of 10 times, but with more trials the probability of flipping tails should get closer to 0.5.

3. There is a 50% chance you will get a favorable outcome.

4. *Sample answer:* picking a 1 out of 1, 2, 3, 4

5. experimental probability; The population is too large to survey every person, so a sample will be used to predict the outcome.

Assignment Guide and Homework Check

Level	Day 1 Activity Assignment	Day 2 Lesson Assignment	Homework Check
Basic	6, 7, 35–37	1–5, 9–23 odd	9, 11, 13, 15, 21
Average	6, 7, 35–37	1–5, 10–14 even, 15–29 odd	10, 12, 17, 23, 29
Advanced	6, 7, 35–37	1–5, 14–34 even	18, 22, 30, 32
Accelerated	1–7, 14–34 even, 35–37		18, 22, 30, 32

Common Errors

- **Exercises 8–11** Students may forget to total all of the trials before writing the experimental probability. They may have an incorrect number of trials in the denominator. Remind them that they need to know the total number of trials when finding the probability.

Practice and Problem Solving

6. $\dfrac{7}{50}$, or 14%

7. $\dfrac{12}{25}$, or 48%

8. $\dfrac{7}{25}$, or 28%

9. $\dfrac{21}{25}$, or 84%

10. $\dfrac{17}{50}$, or 34%

11. 0, or 0%

12. $\dfrac{3}{20}$, or 15%

13. 45 tiles

14. 5 cards

10.3 Record and Practice Journal

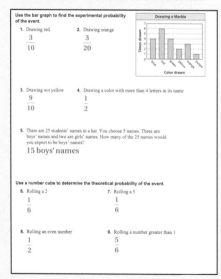

 Vocabulary and Concept Check

1. **VOCABULARY** Describe how to find the experimental probability of an event.

2. **REASONING** You flip a coin 10 times and find the experimental probability of flipping tails to be 0.7. Does this seem reasonable? Explain.

3. **VOCABULARY** An event has a theoretical probability of 0.5. What does this mean?

4. **OPEN-ENDED** Describe an event that has a theoretical probability of $\frac{1}{4}$.

5. **LOGIC** A pollster surveys randomly selected individuals about an upcoming election. Do you think the pollster will use experimental probability or theoretical probability to make predictions? Explain.

 Practice and Problem Solving

Use the bar graph to find the relative frequency of the event.

6. Spinning a 6

7. Spinning an even number

Use the bar graph to find the experimental probability of the event.

① 8. Spinning a number less than 3

9. *Not* spinning a 1

10. Spinning a 1 or a 3

11. Spinning a 7

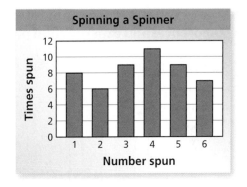

12. **EGGS** You check 20 cartons of eggs. Three of the cartons have at least one cracked egg. What is the experimental probability that a carton of eggs has at least one cracked egg?

② 13. **BOARD GAME** There are 105 lettered tiles in a board game. You choose the tiles shown. How many of the 105 tiles would you expect to be vowels?

14. **CARDS** You have a package of 20 assorted thank-you cards. You pick the four cards shown. How many of the 20 cards would you expect to have flowers on them?

Use the spinner to find the theoretical probability of the event.

③ **15.** Spinning red

16. Spinning a 1

17. Spinning an odd number

18. Spinning a multiple of 2

19. Spinning a number less than 7

20. Spinning a 9

21. **LETTERS** Each letter of the alphabet is printed on an index card. What is the theoretical probability of randomly choosing any letter except Z?

④ **22.** **GAME SHOW** On a game show, a contestant randomly chooses a chip from a bag that contains numbers and strikes. The theoretical probability of choosing a strike is $\frac{3}{10}$. The bag contains 9 strikes. How many chips are in the bag?

23. **MUSIC** The theoretical probability that a pop song plays on your MP3 player is 0.45. There are 80 songs on your MP3 player. How many of the songs are pop songs?

24. **MODELING** There are 16 females and 20 males in a class.

    a. What is the theoretical probability that a randomly chosen student is female?

    b. One week later, there are 45 students in the class. The theoretical probability that a randomly chosen student is a female is the same as last week. How many males joined the class?

The bar graph shows the results of spinning the spinner 200 times. Compare the theoretical and experimental probabilities of the event.

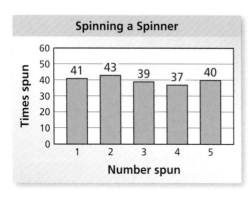

⑤ **25.** Spinning a 4

26. Spinning a 3

27. Spinning a number greater than 4

28. Should you use *theoretical* or *experimental* probability to predict the number of times you will spin a 3 in 10,000 spins?

29. **NUMBER SENSE** The table at the right shows the results of flipping two coins 12 times each.

HH	HT	TH	TT
2	6	3	1

    a. What is the experimental probability of flipping two tails? Using this probability, how many times can you expect to flip two tails in 600 trials?

HH	HT	TH	TT
23	29	26	22

    b. The table at the left shows the results of flipping the same two coins 100 times each. What is the experimental probability of flipping two tails? Using this probability, how many times can you expect to flip two tails in 600 trials?

    c. Why is it important to use a large number of trials when using experimental probability to predict results?

Common Errors

- **Exercises 15–20** Students may write a different probability than what is asked, or forget to include a favorable outcome. For example, in Exercise 15 a student may not realize that there are two red sections and will write the probability as $\frac{1}{6}$ instead of $\frac{1}{3}$. Remind them to read the event carefully and to write the favorable outcomes before finding the probability.
- **Exercise 22** Students may write an incorrect proportion when finding how many strikes are in the bag. Encourage them to write the proportion in words before substituting and solving.

15. $\frac{1}{3}$, or about 33.3%

16. $\frac{1}{6}$, or about 16.7%

17. $\frac{1}{2}$, or 50%   **18.** $\frac{1}{2}$, or 50%

19. 1, or 100%   **20.** 0, or 0%

21. $\frac{25}{26}$, or about 96.2%

22. 30 chips   **23.** 36 songs

24. **a.** $\frac{4}{9}$, or about 44.4%

    b. 5 males

25. theoretical: $\frac{1}{5}$, or 20%;

    experimental: $\frac{37}{200}$, or 18.5%;

    The experimental probability is close to the theoretical probability.

26. theoretical: $\frac{1}{5}$, or 20%;

    experimental: $\frac{39}{200}$, or 19.5%;

    The experimental probability is close to the theoretical probability.

27. theoretical: $\frac{1}{5}$, or 20%;

    experimental: $\frac{1}{5}$, or 20%;

    The probabilities are equal.

28–29. See Additional Answers.

English Language Learners

Group Activity

Set up groups of English learners and English speakers. Have each group predict the number of times a coin will land on *heads* when flipped 30 times. Each group should track their results in a frequency table. Ask groups to present their predictions and results to the class. Then combine the results of all the groups and discuss how the combined results compare to the individual group results.

Practice and Problem Solving

30–32. See *Taking Math Deeper*.

33. a. As the number of trials increases, the most likely sum will change from 6 to 7.

b. As an experiment is repeated over and over, the experimental probability of an event approaches the theoretical probability of the event.

34. See Additional Answers.

Fair Game Review

35. 4%   **36.** 3.5%

37. D

Mini-Assessment

You have three sticks. Each stick has one red side and one blue side. You throw the sticks 10 times and record the results. Use the table to find the experimental probability of the event.

Outcome	Frequency
3 red	3
3 blue	2
2 blue, 1 red	4
2 red, 1 blue	1

1. Tossing 3 blue $\frac{1}{5}$

2. Tossing 2 blue, 1 red $\frac{2}{5}$

3. *Not* tossing all blue $\frac{4}{5}$

Use the spinner to determine the theoretical probability of the event.

4. $P(\text{purple})$ $\frac{1}{6}$ **5.** $P(3)$ $\frac{1}{6}$

6. $P(\text{even})$ $\frac{1}{2}$ **7.** $P(\text{multiple of 3})$ $\frac{1}{3}$

Taking Math Deeper

Exercises 30–32

In this problem, students compare *experimental probabilities* (results of trials), shown in the bar graph, with the *theoretical probabilities* (all possible outcomes), shown in the table.

 Make a list of the different sums and the number of times rolled from the bar graph. Check that the total is 60. Find the experimental probability of each sum.

2: 2/60   3: 4/60   4: 5/60   5: 6/60   6: 13/60   7: 10/60
8: 6/60   9: 8/60   10: 2/60   11: 3/60   12: 1/60

30. No, each sum is not equally likely. The experimental probabilities vary. The most likely sum is 6.

 Make a list of all possible ways to get each sum.

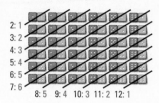

7 is most likely.

2: 1
3: 2
4: 3
5: 4
6: 5
7: 6
8: 5   9: 4   10: 3   11: 2   12: 1

Find the theoretical probability of each sum.

2: 1/36   3: 2/36   4: 3/36   5: 4/36   6: 5/36   7: 6/36
8: 5/36   9: 4/36   10: 3/36   11: 2/36   12: 1/36

31. No, because there is not an equal number of possible ways to get each sum, each sum is not equally likely. So, the theoretical probabilities vary. The most likely sum is 7.

 32. Compare the two types of probabilities with a double bar graph.

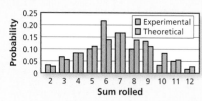

Reteaching and Enrichment Strategies

If students need help. . .	If students got it. . .
Resources by Chapter • Practice A and Practice B • Puzzle Time Record and Practice Journal Practice Differentiating the Lesson Lesson Tutorials Skills Review Handbook	Resources by Chapter • Enrichment and Extension • Technology Connection Start the next section

You roll a pair of number cubes 60 times. You record your results in the bar graph shown.

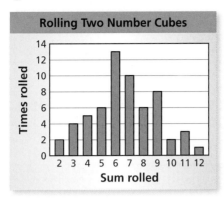

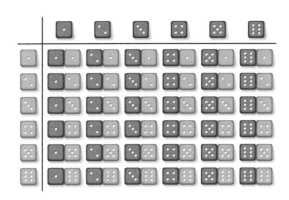

30. Use the bar graph to find the experimental probability of rolling each sum. Is each sum equally likely? Explain. If not, which is most likely?

31. Use the table to find the theoretical probability of rolling each sum. Is each sum equally likely? Explain. If not, which is most likely?

32. PROBABILITIES Compare the probabilities you found in Exercises 30 and 31.

33. REASONING Consider the results of Exercises 30 and 31.

  a. Which sum would you expect to be most likely after 500 trials? 1000 trials? 10,000 trials?

  b. Explain how experimental probability is related to theoretical probability as the number of trials increases.

34. **Project** When you toss a paper cup into the air, there are three ways for the cup to land: *open-end up*, *open-end down*, or *on its side*.

  a. Toss a paper cup 100 times and record your results. Do the outcomes for tossing the cup appear to be equally likely? Explain.

  b. What is the probability of the cup landing open-end up? open-end down? on its side?

  c. Use your results to predict the number of times the cup lands on its side in 1000 tosses.

  d. Suppose you tape a quarter to the bottom of the cup. Do you think the cup will be *more likely* or *less likely* to land open-end up? Justify your answer.

Fair Game Review *What you learned in previous grades & lessons*

Find the annual interest rate. *(Section 6.7)*

35. $I = \$16$, $P = \$200$, $t = 2$ years

36. $I = \$26.25$, $P = \$500$, $t = 18$ months

37. MULTIPLE CHOICE The volume of a prism is 9 cubic yards. What is its volume in cubic feet? *(Section 9.4)*

  Ⓐ 3 ft^3 **Ⓑ** 27 ft^3 **Ⓒ** 81 ft^3 **Ⓓ** 243 ft^3

10.4 Compound Events

Essential Question How can you find the number of possible outcomes of one or more events?

1 ACTIVITY: Comparing Combination Locks

Work with a partner. You are buying a combination lock. You have three choices.

a. This lock has 3 wheels. Each wheel is numbered from 0 to 9.

The least three-digit combination possible is ____.

The greatest three-digit combination possible is ____.

How many possible combinations are there?

b. Use the lock in part (a).

There are ____ possible outcomes for the first wheel.

There are ____ possible outcomes for the second wheel.

There are ____ possible outcomes for the third wheel.

How can you use multiplication to determine the number of possible combinations?

c. This lock is numbered from 0 to 39. Each combination uses three numbers in a right, left, right pattern. How many possible combinations are there?

Probability and Statistics

In this lesson, you will

- use tree diagrams, tables, or a formula to find the number of possible outcomes.
- find probabilities of compound events.

d. This lock has 4 wheels.

Wheel 1: 0–9

Wheel 2: A–J

Wheel 3: K–T

Wheel 4: 0–9

How many possible combinations are there?

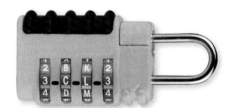

e. For which lock is it most difficult to guess the combination? Why?

Laurie's Notes

Introduction

Applying Mathematical Practices

- **Model with Mathematics:** Mathematically proficient students use visual models to represent problems. Tree diagrams can be used to visualize the possible outcomes of events.

Motivate

- **Acting Time:** If possible, dress as Sherlock Holmes with overcoat, hat, eye monocle, and pipe. Probe students about your identity by sharing a little information:
  - You live on 221b Baker Street in London.
  - You owe your fame to Sir Arthur Conan Doyle (your creator).
  - You were first written about in 1887.
  - You have a friend named Watson and a foe named Professor Moriarty.
- Tell students you have a case that you're working on and they can help you with it. The principal put the teachers' paychecks in a locked box and forgot the combination.
- ❓ Hold up a combination lock and tell the students that you're trying to figure out the combination. "How many different combinations do you think I would need to try in order to open this lock?"
- Tell students that today's activity may help them answer this question.

Activity Notes

Activity 1

- If your school has lockers with combination locks, discuss their operation.
- ❓ "Do any of you have combination locks for your bike, skis, or other possessions?"
- **FYI:** Many local post offices have boxes with combinations. Some post offices now have digital locks where a code must be entered.
- Students are guided with questions in parts (a) and (b) to help them reason about the number of possible outcomes.
- ❓ "If the digits 0–9 are used, how many outcomes are there?" 10
- When students finish part (b), make sure they understand the connection to multiplication. This prepares them for parts (c) and (d), laying the foundation for the Fundamental Counting Principle.
- ❓ "In part (c), how many possible outcomes are there when the dial is turned to the right?" 40
- If students have made sense of parts (a) and (b), they will understand that the number of combinations in part (c) is 40 • 40 • 40.
- It may be helpful for some students if sample combinations are listed. For instance in part (d), possible combinations are 4AK0, 4AK1, 4AK2, etc.
- **Attend to Precision:** Ask a volunteer to explain his or her reasoning for parts (d) and (e).

What Your Students Will Learn

- Use tree diagrams, tables, or a formula to find the number of possible outcomes of events.
- Find probabilities of compound events.

Previous Learning

Students have counted the number of possible outcomes of an event.

Technology for the Teacher

Dynamic Classroom

Lesson Plans
Complete Materials List

10.4 Record and Practice Journal

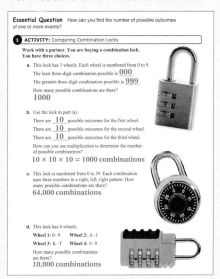

Essential Question How can you find the number of possible outcomes of one or more events?

1 ACTIVITY: Comparing Combination Locks

Work with a partner. You are buying a combination lock. You have three choices.

a. This lock has 3 wheels. Each wheel is numbered from 0 to 9.
The least three-digit combination possible is 000.
The greatest three-digit combination possible is 999.
How many possible combinations are there?
1000

b. Use the lock in part (a).
There are 10 possible outcomes for the first wheel.
There are 10 possible outcomes for the second wheel.
There are 10 possible outcomes for the third wheel.
How can you use multiplication to determine the number of possible combinations?
$10 \times 10 \times 10 = 1000$ combinations

c. This lock is numbered from 0 to 39. Each combination uses three numbers in a right, left, right pattern. How many possible combinations are there?
64,000 combinations

d. This lock has 4 wheels.
Wheel 1: 0–9   Wheel 2: A–J
Wheel 3: K–T   Wheel 4: 0–9
How many possible combinations are there?
10,000 combinations

Visual

When solving counting problems like part (a) in Activity 2, have students write blanks to represent each outcome.

Step 1: Write 4 blanks to show that the password has 4 digits.

____ ____ ____ ____

Step 2: Because there are 10 choices for each of the digits, write:

<u>10</u> <u>10</u> <u>10</u> <u>10</u>

Step 3: Multiply all of the values together to get 10,000 different possible passwords.

10.4 Record and Practice Journal

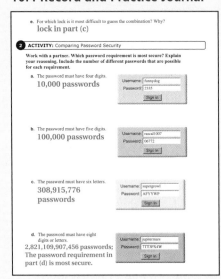

Laurie's Notes

Activity 2

- If students have usernames and passwords for computers or any other aspect of school, be sure to discuss this.
- In this activity, the students are not considering the username, only the password.
- Point out that the passwords are *not* case sensitive, meaning the letters could be either upper- or lower-case.
- **?** If students have difficulty getting started, ask a simpler question such as, "How many passwords are possible using just 2 digits, 0 to 9?"
  100: 00, 01, 02, 03, . . . , 99
- Remind students that the digit "0" is included so there are 10 possible digits when numbers are considered.
- As with Activity 1, the numbers or letters can be repeated.
- For parts (c) and (d), you might have students describe how to find the answer instead of having them actually compute the answer.

What Is Your Answer?

- If time permits, have students convert their answer of minutes to a larger unit of time such as hours, days, and years.

Closure

- **Exit Ticket:** Help the teachers get their paychecks by determining how many different combinations there are for a lock that has 3 dials, each one containing the digits 0 through 4. $5 \cdot 5 \cdot 5 = 125$

ACTIVITY: Comparing Password Security

Work with a partner. Which password requirement is most secure? Explain your reasoning. Include the number of different passwords that are possible for each requirement.

a. The password must have four digits.

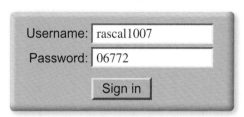

Math Practice

View as Components

What is the number of possible outcomes for each character of the password? Explain.

b. The password must have five digits.

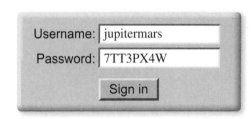

c. The password must have six letters.

Username: supergrowl
Password: AFYYWP

Sign in

d. The password must have eight digits or letters.

Username: jupitermars
Password: 7TT3PX4W

Sign in

What Is Your Answer?

3. **IN YOUR OWN WORDS** How can you find the number of possible outcomes of one or more events?

4. **SECURITY** A hacker uses a software program to guess the passwords in Activity 2. The program checks 600 passwords per minute. What is the greatest amount of time it will take the program to guess each of the four types of passwords?

Practice > Use what you learned about the total number of possible outcomes of one or more events to complete Exercise 5 on page 425.

10.4 Lesson

Key Vocabulary ◀))
sample space, *p. 422*
Fundamental
    Counting Principle,
    p. 422
compound event,
    p. 424

The set of all possible outcomes of one or more events is called the **sample space**.

You can use tables and tree diagrams to find the sample space of two or more events.

EXAMPLE ① **Finding a Sample Space**

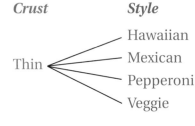

Crust

• Thin Crust
• Stuffed Crust

Style

• Hawaiian
• Mexican
• Pepperoni
• Veggie

You randomly choose a crust and style of pizza. Find the sample space. How many different pizzas are possible?

Use a tree diagram to find the sample space.

Crust	Style	Outcome
Thin	Hawaiian	Thin Crust Hawaiian
	Mexican	Thin Crust Mexican
	Pepperoni	Thin Crust Pepperoni
	Veggie	Thin Crust Veggie
Stuffed	Hawaiian	Stuffed Crust Hawaiian
	Mexican	Stuffed Crust Mexican
	Pepperoni	Stuffed Crust Pepperoni
	Veggie	Stuffed Crust Veggie

∴ There are 8 different outcomes in the sample space. So, there are 8 different pizzas possible.

⬤ **On Your Own**

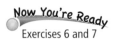
Now You're Ready
Exercises 6 and 7

1. **WHAT IF?** The pizza shop adds a deep dish crust. Find the sample space. How many pizzas are possible?

Another way to find the total number of possible outcomes is to use the **Fundamental Counting Principle**.

Study Tip

The Fundamental
Counting Principle can
be extended to more
than two events.

Key Idea

Fundamental Counting Principle

An event M has m possible outcomes. An event N has n possible outcomes. The total number of outcomes of event M followed by event N is $m \times n$.

◀) Multi-Language Glossary at BigIdeasMath.com

Laurie's Notes

● Introduction

Connect

- **Yesterday:** Students explored how the number of choices on locks and in passwords affected the total number of combinations possible.
- **Today:** Students will find the number of outcomes of compound events.

Motivate

- Display 3 different cups on your desk. I like to use a ceramic cup, a travel mug, and a foam cup. In addition, have a tea bag and hot cocoa mix.
- Pose to students: You are thirsty. You have three different cups to select from and two different beverages.
- **?** "How many different ways can I select a cup and a beverage?" The answer of 6 may or may not be obvious. Hold up the travel mug and tea bag, and then the travel mug and cocoa. Repeat for the other two cups.
- **?** "How many different ways if I add another cup, say a heavy plastic cup?" 8
- **?** "How many different ways if I add another beverage, say coffee?" 9

● Lesson Notes

Discuss

- Today's lesson is about outcomes of one event followed by one or more other events.
- Define sample space. Give examples.

Example 1

- **?** Refer to the pizza shop menu and ask, "How many types of crust are available?" two Write the possible crusts with space between them as shown.
- **?** "For either crust, how many different styles of pizza can you order?" four List the four styles with each crust using the tree diagram as shown.
- **Model with Mathematics:** The tree diagram helps students visualize the 8 outcomes in the sample space.
- **Extension:** Add a size to each (10″ and 14″). Have students determine the number of possible outcomes. 16

On Your Own

- Ask a volunteer to share his or her thinking about the question.

Key Idea

- Write the Fundamental Counting Principle.
- Revisit previous problems to see if the Fundamental Counting Principle gives the same answer.

Goal Today's lesson is using **sample spaces** and the total number of possible outcomes to find probabilities of **compound events**.

Technology for the **Teacher**

Dynamic Classroom

Lesson Tutorials
Lesson Plans
Answer Presentation Tool

Extra Example 1

At a sub shop, you can choose ham, turkey, or roast beef on either white or wheat bread. You randomly choose a meat and bread. Find the sample space. How many subs are possible?

ham on white bread,
ham on wheat bread,
turkey on white bread,
turkey on wheat bread,
roast beef on white bread,
roast beef on wheat bread; 6

● On Your Own

1. thin crust Hawaiian,
   thin crust Mexican,
   thin crust pepperoni,
   thin crust veggie,
   stuffed crust Hawaiian,
   stuffed crust Mexican,
   stuffed crust pepperoni,
   stuffed crust veggie,
   deep dish Hawaiian,
   deep dish Mexican,
   deep dish pepperoni,
   deep dish veggie; 12

Writing

Pair or group English speakers with English learners. Have each group of students write a problem involving the Fundamental Counting Principle. On a separate sheet of paper, have the groups show a detailed solution to their problem. Next, have groups exchange their problems with other groups. Students in each group should work together to solve the problem they receive. Both groups should meet to discuss the problems and solutions.

Extra Example 2

Find the total number of possible outcomes of rolling two number cubes. $6 \times 6 = 36$

Extra Example 3

How many different outfits can you make from 5 T-shirts, 3 pairs of jeans, and 2 pairs of shoes? $5 \times 3 \times 2 = 30$

 On Your Own

2. 20

3. 100

Laurie's Notes

Example 2

- Use a number cube and coin to model this example.
- Work through the example using each method.
- Discuss the efficiency of using the Fundamental Counting Principle instead of using a table. If you only need to know the number of outcomes, the Fundamental Counting Principle should be used. The table, however, shows the sample space instead of just the number of outcomes.
- **?** "Is the answer the same if the coin is flipped first followed by rolling the number cube?" yes; The order changes but the number of outcomes is the same.
- **FYI:** Lists generated from tables and tree diagrams like those in Examples 1 and 2 are examples of organized lists.

Example 3

- Ask a volunteer to tell how many of each type of clothing is shown in the picture.
- In working through this question there will be students who suggest that certain shoes would not be worn with certain jeans, and so on. Remind students that the question is, "How many different outfits can be made, not would you wear the outfit."
- **?** "Would you want to make a tree diagram or table to answer this question? Explain." no; It would take a long time to list the possible outcomes. The Fundamental Counting Principle is more efficient in finding the number of different outfits possible.

On Your Own

- Question 2 could be modeled using a spinner and five pieces of paper.

EXAMPLE 2 **Finding the Total Number of Possible Outcomes**

Find the total number of possible outcomes of rolling a number cube and flipping a coin.

Method 1: Use a table to find the sample space. Let H = heads and T = tails.

	1	**2**	**3**	**4**	**5**	**6**
	1H	2H	3H	4H	5H	6H
	1T	2T	3T	4T	5T	6T

⋮ There are 12 possible outcomes.

Method 2: Use the Fundamental Counting Principle. Identify the number of possible outcomes of each event.

> **Event 1:** Rolling a number cube has 6 possible outcomes.
>
> **Event 2:** Flipping a coin has 2 possible outcomes.
>
> $6 \times 2 = 12$ Fundamental Counting Principle

⋮ There are 12 possible outcomes.

EXAMPLE 3 **Finding the Total Number of Possible Outcomes**

How many different outfits can you make from the T-shirts, jeans, and shoes in the closet?

Use the Fundamental Counting Principle. Identify the number of possible outcomes for each event.

> **Event 1:** Choosing a T-shirt has 7 possible outcomes.
>
> **Event 2:** Choosing jeans has 4 possible outcomes.
>
> **Event 3:** Choosing shoes has 3 possible outcomes.
>
> $7 \times 4 \times 3 = 84$ Fundamental Counting Principle

⋮ So, you can make 84 different outfits.

On Your Own

Now You're Ready
Exercises 8–11

2. Find the total number of possible outcomes of spinning the spinner and choosing a number from 1 to 5.

3. How many different outfits can you make from 4 T-shirts, 5 pairs of jeans, and 5 pairs of shoes?

A **compound event** consists of two or more events. As with a single event, the probability of a compound event is the ratio of the number of favorable outcomes to the number of possible outcomes.

EXAMPLE (4) **Finding the Probability of a Compound Event**

In Example 2, what is the probability of rolling a number greater than 4 and flipping tails?

There are two favorable outcomes in the sample space for rolling a number greater than 4 and flipping tails: 5T and 6T.

$$P(\text{event}) = \frac{\text{number of favorable outcomes}}{\text{number of possible outcomes}}$$

$$P(\text{greater than 4 and tails}) = \frac{2}{12} \qquad \text{Substitute.}$$

$$= \frac{1}{6} \qquad \text{Simplify.}$$

∴ The probability is $\frac{1}{6}$, or $16\frac{2}{3}\%$.

EXAMPLE (5) **Finding the Probability of a Compound Event**

You flip three nickels. What is the probability of flipping two heads and one tails?

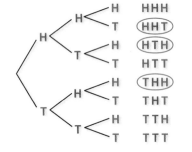

Use a tree diagram to find the sample space. Let H = heads and T = tails.

There are three favorable outcomes in the sample space for flipping two heads and one tails: HHT, HTH, and THH.

$$P(\text{event}) = \frac{\text{number of favorable outcomes}}{\text{number of possible outcomes}}$$

$$P(2 \text{ heads and 1 tails}) = \frac{3}{8} \qquad \text{Substitute.}$$

∴ The probability is $\frac{3}{8}$, or 37.5%.

On Your Own

Now You're Ready
Exercises 15–24

4. In Example 2, what is the probability of rolling at most 4 and flipping heads?

5. In Example 5, what is the probability of flipping at least two tails?

6. You roll two number cubes. What is the probability of rolling double threes?

7. In Example 1, what is the probability of choosing a stuffed crust Hawaiian pizza?

Laurie's Notes

Discuss

- Define a *compound event*. Refer to first three examples and explain that they all involved compound events. Tell students that the next step is to determine probabilities of compound events.
- Define the probability of a compound event: the ratio of the number of favorable outcomes to the number of possible outcomes.

Example 4

- **?** "How many possible outcomes are there when you roll a number cube and flip a coin as in Example 2?" 12
- **?** "How many favorable outcomes are there for rolling a number greater than 4 and flipping tails? Explain." There are 2 favorable outcomes, 5T and 6T.
- So, *P*(rolling greater than 4 and flipping tails) $= \dfrac{2}{12} = \dfrac{1}{6}$.
- Students should now see the value in creating tree diagrams and tables to find sample spaces. When finding probabilities, it can help easily identify the number of favorable outcomes and number of possible outcomes.

Example 5

- **Model with Mathematics:** Read the problem and ask students to draw a tree diagram to find the sample space. The tree diagram makes the number of favorable outcomes and the number of possible outcomes easy to find.
- **?** "If you use the Fundamental Counting Principle, how many possible outcomes are there?" $2 \cdot 2 \cdot 2 = 8$
- Continue to work the problem as shown.

On Your Own

- **Common Error:** In Exercise 6, students often say that there are 12 possible outcomes ($6 + 6$) instead of 36 possible outcomes (6×6).
- **Extension:** Ask students to find the probability of randomly selecting the green T-shirt, green jeans, and green shoes in Example 3. Rather than list the sample space, they should reason that there is only 1 favorable outcome, so the probability is $\dfrac{1}{84}$.

Closure

- **Writing Prompt:** The Fundamental Counting Principle is used to . . .
  Tree diagrams and tables can be used to . . .

Extra Example 4

In Example 2, what is the probability of rolling an even number and flipping heads? 25%

Extra Example 5

You flip three dimes. What is the probability of flipping three heads? 12.5%

On Your Own

4. $\dfrac{1}{3}$, or $33\dfrac{1}{3}\%$

5. $\dfrac{1}{2}$, or 50%

6. $\dfrac{1}{36}$, or about $2\dfrac{7}{9}\%$

7. $\dfrac{1}{8}$, or 12.5%

1. A sample space is the set of all possible outcomes of an event. Use a table or tree diagram to list all the possible outcomes.

2. An event M has m possible outcomes and event N has n possible outcomes. The total number of outcomes of event M followed by event N is $m \times n$.

3. You could use a tree diagram or the Fundamental Counting Principle. Either way, the total number of possible outcomes is 30.

4. *Sample answer:* choosing two marbles from a bag

Practice and Problem Solving

5. 125,000

6. *Sample space:*
Miniature golf 1 P.M.–3 P.M.,
Miniature golf 6 P.M.–8 P.M.,
Laser tag 1 P.M.–3 P.M.,
Laser tag 6 P.M.–8 P.M.,
Roller skating 1 P.M.–3 P.M.,
Roller skating 6 P.M.–8 P.M.;
6 possible outcomes

7. *Sample space:* Realistic Lion, Realistic Bear, Realistic Hawk, Realistic Dragon, Cartoon Lion, Cartoon Bear, Cartoon Hawk, Cartoon Dragon; 8 possible outcomes

8. 21   9. 20

10. 24   11. 60

12. See Additional Answers.

13. The possible outcomes of each question should be multiplied, not added. The correct answer is $2 \times 2 \times 2 \times 2 \times 2 = 32$.

Assignment Guide and Homework Check

Level	Day 1 Activity Assignment	Day 2 Lesson Assignment	Homework Check
Basic	5, 31–33	1–4, 7–25 odd	7, 11, 17, 23, 25
Average	5, 31–33	1–4, 7–13 odd, 14–28 even	7, 11, 18, 22, 26
Advanced	5, 31–33	1–4, 6–12 even, 13, 14–30 even	10, 18, 24, 26, 28
Accelerated	1–5, 6–12 even, 13, 14–30 even, 31–33		10, 18, 24, 26, 28

Common Errors

- **Exercises 8–11** Students may add the number of possible outcomes for each event rather than multiply them. Remind them that the total number of outcomes of two or more events is found by multiplying the number of possible outcomes of each event.

10.4 Record and Practice Journal

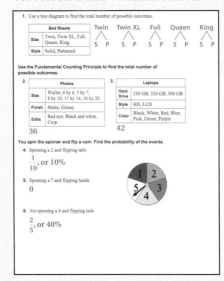

10.4 Exercises

 Vocabulary and Concept Check

1. **VOCABULARY** What is the sample space of an event? How can you find the sample space of two or more events?

2. **WRITING** Explain how to use the Fundamental Counting Principle.

3. **WRITING** Describe two ways to find the total number of possible outcomes of spinning the spinner and rolling the number cube.

4. **OPEN-ENDED** Give a real-life example of a compound event.

 Practice and Problem Solving

5. **COMBINATIONS** The lock is numbered from 0 to 49. Each combination uses three numbers in a right, left, right pattern. Find the total number of possible combinations for the lock.

Use a tree diagram to find the sample space and the total number of possible outcomes.

① 6.

Birthday Party	
Event	Miniature golf, Laser tag, Roller skating
Time	1:00 P.M.–3:00 P.M., 6:00 P.M.–8:00 P.M.

7.

New School Mascot	
Type	Lion, Bear, Hawk, Dragon
Style	Realistic, Cartoon

Use the Fundamental Counting Principle to find the total number of possible outcomes.

② 8.

Beverage	
Size	Small, Medium, Large
Flavor	Root beer, Cola, Diet cola, Iced tea, Lemonade, Water, Coffee

9.

MP3 Player	
Memory	2 GB, 4 GB, 8 GB, 16 GB
Color	Silver, Green, Blue, Pink, Black

③ 10.

Clown	
Suit	Dots, Stripes, Checkers board
Wig	One color, Multicolor
Talent	Balloon animals, Juggling, Unicycle, Magic

11.

Meal	
Appetizer	Nachos, Soup, Spinach dip, Salad, Fruit
Entrée	Chicken, Beef, Spaghetti, Fish
Dessert	Cake, Cookies, Ice cream

12. **NOTE CARDS** A store sells three types of note cards. There are three sizes of each type. Show two ways to find the total number of note cards the store sells.

13. **ERROR ANALYSIS** A true-false quiz has five questions. Describe and correct the error in using the Fundamental Counting Principle to find the total number of ways that you can answer the quiz.

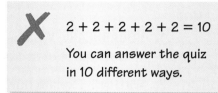

$2 + 2 + 2 + 2 + 2 = 10$
You can answer the quiz in 10 different ways.

14. **CHOOSE TOOLS** You randomly choose one of the marbles. Without replacing the first marble, you choose a second marble.

  a. Name two ways you can find the total number of possible outcomes.

  b. Find the total number of possible outcomes.

You spin the spinner and flip a coin. Find the probability of the compound event.

④ 15. Spinning a 1 and flipping heads

  16. Spinning an even number and flipping heads

  17. Spinning a number less than 3 and flipping tails

  18. Spinning a 6 and flipping tails

  19. *Not* spinning a 5 and flipping heads

  20. Spinning a prime number and *not* flipping heads

You spin the spinner, flip a coin, then spin the spinner again. Find the probability of the compound event.

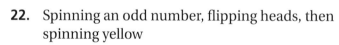

⑤ 21. Spinning blue, flipping heads, then spinning a 1

  22. Spinning an odd number, flipping heads, then spinning yellow

  23. Spinning an even number, flipping tails, then spinning an odd number

  24. *Not* spinning red, flipping tails, then *not* spinning an even number

25. **TAKING A TEST** You randomly guess the answers to two questions on a multiple-choice test. Each question has three choices: A, B, and C.

  a. What is the probability that you guess the correct answers to both questions?

  b. Suppose you can eliminate one of the choices for each question. How does this change the probability that your guesses are correct?

Common Errors

- **Exercises 15–24** Students may find the probability of each individual event. Remind them that they are finding the probability of a *compound event*, so they need to find the ratio of the number of favorable outcomes to the number of possible outcomes using tables, tree diagrams, or the Fundamental Counting Principle.

Practice and Problem Solving

14. **a.** tree diagram or the Fundamental Counting Principle

   b. 12 possible outcomes

15. $\frac{1}{10}$, or 10%   16. $\frac{1}{5}$, or 20%

17. $\frac{1}{5}$, or 20%   18. 0, or 0%

19. $\frac{2}{5}$, or 40%   20. $\frac{3}{10}$, or 30%

21. $\frac{1}{18}$, or $5\frac{5}{9}$%

22. $\frac{1}{9}$, or $11\frac{1}{9}$%

23. $\frac{1}{9}$, or $11\frac{1}{9}$%

24. $\frac{2}{9}$, or $22\frac{2}{9}$%

25. **a.** $\frac{1}{9}$, or about 11.1%

   b. It increases the probability that your guesses are correct to $\frac{1}{4}$, or 25%, because you are only choosing between 2 choices for each question.

26. **a.** $\frac{1}{100}$, or 1%

   b. It increases the probability that your choice is correct to $\frac{1}{25}$, or 4%, because each digit could be 0, 2, 4, 6, or 8.

English Language Learners

Pair Activity

Give pairs of students one or two problems similar to Example 4. Each pair is responsible for reaching a consensus on the answers. This will allow English learners to ask questions or explain concepts to another student.

 Practice and Problem Solving

27. See Additional Answers.

28. See *Taking Math Deeper*.

29. See Additional Answers.

30. 10 ways

 Fair Game Review

31. *Sample answer:* adjacent: ∠*XWY* and ∠*ZWY*, ∠*XWY* and ∠*XWV*; vertical: ∠*VWX* and ∠*YWZ*, ∠*YWX* and ∠*VWZ*

32. *Sample answer:* adjacent: ∠*LJM* and ∠*LJK*, ∠*LJM* and ∠*NJM*; vertical: ∠*KJL* and ∠*PJN*, ∠*PJQ* and ∠*MJL*

33. B

Mini-Assessment

1. Use a tree diagram to find the sample space and the total number of possible outcomes.

Snack	
Fruit	Apple, Banana, Pear
Drink	Water, Iced tea, Milk

AW, AI, AM, BW, BI, BM, PW, PI, PM; 9

2. Use the Fundamental Counting Principle to find the total number of possible outcomes.

Shirt	
Size	S, M, L, XL
Color	White, Blue, Red, Black, Gray
Style	T-shirt, Dress shirt

40

3. What is the probability of randomly choosing a banana and milk in Question 1? $11\frac{1}{9}$%

Taking Math Deeper

Exercise 28

This problem can be used to introduce students to a new mathematical term: *factorial*. If *n* is any whole number, then **n factorial**, written as *n*!, is defined to be

$$n \cdot (n - 1) \cdot (n - 2) \cdot \cdots \cdot 3 \cdot 2 \cdot 1.$$

1 Use the Fundamental Counting Principle.

Number of choices for 1st car: 8
Number of choices for 2nd car: 7
Number of choices for 3rd car: 6
Number of choices for 4th car: 5
Number of choices for 5th car: 4
Number of choices for 6th car: 3
Number of choices for 7th car: 2
Number of choices for 8th car: 1

So, the total number of ways to arrange the cars in the train is

$$8! = 8 \cdot 7 \cdot 6 \cdot 5 \cdot 4 \cdot 3 \cdot 2 \cdot 1 = 40{,}320.$$

That's a lot of ways!

2 Factorials are used to count the number of **permutations** of *n* objects. Later in the study of probability, students will learn that there are *n*! ways to order *n* objects.

3 *Solve A Simpler Problem:* Let's suppose there are only 3 train cars. We can list the 3! = 6 ways as follows.

Reteaching and Enrichment Strategies

If students need help...	If students got it...
Resources by Chapter • Practice A and Practice B • Puzzle Time Record and Practice Journal Practice Differentiating the Lesson Lesson Tutorials Skills Review Handbook	Resources by Chapter • Enrichment and Extension • Technology Connection Start the next section

26. **PASSWORD** You forget the last two digits of your password for a website.

    a. What is the probability that you randomly choose the correct digits?

    b. Suppose you remember that both digits are even. How does this change the probability that your choices are correct?

27. **COMBINATION LOCK** The combination lock has 3 wheels, each numbered from 0 to 9.

    a. What is the probability that someone randomly guesses the correct combination in one attempt?

    b. You try to guess the combination by writing five different numbers from 0 to 999 on a piece of paper. Explain how to find the probability that the correct combination is written on the paper.

28. **TRAINS** Your model train has one engine and eight train cars. Find the total number of ways you can arrange the train. (The engine must be first.)

29. **REPEATED REASONING** You have been assigned a 9-digit identification number.

    a. Why should you use the Fundamental Counting Principle instead of a tree diagram to find the total number of possible identification numbers?

    b. How many identification numbers are possible?

    c. **RESEARCH** Use the Internet to find out why the possible number of Social Security numbers is not the same as your answer to part (b).

30. From a group of 5 candidates, a committee of 3 people is selected. In how many different ways can the committee be selected?

Fair Game Review What you learned in previous grades & lessons

Name two pairs of adjacent angles and two pairs of vertical angles in the figure. *(Section 7.1)*

31.
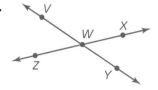

32.
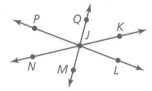

33. **MULTIPLE CHOICE** A drawing has a scale of 1 cm : 1 m. What is the scale factor of the drawing? *(Section 7.5)*

    (A) 1 : 1   **(B)** 1 : 100   **(C)** 10 : 1   **(D)** 100 : 1

Essential Question What is the difference between dependent and independent events?

1 **ACTIVITY: Drawing Marbles from a Bag (With Replacement)**

Work with a partner. You have three marbles in a bag. There are two green marbles and one purple marble. Randomly draw a marble from the bag. Then put the marble back in the bag and draw a second marble.

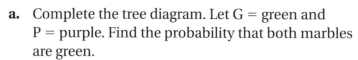

a. Complete the tree diagram. Let G = green and P = purple. Find the probability that both marbles are green.

First draw:

Second draw:

GG

b. Does the probability of getting a green marble on the second draw *depend* on the color of the first marble? Explain.

2 **ACTIVITY: Drawing Marbles from a Bag (Without Replacement)**

Work with a partner. Using the same marbles from Activity 1, randomly draw two marbles from the bag.

Probability and Statistics

In this lesson, you will

• identify independent and dependent events.
• use formulas to find probabilities of independent and dependent events.

a. Complete the tree diagram. Let G = green and P = purple. Find the probability that both marbles are green.

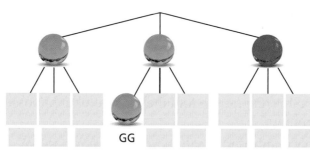

First draw:

Second draw:

GP

Is this event more likely than the event in Activity 1? Explain.

b. Does the probability of getting a green marble on the second draw *depend* on the color of the first marble? Explain.

Laurie's Notes

Introduction

Applying Mathematical Practices

- **Attend to Precision:** Mathematically proficient students try to communicate precisely to others. In the activities today, students should pay attention to whether there is *replacement* after an event. This will be important in defining independent and dependent events in the lesson.

Motivate

- Hand two volunteers a bag containing 2 red and 2 blue cubes (any similar objects will work). The goal is to draw twice and to end up with two of the same color. The only difference is that person A *replaces* the cube after the first draw and person B *does not replace* the cube after the first draw. So, Person A is always drawing 1 out of 4 items. Person B is drawing 1 out of 4 items, and then 1 out of 3 items.
- Have each person do 10 trials and gather data. This should help students start to see a pattern.
- **Big Idea:** The probability on the second draw *depends upon* what is drawn on the first draw when you do *not* replace the cube.
- Draw a tree diagram to show the difference between Person A and Person B. This will help students read the tree diagrams in the activity.

Activity Notes

Activity 1 and Activity 2

- **Model with Mathematics:** The tree diagrams are used in the activities to help compute the theoretical probabilities.
- **Common Error:** Students read the tree diagram incorrectly and think a person is drawing nine times in Activity 1 and six times in Activity 2. Remind students that the first horizontal row of marbles displays the possible outcomes of the first draw and the second horizontal row displays the possible outcomes of the second draw.
- ❓ "On the first draw, you will either draw a green marble or a purple marble. Why does the tree diagram show more than just one of each?" There are two different green marbles that could be drawn and only one purple.
- **Teaching Tip:** Use the labels green marble 1 and green marble 2 to help students recognize why you need two greens and one purple in the tree diagram for the first draw.
- ❓ "When you *do* replace the marble, are the probabilities on the second draw the same as on the first draw? Explain." yes; There are the same marbles in the bag each time.
- ❓ "When you *do not* replace the marble, are the probabilities on the second draw the same as on the first draw? Explain." no; There are only two marbles instead of three to draw from. The probability of the second draw depends upon the first marble drawn.
- ❓ "For Activity 1, what is the probability of drawing 2 purple marbles? for Activity 2?" $\frac{1}{9}$, 0

What Your Students Will Learn

- Identify independent and dependent events.
- Use formulas to find probabilities of independent, dependent, and compound events.

Previous Learning

Students have used samples spaces to find probabilities of compound events.

Technology for the Teacher

Dynamic Classroom

Lesson Plans
Complete Materials List

10.5 Record and Practice Journal

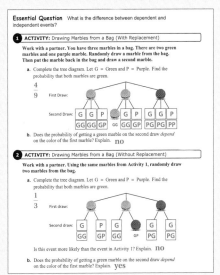

Differentiated Instruction

Kinesthetic

Before class, choose three books to put on your desk. Ask 3 students to each choose a book by writing the title on a piece of paper. Then ask another 3 students to select a book, one at a time, and take them back to their seats. Discuss when the selection of the books involved replacement (The students wrote their choice.) and when the selection of books did not involve replacement (The students took the books to their desks.).

10.5 Record and Practice Journal

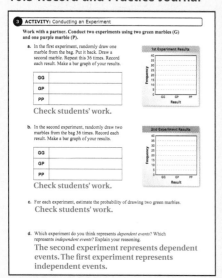

3 ACTIVITY: Conducting an Experiment

Work with a partner. Conduct two experiments using two green marbles (G) and one purple marble (P).

a. In the first experiment, randomly draw one marble from the bag. Put it back. Draw a second marble. Repeat this 36 times. Record each result. Make a bar graph of your results.

GG	
GP	
PP	

1st Experiment Results

Check students' work.

b. In the second experiment, randomly draw two marbles from the bag 36 times. Record each result. Make a bar graph of your results.

GG	
GP	
PP	

2nd Experiment Results

Check students' work.

c. For each experiment, estimate the probability of drawing two green marbles.
Check students' work.

d. Which experiment do you think represents *dependent events*? Which represents *independent events*? Explain your reasoning.
The second experiment represents dependent events. The first experiment represents independent events.

What Is Your Answer?

4. **IN YOUR OWN WORDS** What is the difference between *dependent* and *independent* events? Describe a real-life example of each.
dependent events: occurrence of one event will affect the occurrence of another event
independent events: occurrence of one event will not affect the occurrence of another event

In Questions 5–7, tell whether the events are *independent* or *dependent*. Explain your reasoning.

5. You roll a 5 on a number cube and spin blue on a spinner.
independent

6. Your teacher chooses one student to lead a group, and then chooses another student to lead another group.
dependent

7. You spin red on one spinner and green on another spinner.
independent

8. In Activities 1 and 2, what is the probability of drawing a green marble on the first draw? on the second draw? How do you think you can use these two probabilities to find the probability of drawing two green marbles?
Activity 1: $\frac{2}{3}$; $\frac{2}{3}$; find the product of the probabilities
Activity 2: $\frac{2}{3}$; $\frac{1}{2}$; find the product of the probabilities

Laurie's Notes

Activity 3

- Students will now collect the data for the same events described in Activities 1 and 2. The experimental probabilities should be close to the theoretical probabilities.
- Explain that in part (a), you cannot draw two marbles at once. You draw one, replace it, and draw again. There are three possibilities: two green marbles (GG), one of each color (GP), or two purple marbles (PP).
- In part (b), it does not matter if you draw one marble followed by another, or if you draw two at one time. The result is the same. This drawing procedure eliminates the possibility of drawing two purple marbles (PP).
- Have pairs of students hold up the results of each experiment so that bar graphs can be analyzed visually.
- **? Attend to Precision:** "For the first experiment, compare the lengths of the three bars. What do you notice? Interpret your answer." The GG bar and the GP bar are about the same length. The PP bar is not very tall, maybe about a fourth as tall as either of the other bars. So, you are as likely to draw two greens as you are to draw one of each color, and you are 4 times more likely to draw two greens or one of each color as you are to draw two purples.
- **? Attend to Precision:** "For the second experiment, compare the lengths of the bars. What do you notice? Interpret your answer." The PP bar length is 0 because it is not a possible outcome. The GP bar should be about twice as long as the GG bar, so you are twice as likely to draw one of each color as you are to draw two greens.

What Is Your Answer?

- In Question 4, student language may not be precise at this point. Students may only be comfortable describing the difference in dependent and independent events in terms of purple and green marbles.
- Questions 5–7 will help extend student knowledge of independent and dependent events to other contexts.
- Question 8 is important in helping students discover the multiplication rule for independent and dependent events, which will be formalized in the lesson.

Closure

- You have a deck of playing cards (4 suits, 13 cards in each suit). Give an example of dependent events and independent events using the deck of cards.
  Sample answer:
  Dependent events—draw one card, set it aside, then draw another
  Independent events—draw one card, replace it, then draw another

Work with a partner. Conduct two experiments.

a. In the first experiment, randomly draw one marble from the bag. Put it back. Draw a second marble. Repeat this 36 times. Record each result. Make a bar graph of your results.

b. In the second experiment, randomly draw two marbles from the bag 36 times. Record each result. Make a bar graph of your results.

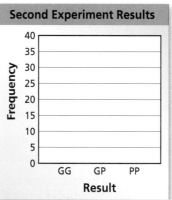

Math Practice

Use Definitions

In what other mathematical context have you seen the terms *independent* and *dependent*? How does knowing these definitions help you answer the questions in part (d)?

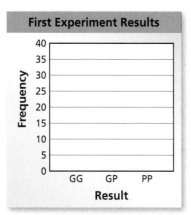

c. For each experiment, estimate the probability of drawing two green marbles.

d. Which experiment do you think represents *dependent events*? Which represents *independent events*? Explain your reasoning.

What Is Your Answer?

4. **IN YOUR OWN WORDS** What is the difference between dependent and independent events? Describe a real-life example of each.

In Questions 5–7, tell whether the events are *independent* or *dependent*. Explain your reasoning.

5. You roll a 5 on a number cube and spin blue on a spinner.

6. Your teacher chooses one student to lead a group, and then chooses another student to lead another group.

7. You spin red on one spinner and green on another spinner.

8. In Activities 1 and 2, what is the probability of drawing a green marble on the first draw? on the second draw? How do you think you can use these two probabilities to find the probability of drawing two green marbles?

Practice

Use what you learned about independent and dependent events to complete Exercises 3 and 4 on page 433.

10.5 Lesson

Check It Out
Lesson Tutorials
BigIdeasMath⩗com

Key Vocabulary 🔊
independent events,
  p. 430
dependent events,
  p. 431

Compound events may be *independent events* or *dependent events*.
Events are **independent events** if the occurrence of one event *does
not* affect the likelihood that the other event(s) will occur.

🗝️ Key Idea

Probability of Independent Events

Words The probability of two or more independent events is
the product of the probabilities of the events.

Symbols $P(A \text{ and } B) = P(A) \cdot P(B)$

$P(A \text{ and } B \text{ and } C) = P(A) \cdot P(B) \cdot P(C)$

EXAMPLE ① **Finding the Probability of Independent Events**

**You spin the spinner and flip the coin. What is the probability of
spinning a prime number and flipping tails?**

The outcome of spinning the spinner does not affect the outcome of
flipping the coin. So, the events are independent.

$$P(\text{prime}) = \frac{3}{5}$$

There are 3 prime numbers (2, 3, and 5).
There is a total of 5 numbers.

$$P(\text{tails}) = \frac{1}{2}$$

There is 1 tails side.
There is a total of 2 sides.

Use the formula for the probability of independent events.

$$P(A \text{ and } B) = P(A) \cdot P(B)$$

$$P(\text{prime and tails}) = P(\text{prime}) \cdot P(\text{tails})$$

$$= \frac{3}{5} \cdot \frac{1}{2} \qquad \text{Substitute.}$$

$$= \frac{3}{10} \qquad \text{Multiply.}$$

∴ The probability of spinning a prime number and flipping tails
is $\frac{3}{10}$, or 30%.

On Your Own

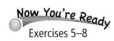
Now You're Ready
Exercises 5–8

1. What is the probability of spinning a multiple of 2 and
flipping heads?

🔊 Multi-Language Glossary at BigIdeasMath⩗com

Laurie's Notes

Introduction

Connect
- **Yesterday:** Students developed an understanding of the difference between independent events and dependent events.
- **Today:** Students will use a formal definition to compute theoretical probabilities of independent events and dependent events.

Motivate
- Tell students that the bag contains slips of paper with each of their names. You will draw two names. The first name you draw will not have homework tonight (or wins a pencil). The second name you draw will get 5 bonus points on the test (or wins a certificate for an ice cream in the cafeteria).
- Draw and reveal the first name. Now slowly put the piece of paper back in the bag and wait for student reaction. They should recognize that they have greater probability of winning if the piece of paper is not replaced.

Lesson Notes

Discuss
- Write the definition of independent events on the board.
- Relate the definition to drawing cards from a deck. If you replace a drawn card before drawing again, the events are independent.

Key Idea
- Write the Key Idea on the board. Note that the probability of independent events is not defined in terms of two events only, it can be extended to three or more events by continuing to multiply.
- The formulas may seem daunting to students because of all the symbols. Be sure to use the words along with the notation.

Example 1
- **Model with Mathematics:** Use a spinner and a coin to model the example. If you do not have a spinner, sketch one on a transparency and use a paper clip, anchored at the center with a pen, as your spinner. You could also use an online simulator.
- ❓ "Are the two events independent? Explain." yes; The outcome of the spinner does not affect the outcome of flipping the coin.
- Students should know that a *prime number* is a number greater than 1 that is divisible by only two numbers, itself and 1, and that a *composite number* is a number having two or more prime factors.
- ❓ "What numbers on the spinner are prime?" 2, 3, and 5
- Write the formula for finding the probability of independent events. You are finding the probability of spinning a prime *and* flipping tails. This is not the same as spinning a prime *or* flipping tails.

On Your Own
- **Connection:** Drawing a tree diagram will help students see the probability of spinning a multiple of 2 and flipping heads.

Goal Today's lesson is identifying and finding probabilities of **independent events** and **dependent events**.

Technology for the **Teacher**

Dynamic Classroom

Lesson Tutorials
Lesson Plans
Answer Presentation Tool

Extra Example 1

In Example 1, what is the probability of spinning a composite number and flipping tails? $\frac{1}{10}$

On Your Own

1. $\frac{1}{5}$

Group Activity

To assist English learners in understanding the concepts and ideas presented in this chapter, provide them with the opportunity to interact, discuss, and listen in a group situation.

Extra Example 2

In Example 2, what is the probability that one of your friends is chosen first, and then you are chosen second? $\frac{1}{1650}$, or about 0.061%

On Your Own

2. $\frac{58}{75}$, or about 77.3%

Laurie's Notes

Discuss

- Write the definition of dependent events on the board.
- Relate the definition to drawing cards from a deck. If you do *not* replace a drawn card before drawing again, the events are dependent.
- Explain that you can find probabilities of dependent events using a formula that is similar to the formula for the probability of independent events, except that the probability of the second event has been affected by the occurrence of the first event.

Key Idea

- Write the Key Idea on the board. Again, be sure to use the words along with the notation.
- Demonstrate the formula using four cards, two with even numbers and two with odd numbers. Ask what the probability of randomly drawing two even-numbered cards would be without replacement. The size of the sample space is 4 for the first draw but only 3 for the second draw. The occurrence of the first event (drawing an even-numbered card) affects the likelihood of the second event (drawing another even-numbered card). So, the probability of drawing both even-numbered cards would be $\frac{2}{4} \cdot \frac{1}{3} = \frac{1}{6}$.

Example 2

- **Representation:** Select five students to be the relatives and six students to be the friends. Give each relative a card marked with an R, and give each friend a card marked with an F. The whole class is the audience. Find the probability of choosing a relative. Then find the probability of choosing a friend after a relative is chosen. Finally, multiply the two probabilities.

On Your Own

? "Of the 100 audience members, how many are *not* you, your relatives, or your friends?" $100 - 12 = 88$

Events are **dependent events** if the occurrence of one event *does* affect the likelihood that the other event(s) will occur.

 **Key Idea**

Probability of Dependent Events

Words The probability of two dependent events A and B is the probability of A times the probability of B after A occurs.

Symbols $P(A \text{ and } B) = P(A) \cdot P(B \text{ after } A)$

EXAMPLE ② **Finding the Probability of Dependent Events**

People are randomly chosen to be game show contestants from an audience of 100 people. You are with 5 of your relatives and 6 other friends. What is the probability that one of your relatives is chosen first, and then one of your friends is chosen second?

Choosing an audience member changes the number of audience members left. So, the events are dependent.

> There are 5 relatives.

$$P(\text{relative}) = \frac{5}{100} = \frac{1}{20}$$

> There is a total of 100 audience members.

> There are 6 friends.

$$P(\text{friend}) = \frac{6}{99} = \frac{2}{33}$$

> There is a total of 99 audience members left.

Use the formula for the probability of dependent events.

$$P(A \text{ and } B) = P(A) \cdot P(B \text{ after } A)$$

$$P(\text{relative and friend}) = P(\text{relative}) \cdot P(\text{friend after relative})$$

$$= \frac{1}{20} \cdot \frac{2}{33} \qquad \text{Substitute.}$$

$$= \frac{1}{330} \qquad \text{Simplify.}$$

∴ The probability is $\frac{1}{330}$, or about 0.3%.

● **On Your Own**

Now You're Ready
Exercises 9–12

2. What is the probability that you, your relatives, and your friends are *not* chosen to be either of the first two contestants?

A student randomly guesses the answer for each of the multiple-choice questions. What is the probability of answering all three questions correctly?

1. In what year did the United States gain independence from Britain?
   A. 1492   **B.** 1776   **C.** 1788   **D.** 1795   **E.** 2000

2. Which amendment to the Constitution grants citizenship to all persons born in the United States and guarantees them equal protection under the law?
   A. 1st   **B.** 5th   **C.** 12th   **D.** 13th   **E.** 14th

3. In what year did the Boston Tea Party occur?
   A. 1607   **B.** 1773   **C.** 1776   **D.** 1780   **E.** 1812

Choosing the answer for one question does not affect the choice for the other questions. So, the events are independent.

Method 1: Use the formula for the probability of independent events.

$$P(\#1 \text{ and } \#2 \text{ and } \#3 \text{ correct}) = P(\#1 \text{ correct}) \cdot P(\#2 \text{ correct}) \cdot P(\#3 \text{ correct})$$

$$= \frac{1}{5} \cdot \frac{1}{5} \cdot \frac{1}{5} \qquad \text{Substitute.}$$

$$= \frac{1}{125} \qquad \text{Multiply.}$$

⋮• The probability of answering all three questions correctly is $\frac{1}{125}$, or 0.8%.

Method 2: Use the Fundamental Counting Principle.

There are 5 choices for each question, so there are $5 \cdot 5 \cdot 5 = 125$ possible outcomes. There is only 1 way to answer all three questions correctly.

$$P(\#1 \text{ and } \#2 \text{ and } \#3 \text{ correct}) = \frac{1}{125}$$

⋮• The probability of answering all three questions correctly is $\frac{1}{125}$, or 0.8%.

On Your Own

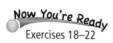
Now You're Ready
Exercises 18–22

3. The student can eliminate Choice A for all three questions. What is the probability of answering all three questions correctly? Compare this probability with the probability in Example 3. What do you notice?

Laurie's Notes

Example 3

- Note that this example involves more than two events.
- **?** "Do you think it is wise to randomly guess on a true-false test or a multiple-choice test?" Answers will vary.
- Explain that this question is an example of guessing on a 3-question multiple-choice test with each question having 5 possible responses.
- You might have students try to answer the questions before finding the probability. B; E; B
- Work through the problem using both methods.
- **Construct Viable Arguments:** To make sure that students understand probabilities of independent and dependent events, ask them why they would or would not use a tree diagram or table to find the probabilities in Examples 1–3.

On Your Own

- The goal of this question is for students to recognize how the probability changes when a choice can be eliminated from each example.
- **Extension:** Ask students which choices they could eliminate from the questions and why.

Closure

- **Exit Ticket:** You and your friend are among 5 volunteers to help distribute workbooks. What is the probability that your teacher randomly selects you and your friend to distribute the workbooks? $\frac{1}{10}$, or 10%

Extra Example 3

A true-false quiz has 4 questions. What is the probability of randomly guessing and answering all of the questions correctly? $\frac{1}{16}$, or 6.25%

On Your Own

3. $\frac{1}{64}$; The probability of getting the correct answer increases when there are only four choices.

Vocabulary and Concept Check

1. What is the probability of choosing a 1 and then a blue chip?; $\frac{1}{15}$; $\frac{1}{10}$

2. For independent events, find the probability of the first event, find the probability of the second event, and then multiply. For dependent events, find the probability of the first event, find the probability of the second event after the first event occurs, and then multiply.

Practice and Problem Solving

3. independent; The outcome of the first roll does not affect the outcome of the second roll.

4. dependent; Your friend's lane number cannot be the same as your lane number. So, your friend's lane number depends on your lane number.

5. $\frac{1}{8}$ 6. $\frac{1}{4}$

7. $\frac{3}{8}$ 8. $\frac{3}{8}$

9. $\frac{1}{42}$ 10. $\frac{1}{14}$

11. $\frac{2}{21}$ 12. $\frac{2}{7}$

13. The two events are dependent, so the probability of the second event is $\frac{1}{3}$.

   $P(\text{red and green}) = \frac{1}{4} \cdot \frac{1}{3} = \frac{1}{12}$

14. dependent; The second draw is affected by the first draw.

Assignment Guide and Homework Check

Level	Day 1 Activity Assignment	Day 2 Lesson Assignment	Homework Check
Basic	3, 4, 27–30	1, 2, 5–25 odd	7, 11, 15, 19
Average	3, 4, 27–30	1, 2, 5–21 odd, 24, 26	7, 11, 15, 19, 24
Advanced	3, 4, 27–30	1, 2, 6–12 even, 13, 14–26 even	10, 16, 22, 24
Accelerated	1–4, 6–12 even, 13, 14–26 even, 27–30		10, 16, 22, 24

Common Errors

- **Exercises 3 and 4** Students may mix up independent and dependent events or may have difficulty determining which type of event it is. Remind them that independent events are where you do two different things or events where you start over before the next trial. Dependent events have at least one less possible outcome after the first event.

- **Exercises 5–12** Students may find the probabilities of each individual event. Remind them that they are finding the probability of a compound event.

10.5 Record and Practice Journal

> **You roll a number cube twice. Find the probability of the events.**
>
> 1. Rolling a 3 twice $\frac{1}{36}$
> 2. Rolling an even number and a 5 $\frac{1}{12}$
> 3. Rolling an odd number and a 2 or a 4 $\frac{1}{6}$
> 4. Rolling a number less than 6 and a 3 or a 1 $\frac{5}{18}$
>
> **You randomly choose a letter from a hat with the letters A through J. Without replacing the first letter, you choose a second letter. Find the probability of the events.**
>
> 5. Choosing an H and then a D $\frac{1}{90}$
> 6. Choosing a consonant and then an E or an I $\frac{7}{45}$
> 7. Choosing a vowel and then an F $\frac{1}{30}$
> 8. Choosing a vowel and then a consonant $\frac{7}{30}$
>
> 9. You have 3 clasp bracelets, 4 watches, and 5 stretch bracelets. You randomly choose two from your jewelry box. What is the probability that you will choose 2 watches? $\frac{1}{11}$
>
> **You flip a coin, and then roll a number cube twice. Find the probability of the event.**
>
> 10. Flipping heads, rolling a 5, and rolling a 2 $\frac{1}{72}$, or about 1.39%
> 11. Flipping tails, rolling an odd number, and rolling a 4 $\frac{1}{24}$, or about 4.17%
> 12. Flipping tails, rolling a 6 or a 1, and rolling a 3 $\frac{1}{36}$, or about 2.8%
> 13. Flipping heads, *not* rolling a 2, and rolling an even number $\frac{5}{24}$, or about 20.8%

Vocabulary and Concept Check

1. **DIFFERENT WORDS, SAME QUESTION** You randomly choose one of the chips. Without replacing the first chip, you choose a second chip. Which question is different? Find "both" answers.

What is the probability of choosing a 1 and then a blue chip?

What is the probability of choosing a 1 and then an even number?

What is the probability of choosing a green chip and then a chip that is *not* red?

What is the probability of choosing a number less than 2 and then an even number?

2. **WRITING** How do you find the probability of two events A and B when A and B are independent? dependent?

 Practice and Problem Solving

Tell whether the events are *independent* or *dependent*. Explain.

3. You roll a 4 on a number cube. Then you roll an even number on a different number cube.

4. You randomly draw a lane number for a 100-meter race. Then your friend randomly draws a lane number for the same race.

You spin the spinner and flip a coin. Find the probability of the compound event.

① 5. Spinning a 3 and flipping heads

6. Spinning an even number and flipping tails

7. Spinning a number greater than 1 and flipping tails

8. *Not* spinning a 2 and flipping heads

You randomly choose one of the tiles. Without replacing the first tile, you choose a second tile. Find the probability of the compound event.

② 9. Choosing a 5 and then a 6

10. Choosing an odd number and then a 20

11. Choosing a number less than 7 and then a multiple of 4

12. Choosing two even numbers

13. **ERROR ANALYSIS** Describe and correct the error in finding the probability.

✗ You randomly choose one of the marbles. Without replacing the first marble, you choose a second marble. What is the probability of choosing red and then green?

$P(\text{red and green}) = \dfrac{1}{4} \cdot \dfrac{1}{4} = \dfrac{1}{16}$

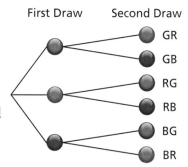

First Draw   Second Draw
GR
GB
RG
RB
BG
BR

14. **LOGIC** A bag contains three marbles. Does the tree diagram show the outcomes for *independent* or *dependent* events? Explain.

15. **EARRINGS** A jewelry box contains two gold hoop earrings and two silver hoop earrings. You randomly choose two earrings. What is the probability that both are silver hoop earrings?

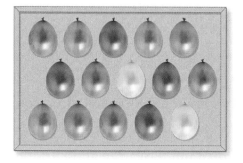

You

16. **HIKING** You are hiking to a ranger station. There is one correct path. You come to a fork and randomly take the path on the left. You come to another fork and randomly take the path on the right. What is the probability that you are still on the correct path?

17. **CARNIVAL** At a carnival game, you randomly throw two darts at the board and break two balloons. What is the probability that both of the balloons you break are purple?

You spin the spinner, flip a coin, then spin the spinner again. Find the probability of the compound event.

③ 18. Spinning a 4, flipping heads, then spinning a 7

19. Spinning an odd number, flipping heads, then spinning a 3

20. Spinning an even number, flipping tails, then spinning an odd number

21. *Not* spinning a 5, flipping heads, then spinning a 1

22. Spinning an odd number, *not* flipping heads, then *not* spinning a 6

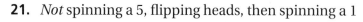

Common Errors

- **Exercise 17** Students may forget to decrease both the number of favorable outcomes and the number of possible outcomes by 1 when finding the probability of the second dart hitting a purple balloon. Remind them that the second throw has 1 less favorable outcome and 1 less possible outcome.
- **Exercises 18–22** Students may find the probabilities of each individual event. Remind them that they are finding the probability of a compound event.

Practice and Problem Solving

15. $\frac{1}{6}$, or about 16.7%

16. $\frac{1}{4}$, or 25%

17. $\frac{2}{35}$

18. $\frac{1}{162}$, or about 0.62%

19. $\frac{5}{162}$, or about 3.1%

20. $\frac{10}{81}$, or about 12.3%

21. $\frac{4}{81}$, or about 4.9%

22. $\frac{20}{81}$, or about 24.7%

23. $\frac{3}{4}$

24. 51.2%

25. **a.** If you and your best friend were in the same group, then the probability that you both are chosen would be 0 because only one leader is chosen from each group. Because the probability that both you and your best friend are chosen is $\frac{1}{132}$, you and your best friend are not in the same group.
    b. $\frac{1}{11}$ **c.** 23

Differentiated Instruction

Vocabulary

Some students may have difficulty understanding how the independence or dependence of events affects probability. Allow students to experiment with bags containing similar small objects, as in the marble activity. Have them draw three objects from the bag with and without replacement and then compare probabilities. Ask volunteers to present their results to the class.

Practice and Problem Solving

26. See *Taking Math Deeper*.

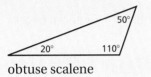 ## Fair Game Review

27. See Additional Answers.

28.

obtuse scalene

29.

acute isosceles

30. C

Mini-Assessment

You spin the spinner and flip a coin. Find the probability of the compound event.

1. Spinning a five and flipping tails $\frac{1}{18}$

2. Spinning an odd number and flipping heads $\frac{5}{18}$

A bag holds six chips numbered 1–6. Without replacing the first chip, you choose a second chip. Find the probability of the compound event.

3. Choosing an even number and then choosing an odd number $\frac{3}{10}$

4. Choosing a number less than 3 and then choosing a number greater than 3 $\frac{1}{5}$

Taking Math Deeper

Exercise 26

This is an example of the strategy *Working Backward*.

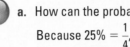

1. Who was the oldest?

    A. Ned   **B.** Yvonne   **C.** Sun Li   **D.** Angel   **E.** Dusty

2. What city was Stacey from?

    A. Raleigh   **B.** New York   **C.** Roanoke   **D.** Dallas   **E.** San Diego

①   **a.** How can the probability of getting both answers correct be 25%?

Because $25\% = \frac{1}{4}$, you need two probabilities whose product is $\frac{1}{4}$.

By eliminating all but 2 of the choices in each question, the probability of randomly guessing the correct answer to both questions is

$$P = \frac{1}{2} \cdot \frac{1}{2} = \frac{1}{4}.$$

②   **b.** How can the probability of getting both answers correct be $8\frac{1}{3}\%$?

Because $8\frac{1}{3}\% = \frac{1}{12}$, you have to find the different ways that you can multiply two probabilities to get $\frac{1}{12}$. As it turns out, there is only one way. You eliminate 1 of the choices to one question and eliminate 2 of the choices to another question. Then, the probability is

$$P = \frac{1}{4} \cdot \frac{1}{3} = \frac{1}{12}.$$

Strategy

③ You might consider using this as an opportunity to talk about test-taking strategies. It should seem clear to students that the more choices you can eliminate, the greater your chance for a higher score.

Reteaching and Enrichment Strategies

If students need help...	If students got it...
Resources by Chapter • Practice A and Practice B • Puzzle Time Record and Practice Journal Practice Differentiating the Lesson Lesson Tutorials Skills Review Handbook	Resources by Chapter • Enrichment and Extension • Technology Connection Start the next section

23. LANGUAGES There are 16 students in your Spanish class. Your teacher randomly chooses one student at a time to take a verbal exam. What is the probability that you are *not* one of the first four students chosen?

24. SHOES Twenty percent of the shoes in a factory are black. One shoe is chosen and replaced. A second shoe is chosen and replaced. Then a third shoe is chosen. What is the probability that *none* of the shoes are black?

25. PROBLEM SOLVING Your teacher divides your class into two groups, and then randomly chooses a leader for each group. The probability that you are chosen to be a leader is $\frac{1}{12}$. The probability that both you and your best friend are chosen is $\frac{1}{132}$.

  a. Is your best friend in your group? Explain.

  b. What is the probability that your best friend is chosen as a group leader?

  c. How many students are in the class?

26. **Structure** After ruling out some of the answer choices, you randomly guess the answer for each of the story questions below.

> **1.** Who was the oldest?
>   **A.** Ned   **B.** Yvonne   **C.** Sun Li   **D.** Angel   **E.** Dusty
>
> **2.** What city was Stacey from?
>   **A.** Raleigh   **B.** New York   **C.** Roanoke   **D.** Dallas   **E.** San Diego

  a. How can the probability of getting both answers correct be 25%?

  b. How can the probability of getting both answers correct be $8\frac{1}{3}$%?

Fair Game Review What you learned in previous grades & lessons

Draw a triangle with the given angle measures. Then classify the triangle. *(Section 7.3)*

27. 30°, 60°, 90°   **28.** 20°, 50°, 110°   **29.** 50°, 50°, 80°

30. MULTIPLE CHOICE Which set of numbers is in order from least to greatest? *(Section 6.2)*

  Ⓐ $\frac{2}{3}$, 0.6, 67%   **Ⓑ** 44.5%, $\frac{4}{9}$, $0.4\overline{6}$

  Ⓒ 0.269, 27%, $\frac{3}{11}$ **Ⓓ** $2\frac{1}{7}$, 214%, $2.\overline{14}$

Check It Out
Lesson Tutorials
BigIdeasMath ✓.com

Key Vocabulary 🔊
simulation, *p. 436*

A **simulation** is an experiment that is designed to reproduce the conditions of a situation or process. Simulations allow you to study situations that are impractical to create in real life.

EXAMPLE **1** **Simulating Outcomes That Are Equally Likely**

HTH	HTT
HTT	HTH
HTT	TTT
(HHH)	HTT
HTT	TTT
HTT	HTH
HTH	(HHH)
HTT	HTT
TTT	HTH
HTH	HTT

A couple plans on having three children. The gender of each child is equally likely. (a) Design a simulation involving 20 trials that you can use to model the genders of the children. (b) Use your simulation to find the experimental probability that all three children are boys.

a. Choose an experiment that has two equally likely outcomes for each event (gender), such as tossing three coins. Let heads (H) represent a boy and tails (T) represent a girl.

b. To find the experimental probability, you need repeated trials of the simulation. The table shows 20 trials.

$$P(\text{three boys}) = \frac{2}{20} = \frac{1}{10}$$

⟵ HHH occurred 2 times.

⟵ There is a total of 20 trials.

∴ The experimental probability is $\frac{1}{10}$, 0.1, or 10%.

EXAMPLE **2** **Simulating Outcomes That Are Not Equally Likely**

Study Tip

In Example 2, the digits 1 through 6 represent 60% of the possible digits (0 through 9) in the tens place. Likewise, the digits 1 and 2 represent 20% of the possible digits in the ones place.

There is a 60% chance of rain on Monday and a 20% chance of rain on Tuesday. Design and use a simulation involving 50 randomly generated numbers to find the experimental probability that it will rain on both days.

Use the random number generator on a graphing calculator. Randomly generate 50 numbers from 0 to 99. The table below shows the results.

Let the digits 1 through 6 in the tens place represent rain on Monday. Let digits 1 and 2 in the ones place represent rain on Tuesday. Any number that meets these criteria represents rain on both days.

```
randInt(0,99,50)
{52 66 73 68 75…
```

(52)	66	73	68	75	28	35	47	48	2
16	68	49	3	77	35	92	78	6	6
58	18	89	39	24	80	(32)	(41)	77	(21)
(32)	40	96	59	86	1	(12)	0	94	73
40	71	28	(61)	1	24	37	25	3	25

$$P(\text{rain both days}) = \frac{7}{50}$$

⟵ 7 numbers meet the criteria.

⟵ There is a total of 50 trials.

∴ The experimental probability is $\frac{7}{50}$, 0.14, or 14%.

🔊 Multi-Language Glossary at BigIdeasMath ✓.com

Laurie's Notes

Introduction

Connect

- **Yesterday:** Students found probabilities of compound events.
- **Today:** Students will perform simulations to find probabilities of compound events.

Motivate

- It is near the end of the year and final exams are coming. Tell students that you heard the history exam will include 10 true-false questions.
- **?** "Assume you don't read the questions. How do you think you could find the probability of randomly guessing the correct answer to *at least* 7 questions?" Students will have a range of ideas about this.
- Explain that this is an event that you probably don't want to perform. Studying is a much better option. However, you can *simulate* the results.

Lesson Notes

Discuss

- Define simulations and give an example. Explain that simulations should be kept simple, allowing you to repeatedly duplicate the results of an event.
- To perform a simulation: (a) state the situation, (b) describe a model that randomly generates appropriate outcomes, (c) use the model repeatedly and record the results, and (d) use the results to make a conclusion.
- Students know that experimental probability approximates theoretical probability as the number of trials increases. So, simulations can be designed to accurately approximate theoretical probabilities.

Example 1

- **?** "How can you simulate an event that has two equally likely outcomes?" Answers will vary, but students will likely mention flipping coins.
- If time permits, students could use coins or an online tool to generate data.
- **?** "If heads (H) represents a boy and tails (T) represents a girl, what outcome represents 3 boys?" HHH
- Students should view the data and interpret the first few results.
- The table shows that 2 out of 20, or 10% of the results represent 3 boys.
- **Extension:** Ask students to use the table to find the experimental probability that all three children are (a) girls and (b) of the same gender.

Example 2

- **?** "Is it possible to simulate a situation with outcomes that are not equally likely?" Students may say no at first. If you tell them to consider a number cube or different colors of marbles in a bag, they may change their minds.
- **Use Appropriate Tools Strategically:** In this simulation, a graphing calculator generates random whole numbers between 0 and 99 using the *randInt* function. The syntax is *randInt(a,b,c)* where *a* and *b* represent the range of numbers and *c* represents how many are generated. You can use the right arrow key to scroll through the results. Commands and syntax may vary for different graphing calculators.

What Your Students Will Learn

- Use simulations to find experimental probabilities.

Goal Today's lesson is performing **simulations** to find probabilities of compound events.

Lesson Tutorials
Lesson Plans
Answer Presentation Tool

Extra Example 1

A couple plans on having four children. The gender of each child is equally likely.

a. Design a simulation involving 25 trials that you can use to model the genders of the children. Toss four coins, letting H = boy and T = girl.

b. Use your simulation to find the experimental probability that all four children have the same gender. Answers will vary, but should be close to 12.5%.

Extra Example 2

There is a 70% chance of snow on Thursday and a 40% chance of snow on Friday. Design and use a simulation involving 50 randomly generated numbers to find the experimental probability that it will snow both days. Randomly generate 50 numbers from 0 to 99 on a graphing calculator, let the digits 1–7 in the tens place represent snow on Thursday and let the digits 1–4 in the ones place represent snow in Friday; Answers will vary, but should be close to 28%.

Record and Practice Journal Extension 10.5 Practice

1–2. See Additional Answers.

Extra Example 3

Each school year, there is a 20% chance that weather causes one or more days of school to be canceled. Design and use a simulation involving 50 randomly generated numbers to find the experimental probability that weather will cause school to be canceled in exactly two of the next five school years. Randomly generate 50 five-digit whole numbers in a spreadsheet, let the digits 1 and 2 represent a school year with a weather cancellation; Answers will vary, but should be close to 20% (the theoretical probability is 20.48%).

Practice

1–4. See Additional Answers.

Mini-Assessment

1. You randomly guess on five true-false questions. Design and use a simulation to find the experimental probability that you correctly answer exactly 3 questions. *Sample answer:* Toss five coins 50 times, letting H = correct and T = incorrect, probability should be close to 31.25%.

2. Suppose 40% of the people in a shopping mall are willing to take a survey. Design and use a simulation to find the experimental probability that out of the next 5 random shoppers, there will be 3 or more in a row that are willing to take the survey. Randomly generate 50 five-digit whole numbers in a spreadsheet, let the digits 1–4 represent a customer willing to take the survey; Answers will vary, but should be close to 11% (the theoretical probability is 11.008%).

Example 2 (continued)

- Discuss the design of the simulation.
  - Randomly generating a digit from 0–9 has 10 possible outcomes, each digit having a 10% probability of being generated.
  - You can simulate rain on Monday (60% chance) using digits from 1–6 (60% probability of being generated).
  - You can simulate rain on Tuesday (20% chance) using digits of 1 or 2 (20% probability of being generated).
  - One way to simulate the weather on both days is to generate random numbers between 0 and 99, letting the tens digit represent Monday and the ones digit represent Tuesday. Results with 1, 2, 3, 4, 5, or 6 in the tens place and 1 or 2 in the ones place represent rain on both days.
- The table shows that 14% of the results represent rain on both days.

Example 3

? "What is the chance that weather causes a school cancellation?" 50%

- **Use Appropriate Tools Strategically:** The Study Tip shows how to randomly generate four-digit numbers in a spreadsheet. This could also be done using a graphing calculator.

? "If we randomly generate 50 numbers from 0 to 9999, what numbers represent school being canceled in at least three of the next four school years?" numbers with at least three of the four digits being 1, 2, 3, 4, or 5

- From the table, 17 out of 50, or 34% of the results represent a cancellation in at least three of the next four school years.
- Make sure students understand that assigning digits 1–5 to represent a school cancellation was arbitrary. They can use other digits, such as 0–4, or odd numbers. Have students repeat Examples 2 and 3 using different digits.
- **Teaching Tip:** Because the two outcomes are equally likely, you could also simulate by flipping coins. Have each student in a group of 4 flip a coin. If heads represents a cancellation and at least 3 heads result, school is cancelled three times in that four-year period. Have groups run several trials. Discuss the results.

Practice

- Have students share their simulation design for Questions 1 and 2.
- **Big Idea:** Students need to understand that as you increase the number of trials in a simulation, the experimental probability better approximates the theoretical probability.
- Question 4 helps students see that simulations can help them approximate theoretical probabilities that are difficult to compute directly. Finding the theoretical probability in Example 3 is difficult, but manageable. However, it would be unmanageable for 5 or more school years.

Closure

- **Exit Ticket:** Design and use a simulation to find the experimental probability that you correctly guess at least 7 of the 10 true-false questions in the Motivate.
- How does a probability of 15% in Example 2 or 25% in Example 3 change the simulation you design? How does it change the probabilities?

Probability and Statistics
In this extension, you will

- use simulations to find experimental probabilities.

Each school year, there is a 50% chance that weather causes one or more days of school to be canceled. Design and use a simulation involving 50 randomly generated numbers to find the experimental probability that weather will cause school to be canceled in at least three of the next four school years.

Use a random number table in a spreadsheet. Randomly generate 50 four-digit whole numbers. The spreadsheet below shows the results.

Let the digits 1 through 5 represent school years with a cancellation. The numbers in the spreadsheet that contain at least three digits from 1 through 5 represent four school years in which at least three of the years have a cancellation.

Study Tip

To create a four-digit random number table in a spreadsheet, follow these steps.

1. Highlight the group of cells to use for your table.
2. Format the cells to display four-digit whole numbers.
3. Enter the formula RAND()*10000 into each cell.

	A	B	C	D	E	F
1	7584	3974	8614	2500	4629	
2	3762	3805	2725	7320	6487	
3	3024	1554	2708	1126	9395	
4	4547	6220	9497	7530	3036	
5	1719	0662	1814	6218	2766	
6	7938	9551	8552	4321	8043	
7	6951	0578	5560	0740	4479	
8	4714	4511	5115	6952	5609	
9	0797	3022	9067	2193	6553	
10	3300	5454	5351	6319	0387	
11						

$$P\left(\begin{array}{l}\text{cancellation in at least three} \\ \text{of the next four school years}\end{array}\right) = \frac{17}{50}$$

17 numbers contain at least three digits from 1 to 5.

There is a total of 50 trials.

∴ The experimental probability is $\frac{17}{50}$, 0.34, or 34%.

Practice

1. **QUIZ** You randomly guess the answers to four true-false questions. (a) Design a simulation that you can use to model the answers. (b) Use your simulation to find the experimental probability that you answer all four questions correctly.

2. **BASEBALL** A baseball team wins 70% of its games. Assuming this trend continues, design and use a simulation to find the experimental probability that the team wins the next three games.

3. **WHAT IF?** In Example 3, there is a 40% chance that weather causes one or more days of school to be canceled each school year. Find the experimental probability that weather will cause school to be canceled in at least three of the next four school years.

4. **REASONING** In Examples 1–3 and Exercises 1–3, try to find the theoretical probability of the event. What do you think happens to the experimental probability when you increase the number of trials in the simulation?

10 Study Help

You can use a **notetaking organizer** to write notes, vocabulary, and questions about a topic. Here is an example of a notetaking organizer for probability.

Write important vocabulary or formulas in this space.

If *P*(event) = 0, the event is *impossible*.

If *P*(event) = 0.25, the event is *unlikely*.

If *P*(event) = 0.5, the event is *equally likely to happen or not happen*.

If *P*(event) = 0.75, the event is *likely*.

If *P*(event) = 1, the event is *certain*.

Probability

A number that measures the likelihood that an event will occur

Can be written as a fraction, decimal, or percent

Always between 0 and 1, inclusive

Write your notes about the topic in this space.

Write your questions about the topic in this space.

How do you find the probability of two or more events?

On Your Own

Make notetaking organizers to help you study these topics.

1. experimental probability

2. theoretical probability

3. Fundamental Counting Principle

4. independent events

5. dependent events

After you complete this chapter, make notetaking organizers for the following topics.

6. sample

7. population

Formulas: Newton = beagle   Notes: Likes dog biscuits
Questions: What is my greatest accomplishment?

I hope it's better than your 200-page essay on bacon.

"I am using a notetaking organizer to plan my autobiography."

Sample Answers

1.

$P(event) = \dfrac{\text{number of times the event occurs}}{\text{total number of trials}}$	**Experimental Probability**
Experiment: an investigation or a procedure that has varying results	Probability that is based on repeated trials of an experiment
Outcomes: the possible results of an experiment	Example: You flip a coin 100 times. You flip heads 52 times and tails 48 times. The experimental probabilites are $P(\text{heads}) = \dfrac{52}{100} = 0.52 = 52\%$, and $P(\text{tails}) = \dfrac{48}{100} = 0.48 = 48\%.$
Event: a collection of one or more outcomes	

How can I find the probability of an event without doing an experiment?

2.

$P(event) = \dfrac{\text{number of favorable outcomes}}{\text{number of possible outcomes}}$	**Theoretical Probability**
Outcomes: the possible results of an experiment	The ratio of the number of favorable outcomes to the number of possible outcomes, when all possible outcomes are equally likely
Event: a collection of one or more outcomes	Example: You flip a coin. The theoretical probability of flipping heads and the theoretical probability of flipping tails is $P(\text{heads}) = \dfrac{1}{2}$, and $P(\text{tails}) = \dfrac{1}{2}.$
Favorable outcomes: the outcomes of a specific event	

What if the possible outcomes are not equally likely?

3–5. Available at *BigIdeasMath.com*.

List of Organizers
Available at *BigIdeasMath.com*

Comparison Chart
Concept Circle
Example and Non-Example Chart
Formula Triangle
Four Square
Idea (Definition) and Examples Chart
Information Frame
Information Wheel
Notetaking Organizer
Process Diagram
Summary Triangle
Word Magnet
Y Chart

About this Organizer

A **Notetaking Organizer** can be used to write notes, vocabulary, and questions about a topic. In the space on the left, students write important vocabulary or formulas. In the space on the right, students write their notes about the topic. In the space at the bottom, students write their questions about the topic. A notetaking organizer can also be used as an assessment tool, in which blanks are left for students to complete.

Technology for the *Teacher*

Editable Graphic Organizer

Answers

1. 2

2. 0

3. 4

4. $\frac{3}{10}$, or 30%

5. $\frac{1}{4}$, or 25%

6. $\frac{3}{4}$, or 75%

7. 0, or 0%

8. $\frac{2}{15}$, or about 13.3%

9. $\frac{11}{30}$, or about 36.7%

10. $\frac{43}{120}$, or about 35.8%

11. 1, or 100%

12. 12

13. 8

14. $\frac{2}{5}$, or 40%

15. $\frac{1}{10}$, or 10%

Alternative Quiz Ideas

100% Quiz	Math Log
Error Notebook	Notebook Quiz
Group Quiz	Partner Quiz
Homework Quiz	Pass the Paper

Group Quiz

Students work in groups. Give each group a large index card. Each group writes five questions that they feel evaluate the material they have been studying. On a separate piece of paper, students solve the problems. When they are finished, they exchange cards with another group. The new groups work through the questions on the card.

Reteaching and Enrichment Strategies

If students need help. . .	If students got it. . .
Resources by Chapter • Practice A and Practice B • Puzzle Time Lesson Tutorials *BigIdeasMath.com*	Resources by Chapter • Enrichment and Extension • Technology Connection Game Closet at *BigIdeasMath.com* Start the next section

10.1–10.5 Quiz

You randomly choose one butterfly. Find the number of ways the event can occur. *(Section 10.1)*

1. Choosing red

2. Choosing brown

3. Choosing *not* blue

6 Green
3 White
4 Red
2 Blue
5 Yellow

You randomly choose one paper clip from the jar. Find the probability of the event. *(Section 10.2)*

4. Choosing a green paper clip

5. Choosing a yellow paper clip

6. *Not* choosing a yellow paper clip

7. Choosing a purple paper clip

Use the bar graph to find the experimental probability of the event. *(Section 10.3)*

8. Rolling a 4

9. Rolling a multiple of 3

10. Rolling a 2 or a 3

11. Rolling a number less than 7

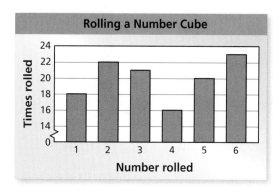

Use the Fundamental Counting Principle to find the total number of possible outcomes. *(Section 10.4)*

12.

	Calculator
Type	Basic display, Scientific, Graphing, Financial
Color	Black, White, Silver

13.

	Vacation
Destination	Florida, Italy, Mexico, England
Length	1 week, 2 weeks

14. BLACK PENS You randomly choose one of the pens shown. What is the theoretical probability of choosing a black pen? *(Section 10.3)*

15. BLUE PENS You randomly choose one of the five pens shown. Your friend randomly chooses one of the remaining pens. What is the probability that you and your friend both choose a blue pen? *(Section 10.5)*

Essential Question How can you determine whether a sample accurately represents a population?

A **population** is an entire group of people or objects. A **sample** is a part of the population. You can use a sample to make an *inference*, or conclusion, about a population.

Identify a population.	Select a sample.	Interpret the data in the sample.	Make an inference about the population.

Population → Sample → Interpretation → Inference

1 ACTIVITY: Identifying Populations and Samples

Work with a partner. Identify the population and the sample.

a.

The students in a school   The students in a math class

b.

The grizzly bears with GPS collars in a park   The grizzly bears in a park

c.

150 quarters   All quarters in circulation

d.

All books in a library   10 fiction books in a library

Probability and Statistics

In this lesson, you will
- determine when samples are representative of populations.
- use data from random samples to make predictions about populations.

2 ACTIVITY: Identifying Random Samples

Work with a partner. When a sample is selected at random, each member of the population is equally likely to be selected. You want to know the favorite extracurricular activity of students at your school. Determine whether each method will result in a random sample. Explain your reasoning.

a. You ask members of the school band.

b. You publish a survey in the school newspaper.

c. You ask every eighth student who enters the school in the morning.

d. You ask students in your class.

Laurie's Notes

Introduction

Applying Mathematical Practices

- **Construct Viable Arguments and Critique the Reasoning of Others:** Mathematically proficient students understand and use stated assumptions and previously established results in constructing arguments. They reason about data and make plausible arguments that take into account the context from which the data arose.

Motivate

- Conduct a quick survey of your class. Ask a couple of fun questions and then ask a math related question.
  - How many of you can roll your tongue?
  - Who likes spicy brown mustard better than yellow mustard?
  - Can you simplify a fraction?
- Discuss each of these questions, who would ask the question, and why they might be asked. Point out the following:
  - Tongue rolling is probably the most commonly used classroom example of a simple genetic trait in humans.
  - This question doesn't allow a person who doesn't like mustard at all to answer.
  - Teachers survey students all the time to help guide instruction.

Discuss

- **Teacher Note:** In a previous course, students answered simple statistical questions about a data set. The flowchart shows one of the basic principles of statistical reasoning, that a result from a sample can be used to make a generalization about the population from which it was selected.
- Define population and sample. In the Motivate, the students in the school can be the *population* and the students in the math class are the *sample*.
- **Construct Viable Arguments and Critique the Reasoning of Others:** Students should be able to give examples of cases where a sample would be representative (eye color) and cases where it would not be representative of a population (favorite musical group/singer).

Activity Notes

Activity 1

- Discuss student responses.

Activity 2

- Remind students that a survey is not the only way you can collect data from a sample, but it is likely the most common method.
- ? "What are the characteristics of a random sample?" Answers will vary.
- ? "Why do think a sample is taken instead of trying to survey an entire population?" You may not have the time, energy, or resources to do so.
- **Construct Viable Arguments and Critique the Reasoning of Others:** When students are finished, ask for volunteers to offer their reasoning about whether the method will result in a random sample.

What Your Students Will Learn

- Determine when samples are representative of populations.
- Use data from random samples to make predictions about populations.

Previous Learning

Students should know how to write and solve a proportion and how to interpret a circle graph.

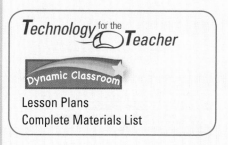

Lesson Plans
Complete Materials List

10.6 Record and Practice Journal

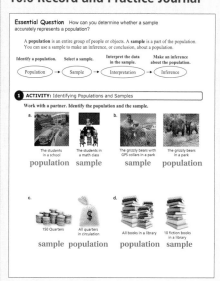

Visual

Ask students to discuss the pros and cons of showing survey results using visual representation. In Question 5, have students predict the types of visuals they would use to display and interpret their survey results.

10.6 Record and Practice Journal

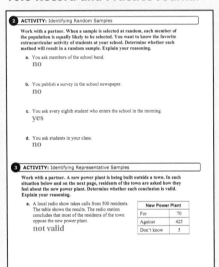

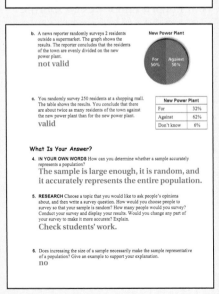

Laurie's Notes

Activity 3

• Give students time to discuss each scenario with their partners.

? "Can the results in part (a) be used to draw a valid conclusion? Explain." No; not everyone in the town may listen to the radio, only those who are really opposed to the new power plant may call the radio show, or it is possible that most of the people who call live close to the power plant.

? "Can the results in part (b) be used to draw a valid conclusion? Explain." No; the sample size is too small.

? "Can the results in part (c) be used to draw a valid conclusion? Explain." Yes; the sample size is large enough and randomly selected, and your conclusion is reasonable based on the results.

• **Reason Abstractly and Quantitatively** and **Construct Viable Arguments and Critique the Reasoning of Others:** Students should recognize that you can have a large sample or a random sample, and still not have one that accurately represents a population. Be sure students recognize the significance of the size of the sample. When the sample size is large enough, random sampling tends to produce representative samples that can be used to draw valid inferences.

What Is Your Answer?

• Ask volunteers to share their ideas for Question 5.

Closure

• **Exit Ticket:** What survey question would you ask to find out what vegetable should be served more often for the hot lunch program at your school? Answers will vary.

There are many different ways to select a sample from a population. To make valid inferences about a population, you must choose a random sample very carefully so that it accurately represents the population.

3 ACTIVITY: Identifying Representative Samples

Work with a partner. A new power plant is being built outside a town. In each situation below, residents of the town are asked how they feel about the new power plant. Determine whether each conclusion is valid. Explain your reasoning.

a. A local radio show takes calls from 500 residents. The table shows the results. The radio station concludes that most of the residents of the town oppose the new power plant.

New Power Plant	
For	70
Against	425
Don't know	5

New Power Plant

b. A news reporter randomly surveys 2 residents outside a supermarket. The graph shows the results. The reporter concludes that the residents of the town are evenly divided on the new power plant.

c. You randomly survey 250 residents at a shopping mall. The table shows the results. You conclude that there are about twice as many residents of the town against the new power plant than for the new power plant.

New Power Plant	
For	32%
Against	62%
Don't know	6%

Math Practice

Understand Quantities

Can the size of a sample affect the validity of a conclusion about a population?

What Is Your Answer?

4. **IN YOUR OWN WORDS** How can you determine whether a sample accurately represents a population?

5. **RESEARCH** Choose a topic that you would like to ask people's opinions about, and then write a survey question. How would you choose people to survey so that your sample is random? How many people would you survey? Conduct your survey and display your results. Would you change any part of your survey to make it more accurate? Explain.

6. Does increasing the size of a sample necessarily make the sample representative of a population? Give an example to support your explanation.

Practice

Use what you learned about populations and samples to complete Exercises 3 and 4 on page 444.

10.6 Lesson

Key Vocabulary))
population, *p. 440*
sample, *p. 440*
unbiased sample,
  p. 442
biased sample, *p. 442*

An **unbiased sample** is representative of a population. It is selected at random and is large enough to provide accurate data.

A **biased sample** is not representative of a population. One or more parts of the population are favored over others.

EXAMPLE **1** **Identifying an Unbiased Sample**

You want to estimate the number of students in a high school who ride the school bus. Which sample is unbiased?

(A) 4 students in the hallway

(B) all students in the marching band

(C) 50 seniors at random

(D) 100 students at random during lunch

Choice A is not large enough to provide accurate data.

Choice B is not selected at random.

Choice C is not representative of the population because seniors are more likely to drive to school than other students.

Choice D is representative of the population, selected at random, and large enough to provide accurate data.

So, the correct answer is (D).

On Your Own

Now You're Ready
Exercises 5–7

1. **WHAT IF?** You want to estimate the number of seniors in a high school who ride the school bus. Which sample is unbiased? Explain.

2. You want to estimate the number of eighth-grade students in your school who consider it relaxing to listen to music. You randomly survey 15 members of the band. Your friend surveys every fifth student whose name appears on an alphabetical list of eighth graders. Which sample is unbiased? Explain.

The results of an unbiased sample are proportional to the results of the population. So, you can use unbiased samples to make predictions about the population.

Biased samples are not representative of the population. So, you should not use them to make predictions about the population because the predictions may not be valid.

Laurie's Notes

Introduction

Connect

- **Yesterday:** Students identified samples and populations.
- **Today:** Students will identify biased and unbiased samples and determine whether a sample can be used to draw conclusions and make predictions about a population.

Motivate

- Tell students about something you read recently that reported, "Four out of five students who responded to the survey said they should have more homework!"
- After students quiet down, restate what you read, omitting a few key words. "Four out of five students said they should have more homework!"
- ❓ "How are the two claims different?" Students should recognize that you could survey 100 people, only 5 respond to the survey and 4 of the 5 answered one way. 95 people did not respond. In the other scenario, 80 of the 100 students answered one way.

Discuss

- Describe unbiased and biased samples. Give a few examples of each type.

Lesson Notes

Example 1

- Work through the problem and discuss why the first three samples are not reasonable.
- ❓ "What other samples might not be reasonable?" Answers will vary.

On Your Own

- **Neighbor Check:** Have students work independently and then have their neighbors check their work. Have students discuss any discrepancies.

Discuss

- ❓ "How do you think the results from an unbiased sample can be used to make predictions about a population?" Listen for students to suggest you can use a proportion.
- State that the results of an unbiased sample are proportional to the results of the population.

Lesson Tutorials
Lesson Plans
Answer Presentation Tool

Extra Example 1

You want to know the number of students in your classroom who do their homework right after school. You survey every third student who arrives in the classroom.

a. What is the population of your survey? your class

b. What is the sample of your survey? every third student who arrives

c. Is the sample unbiased? Explain. Yes. The sample is selected at random, representative of the population, and large enough to provide accurate data.

On Your Own

1. C

2. Your friend's sample is unbiased.

Extra Example 2

You want to know how the residents of your town feel about a ban on texting while driving. Determine whether each conclusion is valid.

a. You survey 200 residents at random. One hundred sixty-four residents support the ban and thirty-six do not. So, you conclude that 82% of the residents of your town support the ban. **The sample is unbiased and the conclusion is valid.**

b. You survey the first 15 residents who drive into your neighborhood. Five support the ban and ten do not.

So, you conclude that $33\frac{1}{3}\%$ of the residents of your town support the ban. **The sample is biased and the conclusion is not valid.**

Extra Example 3

You ask 50 randomly chosen students to name their favorite sport. There are 600 students in the school. Predict the number n of students in the school who would name soccer as their favorite sport. **144 students**

Favorite Sport	Number of Students
Football	8
Soccer	12
Basketball	20
Baseball	10

⬤ On Your Own

3. No, firefighters are more likely to support the new sign.

4. about 384 students

English Language Learners

Vocabulary

English learners may easily grasp the concept of a sample, because it is similar to receiving a small amount of a product. However, the word *population* may be confusing. Relate both words to a sample of food. The sample is what the student tastes, and the population is all of the food.

Laurie's Notes

Example 2

- **Construct Viable Arguments and Critique the Reasoning of Others:** Ask a volunteer to read part (a) and ask whether the conclusion is valid. Students should recognize that the sample is biased because the survey was not random—you only survey nearby residents.
- **Construct Viable Arguments and Critique the Reasoning of Others:** Ask a volunteer to read part (b) and ask whether the conclusion is valid. Students should recognize that the sample is random and large enough so it is an unbiased sample.
- **?** "If there are 2000 residents in the town, then how many would you expect to be in favor of the new sign?" 40% of 2000, or 800 residents

Example 3

- Read and discuss the information given.
- **?** "Why is this sample unbiased?" The sample is representative of the population, the students were selected at random, and the sample is large enough to provide accurate data.
- **?** "How many students were surveyed and how many of them watch one movie each week?" 75 were surveyed; 21 of them watch one movie each week
- Write a proportion to predict the number of students in the school who watch one movie each week.

On Your Own

- **Think-Pair-Share:** Students should read each question independently and then work in pairs to answer the questions. When they have answered the questions, the pair should compare their answers with another group and discuss any discrepancies.

⬤ Closure

- **Exit Ticket:** Describe an unbiased sample. Listen for students to mention the three qualities of an unbiased sample stated at the beginning of the lesson.

EXAMPLE 2 **Determining Whether Conclusions Are Valid**

You want to know how the residents of your town feel about adding a new stop sign. Determine whether each conclusion is valid.

a. You survey the 20 residents who live closest to the new sign. Fifteen support the sign, and five do not. So, you conclude that 75% of the residents of your town support the new sign.

The sample is not representative of the population because residents who live close to the sign are more likely to support it.

⋮⋗ So, the sample is biased, and the conclusion is not valid.

b. You survey 100 residents at random. Forty support the new sign, and sixty do not. So, you conclude that 40% of the residents of your town support the new sign.

The sample is representative of the population, selected at random, and large enough to provide accurate data.

⋮⋗ So, the sample is unbiased, and the conclusion is valid.

EXAMPLE 3 **Making Predictions**

Movies per Week

You ask 75 randomly chosen students how many movies they watch each week. There are 1200 students in the school. Predict the number n of students in the school who watch one movie each week.

The sample is representative of the population, selected at random, and large enough to provide accurate data. So, the sample is unbiased, and you can use it to make a prediction about the population.

Write and solve a proportion to find n.

Sample	Population
$\dfrac{\text{students in survey (one movie)}}{\text{number of students in survey}}$	$=\dfrac{\text{students in school (one movie)}}{\text{number of students in school}}$

$$\frac{21}{75} = \frac{n}{1200} \qquad \text{Substitute.}$$

$$336 = n \qquad \text{Solve for } n.$$

⋮⋗ So, about 336 students in the school watch one movie each week.

● **On Your Own**

Now You're Ready
Exercises 8, 9, and 12

3. In Example 2, each of 25 randomly chosen firefighters supports the new sign. So, you conclude that 100% of the residents of your town support the new sign. Is the conclusion valid? Explain.

4. In Example 3, predict the number of students in the school who watch two or more movies each week.

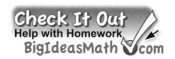
 Vocabulary and Concept Check

1. **VOCABULARY** Why would you survey a sample instead of a population?

2. **CRITICAL THINKING** What should you consider when conducting a survey?

 Practice and Problem Solving

Identify the population and the sample.

3.

Residents of New Jersey   Residents of Ocean County

4.

4 cards   All cards in a deck

Determine whether the sample is *biased* or *unbiased*. Explain.

① 5. You want to estimate the number of students in your school who play a musical instrument. You survey the first 15 students who arrive at a band class.

6. You want to estimate the number of books students in your school read over the summer. You survey every fourth student who enters the school.

7. You want to estimate the number of people in a town who think that a park needs to be remodeled. You survey every 10th person who enters the park.

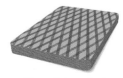

Determine whether the conclusion is valid. Explain.

② 8. You want to determine the number of students in your school who have visited a science museum. You survey 50 students at random. Twenty have visited a science museum, and thirty have not. So, you conclude that 40% of the students in your school have visited a science museum.

9. You want to know how the residents of your town feel about building a new baseball stadium. You randomly survey 100 people who enter the current stadium. Eighty support building a new stadium, and twenty do not. So, you conclude that 80% of the residents of your town support building a new baseball stadium.

Which sample is better for making a prediction? Explain.

10.

Predict the number of students in a school who like gym class.	
Sample A	A random sample of 8 students from the yearbook
Sample B	A random sample of 80 students from the yearbook

11.

Predict the number of defective pencils produced per day.	
Sample A	A random sample of 500 pencils from 20 machines
Sample B	A random sample of 500 pencils from 1 machine

Assignment Guide and Homework Check

Level	Day 1 Activity Assignment	Day 2 Lesson Assignment	Homework Check
Basic	3, 4, 20–24	1, 2, 5–17 odd	7, 9, 11, 13, 17
Average	3, 4, 20–24	1, 2, 5–11 odd, 12–18 even	7, 9, 11, 14, 18
Advanced	3, 4, 20–24	1, 2, 6–18 even, 19	8, 10, 12, 14, 18
Accelerated	1–4, 6–18 even, 19–24		8, 10, 12, 14, 18

Common Errors

- **Exercises 3 and 4** Students may get confused with the words population and sample. Encourage them to think about what it means to eat a sample of something, and then compare the whole to the population.
- **Exercise 5** Students may ignore the fact that the survey is conducted among students arriving at a band class. They may only see that it is a class. Ask them what kind of answers they would expect to get from different classes in the school (such as math), including band class. They should recognize that everyone going to band class will play an instrument, but not everyone in a math class will.
- **Exercises 13–15** Students may not understand why you would want to question the entire population for a survey. Ask them to estimate the population size for each survey, and then ask if it would be reasonable to ask everyone in that population for a response.

10.6 Record and Practice Journal

Determine whether the sample is *biased* or *unbiased*. Explain.

1. You want to estimate the number of students in your school who want a football stadium to be built. You survey the first 20 students who attend a Friday night football game.
biased

2. You want to estimate the number of students in your school who drive their own cars to school. You survey every 8th person who enters the cafeteria for lunch.
unbiased

Determine whether the conclusion is valid. Explain.

3. You want to determine the number of city residents who want to have 38th Street repaved. You randomly survey 15 residents who live on 38th Street. Twelve want the street to be repaved and three do not. So, you conclude that 80% of city residents want the street to be repaved.
no

4. You want to determine how many students consider math to be their favorite school subject. You randomly survey 75 students. Thirty-three students consider math to be their favorite subject and forty-two do not. So, you conclude that 40% of students at your school consider math to be their favorite subject.
yes

Vocabulary and Concept Check

1. Samples are easier to obtain.

2. You should make sure the people surveyed are selected at random and are representative of the population, as well as making sure your sample is large enough.

Practice and Problem Solving

3. Population: Residents of New Jersey
Sample: Residents of Ocean County

4. Population: All cards in a deck
Sample: 4 cards

5. biased; The sample is not selected at random and is not representative of the population because students in a band class play a musical instrument.

6. unbiased; The sample is representative of the population, selected at random, and large enough to provide accurate data.

7. biased; The sample is not representative of the population because people who go to a park are more likely to think that the park needs to be remodeled.

8. yes; The sample is representative of the population, selected at random, and large enough to provide accurate data. So, the sample is unbiased and the conclusion is valid.

9–10. See Additional Answers.

11. Sample A; it is representative of the population.

Practice and Problem Solving

12. 696 students

13. a sample; It is much easier to collect sample data in this situation.

14. a population; There are few enough students in your homeroom to not make the surveying difficult.

15–18. See Additional Answers.

19. See *Taking Math Deeper.*

Fair Game Review

20. 30%   **21.** 140

22. 200   **23.** 3

24. A

Mini-Assessment

1. You want to know the number of students in your school who play a sport. You survey the first 10 students who arrive for lunch.

  a. What is the population of your survey? All students in your school the sample? First 10 students who arrive for lunch

  b. Is the sample *biased* or *unbiased*? Explain. biased; The sample is not large enough to provide accurate data or may not be representative of the population.

2. You ask 120 randomly chosen people at a stadium to name their favorite stadium food. There are about 50,000 people in the stadium. Predict the number of people in the stadium whose favorite stadium food is nachos. about 10,000 people

Favorite Stadium Food	
Nachos	24
Hot Dog	55
Peanuts	16
Popcorn	25

Taking Math Deeper

Exercise 19

This problem may look straightforward. In reality, it is filled with questions that encompass the nature of statistical sampling.

 Straightforward Approach: 75% of the students in the sample said that they plan to go to college. Because 75% of 900 is 675, you can predict that 675 students in the high school plan to go to college.

Here, the guidance counselor assumes that the students who answered "Maybe" plan to attend college. So, because 75% + 5% = 80% and 80% of 900 is 720, the counselor predicts that 720 students in the high school plan to go to college.

Perhaps a better way to predict the number of students is to use the range and say, "675 to 720 students in the school plan to go to college."

75% is 675 and 80% is 720.

 Is the sample large enough to make an accurate prediction?

One of the most surprising results in statistics is that relatively small sample sizes can produce accurate results. *If the sample of 60 students was random, then it is a large enough sample to provide results that are accurate.*

 Once you address the issues of sample size and randomness, there are still many other things to consider. Here are some of them:

- How were the students asked the question? Were they asked in such a way that they felt free to tell the truth?
- Do students in 9th grade really know whether they plan to go to college or not?
- What time of year was the survey taken? If it was in the spring, then the juniors and seniors should have a reasonable idea of whether they are going to college or not.

Project

Select a circle graph from the newspaper, a magazine, or online. Explain the graph. Explain another way that the data could be displayed.

Reteaching and Enrichment Strategies

If students need help. . .	If students got it. . .
Resources by Chapter • Practice A and Practice B • Puzzle Time Record and Practice Journal Practice Differentiating the Lesson Lesson Tutorials Skills Review Handbook	Resources by Chapter • Enrichment and Extension • Technology Connection Start the next section

③ **12. FOOD** You ask 125 randomly chosen students to name their favorite food. There are 1500 students in the school. Predict the number of students in the school whose favorite food is pizza.

Favorite Food	
Pizza	58
Hamburger	36
Pasta	14
Other	17

Determine whether you would survey the population or a sample. Explain.

13. You want to know the average height of seventh graders in the United States.

14. You want to know the favorite types of music of students in your homeroom.

15. You want to know the number of students in your state who have summer jobs.

Theater Ticket Sales	
Adults	**Students**
522	210

16. THEATER You su0.rvey 72 randomly chosen students about whether they are going to attend the school play. Twelve say yes. Predict the number of students who attend the school.

17. CRITICAL THINKING Explain why 200 people with email addresses may not be a random sample. When might it be a random sample?

18. LOGIC A person surveys residents of a town to determine whether a skateboarding ban should be overturned.

    a. Describe how the person could conduct the survey so that the sample is biased toward overturning the ban.

    b. Describe how the person could conduct the survey so that the sample is biased toward keeping the ban.

19. ⚡Reasoning⚡ A guidance counselor surveys a random sample of 60 out of 900 high school students. Using the survey results, the counselor predicts that approximately 720 students plan to attend college. Do you agree with her prediction? Explain.

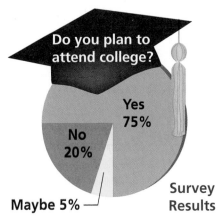

Do you plan to attend college?

Yes 75%

No 20%

Maybe 5%

Survey Results

 Fair Game Review *What you learned in previous grades & lessons*

Write and solve a proportion to answer the question. *(Section 6.3)*

20. What percent of 60 is 18?

21. 70% of what number is 98?

22. 30 is 15% of what number?

23. What number is 0.6% of 500?

24. MULTIPLE CHOICE What is the volume of the pyramid? *(Section 9.5)*

    Ⓐ 40 cm³   Ⓑ 50 cm³

    Ⓒ 100 cm³   Ⓓ 120 cm³

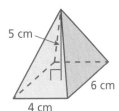

5 cm

6 cm

4 cm

Check It Out
Lesson Tutorials
BigIdeasMath.com

You have already used unbiased samples to make inferences about a population. In some cases, making an inference about a population from only one sample is not as precise as using multiple samples.

1 ACTIVITY: Using Multiple Random Samples

Work with a partner. You and a group of friends want to know how many students in your school listen to pop music. There are 840 students in your school. Each person in the group randomly surveys 20 students.

Step 1: The table shows your results. Make an inference about the number of students in your school who prefer pop music.

Favorite Type of Music			
Country	Pop	Rock	Rap
4	10	5	1

Step 2: The table shows Kevin's results. Use these results to make another inference about the number of students in your school who prefer pop music.

Compare the results of Steps 1 and 2.

Favorite Type of Music			
Country	Pop	Rock	Rap
2	13	4	1

Probability and Statistics
In this extension, you will
• use multiple samples to make predictions about populations.

Step 3: The table shows the results of three other friends. Use these results to make three more inferences about the number of students in your school who prefer pop music.

	Favorite Type of Music			
	Country	Pop	Rock	Rap
Steve	3	8	7	2
Laura	5	10	4	1
Ming	5	9	3	3

Step 4: Describe the variation of the five inferences. Which one would you use to describe the number of students in your school who prefer pop music? Explain your reasoning.

Step 5: Show how you can use all five samples to make an inference.

Practice

1. **PACKING PEANUTS** Work with a partner. Mark 24 packing peanuts with either a red or a black marker. Put the peanuts into a paper bag. Trade bags with other students in the class.

   a. Generate a sample by choosing a peanut from your bag six times, replacing the peanut each time. Record the number of times you choose each color. Repeat this process to generate four more samples. Organize the results in a table.

   b. Use each sample to make an inference about the number of red peanuts in the bag. Then describe the variation of the five inferences. Make inferences about the numbers of red and black peanuts in the bag based on all the samples.

   c. Take the peanuts out of the bag. How do your inferences compare to the population? Do you think you can make a more accurate prediction? If so, explain how.

Laurie's Notes

Introduction

Connect

- **Yesterday:** Students identified biased and unbiased samples and determined whether a sample can be used to draw conclusions and make predictions about a population.
- **Today:** Students will generate multiple samples of data and draw inferences about a population.

Motivate

- ❓ "Are you familiar with different groups that poll large groups of people?" Students may have heard of the Gallup Poll or Rasmussen Reports.
- ❓ "Have any of you or has someone you know ever been asked to participate in a survey, perhaps at the mall or on the telephone?" Answers will vary.
- Explain that in these activities, multiple samples are compared in order to make an inference about a population.

Activity Notes

Activity 1

- Read the introduction and context for the activity. Make sure students understand that more than one random sample is being taken of the 840 students in a school.
- ❓ "How can an unbiased sample be used to make an inference about a population?" Students should describe how a ratio table or proportion can be used to make predictions about the population based upon the unbiased sample.
- Give sufficient time for students to work through the steps.
- There are results from five different random samples. When students have finished Step 3, you could gather the results before students move on to Steps 4 and 5.
- **Construct Viable Arguments and Critique the Reasoning of Others:** Students may have different ways in which they arrive at their conclusions when describing the music preferences of the school. Ask volunteers to explain their reasoning.

Practice

- Tell students to vary the amount of peanuts they mark with red and black markers. It does not have to be 50-50. They can mark as many as they want but you can put a limit on the least amount for each color. So, a student cannot mark only 1 peanut black or all the peanuts red.

What Your Students Will Learn

- Use measures from multiple random samples and simulations to make predictions about populations.

Goal Today's lesson is generating multiple samples of data and drawing inferences about a population.

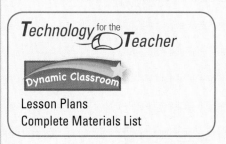

Technology for the Teacher

Dynamic Classroom

Lesson Plans
Complete Materials List

Extension 10.6
Record and Practice Journal

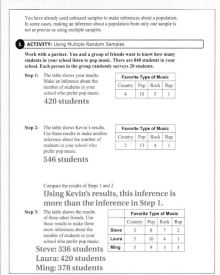

You have already used unbiased samples to make inferences about a population. In some cases, making an inference about a population from only one sample is not as precise as using multiple samples.

1 ACTIVITY: Using Multiple Random Samples

Work with a partner. You and a group of friends want to know how many students in your school listen to pop music. There are 840 students in your school. Each person in the group randomly surveys 20 students.

Step 1: The table shows your results. Make an inference about the number of students in your school who prefer pop music.

Favorite Type of Music			
Country	Pop	Rock	Rap
4	10	5	1

420 students

Step 2: The table shows Kevin's results. Use these results to make another inference about the number of students in your school who prefer pop music.

Favorite Type of Music			
Country	Pop	Rock	Rap
2	13	4	1

546 students

Compare the results of Steps 1 and 2.
Using Kevin's results, this inference is more than the inference in Step 1.

Step 3: The table shows the results of three other friends. Use these results to make three more inferences about the number of students in your school who prefer pop music.

	Country	Pop	Rock	Rap
	Favorite Type of Music			
Steve	3	8	7	2
Laura	5	10	4	1
Ming	5	9	3	3

Steve: 336 students
Laura: 420 students
Ming: 378 students

Extension 10.6
Record and Practice Journal

Step 4: Describe the variation of the five inferences. Which one would you use to describe the number of students in your school who prefer pop music? Explain your reasoning.

Sample answer: The greatest is 546 students. The least is 336 students. 420 is the median and the mode of the data. So, use the inference of 420 students.

Step 5: Show how you can use all five samples to make an inference.
Combine all the samples. 50 students out of the 100 students chose pop music.
Use the proportion $\frac{50}{100} = \frac{x}{840}$ to predict 420 students prefer pop music.

Practice

1. Work with a partner. Mark 24 packing peanuts with either a red or a black marker. Put the peanuts into a paper bag. Trade bags with other students in the class.

   a. Generate a sample by choosing a peanut from your bag six times, replacing the peanut each time. Record the number of times you choose each color. Repeat this process to generate four more samples. Organize the results in a table.
   Check students' work.

   b. Use each sample to make an inference about the number of red peanuts in the bag. Then describe the variation of the five inferences. Make inferences about the numbers of red and black peanuts in the bag based on all the samples.
   Check students' work.

   c. Take the peanuts out of the bag. How do your inferences compare to the population? Do you think you can make a more accurate prediction? If so, explain how.
   Check students' work; Yes, increase the number of random samples.

---

2 ACTIVITY: Using Measures from Multiple Random Samples

Work with a partner. You want to know the mean number of hours students with part-time jobs work each week. You go to 8 different schools. At each school, you randomly survey 10 students with part-time jobs. Your results are shown at the right.

Hours Worked Each Week
1: 6, 8, 6, 6, 7, 4, 10, 8, 7, 8
2: 10, 4, 4, 6, 8, 6, 7, 12, 8, 8
3: 10, 9, 8, 6, 5, 8, 6, 6, 9, 10
4: 4, 8, 4, 4, 5, 4, 4, 6, 5, 6
5: 6, 8, 8, 6, 12, 4, 10, 8, 6, 12
6: 10, 10, 8, 9, 16, 8, 7, 12, 16, 14
7: 4, 5, 6, 6, 4, 5, 6, 6, 4, 4
8: 16, 20, 8, 12, 10, 8, 8, 14, 16, 8

Step 1: Find the mean of each sample.
7, 7.3, 7.7, 5, 8, 11, 5, 12

Step 2: Make a box-and-whisker plot of the sample means.

Mean hours worked each week
5 6 7 8 9 10 11 12

Step 3: Use the box-and-whisker plot to estimate the actual mean number of hours students with part-time jobs work each week. How does your estimate compare to the mean of the entire data set?
Sample answer: 7.5 hours

---

3 ACTIVITY: Using a Simulation

Work with a partner. Another way to generate multiple samples of data is to use a simulation. Suppose 70% of all seventh graders watch reality shows on television.

Step 1: Design a simulation involving 50 packing peanuts by marking 70% of the peanuts with a certain color. Put the peanuts into a paper bag.
Check students' work.

Step 2: Simulate choosing a sample of 30 students by choosing peanuts from the bag, replacing the peanut each time. Record the results. Repeat this process to generate eight more samples. How much variation do you expect among the samples? Explain.

Sample answer: Because the actual percent of students is 70%, expect the number of red peanuts in their sample to be between 60%–80%, or 55%–85%, or 65%–75%, etc.

Step 3: Display your results.
Check students' work.

Practice

2. You want to know whether student-athletes prefer water or sports drinks during games. You go to 10 different schools. At each school, you randomly survey 10 student-athletes. The percents of student-athletes who prefer water are shown.

   60% 70% 60% 50% 80% 70% 30% 70% 80% 40%

   a. Make a box-and-whisker plot of the data.

   Percent that prefer water
   30 40 50 60 70 80

   b. Use the box-and-whisker plot to estimate the actual percent of student-athletes who prefer water. How does your estimate compare to the mean of the data?
   Sample answer: 60%

3. Repeat Activity 2 using the medians of the samples. 7, 7.5, 8, 4.5, 8, 10, 5, 11

   Median hours worked each week *Sample answer:* 7.5
   4.5 6 7.75 9 11
   4 5 6 7 8 9 10 11

4. In Activity 3, how do the percents in your samples compare to the actual percent of seventh graders who watch reality shows on television?
   Check students' work.

5. **REASONING** Why is it better to make inferences about a population based on multiple samples instead of only one sample? What additional information do you gain by taking multiple random samples? Explain.
   The more samples you have, the more accurate your inferences will be.

Activity 2

- Discuss with students the description of the samples. The students all have part-time jobs and are from 8 different schools.
- **Use Appropriate Tools Strategically:** Students can use calculators to quickly find the means.
- The box-and-whisker plot is NOT using all of the data. Only the mean of each sample (from 8 different schools) is used to construct the box-and-whisker plot in Step 2.
- **?** "Can you use your box-and-whisker plot to estimate the actual mean number of hours students work each week? Explain." yes; *Sample answer:* It probably lies somewhere within the "box."
- **?** "How do you think your estimate compares to the mean of the entire data set?" Students should realize that it is a good estimate.

Activity 3

- This is a fun activity for students, if time permits. Make sure you have enough time to complete this simulation.
- The materials do not have to be peanuts. Any congruent shapes that differ in color will work, such as colored tiles.
- In Step 2, there is no wrong answer here. Students are just recognizing variability in their samples and determining a range where they think their percents will fall. They will compare in Question 4.
- Students can use percents or the actual numbers of red peanuts.
- Allow time for different groups to be able to share their method, and their results.

Practice

- Have students discuss Question 5 with their partners and then share their thoughts with the class.

Closure

- Describe how you can generate multiple samples using students in your school to determine their preference for an end-of-the-year field trip from four possible locations.

Hours Worked Each Week
1: 6, 8, 6, 6, 7, 4, 10, 8, 7, 8
2: 10, 4, 4, 6, 8, 6, 7, 12, 8, 8
3: 10, 9, 8, 6, 5, 8, 6, 6, 9, 10
4: 4, 8, 4, 4, 5, 4, 4, 6, 5, 6
5: 6, 8, 8, 6, 12, 4, 10, 8, 6, 12
6: 10, 10, 8, 9, 16, 8, 7, 12, 16, 14
7: 4, 5, 6, 6, 4, 5, 6, 6, 4, 4
8: 16, 20, 8, 12, 10, 8, 8, 14, 16, 8

Work with a partner. You want to know the mean number of hours students with part-time jobs work each week. You go to 8 different schools. At each school, you randomly survey 10 students with part-time jobs. Your results are shown at the left.

Step 1: Find the mean of each sample.

Step 2: Make a box-and-whisker plot of the sample means.

Step 3: Use the box-and-whisker plot to estimate the actual mean number of hours students with part-time jobs work each week.

How does your estimate compare to the mean of the entire data set?

3 **ACTIVITY: Using a Simulation**

Work with a partner. Another way to generate multiple samples of data is to use a simulation. Suppose 70% of all seventh graders watch reality shows on television.

Step 1: Design a simulation involving 50 packing peanuts by marking 70% of the peanuts with a certain color. Put the peanuts into a paper bag.

Step 2: Simulate choosing a sample of 30 students by choosing peanuts from the bag, replacing the peanut each time. Record the results. Repeat this process to generate eight more samples. How much variation do you expect among the samples? Explain.

Step 3: Display your results.

● Practice

2. **SPORTS DRINKS** You want to know whether student-athletes prefer water or sports drinks during games. You go to 10 different schools. At each school, you randomly survey 10 student-athletes. The percents of student-athletes who prefer water are shown.

   60%   70%   60%   50%   80%   70%   30%   70%   80%   40%

   a. Make a box-and-whisker plot of the data.

   b. Use the box-and-whisker plot to estimate the actual percent of student-athletes who prefer water. How does your estimate compare to the mean of the data?

3. **PART-TIME JOBS** Repeat Activity 2 using the medians of the samples.

4. **TELEVISION** In Activity 3, how do the percents in your samples compare to the given percent of seventh graders who watch reality shows on television?

5. **REASONING** Why is it better to make inferences about a population based on multiple samples instead of only one sample? What additional information do you gain by taking multiple random samples? Explain.

Essential Question How can you compare data sets that represent two populations?

1 ACTIVITY: Comparing Two Data Distributions

Work with a partner. You want to compare the shoe sizes of male students in two classes. You collect the data shown in the table.

Male Students in Eighth-Grade Class													
7	9	8	$7\frac{1}{2}$	$8\frac{1}{2}$	10	6	$6\frac{1}{2}$	8	8	$8\frac{1}{2}$	9	11	$7\frac{1}{2}$	$8\frac{1}{2}$

Male Students in Sixth-Grade Class													
6	$5\frac{1}{2}$	6	$6\frac{1}{2}$	$7\frac{1}{2}$	$8\frac{1}{2}$	7	$5\frac{1}{2}$	5	$5\frac{1}{2}$	$6\frac{1}{2}$	7	$4\frac{1}{2}$	6	6

a. How can you display both data sets so that you can visually compare the measures of center and variation? Make the data display you chose.

b. Describe the shape of each distribution.

c. Complete the table.

	Mean	Median	Mode	Range	Interquartile Range (IQR)	Mean Absolute Deviation (MAD)
Male Students in Eighth-Grade Class						
Male Students in Sixth-Grade Class						

d. Compare the measures of center for the data sets.

e. Compare the measures of variation for the data sets. Does one data set show more variation than the other? Explain.

f. Do the distributions overlap? How can you tell using the data display you chose in part (a)?

g. The double box-and-whisker plot below shows the shoe sizes of the members of two girls basketball teams. Can you conclude that at least one girl from each team has the same shoe size? Can you conclude that at least one girl from the Bobcats has a larger shoe size than one of the girls from the Tigers? Explain your reasoning.

Probability and Statistics

In this lesson, you will

- use measures of center and variation to compare populations.
- use random samples to compare populations.

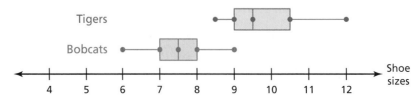

Laurie's Notes

What Your Students Will Learn
- Use measures of center, variations, and random samples to compare populations.

Previous Learning
Students have found measures of center and measures of variation.

Introduction

Applying Mathematical Practices
- **Reason Abstractly and Quantitatively:** Mathematically proficient students make sense of quantities and their relationships in problem situations. The overlap between two data sets can be compared visually using various data displays.

Motivate
- Display the double box-and-whisker plot showing the electricity produced (in kilowatt-hours, kWh) each day in July from solar panels on two houses.

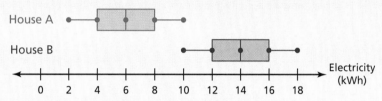

- Ask questions comparing the box-and-whisker plots.
- **?** "Were there days in which House A generated more electricity than House B?" no
- **?** "Could there have been days in which both houses generated the same amount of electricity?" yes

Activity Notes

Activity 1
- **FYI:** This first activity reviews statistical measures that students learned in a previous course as well as shapes of distributions. Students may need more guidance or probing questions for this activity than usual.
- Students should work with their partners to complete the activity.
- **?** "If we were to collect shoe size data from the boys in this class, what would the data look like?" Answers vary, but students should mention whole and half sizes.
- **?** "What data displays can you use to visually compare two data sets?" Students may say stem-and-leaf plots, box-and-whisker plots, or dot plots.
- Students may need to be reminded about measures of center, measures of variation, and shapes of distributions. You may need to review how to find the IQR and the MAD. This was all taught in a previous course.
- Discuss the measures of center and variation for the data sets.
- **?** "Do the data sets represent samples or populations?" Because you are comparing shoe sizes of male students in two classes, not your grade or your school, these represent populations.
- If desired, have students draw dotted lines vertically through the measures of center as one way to visualize overlap.
- **Reason Abstractly and Quantitatively:** In Activity 1(g), students should conclude that just because two data sets overlap slightly doesn't necessarily mean they contain one or more of the same data values.

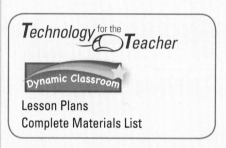

Technology for the Teacher

Dynamic Classroom

Lesson Plans
Complete Materials List

10.7 Record and Practice Journal

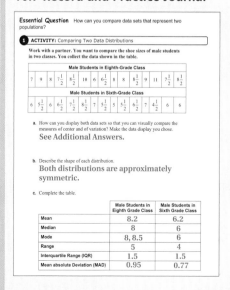

Visual

Use a diagram of a generic box-and-whisker plot on an overhead as a visual aid for English learners. Have students identify the parts of the box-and-whisker plot: *median, first quartile, third quartile, least value,* and *greatest value.* Make sure students understand that they can interpret a box-and-whisker plot that does not have a scale.

10.7 Record and Practice Journal

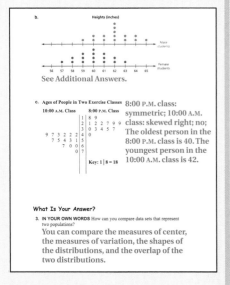

d. Compare the measures of center for the data sets.
The mean and median for the eighth-grade class is 2 more than the mean and median for the sixth-grade class.

e. Compare the measures of variation for the data sets. Does one data set show more variation than the other? Explain.
yes

f. Do the distributions overlap? How can you tell using the data display you chose in part (a)?
yes

g. The double box-and-whisker plot below shows the shoe sizes of the members of two girls basketball teams. Can you conclude that at least one girl from each team has the same shoe size? Can you conclude that at least one girl from the Bobcats has a larger shoe size than one of the girls from the Tigers? Explain your reasoning. no; yes

2 ACTIVITY: Comparing Two Data Distributions

Work with a partner. Compare the shapes of the distributions. Do the two data sets overlap? Explain. If so, use measures of center and the least and the greatest values to describe the overlap between the two data sets.

a.

See Additional Answers.

b.

See Additional Answers.

c. Ages of People in Two Exercise Classes

8:00 P.M. class: symmetric; 10:00 A.M. class: skewed right; no; The oldest person in the 8:00 P.M. class is 40. The youngest person in the 10:00 A.M. class is 42.

Key: 1 | 8 = 18

What Is Your Answer?

3. IN YOUR OWN WORDS How can you compare data sets that represent two populations?
You can compare the measures of center, the measures of variation, the shapes of the distributions, and the overlap of the two distributions.

Laurie's Notes

Activity 2

- Once students have discussed Activity 1 thoroughly, Activity 2 should take much less time.
- This activity is a good review of the three data displays shown.
- **Reason Abstractly and Quantitatively** and **Use Appropriate Tools Strategically:** The analysis of each plot in this activity will give good insights into student understanding of each type of data display.
- If time permits, ask students to summarize what is known about each data set in each of the three parts.

? "Do you know how many students are represented in the box-and-whisker plots?" no

? "Do you know how many students are represented in the dot plots?" yes

? "Do you know how many people are represented in the stem-and-leaf plots?" yes

What Is Your Answer?

- Ask volunteers to share their answers.

Closure

- Refer back to the box-and-whisker plots in the Motivate. Write a summary that compares the data sets.

2 ACTIVITY: Comparing Two Data Distributions

Work with a partner. Compare the shapes of the distributions. Do the two data sets overlap? Explain. If so, use measures of center and the least and the greatest values to describe the overlap between the two data sets.

a.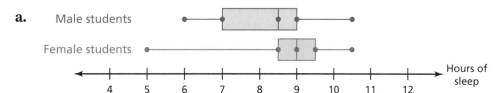

b.

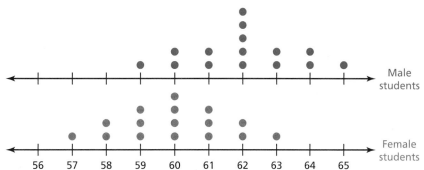

Math Practice

Recognize Usefulness of Tools

How is each type of data display useful? Which do you prefer? Explain.

c. **Ages of People in Two Exercise Classes**

10:00 A.M. Class		8:00 P.M. Class
	1	8 9
	2	1 2 2 7 9 9
	3	0 3 4 5 7
9 7 3 2 2 2	4	0
7 5 4 3 1	5	
7 0 0	6	
0	7	

Key: 1 | 8 = 18

What Is Your Answer?

3. **IN YOUR OWN WORDS** How can you compare data sets that represent two populations?

Practice Use what you learned about comparing data sets to complete Exercise 3 on page 452.

Section 10.7   Comparing Populations   **449**

Recall that you use the mean and the mean absolute deviation (MAD) to describe symmetric distributions of data. You use the median and the interquartile range (IQR) to describe skewed distributions of data.

To compare two populations, use the mean and the MAD when both distributions are symmetric. Use the median and the IQR when either one or both distributions are skewed.

EXAMPLE 1 **Comparing Populations**

The double dot plot shows the time that each candidate in a debate spent answering each of 15 questions.

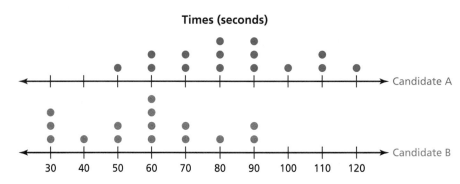

> **Study Tip**
>
> You can more easily see the visual overlap of dot plots that are aligned vertically.

a. Compare the populations using measures of center and variation.

Both distributions are approximately symmetric, so use the mean and the MAD.

Candidate A

$$\text{Mean} = \frac{1260}{15} = 84$$

$$\text{MAD} = \frac{244}{15} \approx 16$$

Candidate B

$$\text{Mean} = \frac{870}{15} = 58$$

$$\text{MAD} = \frac{236}{15} \approx 16$$

> **Study Tip**
>
> When two populations have similar variabilities, the value in part (b) describes the visual overlap between the data. In general, the greater the value, the less the overlap.

∴ So, the variation in the times was about the same, but Candidate A had a greater mean time.

b. Express the difference in the measures of center as a multiple of the measure of variation.

$$\frac{\text{mean for Candidate A} - \text{mean for Candidate B}}{\text{MAD}} = \frac{26}{16} \approx 1.6$$

∴ So, the difference in the means is about 1.6 times the MAD.

On Your Own

1. **WHAT IF?** Each value in the dot plot for Candidate A increases by 30 seconds. How does this affect the answers in Example 1? Explain.

Laurie's Notes

Introduction

Connect

- **Yesterday:** Students explored overlap between data sets.
- **Today:** Students will compare two populations using measures of center, measures of variation, and overlap.

Motivate

- Recall that in a previous course you used the mean and the MAD to describe symmetric distributions of data. You used the median and the IQR to describe skewed distributions of data.
- Tell students that when comparing two populations, they will use the mean and MAD when *both* distributions are symmetric. If either one or both of the distributions are skewed, they will use the median and IQR.

Lesson Notes

Example 1

- Refer to the double dot plot and ask students to describe what the dots represent. They should explain that each dot represents the number of seconds a candidate spent answering a question in a debate.
- **?** "How can you describe the distributions shown in the display?" Candidate A's responses were from 50 seconds to 120 seconds and the distribution is approximately symmetric. Candidate B's responses were from 30 seconds to 90 seconds and the distribution is approximately symmetric.
- **?** "How do you describe the centers and variation of symmetric distributions?" using the mean and the MAD
- **Use Appropriate Tools Strategically:** Students should use a calculator to quickly compute the mean and the MAD. Split the class with each half computing one of the two means and the associated MAD.
- **Attend to Precision:** Ask students to interpret the results in part (a) and to reference the visual display. The dot plots show two symmetric distributions with similar variabilities but different centers. Candidate A's graph is shifted to the right, meaning longer responses.
- **?** "Do the dot plots overlap? If so, do you think we can *measure* the overlap?" yes; Students may think it can be measured but are unsure what it means.
- **Make Sense of Problems and Persevere in Solving Them:** In part (b), students need to find the difference in the means as a multiple of the MAD, so they must divide by the MAD. Make sure students understand the meaning of this number, as described in the Study Tip.
- Work to finish the problem and interpret the result.

On Your Own

- **Big Idea:** Increasing each data value by 30 seconds affects the mean but not the variability. The difference in the measures of center will be greater and the visual overlap will be much less. The variabilities are still similar, so the value in part (b) will increase, indicating less overlap.

Differentiated Instruction

Vocabulary

Key words in this lesson were introduced in a previous grade. Have students add the words and acronyms, *mean, mean absolute deviation (MAD), symmetric distribution, median, interquartile range (IQR),* and *skewed distribution,* to their math notebook glossaries.

Extra Example 1

The data sets below give the times (in seconds) that the candidates in Example 1 spent answering each of 15 questions in a second debate.

Candidate A: 40, 50, 50, 50, 50, 50, 60, 60, 60, 70, 70, 70, 80, 80, 90

Candidate B: 40, 50, 60, 70, 70, 70, 80, 80, 80, 80, 90, 90, 90, 90, 90

a. Compare the populations using measures of center and variation. same variation; Candidate B had a greater median time.

b. Express the difference in the measures of center as a multiple of the measure of variation. difference in medians is 1 times the IQR

On Your Own

1. Part (a): Candidate A's mean increases by 30 to 114 and the MAD does not change, so Candidate A still has a greater mean time.

   Part (b): The difference in the means is now 3.5 times the MAD. The number is greater, indicating less overlap in the data.

Extra Example 2

You want to compare the costs of speeding tickets in State C to the costs in States A and B in Example 2.

a. The box-and-whisker plot shows a random sample of 10 speeding tickets in State C. Compare the samples using measures of center and variation. Can you use this to make a valid comparison about speeding tickets in the three states? Explain.

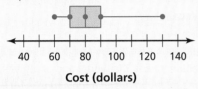

Cost (dollars)

The median, 80, is the same as State B but greater than State A; The IQR, 20, is the same as State A but less than State B; No, the sample size is too small and the variability too great.

b. The box-and-whisker plot shows the medians of 100 random samples of 10 speeding tickets in State C. Compare the variability of the sample medians to the variability of the sample costs in part (a).

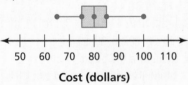

Cost (dollars)

Sample medians vary much less than the sample costs.

c. Make a conclusion about the costs of speeding tickets in the three states. Speeding tickets generally cost less in State A than in States B and C, where the costs are about the same.

Discuss

- Explain that you do not need to have all of the data from two populations to make comparisons. You can use random samples to make comparisons.

Example 2

- Ask a volunteer to read part (a).
- **?** "What do you notice about the distributions?" They are both skewed right.
- **?** "How do you describe the centers and variation of skewed distributions?" using the median and the IQR
- Give students time to find the median and the IQR.
- **Attend to Precision:** Discuss part (a). The samples have different medians and different IQRs. The median and IQR for State A are each less than the median and IQR for State B. However, the sample size is too small and the variability is too great to make comparisons about the populations.
- In part (b), each box-and-whisker plot represents the medians of 100 random samples of 10 speeding tickets. The sample size is no longer small.
- Make sure students see the distinction between parts (a) and (b). In part (a), one random sample is taken (in each state), and the individual speeding ticket costs are used to make the box-and-whisker plot. In part (b), 100 random samples are taken, and the 100 *medians* of the samples are used to make the box-and-whisker plot.
- If students are confused about what this means, refer to part (a). The median, 70, for State A represents one of the 100 values for State A in part (b). A similar statement can be made for State B.
- **?** "What can you say about the variation of the sample medians compared to the variation of the sample costs?" The sample medians vary much less than the sample costs.
- **Construct Viable Arguments** and **Attend to Precision:** Students should conclude from parts (a) and (b) that it is reasonable to assume that speeding tickets generally cost more in State B than in State A.

On Your Own

- **Neighbor Check:** Have students work independently and then have their neighbors check their work. Have students discuss any discrepancies.

Closure

- **Writing Prompt:** To compare two populations . . .

On Your Own

2. No, the sample size is too small to make a conclusion about the population.

You do not need to have all the data from two populations to make comparisons. You can use random samples to make comparisons.

EXAMPLE ② **Using Random Samples to Compare Populations**

You want to compare the costs of speeding tickets in two states.

a. **The double box-and-whisker plot shows a random sample of 10 speeding tickets issued in two states. Compare the samples using measures of center and variation. Can you use this to make a valid comparison about speeding tickets in the two states? Explain.**

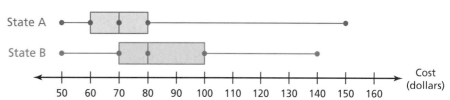

Both distributions are skewed right, so use the median and the IQR.

⁖ The median and the IQR for State A, 70 and 20, are less than the median and the IQR for State B, 80 and 30. However, the sample size is too small and the variability is too great to conclude that speeding tickets generally cost more in State B.

b. **The double box-and-whisker plot shows the medians of 100 random samples of 10 speeding tickets for each state. Compare the variability of the sample medians to the variability of the sample costs in part (a).**

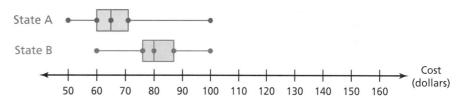

The IQR of the sample medians for each state is about 10.

⁖ So, the sample medians vary much less than the sample costs.

c. **Make a conclusion about the costs of speeding tickets in the two states.**

The sample medians show less variability. Most of the sample medians for State B are greater than the sample medians for State A.

⁖ So, speeding tickets generally cost more in State B than in State A.

● **On Your Own**

Exercise 8

2. WHAT IF? A random sample of 8 speeding tickets issued in State C has a median of $120. Can you conclude that a speeding ticket in State C costs more than in States A and B? Explain.

Vocabulary and Concept Check

1. **REASONING** When comparing two populations, when should you use the mean and the MAD? the median and the IQR?

2. **WRITING** Two data sets have similar variabilities. Suppose the measures of center of the data sets differ by 4 times the measure of variation. Describe the visual overlap of the data.

Practice and Problem Solving

3. **SNAKES** The tables show the lengths of two types of snakes at an animal store.

Garter Snake Lengths (inches)					
26	30	22	15	21	24
28	32	24	25	18	35

Water Snake Lengths (inches)					
34	25	24	35	40	32
41	27	37	32	21	30

   a. Find the mean, median, mode, range, interquartile range, and mean absolute deviation for each data set.

   b. Compare the data sets.

4. **HOCKEY** The double box-and-whisker plot shows the goals scored per game by two hockey teams during a 20-game season.

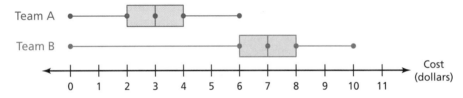

   a. Compare the populations using measures of center and variation.

   b. Express the difference in the measures of center as a multiple of the measure of variation.

5. **TEST SCORES** The dot plots show the test scores for two classes taught by the same teacher.

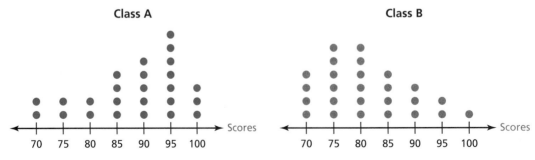

   a. Compare the populations using measures of center and variation.

   b. Express the difference in the measures of center as a multiple of each measure of variation.

Assignment Guide and Homework Check

Level	Day 1 Activity Assignment	Day 2 Lesson Assignment	Homework Check
Basic	3, 10–14	1, 2, 4–6	1, 4, 5, 6
Average	3, 10–14	1, 2, 4–8	1, 4, 5, 6, 7
Advanced	3, 10–14	1, 2, 4–9	1, 5, 6, 7, 8
Accelerated	1–14		1, 5, 6, 7, 8

Common Errors

- **Exercises 4–6** Students may use the wrong measures of center and variation when comparing populations. Remind them to use the mean and the MAD when *both* distributions are symmetric. Otherwise, use the median and the IQR.

10.7 Record and Practice Journal

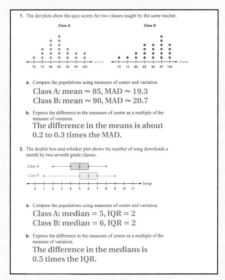

Vocabulary and Concept Check

1. When comparing two populations, use the mean and the MAD when each distribution is symmetric. Use the median and the IQR when either one or both distributions are skewed.

2. There will probably be little or no visual overlap of the data. The core (center) portions of the data are too far apart.

Practice and Problem Solving

3. **a.** garter snake: mean = 25, median = 24.5, mode = 24, range = 20, IQR = 7.5, MAD ≈ 4.33
   water snake: mean = 31.5, median = 32, mode = 32, range = 20, IQR = 10, MAD ≈ 5.08

   b. The water snakes have greater measures of center because the mean, median, and mode are greater. The water snakes also have greater measures of variation because the interquartile range and mean absolute deviation are greater.

4. **a.** Team A: median = 3, IQR = 2
   Team B: median = 7, IQR = 2
   The variation in the goals scored is the same, but Team B usually scores about 4 more goals per game.

   b. The difference in the medians is 2 times the IQR.

5. See Additional Answers.

6. See Additional Answers.

7. See *Taking Math Deeper*.

8–9. See Additional Answers.

 Fair Game Review

10.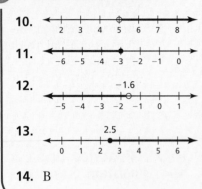

11.

12.

13.

14. B

Mini-Assessment

The data sets below give the final grades of the females in a 7th grade math class and the females in an 8th grade math class.

Grade 7: 78, 82, 84, 87, 88, 89, 89, 90, 93, 100

Grade 8: 77, 80, 81, 84, 86, 87, 88, 90, 91, 96

1. Compare the populations using measures of center and variation. The variation was about the same, but Grade 7 females had a greater mean final grade.

2. Express the difference in the measures of center as a multiple of the measure of variation. difference in means is about 0.5 times the MAD

3. Now you want to compare the final grades of all females in 7th and 8th grade math classes. Can you conclude that the grades are better in Grade 7? Explain. No, sample size is too small.

Taking Math Deeper

Exercise 7

The values found in Exercises 4(b), 5(b), and 6(b) measure the overlap of the data sets. Begin by displaying the data so that you can visualize the overlap.

1 Describe the visual overlap.

Looking at the double box-and-whisker plot in Exercise 4, you can see that there is *some* overlap in the data.

Arrange the dot plots in Exercise 5 vertically. There is *a lot* of overlap.

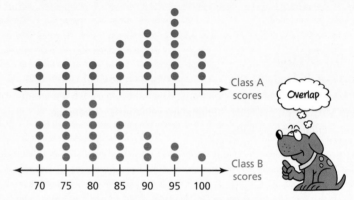

Make a double box-and-whisker plot in Exercise 6. There is *no* overlap.

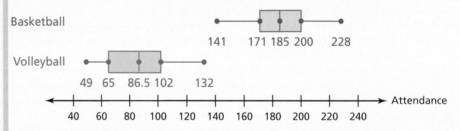

2 Compare the values in Exercises 4(b), 5(b), and 6(b): 2, 0.8 to 1, and 5.1 to 5.6. The value in Exercise 6(b) is the greatest, indicating that these data have less overlap than the data in Exercises 4 and 5.

3 Notice that greater numbers indicate less overlap, and lesser numbers indicate more overlap. The value in Exercise 5(b) is the least, indicating that these data have more overlap than the data in Exercises 4 and 6.

Reteaching and Enrichment Strategies

If students need help. . .	If students got it. . .
Resources by Chapter • Practice A and Practice B • Puzzle Time Record and Practice Journal Practice Differentiating the Lesson Lesson Tutorials Skills Review Handbook	Resources by Chapter • Enrichment and Extension • Technology Connection Start the next section

6. **ATTENDANCE** The tables show the attendances at volleyball games and basketball games at a school during the year.

Volleyball Game Attendance						
112	95	84	106	62	68	53
75	88	93	127	98	117	60
49	54	85	74	88	132	

Basketball Game Attendance						
202	190	173	155	169	188	195
176	141	152	181	198	214	179
163	186	184	207	219	228	

   a. Compare the populations using measures of center and variation.

   b. Express the difference in the measures of center as a multiple of each measure of variation.

7. **NUMBER SENSE** Compare the answers to Exercises 4(b), 5(b), and 6(b). Which value is the greatest? What does this mean?

② 8. **MAGAZINES** You want to compare the number of words per sentence in a sports magazine to the number of words per sentence in a political magazine.

   a. The data represent random samples of 10 sentences in each magazine. Compare the samples using measures of center and variation. Can you use this to make a valid comparison about the magazines? Explain.

      Sports magazine: 9, 21, 15, 14, 25, 26, 9, 19, 22, 30

      Political magazine: 31, 22, 17, 5, 23, 15, 10, 20, 20, 17

   b. The double box-and-whisker plot shows the means of 200 random samples of 20 sentences. Compare the variability of the sample means to the variability of the sample numbers of words in part (a).

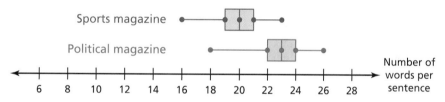

   c. Make a conclusion about the numbers of words per sentence in the magazines.

9. **Project** You want to compare the average amounts of time students in sixth, seventh, and eighth grade spend on homework each week.

   a. Design an experiment involving random sampling that can help you make a comparison.

   b. Perform the experiment. Can you make a conclusion about which students spend the most time on homework? Explain your reasoning.

Fair Game Review What you learned in previous grades & lessons

Graph the inequality on a number line. *(Section 4.1)*

10. $x > 5$ 11. $b \le -3$ 12. $n < -1.6$ 13. $p \ge 2.5$

14. **MULTIPLE CHOICE** The number of students in the marching band increased from 100 to 125. What is the percent of increase? *(Section 6.5)*

   Ⓐ 20%   Ⓑ 25%   Ⓒ 80%   Ⓓ 500%

1. Which sample is better for making a prediction? Explain. *(Section 10.6)*

Predict the number of students in your school who play at least one sport.	
Sample A	A random sample of 10 students from the school student roster
Sample B	A random sample of 80 students from the school student roster

2. **GYMNASIUM** You want to estimate the number of students in your school who think the gymnasium should be remodeled. You survey 12 students on the basketball team. Determine whether the sample is *biased* or *unbiased*. Explain. *(Section 10.6)*

3. **TOWN COUNCIL** You want to know how the residents of your town feel about a recent town council decision. You survey 100 residents at random. Sixty-five support the decision, and thirty-five do not. So, you conclude that 65% of the residents of your town support the decision. Determine whether the conclusion is valid. Explain. *(Section 10.6)*

4. **FIELD TRIP** Of 60 randomly chosen students surveyed, 16 chose the aquarium as their favorite field trip. There are 720 students in the school. Predict the number of students in the school who would choose the aquarium as their favorite field trip. *(Section 10.6)*

5. **FOOTBALL** The double box-and-whisker plot shows the points scored per game by two football teams during the regular season. *(Section 10.7)*

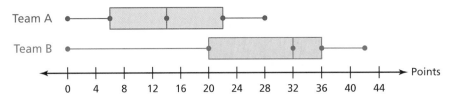

   a. Compare the populations using measures of center and variation.

   b. Express the difference in the measures of center as a multiple of the measure of variation.

6. **SUMMER CAMP** The dot plots show the ages of campers at two summer camps. *(Section 10.7)*

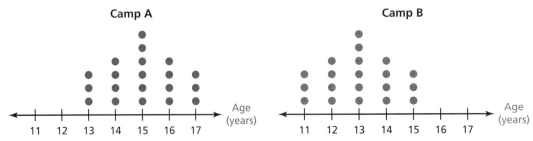

   a. Compare the populations using measures of center and variation.

   b. Express the difference in the measures of center as a multiple of the measure of variation.

Alternative Assessment Options

Math Chat   **Student Reflective Focus Question**
Structured Interview   Writing Prompt

Student Reflective Focus Question

Ask students to summarize the similarities and differences between samples and populations. Be sure that they include examples. Select students at random to present their summary to the class.

Study Help Sample Answers

Remind students to complete Graphic Organizers for the rest of the chapter.

6.

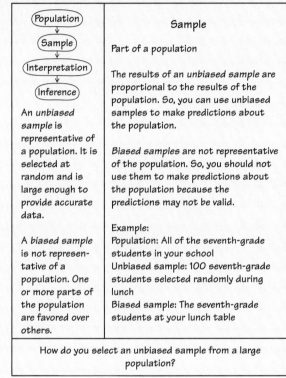

7. Available at *BigIdeasMath.com*.

Answers

1. Sample B because the sample size is larger.

2. biased; The sample is not selected at random and is not representative of the population because students on the basketball team use the gymnasium regularly when practicing.

3. yes; The sample is representative of the population, selected at random, and large enough to provide accurate data. So, the sample is unbiased and the conclusion is valid.

4. 192 students

5. a. Team A:
   median = 14, IQR = 16;
   Team B:
   median = 32, IQR = 16;
   The variation in the points scored is the same, but Team B generally has a greater score.

   b. The difference in the medians is 1.125 times the IQR.

6. a. Camp A:
   mean = 15, MAD = 1;
   Camp B:
   mean = 13, MAD = 1;
   The variation in the ages is the same, but Camp A has a greater age.

   b. The difference in the means is 2 times the MAD.

Reteaching and Enrichment Strategies

If students need help. . .	If students got it. . .

Resources by Chapter
  • Practice A and Practice B
  • Puzzle Time
Lesson Tutorials
BigIdeasMath.com
	Resources by Chapter
	  • Enrichment and Extension
	  • Technology Connection
	Game Closet at *BigIdeasMath.com*
	Start the Chapter Review

Technology for the *Teacher*

Online Assessment
Assessment Book
ExamView® Assessment Suite

For the Teacher
Additional Review Options
- *BigIdeasMath.com*
- Online Assessment
- Game Closet at *BigIdeasMath.com*
- Vocabulary Help
- Resources by Chapter

Answers

1. **a.** 2
   b. 1 green, 1 purple

2. **a.** 3
   b. 3 orange, 3 blue, 3 purple

3. **a.** 5
   b. 1 green, 1 purple, 3 orange, 3 blue, 3 purple

4. **a.** 3
   b. 2 blue, 2 orange, 2 green

5. **a.** 8
   b. 1 green, 1 purple, 2 blue, 2 orange, 2 green, 3 orange, 3 blue, 3 purple

6. **a.** 5
   b. 1 green, 1 purple, 2 blue, 2 orange, 2 green

Review of Common Errors

Exercises 1–6
- Students may forget to include, or include too many, favorable outcomes. Encourage them to write out all of the possible outcomes and then circle the favorable outcomes for the given event.

10 Chapter Review

Check It Out
Vocabulary Help
BigIdeasMath ✓com

Review Key Vocabulary

experiment, *p. 402*
outcomes, *p. 402*
event, *p. 402*
favorable outcomes, *p. 402*
probability, *p. 408*
relative frequency, *p. 412*

experimental probability, *p. 414*
theoretical probability, *p. 415*
sample space, *p. 422*
Fundamental Counting Principle, *p. 422*
compound event, *p. 424*

independent events, *p. 430*
dependent events, *p. 431*
simulation, *p. 436*
population, *p. 440*
sample, *p. 440*
unbiased sample, *p. 442*
biased sample, *p. 442*

Review Examples and Exercises

10.1 Outcomes and Events *(pp. 400–405)*

You randomly choose one toy race car.

a. **In how many ways can choosing a green car occur?**

b. **In how many ways can choosing a car that is *not* green occur? What are the favorable outcomes of choosing a car that is *not* green?**

a. There are 5 green cars. So, choosing a green car can occur in 5 ways.

b. There are 2 cars that are *not* green. So, choosing a car that is *not* green can occur in 2 ways.

green	*not* green
green, green, green, green, green	blue, red

∴ The favorable outcomes of the event are blue and red.

Exercises

You spin the spinner. (a) Find the number of ways the event can occur. (b) Find the favorable outcomes of the event.

1. Spinning a 1

2. Spinning a 3

3. Spinning an odd number

4. Spinning an even number

5. Spinning a number greater than 0

6. Spinning a number less than 3

10.2 Probability *(pp. 406–411)*

You flip a coin. What is the probability of flipping tails?

$$P(\text{event}) = \frac{\text{number of favorable outcomes}}{\text{number of possible outcomes}}$$

$$P(\text{tails}) = \frac{1}{2} \quad \leftarrow \boxed{\text{There is 1 tails.}}$$
$$\quad\quad\quad\quad\quad \leftarrow \boxed{\text{There is a total of 2 sides.}}$$

The probability of flipping tails is $\frac{1}{2}$, or 50%.

Exercises

7. You roll a number cube. Find the probability of rolling an even number.

10.3 Experimental and Theoretical Probability *(pp. 412–419)*

a. The bar graph shows the results of spinning the spinner 70 times. What is the experimental probability of spinning a 2?

The bar graph shows 12 twos. So, the spinner landed on two 12 times in a total of 70 spins.

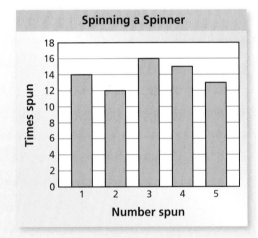

Spinning a Spinner

$$P(\text{event}) = \frac{\text{number of times the event occurs}}{\text{total number of trials}}$$

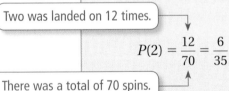

$$\boxed{\text{Two was landed on 12 times.}}$$
$$P(2) = \frac{12}{70} = \frac{6}{35}$$
$$\boxed{\text{There was a total of 70 spins.}}$$

The experimental probability is $\frac{6}{35}$, or about 17%.

b. The theoretical probability of choosing a purple grape from a bag is $\frac{2}{9}$. There are 8 purple grapes in the bag. How many grapes are in the bag?

$$P(\text{purple}) = \frac{\text{number of purple grapes}}{\text{total number of grapes}}$$

$$\frac{2}{9} = \frac{8}{g} \quad\quad \text{Substitute. Let } g \text{ be the total number of grapes.}$$

$$g = 36 \quad\quad \text{Solve for } g.$$

So, there are 36 grapes in the bag.

Review of Common Errors (continued)

Exercise 7

- Students may write the probability as the ratio of the number of favorable outcomes to the number of unfavorable outcomes. Remind them that the probability of an event is the ratio of the number of favorable outcomes to the number of possible outcomes.

7. $\frac{1}{2}$, or 50%

8. $\frac{8}{35}$, or about 22.9%

9. $\frac{43}{70}$, or about 61.4%

10. $\frac{57}{70}$, or about 81.4%

11. $\frac{2}{5}$, or 40%

12. $\frac{1}{4}$, or 25%

13. $\frac{3}{8}$, or 37.5%

14. $\frac{5}{8}$, or 62.5%

15. $\frac{1}{8}$, or 12.5%

16. 12

17. 90

18. $\frac{1}{8}$, or 12.5%

Review of Common Errors (continued)

Exercises 8–11

- Students may forget to total all of the trials before writing the experimental probability. They may have an incorrect number of trials in the denominator. Remind them that they need to know the total number of trials when finding the probability.

- Students may find the theoretical probability of the event instead of using the data to find the experimental probability. Remind them that they are using the experimental probability and assuming that this trend will continue to predict the outcome of an event.

Exercises 12–15

- Students may write a different probability than what is asked, or forget to include a favorable outcome. For example, in Exercise 13 a student may not realize that there are three "1" sections and will write the probability as $\frac{1}{8}$ or $\frac{1}{4}$ instead of $\frac{3}{8}$. Remind them to read the event carefully and to write the favorable outcomes before finding the probability.

Exercises 17 and 18

- Students may try to use a tree diagram to solve these problems. Although it is possible and not incorrect to do so, recommend that students use the Fundamental Counting Principle as a much less time consuming alternative.

Exercises 19–22

- Students may mix up independent and dependent events or may have difficulty determining which type of event it is. Remind them that independent events are where you do two different things or events where you start over before the next trial. Dependent events have at least one less possible outcome after the first draw, roll, or flip.

Exercise 23

- Students may ignore the fact that the survey is conducted among students arriving at a biology club meeting. They may only see that it is a meeting at school. Ask them what kind of answers they would expect to get from different club meetings (such as math club). They should recognize that everyone going to a biology club meeting likes biology, but not everyone in the math club will.

Exercise 24

- Students may use the wrong measures of center and variation when comparing populations. Remind them to use the mean and the MAD when *both* distributions are symmetric. Otherwise, use the median and the IQR.

Exercises

Use the bar graph on page 456 to find the experimental probability of the event.

8. Spinning a 3

9. Spinning an odd number

10. *Not* spinning a 5

11. Spinning a number greater than 3

Use the spinner to find the theoretical probability of the event.

12. Spinning blue

13. Spinning a 1

14. Spinning an even number

15. Spinning a 4

16. The theoretical probability of spinning an even number on a spinner is $\frac{2}{3}$. The spinner has 8 even-numbered sections. How many sections are on the spinner?

10.4 ## Compound Events *(pp. 420–427)*

a. How many different home theater systems can you make from 6 DVD players, 8 TVs, and 3 brands of speakers?

$$6 \times 8 \times 3 = 144 \qquad \text{Fundamental Counting Principle}$$

⋮ So, you can make 144 different home theater systems.

b. You flip two pennies. What is the probability of flipping two heads?

Use a tree diagram to find the probability. Let H = heads and T = tails.

There is one favorable outcome in the sample space for flipping two heads: HH.

$$P(\text{event}) = \frac{\text{number of favorable outcomes}}{\text{number of possible outcomes}}$$

$$P(2 \text{ heads}) = \frac{1}{4} \qquad \text{Substitute.}$$

⋮ The probability is $\frac{1}{4}$, or 25%.

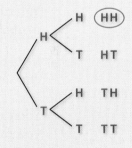

Exercises

17. You have 6 bracelets and 15 necklaces. Find the number of ways you can wear one bracelet and one necklace.

18. You flip two coins and roll a number cube. What is the probability of flipping two tails and rolling an even number?

Independent and Dependent Events *(pp. 428–437)*

You randomly choose one of the tiles and flip the coin. What is the probability of choosing a vowel and flipping heads?

Choosing one of the tiles does not affect the outcome of flipping the coin. So, the events are independent.

$P(\text{vowel}) = \dfrac{2}{7}$ ← There are 2 vowels (A and E).

← There is a total of 7 tiles.

$P(\text{tails}) = \dfrac{1}{2}$ ← There is 1 tails side.

← There is a total of 2 sides.

Use the formula for the probability of independent events.

$$P(A \text{ and } B) = P(A) \cdot P(B)$$

$$= \dfrac{2}{7} \cdot \dfrac{1}{2} = \dfrac{1}{7}$$

∴ The probability of choosing a vowel and flipping heads is $\dfrac{1}{7}$, or about 14%.

Exercises

You randomly choose one of the tiles above and flip the coin. Find the probability of the compound event.

19. Choosing a blue tile and flipping tails

20. Choosing the letter G and flipping tails

You randomly choose one of the tiles above. Without replacing the first tile, you randomly choose a second tile. Find the probability of the compound event.

21. Choosing a green tile and then a blue tile

22. Choosing a red tile and then a vowel

Samples and Populations *(pp. 440–447)*

You want to estimate the number of students in your school whose favorite subject is math. You survey every third student who leaves the school. Determine whether the sample is *biased* or *unbiased*.

The sample is representative of the population, selected at random, and large enough to provide accurate data.

∴ So, the sample is unbiased.

Review Game

Making Predictions

Materials per team:
- deck of cards
- paper
- pencil
- calculator

Directions:

Each team shuffles their deck of 52 cards. The first 39 cards are flipped over and placed in a discard pile. Teams are to keep track of what cards are discarded and what cards are left in the deck. The 13 cards remaining in the deck are shuffled by the teacher.

With the whole class, teams take turns predicting the next card to be flipped over from their deck. Predictions can include black card, red card, face card, single number card (if they are brave), etc. The team calculates the probability of the prediction on the board for the class to see.

Each team will do this for 5 cards. For each correct prediction, the team gets a point. The teacher holds onto each team's remaining cards until the game is over, in case of a tie.

Who Wins?

The team with the most points wins. If there is a tie, a one card draw will be the tie breaker. With the remaining cards in the team's deck, each team will make a prediction. The team whose prediction is correct wins.

For the Student
Additional Practice
- Lesson Tutorials
- Multi-Language Glossary
- Self-Grading Progress Check
- *BigIdeasMath.com*
  Dynamic Student Edition
  Student Resources

Answers

19. $\frac{2}{7}$, or about 28.6%

20. $\frac{1}{14}$, or about 7.1%

21. $\frac{4}{21}$, or about 19.0%

22. $\frac{1}{21}$, or about 4.8%

23. biased; The sample is not selected at random and is not representative of the population because students in the biology club like biology.

24. **a.** Class A:
    median = 88, IQR = 6;
    Class B:
    median = 91, IQR = 9;
    In general, Class B has greater scores than Class A. Class A has less variation than Class B.

    b. The difference in the medians is about 0.3 to 0.5 times the IQR.

My Thoughts on the Chapter

What worked. . .

What did not work. . .

What I would do differently. . .

23. You want to estimate the number of students in your school whose favorite subject is biology. You survey the first 10 students who arrive at biology club. Determine whether the sample is *biased* or *unbiased*. Explain.

10.7 **Comparing Populations** *(pp. 448–453)*

The double box-and-whisker plot shows the test scores for two French classes taught by the same teacher.

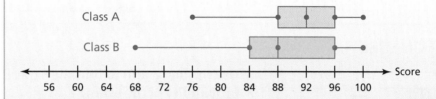

a. **Compare the populations using measures of center and variation.**

Both distributions are skewed left, so use the median and the IQR.

⸪ The median for Class A, 92, is greater than the median for Class B, 88. The IQR for Class B, 12, is greater than the IQR for Class A, 8. The scores in Class A are generally greater and have less variability than the scores in Class B.

b. **Express the difference in the measures of center as a multiple of each measure of variation.**

$$\frac{\text{median for Class A} - \text{median for Class B}}{\text{IQR for Class A}} = \frac{4}{8} = 0.5$$

$$\frac{\text{median for Class A} - \text{median for Class B}}{\text{IQR for Class B}} = \frac{4}{12} = 0.3$$

⸪ So, the difference in the medians is about 0.3 to 0.5 times the IQR.

24. **SPANISH TEST** The double box-and-whisker plot shows the test scores of two Spanish classes taught by the same teacher.

a. Compare the populations using measures of center and variation.

b. Express the difference in the measures of center as a multiple of each measure of variation.

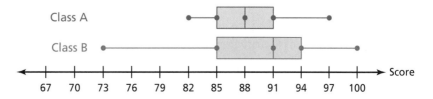

You randomly choose one game piece. (a) Find the number of ways the event can occur. (b) Find the favorable outcomes of the event.

1. Choosing green

2. Choosing *not* yellow

3. Use the Fundamental Counting Principle to find the total number of different sunscreens possible.

Sunscreen	
SPF	10, 15, 30, 45, 50
Type	Lotion, Spray, Gel

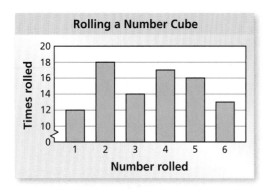

Rolling a Number Cube

Use the bar graph to find the experimental probability of the event.

4. Rolling a 1 or a 2

5. Rolling an odd number

6. *Not* rolling a 5

Use the spinner to find the theoretical probability of the event(s).

7. Spinning an even number

8. Spinning a 1 and then a 2

Knight Queen King Bishop Rook Pawn

You randomly choose one chess piece. Without replacing the first piece, you randomly choose a second piece. Find the probability of choosing the first piece, then the second piece.

9. Bishop and bishop

10. King and queen

11. LUNCH You want to estimate the number of students in your school who prefer to bring a lunch from home rather than buy one at school. You survey five students who are standing in the lunch line. Determine whether the sample is *biased* or *unbiased*. Explain.

12. AGES The double box-and-whisker plot shows the ages of the viewers of two television shows in a small town.

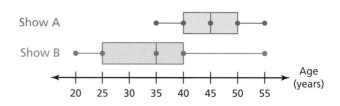

a. Compare the populations using measures of center and variation.

b. Express the difference in the measures of center as a multiple of each measure of variation.

Test Item References

Chapter Test Questions	Section to Review
1, 2	10.1
4–8	10.2
4–8	10.3
3	10.4
9, 10	10.5
11	10.6
12	10.7

Test-Taking Strategies

Remind students to quickly look over the entire test before they start so that they can budget their time. There is a lot of vocabulary in this chapter, so students should have been making flash cards as they worked through the chapter. Words that get mixed up should be jotted on the back of the test before they start. Students need to use the **Stop** and **Think** strategy before they answer a question.

Common Errors

- **Exercises 1 and 2** Students may forget to include, or include too many, favorable outcomes. Encourage them to write out all of the possible outcomes and then circle the favorable outcomes for the given event.
- **Exercise 3** Students may use a tree diagram to solve this problem. Although it is possible to do so, point out that the directions state to use the Fundamental Counting Principle. Also, point out that using the Fundamental Counting Principle is a much less time consuming method.
- **Exercises 4–6** Students may forget to total all of the trials before writing the experimental probability. Remind them that they need to know the total number of trials when finding the probability.
- **Exercises 9 and 10** Students may forget to subtract one from the total number of possible outcomes when finding the probability of choosing the second chess piece. Remind them that the second draw has one less possible outcome because they have removed one of the chess pieces.

Reteaching and Enrichment Strategies

If students need help...	If students got it...
Resources by Chapter • Practice A and Practice B • Puzzle Time Record and Practice Journal Practice Differentiating the Lesson Lesson Tutorials *BigIdeasMath.com* Skills Review Handbook	Resources by Chapter • Enrichment and Extension • Technology Connection Game Closet at *BigIdeasMath.com* Start Cumulative Assessment

Answers

1. **a.** 1   **b.** green

2. **a.** 5

   b. red, blue, red, green, blue

3. 15

4. $\frac{1}{3}$, or about 33.3%

5. $\frac{7}{15}$, or about 46.7%

6. $\frac{37}{45}$, or about 82.2%

7. $\frac{4}{9}$, or about 44.4%

8. $\frac{1}{81}$, or about 1.2%

9. $\frac{1}{120}$, or about 0.8%

10. $\frac{1}{240}$, or about 0.4%

11. biased; The sample size is too small and students standing in line are more likely to say they prefer to buy their lunches at school.

12. **a.** Show A:
    median = 45, IQR = 10;
    Show B:
    median = 35, IQR = 15;
    Show B generally has a younger audience and more variation in ages than Show A.

    b. The difference in the medians is about 0.7 to 1 times the IQR.

Technology for the *Teacher*

Online Assessment
Assessment Book
ExamView® Assessment Suite

After Answering Easy Questions, Relax
Answer Easy Questions First
Estimate the Answer
Read All Choices before Answering
Read Question before Answering
Solve Directly or Eliminate Choices
Solve Problem before Looking at
   Choices
Use Intelligent Guessing
Work Backwards

About this Strategy

When taking a multiple choice test, be sure to read each question carefully and thoroughly. Sometimes you don't know the answer. So . . . guess intelligently! Look at the choices and choose the ones that are possible answers.

Answers

1. C

2. $\frac{1}{5}$, or 0.2

Item Analysis

1. **A.** The student does not understand the concepts of certainty and likelihood.

   B. The student does not understand the difference between likely and unlikely.

   C. Correct answer

   D. The student does not understand that even a highly unlikely event is not impossible.

2. **Gridded Response:** Correct answer: $\frac{1}{5}$, or 0.2

   Common Error: The student only considers that Sunday is one day of the week and gets an answer of $\frac{1}{7}$.

1. A school athletic director asked each athletic team member to name his or her favorite professional sports team. The results are below:

- D.C. United: 3
- Florida Panthers: 8
- Jacksonville Jaguars: 26
- Jacksonville Sharks: 7
- Miami Dolphins: 22
- Miami Heat: 15
- Miami Marlins: 20
- Minnesota Lynx: 4
- New York Knicks: 5
- Orlando Magic: 18
- Tampa Bay Buccaneers: 17
- Tampa Bay Lightning: 12
- Tampa Bay Rays: 28
- Other: 6

Test-Taking Strategy
Use Intelligent Guessing

What's the probability of drawing 1 hyena out of a bag with 2 hyenas and 3 mice?

Ⓐ -10% Ⓑ 40% Ⓒ 60% Ⓓ 500%

40% < 60% I'm hoping 40%.

"You know it can't be -10% or 500%. So, you can intelligently guess between 40% and 60%."

One athletic team member is picked at random. What is the likelihood that this team member's favorite professional sports team is *not* located in Florida?

A. certain

B. likely, but not certain

C. unlikely, but not impossible

D. impossible

2. Each student in your class voted for his or her favorite day of the week. Their votes are shown below:

Favorite Day of the Week

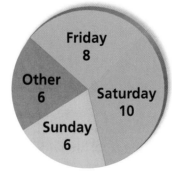

Friday 8

Other 6

Saturday 10

Sunday 6

A student from your class is picked at random. What is the probability that this student's favorite day of the week is Sunday?

3. How far, in millimeters, will the tip of the hour hand of the clock travel in 2 hours? (Use $\frac{22}{7}$ for π.)

F. 44 mm

G. 88 mm

H. 264 mm

I. 528 mm

4. Nathaniel solved the proportion in the box below.

$$\frac{16}{40} = \frac{p}{27}$$

$$16 \cdot p = 40 \cdot 27$$

$$16p = 1080$$

$$\frac{16p}{16} = \frac{1080}{16}$$

$$p = 67.5$$

What should Nathaniel do to correct the error that he made?

A. Add 40 to 16 and 27 to p.

B. Subtract 16 from 40 and 27 from p.

C. Multiply 16 by 27 and p by 40.

D. Divide 16 by 27 and p by 40.

5. A North American hockey rink contains 5 face-off circles. Each of these circles has a radius of 15 feet. What is the total area, in square feet, of all the face-off circles? (Use 3.14 for π.)

F. 706.5 ft^2

G. 2826 ft^2

H. 3532.5 ft^2

I. 14,130 ft^2

Item Analysis (continued)

Answers

3. G

4. C

5. H

3. **F.** The student does not multiply the radius by two to find the circumference but then correctly divides by 6, or the student correctly finds the circumference but then divides by 12 to find how far the tip of the hour hand will travel in 1 hour.

   G. Correct answer

   H. The student does not multiply the radius by two to find the circumference and then uses the incorrect circumference to find how far the tip of the hour hand will travel in 12 hours.

   I. The student finds how far the tip of the hour hand will travel in 12 hours.

4. **A.** The student incorrectly thinks that proportions involve addition.

   B. The student incorrectly thinks that proportions involve subtraction.

   C. Correct answer

   D. The student switches the 40 and the 27 in the proportion, resulting in a proportion that is not equivalent to the original proportion.

5. **F.** The student finds the area of only 1 circle.

   G. The student finds the area of only 1 circle and uses the diameter of the circle instead of the radius, finding $\pi \cdot 30^2$.

   H. Correct answer

   I. The student uses the diameters of the circles instead of the radii, finding $\pi \cdot 30^2$ for each of the circles.

Answers

6. $\frac{1}{16}$, or 0.0625

7. C

8. H

9. *Part A* independent

   Part B favorable outcomes: 3
   possible outcomes: 6

   Part C $\frac{1}{4}$, or 0.25

Item Analysis (continued)

6. **Gridded Response:** Correct answer: $\frac{1}{16}$, or 0.0625

   Common Error: The student does not realize the compound nature of the event and gets an answer of $\frac{1}{4}$ or 0.25 or equivalent.

7. **A.** The student finds the area of one lateral face.

   B. The student finds the area of the four lateral faces.

   C. Correct answer

   D. The student forgets to multiply the area of each triangular lateral face by $\frac{1}{2}$.

8. **F.** The student finds what percent $15.00 is of $6.00.

   G. The student subtracts $6.00 from $15.00 to get $9.00 and thinks that this means 90%.

   H. Correct answer

   I. The student finds what percent $6.00 is of $15.00.

9. **2 points** The student demonstrates a thorough understanding of determining probability. In Part A, the student determines the events are independent. In Part B, the student finds the number of possible outcomes is 6 and the number of favorable outcomes is 3. In Part C, the student gets an answer of $\frac{1}{4}$, or 0.25.

   1 point The student demonstrates a partial understanding of determining probability. The student gets a correct answer for Part A and Part B but the answer for Part C is incorrect.

   0 points The student provided no response, a completely incorrect or incomprehensible response, or a response that demonstrates insufficient understanding of probability.

6. A spinner is divided into eight congruent sections, as shown below.

You spin the spinner twice. What is the probability that the arrow will stop in a yellow section both times?

7. What is the surface area, in square inches, of the square pyramid?

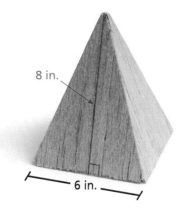

8 in.

6 in.

A. 24 in.²

B. 96 in.²

C. 132 in.²

D. 228 in.²

8. The value of one of Kevin's baseball cards was $6.00 when he first got it. The value of this card is now $15.00. What is the percent increase in the value of the card?

F. 40%

G. 90%

H. 150%

I. 250%

9. You roll a number cube twice. You want to roll two even numbers.

Part A Determine whether the events are independent or dependent.

Part B Find the number of favorable outcomes and the number of possible outcomes of each roll.

Part C Find the probability of rolling two even numbers. Explain your reasoning.

Appendix A
My Big Ideas Projects

A.1   Literature Project
The Mathematics of Jules Verne

A.2   History Project
Mathematics in Ancient Greece

About the Appendix

- The interdisciplinary projects can be used anytime throughout the year.
- The projects offer students an opportunity to build on prior knowledge, to take mathematics to a deeper level, and to develop organizational skills.
- Students will use the Essential Questions to help them form "need to knows" to focus their research.

Essential Question

- **Literature Project**
  How does the knowledge of mathematics influence science fiction writing?
- **History Project**
  How do you use mathematical knowledge that was originally discovered by the Greeks?

Additional Resources

BigIdeasMath.com

Essential Question

- **Art Project**
  How have circles influenced ancient and modern art?
- **Science Project**
  How does the classification of living organisms help you understand the similarities and differences of animals?

My Big Ideas Projects

A.3   Art Project
      Circle Art

A.4   Science Project
      Classifying Animals

The Mathematics of Jules Verne

1 Project Overview

Jules Verne (1828–1905) was a famous French science fiction writer. He wrote about space, air, and underwater travel before aircraft and submarines were commonplace, and before any means of space travel had been devised.

For example, in his 1865 novel *From the Earth to the Moon*, he wrote about three astronauts who were launched from Florida and recovered through a splash landing. The first actual moon landing wasn't until 1969.

Essential Question How does the knowledge of mathematics influence science fiction writing?

Read one of Jules Verne's science fiction novels. Then write a book report about some of the mathematics used in the novel.

Sample: A league is an old measure of distance. It is approximately equal to 4 kilometers. You can convert 20,000 leagues to miles as follows.

$$20{,}000 \text{ leagues} \cdot \frac{4 \text{ km}}{1 \text{ league}} \cdot \frac{1 \text{ mile}}{1.61 \text{ km}} \approx 50{,}000 \text{ miles}$$

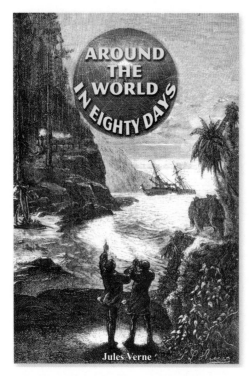

Project Notes

Introduction

For the Teacher

- **Goal:** Students will read one of Jules Verne's science-fiction novels and write a book report about some of the mathematics used in the novel. A sample for *Twenty Thousand Leagues Under the Sea* is included below. Samples for *Journey to the Center of the Earth* and *Around the World in Eighty Days* can be found at *BigIdeasMath.com*.
- **Management Tip:** You may want to have students work together in groups.

Essential Question

- How does the knowledge of mathematics influence science fiction writing?

Things to Think About

Summary of *Twenty Thousand Leagues Under the Sea*

- In 1866, several ships around the world had contact with a strange sea creature. The United States sent the Abraham Lincoln, a frigate capable of great speed, to pursue the monster. Monsieur Pierre Aronnax, Professor of Natural History at the Museum of Paris, was asked to join the expedition. Aronnax was accompanied by his devoted servant Conseil, who followed his master anywhere. Ned Land, a Canadian whaler, also joined the expedition.
- They searched for months for the monster but could not find it. Just as they were about to give up their search, it was spotted. After giving chase, there was a collision, and the three main characters fell into the sea. As it turns out, the "monster" was actually the Nautilus, a spectacular submarine that was captained by Nemo. Aronnax, Conseil, and Land were rescued by Captain Nemo, but became his prisoners. They had free run of the submarine, were well fed, and were able to take part in exciting activities, but they were not allowed to leave the submarine permanently.
- The submarine was powered by electricity. It had moving covers on the side windows, which allowed everyone to see what was happening in the ocean. The crew used deep sea diving suits to walk on the ocean floor in pursuit of pearls, volcanoes, and sea creatures. They traveled to the South Pole and fought off gigantic octopi.
- In less than 10 months, the crew traveled twenty thousand leagues. All the while, Ned Land planned their escape. Finally, the Nautilus got caught in "the maelstrom," a whirlpool off the coast of Norway from which no ship had ever escaped. At the same time, the three prisoners were trying to get away. They were able to escape, but they had no knowledge of the fate of the Nautilus.

References

Go to *BigIdeasMath.com* to access links related to this project.

Meet with a reading or language arts teacher and review curriculum maps to identify whether students have read *20,000 Leagues Under the Sea*, *Journey to the Center of the Earth*, or *Around the World in Eighty Days*. If the books have already been read, you may want to discuss the work students have completed and then review the books with them. If the books have not been read, perhaps you can both work simultaneously and share notes. Or, you may want to explore activities that the reading or language arts teacher has done in the past to support student learning in this particular area.

Project Notes

Mathematics Used in the Story

- 20,000 leagues was the distance traveled during the 10 months, not the depth below the surface. One league is about 4 kilometers, so the journey was about 50,000 miles.
- They interchanged the units of distance frequently; using inches, feet, yards, meters, fathoms (1 fathom = 6 feet), miles, and of course, leagues.
- Other units of measure used include knots (1 nautical mile per hour, approximately 6076 feet per hour), miles per hour, revolutions per second, pounds per square inch, atmospheres, horsepower, hundredweight, degrees, and cubic yards.
- One of the first ships to collide with the "monster" ended up with a hole that was in the shape of an isosceles triangle.
- One of the weapons was described as being able to "throw with ease a conical projectile of nine pounds a mean distance of ten miles."
- It was mentioned that water covers $\frac{7}{10}$ of the Earth's surface.
- Proportions were used frequently. For example, the common narwhal, or unicorn of the sea, was compared to the "monster." "Increase its size fivefold or tenfold, give it strength proportionate to its size, lengthen its destructive weapons, and you obtain the animal required."

Scientific Predictions that Have Happened

- submarines
- deep sea diving gear with metal helmets receiving air by use of pumps and regulators
- traveling underwater to the South Pole
- depth gauges
- electric reflectors to light the sea
- compressed air guns
- electric lanterns
- handrails made of metal and charged with electricity

Journey to the Center of the Earth and *Around the World in Eighty Days*

Go to *BigIdeasMath.com* for sample reports.

Closure

- **Rubric** An editable rubric for this project is available at *BigIdeasMath.com*.
- Students may present their reports to the class or school as a television report or public information broadcast.

② Things to Include

- Describe the major events in the plot.

- Write a brief paragraph describing the setting of the story.

- List and identify the main characters. Explain the contribution of each character to the story.

- Explain the major conflict in the story.

- Describe at least four examples of mathematics used in the story.

- Which of Jules Verne's scientific predictions have come true since he wrote the novel?

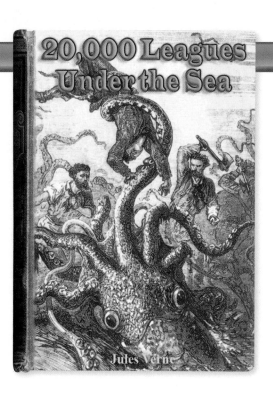

Jules Verne (1828–1905)

③ Things to Remember

- You can download one of Jules Verne's novels at *BigIdeasMath.com*.

- Add your own illustrations to your project.

- Organize your report in a folder, and think of a title for your report.

Mathematics in Ancient Greece

1 Getting Started

The ancient Greek period began around 1100 B.C. and lasted until the Roman conquest of Greece in 146 B.C.

The civilization of the ancient Greeks influenced the languages, politics, educational systems, philosophy, science, mathematics, and arts of Western Civilization. It was a primary force in the birth of the Renaissance in Europe between the 14th and 17th centuries.

Corinthian Helmet

Essential Question How do you use mathematical knowledge that was originally discovered by the Greeks?

Sample: Ancient Greek symbols for the numbers from 1 through 10 are shown in the table.

I	II	III	IIII	Γ	ΓI	ΓII	ΓIII	ΓIIII	△
1	2	3	4	5	6	7	8	9	10

These same symbols were used to write the numbers between 11 and 39. Here are some examples.

$$\triangle\ \Gamma\ ||| = 18 \qquad \triangle\triangle\triangle\Gamma = 35 \qquad \triangle\ \triangle\ |||| = 24$$

Alexander the Great

Parthenon

Project Notes

Introduction

For the Teacher

- **Goal:** Students will discover how ancient Greeks used and applied mathematics in many areas of life.
- **Management Tip:** Students can work in groups to create reports about the mathematics used in ancient Greece.

Essential Question

- How do you use mathematical knowledge that was originally discovered by the Greeks?

Things to Think About

? **How have the following Greeks contributed to the field of mathematics?**

- **Pythagoras (c. 570 B.C.–490 B.C.):** Most students will recognize the name Pythagoras as being the person after whom the Pythagorean Theorem was named. The theorem states that the sum of the squares of the lengths of the legs of a right triangle equals the square of the length of the hypotenuse. The Babylonians discovered this theorem 1000 years before Pythagoras, but he is credited with proving the theorem which bears his name. He also founded the idea of irrational numbers.

- **Aristotle (c. 384 B.C.–322 B.C.):** Aristotle is famous for being a philosopher, an astronomer, and a scientist. He is also responsible for developing the study of logic, which he called analytics. He believed that all mathematical ideas needed to be proven, and so he created ways to do this, which were later called proofs. Mathematicians still study his logic and syllogism today.

- **Euclid (c. 300 B.C.):** Euclid is known for his discovery of geometry, which he called *The Elements*. *The Elements* consisted of plane geometry, number theory, irrational numbers, and solid geometry. Within the chapters of number theory, Euclid included a way to find the greatest common divisor (GCD) of two integers, without factoring the integers. Mathematicians often refer to this as Euclid's algorithm.

- **Archimedes (c. 287 B.C.–212 B.C.):** One of the most famous stories about Archimedes was his discovery that an object placed in water becomes lighter by the amount equal to the weight of the water it displaces. He supposedly made this discovery when he got into his bathtub, and then shouted "Eureka!" as he ran through the town. Archimedes is responsible for calculating an approximation of pi and creating a way to find the areas and volumes of solid figures.

- **Eratosthenes (c. 276 B.C.–194 B.C.):** In addition to being a mathematician, Eratosthenes was a geographer and an astronomer. His most famous mathematical discovery is called the Sieve of Eratosthenes. It is used to find prime numbers. He also discovered an accurate method to measure the circumference of Earth, the distance from Earth to the sun, and the distance from Earth to the moon.

References

Go to *BigIdeasMath.com* to access links related to this project.

Cross-Curricular Instruction

Meet with a history teacher and review curriculum maps to identify whether students have covered the ancient Greeks. If the topic has been covered, you may want to discuss the work students have completed and then review prior knowledge with them. If the history teacher has not discussed these concepts, perhaps you can both work simultaneously on these concepts and share notes. Or, you may want to explore activities that the history teacher has done in the past to support student learning in this particular area.

Project Notes

? Who taught Alexander the Great?
- Alexander the Great attended the School of Royal Pages when he was thirteen. The school hired Aristotle to be a tutor for him and about 50 other boys his age. Aristotle taught them philosophy, politics, ethics, geography, and marine biology.

? How did the ancient Greeks represent fractions?
- The ancient Greeks did not use the fraction bar to separate the numerator from the denominator. Instead, they used a ′ (which is called a diacritical mark) to show the denominator.

$$\frac{1}{2} = \beta'$$

- In more complex fractions the numerator was written with an overbar.

$$\frac{17}{28} = \overline{\iota\varsigma}\kappa\eta$$

(**Note:** By the Alexandrian Age, the Greek Attic system of enumeration, shown on page A4 in the pupil's edition, was being replaced by the Ionian Greek system of enumeration as shown above.)

? How did the ancient Greeks use mathematics?
- Unlike the ancient Egyptians, who used mathematics to measure the depth of the Nile River or to build the great pyramids, the ancient Greeks used mathematics to expand their knowledge and philosophy.
- They used arithmetic (then called logistic) in business and in war. A leader needed to know how to use numbers in order to line up his troops for battle.
- Philosophers used number theory, which was called arithmetic, to prove what was previously accepted without proof.
- Some specific mathematical uses came from discoveries by Pythagoras and Archimedes. Pythagoras is famous for proving the Pythagorean Theorem. Archimedes is famous for approximating the value of pi so that the circumferences of circles could be computed. In addition to Archimedes' discovery of the law of displacement, he created a system of levers and pulleys to move heavy objects, such as ships. He also invented an object, which was later called the Archimedes Screw, that helped farmers irrigate their fields by raising the level of the water in the rivers.

Closure

- **Rubric** An editable rubric for this project is available at *BigIdeasMath.com*.
- You may hold a class debate where students can compare, defend, and discuss their findings with other students.

- Describe at least one contribution that each of the following people made to mathematics.

  Pythagoras (c. 570 B.C.–c. 490 B.C.)

  Aristotle (c. 384 B.C.–c. 322 B.C.)

  Euclid (c. 300 B.C.)

  Archimedes (c. 287 B.C.–c. 212 B.C.)

  Eratosthenes (c. 276 B.C.–c. 194 B.C.)

- Which of the people listed above was the teacher of Alexander the Great? What subjects did Alexander the Great study when he was in school?

- How did the ancient Greeks represent fractions?

- Describe how the ancient Greeks used mathematics. How does this compare with the ways in which mathematics is used today?

A α	alpha	N ν	nu
B β	beta	Ξ ξ	xi
Γ γ	gamma	O o	omicron
Δ δ	delta	Π π	pi
E ε	epsilon	P ρ	rho
Z ζ	zeta	Σ σ	sigma
H η	eta	T τ	tau
Θ θ	theta	Υ υ	upsilon
I ι	iota	Φ φ	phi
K κ	kappa	X χ	chi
Λ λ	lambda	Ψ ψ	psi
M μ	mu	Ω ω	omega

3 **Things to Remember**

- Add your own illustrations to your project.

- Try to include as many different math concepts as possible. Your goal is to include at least one concept from each of the chapters you studied this year.

- Organize your report in a folder, and think of a title for your report.

Greek Pottery

Trireme Greek Warship

A.3 Art Project

Circle Art

1 Getting Started

Circles have been used in art for thousands of years.

Essential Question How have circles influenced ancient and modern art?

Find examples of art in which circles were used. Describe how the artist might have used properties of circles to make each piece of art.

Sample: Here is a technique for making a pattern that uses circles in a stained glass window.

1. Use a compass to draw a circle.

2. Without changing the radius, draw a second circle whose center lies on the first circle.

3. Without changing the radius, draw a third circle whose center is one of the points of intersection.

4. Without changing the radius, continue to draw new circles at the points of intersection.

5. Color the design to make a pattern.

Mexican Tile

Ancient Roman Mosaic

Project Notes

Introduction

For the Teacher
- **Goal:** Students will discover how circles are used in art and architecture.
- **Management Tip:** Students may work in groups to find mosaic tile patterns. They can brainstorm as a group to come up with a way to measure the angles contained in the tile patterns.

Essential Question
- How have circles influenced ancient and modern art?

Things to Think About

? Where have artists used circles?
- Artists have used the circle as a basis for paintings, jewelry, wrought iron designs, architecture, and religious beliefs.
- Stonehenge on the Salisbury Plain in southwestern England is made of large upright stones set in a circular pattern. It is estimated that Stonehenge was built around 2000 B.C.

? How significant is the history of one of the most used circles, the ring?
- In ancient Greece, the *signet ring* was used to make an impression indicating authenticity.
- In Rome, citizens could wear rings of iron but slaves were forbidden to wear rings.
- Some churches use rings to signify the office of the wearer.
- The Egyptians wore rings of jasper or bloodstone for success in battle.
- Some cultures believe a magic ring will provide the wearer with power over others.
- The circle is a symbol of eternity because it has no beginning or end.
- The circle is also a symbol of mobility because of the wheel.
- The wheel is associated with chance and fortune.

References

Go to *BigIdeasMath.com* to access links related to this project.

Cross-Curricular Instruction

Meet with an art teacher and review curriculum maps to identify whether students have covered circles in art. If the topic has been covered, you may want to discuss the work students have completed and review prior knowledge with them. If the art teacher has not discussed these concepts, perhaps you can both work simultaneously on these concepts and share notes. Or, you may want to explore activities that the art teacher has done in the past to support student learning in this particular area.

Project Notes

? How did ancient artists draw circles?
- To draw a circle, ancient artists could have used two sticks connected by a length of rope or string. One stick would be placed at the center of the desired circle. The rope, which represented the radius, would be stretched out and the other stick would be used to draw the circle.

? What are some examples of how circles are used in mosaic tile patterns?

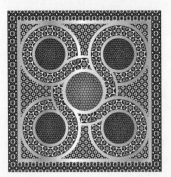

? What is an example of how circles are used in modern art or architecture?
- The Rotunda at the United States Capital is a circular room 96 feet in diameter and 180 feet in height.

Closure

- **Rubric** An editable rubric for this project is available at *BigIdeasMath.com*.
- Students may present their reports to a parent panel or community members.

2 Things to Include

- Describe how ancient artists drew circles.

- Describe the symbolism of circles in ancient art.

- Find examples of how circles are used to create mosaic tile patterns.

- Measure the angles that are formed by the patterns in the circle art you find. For instance, you might describe the angles formed by the netting in the Native American dreamcatcher.

- Find examples of the use of circles in modern art.

- Use circles to create your own art. Describe how you used mathematics and the properties of circles to make your art.

Native American Dreamcatcher

3 Things to Remember

- Add your own illustrations to your project.

- Try to include as many different math concepts as possible. Your goal is to include at least one concept from each of the chapters you studied this year.

- Organize your report in a folder, and think of a title for your report.

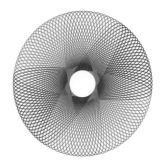

Classifying Animals

1 Getting Started

Biologists classify animals by placing them in phylums, or groups, with similar characteristics. Latin names, such as Chordata (having a spinal cord) or Arthropoda (having jointed limbs and rigid bodies) are used to describe these groups.

Biological classification is difficult, and scientists are still developing a complete system. There are seven main ranks of life on Earth; kingdom, phylum, class, order, family, genus, and species. However, scientists usually use more than these seven ranks to classify organisms.

Essential Question How does the classification of living organisms help you understand the similarities and differences of animals?

Write a report about how animals are classified. Choose several different animals and list the phylum, class, and order of each animal.

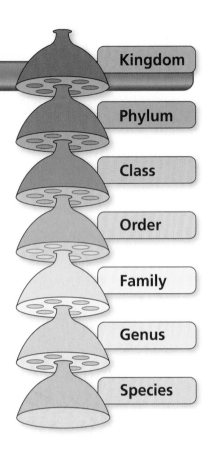

Kingdom

Phylum

Class

Order

Family

Genus

Species

Sample: A bat is classified as an animal in the phylum Chordata, class Mammalia, and order Chiroptera. *Chiroptera* is a Greek word meaning "hand-wing."

Wasp
Phylum: Arthropoda
Class: Insecta
Order: Hymenoptera
(membranous wing)

Bat
Phylum: Chordata
Class: Mammalia
Order: Chiroptera
(hand-wing)

Monkey
Phylum: Chordata
Class: Mammalia
Order: Primate (large brain)

Kangaroo
Phylum: Chordata
Class: Mammalia
Order: Diprotodontia
(two front teeth)

Project Notes

Introduction

For the Teacher

- **Goal:** Students will discover how to classify animals.
- **Management Tip:** You may want students to work together in groups.

Essential Question

- How does the classification of living organisms help you understand the similarities and differences of animals?

Things to Think About

? How are living organisms classified?

- All living things are classified into groups based on their similarities and differences. Here is a mnemonic that may help: King Phillip Came Over From Great Spain – Kingdom, Phylum, Class, Order, Family, Genus, and Species.

? How many kingdoms are there?

- The kingdom that categorizes animals is called Animalia. The other kingdoms categorize plants, fungi, bacteria, and protists.

? What are some examples of phyla in the animal kingdom?

- There are many phyla in the Animalia Kingdom, a few of them are: Chordata (all animals with backbones such as fish, birds, mammals, reptiles, and amphibians), Arthropoda (insects, spiders, and crustaceans), Mollusca (snails, squid, and clams), Annelida (segmented worms), and Echinodermata (starfish and sea urchins).

? What are some examples of classes in the animal phyla?

- The phylum Chordata is broken into several classes including Mammalia (mammals), Aves (birds), Reptilia (reptiles), and Amphibia (amphibians).

? What are some examples of orders in the animal classes?

- The class Mammalia is broken into many orders. A few of them are: Rodentia (rats and mice), Primates (Old- and New-World monkeys), Chiroptera (bats), Carnivora (dogs and cats), Perissodactyla (horses and zebras), and Proboscidea (elephants).

? What are some examples of families in the animal orders?

- A few of the families for the order Carnivora are: Canidae (dogs), Felidae (cats), Ursidae (bears), Hyaenidae (hyaena and aardwolves), and Mustelidae (weasels and wolverines).

? What are some examples of genera in the animal families?

- The Felidae family is divided into: Acinonyx (cheetah), Panthera (lion and tiger), Neofelis (clouded leopard), and Felis (domestic cats).

? What are some examples of species in the animal genera?

- When naming the species, scientists place the genus in front of the species, so that Panthera is divided into Panthera leo (lion) and Panthera tigris (tiger).

References

Go to *BigIdeasMath.com* to access links related to this project.

Cross-Curricular Instruction

Meet with a science teacher and review curriculum maps to identify whether students have covered animal classification. If the topic has already been covered, you may want to discuss the work students have completed and then review prior knowledge with them. If the science teacher has not discussed these concepts, perhaps you can both work simultaneously on these concepts and share notes. Or, you may want to explore activities that the science teacher has done in the past to support student learning in this particular area.

Project Notes

? How can you use a graphic organizer to help classify animals?
- You can use many different graphic organizers. Examples are shown.
- If you are trying to remember pieces of information about a particular animal, you may want to use a four square shown below.

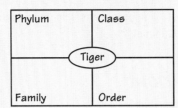

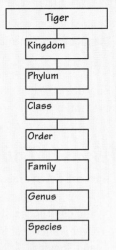

? Can animals be divided into vertebrates and invertebrates?
- All vertebrates are from the phylum Chordata. Invertebrates may be from different phyla.

Invertebrates	Vertebrates
Protozoa	Fish
Echinoderms (starfish)	Amphibians (frogs)
Annelids (earthworms)	Reptiles (alligators)
Mollusks (octopus)	Birds
Arthropods (crabs)	Mammals
Crustaceans (crabs)	Marsupials (kangaroos)
Arachnids (spiders)	Primates (gorillas)
Insects	Rodents (mice)
	Cetaceans (whales, dolphins)
	Animals such as seals

? What percent of the animal species are invertebrates?
- 97% of the animal species in the world are invertebrates. Even though vertebrates represent a much smaller portion of the animal kingdom, they dominate their environments because of their size and mobility.

Closure

- **Rubric** An editable rubric for this project is available at *BigIdeasMath.com*.
- Students may present their reports to the class or compare their reports with other students' reports.

② Things to Include

- List the different classes of phylum Chordata. Have you seen a member of each class?

- List the different classes of phylum Arthropoda. Have you seen a member of each class?

- Show how you can use graphic organizers to help classify animals. Which types of graphic organizers seem to be most helpful? Explain your reasoning.

- Summarize the number of species in each phlyum in an organized way. Be sure to include fractions, decimals, and percents.

Parrot
Phylum: Chordata
Class: Aves
Order: Psittaciformes
(strong, curved bill)

Frog
Phylum: Chordata
Class: Amphibia
Order: Anura (no tail)

Spider
Phylum: Arthropoda
Class: Arachnida
Order: Araneae (spider)

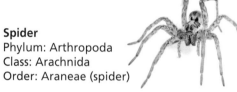

③ Things to Remember

- Organize your report in a folder, and think of a title for your report.

Lobster
Phylum: Arthropoda
Class: Malacostraca
Order: Decopada (ten footed)

Crocodile
Phylum: Chordata
Class: Reptilia
Order: Crocodilia
(pebble-worm)

Cougar
Phylum: Chordata
Class: Mammalia
Order: Carnivora (meat eater)

Key Vocabulary Index

Mathematical terms are best understood when you see them used and defined *in context*. This index lists where you will find key vocabulary. A full glossary is available in your Record and Practice Journal and at *BigIdeasMath.com*.

absolute value, 4
additive inverse, 10
adjacent angles, 272
biased sample, 442
center, 318
circle, 318
circumference, 319
complementary angles, 278
complex fraction, 165
composite figure, 326
compound event, 424
congruent angles, 272
congruent sides, 284
constant of proportionality, 200
cross products, 173
cross section, 388
dependent events, 431
diameter, 318
direct variation, 200
discount, 248
equivalent equations, 98
event, 402
experiment, 402
experimental probability, 414
factoring an expression, 92
favorable outcomes, 402
Fundamental Counting Principle, 422
graph of an inequality, 127
independent events, 430
inequality, 126
integer, 4
interest, 254
kite, 294
lateral surface area, 358
like terms, 82
linear expression, 88
markup, 248
opposites, 10

outcomes, 402
percent of change, 242
percent of decrease, 242
percent error, 243
percent of increase, 242
pi, 319
population, 440
principal, 254
probability, 408
proportion, 172
proportional, 172
radius, 318
rate, 164
ratio, 164
rational number, 46
regular pyramid, 364
relative frequency, 412
repeating decimal, 46
sample, 440
sample space, 422
scale, 300
scale drawing, 300
scale factor, 301
scale model, 300
semicircle, 320
simple interest, 254
simplest form, 82
simulation, 436
slant height, 364
slope, 194
solution of an inequality, 126
solution set, 126
supplementary angles, 278
terminating decimal, 46
theoretical probability, 415
unbiased sample, 442
unit rate, 164
vertical angles, 272

Student Index

This student-friendly index will help you find vocabulary, key ideas, and concepts. It is easily accessible and designed to be a reference for you whether you are looking for a definition, real-life application, or help with avoiding common errors.

A

Absolute value, 2–7
defined, 4
error analysis, 6
real-life application, 5
Addition
of expressions
linear, 86–91
modeling, 86–87
of integers, 8–15
with different signs, 8–11
error analysis, 12
with the same sign, 8–11
Property
of Equality, 98
of Inequality, 132
real-life application, 99
of rational numbers, 50–55
error analysis, 54
real-life application, 53
writing, 54, 55, 65
to solve inequalities, 130–135
Addition Property of Equality, 98
real-life application, 99
Addition Property of Inequality,
132
Additive inverse
defined, 10
writing, 12
Additive Inverse Property, 10
Adjacent angles
constructions, 270–275
defined, 272
Algebra
equations
equivalent, 98
modeling, 96–97, 102,
108–109
solving, 96–113
two-step, 108–113
writing, 100, 106, 112
expressions
linear, 86–91
modeling, 86–87, 91
simplifying, 80–85
writing, 84, 90
formulas, *See* Formulas
Algebra tiles
equations, 96–97, 102, 108–109
expressions, 86–87

Angle(s)
adjacent
constructions, 270–275
defined, 272
classifying, 277–278
triangles by, 284
complementary
constructions, 276–281
defined, 278
congruent
defined, 272
reading, 284
measures
of a quadrilateral, 295
of a triangle, 288–289
naming, 272
error analysis, 274
sums
for a quadrilateral, 295
for a triangle, 288–289
supplementary
constructions, 276–281
defined, 278
vertical
constructions, 270–275
defined, 272
Area, *See also* Surface area
of a circle, 332–337
formula, 334
of a composite figure, 338–343

B

Biased sample(s), defined, 442

C

Center, defined, 318
Choose Tools, *Throughout. For
example, see:*
angles, 271
circles and circumference, 321
probability, 406, 426
Circle(s)
area of, 332–337
formula, 334
center of, 318
circumference and, 316–323
defined, 319
formula, 319
research, 323
defined, 318

diameter of, 318
pi, 316, 319
semicircle, 320
Circumference
and circles, 316–323
defined, 319
formula, 319
research, 323
Common Error
inequalities, 141
percent equations, 234
Complementary angles
constructing, 276–281
defined, 278
Complex fraction, defined, 165
Composite figure(s)
area of, 338–343
defined, 326
perimeters of, 324–329
error analysis, 328
Compound event(s), *See also*
Events, Probability
defined, 424
error analysis, 426
writing, 425
Congruent angles
defined, 272
reading, 284
Congruent sides
defined, 284
reading, 284
Connections to math strands
Algebra, 13, 19, 27, 32
Geometry, 91, 101, 113, 175, 245,
366
Constant of proportionality,
defined, 200
Constructions
angles
adjacent, 270–275
vertical, 270–275
quadrilaterals, 292–297
triangles, 282–289
error analysis, 286
Critical Thinking, *Throughout. For
example, see:*
circles, area of, 337
composite figures
area of, 343
perimeter of, 329
cylinders, 372
discounts, 251

inequalities, 129
interest, 257
percent
    of change, 245
    of decrease, 245
    of increase, 245
prisms, 381
probability, 411
proportions, 175, 191
    direct variation, 203
pyramids, 366
rates, 169
rational numbers, 49
    dividing, 69
ratios, 168, 175
samples, 444, 445
scale drawing, 303
scale factor, 304
simplifying expressions, 85
slant height, 366
slope, 197
solving equations, 101, 107
surface area, 366, 372, 380
triangles, 287
volume, 380, 381
Cross products, defined, 173
Cross Products Property, 173
Cross section(s)
    defined, 388
    of three-dimensional figures,
        388–389
Cube(s), surface area of, 358
Cylinder(s)
    cross section of, 388–389
    surface area of, 368–373
        error analysis, 372
        formula, 370
        real-life application, 371

D

Decimal(s)
    comparing
        fractions with, 220–225
        percents with, 220–225
        real-life application, 222
    ordering
        with fractions, 220–225
        with percents, 220–225
        real-life application, 223
    percents as, 214–219
        error analysis, 218
        modeling, 214–215
        real-life application, 217
    repeating
        defined, 46
        error analysis, 48
        writing, 48

terminating
    defined, 46
    writing, 48
writing
    as fractions, 47
    as percents, 215–216
    rational numbers as, 46
Dependent events, *See also* Events,
        Probability
    defined, 431
    error analysis, 434
    formula, 431
    writing, 433
Diameter, defined, 318
Different Words, Same Question,
        Throughout. For example,
        see:
    adding rational numbers, 54
    area of a circle, 336
    constructing triangles, 286
    direct variation, 202
    inequalities, 128
    linear expressions, 90
    percent equations, 236
    probability, 433
    subtracting integers, 18
    surface area of a prism, 359
Direct variation, 198–203
    constant of proportionality, 200
    defined, 200
    error analysis, 202
    modeling, 203
    real-life application, 201
    writing, 202
Discount(s), 246–251
    defined, 248
    writing, 250
Division
    equations
        modeling, 102
        real-life application, 105
        solving by, 102–107
    of integers, 28–33
        with different signs, 28–30
        error analysis, 32
        reading, 32
        real-life application, 31
        with the same sign, 28–30
        writing, 32
    Property
        of Equality, 104
        of Inequality, 140–141
        real-life application, 105
    of rational numbers, 64–69
        error analysis, 68
        real-life application, 67
        writing, 65, 68
    to solve inequalities, 138–145

Division Property of Equality, 104
Division Property of Inequality,
        140–141
    error analysis, 144

E

Equality
    Addition Property of, 98
    Division Property of, 104
    Multiplication Property of, 104
    Subtraction Property of, 98
Equation(s)
    addition, 96–101
        error analysis, 100
        modeling, 96–97
        writing, 100
    division, 102–107
        modeling, 102
        real-life application, 105
    equivalent, 98
    multiplication, 102–107
        error analysis, 106
        writing, 106
    percent, 232–237
        error analysis, 236
        real-life application, 235
    percent proportions, 226–231
        error analysis, 230
        real-life application, 229
        writing, 230
    subtraction, 96–101
        modeling, 96–97
        real-life application, 99
    two-step
        error analysis, 112
        modeling, 108–109
        real-life application, 111
        solving, 108–113
        writing, 112, 113
Equivalent equations
    defined, 98
    writing, 100
Error Analysis, *Throughout. For*
        example, see:
    absolute value, 6
    classifying triangles, 286
    decimals, as percents, 218
    direct variation, 202
    equations
        solving, 100, 106
        two-step, 112
    expressions
        like terms of, 84
        linear, 91
    inequalities
        solving, 134, 143, 144, 150
        writing, 128

integers
  adding, 12
  dividing, 32
  multiplying, 26
  subtracting, 18
naming angles, 274
outcomes, 404
percent
  equations, 236
  of increase, 244
  proportions, 230
perimeter of a composite figure,
    328
prisms
  surface area, 360
  volume of, 380
probability, 410
  dependent events, 434
  Fundamental Counting
    Principle, 426
  outcomes, 404
proportions
  direct variation, 202
  percent, 230
  solving, 190
  writing, 182
rational numbers
  adding, 54
  decimal representation of, 48
  dividing, 68
  multiplying, 68
  subtracting, 62
scale drawings, 303
simple interest, 256
slope, 196
surface area
  of a cylinder, 372
  of a prism, 360
volume of a prism, 380
Event(s), *See also* Probability
  compound, 420–427
    defined, 424
  defined, 402
  dependent, 428–435
    defined, 431
    error analysis, 434
    writing, 433
  independent, 428–435
    defined, 430
    writing, 433
  outcomes of, 402
  probability of, 406–411
    defined, 408
    error analysis, 410
Example and non-example chart,
    290

Experiment(s)
  defined, 402
  outcomes of, 400–405
  project, 453
  reading, 402
  simulations, 436–437
    defined, 436
Experimental probability, 412–419
  defined, 414
  formula, 414
Expression(s)
  algebraic
    error analysis, 84
    like terms, 82
    modeling, 85
    real-life application, 83
    simplest form of, 82
    simplifying, 80–85
    writing, 84
  factoring, 92–93
    defined, 92
  linear
    adding, 86–91
    defined, 88
    error analysis, 91
    modeling, 86–87, 91
    real-life application, 89
    subtracting, 86–91
    writing, 90

F

Factoring an expression, 92–93
  defined, 92
Favorable outcome(s), defined,
    402
Formulas
  angles
    sum for a quadrilateral, 295
    sum for a triangle, 288
  area
    of a circle, 334
    of a parallelogram, 341
    of a rectangle, 341
    of a semicircle, 341
    of a triangle, 341
  circumference, 319
  pi, 316
  probability
    dependent events, 431
    of an event, 408
    experimental, 414
    independent events, 430
    relative frequency, 412
    theoretical, 415
  surface area
    of a cube, 358
    of a cylinder, 370

  of a prism, 357
  of a pyramid, 364
  of a rectangular prism, 356
  volume
    of a cube, 378
    of a prism, 378
    of a pyramid, 384
Four square, 94
Fraction(s)
  comparing
    decimals with, 220–225
    percents with, 220–225
    real-life application, 222
  complex, defined, 165
  decimals as
    error analysis, 48
    writing, 47
  ordering
    with decimals, 220–225
    with percents, 220–225
    real-life application, 223
Fundamental Counting Principle
  defined, 422
  error analysis, 426
  writing, 425

G

Geometry
  angles
    adjacent, 270–275
    complementary, 276–281
    congruent, 272
    constructing, 270–281
    error analysis, 274
    supplementary, 276–281
    vertical, 270–275
  circles
    area of, 332–337
    center of, 318
    circumference, 316–323
    diameter, 318
    radius, 318
    semicircles, 320
  composite figures
    area of, 338–343
    perimeter of, 324–329
  constructions and drawings,
    270–289, 292–305
  pi, 316
  quadrilaterals, 294
    constructing, 292–297
    defined, 292
    kite, 294
    parallelogram, 294, 341
    rectangle, 294, 341
    rhombus, 294

square, 294
    trapezoid, 294
surface area
    of a cylinder, 368–373
    of a prism, 354–361
    of a pyramid, 362–367
    of a rectangular prism, 356
trapezoids, 294
triangles
    area of, 341
    classifying, 284
    congruent sides, 284
    constructing, 282–287
volume of a prism, 376–381
Graph of an inequality, defined, 127
Graphic Organizers
    example and non-example chart, 290
    four square, 94
    idea and examples chart, 20
    information frame, 374
    information wheel, 184
    notetaking organizer, 438
    process diagram, 56
    summary triangle, 238
    word magnet, 330
    Y chart, 136
Graphing
    inequalities, 124–129
        defined, 127
    proportional relationships, 176–177

I

Idea and examples chart, 20
Independent events, *See also*
    Events, Probability
    defined, 430
    formula, 430
    writing, 433
Inequality
    Addition Property of, 132
    defined, 126
    Division Property of, 140–141
    graphing, 124–129
        defined, 127
    Multiplication Property of, 140–141
    solution of
        defined, 126
        reading, 133
    solution set of
        defined, 126
    solving two-step, 146–151
        error analysis, 150

real-life application, 149
    writing, 150
solving using addition, 130–135
    error analysis, 134
    real-life application, 133
    writing, 130
solving using division, 138–145
    error analysis, 144
    project, 145
    writing, 139
solving using multiplication, 138–145
    error analysis, 143
    project, 145
    writing, 139, 143
solving using subtraction, 130–135
    error analysis, 134
    writing, 130
Subtraction Property of, 132
symbols, 126
    reading, 133
writing, 124–129
    error analysis, 128
    modeling, 129
Information frame, 374
Information wheel, 184
Integer(s)
    absolute value of, 2–7
        error analysis, 6
        real-life application, 5
    adding, 8–15
        with different signs, 8–11
        error analysis, 12
        with the same sign, 8–11
    additive inverse of, 10
        writing, 12
    defined, 4
    dividing, 28–33
        with different signs, 28–30
        error analysis, 32
        reading, 32
        real-life application, 31
        with the same sign, 28–30
        writing, 32
    multiplying, 22–27
        with different signs, 22–24
        error analysis, 26
        modeling, 27
        real-life application, 25
        with the same sign, 22–24
        writing, 26
    subtracting, 14–19
        error analysis, 18
        real-life application, 17
        writing, 18

Interest
    defined, 254
    principal, 254
    simple, 252–257
        defined, 254
        error analysis, 256
    writing, 256

K

Kite, defined, 294

L

Lateral surface area, defined, 358
Like terms
    defined, 82
    error analysis, 84
    writing, 84
Linear expression(s)
    adding, 86–91
        modeling, 86–87, 91
        writing, 90
    factoring, 92–93
    subtracting, 86–91
        error analysis, 91
        modeling, 87
        real-life application, 89
        writing, 90
Logic, *Throughout. For example, see:*
    absolute value, 3
    angles, 281
    circles, 323
    circumference, 323
    constructing a triangle, 287
    inequalities
        solving, 145
        writing, 129
    linear expression, 91
    percent equations, 237
    prisms, 381
    probability
        dependent events, 434
        experimental, 417
        independent events, 434
        theoretical, 417
    samples, 445

M

Markup(s), 246–251
    defined, 248
    writing, 250
Meaning of a Word
    adjacent, 270
    opposite, 10
    proportional, 170

rate, 162
rational, 44
Mental Math, *Throughout. For example, see:*
  integers
    adding, 13
    subtracting, 19
  proportions, 181
Modeling, *Throughout. For example, see:*
  decimals as percents, 214–215
  direct variation, 203
  expressions
    algebraic, 85
    linear, 86–87, 91
  inequalities, 129
  multiplying integers, 27
  percents as decimals, 214–215
  probability, 418
  scale models, 305
  solutions of equations, 96–97, 102
Multiplication
  equations, solving by, 102–107
  inequalities, solving by, 138–145
  of integers, 22–27
    with different signs, 22–24
    error analysis, 26
    modeling, 27
    real-life application, 25
    with the same sign, 22–24
    writing, 26
  Property
    of Equality, 104
    of Inequality, 140–141
  of rational numbers, 64–69
    error analysis, 68
    writing, 65, 68
Multiplication Property of Equality, 104
Multiplication Property of Inequality, 140–141
  error analysis, 143

N

Notetaking organizer, 438
Number Sense, *Throughout. For example, see:*
  adding rational numbers, 55
  fractions
    comparing percents with, 224
    ordering percents and, 225
  inequalities, 135, 145
  integers
    adding, 12
    dividing, 33
    multiplying, 27

percents
  comparing fractions with, 224
  of increase, 244
  ordering fractions and, 225
  proportions, 230
probability, 418
proportions, 190, 230
reciprocals, 68
slope, 197
solving equations, 107
statistics, 453
surface area of a prism, 361

O

Open-Ended, *Throughout. For example, see:*
  absolute value, 7
  angles, 275, 281
  area of a composite figure, 342
  composite figures, 328
  inequalities, 143, 150
  integers, 7
    adding, 13
    dividing, 32
    multiplying, 26
    opposite, 18
    subtracting, 19
  inverse operations, 106
  percents as decimals, 218
  probability, 410
    compound events, 425
    theoretical, 417
  proportions
    solving, 190
    writing, 174, 182
  rates, 167
  rational numbers
    adding, 54
    decimal representation of, 49
    multiplying, 69
    subtracting, 63
  ratios, 174
  scale factor, 304
  solving equations, 107
  surface area of a prism, 360
  volume
    of a prism, 381
    of a pyramid, 386, 387
Opposites, defined, 10
Outcomes, *See also* Events, Probability
  counting, 403
    error analysis, 404
  defined, 402
  experiment, 402
  favorable, 402

reading, 402
writing, 404

P

Parallelogram(s)
  area of, 341
  defined, 294
Patterns, *Throughout. For example, see:*
  dividing integers, 33
Percent(s)
  of change, 242
  comparing
    decimals with, 220–225
    fractions with, 220–225
    real-life application, 222
  as decimals, 214–219
    error analysis, 218
    modeling, 214–216
    real-life application, 217
  of decrease, 240–245
  equation, 232–237
    error analysis, 236
    real-life application, 235
  error, 243
  of increase, 240–245
  ordering
    with decimals, 220–225
    with fractions, 220–225
    real-life application, 223
  proportions, 226–231
    error analysis, 230
    real-life application, 229
    writing, 230
  research, 219
Percent of change
  defined, 242
  formula, 242
Percent of decrease
  defined, 242
  formula, 242
  writing, 244
Percent error
  defined, 243
  formula, 243
Percent of increase
  defined, 242
  error analysis, 244
  formula, 242
Perimeter(s)
  of composite figures, 324–329
    error analysis, 328
Pi
  defined, 319
  formula, 316

Polygon(s)
  kite, 294
  parallelogram, 294
  quadrilateral, 292–297
  rectangle, 294, 341
  rhombus, 294
  square, 294
  trapezoid, 294
  triangle, 282–289
Population(s), 440–453
  comparing, 448–453
    project, 453
    writing, 452
  defined, 440
  research, 441
  samples, 440–447
    biased, 442
    defined, 440
    unbiased, 442
Precision, *Throughout. For
    example, see:*
  constructing angles, 275, 281
  inequalities
    graphing, 128
    solving, 143
  ordering fractions, decimals,
    and percents, 225
  outcomes, 405
  percents
    of change, 245
  prisms, 381
  rates, 169
  rational numbers
    dividing, 69
    multiplying, 65
Prices
  discounts and markups, 246–251
    defined, 248
    writing, 250
Principal, defined, 254
Prism(s)
  cross section of, 388
  surface area of, 354–361
    error analysis, 360
    formula, 357
    real-life application, 358
    rectangular, 356
    writing, 359
  volume of, 376–381
    error analysis, 380
    formula, 378
    real-life application, 379
    rectangular, 376
    writing, 386

Probability, 406–419
  defined, 408
  error analysis, 410
  events, 400–405
    compound, 420–427
    dependent, 428–435
    error analysis, 434
    independent, 428–435
    writing, 433
  of events
    defined, 408
    formula for, 408
  experimental, 412–419
    defined, 414
    formula, 414
  experiments
    defined, 402
    reading, 402
    simulations, 436–437
  Fundamental Counting
      Principle, 422
    error analysis, 426
  outcomes, 400–405
    defined, 402
    error analysis, 404
    favorable, 402
    reading, 402
    writing, 404
  project, 419
  research, 427
  sample space, 422
  theoretical, 412–419
    defined, 415
    formula, 415
    modeling, 418
  writing, 425
Problem Solving, *Throughout. For
    example, see:*
  adding integers, 13
  angles, 281
  circumference, 323
  decimals as percents, 219
  equations, two-step, 113
  inequalities, two-step, 151
  percent proportions, 231
  probability, 427, 435
  proportions, 191
  rational numbers, 49
  simple interest, 257
  surface area, 367
  volume, 387
Process diagram, 56
Properties
  Addition Property of Equality, 98
    real-life application, 99

  Addition Property of Inequality,
      132
  Additive Inverse Property, 10
  Cross Products Property, 173
  Division Property of Equality,
      104
  Division Property of Inequality,
      140–141
    error analysis, 144
  Multiplication Property of
      Equality, 104
  Multiplication Property of
      Inequality, 140–141
    error analysis, 143
  Subtraction Property of Equality,
      98
  Subtraction Property of
      Inequality, 132
    real-life application, 133
Proportion(s), 170–175
  cross products, 173
  Cross Products Property, 173
  defined, 173
  percent, 226–231
    error analysis, 230
    real-life application, 229
    writing, 230
  scale and, 300
  solving, 186–191
    error analysis, 190
    real-life application, 189
    writing, 190
  writing, 178–183
    error analysis, 182
Proportional
  defined, 172
  relationship
    constant of proportionality,
        200
    direct variation, 198–203
    error analysis, 202
    graphing, 176–177
    modeling, 203
    reading, 172
    real-life application, 201
    writing, 202
Pyramid(s)
  cross section of, 388
  regular, 364
  slant height, 364
  surface area of, 362–367
    formula, 364
    real-life application, 365
  volume of, 382–387
    formula, 384
    real-life application, 385
    writing, 386

Q

Quadrilateral(s), 292–297
  classifying, 294
  constructions, 292–297
    reading, 294
  defined, 292
  kite, 294
  parallelogram, 294, 341
  rectangle, 294, 341
  rhombus, 294
  square, 294
  sum of angle measures, 295
  trapezoid, 294

R

Radius, defined, 318
Rate(s)
  defined, 164
  ratios and, 162–169
  research, 169
  unit rate
    defined, 164
    writing, 167
Ratio(s), *See also* Proportions, Rates
  complex fraction, 165
  defined, 164
  proportions and, 170–189
    cross products, 173
    Cross Products Property, 173
    error analysis, 182, 190
    proportional, 172
    real-life application, 189
    solving, 186–191
    writing, 178–183
  rates and, 162–169
    complex fractions, 165
    writing, 167
  scale and, 300
  slope, 192–197
    error analysis, 196
Rational number(s), *See also*
    Fractions, Decimals
  adding, 50–55
    error analysis, 54
    real-life application, 53
    writing, 54, 55, 65
  defined, 44, 46
  dividing, 64–69
    error analysis, 68
    real-life application, 67
    writing, 65, 68
  multiplying, 64–69
    error analysis, 68
    writing, 65, 68
  ordering, 47

repeating decimals
  defined, 46
  error analysis, 48
  writing, 46
subtracting, 58–63
  error analysis, 62
  real-life application, 61
  writing, 62
terminating decimals
  defined, 46
  writing, 46
writing as decimals, 46
Reading
  congruent angles, 284
  congruent sides, 284
  experiments, 402
  inequalities
    solving, 133
    symbols of, 126
  outcomes, 402
  proportional relationship, 172
  quadrilaterals, 294
Real-Life Applications, *Throughout.*
    For example, see:
  absolute value, 5
  decimals, 217, 222, 223
  direct variation, 201
  expressions
    linear, 89
    simplifying, 83
  fractions, 217, 222, 223
  integers
    dividing, 31
    multiplying, 25
    subtracting, 17
  percent, 217, 222, 223
    equations, 235
    proportions, 229
  proportions, 189, 229
  rational numbers
    adding, 53
    dividing, 67
    subtracting, 61
  solving equations, 99, 105, 111
  solving inequalities, 133, 149
  surface area
    of a cylinder, 371
    of a prism, 358
    of a pyramid, 365
  volume
    of a prism, 379
    of a pyramid, 385
Reasoning, *Throughout. For*
    example, see:
  absolute value, 7
  angles
    adjacent, 280
    complementary, 280, 281

    congruent, 275
    constructing, 275
    supplementary, 280, 281
  circles, 322
  circumference, 322
  composite figures
    area of, 342
    perimeter of, 328
  cones, 389
  cross sections
    of a cone, 389
    of a prism, 388
  direct variation, 203
  expressions
    linear, 91
    simplifying, 81, 84, 85
  inequalities, 128
    solving, 134, 135, 151
  integers
    multiplying, 27
    subtracting, 19
  markups, 250
  percents
    discount and markups, 250
    equations, 236
    of increase, 245
    proportions, 231
  probability, 410
    experimental, 417, 419
    simulations, 437
    theoretical, 419
  proportions
    proportional relationships,
      177, 203
    solving, 191
    writing, 183
  quadrilaterals, 293, 296
  rational numbers
    adding, 55
    subtracting, 63
  scale factor, 303
  scale models, 304
  simple interest, 256, 257
  slope, 196
  solving equations, 107, 113
  statistics
    comparing populations, 452
    samples, 445, 447
  surface area, 367
    of a cylinder, 373
    of a prism, 355
    of a pyramid, 363
  triangles
    angle measures of, 289
    constructions, 283
  volume
    of a prism, 387
    of a pyramid, 386, 387

Rectangle(s)
  area of, formula, 341
  defined, 294
Regular pyramid, defined, 364
Relative frequency, *See also*
      Probability
  defined, 412
  formula, 412
Repeated Reasoning, *Throughout.*
      For example, see:
  adding rational numbers, 55
  circles
      area, 337
      circumference, 337
  cylinders, 373
  probability, 427
  slope, 193
  solving equations, 109
  surface area, 373
Repeating decimal(s)
  defined, 46
  error analysis, 48
  writing, 48
Rhombus, defined, 294

Ⓢ

Sample(s), 440–447
  biased, 442
  defined, 440
  research, 441
  unbiased, 442
Sample space, defined, 422
Scale
  defined, 300
  research, 299
Scale drawing, 298–305
  defined, 300
  error analysis, 303
  research, 299
  scale, 300
  scale factor, 301
Scale factor
  defined, 301
  finding, 301
Scale model
  defined, 300
  finding distance in, 301
  modeling, 305
Semicircle(s)
  area of, 341
  defined, 320
Side(s)
  classifying triangles by, 284
  congruent, 284
      defined, 284
      reading, 284

Simple interest, 252–257
  defined, 254
  error analysis, 256
  writing, 256
Simplest form, defined, 82
Simulation(s), 436–437
  defined, 436
Slant height, defined, 364
Slope, 192–197
  defined, 194
  error analysis, 196
Solids, *See also:* Surface area,
      Three-dimensional figures,
      Volume
  cross section, 388–389
  cube, 358
  cylinder, 368–373
  prism, 354–361, 376–381
  pyramid, 362–367, 382–387
Solution of an inequality, defined,
      126
Solution set, defined, 126
Square, defined, 294
Statistics, *See also* Probability
  populations, 440–453
      comparing, 448–453
      defined, 440
      project, 453
      research, 441
      writing, 452
  samples, 440–447
      biased, 442
      defined, 440
      research, 441
      unbiased, 442
Structure, *Throughout. For*
      example, see:
  angles, 275
  composite figures, 343
  direct variation, 199
  expressions
      factoring, 93
      linear, 91
      simplifying, 85
  integers
      adding, 9
      dividing, 29
      multiplying, 23
      subtracting, 15
  parallelograms, 297
  percent proportions, 231
  probability, 435
  solving equations, 97
  subtracting rational numbers, 63
  surface area of a prism, 361
  volume
      of a prism, 377
      of a pyramid, 383

Study Tip
  additive inverse, 11
  amount of error, 243
  circumference, 319
  comparing populations, 450
  constructing triangles, 285
  cross products, 173
  decimal points, 216
  direct variation, 200
  discounts, 248
  exponents, 25
  expressions
      linear, 89
      simplifying, 82
  inequalities
      graphing, 127
      solving, 132
  opposites, 11
  ordering decimals and percents,
      222
  percent proportions, 228
  powers, 25
  prisms, 378
  probability, 408
      Fundamental Counting
          Principle, 422
      simulating outcomes, 436,
          437
  proportional relationships, 177
  pyramids, 384
      volume of, 384
  quadrilaterals, 294
  radius, 319
  rational numbers, 47, 52
  ratios, 302
  scale, 300
  scale drawings, 302
  simplifying equations, 110
  slope, 194
  squares, 294
  volume
      of a prism, 378
      of a pyramid, 384
Subtraction
  of expressions
      error analysis, 91
      linear, 86–91
      modeling, 87
      real-life application, 89
  of integers, 14–19
      error analysis, 18
      real-life application, 17
      writing, 18
  Property
      of Equality, 98
      of Inequality, 132

of rational numbers, 58–63
  error analysis, 62
  real-life application, 61
  writing, 62
to solve inequalities, 130–135
Subtraction Property of Equality, 98
Subtraction Property of Inequality,
      132
  real-life application, 133
Summary triangle, 238
Supplementary angles
  constructing, 276–281
  defined, 278
Surface area
  of a cylinder, 368–373
    error analysis, 372
    formula, 370
    real-life application, 371
  formula, 357
    cylinders, 370
    prism, 357
    pyramid, 364
    rectangular prism, 356
  lateral, defined, 358
  of a prism, 354–361
    cube, 358
    error analysis, 360
    formula, 357
    real-life application, 358
    rectangular, 356
    triangular, 357
    writing, 359
  of a pyramid, 362–367
    formula, 364
    real-life application, 365
Symbols of inequality, 126

T

Term(s)
  like, 82
    error analysis, 84
  writing, 84
Terminating decimal
  defined, 46
  writing, 48
Theoretical probability, *See also*
      Probability
  defined, 415
  formula, 415
  modeling, 418
Three-dimensional figure(s),
      388–389
  cross sections of, 388
Trapezoid, defined, 294

Triangle(s)
  acute, 284
  area of, formula, 341
  classifying
    by angles, 284
      error analysis, 286
    by sides, 284
      writing, 286
  congruent sides, 284
  constructing, 282–287
  equiangular, 284
  equilateral, 284
  isosceles, 284
  obtuse, 284
  right, 284
  scalene, 284
  sum of angle measures, 288–289
Two-step inequalities, *See*
      Inequality

U

Unbiased sample(s), defined, 442
Unit rate
  defined, 164
  writing, 167

V

Vertical angles
  constructions, 270–275
  defined, 272
  error analysis, 274
Volume
  of a prism, 376–381
    error analysis, 380
    formula, 378
    real-life application, 379
    rectangular, 376
  writing, 386
  of a pyramid, 382–387
    formula, 384
    real-life application, 385
  writing, 386

W

Which One Doesn't Belong?,
      Throughout. For example,
      see:
  absolute value, 6
  circles, 321
  comparing fractions, decimals,
      and percents, 218, 224
  dividing integers, 32
  inequalities, 134

percent proportions, 230
pyramids, 366
quadrilaterals, 296
rational numbers, 62
ratios, 174
solving equations, 100
Word magnet, 330
Writing, *Throughout. For example,*
      see:
  comparing populations, 452
  decimals
    repeating, 48
    terminating, 48
  discounts, 250
  equations
    equivalent, 100
    solving, 106
    two-step, 112, 113
  expressions
    algebraic, 84
    linear, 90
  integers
    additive inverse of, 12
    dividing, 32
    multiplying, 26
    subtracting, 18
  interest rate, 256
  markups, 250
  percents
    of decrease, 244
    proportions, 230
  probability
    events, 433
    Fundamental Counting
      Principle, 425
    outcomes, 404, 425
  proportions, 182
    solving, 190
  rates, 167
  rational numbers
    adding, 54, 55, 65
    dividing, 65, 68
    multiplying, 65, 68
    subtracting, 62, 65
  solving inequalities, 139, 143,
      150
  surface area of a prism, 359
  triangles, classifying, 286
  volume
    of a prism, 386
    of a pyramid, 386

Y

Y chart, 136

Additional Answers

Chapter 1

Try It Yourself

1. $11 + b$;
$$3 + (b + 8) = 3 + (8 + b) \qquad \text{Comm. Prop. of Add.}$$
$$= (3 + 8) + b \qquad \text{Assoc. Prop. of Add.}$$
$$= 11 + b \qquad \text{Add 3 and 8.}$$

2. $d + 10$;
$$(d + 4) + 6 = d + (4 + 6) \qquad \text{Assoc. Prop. of Add.}$$
$$= d + 10 \qquad \text{Add 4 and 6.}$$

3. $30p$;
$$6(5p) = (6 \cdot 5)p \qquad \text{Assoc. Prop. of Mult.}$$
$$= 30p \qquad \text{Multiply 6 and 5.}$$

4. 0;
$$13 \cdot m \cdot 0 = 13 \cdot 0 \cdot m \qquad \text{Comm. Prop. of Mult.}$$
$$= (13 \cdot 0) \cdot m \qquad \text{Assoc. Prop. of Mult.}$$
$$= 0 \cdot m \qquad \text{Mult. Prop. of Zero}$$
$$= 0 \qquad \text{Mult. Prop. of Zero}$$

5. $29x$;
$$1 \cdot x \cdot 29 = 1 \cdot 29 \cdot x \qquad \text{Comm. Prop. of Mult.}$$
$$= (1 \cdot 29) \cdot x \qquad \text{Assoc. Prop. of Mult.}$$
$$= 29x \qquad \text{Mult. Prop. of One}$$

6. $n + 14$;
$$(n + 14) + 0 = n + (14 + 0) \qquad \text{Assoc. Prop. of Add.}$$
$$= n + 14 \qquad \text{Add. Prop. of Zero}$$

Record and Practice Journal
Fair Game Review

6. $32k$;
$$8(4k) = (8 \cdot 4)k \qquad \text{Assoc. Prop. of Mult.}$$
$$= 32k \qquad \text{Multiply 8 and 4.}$$

7. 0;
$$13 \cdot 0 \cdot p = (13 \cdot 0)p \qquad \text{Assoc. Prop. of Mult.}$$
$$= 0 \cdot p = 0 \qquad \text{Mult. Prop. of Zero}$$

8. 0;
$$7 \cdot z \cdot 0 = 7 \cdot 0 \cdot z \qquad \text{Comm. Prop. of Mult.}$$
$$= (7 \cdot 0) \cdot z \qquad \text{Assoc. Prop. of Mult.}$$
$$= 0 \cdot z = 0 \qquad \text{Mult. Prop. of Zero}$$

9. $2.5w$;
$$2.5 \cdot w \cdot 1 = 2.5 \cdot (w \cdot 1) \qquad \text{Assoc. Prop. of Mult.}$$
$$= 2.5 \cdot w \qquad \text{Mult. Prop. of One}$$
$$= 2.5w$$

10. $19x$;
$$1 \cdot x \cdot 19 = 1 \cdot 19 \cdot x \qquad \text{Comm. Prop. of Mult.}$$
$$= (1 \cdot 19) \cdot x \qquad \text{Assoc. Prop. of Mult.}$$
$$= 19x \qquad \text{Mult. Prop. of One}$$

11. $t + 3$;
$$(t + 3) + 0 = t + (3 + 0) \qquad \text{Assoc. Prop. of Add.}$$
$$= t + 3 \qquad \text{Add. Prop. of Zero}$$

12. $4 + g$;
$$0 + (g + 4) = 0 + (4 + g) \qquad \text{Comm. Prop. of Add.}$$
$$= (0 + 4) + g \qquad \text{Assoc. Prop. of Add.}$$
$$= 4 + g \qquad \text{Add. Prop. of Zero}$$

Section 1.2
Practice and Problem Solving

48. a. point C; E is $15 + (-13) = 2$ higher than C, so C is deeper.

b. point B; D is $-18 + 15 = -3$ from B, so D is 3 units lower than B.

Section 1.5
Fair Game Review

42.

43.

44.

Chapter 2

Record and Practice Journal
Fair Game Review

11. $\dfrac{47}{30}$

12. $\dfrac{1}{3}$

13. $\dfrac{2}{35}$

14. $\dfrac{5}{27}$

15. $\dfrac{2}{5}$

16. $\dfrac{14}{11}$

17. $\dfrac{3}{4}$

18. $7\dfrac{1}{12}$ cups

Section 2.1
Practice and Problem Solving

40. $-5\dfrac{3}{11} < -5.\overline{2}$

41. $-2\dfrac{13}{16} < -2\dfrac{11}{14}$

42. *Sample answer:* $-0.4, -0.\overline{45}$

43. Michelle

44. math quiz

45. No; The base of the skating pool is at -10 feet, which is deeper than $-9\dfrac{5}{6}$ feet.

47. a. when a is negative

b. when a and b have the same sign, $a \neq 0 \neq b$

Section 2.2
Practice and Problem Solving

24. The sum will be positive when the addend with the greater absolute value is positive. The sum will be negative when the addend with the greater absolute value is negative. The sum will be zero when the numbers are opposites.

Section 2.4
Practice and Problem Solving

46. b. 0.03 in.;

$$\frac{-0.05 + 0.09 + (-0.04) + (-0.08) + 0.03}{5} = -0.01$$

Chapter 3

Section 3.1
Practice and Problem Solving

24. Solution B is correct.

Solution A: The expression inside the parentheses was simplified first, but 2 and $-5x$ are not like terms.

Solution C: The solution evaluated the expression from left to right by subtracting 4 from 6, instead of using order of operations and distributing the -4 first.

Solution D: The solution did not distribute the negative sign to both of the terms inside the parentheses.

25. $(9 + 3x)$ ft^2

26. $(20x + 25y)$ dollars

27. *Sample answer:*

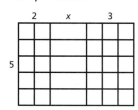

$5x + 25$

28. *Sample answer:*

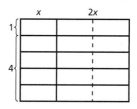

$15x$

Section 3.2
Record and Practice Journal

5. Add—combine like tiles and remove zero pairs

Subtract—remove like tiles, add zero pairs to the first expression as necessary, then remove like tiles

Practice and Problem Solving

25. The -3 was not distributed to both terms inside the parentheses.

$$(4m + 9) - 3(2m - 5) = 4m + 9 - 6m + 15$$
$$= 4m - 6m + 9 + 15$$
$$= -2m + 24$$

Extension 3.2
Practice

16. a. *Sample answer:*

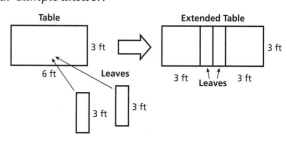

b. $6 + 2x$; shorter dimension of each leaf

Section 3.5
Record and Practice Journal

5. Add enough 1 or -1 tiles to each side to remove the zero pairs and leave only variable tiles on one side. Form as many equal groups of 1 or -1 tiles as there are variable tiles. One of these groups shows the value of the variable.

Chapter 4

Section 4.2

Practice and Problem Solving

12. $z \geq 3.1$;

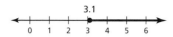

13. $-2.8 < d$;

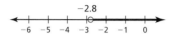

14. $-\dfrac{4}{5} > s$;

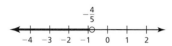

15. $\dfrac{3}{4} \geq m$;

16. $r < -0.9$;

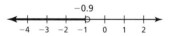

17. $h \leq -2.4$;

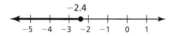

18. The 2 should have been subtracted rather than added.

$$x - 7 > -2$$
$$\underline{+7 \quad +7}$$
$$x > 5$$

19. The wrong side of the number line is shaded.

20. **a.** $15 + p \leq 44$; $p \leq 29$ passengers

   b. no; Only 29 more passengers can board the plane.

Section 4.3

Record and Practice Journal

5. If you multiply or divide each side of an inequality by the same positive number, the inequality remains true.

   If you multiply or divide each side of an inequality by the same negative number, the direction of the inequality symbol must be reversed for the inequality to remain true.

Practice and Problem Solving

17. $y \leq -3$;

18. $b < 48.59$;

19. The inequality sign should not have been reversed.

$$\frac{x}{3} < -9$$
$$3 \cdot \frac{x}{3} < 3 \cdot (-9)$$
$$x < -27$$

20. $\dfrac{x}{4} \leq 5$; $x \leq 20$ **21.** $\dfrac{x}{7} < -3$; $x < -21$

22. $6x \geq -24$; $x \geq -4$ **23.** $-2x > 30$; $x < -15$

35. $b > 6$;

36. They forgot to reverse the inequality symbol.

$$-3m \geq 9$$
$$\frac{-3m}{-3} \leq \frac{9}{-3}$$
$$m \leq -3$$

37. $-2.5x < -20$; $x > 8$ h

38. **a.** $27x \leq 150$; $x \leq \dfrac{50}{9}$, or $5\dfrac{5}{9}$

   b. no; The maximum height allowed is 150 inches, and 6 boxes has a height of 162 inches.

39. $10x \geq 120$; $x \geq 12$ cm

46. $x \geq 3$;

47. $s < 14$;

Section 4.4
Practice and Problem Solving

12. $g > -1$;

13. $w \le 3$;

14. $k \ge -18$;

15. $d > -9$;

16. $n < -0.6$;

17. $c \ge -1.95$;

Chapter 5

Record and Practice Journal
Fair Game Review

7. no

8. yes

9. yes

10. no

11. $\dfrac{6}{29}$

12. $d = -48$

13. $x = 21$

14. $n = 40$

15. $a = -9$

16. $k = -5$

17. $y = -3$

18. $w = 18$

19. $z = 90$

20. $4p = 35$; $p = \$8.75$

Section 5.1
Record and Practice Journal

1.

Description	Verbal Rate	Numerical
Your running rate in a 100-meter dash	meters per second	$\dfrac{8 \text{ m}}{\text{sec}}$; $\dfrac{80 \text{ m}}{\text{sec}}$
The fertilization rate for an apple orchard	pounds per acre	$\dfrac{150 \text{ lb}}{\text{acre}}$; $\dfrac{1 \text{ lb}}{\text{acre}}$
The average pay rate for a professional athlete	dollars per year	$\dfrac{\$3{,}000{,}000}{\text{yr}}$; $\dfrac{\$3000}{\text{yr}}$
The average rainfall rate in a rain forest	inches per year	$\dfrac{100 \text{ in.}}{\text{yr}}$; $\dfrac{5 \text{ in.}}{\text{yr}}$

Section 5.2
Practice and Problem Solving

32. yes; Because Ratio A is equivalent to Ratio B, Ratios A and B simplify to the same ratio. Because Ratio B is equivalent to Ratio C, Ratios B and C simplify to the same ratio. Ratios A and C simplify to the same ratio, so they are equivalent.

Extension 5.2
Practice

3. (0, 0): You earn $0 for working 0 hours.

(1, 15): You earn $15 for working 1 hour; unit rate: $\dfrac{\$15}{1 \text{ h}}$

(4, 60): You earn $60 for working 4 hours; unit rate: $\dfrac{\$60}{4 \text{ h}} = \dfrac{\$15}{1 \text{ h}}$

4. (0, 0): The balloon rises 0 feet in 0 seconds.

(1, 5): The balloon rises 5 feet in 1 second; unit rate: $\dfrac{5 \text{ ft}}{1 \text{ sec}}$

(6, 30): The balloon rises 30 feet in 6 seconds; unit rate: $\dfrac{30 \text{ ft}}{6 \text{ sec}} = \dfrac{5 \text{ ft}}{1 \text{ sec}}$

5. yes; 5 ft/h

6. no

7. $y = \dfrac{4}{3}$

8. a. You

Days	1	2	3	4	5
Cost (dollars)	1.50	2	2.50	3	3.50

Friend

Days	1	2	3	4	5
Cost (dollars)	1.25	2.50	3.75	5	6.25

b. your friend

Record and Practice Journal Practice

1.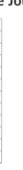
no

2.
yes

3.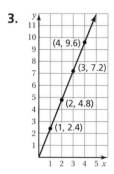
yes

4.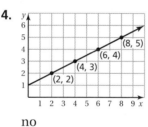
no

5. (0, 0): The car travels 0 miles in 0 hours.
(1, 60): The car travels 60 miles in 1 hour.
(2, 120): The car travels 120 miles in 2 hours.

6. (0, 0): 0 pounds of shrimp costs \$0.
(4, 40): 4 pounds of shrimp costs \$40.
(7, 70): 7 pounds of shrimp costs \$70.

7. (0, 0): You receive 0 emails in 0 days.
(3, 45): You receive 45 emails in 3 days.
(4, 60): You receive 60 emails in 4 days.

8. (0, 0): There are 0 cups of blueberries in 0 pies.
(2, 12): There are 12 cups of blueberries in 2 pies.
(4, 24): There are 24 cups of blueberries in 4 pies.

Section 5.5

Record and Practice Journal

1. c. mi/h to ft/sec: Multiply by 5280 to convert miles to feet and divide by 3600 to convert hours to seconds.

ft/sec to mi/h: Divide by 5280 to convert feet to miles and multiply by 3600 to convert seconds to hours.

Practice and Problem Solving

12. The change in y should be in the numerator. The change in x should be in the denominator.
Slope $= \dfrac{5}{4}$

13.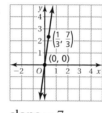
slope $= 7$

14.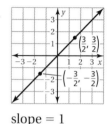
slope $= 1$

15.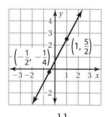
slope $= \dfrac{11}{6}$

17. a.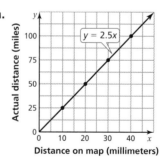

b. 2.5; Every millimeter represents 2.5 miles.

c. 120 mi

d. 90 mm

Section 5.6

On Your Own

4. no; The equation cannot be written as $y = kx$.

5. yes; The equation can be written as $y = kx$.

6. no; The equation cannot be written as $y = kx$.

Practice and Problem Solving

14. yes; The equation can be written as $y = kx$; $k = 1$

15. yes; The equation can be written as $y = kx$; $k = \frac{1}{2}$

16. no; The equation cannot be written as $y = kx$.

17. no; The equation cannot be written as $y = kx$.

18. The line does not pass through the origin, so x and y do not show direct variation.

19.

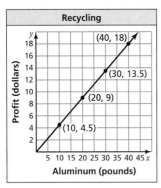

yes; $y = 0.45x$

Chapter 5 Test

12.

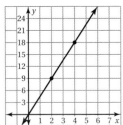 slope $= \frac{9}{2}$

13. no; The equation cannot be written as $y = kx$.

14. no; The equation cannot be written as $y = kx$.

15. yes; The equation can be written as $y = kx$.

16. $58

17. a. The number of cycles during the day is greater than the number of cycles during the night.

   b. Day: 40; Night: 30
   The crosswalk cycles 40 times per hour during the day and 30 times per hour during the night.

Chapter 6

Section 6.2

Practice and Problem Solving

20.

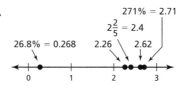

Section 6.3

Practice and Problem Solving

19. 34 represents the part, not the whole.

$$\frac{a}{w} = \frac{p}{100}$$

$$\frac{34}{w} = \frac{40}{100}$$

$$w = 85$$

29. a. a scale along the vertical axis

   b. 6.25%; *Sample answer:* Although you do not know the actual number of votes, you can visualize each bar as a model with the horizontal lines breaking the data into equal parts. The sum of all the parts is 16. Greg has the least parts with 1, which is $100\% \div 16 = 6.25\%$.

   c. 31 votes

Section 6.5

Practice and Problem Solving

24. Increasing 20 to 40 is the same as increasing 20 by 20. So, it is a 100% increase. Decreasing 40 to 20 is the same as decreasing 40 by one-half of 40. So, it is a 50% decrease.

25. a. about 16.95% increase

   b. 161,391 people

29. less than; *Sample answer:* Let x represent the number. A 10% increase is equal to $x + 0.1x$, or $1.1x$. A 10% decrease of this new number is equal to $1.1x - 0.1(1.1x)$, or $0.99x$. Because $0.99x < x$, the result is less than the original number.

Chapter 7

Record and Practice Journal Fair Game Review

7.
80°

8.
35°

9.
100°

10.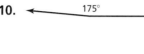
175°

Section 7.1

Practice and Problem Solving

15.
85°
85°

16.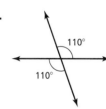
110°
110°

17.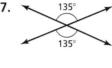
135°
135°

18. 43

19. a. *Sample answer:*

b. *Sample answer:*
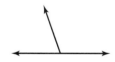

c. *Sample answer:*

$x°$
$(135 - x)°$

20. *Sample answer:*
1) Draw one angle, then draw the other angle using a side of the first angle.
2) Draw one angle that is the sum of the two angles, then draw the shared side.

21. never   **22.** always

23. sometimes   **24.** always

25.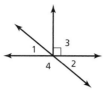
1
3
4
2

Section 7.2

Practice and Problem Solving

17.
20°

18.
35°

19.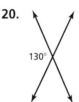
80°

20.
130°

21. *Sample answer:* 1) Draw one angle, then draw the other using a side of the first angle; 2) Draw a right angle, then draw the shared side.

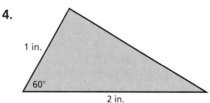
30°

22. *Sample answer:* 120°; It is supplementary with a 60° angle, but it is greater than 90°, so it cannot be complementary with another angle.

23. a. 25°   **b.** 65°

Section 7.3

On Your Own

4.
1 in.
60°
2 in.

right scalene triangle

Additional Answers   **A27**

Practice and Problem Solving

3. *Sample answer:*

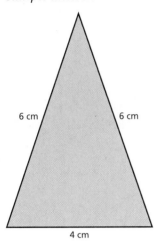

6 cm   6 cm

4 cm

4. *Sample answer:*

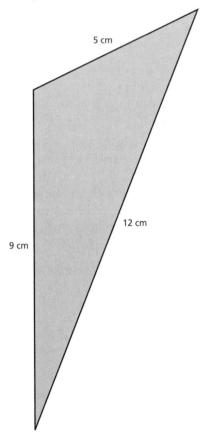

5 cm

12 cm

9 cm

5. *Sample answer:*

55°

65°   60°

12. The triangle is not an acute triangle because acute triangles have 3 angles less than 90°. The triangle is an obtuse scalene triangle because it has one angle greater than 90° and no congruent sides.

13. acute isosceles

14. 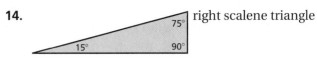 right scalene triangle

75°

15°   90°

15. 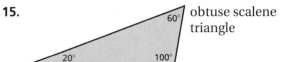 obtuse scalene triangle

60°

20°   100°

16. 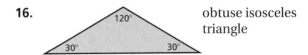 obtuse isosceles triangle

120°

30°   30°

17.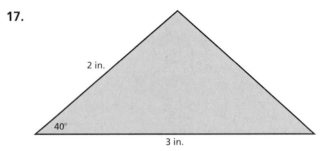

2 in.

40°

3 in.

18.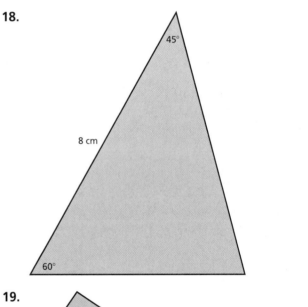

45°

8 cm

60°

19.

20. not necessarily; Just because one angle is acute doesn't mean it will be an acute triangle. The classification depends on the third side. It could form a right angle or an obtuse angle.

23. many; You can change the angle formed by the two given sides to create many triangles.

24. one; Only one line segment can be drawn between the end points of the two given sides.

25. no; The sum of any two side lengths must be greater than the remaining length.

26. no; An equilateral triangle cannot have a right angle.

27. a. green: 65; purple: 25; red: 45

b. The angles opposite the congruent sides are congruent.

c. An isosceles triangle has at least two angles that are congruent.

Extension 7.3

Practice

16. If two angle measures of a triangle were each greater than or equal to 90°, the sum of three angle measures would be greater than 180°, which is not possible.

17. a. 72

b. You can change the distance between the bottoms of the two upright cards; yes; x must be greater than 60 and less than 90; If x were less than or equal to 60, the two upright cards would have to be exactly on the edges of the base card or off the base card. It is not possible to stack cards at these angles. If x where equal to 90, then the two upright cards would be vertical, which is not possible. The card structure would not be stable. In practice, the limits on x are probably closer to $70 < x < 80$.

Record and Practice Journal Practice

1. $x = 60$; acute, isosceles

2. $x = 30$; obtuse, isosceles

3. $x = 31$; right, scalene

4. $x = 28$; acute, scalene

5. $x = 35$ **6.** yes

7. no; 54.5°   **8.** no; 86°

9. no; 28°   **10.** yes

11. yes

7.1–7.3 Quiz

9.

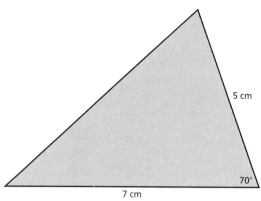

Section 7.4

Record and Practice Journal

1. Sample answers are given.

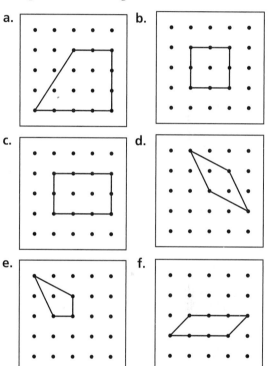

Practice and Problem Solving

25.

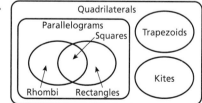

26. a. $x = 125$; $y = 55$

b. Opposite angles of a parallelogram are equal.

c. Consecutive interior angles of a parallelogram are supplementary.

Section 7.5

Record and Practice Journal

6. Painting:

	Actual Object	Original Drawing	Your Drawing
Perimeter	192 in.	24 units	8 units
Area	2304 in.^{2}	36 units^2	4 units^2

Check students' work. Students should conclude that the ratio of the perimeters for the original drawings to the actual objects is equal to the ratios they found in Activities 1(c) and 3(c). The ratio of the areas for the original drawings to the actual objects is equal to the square of those ratios. Students can find these ratios for their own drawings to the actual object and see the same relationships.

On Your Own

4. c. The scale factor and the ratio of the perimeters change to $\frac{10}{3}$, and the ratio of the areas changes to $\left(\frac{10}{3}\right)^2$; The change in scale results in a change to each of these three values, but the ratio of the perimeters is still the same as the scale factor, and the ratio of the areas is still the same as the square of the scale factor.

Practice and Problem Solving

25.

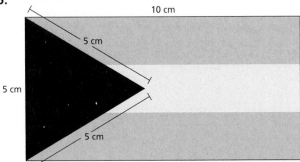

Not actual size

7.4–7.5 Quiz

6.

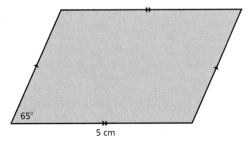

Chapter 7 Review

6.

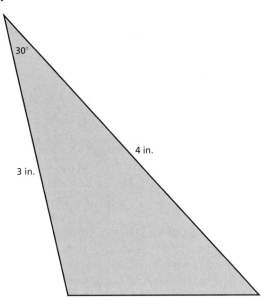

Chapter 7 Test

7.

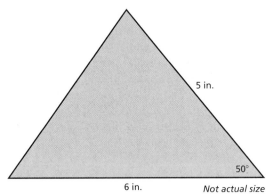

Not actual size

15.

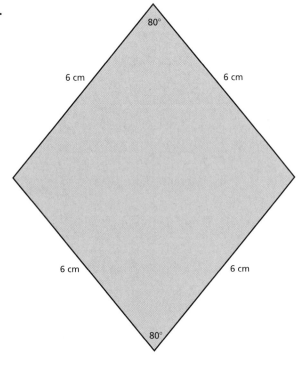

Chapter 8

Section 8.1

Practice and Problem Solving

25. b. about 320.28 mm; Subtract $\frac{1}{12}$ the circumference of a circle whose radius is the length of the hour hand from the circumference of a circle whose radius is the length of the minute hand.

Section 8.2

Practice and Problem Solving

18. The starting points are staggered so that each runner can run the same distance and use the same finish line. This is necessary because the circumference is different for each lane. The diagram shows this because the diameter is greater in the outer lanes.

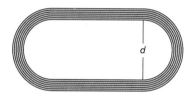

Section 8.3

Practice and Problem Solving

13. about 628 cm²

14. about 226.08 in.²

15. about 1.57 ft²

16. a.

2	4π in.	4π in.²
4	8π in.	16π in.²
8	16π in.	64π in.²
16	32π in.	256π in.²

    b. The circumference doubles; The area becomes four times as great.

    c. The circumference triples; The area becomes 9 times as great.

19. about 9.8125 in.²; The two regions are identical, so find one-half the area of the circle.

20. about 17.415 m²; Subtract the area of a circle with a 9-meter diameter from the area of the square.

21. about 4.56 ft²; Find the area of the shaded regions by subtracting the areas of both unshaded regions from the area of the quarter-circle containing them. The area of each unshaded region can be found by subtracting the area of the smaller shaded region from the semicircle. The area of the smaller shaded region can be found by drawing a square about the region.

Subtract the area of a quarter-circle from the area of the square to find an unshaded area. Then subtract both unshaded areas from the square's area to find the shaded region's area.

Section 8.4

Practice and Problem Solving

12. 89 m²

13. 23.5 in.²

14. about 21.87 ft²

15. 24 m²

Chapter 9

Section 9.1

Practice and Problem Solving

26. The dimensions of the red prism are $\frac{3}{2}$ times the dimensions of the blue prism. The surface area of the red prism is $\frac{9}{4}$ times the surface area of the blue prism.

27. a. 0.125 pint

    b. 1.125 pints

    c. red and green: The ratio of the paint amounts (red to green) is 4 : 1 and the ratio of the side lengths is 2 : 1.

       green and blue: The ratio of the paint amounts (blue to green) is 9 : 1 and the ratio of the side lengths is 3 : 1.

       The ratio of the paint amounts is the square of the ratio of the side lengths.

28. a. 0.5 in.

    b. 13.5 in.²

Section 9.2

Record and Practice Journal

1. a. $S = 85,560$ m^2 **b.** $S = 1404$ m^2

   c. $S = 1960$ m^2 **d.** $S = 1276$ m^2

2. b.

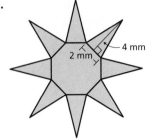

Practice and Problem Solving

18. The slant height is greater. The height is the distance between the top and the point on the base directly beneath it. The distance from the top to any other point on the base is greater than the height.

19. 124 cm^2

20. greater than; If it is less than or equal to, then the lateral face could not meet at a vertex to form a solid.

Section 9.3

Practice and Problem Solving

12. The area of only one base is added. The first term should have a factor of 2;

$$S = 2\pi r^2 + 2\pi rh$$
$$= 2\pi (5)^2 + 2\pi (5)(10.6)$$
$$= 50\pi + 106\pi$$
$$= 156\pi \approx 489.8 \text{ yd}^2$$

18. a. 4 times greater; 9 times greater; 25 times greater; 100 times greater

   b. When both dimensions are multiplied by a factor of k, the surface area increases by a factor of k^2; 400 times greater

Section 9.4

Practice and Problem Solving

19. 1728 in.^{3}

$1 \times 1 \times 1 = 1$ ft^3 $12 \times 12 \times 12 = 1728$ in.^{3}

Section 9.5

Practice and Problem Solving

17. 12,000 in.^{3}; The volume of one paperweight is 12 cubic inches. So, 12 cubic inches of glass is needed to make one paperweight. So, it takes $12 \times 1000 = 12,000$ cubic inches to make 1000 paperweights.

Chapter 10

Section 10.3

Record and Practice Journal

6. *Sample answer:* Most likely this is true because there are only 20 chips in the bag and you did not select an orange chip in 50 tries. However, you cannot say this for certain because there may be 1 orange chip in the bag and you never selected it.

Practice and Problem Solving

28. theoretical

29. a. $\dfrac{1}{12}$; 50 times

   b. $\dfrac{11}{50}$; 132 times

   c. A larger number of trials should result in a more accurate probability, which gives a more accurate prediction.

34. a. Check students' work. The cup should land on its side most of the time.

   b. Check students' work.

   c. Check students' work.

   d. more likely; Due to the added weight, the cup will be more likely to hit open-end up and thus more likely to land open-end up. Some students may justify by performing multiple trials with a quarter taped to the bottom of the cup.

Section 10.4

Practice and Problem Solving

12. Tree Diagram:

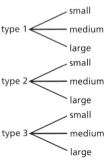

Fundamental Counting Principle: $3 \cdot 3 = 9$

27. a. $\dfrac{1}{1000}$, or 0.1%

b. There are 1000 possible combinations. With 5 tries, someone would guess 5 out of the 1000 possibilities. So, the probability of getting the correct combination is $\dfrac{5}{1000}$, or 0.5%.

29. a. The Fundamental Counting Principle is more efficient. A tree diagram would be too large.

b. 1,000,000,000 or one billion

c. *Sample answer:* Not all possible number combinations are used for Social Security Numbers (SSN). SSNs are coded into geographical, group, and serial numbers. Some SSNs are reserved for commercial use and some are forbidden for various reasons.

Section 10.5

Fair Game Review

27.

; right scalene

Extension 10.5

Practice

1. a. *Sample answer:* Roll four number cubes. Let an odd number represent a correct answer and an even number represent an incorrect answer. Run 40 trials.

b. Check students' work. The probability should be "close" to 6.25% (depending on the number of trials, because that is the theoretical probability).

2. *Sample answer:* Place 7 green and 3 red marbles in a bag. Let the green marbles represent a win and the red marbles represent a loss. Randomly pick one marble to simulate the first game. Replace the marble and repeat two more times. This is one trial. Run 30 trials. Check students' work. The probability should be close to 34.3% (depending on the number of trials, because that is the theoretical probability).

3. *Sample answer:* Using the spreadsheet in Example 3 and using digits 1–4 as successes, the experimental probability is 16%.

4. Example 1: $\dfrac{1}{2} \cdot \dfrac{1}{2} \cdot \dfrac{1}{2} = \dfrac{1}{8}$

Example 2: $0.6 \times 0.2 = 0.12$, or 12%

Example 3: Students likely will not be able to find this theoretical probability. However, it can be found by examining the favorable outcomes and using logic:

$$P(1, 2, 3, 4) = \dfrac{1}{2} \cdot \dfrac{1}{2} \cdot \dfrac{1}{2} \cdot \dfrac{1}{2} = \dfrac{1}{16}$$

$$P(1, 2, 3, \text{not } 4) = \dfrac{1}{2} \cdot \dfrac{1}{2} \cdot \dfrac{1}{2} \cdot \dfrac{1}{2} = \dfrac{1}{16}$$

$$P(1, 2, \text{not } 3, 4) = \dfrac{1}{2} \cdot \dfrac{1}{2} \cdot \dfrac{1}{2} \cdot \dfrac{1}{2} = \dfrac{1}{16}$$

$$P(1, \text{not } 2, 3, 4) = \dfrac{1}{2} \cdot \dfrac{1}{2} \cdot \dfrac{1}{2} \cdot \dfrac{1}{2} = \dfrac{1}{16}$$

$$P(\text{not } 1, 2, 3, 4) = \dfrac{1}{2} \cdot \dfrac{1}{2} \cdot \dfrac{1}{2} \cdot \dfrac{1}{2} = \dfrac{1}{16}$$

So, the theoretical probability is $5\left(\dfrac{1}{16}\right) = \dfrac{5}{16} = 31.25\%$.

Or, they could realize that there are 5 favorable outcomes out of 16 (from the Fundamental Counting Principle), and all outcomes are equally likely, so the probability is $\dfrac{5}{16}$.

Exercise 1: $\dfrac{1}{2} \cdot \dfrac{1}{2} \cdot \dfrac{1}{2} \cdot \dfrac{1}{2} = \dfrac{1}{16}$

Exercise 2: $0.7^3 = \dfrac{343}{1000}$, or 34.3%

Exercise 3: Students likely will not be able to find this theoretical probability. However, it can be found by examining the favorable outcomes and using logic:

$$P(1, 2, 3, 4) = \dfrac{2}{5} \cdot \dfrac{2}{5} \cdot \dfrac{2}{5} \cdot \dfrac{2}{5} = \dfrac{16}{625}$$

$$P(1, 2, 3, \text{not } 4) = \dfrac{2}{5} \cdot \dfrac{2}{5} \cdot \dfrac{2}{5} \cdot \dfrac{3}{5} = \dfrac{24}{625}$$

$$P(1, 2, \text{not } 3, 4) = \dfrac{2}{5} \cdot \dfrac{2}{5} \cdot \dfrac{3}{5} \cdot \dfrac{2}{5} = \dfrac{24}{625}$$

$$P(1, \text{not } 2, 3, 4) = \dfrac{2}{5} \cdot \dfrac{3}{5} \cdot \dfrac{2}{5} \cdot \dfrac{2}{5} = \dfrac{24}{625}$$

$$P(\text{not } 1, 2, 3, 4) = \dfrac{3}{5} \cdot \dfrac{2}{5} \cdot \dfrac{2}{5} \cdot \dfrac{2}{5} = \dfrac{24}{625}$$

So, the theoretical probability is $\dfrac{16}{625} + 4\left(\dfrac{24}{625}\right) = \dfrac{112}{625} = 17.92\%$.

Notice that you still have 5 favorable outcomes and 16 possible outcomes but the outcomes are not equally likely, so the probability is not $\dfrac{5}{16}$ as before.

When you increase the number of trials in a simulation, the experimental probability approaches the theoretical probability of the event that you are simulating.

Record and Practice Journal Practice

1. **a.** *Sample answer:* Roll four number cubes. Let an odd number represent a *no* answer and an even number represent a *yes* answer. Run 40 trials.

   b. Check students' work. The probability should be close to 6.25% (depending on the number of trials, because that is the theoretical probability).

2. **a.** Place 7 green and 3 red marbles in bag. Let the green marbles represent a snowy day and the red marbles represent a snowless day. Randomly pick one marble to simulate the today. Replace the marble and repeat one more time. This is one trial. Run 50 trials.

   b. Check students' work. The probability should be close to 42% (depending on the number of trials, because that is the theoretical probability).

Section 10.6

Practice and Problem Solving

9. no; The sample is not representative of the population because people going to the baseball stadium are more likely to support building a new baseball stadium. So, the sample is biased and the conclusion is not valid.

10. Sample B because it is a larger sample.

15. sample; It is much easier to collect sample data in this situation.

16. 1260 students

17. Not everyone has an email address, so the sample may not be representative of the entire population. *Sample answer:* When the survey question is about technology or which email service you use, the sample may be representative of the entire population.

18. **a.** *Sample answer:* The person could ask, "Do you agree with the town's unfair ban on skateboarding on public property?"

    b. *Sample answer:* The person could ask, "Do you agree that the town's ban on skateboarding on public property has made the town quieter and safer?"

Section 10.7

Record and Practice Journal

1. **a.** *Sample answer:* double box-and-whisker plot or double dot plot

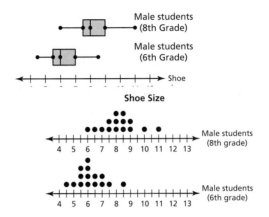

2. **a.** male students: symmetric; female students: skewed left; yes; *Sample answer:* The data set for the female students completely overlaps the data set for the male students. The overlaps between the centers and between the extreme values are shown.

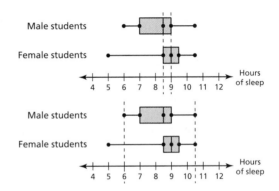

   b. male students: symmetric; female students: symmetric; yes; *Sample answer:* The overlaps between the centers and between the extreme values are shown.

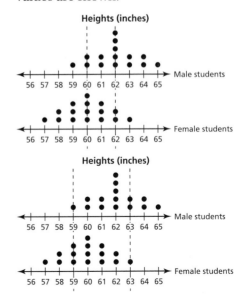

Practice and Problem Solving

5. a. Class A: median = 90, IQR = 12.5
Class B: median = 80, IQR = 10
The variation in the test scores is about the same, but Class A has greater test scores.

  b. The difference in the medians is 0.8 to 1 times the IQR.

6. a. volleyball: mean = 86, MAD = 19.6
basketball: mean = 185, MAD = 17.7
The variation in the attendances is about the same, but basketball has a greater attendance.

  b. The difference in the means is about 5.1 to 5.6 times the MAD.

8. a. The mean and MAD for the sports magazine, 19 and 5.8, are close to the mean and MAD for the political magazine, 18 and 5.2. However, the sample size is small and the variability is too great to conclude that the number of words per sentence is about the same.

  b. The sample means vary much less than the sample numbers of words per sentence.

  c. The number of words per sentence is generally greater in the political magazine than in the sports magazine.

9. a. Check students' work. Experiments should include taking many samples of a manageable size from each grade level. This will be more doable if the work of sampling is divided among the whole class, and the results are pooled together.

  b. Check students' work. The data may or may not support a conclusion.

Photo Credits

Chapter 8

314 ©iStockphoto.com/Michael Flippo, ©iStockphoto.com/Ann Marie Kurtz; **316** ©iStockphoto.com/HultonArchive; **317** Sir Cumference and the First Round Table; text copyright © 1997 by Cindy Neuschwander; illustrations copyright © 1997 by Wayne Geehan. Used with permission by Charlesbridge Publishing, Inc., 85 Main Street, Watertown, MA 02472; 617-926-0329; www.charlesbridge.com. All rights reserved.; **321** *Exercises 3 and 4* ©iStockphoto.com/zentilia; *Exercise 5* ©iStockphoto.com/Dave Hopkins; *Exercise 6* ©iStockphoto.com/Boris Yankov; *Exercise 7* ©iStockphoto.com/ALEAIMAGE; *Exercise 8* ©iStockphoto.com/iLexx; *Exercise 9–11* ©iStockphoto.com/alexander mychko; **323** *center left* ©iStockphoto.com/HultonArchive; *center right* ©iStockphoto.com/winterling; **329** ©iStockphoto.com/Scott Slattery; **331** *Exercise 6* ©iStockphoto.com/DivaNir4A; *Exercise 7* ©iStockphoto.com/Stacey Walker; *Exercise 8* ©iStockphoto.com/Creativeye99; *Exercise 11* Kalamazoo (Michigan) Public Library; **335** ©iStockphoto.com/Brian Sullivan; **336** *Exercise 3* ©iStockphoto.com/zentilia; *Exercises 4 and 5* ©iStockphoto.com/PhotoHamster; *Exercise 6* ©iStockphoto.com/subjug; *Exercise 7* ©iStockphoto.com/Dave Hopkins; *Exercise 8* ©iStockphoto.com/7nuit; **342** Big Ideas Learning, LLC; **344** *Exercise 4* ©iStockphoto.com/Mr_Vector; *Exercises 5 and 6* ©iStockphoto.com/sndr

Chapter 9

352 stephan kerkhofs/Shutterstock.com, Cigdem Sean Cooper/Shutterstock.com, ©iStockphoto.com/Andreas Gradin; **355** ©iStockphoto.com/Remigiusz Załucki; **358** *center left* Bob the Wikipedian / CC-BY-SA-3.0; *center right* ©iStockphoto.com/Sherwin McGehee; **360** Yasonya/Shutterstock.com; **T-361** Yasonya/Shutterstock.com; **361** Stankevich/Shutterstock.com; **362** *top left* ©iStockphoto.com/Luke Daniek; *top right* ©iStockphoto.com/Jeff Whyte; *bottom left* ©Michael Mattox. Image from BigStockPhoto.com; *bottom right* ©iStockphoto.com/Hedda Gjerpen; **363** ©iStockphoto.com/josh webb; **372** ©iStockphoto.com/Tomasz Pietryszek; **T-373** ©iStockphoto.com/scol22; **373** *center left* Newcastle Drum Centre; *center right* ©iStockphoto.com/scol22; **381** *top left* ©iStockphoto.com/david franklin; *top right* ©Ruslan Kokarev. Image from BigStockPhoto.com; *center right* ©iStockphoto.com/Lev Mel, ©iStockphoto.com/Ebru Baraz; **382** *bottom left* ©iStockphoto.com/Jiri Vatka; *bottom right* Patryk Kosmider/Shutterstock.com; **T-384** Christophe Testi/Shutterstock.com **386** ©iStockphoto.com/ranplett, Image © Courtesy of Museum of Science, Boston; **T-387** ©iStockphoto.com/Yails; **387** *center right* ©iStockphoto.com/James Kingman; *center left* ©iStockphoto.com/Yails; **389** *Exercise 12* ©iStockphoto.com/AlexStar; *Exercise 13* Knartz/Shutterstock.com; *Exercise 14* SOMMAI/Shutterstock.com

Chapter 10

398 infografick/Shutterstock.com, ©iStockphoto.com/Ann Marie Kurtz; **401** ryasick photography/Shutterstock.com; **402** *center* ©iStockphoto.com/sweetym; *bottom left* Big Ideas Learning, LLC; **403** ©iStockphoto.com/sweetym; **404** ©iStockphoto.com/Joe Potato Photo, ©iStockphoto.com/sweetym; **405** *top left* ©iStockphoto.com/Jennifer Morgan; *top center* United States coin image from the United States Mint; **409** *top left* design56/Shutterstock.com; *center right* ©iStockphoto.com/spxChrome; **410** *center right* Daniel Skorodyelov/Shutterstock.com; *bottom left* Mitrofanova/Shutterstock.com; **411** Jamie Wilson/Shutterstock.com; **412** *top right* James Steidl/Shutterstock.com; *bottom right* traudl/Shutterstock.com; **413** Warren Goldswain/Shutterstock.com; **414** ©iStockphoto.com/Frank van de Bergh; **415** ©iStockphoto.com/Eric Ferguson; **419** Feng Yu/Shutterstock.com; **420** *top right* John McLaird/Shutterstock.com; *center right* Robert Asento/Shutterstock.com; *bottom right* Mark Aplet/Shutterstock.com; **425** ©iStockphoto.com/Justin Horrocks; **426** ©iStockphoto.com/sweetym; **T-427** tele52/Shutterstock.com; **427** *top right* John McLaird/Shutterstock.com; *center* tele52/Shutterstock.com; **428 and 429** ©iStockphoto.com/Joe Potato Photo, ©iStockphoto.com/sweetym; **431** Univega/Shutterstock.com; **432** James Steidl/Shutterstock.com; **434** *top left* ©iStockphoto.com/sweetym; *center right* ©iStockphoto.com/Andy Cook; **435** -Albachiaraa-/Shutterstock.com; **439** ©iStockphoto.com/Doug Cannell; **440** *Activity 1a left* ©iStockphoto.com/Shannon Keegan; *Activity 1a right* ©iStockphoto.com/Lorelyn Medina; *Activity 1b left* Joel Sartore/joelsartore.com; *Activity 1b right* Feng Yu/Shutterstock.com; *Activity 1c left* ©iStockphoto.com/kledge; *Activity 1c right* ©iStockphoto.com/spxChrome; *Activity 1d* ©iStockphoto.com/Alex Slobodkin; **442** ©iStockphoto.com/Philip Lange; **444** ©iStockphoto.com/blackwaterimages, ©iStockphoto.com/Rodrigo Blanco, ©iStockphoto.com/7nuit; **447** Feng Yu/Shutterstock.com; **451** artis777/Shutterstock.com; **457** GoodMood Photo/Shutterstock.com; **458** Asaf Eliason/Shutterstock.com; **460** *top right* ©iStockphoto.com/Frank Ramspott; *bottom left* ©iStockphoto.com/7nuit; **462** mylisa/Shutterstock.com; **463** ra-design/Shutterstock.com

Appendix A

A0 *background* ©iStockphoto.com/Björn Kindler; *top left* ©iStockphoto.com/MichaelMattner; *top right* AKaiser/Shutterstock.com; **A1** *top right* ©iStockphoto.com/daver2002ua; *bottom left* ©iStockphoto.com/Eric Isselée; *bottom right* ©iStockphoto.com/Victor Paez; **A2** sgame/Shutterstock.com; **A3** *top right* AKaiser/Shutterstock.com; *bottom left* Photononstop/SuperStock, ©iStockphoto.com/Adam Radosavljevic; *bottom right* ©iStockphoto.com/MichaelMattner; **A4** ©iStockphoto.com/Duncan Walker; **A5** *top right* ©iStockphoto.com/DNY59; *bottom left* ©iStockphoto.com/Boris Katsman; *bottom right* © User:MatthiasKabel/Wikimedia Commons / CC-BY-SA-3.0 / GFDL; **A6** *top right* ©iStockphoto.com/Keith Webber Jr.; *bottom left* ©iStockphoto.com/Eliza Snow; *bottom right* ©iStockphoto.com/David H. Seymour; **A7** *top right* ©iStockphoto.com/daver2002ua; *bottom left* ©iStockphoto.com/Victor Paez; **A8** *wasp* ©iStockphoto.com/arlindo71; *bat* ©iStockphoto.com/Alexei Zaycev; *monkey and kangaroo* ©iStockphoto.com/Eric Isselée; **A9** *macaw* ©iStockphoto.com/Eric Isselée; *frog* ©iStockphoto.com/Brandon Alms; *spider* ©iStockphoto.com/arlindo71; *puma* ©iStockphoto.com/Eric Isselée; *lobster* ©iStockphoto.com/Joan Vicent Cantó Roig; *crocodile* ©iStockphoto.com/Mehmet Salih Guler

Cartoon illustrations Tyler Stout

Learning Progression

Kindergarten

Counting and Cardinality	– Count to 100 by Ones and Tens; Compare Numbers
Operations and Algebraic Thinking	– Understand and Model Addition and Subtraction
Number and Operations in Base Ten	– Work with Numbers 11–19 to Gain Foundations for Place Value
Measurement and Data	– Describe and Compare Measurable Attributes; Classify Objects into Categories
Geometry	– Identify and Describe Shapes

Grade 1

Operations and Algebraic Thinking	– Represent and Solve Addition and Subtraction Problems
Number and Operations in Base Ten	– Understand Place Value for Two-Digit Numbers; Use Place Value and Properties to Add and Subtract
Measurement and Data	– Measure Lengths Indirectly; Write and Tell Time; Represent and Interpret Data
Geometry	– Draw Shapes; Partition Circles and Rectangles into Two and Four Equal Shares

Grade 2

Operations and Algebraic Thinking	– Solve One- and Two-Step Problems Involving Addition and Subtraction; Build a Foundation for Multiplication
Number and Operations in Base Ten	– Understand Place Value for Three-Digit Numbers; Use Place Value and Properties to Add and Subtract
Measurement and Data	– Measure and Estimate Lengths in Standard Units; Work with Time and Money
Geometry	– Draw and Identify Shapes; Partition Circles and Rectangles into Two, Three, and Four Equal Shares

Grade 3

Operations and Algebraic Thinking — Represent and Solve Problems Involving Multiplication and Division; Solve Two-Step Problems Involving Four Operations

Number and Operations in Base Ten — Round Whole Numbers; Add, Subtract, and Multiply Multi-Digit Whole Numbers

Number and Operations— Fractions — Understand Fractions as Numbers

Measurement and Data — Solve Time, Liquid Volume, and Mass Problems; Understand Perimeter and Area

Geometry — Reason with Shapes and Their Attributes

Grade 4

Operations and Algebraic Thinking — Use the Four Operations with Whole Numbers to Solve Problems; Understand Factors and Multiples

Number and Operations in Base Ten — Generalize Place Value Understanding; Perform Multi-Digit Arithmetic

Number and Operations— Fractions — Build Fractions from Unit Fractions; Understand Decimal Notation for Fractions

Measurement and Data — Convert Measurements; Understand and Measure Angles

Geometry — Draw and Identify Lines and Angles; Classify Shapes

Grade 5

Operations and Algebraic Thinking — Write and Interpret Numerical Expressions

Number and Operations in Base Ten — Perform Operations with Multi-Digit Numbers and Decimals to Hundredths

Number and Operations— Fractions — Add, Subtract, Multiply, and Divide Fractions

Measurement and Data — Convert Measurements within a Measurement System; Understand Volume

Geometry — Graph Points in the First Quadrant of the Coordinate Plane; Classify Two-Dimensional Figures

Mathematics Reference Sheet

Conversions

U.S. Customary
1 foot = 12 inches
1 yard = 3 feet
1 mile = 5280 feet
1 acre = 43,560 square feet
1 cup = 8 fluid ounces
1 pint = 2 cups
1 quart = 2 pints
1 gallon = 4 quarts
1 gallon = 231 cubic inches
1 pound = 16 ounces
1 ton = 2000 pounds
1 cubic foot ≈ 7.5 gallons

U.S. Customary to Metric
1 inch = 2.54 centimeters
1 foot ≈ 0.3 meter
1 mile ≈ 1.61 kilometers
1 quart ≈ 0.95 liter
1 gallon ≈ 3.79 liters
1 cup ≈ 237 milliliters
1 pound ≈ 0.45 kilogram
1 ounce ≈ 28.3 grams
1 gallon ≈ 3785 cubic centimeters

Time
1 minute = 60 seconds
1 hour = 60 minutes
1 hour = 3600 seconds
1 year = 52 weeks

Temperature
$$C = \frac{5}{9}(F - 32)$$

$$F = \frac{9}{5}C + 32$$

Metric
1 centimeter = 10 millimeters
1 meter = 100 centimeters
1 kilometer = 1000 meters
1 liter = 1000 milliliters
1 kiloliter = 1000 liters
1 milliliter = 1 cubic centimeter
1 liter = 1000 cubic centimeters
1 cubic millimeter = 0.001 milliliter
1 gram = 1000 milligrams
1 kilogram = 1000 grams

Metric to U.S. Customary
1 centimeter ≈ 0.39 inch
1 meter ≈ 3.28 feet
1 kilometer ≈ 0.62 mile
1 liter ≈ 1.06 quarts
1 liter ≈ 0.26 gallon
1 kilogram ≈ 2.2 pounds
1 gram ≈ 0.035 ounce
1 cubic meter ≈ 264 gallons

Number Properties

Commutative Properties of Addition and Multiplication
$$a + b = b + a$$
$$a \cdot b = b \cdot a$$

Associative Properties of Addition and Multiplication
$$(a + b) + c = a + (b + c)$$
$$(a \cdot b) \cdot c = a \cdot (b \cdot c)$$

Addition Property of Zero
$$a + 0 = a$$

Multiplication Properties of Zero and One
$$a \cdot 0 = 0$$
$$a \cdot 1 = a$$

Distributive Property:
$$a(b + c) = ab + ac$$
$$a(b - c) = ab - ac$$

Properties of Equality

Addition Property of Equality
If $a = b$, then $a + c = b + c$.

Subtraction Property of Equality
If $a = b$, then $a - c = b - c$.

Multiplication Property of Equality
If $a = b$, then $a \cdot c = b \cdot c$.

Multiplicative Inverse Property
$$n \cdot \frac{1}{n} = \frac{1}{n} \cdot n = 1, n \neq 0$$

Division Property of Equality
If $a = b$, then $a \div c = b \div c, c \neq 0$.

Properties of Inequality

Addition Property of Inequality
If $a > b$, then $a + c > b + c$.

Subtraction Property of Inequality
If $a > b$, then $a - c > b - c$.

Multiplication Property of Inequality
If $a > b$ and c is positive, then $a \cdot c > b \cdot c$.
If $a > b$ and c is negative, then $a \cdot c < b \cdot c$.

Division Property of Inequality
If $a > b$ and c is positive, then $a \div c > b \div c$.
If $a > b$ and c is negative, then $a \div c < b \div c$.

Circumference and Area of a Circle

$C = \pi d$ or $C = 2\pi r$

$A = \pi r^2$

$\pi \approx \dfrac{22}{7}$, or 3.14

Angles of Polygons

Sum of the Angle Measures of a Triangle

$x + y + z = 180$

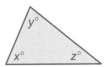

Sum of the Angle Measures of a Quadrilateral

$w + x + y + z = 360$

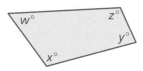

Surface Area
Prism

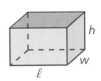

$S = 2\ell w + 2\ell h + 2wh$

$S = $ areas of bases
+ areas of lateral faces

Pyramid

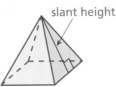

slant height

$S = $ area of base
+ areas of lateral faces

Cylinder

$S = 2\pi r^2 + 2\pi rh$

Volume
Prism

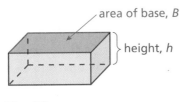

area of base, B

height, h

$V = Bh$

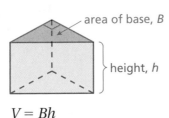

area of base, B

height, h

$V = Bh$

Pyramid

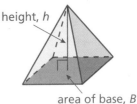

height, h

area of base, B

$V = \dfrac{1}{3}Bh$

Simple Interest

Simple interest formula
$I = Prt$